THERMOFLUID DYNAMICS

Thermofluid Dynamics

A. J. Reynolds
Department of Mechanical Engineering,
Brunel University, London, England

WILEY-INTERSCIENCE
a division of John Wiley & Sons Ltd.
London · New York · Sydney · Toronto

HN

Library of Congress Catalog card No. 78–149580

ISBN 0 471 71780 0

Printed in Hungary
by Franklin Printing House, Budapest

Contents

Preface

These remarks are addressed primarily to teachers, but may also prove helpful to students.

This book is intended for students of mechanical, aeronautical and chemical engineering during the first part of their degree courses. It is difficult to be more definite concerning its place in the overall scheme of studies, for in combining certain aspects of the two subjects of thermodynamics and fluid mechanics this book departs from the standard pattern of university engineering courses, and teachers who wish to adopt this approach will have to fit it into existing schemes. To make this possible in varied circumstances, I have assumed no prior knowledge of thermodynamics and fluid mechanics, and have provided introductory chapters dealing in turn with elementary aspects of the two subjects. Students who have some background in one or both will be able to move quickly through portions of these chapters.

In writing this book I have assumed that the reader has some background in elementary physics and chemistry, in particular, the elementary ideas of rigid-body mechanics and "heat", and of atomic and molecular structure, and an acquaintance with the perfect-gas laws. I have also assumed that the student is familiar with at least one system of units involving absolute and gravitational force units, and have therefore not discussed unit systems in detail. The methods of mathematical analysis have been kept as simple as is consistent with the aim of the book. To this end, the main line of development relates to one-dimensional flow. However, in introducing the basic ideas of thermodynamics, it has been assumed that the student has some acquaintance with the notation and elementary concepts of functions of several variables. In fact, the requisite background could be supplied, for students who have studied only the calculus of functions of a single variable, in a few hours of supplementary lectures. At other points, simple partial differential equations emerge from the analysis, but these do not play a central role in the development. Frequent use has been made of series expansions of functions, and in the examples a variety of approximate and iterative techniques have been demonstrated. In my experience, students require a good deal of practice in the use of these methods before developing confidence in them, and I have therefore laid some stress here.

The aim of this book is to develop the basic ideas of the component subjects, thermodynamics and fluid mechanics, and to illustrate their application to a variety of systems of practical importance. The approach adopted is as practical as is consistent with this basic intention, and it is hoped that through the examples and problems the student will develop confidence in applying the fundamental concepts to obtain numerical results pertaining to realistic systems. By way of illustration, the workings of some typical pieces of engineering equipment have been discussed, and some of the restrictions which the designer must observe have been brought out. These insights into engineering practice are, in a sense, fortunate by-products of the basic line of development, for it has not been possible in this introductory treatment to look systematically into these matters.

I shall not rehearse the arguments for introducing thermodynamics and fluid mechanics to engineering students in combination. I hope that this book, with others of its genre, will demonstrate that this procedure is a viable alternative to the traditional separate treatment of the two subjects. Not every aspect of the component subjects falls within the compass of a unified treatment, and it is envisaged that this introduction to the field will be supplemented towards the end of a student's degree course by more specialized work. Some of the fluid mechanical topics which normally find a place in a mechanical engineering course, but which have not been included in this book, are a discussion of a developing boundary layer and of other two- and three-dimensional flows, a more detailed treatment of flow within turbomachines, and perhaps further consideration of open-channel flows. Thermodynamic topics which will probably be given further attention include the operation of vapour-cycle and internal-combustion plants, and more thorough treatments of combustion and heat transfer. For aeronautical and chemical engineering students, other lines of specialization will need to be developed.

The topics discussed in this book are arranged in the order which seems to me the most convenient for a student to follow. However, the last five chapters are to a large extent independent of one another, and students with a grasp of the work of the first two chapters should be able to approach the later work in a variety of ways, or to omit considerable portions of it, as dictated by their specific needs and the time available. Some sections whose results are used more widely in other chapters should be kept in mind when selecting an alternative order of presentation: Section 3.2, which introduces sound-wave propagation, the speed of sound and Mach number; Section 4.2, which deals with isentropic flow with varying area; Section 5.2, which introduces the techniques of dimensional analysis; Section 4.4, which introduces nozzle and diffuser efficiencies; and Section 6.2, which deals with the efficiencies of flow machines.

A comment on the choice of unit systems may be appropriate. Any writer of a textbook in the United Kingdom is under considerable pressure to concentrate on SI units. I have avoided doing this. There is still room for debate as to the most useful procedure in teaching students in this country. However, the result of this debate is not decisive in selecting units for a text book which will, it is hoped, prove useful to students in countries which have not opted for the SI units. It is hoped that sufficient use has been made of both metric units and the traditional units of British and American engineers to familiarize the student with whichever of the two systems is the more important for him.

A. J. Reynolds

Introduction

Most of the devices which engineers design, manufacture and use interact in one way or another with flowing fluids. Sometimes the motion of the fluid is the aim of the device, as in cross-country gas pipelines, irrigation canals, and the ventilating systems of automobiles and buildings. In other cases, the fluid motion, though not sought, is nevertheless a vital part of the process, for instance, in driving forward an aeroplane, ship or automobile. In yet another class of situations, the fluid is introduced as a means of carrying energy from place to place; in a conventional boiler water and steam are used in this way; in a nuclear reactor the heat-transfer fluid may be carbon dioxide or liquid sodium; and a domestic hair dryer uses simply atmospheric air. The interchange of energy with a fluid can also be accomplished in flow machines—pumps, turbines and compressors—which interact with liquids, vapours and gases. Fluid flows are also involved in basic energy-releasing processes in which fuels are burned in a flow of air, for example, in a jet engine, diesel engine or gas cooking stove. Nor should it be forgotten that natural flows in the atmosphere, in rivers and in the sea are often of importance to engineers, since they apply forces to engineering structures, and are also used in carrying away wastes, including waste heat, from cities and industrial plants. Biological processes also involve fluid motions, and the flows of blood within the body and of air in the respiratory system have attracted considerable attention. Finally, in many machines there are small-scale flows which nevertheless contribute vitally to the overall operation. The basic operation of a rotary ore-crushing mill, for instance, does not involve a flowing fluid, but the successful operation of the mill does depend on the proper supply of lubricating oil to its bearings. Again, the steady operation of an electric motor requires that the energy dissipated within it be removed, and this is usually done by passing a fluid through the motor.

In many of the situations mentioned above as being of interest to engineers, a flow of energy is associated with the fluid motion. In analysing such situations the engineer must use ideas and methods belonging to the two subjects of fluid mechanics and thermodynamics. This book aims to look at both the thermodynamic and fluid mechanical aspects of these processes, and to show how they fit together and complement one another. This unified survey of the two fields forms a basis for later, more specialized studies of thermodynamics and fluid mechanics, in which the interaction between the two is not of vital importance.

The subject matter of this book provides the basis for the analysis of a considerable fraction of the problems which the engineering profession has to solve. There are, of course, many problems associated with the strength and motion of solid bodies which do not fall within our terms of reference. However, even problems of this class may require a consideration of the temperature distribution and forces imposed by flowing fluids. Thus, for example, the mechanical design of an aircraft wing or a turbine rotor depends on the temperature of the fluid passing over the surface and the forces applied by the flowing fluid.

In order to maintain a reasonable length, certain restrictions must be imposed in this introductory survey of one whole side of engineering analysis. The restrictions are of two kinds. The first relates to the analytical methods which will be used. We shall confine our attention for the most part to flows which can be represented with reasonable accuracy by a single vein of fluid whose geometry is known in advance, and across which the velocity is sensibly constant. This idealization is referred to as the one-dimensional flow model. Using it we might, for example, discuss the gross features of the flow through a propeller, and see in general terms how thrust is generated. However, the details of the flow around the individual blades and in the vortices trailing downstream from the blade tips cannot be discussed using the one-dimensional model. Similar restrictions apply in many of the situations we shall discuss. We can look at the overall behaviour of a system and see in general terms how it operates, but cannot always examine the details of the processes by which the ultimate effect is achieved.

The second major restriction imposed to keep this book within reasonable bounds relates to the selection of subject matter. Particular attention will be given to those situations which illustrate the interaction between fluid motion and thermodynamics, that is, between momentum changes, forces and pressures on the one hand, and temperature changes and energy flows on the other. The jet engine falls into this category. Its operation cannot be analysed without referring both to the burning of fuel within the combustion chambers and to the high-speed jet emerging from it. Similarly, an understanding of the operation of a steam turbine requires an analysis of the change in the thermodynamic state of the fluid, and of the forces applied to the turbine blades. In many processes, however, either the thermodynamics or the mechanics of the flow is relatively uninteresting, and attention is naturally focussed on the other aspect. Consider, for example, the operation of a diesel engine. This involves several flow processes: air is drawn into the cylinders and compressed, fuel is injected into the compressed air and burned, and hot combustion products are exhausted from the engine. However, the thermodynamic aspect of the process is dominant, and a detailed discussion of a diesel engine is usually taken to fall within the confines of thermodynamics. Consider, by way of contrast, the flow of water along an irrigation ditch. The fluid is subject to certain thermodynamic processes, most notably, evaporation at the surface and frictional dissipation. However, a study of the flow along the ditch and through control devices installed to adjust the flow rate will use methods commonly included in the subject of fluid mechanics. Nevertheless, the processes in a jet engine, in a diesel engine and in an an irrigation ditch are all governed by the same fundamental mechanical and thermodynamic principles. It is the aim of this book to show how these processes are related, and to indicate why, when and how it is possible to neglect one aspect of certain processes.

It will be helpful in using this book to keep in mind its organization. The first two chapters present the basic ideas and methods of thermodynamics and fluid mechanics or, to be precise, those ideas and methods which will be used in this book. For some students, parts of these chapters will serve as a review of earlier work. If this is the case, they can be taken fairly rapidly, serving to introduce the symbols and terminology of this book, and perhaps to provide another slant on the fundamental ideas of the two subjects. The discussion of fluid motions in the second chapter is confined for the most part to isenergic flows, those which do not exchange energy with their surroundings. Even with this restriction, a flowing fluid proves to be capable of a surprising variety of behaviour, and this is explored in three further chapters concerning isenergic flow. These deal in turn with wave propagation in a compressible fluid, with ideal flows in which friction is unimportant, and finally with mechanisms of friction. Each of these three chapters opens the way to the analysis of important classes of problems, for example, the formation of shock waves around high-speed aircraft, the nearly ideal flows around the blades of turbomachines, and flows through systems of pipes and ducts. The final two chapters of the book consider the effects of energy addition, first through heat flow into or out of the fluid, and then through work transfer in turbomachines. The results of these two chapters are fundamental to the design of a wide range of engineering equipment—heat exchangers, combustion chambers, pumps and turbines.

The arrangement of the problems located at the end of each chapter should be noted. Within each of the three groups into which the problems are divided the ordering corresponds to the sequence in which topics arise in the chapter. The general level of difficulty increases from Group A to Group C. The problems of Group A are fairly straightforward; they allow the reader to test his understanding, to develop confidence, and to become familiar with units and orders of magnitude. Group B contains somewhat more challenging problems, some requiring a measure of independent thought, others calling for more extensive calculations and manipulations. The problems of Group C deal in more detail with matters mentioned in the text, and indicate directions in which the subject can be developed. It is not intended that any student should attempt all of these problems, but only a selection appropriate to his background and the intention of the course he is following. Students requiring further guidance in problem solving would do well to consult "A, B, and C: The Human Element in Mathematics" in "Literary Lapses" by Stephen Leacock.

1

Thermodynamics

Although it is the aim of this book to give a unified view of the mechanics and thermodynamics of flowing fluids, little mention will be made of fluid motions in the first chapter. It has seemed better to introduce the basic concepts of thermodynamics without the complication of mechanical effects.

It may seem that we should introduce this chapter by defining thermodynamics. In fact this could be done only by using terms whose meaning is, at this stage, somewhat obscure. Let us adopt the alternative procedure of looking briefly at the basic concepts of the subject, in this way building up a definition for it. The basic ideas are few in number, although their precise expression requires careful definition and analysis. In the following paragraph they will be outlined briefly and illustrated with a few examples, in order that their essential simplicity will not be forgotten when they are later expressed in mathematical form.

The first important idea is that of an *equation of state*. This is a relationship, which may or may not take the form of an explicit equation, between the properties of a substance; for example, between its pressure, temperature and density. Thus if the pressure and temperature of the steam entering a turbine are known, its density is given by the equation of state for water vapour. The second major concept is that of *energy conservation*. Two modes of energy addition will be defined, work and heat; only by these means can the intrinsic energy of a body or a mass of fluid be changed. Overall, there is conservation of energy, since what is gained by the body is lost by its surroundings. The last fundamental idea is that of *irreversibility*. There is a preferred direction in natural energy flows. If you place your hands on a warm radiator, they will be warmed; you need have no fear that heat will flow from your hands to the already warm radiator. Simple though such ideas associated with the direction of energy flow may seem, considerable effort is required to formulate a statement of the principle of irreversibility which encompasses the varied processes undergone by a flowing fluid.

This chapter is taken up with the mathematical expression of these three basic ideas and their consequences. But these concepts cannot be fully elucidated in a single chapter. In a sense it would be true to say that the whole book is devoted to this task. A real understanding of the basic concepts of thermodynamics is achieved only when one has seen how they apply in a variety of situations. Only then does their generality and power become fully apparent.

1.1 THE SCOPE OF THERMODYNAMICS

Matter can be visualized in two ways. On the microscopic scale it consists of enormous numbers of interacting atoms and molecules. The nature of the interaction differs from substance to substance; the molecules of a gas move about randomly, only occasionally colliding, while the atoms of a solid may only vibrate about their mean positions in a crystal lattice. The macroscopic point of view is that of the human senses and simple measuring devices, such as a mercury-bulb thermometer. These see matter as continuous, taking the form of discrete bodies or homogeneous substances, such as water or air.

The engineer usually wants to know how large pieces of matter are going to behave; he must ultimately adopt a macroscopic description of matter. In principle, one could proceed by first looking at the molecular scale, then summing the effects of microscopic events to predict what will happen on the macroscopic level. In practice, this procedure proves to be impossibly difficult for all but the simplest situations. Hence the engineer has no option: he must try to describe the behaviour of matter in the large, without direct reference to the underlying molecular structure. If, for example, a one-kilowatt electric heater were brought into the room in which you are sitting, you would be ill-advised to attempt to predict the temperature in one corner of the room by studying individual molecular migrations. Even though one knows that molecular motions are ultimately the means by which the energy from the heater spreads through the room, a useful discussion must be based on macroscopic properties such as temperature, density and thermal conductivity.

Corresponding to the two ways of visualizing matter, there are two branches of thermodynamics. Statistical thermodynamics looks at matter on the microscopic scale, and then tries to predict the behaviour of macroscopic systems. This procedure is capable of providing deep insight into the behaviour of matter. But it is the macroscopic branch of thermodynamics which will be followed in this book. This is called either classical thermodynamics, by virtue of its place in the history of science, or engineering thermodynamics, since it is now of interest primarily to applied scientists.

Classical thermodynamics is necessarily a somewhat abstract subject, since it is an attempt to describe the behaviour of matter without specifying its underlying structure. The generalized analysis can often be made more meaningful by keeping in mind the microscopic processes represented by the macroscopic properties. The concept of temperature, for instance, is somewhat nebulous, even though we use it every day and know how to measure temperature using thermometers. However, one can get some feeling for the significance of the property by considering the particular case of a gas, for which the temperature is a measure of the kinetic energy of the randomly moving molecules.

The underlying molecular processes are often conceptually simpler than the macroscopic properties which represent them. On the microscopic level, the pressure of a gas against a surface can be ascribed to the momentum change in molecules striking and rebounding from the surface. Thus pressure and temperature are both seen to be simple manifestations of molecular activity. The microscopic point of view has revealed a unity lacking at the macroscopic level, where pressure and temperature appear to be quite distinct entities, and are measured using very different instruments.

Thermodynamics relates the properties of systems in states of equilibrium, and describes processes and cycles between equilibrium states. This statement is important for two reasons. First, it states restrictions which must be kept in mind during every thermodynamic calculation. Secondly, it uses six fairly common words in a specialized technical sense. Before going further we must define these terms and introduce some others which will be needed later.

System

In this introductory chapter the system chosen for study will be a fixed amount of matter. Such a system of *fixed contents* is often called a *closed system*, that is, closed to the entry and exit of matter. The word *isolated* is used to describe a closed system on which no work is done and into which no heat flows.

In later chapters devoted to the study of flowing fluids it will often be convenient to focus attention on a fixed volume, into and out of which the fluid moves. This is called an *open system* or *control volume*. These two kinds of system, closed and open, are combined in the comprehensive definition of a system as any clearly defined region in space, perhaps fixed, perhaps moving so that it always contains the same material, or perhaps satisfying neither of these requirements. The word *surroundings* is used for everything outside the system considered.

Some of the systems which might be considered in studying the performance of a rocket are shown in Figure 1.1. We might consider the structural shell of the rocket; neither the contents nor the shape of this system change

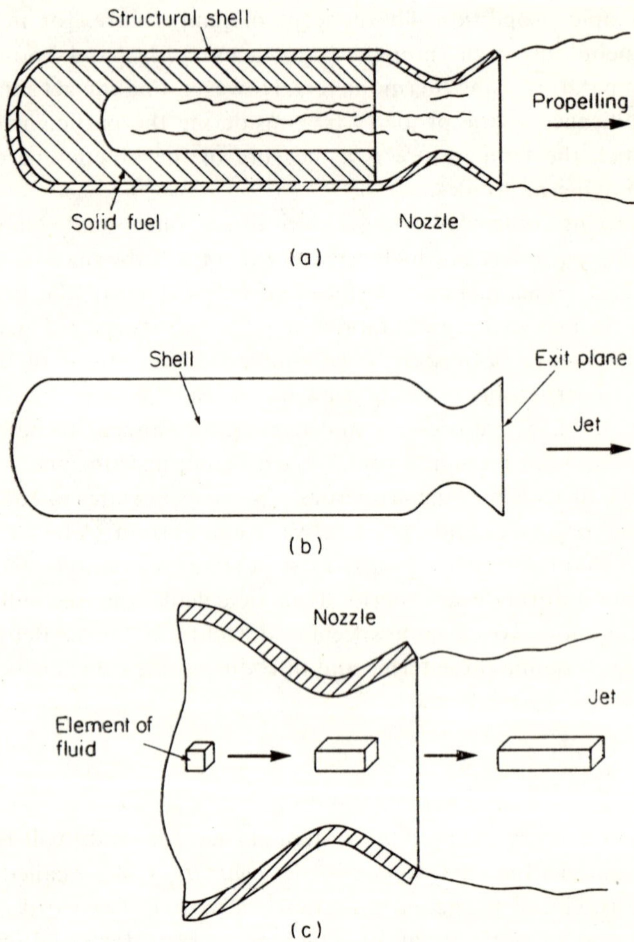

Figure 1.1 Examples of system definition. (a) Section through a solid-fuel rocket. The shell is a system of fixed contents and shape. (b) External view of the rocket. The system bounded by the shell and exit plane has fixed shape, but varying contents. (c) Detail of flow through nozzle. The system is an element of moving fluid, with fixed contents but varying shape

as time passes. A second system which could be examined is the rocket and its varying contents. This system could be defined as the region bounded by the shell and closed by an imaginary surface across the exit plane of the nozzle; the shape of this system does not change in time, but its contents do. Yet another system which might be studied is an element of the propelling gases. Consider a small volume moving with the fluid so that it always contains the same material. This system moves with the discharged gases through the nozzle and out into the jet; its shape changes as it moves, even though its contents remain the same.

In the analysis of any situation perhaps the most important step, and certainly the first step, is the selection of the system to be studied. The choice is to some extent arbitrary, as the discussion above indicates. However, the effort required in the analysis is vitally dependent on this choice; indeed, success or failure may depend on it. At several points in the following chapters different systems will be used to examine the same process, in order that various approaches may be compared.

Property

A property of a system is a quantity describing it. A macroscopic description of the air in a room, for example, would include the length, breadth and height of the room, the mass of each of the constituent gases, the rate at which air enters and leaves through windows and doors, and the pressure and temperature. Many other macroscopic properties could be included, and also data relating to the microscopic structure; even the positions of the individual molecules would be included in the ultimate description. All of this information will not be needed to solve a particular problem, even if it could be obtained. Some of the properties depend upon others and need not be separately prescribed. Moreover, the answer to a particular question concerning the air in a room will be independent of many of its properties. For example, the estimation of the size of heater required to maintain a comfortable temperature will not depend on a detailed knowledge of the microscopic state, nor even require a very precise knowledge of the pressure of the air.

Obviously, a vital step in analysing any one aspect of a system's behaviour is the selection of relevant properties. Macroscopic thermodynamics goes a long way towards making this selection, by assuming that the innumerable microscopic properties can be represented by a relatively few macroscopic properties. However, one has still the considerable task of deciding which macroscopic properties are relevant to a particular situation.

The macroscopic properties can be interpreted as summations of random activity at microscopic scales. At the microscopic level, the state of a system is continually changing. However, when one averages over macroscopic times and distances, the random molecular processes often merge into a stable mean value. A clearly defined mean value will be obtained only if the averaging is carried out over very many random microscopic events. It is to be expected that the mean properties of a few randomly moving molecules will fluctuate wildly. The probability that the mean properties of a very large number will do so is vanishingly small.

In practice, the determination of macroscopic properties is not carried out by mathematical averaging of microscopic events. Instead, use is made of macroscopic instruments which average over time and length scales large compared to the molecular scales. Consider, for example, a mercury thermometer immersed in a gas. The instrument averages the influence of the many molecules which come into contact with it, and the resulting mean temperature is a macroscopic property corresponding in some way to the mean energy of the molecules. Of course, there are situations in which macroscopic instruments do not provide a stable mean value. For instance, a student of cosmic rays may have to wait for some time before his sensors detect a single specimen. Again, if one makes measurements at progressively higher altitudes, the probability of detecting molecular fluctuations will increase as the atmosphere becomes more rarefied. However, for most of the processes to be discussed in this book, and for a wide range of engineering practice, the assumption of the existence of stable mean properties is entirely justified.

State

The state of a system is its condition as defined by all its properties. For many purposes a few properties provide a sufficiently precise definition of state. To describe the air within a room, for example, one may be satisfied with the volume of the room and the pressure and temperature, together with the statement that it is 'normal' atmospheric air. It is not difficult to find systems whose state requires more detailed specification. The description of the state of a piece of steel may involve not only its temperature and the forces applied to it, but also its detailed composition, and a description of its grain structure.

It is easiest to describe the state of a *homogeneous system*, one such that all macroscopic samples taken from it at the same time have the same properties. Of course, if one takes a small enough sample, no real system will be homogeneous. The assumption of homogeneity is particularly appropriate for fluids, since mixing, both microscopic and macroscopic, takes place with relative ease and smooths out variations in the macroscopic properties.

Equilibrium

A system is said to be in thermodynamic equilibrium when changes of state, as measured by variations in time of macroscopic properties, have ceased to occur. In an equilibrium state no macroscopic flows of matter or energy occur, and the chemical composition remains constant. Accordingly, the description of a system is simplest for an equilibrium state.

The distinction between equilibrium and non-equilibrium states is clarified by a consideration of a familiar situation, a pot of water heated from below by a flame. The flame raises the temperature of the water at the bottom of the pot and also reduces the density there. Convection currents of matter and energy are established and a complex varying temperature distribution is set up in the water. However, if the flame is removed, the water will gradually return to an equilibrium state, with nearly uniform properties. This example suggests that the only homogeneous systems are those in equilibrium. The converse is not true. The system comprising a table and the air around it can be in equilibrium, even though it is not homogeneous.

Process

A process is the transition from one state to another. Strictly speaking, a system cannot be at equilibrium during any process, that is, while its state is changing. Classical thermodynamics does not attempt to describe a system undergoing a process, when it is not at equilibrium, but can relate initial and final equilibrium states. Classical thermodynamics cannot, for example, describe the details of an explosion of fuel within the cylinder of an automobile engine, but might be used to predict the equilibrium temperature of the products of combustion.

If this restriction on its application were strictly interpreted, thermodynamics would be of little practical use, for real systems are never exactly in equilibrium. However, during a process which is slow enough, the state of a system may not differ greatly from an equilibrium state. The system may therefore be assumed with reasonable accuracy to pass through a series of equilibrium states during the process.

Consider once again the situation discussed above, a pot of water heated by a flame. As was pointed out earlier, rapid heating will give rise to large temperature gradients and convection currents. However, with a very low flame, it is possible to change the state more slowly, so that molecular conduction and small convection currents have time to maintain fairly

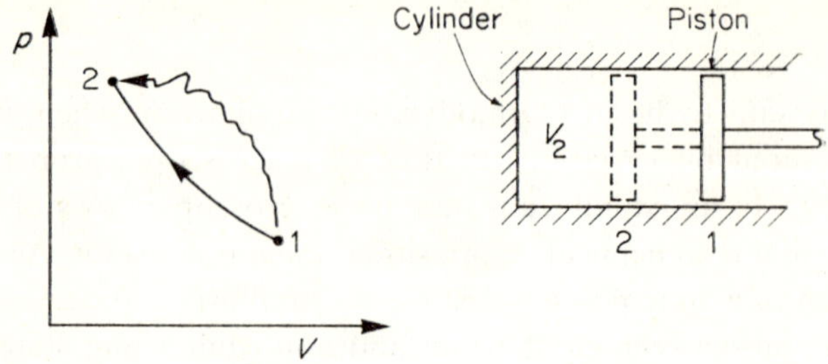

(a)

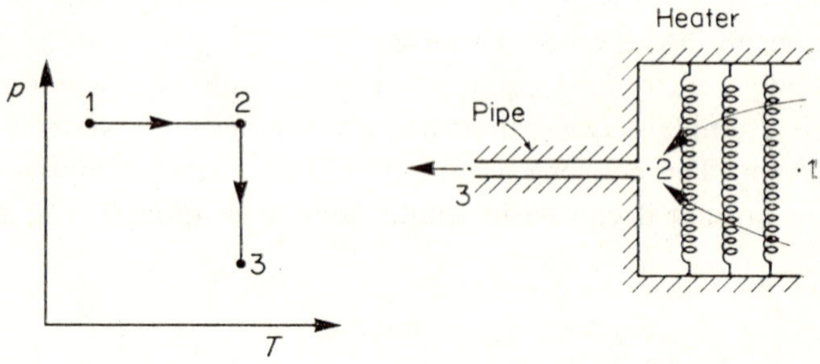

(b)

Figure 1.2 Examples of processes. (a) Non-flow compression within a cylinder for the two cases of slow and abrupt motion of the piston. The plot represents pressure and volume changes. (b) Steady-flow processes in a heater and pipe. The plot represents pressure and temperature changes

uniform conditions throughout the mass of water. During this slower process the system remains nearly in equilibrium.

It is often convenient to represent equilibrium states graphically, by plotting one property of a system against another. Figure 1.2(a) relates the pressure and volume of a gas contained within a cylinder. The states labelled 1 and 2 are equilibrium states. If the piston is moved slowly inwards, so that the pressure remains nearly uniform throughout the cylinder, states intermediate to 1 and 2 will be nearly in equilibrium, and a continuous process path may be drawn. If the piston were rammed in very rapidly, there would exist large pressure differences within the cylinder during the process. Since no single pressure could be found to describe the system, a continuous process path could not be constructed between the initial and final equilibrium states. A discontinuous transition must then be shown on the plot of pressure against volume.

Some of the names used to identify particular kinds of processes may be noted here. The prefix iso- is often used in thermodynamics to indicate that some property is constant during a process. Thus *isothermal*, *isobaric*, *isentropic* and *isenthalpic* denote processes in which temperature, pressure entropy and enthalpy remain constant. (Entropy and enthalpy are properties which you may not have encountered before; they will be introduced later.) Another pair of labels denoting particular processes are *adiabatic* and *isenergic*. The former denotes a process in which no heat flows, and the latter one in which neither heat nor work enters a flowing fluid. The significance of these terms will become clearer when the terms heat and work are defined precisely.

The terms *flow process* and *non-flow process* may also be encountered. In the first, fluid enters and leaves the system, while in the second the contents of the system are fixed. The particular case of a *steady-flow process* should be noted; here the macroscopic state within the system is the same at each instant, although the material within the system is steadily changing. This kind of process provides a useful approximation to many practically important flows. Consider the operation of an aircraft jet engine. There are brief intervals when the flow is not steady: for example, when starting, during violent manoeuvres, when the geometry of the inlet and outlet ducts is changed, or when the fuel flow is suddenly altered. But during a very large part of its operating life, the flow through the engine is essentially steady.

Figure 1.2(b) represents a steady-flow process which can be broken down into two simpler processes. Air flows slowly, first through a heater, then through a pipe. In the heater it undergoes a nearly isobaric process; in the pipe the flow is essentially isothermal. The second part of the flow may also be described as adiabatic or isenthalpic.

Cycle

A cycle is a process or series of processes which returns a system to its initial state. Some of the most familiar cyclic processes are those within the cylinders of piston engines and reciprocating compressors. Here the fluid does not experience a cyclic process, for it ends up with an altered state. The system for which the process is cyclic is the cylinder itself.

Figure 1.3 illustrates a simple cycle which could be achieved using the device shown. It will be instructive to trace the process path around the cycle, deciding how the movements of the piston and valve must be coordinated, and how the pressure beyond the valve must vary.

This completes the definition of the terms contained in the sentence stating

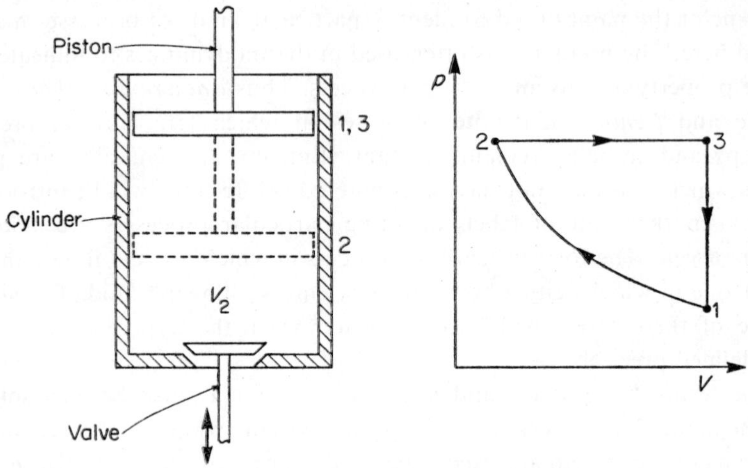

Figure 1.3 A simple cyclic process. The plot represents pressure and volume changes

the fundamental restriction: thermodynamics relates the properties of systems in states of equilibrium, and describes processes and cycles between equilibrium states. Let us now consider the implications of this restriction for a flowing fluid. Strictly, classical thermodynamics does not apply to any of the situations we wish to discuss, for the state of a flowing fluid is continually changing, and it is never precisely in equilibrium. However, the rate of molecular interaction is usually so rapid that the fluid has time to adjust to the relatively slow changes in macroscopic properties. Thus a near-equilibrium condition is established within each small element of fluid.

While the assumption that a flowing fluid passes through a succession of equilibrium states is realistic for most of the situations which engineers are called upon to analyse, it does sometimes break down. For example, in a rarefied gas in the upper atmosphere, the mean time between molecular collisions becomes large compared to the time scale of the events which are to be studied. Other flows in which equilibrium conditions are not established involve ionized gases. This case does not lie beyond the range of engineering practice. The air around a vehicle re-entering the atmosphere from space ionizes on being heated to very high temperatures as it absorbs the vehicle's enormous kinetic energy. Also, efforts are being made to develop plasma generators, which will use ionized gases for power production. A final example of a non-equilibrium state is that within a shock wave formed by a supersonic aeroplane. Later in this book the states upstream and downstream of a shock wave will be related using the principles of classical thermodynamics. But these principles do not apply to the state of the fluid within the wave itself, where the temperature and velocity gradients are very large.

Example 1.1 This simple numerical example illustrates the tendency of a random small-scale process to generate a macroscopic mean value which is unlikely to vary beyond a narrow band of values.

Consider the random selection of a sample of red and white balls from a very large collection of the two species. The selection of groups of four balls will indicate the nature of the probability distribution. The probability of choosing

all red is $(\frac{1}{2})^4 = 1/16$
three red and one white is $4\times(\frac{1}{2})^4 = 4/16$
two red and two white is $6\times(\frac{1}{2})^4 = 6/16$
one red and three white is 4/16
all white is 1/16

Note that the numerators of these fractions are the coefficients of a binomial expansion; this suggests that the probabilities of selection for larger groups of balls will also correspond to sets of binomial coefficients. Thus the probabilities for groups of five balls containing 0, 1, 2, 3, 4 and 5 red balls are 1/32, 5/32, 10/32, 10/32, 5/32 and 1/32; and the probabilities of groups of ten balls containing 0, 1, 2,. . . 10 red balls are 1/1024, 10/1024, 45/1024, 120/1024, 210/1024, 252/1024, 210/1024. etc.

When these two sets are plotted as functions of the concentration of red balls (do this), it is seen that in both cases a 'pink' mixture—one with roughly equal numbers of red and white balls—is highly probable. If the group over which the average is taken changes continually in a random manner, very high and very low concentrations will appear only occasionally. Most of the time the mixture will be 'pink'. Note also that the probability distribution is more peaked for the larger group size. Hence significant deviations from the mean are less likely when the concentration is obtained by averaging over a large group.

1.2 EQUATIONS OF STATE

1.2.1 The Equation of State of a Perfect Gas

An equation of state is a relationship between the equilibrium properties of a system. There is, in general, a different equation for each and every system. However, equations of state do have certain common features, and this discussion of the particular equation for a perfect gas will not only produce results applicable to a gas, but will also reveal something about a wider class of equations of state. The perfect gas is selected for this initial study since its equation has a simple form which is familiar to every student from earlier studies of physics and chemistry.

It is first necessary to decide how many properties and which properties of a perfect gas are to be related in its equation of state. Consider a quantity of a gas, say, oxygen, held within the container shown in Figure 1.4. There are two external means by which the state of the gas can be changed; the force on the piston can be altered, and a flame can be applied to the container. The effect of the changes in the applied force will be considered first.

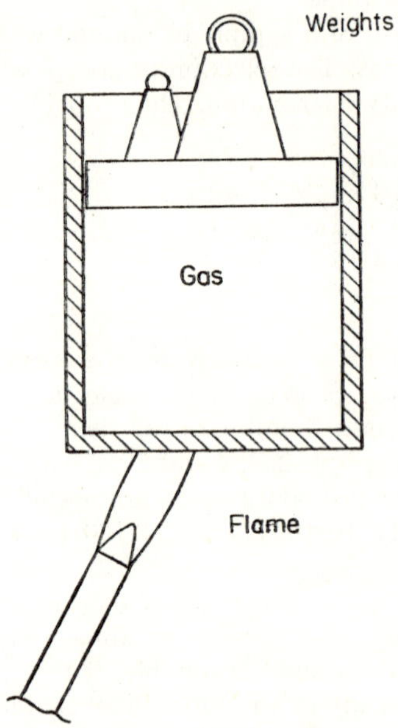

Figure 1.4 External influences on a gas within a cylinder

If one increases the force, and hence the pressure applied to the gas by the piston, the volume of the gas will be reduced. Two aspects of *pressure* must now be distinguished. One is the applied pressure, the external force divided by the piston area; the other is the internal pressure, the outwards push of the gas against the containing walls. When mechanical and thermodynamic equilibrium is established, the two pressures are equal. However, there is a crucial difference between them. While the external pressure is defined by the external force, the internal pressure is spontaneously developed in response to volume changes. Thus it appears that the equation of state should relate the volume of the system to its internal pressure.

Now consider the effect of the flame on the system shown in Figure 1.4. If the applied force is kept constant, the gas will expand. In order to relate this expansion to a change in the properties of the system, a thermodynamic property directly responsive to heat addition must be introduced. The most easily measured property of this kind is *temperature*, and it will therefore be taken to be the fundamental thermodynamic property. It is to be expected that the equation of state will relate the temperature of the system to its volume and, since volume and pressure are linked, to the pressure as well.

Temperature, and the ways of measuring it, are so familiar that it is easy to forget the assumptions which lie behind its use. These assumptions must be made explicit here, in order that the concept of temperature can be used with precision. First it is necessary to state what is meant by equality of temperature. This is done by the *zeroth law of thermodynamics:* the temperatures of two systems are equal if the systems remain in equilibrium while held in contact. This principle is used whenever temperature is measured; when the reading on a thermometer ceases to change it is taken to be in equilibrium with its surroundings, and to have the same temperature as its surroundings. The zeroth law also implies that there is a single temperature scale applicable to all systems. It may appear that there are several temperature scales in common use. But these are in fact rigidly linked, and are merely different ways of describing points on a unique scale of temperature. For example, changes in temperature on the Centigrade and Fahrenheit scales are rigidly connected by the relation 1 °C = 1·8 °F. It is still necessary to define the unique temperature scale or, in other words, to select one kind of thermometer to act as a standard. It will be seen shortly that a convenient scale is provided by the equation linking the properties of gases.

The preceding discussion has suggested which properties will be related by an equation of state for a gas. An explicit relationship amongst these properties will now be examined. It applies with good accuracy to the so-called permanent gases, for example, oxygen, nitrogen and helium, at temperatures and pressures not very different from normal atmospheric values. When a mass m of a gas with molecular weight MW is confined in a container, its volume V, pressure p, and temperature T are found to be related by the single equation

$$pV = \mathbf{R}\frac{m}{MW}T \tag{1.2.1}$$

The same constant **R** is applicable to all the gases; it is called the universal gas constant.

To obtain the simple equation (1.2.1) one must measure the temperature T in a particular way, using a datum level near −273 °C or −460 °F; this datum is very nearly the same for the various gases. The temperature measured from this datum is called the *absolute temperature.* The gas law predicts that the density of the gas tends to infinity as the temperature approaches the absolute zero with the pressure maintained constant. Of course, the perfect-gas law ceases to apply long before absolute zero can be approached, but this prediction does suggest that some very odd things will occur at temperatures near absolute zero. This turns out to be the case, although these complex low-temperature effects will not be discussed in this book. Since the absolute zero is of fundamental interest, and since the much-used gas laws take a convenient form when the absolute temperature is used in them, the perfect-gas scale of temperature is adopted as the universal scale defined by the zeroth law of thermodynamics.

By considering briefly the molecular aspect of matter, we can see why the behaviour of a number of gases is described by a single equation (1.2.1). Consider a number of samples of different gases, each sample having the same volume, pressure and temperature. The gas law then indicates that the value m/MW will be the same for each sample. But the molecular weights are, by definition, proportional to the masses of the constituent molecules. Hence each sample must contain the same number of molecules. This conclusion has far-reaching consequences when one comes to consider reactions between gases.

Equation (1.2.1) also suggests that it may sometimes be advantageous to measure the mass of a gas in units proportional to its molecular weight. This is frequently done by chemists and physicists, and the engineer must be familiar with this convention in order to use physical data presented in this way. These special mass units are called moles, and are simply the molecular weight times the mass unit of the basic unit system which is used. Thus, for oxygen with $MW = 32$, a kilogram–mole is given by 1 kg mol= 32 kg, and a pound–mole is given by 1 lb mol = 32 lb.

The ideal-gas equation can be rewritten in the forms

$$pv = RT \quad \text{and} \quad p = \varrho RT \tag{1.2.2}$$

Here $v = V/m$ is the specific volume of the gas sample, and $\varrho = m/V$ is its density. The constant $R = \mathbf{R}/MW$ applies for only one gas or mixture of gases. The rewriting has fundamentally changed the significance of the gas law. It does not describe the properties of a particular mass, but now relates the properties of any sample of the gas. No longer do we have the equation of state of a system, but rather the equation of state of a substance, relating the properties of that substance wherever it may be found. (The implicit restriction to equilibrium states should, however, be kept in mind.) If, for example, the low temperature and pressure existing at the top of Mount Everest were to be reproduced in a sample of air in a laboratory, the density of the laboratory air would match that at the top of the mountain.

So far little has been said about the units of the quantities appearing in the gas law. Students, and even experienced workers, often have difficulty in manipulating these units, the difficulty arising in part from the multiplicity of unit systems in common use. Nevertheless, since one inevitably encounters a variety of units in reading and in engineering practice, it is unrealistic to concentrate on a single unit system. For this reason the examples, data and problems in this book make use of a mixture of unit systems.

Table 1.1 gives the units which engineers commonly use for the quantities appearing in the gas law. The units given in each column form a consistent set; if three quantities measured in these units are introduced into equations (1.2.2), the fourth quantity will automatically be measured in the corresponding consistent unit. Obviously, it is advantageous to carry out a particular calculation using one consistent set of units, and therefore all the data should first be converted to one system. With experience and confidence this rule may be relaxed, but a beginner will be well advised to follow it.

Table 1.1. Consistent units for the gas law

Property	English system		Système International (SI)
	Engineering	Absolute Engineering	
V	ft³	ft³	m³
T	°R	°R	°K
p	lbf/ft²	lbf/ft²	N/m²
m	lb	sl	kg
$\varrho = 1/v$	lb/ft³	sl/ft³	kg/m³
R	ft lbf/lb °R	ft lbf/sl °R	J/kg °K

Other unit systems may be met: the English absolute system with forces measured in poundals; the CGS absolute and gravitational systems with forces in dynes and grams, respectively; the MKS gravitational system with forces in kilograms; and systems using moles. One must be prepared to deal with data in any of these systems. When possible, however, preference should be given to the SI units, for by international agreement this will in the future become the standard system. Moreover, since these units have been devised on the basis of more extensive experience, they should prove easier to use once they become familiar.

Example 1.2 In practice the information available is often specified in a mixture of unit systems. For example, the data issued by the weather office may take the form: pressure, 980 mbar; temperature, 70 °F. Suppose that the density is required, in lb/ft³.

Under normal atmospheric conditions air behaves much like a perfect gas, and the density can be calculated as $\varrho = p/RT$. Values of R are given in several kinds of units in Appendix II. The most convenient set of units for the present calculation seems to be that for which $R = 53{\cdot}3$ ft lbf/lb °R. In these units

$$T = 70+460 = 530\ °\text{R},$$

and

$$p = 980 \text{ mbar} = 0{\cdot}980 \times 14{\cdot}50 \times 144 = 2040 \text{ lbf/ft}^2$$

Then

$$\varrho = \frac{2040}{53{\cdot}3 \times 530} = 0{\cdot}0724 \text{ lb/ft}^3$$

If one had to do a large number of calculations of this kind, the best solution would be to derive a special equation for the units in which the data is supplied. Thus

$$R = 53{\cdot}3\ (\text{lbf/ft}^2)/(\text{lb/ft}^3)\ °\text{R}$$

$$= 53{\cdot}3 \times \frac{1000}{14{\cdot}50 \times 144} \text{ mbar/(lb/ft}^3)\ °\text{R}$$

$$= 25{\cdot}5 \text{ mbar ft}^3/\text{lb } °\text{R}$$

The working equation for this calculation is then

$$\varrho = \frac{p}{25{\cdot}5(T+460)}$$

Example 1.3 One important application of the perfect-gas law is in relating variations of pressure, temperature and density through the atmosphere. Here we shall consider pressure and density changes within the stratosphere, the nearly isothermal layer which is usually found between the altitudes of 11,000 m and 20,000 m. The precise location of this isothermal layer depends upon geographical position, season of the year and weather conditions; in any case there are significant variations in temperature both below and above the stratosphere.

It is required to find the properties of the air at 20,000 m, given that the temperature and pressure at 11,000 m are 0·227 bar and 217 °K. The universal gas constant may be taken to be 1545 ft lbf/lb mol °R.

Let us first determine the laws governing the variations in pressure and density with altitude. The two are linked by the condition of vertical equilibrium, which states that the pressure variation is just sufficient to support the weight of the air. Consider a thin horizontal layer of height dz and plan area unity. This element weighs $\varrho g\,dz$; the pressure force on its upper face is $p+dp$ acting downwards, and that on its lower face is p acting upwards. Then for static equilibrium

$$\varrho g\,dz = p-(p+dp) = -dp$$

Now the equation of state comes into play: $p = \varrho RT$. Combined with the condition of vertical equilibrium, it gives

$$\frac{dp}{p} = -\frac{g}{RT}\,dz$$

Since the temperature and acceleration of gravity may be assumed constant with reasonable accuracy, an explicit integration is possible:

$$\log_e \frac{p}{p_1} = -\frac{g}{RT}(z-z_1)$$

where p_1 is the pressure at height z_1. Then

$$p = p_1 \exp\left[-\frac{g}{RT}(z-z_1)\right]$$

and

$$\varrho = \varrho_1 \exp\left[-\frac{g}{RT}(z-z_1)\right]$$

give the pressure and density variations through the stratosphere.

We are now able to carry out the required calculations. First the gas constant must be converted to suitable units. The air of the stratosphere may be taken to be 23·3 per cent oxygen and 76·7 per cent nitrogen by weight, with $MW = 32{\cdot}0$ and 28·0 for the two diatomic gases. Hence the mean molecular weight is

$$MW = 0{\cdot}233 \times 32{\cdot}0 + 0{\cdot}767 \times 28{\cdot}0 = 29{\cdot}0$$

and the appropriate value of the gas constant is

$$R = \frac{\mathbf{R}}{MW} = \frac{1545}{29{\cdot}0} = 53{\cdot}3 \text{ ft lbf/lb °R}$$

$$= 53{\cdot}3 \times 4{\cdot}45 \times 2{\cdot}205 \times \frac{1{\cdot}8}{3{\cdot}28} = 287 \text{ J/kg °K}$$

The density at 11,000 m is determined by the pressure and temperature there

$$\varrho_1 = \frac{22{,}700}{287 \times 217} = 0{\cdot}365 \text{ kg/m}^3$$

The factor relating pressures and densities at 11,000 and 20,000 m is

$$\exp\left[-\frac{9{\cdot}81 \times 9000}{287 \times 217}\right] = 0{\cdot}242$$

Finally, the pressure and density at 20,000 m are

$$p = 0{\cdot}227 \times 0{\cdot}242 = 0{\cdot}0550 \text{ bar}$$

and

$$\varrho = 0{\cdot}365 \times 0{\cdot}242 = 0{\cdot}0884 \text{ kg/m}^3$$

The calculation of the properties of the International Standard Atmosphere makes use of a description of the stratosphere very much like that used above. Hence the changes between 11,000 m and 20,000 m found here are very nearly those given in the standard-atmosphere tables of Appendix VI.

1.2.2 General Equations of State

When other substances are subjected to pressure changes or absorb heat, their response may be expected to be generally like that of a gas. No matter what the substance within the system shown in Figure 1.4, its equation of state will have the form

$$p = p(v, T) \tag{1.2.3}$$

This relationship could equally well be written

$$v = v(p, T) \quad \text{or} \quad f(p, v, T) = 0 \tag{1.2.3}$$

using other standard notations for functions of several variables. These equations state that whenever particular values of two of the properties are reproduced, in a system at equilibrium, the third will always take on the same value.

Since each property depends on two others, small changes between equilibrium states will be related by equations such as

$$\mathrm{d}v = \left(\frac{\partial v}{\partial p}\right)_T \mathrm{d}p + \left(\frac{\partial v}{\partial T}\right)_p \mathrm{d}T \tag{1.2.4}$$

Note that, when used in thermodynamics, partial derivatives are customarily enclosed in brackets and followed by a subscript which indicates the property remaining constant during the differentiation. This convention is not strictly necessary at this stage, since only three inter-related properties have been introduced, but it becomes vital when additional properties are considered.

In fact additional properties have already been introduced, since derivatives of properties, like those of equation (1.2.4), are themselves properties. This point becomes clearer if the derivatives are worked out for the particular case of a perfect gas. Using equation (1.2.2) one gets

$$\begin{aligned} \left(\frac{\partial v}{\partial p}\right)_T &= RT\,\mathrm{d}\left(\frac{1}{p}\right)\Big/\mathrm{d}p = -\frac{RT}{p^2} = -\frac{v}{p} \\ \left(\frac{\partial v}{\partial T}\right)_p &= \frac{R}{p} = \frac{v}{T} \end{aligned} \tag{1.2.5}$$

The gas constant R is constant for both differentiations, as is T for one, and p for the other.

Each of the derivatives calculated in equations (1.2.5) takes on a unique value when two properties are specified. In this respect the derivatives are like the properties related in equations (1.2.3). The kind of calculation carried out in equations (1.2.5) could, in principle, be performed for any equation of state. Thus equation (1.2.3), which links three properties, requires that other macroscopic properties derivable from them are also determined once two properties have been specified.

A substance with equation of state (1.2.3) is called a *simple compressible substance*, and a system whose volume V is related to its pressure and temperature through

$$p = p(V, T)$$

is called a *simple system*. For such systems the following *state postulate* is made: the state of a simple system is determined when two independent equilibrium properties have been specified. It has been seen that this postulate is valid for properties related by a three-property equation of state, and for properties derived from them. It will be assumed to hold for every thermodynamic property of a simple system. Thus a *thermodynamic property* of a simple system may be defined as a quantity which can be expressed as a function of two independent thermodynamic properties. It follows that there must be an equation of state linking any three thermodynamic properties of a simple system or of a simple compressible substance. This conclusion will prove to be of vital importance in several of the arguments which follow.

It must not be thought that the properties of every system can be represented by an equation of the form $p = p(V, T)$. The only way in which this kind of system can respond to applied forces and energy addition is through uniform increases in pressure and temperature. It is not difficult to imagine systems capable of different responses. A magnetic substance, for instance, can interact with its surroundings in another way, through magnetic forces and energy. A bridge truss, made up of a number of girders, each carrying a different load, can resist applied loads and absorb energy in a variety of ways. If a chemical reaction were to occur within the system, the equation of state of the products would, in general, differ from that of the mixture before the reaction occurred. None of these effects is taken into account in the simple equation of state which has been considered here.

It must be concluded that the equation $p = p(V, T)$ applies only to systems essentially similar to that considered at the beginning of this discussion of equations of state: a homogeneous mass of fluid within a container. This is in fact very like the situation in which small elements of a flowing fluid find themselves. Figure 1.5 shows a sample of fluid contained by a piston, a second sample within a stationary mass of fluid, and a third sample passing through a nozzle. In each case an external pressure acts on the system; in the first case it is applied by the piston and the other walls of the container; in the other cases the surrounding fluid applies the containing pressure. In each case, a uniform internal pressure is developed to balance the pressure applied from without. It will be assumed that the fluid in each situation obeys the same equation of state, subject to the requirement of a near-equilibrium state in each case.

Any equation of the form (1.2.3) can be represented as a surface in the

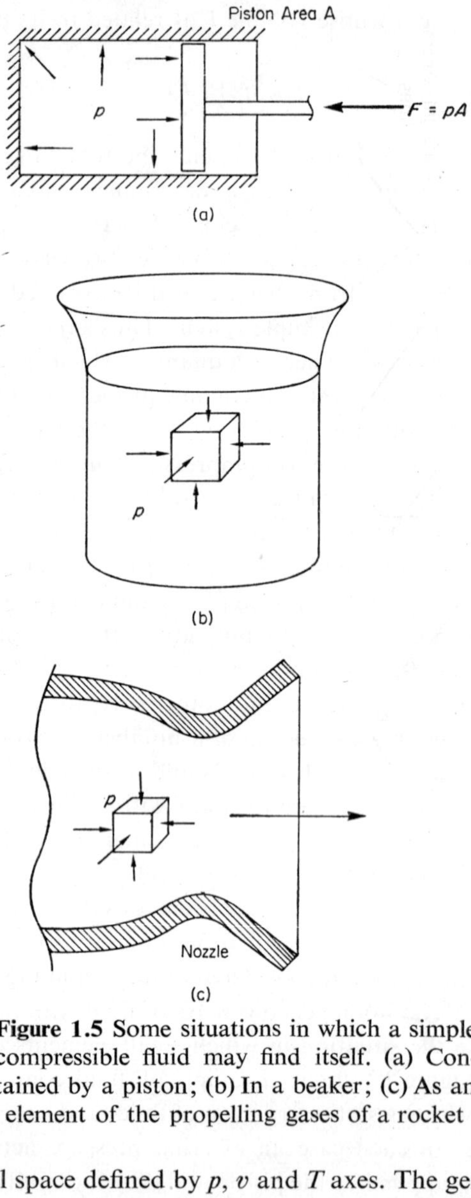

Figure 1.5 Some situations in which a simple compressible fluid may find itself. (a) Contained by a piston; (b) In a beaker; (c) As an element of the propelling gases of a rocket

three-dimensional space defined by p, v and T axes. The general form of the *surface of state* for a pure substance is shown in Figure 1.6(a). For comparison, the surface of state defined by equation (1.2.2) is shown in Figure 1.6(b). A *pure substance* is defined as one whose chemical composition remains, on the macroscopic scale, invariable and homogeneous when its state changes slightly. Hence water and ice in equilibrium make up a pure substance, for the formation of more ice does not alter the chemical compo-

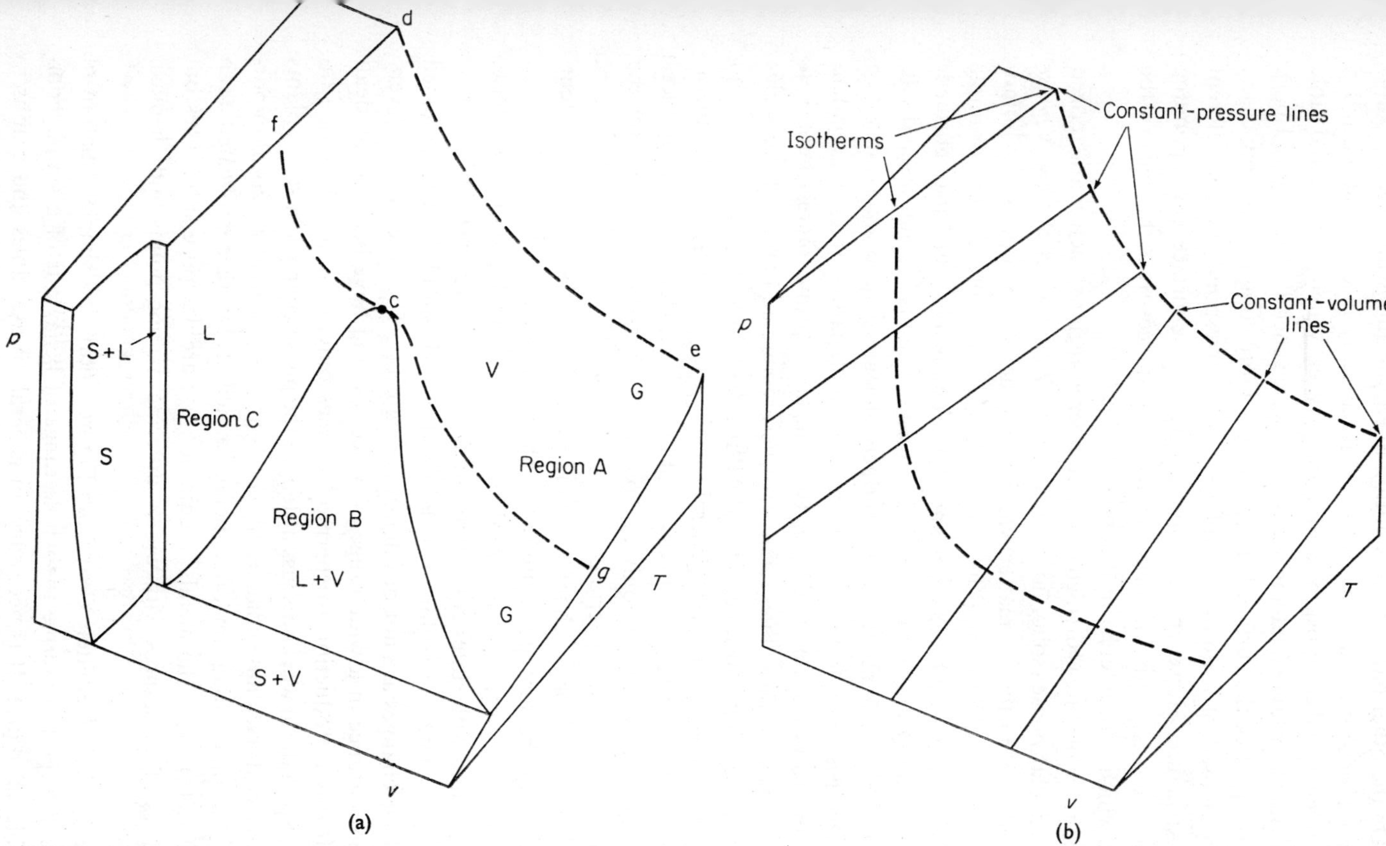

Figure 1.6 Surfaces of state in p, v, T-space. (a) Surface of state for a pure substance. The dashed lines are isotherms. The solid lines mark the limits of phases. Phases are indicated by: S-solid, L-liquid, V-vapour, G-gas. (b) Surface of state defined by the perfect-gas law

sition. On the other hand, a salt solution in equilibrium with a salt crystal is not a pure substance, since there is a discontinuity in composition and a change in chemical composition with state. Oxygen, water, mercury and table salt are pure substances, having surfaces of state generally like Figure 1.6(a). The lower part of the surface is in fact rather more complex in many cases, being composed of a number of planes, each corresponding to a different crystal structure. However, the upper portion of the surface, corresponding to fluid states, is of similar form for many substances. It is this part of the surface which is of interest in this book.

The surface of state is divided into a number of regions, each characterized by a particular phase (solid, liquid or vapour) or mixture of phases. A *phase* is defined as a portion of matter which is, on the macroscopic scale, homogeneous in chemical composition and physical structure. In the example used above (water and ice, water and salt), four phases were distinguished: two liquid, water and a solution of salt in water; and two solid, ice and salt, The lines of demarcation on the surface of state denote states at which phase changes begin to take place: melting, sublimation, freezing, condensation and vaporization. The latter two will be of prime interest here. It is in the location of these lines relative to the pressure and temperature scales along the axes that substances differ. Although oxygen, water, mercury and salt have surfaces of the same general form, normal pressures and temperatures define points on different regions of the surface. Hence these substances are usually encountered in only one phase. Water is the exception, for the normal range of atmospheric conditions falls in the part of the surface of state where many phase changes occur. Hence only small changes in temperature and pressure are required to change its phase.

Most of the processes discussed in this book involve only one phase, either a liquid with states corresponding to points on region C, or a vapour or gas with states on region A of Figure 1.6(a). It should not be assumed that phase changes are not of interest to engineers. They are in fact of very great importance in a wide range of processes, and have been lightly dealt with in this introduction to thermodynamics only in the interests of simplicity. Specialized works discuss at length the phase changes in steam plants and refrigerators. Phase changes in metals are also of great practical importance, the phase diagram often being immensely more complicated than Figure 1.6(a), since the metals in common use are not pure substances but are alloys of substances with widely differing melting points and chemical properties.

One fact is immediately clear from Figure 1.6(a)—no simple equation of state can hope to describe the whole range of behaviour of a simple compressible substance. It is necessary, in general, to use tables and graphs to represent the equation of state. The most widely used tabular presentations of an equation of state are those for water and water vapour, known as *steam tables*. Appendix IV contains steam tables in English and SI units. Comparable tables are available for many other substances. This book contains a number of discussions of processes in steam; they are of some importance in themselves, and have further importance in illustrating methods of calculation which can be applied to other substances whose properties are available in graphs and tables.

The steam tables relate pressure, temperature, specific volume and other properties which will be introduced later as measures of the energy of a substance: specific internal energy, enthalpy and entropy. Two kinds of data are presented in the steam tables. First, there are the properties of saturated water and vapour, that is, of vapour and liquid phases in equilibrium together; these correspond to points on region B in Figure 1.6(a). Secondly, there are the properties of the vapour phase alone, corresponding to points on region A; this is referred to as superheated steam, that is, steam whose temperature is higher than the boiling temperature for its pressure. Consideration of Figure 1.6(a) shows that these two kinds of data are sufficient to describe the whole surface of state. The several regions making up the complex surface are of only two types: single-phase regions, such as A and C, with equations of the form $v = v(p, T)$, and mixed-phase regions, such as B, with equations of the form $v = v(p)$ or $v = v(T)$. Hence the forms of presentation used in steam tables may be used to describe all the regions of the surface of state.

The data for saturated liquid and vapour provide, in effect, the equation of the *saturation line* bounding regions A and C. They also provide the means of describing the intermediate, mixed-phase states of region B. The two phases may exist in equilibrium with one another in any proportion. Thus if the fraction which is liquid is y, the average specific volume of the mixture is

$$v = yv_f + (1-y)v_g = v_g - yv_{fg} \tag{1.2.6}$$

where v_f and v_g are the specific volumes of the liquid and vapour phases at the appropriate pressure and temperature. The fraction y is called the wetness fraction of the steam. One may also introduce the dryness fraction x $(= 1-y)$, in terms of which the average specific volume is

$$v = xv_g + (1-x)v_f = v_f + xv_{fg} \tag{1.2.6}$$

As the water vaporizes, its specific volume increases linearly from v_f to v_g, the net expansion being $v_{fg} = v_g - v_f$. The properties to be introduced later, measures of the energy of the substance, vary in a similar way within the mixed-phase region.

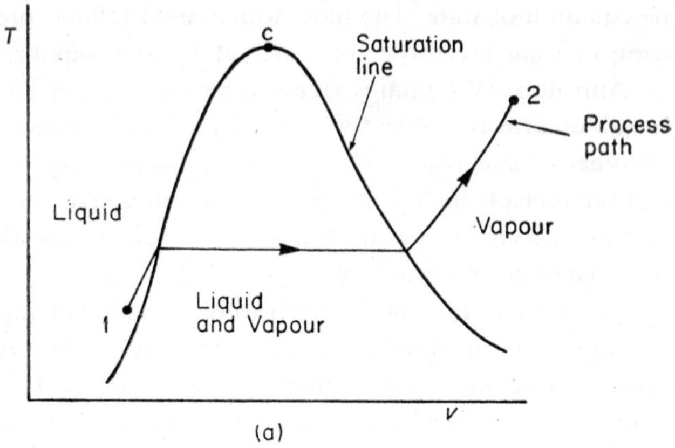

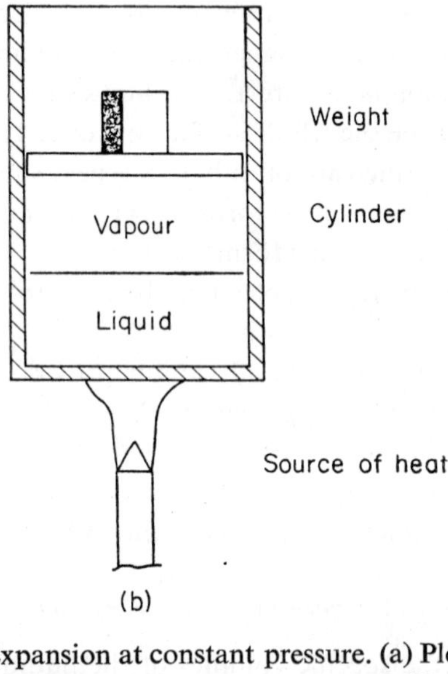

Figure 1.7 Expansion at constant pressure. (a) Plot relating changes in temperature and average specific volume. (b) Apparatus in which the process might take place

Figure 1.7(a) shows how the average specific volume of a substance varies with temperature, pressure being kept constant, for a range of states extending across the mixed-phase region. This transition could be achieved in the apparatus sketched in Figure 1.7(b). Figure 1.7(a) may be thought of as a projection of the surface of state shown in Figure 1.6(a) onto a plane parallel to the v and T axes.

The properties of steam are also presented graphically in Appendix IV. The presentation adopted, known as a *Mollier diagram*, uses the properties enthalpy and entropy as the fundamental variables; this form is more convenient for the solution of a range of practical problems. Note that the values of temperature and pressure are also indicated on the chart. Charts of this kind have considerable utility in providing numerical answers to problems, but are perhaps even more useful as an aid to visualizing the processes under analysis. Tracing a series of processes over a diagram gives a better insight than merely working through the tabled values.

The abrupt transition shown on the Mollier chart and in the steam tables between vapour, liquid and mixed-phase states is somewhat misleading. The sharp transition does indeed represent the equilibrium properties, but during a rapid process a substance may depart from equilibrium. Thus when the pressure and temperature of steam drop rapidly, for example, as steam rushes through a turbine, drops of water vapour may not appear instantaneously when the saturation line is reached. Instead, the steam may be supercooled to a metastable state beyond the saturation line which represents true equilibrium conditions. Whether or not a vapour becomes supercooled depends on the presence of impurities to provide sites at which drops can form, and also on the speed of the process, since the growth of drops is not instantaneous.

Example 1.4 One way of convincing oneself that a tabular equation of state is equivalent to an explicit equation linking the properties is to demonstrate that the same operations can be performed using the two kinds of relationship, and that the results are nearly the same. Here the steam tables will be used to determine the derivatives of equation (1.2.4) for steam at 500 °C and 5 bar. These results will be compared with the values suggested by equations (1.2.5), which are strictly applicable to a perfect gas.

Since temperature and pressure are constant along the rows and columns of the steam tables, the derivatives can be estimated by taking small steps along them. Using data for 500 °C and pressures of 5 and 6 bar, one gets

$$\left(\frac{\Delta v}{\Delta p}\right)_T = \frac{0{\cdot}5919-0{\cdot}7108}{100} = -1{\cdot}19\times10^{-3}\ \mathrm{m^5/kN\ kg}$$

Using data for 5 bar and temperatures of 450 °C and 500 °C, one gets

$$\left(\frac{\Delta v}{\Delta T}\right)_p = \frac{0{\cdot}7108-0{\cdot}6641}{50} = 9{\cdot}35\times10^{-4}\ \mathrm{m^3/kg\ ^\circ K}$$

The steps taken here are not in fact small, and the derivatives will not be very accurately determined. The accuracy could be improved by considering steps in the opposite direction (to 4 bar and 550 °C), the two estimates being averaged.

If a perfect gas had the properties 500 °C, 5 bar, and 0·711 m³/kg, the derivatives of equation (1.2.4) would be, according to equations (1.2.5)

$$\left(\frac{\partial v}{\partial p}\right)_T = -\frac{v}{p} = -\frac{0\cdot711}{500} = -1\cdot42\times10^{-3}\ \text{m}^5/\text{kN kg}$$

$$\left(\frac{\partial v}{\partial T}\right)_p = \frac{v}{T} = \frac{0\cdot711}{773} = 9\cdot2\times10^{-4}\ \text{m}^3/\text{kg °K}$$

Note that the absolute temperature has been used in this calculation, since the results were derived from the ideal-gas law. In steam-table calculations there is little advantage in using absolute scales, and temperatures are given in the tables using the scales on which they are commonly measured.

The values found above using perfect-gas results are fairly close to the estimates made for steam. The differences might arise either from errors of approximation or from the departure of steam from the perfect-gas law. In fact, the former cause is the more important; steam does behave rather like a perfect gas under these conditions. This may be seen by comparing the ratio pv/T obtained using steam-table values, with $\mathbf{R}/MW$ for water vapour. With $MW = 18$ for water

$$R = \frac{\mathbf{R}}{MW} = \frac{8\cdot31}{18} = 0\cdot461\ \text{kJ/kg °K}$$

while steam-table values give

$$\frac{pv}{T} = 500\times\frac{0\cdot711}{773} = 0\cdot460\ \text{kJ/kg °K}$$

Example 1.5 One of the most important applications of the thermodynamic properties of steam is in the analysis of the cycle undergone by steam supplying energy to a turbine. This cycle is the indirect source of much of the electricity used

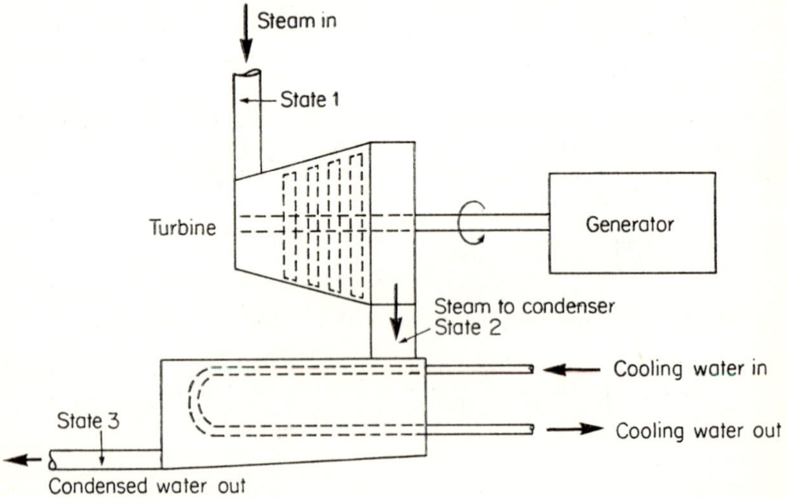

Figure 1.8 Turbine and condenser of an electricity-generating plant

in the world. Many of the examples of steam flow to be discussed in this book will relate to this practically important process.

Consider the part of the cycle which takes place in the turbine and condenser shown in Figure 1.8. Hot steam at high pressure ($T_1 = 1000$ °F, $p_1 = 500$ lbf/in^2) enters the turbine. Two outlet conditions will be considered:

(a) $p_2 = 10$ lbf/in^2, $T_2 = 200$ °F, and
(b) $p_2 = 10$ lbf/in^2, but with 4 per cent of the fluid condensed to liquid water.

In the condenser the steam is wholly converted to water, the pressure remaining constant at p_2. It is required to find the temperature variation through this part of the cycle, and also to study the variations in volume of the steam.

The best way of visualizing the processes defined above is to make use of the Mollier chart. Evidently the first outlet condition (a) is one of dry steam.

(a) Dry Steam at State 2

The only temperature remaining to be found is T_3, which is the saturation temperature for $p = 10$ lbf/in^2. The steam tables give $T_3 = 193$ °F.

Now consider the volume changes. The expansion ratio across the turbine follows from steam-table values:

$$\frac{v_2}{v_1} = \frac{38{\cdot}85}{1{\cdot}700} = 22{\cdot}8$$

After complete condensation, the specific volume is $v_3 = 0{\cdot}01659$ ft^3/lb. The ratio by which the volume of the steam has contracted is

$$\frac{v_2}{v_3} = \frac{38{\cdot}85}{0{\cdot}01659} = 2340$$

and the overall contraction ratio is

$$\frac{v_1}{v_3} = 102{\cdot}6$$

(b) Wet Steam at State 2

The turbine outlet temperature is now the saturation temperature for $p = 10$ lbf/in^2. Then

$$T_2 = T_3 = 193 \text{ °F}$$

Here one must use the relationships (1.2.6) to find the average specific volume of the outlet mixture:

$$v_2 = 0{\cdot}04 \times 0{\cdot}01659 + 0{\cdot}96 \times 38{\cdot}42 = 36{\cdot}9 \text{ ft}^3\text{/lb}$$

Then the expansion ratio across the turbine is

$$\frac{v_2}{v_1} = \frac{36{\cdot}9}{1{\cdot}700} = 21{\cdot}7$$

The contraction in the condenser is given by

$$\frac{v_2}{v_3} = \frac{36{\cdot}9}{0{\cdot}01659} = 2220$$

The overall contraction ratio is the same as that found in (a).

The large changes in fluid density calculated above have important implications for the design of the flow passages in the circuit. In order that the velocities be kept to a reasonable level, the ducts carrying steam must be much larger than those conveying water. Moreover, as there is a large volume change across the turbine, it is necessary to increase the size of the flow passages through the turbine. This is indicated schematically in Figure 1.8, which shows the low-pressure end of the turbine with a greater diameter than the high-pressure end.

1.2.3 Equations of Limited Validity

Although one cannot obtain an equation describing the whole of the surface of state, it is possible to obtain reasonably simple equations describing parts of the surface. The perfect-gas law is one such equation. Comparison of Figures 1.6(a) and (b) shows that the surfaces drawn have the same shape well away from the saturation line. In the high-temperature region of Figure 1.6(a) isobars are very nearly straight lines, and isotherms such as de are nearly hyperbolic. These statements must be qualified, for gases dissociate at very high temperatures. For example, around 3000 °K oxygen will not consist solely of O_2 molecules, but will also contain O atoms. At even higher temperatures the gas will ionize. Once dissociation begins, the gas will cease to follow the pattern for a gas of constant molecular weight.

Since the perfect-gas law provides a useful representation of one part of the surface of state, it is natural to seek representations of other parts. For our purposes equations governing vapours and liquids are particularly useful. At this point it should be noted that there are no absolutely clear-cut distinctions between the various phases of matter. It is immediately clear from the surface of state that there is no real distinction between vapour and gas. Sometimes they are distinguished using the critical temperature, that at point c of Figure 1.6(a): vapour below this temperature, gas above. Nor is there an absolutely clear distinction between a vapour and a liquid. Under normal atmospheric conditions the formation of a surface characterizes the liquid phase, and one may distinguish between the two kinds of fluid, vapour and liquid, by describing their behaviour on being introduced into a closed container. A vapour or gas will expand to fill the container, while a liquid will form a pool at the bottom or perhaps remain partially disseminated in drops. Hence the liquid phase is generally much denser than the gaseous or vapour phase. The transition from vapour to liquid usually involves the creation of a surface; this process takes place near the saturation line but, as was explained earlier, not always exactly on it. Nevertheless, it is possible to imagine a compression process which converts a gas to a liquid without the formation of a surface. The process starts from a state in region A. As the pressure is increased, the temperature must also increase, so that the process path does not touch the saturation line. Once the critical point is passed, the temperature may be reduced, while the pressure increases still further to achieve a high-density state in region C. There has thus been a continuous transition from gas to liquid. It appears that equations devised to describe a single phase will fail not only at the saturation line, where a different phase will appear, but also near the critical point, where the phases are indistinguishable.

Now to return to the problem of representing the behaviour of vapours. The perfect-gas law provides a useful starting point, but an equation governing a vapour or imperfect gas will necessarily be more complicated. One of the best-known generalizations of the gas law is *van der Waals' equation:*

$$\left(p+\frac{a}{v^2}\right)(v-b) = RT \tag{1.2.7}$$

This was derived by considering a simple model of the molecules of a gas, the correction terms, a/v^2 and b, being introduced to account for cohesive forces between molecules and for the finite volume occupied by the molecules themselves. The form of the equation suggests that it is the fluid density which is the most important factor in determining departures from the perfect-gas model.

The constants a and b in van der Waals' equation are chosen to fit measured characteristics of particular gases, even though the structure of the equation was suggested on theoretical grounds. The provision of two addition parameters allows an improved description of the surface of state. However, since the equation is based on a single-phase model, it cannot predict condensation. Figure 1.9 is a sketch of the isotherms given by van der Waals' equation. The upper isotherms can be roughly identified with those shown in Figure 1.6(a), the dotted lines de and fcg. The lower isotherms display a characteristic never to be found on a real surface of state: $(\partial p/\partial v)_T > 0$. This corresponds to an unstable material, one whose volume decreases when the pressure is decreased.

In seeking a description of the liquid phase we must start afresh. The basis of the equation is the observation that the specific volume varies only slightly with pressure and temperature. Equation (1.2.4) relates volume changes to changes in pressure and temperature. It can be rewritten as

$$\frac{dv}{v} = -\varkappa \, dp+\beta \, dT \tag{1.2.8}$$

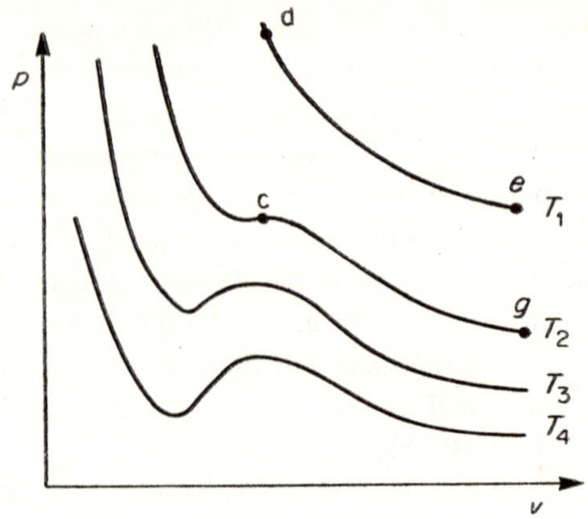

Figure 1.9 Isotherms defined by van der Waals equation

with

$$\varkappa = -\frac{1}{v}\left(\frac{\partial v}{\partial p}\right)_T \quad \text{and} \quad \beta = \frac{1}{v}\left(\frac{\partial v}{\partial T}\right)_p \tag{1.2.8}$$

The property $\varkappa$ specifies the proportional decrease in volume when pressure increases. It is called the coefficient of compressibility, its reciprocal being the bulk modulus of elasticity. The other property β gives the proportional increase in volume when the temperature increases. It is called the coefficient of cubical expansion. Another pair of names for these quantities, which make their relationship clearer, is *isothermal and isobaric compressibilities.*

The compressibilities of many substances are available in graphs and tables covering wide ranges of temperature and pressure. In essence, these provide forms of the equation of state which are suited to the solution of a particular class of problems. They have not, however, the general utility of comprehensive tables like the steam tables, which by relating a number of properties make possible a great variety of calculations. Indeed, Example 1.4 demonstrates how it is possible to obtain the compressibilities themselves from the steam tables.

For many liquids and solids the two compressibilities, $\varkappa$ and β, have the happy characteristic of varying only moderately over considerable ranges of temperature and pressure. In this they are unlike gases, for which equations (1.2.5) give $\varkappa = 1/p$ and $\beta = 1/T$. Of even greater importance is the typically small magnitude of $\varkappa$ and β for liquids and solids. The data for water given in Appendix III indicate that, at normal temperature and pressure, the two compressibilities have values of the order

$$\varkappa \simeq 3\times10^{-6}\ \text{in}^2/\text{lbf} \quad \text{or} \quad 5\times10^{-5}/\text{bar}$$

$$\beta \simeq 2\times10^{-4}/^\circ\text{C} \quad \text{or} \quad 10^{-4}/^\circ\text{F}$$

Hence a proportional change in volume of 1/1000 corresponds to a temperature change in the order of 10 °F or 5 °C, and to a pressure change in the order of 300 lbf/in² or 20 bar. This is consistent with everyday experience, namely, that in a wide variety of situations the densities of liquids and solids are nearly constant. Of course, if a phase change occurs, considerable changes in density must be expected.

Equation (1.2.8) can be integrated formally to give an algebraic equation of state

$$\frac{v}{v_1} = \exp\left[\int_p^{p_1} \varkappa \, dp + \int_{T_1}^{T} \beta \, dT\right]$$

with p_1, v_1, T_1 the properties at some reference state. In principle, one could introduce tabulated or calculated values of $\varkappa$ and β, and so determine the changes in the properties of any substance from this equation. In practice, this technique is not very useful unless $\varkappa$ and β are essentially constant. Then an explicit integration is possible

$$\frac{v}{v_1} = \exp\left[\varkappa(p_1 - p) + \beta(T - T_1)\right] \tag{1.2.9}$$

In many applications either the pressure change or the temperature change is insignificant, and the equation can be simplified even more.

Example 1.6 The van der Waals equation provides a means of investigating the validity of the perfect-gas law and also a means of studying its own validity. For most engineering purposes one might feel that the behaviour of a fluid was adequately represented by the perfect-gas law if a/pv^2, $b/v < 0{\cdot}001$, and that the van der Waals equation itself gave useful results if a/pv^2, $b/v < 0{\cdot}1$. It is easy to see how these criteria apply to steam, since tables of its properties are available. The van der Waals constants for steam are $a = 9{\cdot}5 \times 10^6$ (lbf/ft²) (ft³/sl)² and $b = 0{\cdot}87$ ft³/sl.

Table 1.2 lists the properties and the values of the correction terms for a number of points on the saturation line and for one superheated condition.

Table 1.2. Error terms in the van der Waals equation

State	p (lbf/in²)	T (°F)	v (ft³/lb)	a/pv^2 (unit-free)	b/v (unit-free)
Critical	3206	705	0·0503	7·8	0·54
Condensing vapour	1000	545	0·4456	0·32	0·061
Superheated vapour	1000	1000	0·8294	0·093	0·033
Condensing vapour	14·7	212	26·8	0·0060	0·0010
Condensing vapour	1	102	333·6	0·0006	0·00008
Evaporating liquid	1	102	0·01614	240,000	1·7

As expected, the van der Waals equation fails near the critical point. Moreover, although the isotherms shown in Figure 1.9 have something like the correct shape in the liquid region, the equation is in fact seen to be quite meaningless there. The van der Waals model appears to be reasonably satisfactory for steam with $p < 1000\ \text{lbf/in}^2$ or 70 bar, and the perfect-gas model for steam at sub-atmospheric pressures.

Example 1.7 Volume changes in solids and liquids as a result of temperature changes are sometimes important. Consider the tail pipe of a jet engine, a cylinder of metal 3 m in length, within which 'afterburning' sometimes takes place. This is an additional burning of fuel behind the turbine of the engine, inefficient, but sometimes worthwhile to provide extra thrust for brief periods. Suppose that the temperature of the tail pipe rises from the 500 °C maintained with normal engine operation to 1000 °C when afterburning takes place. By how much would the pipe lengthen, if the metal of which it is made has a coefficient of cubical expansion of about $50\times10^{-6}/°\text{K}$ over this range of temperature?

On the reasonable assumption that the pipe is free to expand diametrically, volumetric and linear expansions will be related by

$$\frac{\Delta V}{V} = \frac{3\Delta L}{L}$$

for the small expansions which will occur in a solid. Then the linear expansion is given by

$$\Delta L = \frac{\beta L\,\Delta T}{3} = 50\times10^{-6}\times300\times\frac{500}{3} = 2{\cdot}5\ \text{cm}$$

Example 1.8 In order to demonstrate the way in which a simplified equation of state, such as equation (1.2.9), can be applied, the work done in isothermal compression of a substance for which $\varkappa$ is constant will be calculated. It will be seen later that the results have a wider applicability than that suggested by this rigid specification.

Consider a homogeneous mass m contained within a cylinder by a piston, as shown in Figure 1.10(a). In the initial state the specific volume is v_1 and the pressure is p_1. The rear face of the piston is exposed to the pressure p_1 throughout the compression. Then the increment of work done when the volume of the material is decreased by $\mathrm{d}V$ is

$$-(p-p_1)\,\mathrm{d}V$$

and the work done per unit mass is given by

$$w = \frac{W}{m} = -\frac{1}{m}\int_1 (p-p_1)\,\mathrm{d}V = -\int_1 (p-p_1)\,\mathrm{d}v$$

To perform the integration we need to know $p = p(v)$. This is easy to find by virtue of the assumptions made concerning the process and the nature of the material. For isothermal compression, with $\varkappa$ constant, equation (1.2.9) gives

$$\frac{v}{v_1} = \exp\,[-\varkappa(p-p_1)]$$

Since

$$dv = -v_1\varkappa \exp[-\varkappa(p-p_1)]\,dp$$

the work integral can be written

$$w = v_1\varkappa \int_0^{p'} p' \exp(-\varkappa p')\,dp'$$

with $p' = p-p_1$ the rise in pressure.

The integral found above is of standard form. Tables of integrals or integration by parts give

$$w = -v_1\left[p'e^{-\varkappa p'} + \frac{(e^{-\varkappa p'}-1)}{\varkappa}\right]$$

This result is rather cumbersome, and it is natural to look for a simpler formula valid under restricted circumstances. If $\varkappa p' \ll 1$, a few terms of the expansion

$$e^{-\varkappa p'} = 1-\varkappa p'+\tfrac{1}{2}(\varkappa p')^2- \ldots$$

will provide an adequate approximation. Introducing the series expansion, one gets

$$w = -v_1\left[p'\left(1-\varkappa p'+\frac{1}{2}\varkappa^2p'^2\right) + \frac{(1-\varkappa p'+\frac{1}{2}\varkappa^2p'^2-1)}{\varkappa}\right]$$

$$= \frac{1}{2}\varkappa v_1 p'^2[1+\text{terms of higher order in } \varkappa p'].$$

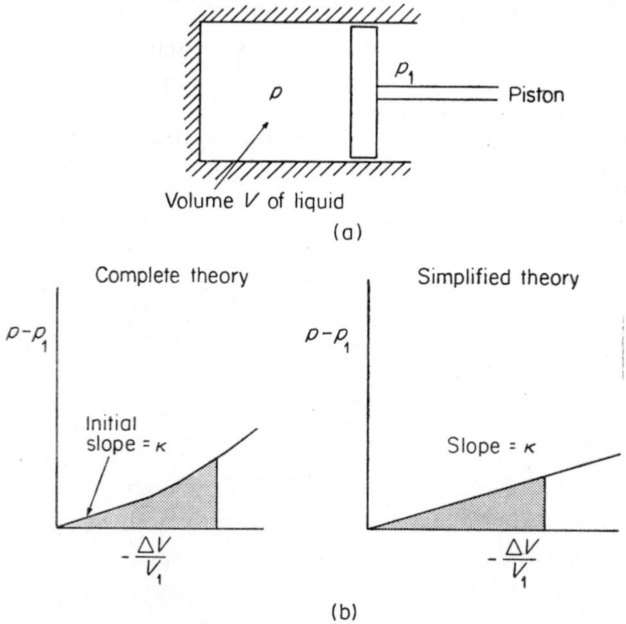

Figure 1.10 Compression of a liquid. (a) Compressing device; (b) Relationship between pressure and volume according to the complete and linearized formulae. The shaded areas represent work done in the compression

If only this simple result had been required, the approximation could have been introduced at an earlier stage. One might have taken $e^{-\varkappa p'} \simeq 1$ in the work integral to get

$$w \simeq v_1 \varkappa \int_0^{p'} p' \, dp'$$

or might even have approximated in equation (1.2.8) to get

$$dv \simeq -v_1 \varkappa \, dp$$

Figure 1.10(b) indicates the relationship between the full equation of state and this approximation to it.

It is reasonable to wonder whether the complete formula will be useful in any case, since pressure changes large enough to justify its use, say $p' > 1/10\varkappa$, will probably correspond to significant changes in $\varkappa$.

1.3 THE CONSERVATION OF ENERGY

1.3.1 Work and Energy

The concept of energy is important both in engineering practice and in the theoretical development of thermodynamics. Much of engineering is devoted to the conversion of energy from one form to another. For example, in a steam-driven electricity generating plant the chemical energy of the fuel is converted into thermal energy in the boiler, then into mechanical power in the turbine, and finally into electrical power in the dynamo. Some of these applications of the concept of energy will be discussed later. For the moment, however, attention will be focussed on the role of energy in the fundamental principles governing natural processes.

The idea that matter possesses energy which is interchangeable with mechanical work has provided some of the most useful ways of describing natural events. The idea was first clearly stated for the motion of solid bodies. This step required some centuries of thought. However, the concept of energy has since been extended to more general systems, by introducing new forms of energy to characterize matter undergoing a wider and wider class of processes. Of course, the definition of each new kind of energy must be consistent with those introduced earlier. In the following pages, we shall follow this historical pattern, first introducing work, kinetic energy and potential energy, and then measures of the energy of a thermodynamic system.

The relationship between work and mechanical energy is revealed by studying a solid body moving vertically, acted on by its own weight, $-mg$, and by an additional vertical force, F. Figure 1.11 shows this situation. The

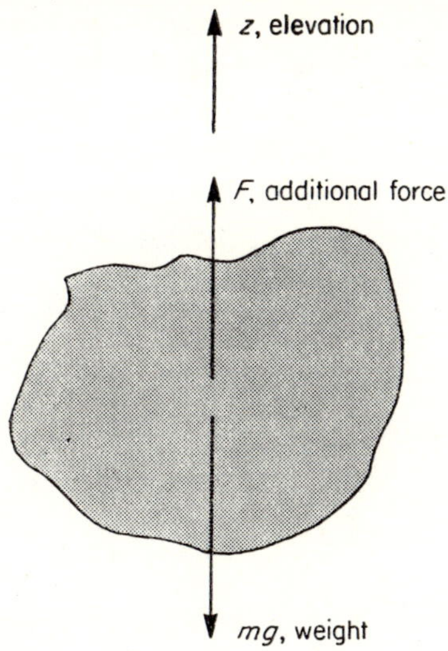

Figure 1.11 A rigid body in vertical motion. Forces, displacement, velocity and acceleration are all measured as positive when upwards

source and direction of the force F are not relevant to this discussion. However, for definiteness, it may be thought of as drag applied by the air through which the body moves.

The equation governing the motion is obtained by applying *Newton's second law*

$$m\frac{d^2z}{dt^2} = -mg+F$$

The mechanical energy equation for this system will be derived directly from this equation of motion. Hence the mechanical energy principle is merely a restatement of Newton's law of motion, and is not an independent principle. The work corresponding to each of the terms can be found by multiplying through by dz, and then integrating each term. This gives first

$$\frac{dz}{dt}\,d\left(\frac{dz}{dt}\right) = \frac{1}{2}m\,d\left(\frac{dz}{dt}\right)^2 = -mg\,dz+F\,dz$$

and then

$$\frac{1}{2}m\left(\frac{dz}{dt}\right)^2+mgz = \int F\,dz+\text{constant} \tag{1.3.1}$$

In the preceding equation the sum of the terms on the left is found to change only as a result of the work done by the force F. The first term is a property of the body, depending only on its mass and velocity. This is called the *kinetic energy* of the body; it has the dimensions of work, although it no longer appears as the product of force and distance. The second term on the left is simply the negative of the work done by the weight of the body during the vertical displacement. It is conventional to treat this quantity as a property of the body; the name *potential energy* is given to it since this amount of energy is potentially convertible to kinetic energy if the body were to fall through distance z.

Using these two kinds of energy, one can write equation (1.3.1) as

$$KE+PE = W+\text{constant}$$

where W is the work done by the additional force. Alternatively,

$$\Delta(KE+PE) = W \tag{1.3.2}$$

states that the change in the total mechanical energy of the body is equal to the work done on it. Work produces changes in energy, rather as force produces changes in momentum; this follows from the origin of the energy relationship in Newton's law. If no force but gravity acts on the body

$$KE+PE = \text{constant} \tag{1.3.3}$$

This states that the total mechanical energy remains constant during the motion. The work which is in fact done against gravity has been accounted for by introducing the potential energy. The simplicity of the conservation law (1.3.3) is the justification for introducing this additional form of energy.

1.3.2 Work and Heat

It has been seen that the energy of a moving solid body changes only as a result of work done on the body. In discussing the energy of a thermodynamic system it is also necessary to take account of the work done on the system. However, in the mechanical example discussed above all the work could be expressed as a product of a force and the distance through which that force acted. For a thermodynamic system other, essentially equivalent, ways of calculating work must be used; for example, the product of the torque on a shaft and the angle through which the shaft turns, and the product of the potential difference and the electric charge moved. One form of work which is of particular importance for a compressible system is the work of *bulk compression*, that done in slowly reducing the volume of the system against a uniform internal pressure. The work of bulk compression is measured as the product of the pressure and the change in volume. Example 1.8 illustrates the calculation of this kind of work.

It is matter of everyday experience that there is another mode of energy transfer—heat. The energy of a thermodynamic system depends on both work and heat flow. But why does the preceding analysis, based on Newton's law, not suggest the possibility of heat flow? The answer to this question can be obtained by considering events on the microscopic level. This is one of the points at which macroscopic thermodynamics is unconvincing unless supported by reference to the underlying reality of microscopic activity.

Consider a balloon filled with air, in equilibrium with the surrounding air, which has the same temperature but a lower pressure. Suppose now that the pressure outside the balloon is raised slightly. In moving to a new equilibrium condition the balloon will be slightly reduced in size. The macroscopic work done on it by the surrounding air can easily be calculated; it is simply the product of the small volume change and the pressure differential across the balloon wall.* In reality, the work done in compressing the balloon is the sum of the work done by many molecules striking the surface of the balloon. Nevertheless, this work can easily be calculated in terms of macroscopic properties. Next suppose that the temperature of the air outside the balloon is raised slightly. Then the molecules striking the outside surface will be more energetic than those inside, and a net transfer of energy will occur. This flow of energy can take place even if the balloon does not expand or contract. It cannot be represented in terms of macroscopic work.

This example leads to the conclusion that certain microscopic energy transfers have no counterpart as macroscopic work. These energy flows involve molecular collisions and vibrations, and electromagnetic radiation. Nevertheless, macroscopic thermodynamics must account for these energy flows, and a second mode of macroscopic energy transfer—heat—is introduced to do this. When the term work is used henceforth, it will be understood to mean macroscopic work, that is, work which can be calculated in terms of macroscopic forces and displacements. Heat is the residual energy transfer which cannot be calculated in this way.

We are not yet in a position to define heat; we have decided what it is not, but not what it is. This is sufficient for the discussions which follow immediately, for they relate to processes in which the heat addition is assumed to be zero. Such a process is termed adiabatic, and the system in which it takes place may be said to be insulated. In practice, it is possible to render many processes effectively adiabatic by surrounding the system with insulating materials.

The arbitrary nature of the distinction between heat and work may be demonstrated by considering the system illustrated in Figure 1.12. A volume

* The strain energy stored by elastic forces in the balloon fabric is taken to be negligible.

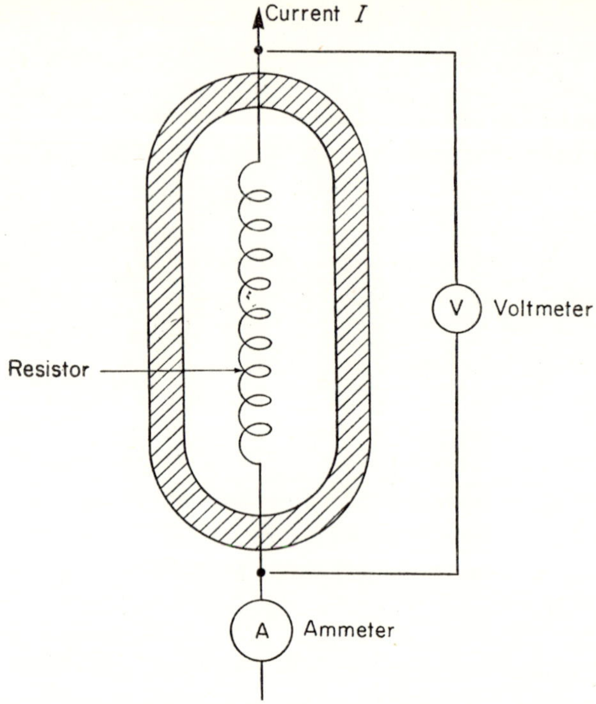

Figure 1.12 Addition of energy to a system by an electrical heating element. Work or heat?

of gas is contained within a rigid, insulated container. The only way of supplying energy to the gas is by passing an electric current through a thin wire within the container. This situation will be looked at in two ways, first by considering the system made up of the gas and the heater wire, and then by considering only the gas. For the first system, the power input can be measured using external meters, which give readings which can be related to macroscopic forces. Thus the energy addition may be thought of as macroscopic work done at the rate VI, where V and I are the potential difference and electric current. For the second system, the gas bounded by the container walls and surrounding the heater wire, the energy input is very nearly that found above. But it now appears as heat transferred from the wire to the gas by complex processes of conduction and convection.

1.3.3 Internal Energy

It is only necessary to rub your hands together to realize that the properties of a system change when work is done on it. Our next task is to relate this change in state to the work done. Since heat has so far been defined only in a negative sense, it is necessary to confine the discussion to a system insulated against heat flow. However, the macroscopic work can take a variety of forms. Figure 1.13 illustrates three of them; the material in the cylinder can be compressed by the piston, stirred with the rotor, or can absorb energy from an electric current. In each case the work addition can be measured in macroscopic terms.

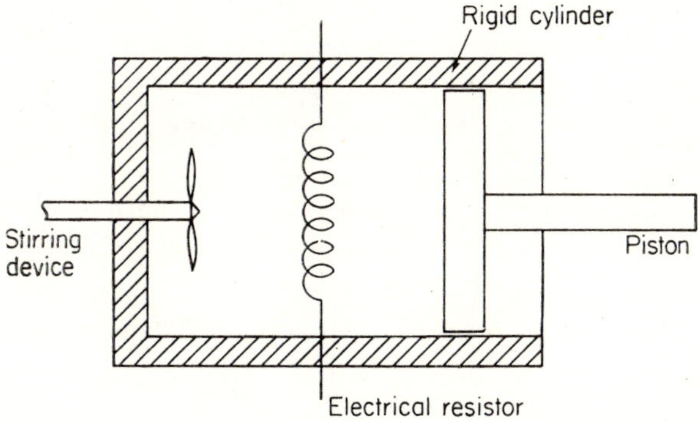

Figure 1.13 A system with three means of energy addition

Equation (1.3.2) indicates that work done on a mechanical system acts to increase the energy of the system. This idea will be carried over into thermodynamics by introducing a new thermodynamic property, the internal energy. The change in internal energy in any adiabatic process is set equal to the macroscopic work done on the system

$$\Delta E = W \tag{1.3.4}$$

This statement implies that, overall internal energy is conserved whenever work is done. Since the work done *on* any system corresponds to an equal amount of work done *by* the surroundings

$$\Delta E_s = -W$$

for the surrounding system. Then

$$\Delta E + \Delta E_s = 0$$

The nature of the internal energy for particular systems can be explained with some precision. For a monatomic gas the internal energy is the kinetic energy associated with molecular translational velocities. For more complex gas molecules, the energy associated with molecular rotation and vibration is included. For a solid, the internal energy is associated with binding forces between atoms. For a complex mechanical system, it might be the energy stored in stretched springs or bent beams. However, one does not need to identify the internal energy-storing mechanisms to calculate the net change in internal energy. Equation (1.3.4) gives this change in terms of a measurable quantity—the macroscopic work done on the insulated system.

In some cases one does know enough about the internal structure of a system to be able to calculate its internal energy. It is then possible to work backwards through the definition (1.3.4) to determine the work which would be done in some specified process. There is no fundamental contradiction to the statements made in the previous paragraph. In developing a subject one must provide a systematic set of definitions and hypotheses. However, once the structure of the subject is complete, it is possible to work backwards and forwards through these relationships, making use of whatever information is available.

The assumption that the internal energy E is a thermodynamic property has important implications. According to the state postulate for a simple system, any property is determined when two other independent properties are specified. Then there must exist equations of state with forms such as

$$E = E(p, T), \quad E = E(V, T) \quad \text{and} \quad p = p(E, T) \qquad (1.3.5)$$

These relationships can be determined empirically by systematically measuring the pressure, temperature and volume changes associated with a measured amount of work. Thus for any substance it is possible, in principle, to construct a surface of state, similar to Figure 1.6(a), but relating the internal energy to the other properties. Tables similar to the steam tables can also be constructed. Since only changes in internal energy are specified by equation (1.3.4), it is necessary to introduce an arbitrary datum level from which E is measured. For example, the values of internal energy tabulated in the steam tables are measured relative to the value at the triple point. This is the unique state at which vapour, solid and liquid can exist in equilibrium together; for water, the triple point is specified by 0·01 °C and 0·006112 bar.

If E is a property, the change in internal energy of a system between two specified states must be the same, no matter what the process of transition. But, according to equation (1.3.4), this change is, in any adiabatic process, equal to the work done on the system. Hence it is possible to check experimentally whether the internal energy is in fact a thermodynamic property. Consider, for example, the system shown in Figure 1.13. This could move between two states in a variety of ways: bulk compression by the piston, followed by stirring; or energy addition by the electric element, followed by bulk compression. These two processes are shown in Figure 1.14, but many other process paths could be drawn in. Only if the total work done is the same in every case will E be a thermodynamic property. Experiments with various adiabatic work inputs have not in fact been used to check the assumptions of thermodynamics, but evidence from other careful experiments

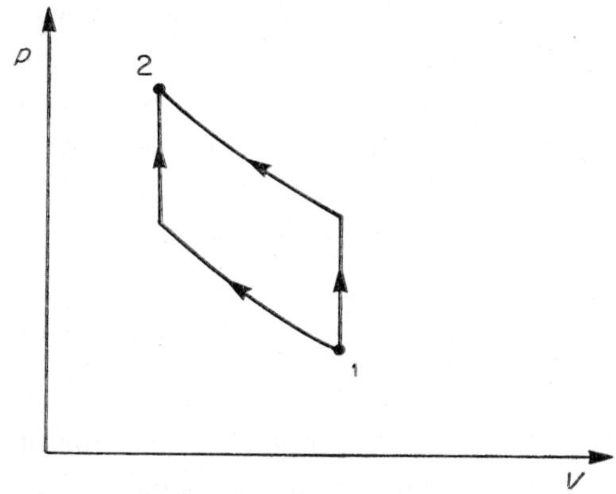

Figure 1.14 Alternative paths between the same pair of terminal states

indicates that such measurements would be consistent with the state postulate if they were carried out.

Example 1.9 To illustrate methods of calculating work and internal energy change, consider a particular set of energy inputs to the system sketched in Figure 1.13. Let us find the increase in internal energy of 10 lb of steam at $T_1 = 500$ °F and $p_1 = 100$ lbf/in^2, when energy is added through all of the mechanisms shown in the sketch—volume change, stirring, and the flow of an electric current. The volume is slowly reduced by 1/10 ft^3, the stirring device is driven for 30 min at 2000 rev/min against a torque of $L = 100$ gf cm, and an electric current of $I = 1$ A flows for 1/20 sec against a potential difference of 30,000 V.

It will be shown later (in Example 1.10) that these energy inputs produce only very small changes in the properties of the system. In particular, the pressure change is small, and the work of bulk compression can be calculated as

$$W_c = -p_1 \Delta V = 100 \times 144 \times 0{\cdot}1 = 1440 \text{ ft lbf} = 1950 \text{ J}$$

The stirring work can be calculated as

$$\begin{aligned} W_s = L\omega \, \Delta t_s &= 100 \times 2000 \times (2\pi/60) \times 1800 \text{ gf cm/sec} \\ &= 120\pi \text{ kgf m/sec} \\ &= 3700 \text{ J} = 2730 \text{ ft lbf} \end{aligned}$$

The electrical work can be calculated as

$$W_e = VI \, \Delta t_e = 30{,}000 \times 1{\cdot}00 \times 0{\cdot}05 = 1500 \text{ J} = 1100 \text{ ft lbf}$$

Then

$$\Delta E = W_c + W_s + W_e = 5270 \text{ ft lbf} = 7150 \text{ J}$$

1.3.4 The First Law

We are now able to discuss heat in a more rigorous way. Consider a system whose boundaries do allow heat to flow, for example, that shown in Figure

1.13, with the insulation removed. During a process in which work W is done on the system, heat Q flows into it. It has been seen earlier that, although heat cannot be expressed in terms of macroscopic forces and displacements, it is nevertheless a mechanism of energy transfer equivalent to macroscopic work. Hence the internal energy change during the process is

$$\Delta E = W + Q \tag{1.3.6}$$

This is the first law of thermodynamics; it applies to a system of fixed contents. Since the internal energy change for the surroundings is given by

$$\Delta E_s = -W - Q$$

it follows, as before, that there is overall a conservation of energy.

Heat has not yet been defined from the point of view of macroscopic thermodynamics. The first law provides a definition, since heat is the only quantity appearing in it for which a method of measurement has not been proposed. For a particular process, W can be measured, while ΔE follows from the equations of state (1.3.5), evaluated at the end points of the process. Hence, in macroscopic terms, heat is simply the difference between the net change in internal energy, and that part of the change produced by macroscopic work. In practice, it is possible to measure and predict heat flows without making direct use of the first law. Once the relationship between work, heat and internal energy is established, it can be used in a variety of ways.

It should be noted that heat and work are not properties of a system. They do not describe a state, but are characteristics of a process by which a system changes from one state to another. Indeed, as has been shown, it is not possible to make a rigorous distinction between work and heat. It is their sum—the total input of energy—which is directly linked to the change in internal energy.

Since heat and work are interchangeable, with respect to their effect on the internal energy, it is possible to determine equations of state involving E by experiments in which heat is added. This is the method used in practice, since large changes in internal energy are difficult to induce in some substances using only macroscopic work. It is common practice to measure heat flows by observing the changes in the temperature of water; these reflect changes in the internal energy of the water, and also changes in the internal energy of a system with which the water has been in contact. Hence it is natural to measure heat and internal energy using units related to the thermal properties of water. Two of these are the kilogram–calorie (approximately the energy required to increase the temperature of a kilogram of liquid water by 1 °C), and the British thermal unit (approximately the energy required to raise the temperature of one pound of liquid water by 1 °F).

The units of internal energy, specified by equation (1.3.4), are those of mechanical work; according to equation (1.3.6), heat has the same units. Thus no additional energy or heat unit is necessary. The thermal units, such as kcal and Btu, are superfluous, being linked rigidly to the mechanical energy units by

$$\text{thermal energy unit} = \text{J} \times \text{mechanical energy unit}$$

with J a conversion factor, often called the *mechanical equivalent of heat*. Although the steam tables in English units give energy in Btu, the thermal unit, the SI tables use the mechanical unit kJ, rather than kcal. Thus calculations using the SI tables often prove to be simpler.

Some writers make the relationship between mechanical and thermal energy units explicit in their equations. For example, they write the first law as

$$\Delta E = \frac{W}{\text{J}} + Q$$

This equation is consistent with measurement of W in mechanical units (say, ft lbf), and ΔE and Q in thermal units (say, Btu); for these units, J = = 778 ft lbf/Btu. This practice of explicitly introducing conversion factors has not been followed in this book. Hence each of ΔE, W, and Q must be converted to a consistent set of units before or on being used in an equation.

For a homogeneous system, it is useful to write the first law in the form

$$\Delta e = w + q$$

with $e = E/m$, the *specific internal energy*, and $w = W/m$, $q = Q/m$ the work done on and heat flow to unit mass of the system. All of these quantities are measured in the same units of energy per unit mass.

Specific internal energy is, like specific volume which was introduced in equations (1.2.2), independent of the mass of the homogeneous system it describes. Properties with this characteristic and, more will be introduced later, are called *intensive properties*, the corresponding total properties E and V being termed *extensive properties*. The defining characteristic of an intensive property is its constancy when the system is divided into macroscopic parts. By this test, pressure and temperature are intensive properties of any simple system in an equilibrium state. If, for example, a partition is erected across the middle of a room, the volume and internal energy are halved, but the pressure and temperature remain the same. For this homogeneous system the volume per unit mass and the internal energy per unit mass remain the same too, and they are properly called intensive properties. For an inhomogeneous system the situation is not so simple. Consider a mixture of water and ice in equilibrium. The pressure and temperature are still intensive properties. However, volume per unit mass is not. One can in fact distinguish three different specific volumes, one for the solid phase, one for the liquid phase, and an average for the whole system (see equations (1.2.6)). The first two are intensive properties of their respective phases, but none of the three is an intensive property of the system.

Some of the units commonly used for energy and related quantities are given in Table 1.3. The mechanical units are consistent with the corresponding columns of Table 1.1. The Centigrade heat unit (Chu = 1·8 Btu), which has been much used in the gas-turbine field, is unique in combining English and metric units. It is (approximately) the energy required to raise the temperature of one pound of liquid water through 1 °C.

Table 1.3. Energy Units

Quantity	English System Mechanical	English System Thermal	Système International Mechanical	Système International Thermal
Work, heat or energy	ft lbf	Btu *or* Chu	J = Nm *or* kJ = kN m	kcal
Power (fundamental)	ft lbf/sec		W = J/sec	
Power (practical)	hp	Btu/hr	kW *or* MW	
Specific Energy	ft lbf/lb *or* ft lbf/sl	Btu/lb *or* Btu/sl	J/kg *or* kJ/kg	kcal/kg

Example 1.10 The determination of the properties of the steam at the end of the process discussed in Example 1.9 provides illustrations of the manipulations of energy units.

Using the conversion factors of Appendix I, one can convert the energy change to thermal units:

$$\Delta E = \frac{5270}{778} = 6{\cdot}78 \text{ Btu}$$

$$= \frac{7{\cdot}15}{4{\cdot}187} = 1{\cdot}71 \text{ kcal}$$

The steam tables do not in fact give values of internal energy, but of specific enthalpy, a property related to the specific internal energy by

$$h = e + pv$$

The significance of this property, and the reason why it is given in the tables rather than internal energy, will be explained later. For the moment, it is of interest solely because it provides a way of determining the internal energy.

For 500 °F and 100 lbf/in², the specific internal energy is

$$e_1 = h_1 - p_1 v_1$$
$$= 1279{\cdot}1 - 100 \times 144 \times \frac{5{\cdot}59}{778} = 1175{\cdot}9 \text{ Btu/lb}$$

Note that the product $p_1 v_1$ has been changed to the units which specific enthalpy h has in the tables. The final specific internal energy can now be found

$$e_2 = e_1 + \frac{\Delta E}{m} = 1175{\cdot}9 + \frac{6{\cdot}8}{10} = 1176{\cdot}6 \text{ Btu/lb}$$

since the system contains 10 lb of steam.

These changes in state are so small that tables and plots are of little help in finding changes in other properties, such as pressure. However, there is an approximate relationship amongst the properties of superheated steam which is very useful in situations of this kind:

$$pv = 1{\cdot}247(h - 835) \quad \text{for English steam-table units}$$

This result—known as Callendar's equation—may be expected to give an adequate prediction of a small change in state. It gives for small changes

$$\Delta(pv) = 1{\cdot}247\ \Delta h \quad \text{for English steam-table units}$$

In introducing the specific internal energy, whose change is known in this case, care must be taken with the units:

$$\Delta(pv) = 1{\cdot}247\left[\Delta e + \Delta(pv) \times \frac{144}{778}\right]$$
$$= 1{\cdot}247\ \Delta e + 0{\cdot}230\ \Delta(pv)$$

Then

$$\Delta(pv) = 1{\cdot}62\ \Delta e \quad \text{in English steam-table units}$$

Conversions of this kind are not so difficult when the SI units are used. However, the units given in the SI steam tables are not quite consistent; powers of 10 must be introduced to achieve consistency.

The last result can be used to find the change in pressure, by rewriting it as

$$p_2 - p_1 = \frac{p_1\ \Delta v + 1{\cdot}62\ \Delta e}{v_2}$$
$$= \frac{100/100 + 1{\cdot}62 \times 0{\cdot}678}{5{\cdot}58} = 0{\cdot}38 \text{ lbf/in}^2$$

The small size of this change justifies the assumption of constant pressure made in Example 1.9 when calculating the work done in compressing the steam.

1.3.5 General Principles of Energy Conservation

The forms of energy which were considered first—kinetic and potential energies of a mechanical system—have not yet been related to the discussion of energy in the context of thermodynamics. It has tacitly been assumed that the thermodynamic systems moved very little during a process and were stationary at the beginning and end of the process. These restrictions must be relaxed when flowing fluids are examined, since changes in kinetic and potential energies are often of great importance. Moreover, it often proves convenient to select system boundaries through which material flows; the energy principles which have been considered so far do not allow for this.

The two vital preliminaries to the generalization of the energy principle (1.3.6) are a reconsideration of the forms which energy can take, and a reconsideration of the ways in which the energy of a system can be altered. Energy is any quantity with the units of work which is considered to be a property of a system. Let us look in turn at the three forms of energy which have been distinguished: kinetic, internal and potential.

Kinetic energy is associated with the macroscopic or bulk velocity of the material of a system. In practice this velocity is found simply by using macroscopic measuring instruments. In principle it could be defined in terms of the average mass flux associated with the underlying molecular motion. If one were to average over very many molecular motions, the random or thermal contributions would cancel out; the residual motion is the bulk velocity. The method of calculating the kinetic energy was indicated by the derivation of equation (1.3.1). Integration of Newton's second law gives the kinetic energy as

$$KE = \tfrac{1}{2}mu^2$$

for a mass m moving with uniform velocity u. More generally, for a system occupying volume V, the kinetic energy can be obtained by integrating over the elements of mass making up the system:

$$KE = \tfrac{1}{2}\int_V \varrho u^2 \,\mathrm{d}V$$

Internal energy has been introduced in an unambiguous manner for a stationary system. However, for a system whose contents have a macroscopic velocity, it is necessary to distinguish carefully between the internal and kinetic energies. It has already been noted that, in some instances, the internal energy of a gas is simply the sum of the kinetic energies associated with the random motions of the constituent molecules. It is the word random which distinguishes this energy from that which was above called kinetic energy. The random motion is that which would be seen by an observer moving with the macroscopic velocity of the material. An example will help to clarify this distinction. The air in the cabin of an aeroplane standing on the runway has internal energy, but no macroscopic velocity or macroscopic kinetic energy.

However, when the aeroplane is moving down the runway, the air has a macroscopic velocity and an associated kinetic energy. The internal energy is unchanged.

Potential energy is associated with position in a macroscopic force field. It is the negative of the work done in moving through the field, as seen in Section 1.3.1 for the special case of gravitational potential energy. Hence, when the potential energy associated with a particular force field is included in the energy equation, one must not include the corresponding work term. To do so would be to account twice for the same force. For a mass m in a uniform gravitational field

$$PE = mgz$$

with z a coordinate measured upwards with respect to the field. For a system of varying density and specific weight

$$PE = \int_V \varrho gz \, \mathrm{d}V$$

Other force fields, for example, radially symmetric gravitational or electrostatic fields, can be dealt with in a broadly similar way. In this book, however, the only force field which will be considered is gravity.

The *total energy* of a system is the sum of the various forms of energy distinguished above:

$$TE = E + KE + PE$$

For a uniform gravitational field

$$te = e + \tfrac{1}{2}u^2 + gz \tag{1.3.7}$$

gives the total energy per unit mass. In particular situations the changes in one or more of these contributions may be negligible, and it proves convenient to calculate the total energy using only the significant contributions. For example, changes in the potential energy of gases are insignificant in many situations, and it is possible to drop the potential energy term from the total energy. Also, in low-velocity flows, the kinetic energy term may be negligible.

Now consider ways in which the energy of a system can be altered. Work and heat have already been introduced. However, when material flows into or out of a system a third mechanism of energy transfer will act—convection. The energy of the material will be carried in and out with it. The net amount of energy added to a system will be called the *convected energy*. If a homogeneous mass m' enters a system, the convected energy is

$$CE = m'te$$

with *te* the total energy per unit mass of the added material. For a uniform gravitational field

$$CE = m'(e+\tfrac{1}{2}u^2+gz) \tag{1.3.8}$$

Note that the sign of the convected-energy contribution depends upon the sign of m'; when material leaves the system, $m' < 0$, and the convected energy is negative. The net convected energy for a complicated process must be calculated by summing, with proper attention to inflows and outflows, the energy fluxes through every part of the surface enclosing the system. For the cases to be considered in this book, the result (1.3.8) will be sufficient.

Having listed the various forms of energy and the ways in which the energy of a system can be changed, we are now able to establish an overall balance between energy changes and energy transfers:

$$\Delta(TE) = \Delta(E+KE+PE) = W+Q+CE \tag{1.3.9}$$

This *overall energy equation* may be thought of as a generalization of the first law of thermodynamics (1.3.6) to include systems of varying contents and finite kinetic and potential energy. However, the overall energy balance may be looked at in another way. The study of a particular case in Section 1.3.1 suggests that any process will be governed by a *mechanical energy equation*, derived from Newton's law, of the form

$$\Delta(KE+PE) = W_{\mathrm{m}}+CE_{\mathrm{m}} \tag{1.3.10}$$

where W_{m} is the mechanical work, that giving rise to macroscopic accelerations and motion through a macroscopic force field, and CE_{m} is the mechanical energy convected into the system. The process is also governed by the *thermodynamic energy equation*, or first law of thermodynamics

$$\Delta E = W_{\mathrm{t}}+Q+CE_{\mathrm{t}} \tag{1.3.11}$$

where W_{t} is the thermodynamic work, that contributing to the change in internal energy of the system, and CE_t is the net convection of internal energy into the system. By adding corresponding terms in equations (1.3.10, 11), one obtains the overall energy balance (1.3.9).

It has been seen that the overall energy equation can be thought of as the summation of the mechanical and thermodynamic energy equations. Only two of the three energy statements (1.3.9, 10, 11) are independent; the other can be obtained by addition or subtraction. Since the mechanical energy equation can be obtained from Newton's second law, it follows that only one independent energy equation can be introduced into an analysis in which the momentum principle is used.

Consideration of a simple example will clarify the ideas of the last paragraphs. The mechanical energy equation for a body moving vertically has been obtained, equation (1.3.1). But as the body moves through the air it may be heated by the rays of the sun. No account is taken of this energy input in the mechanical energy equation. However, the thermodynamic energy equation relates this heat input to a rise in integral energy of the moving body. Thus the total energy of the body changes as a result of both work done on the body and heat flow. In this particular case, the work alters only the mechanical energy components, while the heat affects only the internal energy.

Later these ideas will be applied to a flowing fluid, and it will be found that the two energy equations—mechanical and thermal—do not separate out so easily. For a compressible fluid, the responses to work and heat addition are closely related. It is for this reason that the thermodynamics and fluid mechanics must be examined together.

Example 1.11 The distinction between the work terms introduced in equations (1.3.9, 10, 11) can most easily be seen by examining simple mechanical models. Consider the system shown in Figure 1.15. It consists of two masses, M suspended from a spring of rate k, and m sliding in a vertical slot cut into M. A constant friction force f acts between M and m, taken to be positive upwards on m. The system starts from rest with the centres of gravity of M and m at the same elevation z_0.

Newton's second law applied to the two masses gives

$$m\frac{d^2z}{dt^2} = -mg+f$$

and

$$M\frac{d^2Z}{dt^2} = -Mg+k(z_0-Z)-f$$

Integrated, and with the initial conditions

$$z = Z = z_0 \quad \text{and} \quad \frac{dz}{dt} = \frac{dZ}{dt} = 0$$

satisfied, these give the energy equations

$$\frac{1}{2}m\left(\frac{dz}{dt}\right)^2 = -mg(z-z_0)+f(z-z_0)$$

and

$$\frac{1}{2}M\left(\frac{dZ}{dt}\right)^2 = -Mg(Z-z_0)-\frac{1}{2}k(Z-z_0)^2-f(Z-z_0)$$

The mechanical energy equation for the two-mass system is obtained by summing:

$$\Delta(KE+PE) = W_m = -\tfrac{1}{2}k(Z-z_0)^2-f(Z-z)$$

From the overall point of view, the only work done on the two-mass system is that of the external spring force (the gravitational components are included in the

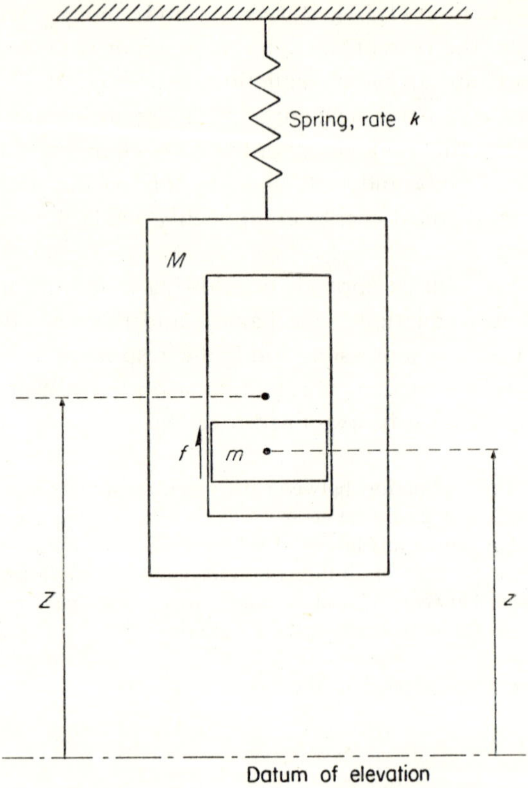

Figure 1.15 A mechanical system composed of two masses

PE and must not be put in again). Then the overall energy equation is

$$\Delta(KE+PE+E) = W$$
$$= -\tfrac{1}{2}k(Z-z_0)^2$$

Subtraction of the mechanical elements gives the thermal equation

$$\Delta E = W_t = f(Z-z)$$

Incidentally, it would be possible to consider the quantity $-W$ to be a form of internal energy. If this were done, the overall energy equation would become simply a conservation statement.

1.4 RESPONSE TO ENERGY ADDITION

1.4.1 Specific Heats

We now have available two basic tools of thermodynamics: a means of describing a substance, its equation of state; and a means of relating energy

Even though the specific heats were introduced by examining simple processes of heat addition, they relate the changes in temperature, internal energy and enthalpy which occur in other constant-volume and constant-pressure processes. Consider a constant-volume, adiabatic process, in which work is done by stirring, as indicated in Figure 1.13. For this process, $\mathrm{d}w = \mathrm{d}e$, and the temperature change for a small energy addition is $\mathrm{d}T = \mathrm{d}e/c_v = \mathrm{d}w/c_v$.

If the dependence of internal energy and enthalpy on pressure and density is small, so that they are essentially functions of temperature, the specific heats relate changes in temperature, internal energy and enthalpy in any process. This can be shown as follows. If $e = e(T)$, $h = h(T)$, the changes in internal energy and enthalpy in any small change of state are given by

$$\mathrm{d}h = \left(\frac{\partial h}{\partial T}\right)_p \mathrm{d}T + \left(\frac{\partial h}{\partial p}\right)_T \mathrm{d}p = c_p\, \mathrm{d}T$$

$$\mathrm{d}e = \left(\frac{\partial e}{\partial T}\right)_v \mathrm{d}T + \left(\frac{\partial e}{\partial v}\right)_T \mathrm{d}v = c_v\, \mathrm{d}T$$

so that

$$\frac{\mathrm{d}h}{\mathrm{d}e} = \frac{c_p}{c_v}$$

Since the pressure and density dependence of the energies of real substances are often found to be small, these results provide useful approximations in practice.

It is found that $c_p > c_v$ for almost all substances. In other words, a specified rise in temperature will produce a larger increase in enthalpy than in internal energy. The simple processes considered above show why this is so. Additional energy must be supplied during the constant-pressure process to provide the work done by the system in expanding against the force of the piston. This matter can also be investigated analytically. The internal energy may be taken to be a function of volume and temperature. Then

$$\mathrm{d}e = \left(\frac{\partial e}{\partial v}\right)_T \mathrm{d}v + \left(\frac{\partial e}{\partial T}\right)_v \mathrm{d}T$$

But changes in enthalpy and internal energy are related by

$$\mathrm{d}h = \mathrm{d}e + p\, \mathrm{d}v + v\, \mathrm{d}p$$

Combining these results, one has

$$\frac{\mathrm{d}h}{\mathrm{d}T} = \left[p + \left(\frac{\partial e}{\partial v}\right)_T\right] \frac{\mathrm{d}v}{\mathrm{d}T} + \left(\frac{\partial e}{\partial T}\right)_v + v \frac{\mathrm{d}p}{\mathrm{d}T}$$

When restricted to a constant-pressure process, this gives

$$c_p - c_v = \left[p + \left(\frac{\partial e}{\partial v}\right)_T\right]\left(\frac{\partial v}{\partial T}\right)_P \tag{1.4.4}$$

For many materials the internal energy does not vary much when the volume changes while the temperature remains constant: that is, $(\partial e/\partial v)_T \simeq 0$. Hence

$$c_p - c_v = p\left(\frac{\partial v}{\partial T}\right)_p \tag{1.4.4}$$

is often a useful approximation.

The last result indicates that $c_p > c_v$ for any material which expands when its temperature rises. It is only for a material of constant density that $c_p = c_v$; in this case only one specific heat is required to describe the response to energy addition. Since liquids and solids do not expand much when the temperature rises, it is often possible to assume that a liquid or solid has a single specific heat at any one state.

For gases and vapours it is never realistic to assume equality of the two specific heats. For a perfect gas this can be demonstrated using the second of equations (1.4.4). For $pv = RT$, the derivative $(\partial v/\partial T)_p = R/p$. Then

$$c_p - c_v \simeq R \tag{1.4.5}$$

It is usually assumed that a perfect gas obeys the additional restriction $(\partial e/\partial v)_T = 0$. For such a gas, the last relationship is exact.

Another means of characterizing the response of materials to energy addition is the unit-free *ratio of specific heats:*

$$\gamma = \frac{c_p}{c_v} \tag{1.4.6}$$

It is found that $\gamma > 1$ for very nearly all substances. For liquids and solids, $\gamma \simeq 1$, and for some purposes it may be assumed that $\gamma = 1$. However, values around $\gamma < 1{\cdot}67$ are found for monatomic gases; for other gases and vapours, $\gamma = 1{\cdot}67$, but is still significantly greater than unity.

For many substances tables are available relating the specific heats to the easily measured properties, pressure and temperature. Examples are to be found in Appendices II and III. These tables, like those of the compressibilities $\varkappa$ and β, are in effect equations of state in a form particularly suited to the solution of certain important classes of problems. Specific heats are particularly useful when they do not vary rapidly as the state changes and when, as is often the case, their variation is primarily dependent on temperature.

So far the effect of heat addition on temperature has been of prime interest. But what happens if the temperature does not change? This can happen when heat is added to a substance in a mixed-phase state, such as those represented by region B in Figure 1.6(a). Here heat addition at constant temperature requires that the pressure be maintained constant. The response of the substance will take the form of vaporization of a further portion of the liquid, and the mixture will expand. For an elemental process

$$dQ = dE + p\,dV = dH - V\,dp$$

Then for the whole constant-pressure heat addition

$$Q = \Delta H$$

The amount of liquid vaporized is determined by

$$\Delta H = m(h_g - h_f) = mh_{fg}$$

where h_f, h_g, and h_{fg} are values appropriate to saturated liquid and vapour at the pressure at which the process takes place. The enthalpy difference $h_{fg} = h_g - h_f$ is called the *latent heat of vaporization.*

The process described above is a good approximation to that occurring in a boiling kettle. The rate at which water is boiled away is given by

$$\dot{m} = \frac{\dot{Q}}{h_{fg}}$$

where $\dot{Q}$ is the rate of heat input. For atmospheric pressure, the steam tables give

$$\frac{\dot{Q}}{\dot{m}} = h_{fg} = 2260 \text{ kW/(kg/sec)} = 630 \text{ W/(kg/hr)}$$

Constant-volume heat addition to a mixed-phase system is more difficult to analyse, since temperature, pressure, internal energy and enthalpy all change, only the average specific volume remaining constant.

Example 1.12 This example illustrates the method of calculating internal energy changes in solids, and also provides a simple illustration of an energy balance for a system whose contents are continually changing. Consider the power requirements for a paint drier on the assembly line of an automobile factory. After being sprayed with paint the body shells pass through a tunnel lined with infrared heaters. The shells weigh 200 kg, and pass along the line at the rate of one per minute. Their temperature on entering the tunnel is 20 °C, and is to be 300 °C on exit. The specific heat of the shells may be taken to be 0·1 cal/g °C, a little below that of iron, and their temperature may be assumed to be uniform, since they are thin metal sheets exposed to uniform radiation.

Consider the energy balance for the region inside the tunnel. Into it flow electrical energy and cold automobile bodies; out of it flow hot automobile bodies. The total energy within the region can be taken to be constant in time, once the unit has warmed up. Then equation (1.3.9) reduces to

$$0 = W+Q+CE$$

where $W+Q$ is the energy input over some period of time Δt, and CE is the net convected energy in that time.

Since the power required by the heaters is to be found, the appropriate result from the energy equation is

$$P = \frac{W+Q}{\Delta t} = -\frac{CE}{\Delta t}$$

The only part of the shells' energy which changes significantly is their internal energy, which can be measured in terms of the change in their temperature. Thus when mass Δm passes through the system

$$CE = \Delta m(e_1-e_2) = c\,\Delta m(T_1-T_2)$$

where c is the specific heat of the shells. The power absorbed is then

$$P = c\frac{\Delta m}{\Delta t}(T_2-T_1)$$

$$= 0{\cdot}1\times\frac{200}{60}\times 280 \text{ kcal/sec} = 390 \text{ kW}$$

The power required will in fact be a good deal higher than this, since some energy will be convected away from the heaters, and some will be radiated through the ends of the tunnel.

Example 1.13 Two competitive designs have been proposed for a nuclear power station. In each, a fluid is circulated through the reactor and through a turbine. The fluid absorbs energy in the reactor and conveys it to the turbine. The portion of the plant which is of interest here is shown in Figure 1.17.

In each of the two designs the pressure in the reactor core is 20 bar, and the temperatures of the fluid entering and leaving the core are 250 °C and 600 °C. However, the fluid circulating through the reactor is carbon dioxide in one design and steam in the other. It is required to compare the energy-carrying capacity of the two fluids, in order to see how rapidly they will have to be circulated.

The pressure may be taken to be constant in the core. Then for unit mass of heat-transfer fluid

$$dq = dh - v\,dp = dh$$

the heat absorbed by the fluid being measured by the change in its enthalpy. Hence the comparison of the two fluids comes down to a comparison of the changes in their enthalpy across the core. The properties of steam are available in the steam tables. For carbon dioxide, the specific heat may be assumed to be independent of pressure; its temperature dependence is given by

$$c_p = 6{\cdot}21+(1{\cdot}040\times10^{-2})T-(3{\cdot}545\times10^{-6})T^2 \text{ cal/g mol °K}$$

with $MW = 44{\cdot}0$.

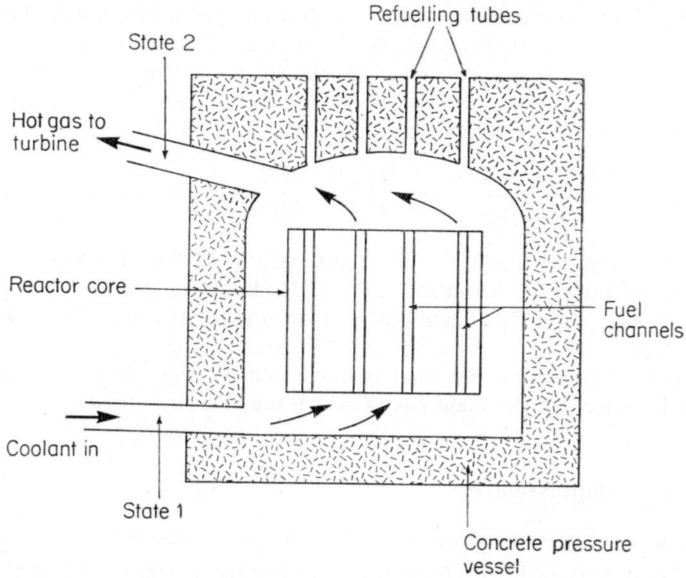

Figure 1.17 Section through a nuclear reactor showing schematically the flow path of the heat-transfer fluid

The tables for steam give immediately

$$\Delta h = 3690-2904 = 786 \text{ kJ/kg}$$

The average value of the specific heat is $\Delta h/\Delta T = 786/350 = 2{\cdot}24$ kJ/kg °K. Note that the values of h do not vary rapidly along the columns of constant temperature. Hence enthalpy is primarily a function of temperature; this is typical of vapours and gases.

For carbon dioxide, another approach is required:

$$\Delta h = \int_1^2 c_p \, dT = 6{\cdot}21\,\Delta T+(1{\cdot}040\times 10^{-2})\frac{T_1+T_2}{2}\,\Delta T$$

$$-(3{\cdot}545\times 10^{-6})\frac{T_1^2+T_1T_2+T_2^2}{3}\,\Delta T$$

With $T_1 = 523$°K, $T_2 = 873$°K, and $\Delta T = 350$°K

$$\Delta h = 11{\cdot}71\,\Delta T = 4100 \text{ kcal/kg mol}$$

$$= 4100\times\frac{4{\cdot}187}{44{\cdot}0} = 390 \text{ kJ/kg}$$

The average specific heat is $\Delta h/\Delta T = 390/350 = 1{\cdot}12$ kJ/kg °K. The much higher specific heat of the steam allows it to absorb a greater amount of energy (per unit mass) for a specified rise in temperature. Hence the mass flow of carbon dioxide must be greater.

The energy-carrying capacity of the two fluids may also be compared on a volumetric basis; this comparison is relevant to the velocities required in the circulating fluid. For the steam at outlet, the specific volume is 0·1995 m^3/kg, and the carrying capacity is 786/0·1995 = 3940 kJ/m^3. For the carbon dioxide, the specific volume at outlet may be estimated as

$$v = \frac{(\mathrm{R}/MW)T}{p} = \frac{8{\cdot}314}{44{\cdot}0} \times \frac{873}{2000} = 0{\cdot}0825 \text{ m}^3/\text{kg}$$

The energy-carrying capacity is then 390/0·0825 = 4730 kJ/m^3. On a volumetric basis the advantage of the steam vanishes. Indeed, a somewhat higher volume flow of steam is required to extract the same amount of energy from the reactor core.

The choice of heat-transfer fluid is influenced by other factors, in particular, its corrosive effect on the materials of which the reactor is made.

1.4.2 Bulk Compression

The specific heats were introduced primarily to describe the response of a substance to heat addition. Now the response to adiabatic bulk compression will be considered. Since the specific heats describe the response to any energy addition, they can be used to describe certain aspects of this kind of process. The change in specific internal energy is equal to the work done on unit mass:

$$\mathrm{d}e = -p \, \mathrm{d}v$$

Since $e = e(v, T)$, this change is in general given by

$$\mathrm{d}e = c_v \mathrm{d}T + \left(\frac{\partial e}{\partial v}\right)_T \mathrm{d}v$$

When the preceding results are combined, it is found that

$$\mathrm{d}e\left[1 + \frac{1}{p}\left(\frac{\partial e}{\partial v}\right)_T\right] = c_v \, \mathrm{d}T$$

For most substances, the second of the terms enclosed by brackets is negligible, so that $e = e(T)$ essentially. Hence

$$\mathrm{d}e = -p \, \mathrm{d}v = c_v \, \mathrm{d}T \tag{1.4.7}$$

is an acceptable approximation for most cases of adiabatic bulk compression. This result indicates that, in general, work done on a compressible substance will lead to an increase in its temperature.

The change in volume in any infinitesimal process is given by

$$\frac{\mathrm{d}v}{v} = -\varkappa \, \mathrm{d}p + \beta \, \mathrm{d}T \tag{1.2.8}$$

The net volume change is expressed as the sum of contributions from pressure change and from temperature change. Adiabatic bulk compression may certainly be expected to increase the pressure. Equations (1.4.7) indicate that the temperature will rise as well, since work produces an internal energy change, and the latter is closely linked to temperature. But how important is this temperature change compared to the 'direct' effect of pressure change? The answer can be found by comparing the total volume change with that associated with temperature change. The ratio of the two volume changes is

$$r = \left| \frac{\beta \, dT}{dv/v} \right|$$

When the relationships (1.4.7) are introduced

$$r = \frac{\beta}{c_v} pv \tag{1.4.8}$$

It is now necessary to consider particular cases.

For a perfect gas, $pv = RT$, and

$$\beta = \frac{1}{v} \left(\frac{\partial v}{\partial T} \right)_p = \frac{1}{T}$$

according to one of equations (1.2.5). Then

$$r = \frac{R}{c_v}$$

Equation (1.4.5) indicates that $R \simeq c_p - c_v = c_v(\gamma - 1)$. Then

$$r \simeq \gamma - 1 \quad \text{for a perfect gas.}$$

The ratio of specific heats of most gases is in the range 1·7 to 1·2 so that

$$r = 0{\cdot}7 \text{ to } 0{\cdot}2$$

This result indicates that it is quite unrealistic to neglect the temperature rise in a gas which is adiabatically compressed. This may be expected to be true for vapours as well.

For liquids and solids no simple formulae are available. However, a consideration of a particular case suggests the general nature of their response. An important special case is water at atmospheric pressure and temperature, for which $\beta \simeq 2 \times 10^{-4}/°\text{C}$, $c_v \simeq 1$ kcal/kg °C $\simeq 4000$ N m/kg °C, $\varrho \simeq 1000$ kg/m³ and $p \simeq 10^5$ N/m².

When introduced into equation (1.4.8), these values give

$$r \simeq \frac{1}{200{,}000}$$

The contribution of temperature change to the volume change during the adiabatic compression of a liquid is thus seen to be negligible. Hence the equation (1.2.8) can be simplified to

$$\frac{dv}{v} = -\varkappa \, dp \tag{1.4.10}$$

if it is required to predict the response of a liquid to adiabatic compression.

In Example 1.8, the work done in isothermally compressing a substance obeying equation (1.2.8) was calculated, on the assumption that $\varkappa$ = constant. These results are now seen to apply very nearly to adiabatic compression as well. An isothermal compression can be achieved by performing the operation slowly, so that thermal equilibrium with the environment can be maintained throughout. Adiabatic compression can be achieved in an insulated cylinder, and by performing the compression more rapidly. Actual compression processes will often be intermediate to these two, and hence are quite accurately described by equation (1.4.10).

It will often prove convenient to have a simple, yet fairly realistic representation of the behaviour of a liquid. For this purpose we shall consider a fluid whose near-adiabatic and near-isothermal changes of state are governed by equation (1.4.10), with $\varkappa$ = constant. This hypothetical substance will be called an *ideal liquid.*

Example 1.14 It has been stated several times that for real substances

$$\frac{1}{p}\left(\frac{\partial e}{\partial v}\right)_T \ll 1, \quad \text{and} \quad e = e(T) \text{ very nearly.}$$

These statements can be checked very easily for steam. Steam-table values of e and h change only slowly down the constant-temperature columns, while very considerable changes occur along the rows.

A more precise idea of the magnitude of the deviations from the simple pattern can be obtained by evaluating the ratio $(1/p)(\partial e/\partial v)_T$ using steam-table values. Table 1.4 gives values of this ratio, approximated by

$$\frac{2}{p_1+p_2} \cdot \frac{e_2-e_1}{v_2-v_1}$$

for some widely differing states.

Except for the last, near-critical condition, the error introduced by neglecting the change of internal energy with volume is only a few per cent. Hence this assumption is useful if one wishes to obtain a general idea of the behaviour of vapours. If, on the other hand, accurate predictions of changes of state are required, variations of internal energy with density cannot be neglected, especially in high-density states. For this reason one must use tables to determine changes of state in steam, rather than relying on formulae derived from the perfect-gas model.

These results display the same general trends as those of Example 1.6, where van der Waals equation was used to investigate the appropriateness of the perfect-gas model.

Table 1.4. Dependence of internal energy upon specific volume

T (°C)	p_1 (bar)	p_2	$(1/p)(\partial e/\partial v)_T$ (unit-free)
150	1	1·5	0·04
350	6	8	0·03
600	50	60	0·10
375	210	220	3·8

1.4.3 The Perfect Gas

The perfect-gas model is of interest not only because it accurately describes certain real gases, but also because it indicates in a general way how a wider range of highly compressible substances behave. The ideal gas was introduced in equations (1.2.2) as one whose properties were related by

$$pv = RT \quad \text{or} \quad p = \varrho RT$$

with R a constant. It has been seen that a number of simple results can be obtained for this equation of state. Certain of these results required the additional restriction

$$\left(\frac{\partial e}{\partial v}\right)_T = 0$$

Henceforth, a gas satisfying these two requirements will be called semi-perfect, since an even more restricted species of perfect gas will be introduced shortly.

The properties of a *semi-perfect gas* are linked by a number of other special relationships. The internal energy and enthalpy are functions of temperature alone:

$$\begin{aligned} e &= e(T) \\ h &= e(T)+RT = h(T) \end{aligned} \tag{1.4.11}$$

Hence the specific heats (1.4.3) are also functions of temperature alone:

$$\begin{aligned} c_v &= c_v(T) \\ c_p &= c_p(T) \\ \gamma &= \gamma(T) \end{aligned} \tag{1.4.11}$$

However, the difference between the specific heats is constant, as shown by equations (1.4.4, 5):

$$c_p - c_v = R$$

so that

$$c_p = \frac{\gamma}{\gamma-1} R \qquad c_v = \frac{1}{\gamma-1} R \tag{1.4.11}$$

These results provide a good approximation to the behaviour of real gases over a considerable range of temperature and pressure.

For a strictly perfect-gas, the additional restriction is made that the specific heats are constant. With

$$c_p,\ c_v,\ \gamma \quad \text{constants} \tag{1.4.12}$$

it follows that

$$e = \int c_v \, \mathrm{d}T = c_v T + \text{constant}$$

$$h = \int c_p \, \mathrm{d}T = c_p T + \text{constant}$$

For this restricted case there is available a convenient datum level for energy. The internal energy and enthalpy of a perfect gas are usually measured relative to their values at absolute zero. Thus they are related to the absolute temperature by

$$e = c_v T \quad \text{and} \quad h = c_p T \tag{1.4.12}$$

Naturally, the last three of the results (1.4.11) apply in this more restricted case.

The name *calorically perfect-gas* is sometimes given to a gas with constant specific heats, for which the results (1.4.12) are valid. When this term is used, the word perfect is used for the less rigidly specified fluid, which was referred to above as semi-perfect. Thus the labels calorically perfect and semi-perfect are unambiguous, but the single word perfect may be used in two quite different senses.

The range of application of the calorically perfect gas is narrower than that of the semi-perfect model. For a monatomic gas, such as helium or argon, it is found that γ, the ratio of specific heats, is a constant (near $5/3 = 1{\cdot}67$) over a wide range of temperature. For a diatomic gas, such as hydrogen, oxygen, nitrogen or air, γ is constant (near $7/5 = 1{\cdot}4$) at temperatures near atmospheric, but falls at high temperatures, above say 800 °R or 450 °K. For polyatomic gases, such as carbon dioxide, methane and steam, γ varies considerably in the normal range of temperatures. This variability in behaviour can be explained in terms of microscopic motions. Monatomic gases are able to absorb energy only by an increase in the mean kinetic energy of their molecules. However, gases with more complex molecules can store energy in vibrations of individual atoms within the molecule and by rotation of the whole molecule. As the temperature increases, further vibrational modes become active, and the energy-storing properties of the gas change. At very high temperatures (> 3000 °K for oxygen, for example), dissociation and ionization occur and further complicate the response to energy addition.

Despite these limitations, the calorically perfect gas gives a good representation of some real gases over a considerable temperature range, and for other gases provides a useful approximation over a narrower band of temperatures. Some particularly useful results can be found for adiabatic bulk compression. For a perfect-gas, equations (1.4.7) relate volume and temperature changes precisely:

$$-p\,dv = c_v\,dT$$

Differentiation of the perfect-gas law (1.2.2) gives

$$p\,dv + v\,dp = R\,dT$$

When the temperature change dT is eliminated

$$(R + c_v)p\,dv + c_v v\,dp = 0$$

or

$$\gamma p\,dv + v\,dp = 0$$

This equation can be integrated to give

$$pv^{\gamma} = \text{constant}$$

or

$$p \propto \varrho^{\gamma} \qquad (1.4.13)$$

These results prescribe the relationship between pressure and density during an adiabatic compression in which the only work done involves uniform change of density. A particular process path is defined when any one state on it is known, so that the constant of integration can be evaluated. Thus

$$pv^{\gamma} = p_1 v_1^{\gamma}$$

The variations of the other properties can be obtained using the perfect-gas law $p = RT/v = \varrho R T$, in conjunction with the results (1.4.13). Thus

$$p \propto T^{\frac{\gamma}{\gamma-1}}$$

$$\varrho \propto T^{\frac{1}{\gamma-1}} \qquad (1.4.13)$$

$$v \propto T^{-\frac{1}{\gamma-1}}$$

The results (1.4.13) are usually called the *isentropic gas relations* for, as will be seen later, the restrictions imposed in their derivation imply that a property called the entropy remains constant during the process. Here they have been derived for adiabatic bulk compression of a calorically perfect gas.

Example 1.15 In Example 1.8 the work done in compressing an ideal liquid was found. The first of the results (1.4.13) allows a similar calculation to be carried out for another special case—adiabatic bulk compression of a perfect gas. In some compressors, gases actually undergo processes very like the ideal one studied here.

The pressure and volume at an intermediate stage of compression are related to the initial pressure and volume by

$$pV^{\gamma} = p_1 V_1^{\gamma}$$

Differentiation gives

$$\frac{\mathrm{d}p}{p} + \gamma \frac{\mathrm{d}V}{V} = 0$$

(Note the simple relationship between small changes in quantities connected by a power law. It is most easily obtained by taking logarithms and then differentiating. Hence quantities such as $\mathrm{d}p/p$ are termed logarithmic derivatives, and the whole process is called logarithmic differentiation. It is possible, however, to omit the intervening steps and to write down the result directly.

Thus

$$x^a y^b z^c = \text{constant}$$

gives

$$a\frac{\mathrm{d}x}{x} + b\frac{\mathrm{d}y}{y} + c\frac{\mathrm{d}z}{z} = 0)$$

The element of work done on the gas can now be written

$$-p\,\mathrm{d}V = \left(\frac{1}{\gamma}\right) V\,\mathrm{d}p = \left(\frac{1}{\gamma}\right) V_1 \left(\frac{p_1}{p}\right)^{\frac{1}{\gamma}} \mathrm{d}p$$

and integration gives

$$W = -\int_1^2 p\,\mathrm{d}V = \left(\frac{1}{\gamma}\right) V_1 p_1^{\frac{1}{\gamma}} \int_1^2 p^{-\frac{1}{\gamma}}\,\mathrm{d}p$$

$$= \frac{p_1 V_1}{\gamma - 1}\left[\left(\frac{p_2}{p_1}\right)^{\frac{\gamma-1}{\gamma}} - 1\right]$$

This result, like that of Example 1.8, specifies the work done in terms of the pressure change. For a small pressure change, an approximate result can be found using a binomial expansion:

$$W = \frac{p_1 V_1}{\gamma - 1}\left[\left(1 + \frac{\Delta p}{p_1}\right)^{\frac{\gamma-1}{\gamma}} - 1\right]$$

$$= \frac{p_1 V_1}{\gamma - 1}\left[\frac{\gamma - 1}{\gamma}\frac{\Delta p}{p_1} + \text{higher-order terms in } \left(\frac{\Delta p}{p_1}\right)\right]$$

$$\simeq \left(\frac{1}{\gamma}\right) V_1\,\Delta p$$

as would be expected.

Another not unexpected result can be obtained from the work integral. Taking

$$p_1V_1 = mRT_1 \quad \text{and} \quad \left(\frac{p_2}{p_1}\right)^{\frac{\gamma-1}{\gamma}} = \frac{T_2}{T_1}$$

one gets

$$W = \frac{mR}{\gamma-1}(T_2-T_1)$$

But for a perfect gas

$$c_v = \frac{R}{\gamma-1} \quad \text{and} \quad E = mc_vT$$

so that

$$W = mc_v(T_2-T_1) = E_2-E_1$$

This result is to be expected, since in an adiabatic process the only means of changing the internal energy is through work.

We can see now that in this case there is an easier way of relating the work of compression to the change in pressure or the change in volume. The integration can be avoided by starting with

$$W = mc_v(T_2-T_1)$$

and using the isentropic gas relations (1.4.13) to express the work in other ways.

1.4.4 The Constant-density Substance

It has been seen that liquids respond to energy addition in a very different way from vapours and gases. Some aspects of the behaviour of nearly incompressible substances are revealed by the limiting case of strictly constant density. Indeed, certain results obtained for constant density can be applied with good accuracy in predicting the behaviour of real liquids and solids.

One of the singular features of a constant-density substance has already been indicated by equations (1.4.4):

$$c_p = c_v \tag{1.4.14}$$

When a material undergoes adiabatic bulk compression, the internal energy change is given by

$$de = -p\,dv$$

For an incompressible substance, $de = 0$; since the substance does not 'give', no work can be done on it by uniform compression. For a compressible substance, the change in internal energy for a specified volume change depends upon the pressure. For the incompressible substance, there is no change in volume or internal energy, no matter what the pressure may be. Thus it appears that

$$e = e(T) \quad \text{for a constant-density substance} \tag{1.4.14}$$

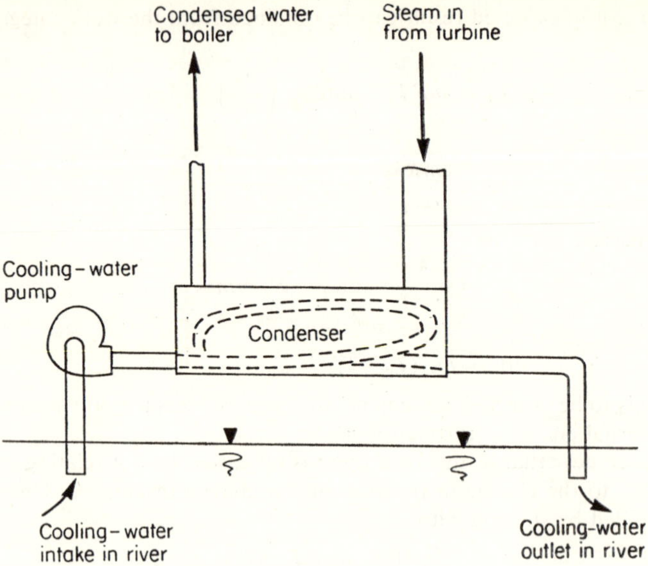

Figure 1.18 Flow of cooling water through a pump and condenser

The thermodynamic properties, which for a compressible substance are closely related, are now separated into two groups. The 'thermal' properties, temperature and internal energy, are interdependent. However, pressure and the constant density do not behave like thermodynamic properties; the latter never changes, while changes in the pressure do not influence the 'thermal' properties. Thus, for this special kind of substance, the pressure and density are not really thermodynamic properties. The only property which depends on both temperature and pressure is the enthalpy:

$$h = e(T) + \frac{p}{\varrho_c} = e(T) + v_c p \tag{1.4.14}$$

The division of the properties into two independent groups has important implications for the study of the flow of a liquid to which heat is supplied. Seemingly it should be possible to treat each of the fluid mechanics and the thermodynamics without reference to the other. This is very nearly what is done in practice. Consider the situation shown in Figure 1.18. Cooling water for a steam plant is pumped from a river, absorbs heat in a condenser, and is returned to the river. (It is the steam which condenses in the condenser; the cooling water remains in the liquid form throughout.) In analysing the flow of cooling water, it is usual to take its density to be constant. The pressure drop through the system is calculated without reference to the heat addition; the viscosity of the water does in fact depend quite strongly upon temperature, but in realistic situations the change in viscosity has only a small effect on the pressure drop. The heat-transfer calculations for the cooling-water circuit make no reference to the pressure of the water. So long as the pressure is high enough to prevent boiling at the temperatures which are achieved, it has no influence on the energy-transfer processes.

Example 1.16 Consider two alternative proposals for the nuclear reactor considered in Example 1.13, using steam as coolant fluid in each. The enthalpy change is to be found for the same core pressure, 20 bar, and outlet temperature $T_2 = 600$ °C, for which $h_2 = 3690$ kJ/kg, but with two different inlet conditions:

(a) 50 per cent wet steam, and
(b) water at 150 °C.

(a) Wet Steam

The average specific enthalpy of a mixture of vapour and liquid can be calculated in the same way as the average specific volume in equations (1.2.6). Thus

$$\begin{aligned} h_1 &= yh_f+(1-y)h_g \\ &= 0{\cdot}50\times909+0{\cdot}50\times2799 \\ &= 1854 \text{ kJ/kg} \end{aligned}$$

and the enthalpy rise across the core is

$$\Delta h = 3690-1854 = 1846 \text{ kJ/kg}$$

It is no longer meaningful to take $\Delta h/\Delta T$ as a measure of the average specific heat. In this case, half of the enthalpy change takes place at constant temperature.

(b) Water at 150 °C

The only properties of liquid water given in the steam tables are those for liquid in equilibrium with vapour. However, these can be used to estimate the properties of water beyond the saturation line. Since the density changes only slightly for pressure and temperature changes of the order considered here, the constant-density model provides a reasonably accurate representation. Thus

$$h_1 = e_f(T_1)+\frac{p_1}{\varrho_c}$$

For saturation at the same temperature

$$h_f = e_f(T_1)+\frac{p_s}{\varrho_c}$$

Then

$$h_1 = h_f+\frac{p_1-p_s}{\varrho_c}$$

For $T_1 = T_s = 150$ °C, interpolation in the steam tables gives $h_f = 632$ kJ/kg, $p_s = 4{\cdot}77$ bar, $v_s = 0{\cdot}001091$ m³/kg. The enthalpy of the water at inlet may then be estimated as

$$h_1 = 632+(20-4{\cdot}77)\times 0{\cdot}1091$$
$$= 634 \text{ kJ/kg}$$

The contribution of the pressure term to the enthalpy is found to be small; for many purposes sufficient accuracy would be obtained by assuming $h(T_1) = h_f(T_1)$. The enthalpy change across the reactor is

$$\Delta h = 3690-634 = 3056 \text{ kJ/kg}$$

Comparison of the enthalpy changes (a) and (b) with that found for steam in Example 1.13, $\Delta h = 786$ kJ/kg for $T_1 = 250$ °C, shows that a very great increase in the energy-carrying capacity of the steam is achieved with rather small changes in the inlet temperature. This is the consequence of the large absorption of energy in boiling, with no corresponding increase in temperature. This example makes clear the reason why liquid/vapour cycles are often selected for engineering equipment in which it is desired to transfer large amounts of heat.

1.5 THE DEGRADATION OF ENERGY

1.5.1 Irreversibility

Energy and heat are measured in the same units as mechanical work, and are in some sense interchangeable with it. But experience in everyday life reveals that there are restrictions on the interchange between heat, work, and various forms of energy. Natural energy flows display a preferred direction: heat flows from a hot body to a cold one, and kinetic energy is converted into internal energy by friction. Local reversals of these dominant trends do occur, both naturally and through human intervention. However, a careful examination of a larger, enveloping system shows that there are universally valid rules governing the direction of energy conversion. Consider, for example, water flowing from a reservoir over a dam and mixing with the river below. If water is pumped back into the reservoir, the process has in a sense been reversed; the reservoir looks the same as it did before the water left it. However, an examination of the pumping process shows that more mechanical energy is used by the pumps than is gained by the water. There is then, overall, a conversion of mechanical energy to internal energy. In such a process, energy is said to be degraded.

It has been noted that natural processes are always irreversible in some measure. For use in thermodynamics, the word 'irreversible' must be defined with precision. It will be defined in terms of an isolated system, one of fixed contents, through whose boundaries neither heat nor work passes. If a process within such a system can proceed in only one direction, the process is said to be irreversible. If a process can proceed in either direction, returning the system to its original state, it is reversible.

The simple system illustrated in Figure 1.19 provides a number of examples of irreversible processes. This isolated system contains two compartments separated by an isolating barrier. The compartments are initially filled with homogeneous, simple compressible substances, with an equilibrium state in each compartment. For ease of visualization, the substances may be thought of as gases.

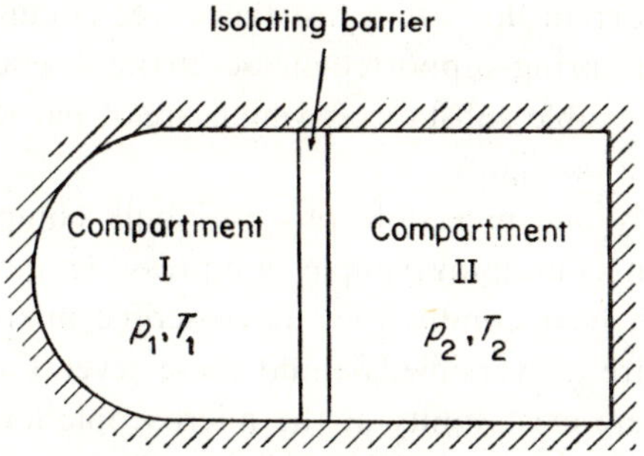

Figure 1.19 An isolated system containing two compartments

Imagine that a small hole is opened in the barrier within the system. If the pressures in the two compartments are initially different, one gas will rush through the hole and will mix with the fluid in the other compartment; ultimately equilibrium will be restored with uniform pressure throughout. If there are initially different species of gas in the two compartments, diffusion through the hole will ultimately produce a mixture which is homogeneous through both compartments. If the temperatures in the two compartments differ initially, again diffusion will in time produce uniformity. Indeed, in this case, the 'hole' between the compartments need not allow the passage of matter, but only of heat. Finally, imagine that the substances separated by the barrier take part in a chemical reaction when they come into contact, so that the ultimate equilibrium state contains new kinds of molecules. Evidently all of the processes described are irreversible. The system, once disturbed, moves into a new equilibrium condition and does not move away from it again.

Another kind of process which can be imagined to occur within the isolated system of Figure 1.19 is one in which the barrier moves along the container, as would a closely fitting piston or a flexible membrane. If the pressures are initially different, say, $p_1 > p_2$, the barrier will first move to the right, the fluid in compartment II being compressed. In the absence of friction at the walls and within the fluid, an oscillatory motion of the barrier can be established. Since the process within the isolated system spontaneously reverses direction, it can be termed reversible. In reality, there will be some friction and the barrier will ultimately come to rest near the position of equalized pressures. Nevertheless, a process has been found—uniform compression and expansion of the bulk of the material—which could in principle be reversible, and can in practice be oscillatory with slow decay of amplitude.

One can imagine other reversible processes within other kinds of system. For instance, the current flows in a resistance-free circuit can be reversible, as can oscillations of spring-supported masses in the absence of friction. However, simple compressible substances, which are of particular interest here, do not contain such oscillators.

In summary, only one reversible process—bulk compression—has been found for a system containing simple compressible substances. Processes involving diffusion, heat conduction, macroscopic mixing, chemical reactions and friction are irreversible. Behind these several macroscopic causes of irreversibility there exists unity at the microscopic level. Ultimately it is movement and interaction at molecular scales which give rise to the various macroscopic manifestations of irreversibility: mass diffusion is the macroscopic counterpart of molecular migration and mingling; heat conduction corresponds to work at the molecular level; fluid friction is the result of momentum interchange among molecules; macroscopic mixing produces small-scale discontinuities that are removed by diffusion and conduction.

An essential feature of molecular events is their randomness. Hence they tend to smooth out every kind of macroscopic discontinuity, and there is a universal tendency for a highly ordered state to be replaced by a disordered one. This can be seen by putting some red beads into a glass jar, adding some white ones on top, and then shaking the jar. The initial ordered state, white over red, rapidly disappears. Many individual patterns are set up as shaking continues, but the original, highly ordered pattern is unlikely to reappear. Seen from a distance the material in the jar always appears to be pink. Thus the random small-scale processes which are responsible for macroscopic irreversibility are also responsible for the existence of stable macroscopic properties.

So far the discussion of irreversibility has referred to an isolated system. How does it apply to a system of the kind which we usually have to examine, one exchanging work and heat with its surroundings? The word reversible still seems to be applicable to the process of bulk compression. If a system is compressed adiabatically, the work done on it being $dW =$

$-p\,dV$, the system can do the same amount of work on the surroundings if it expands to its original volume. For other modes of energy addition, heat and work involving friction and diffusion, it is not so obvious how the ideas of irreversibility in an isolated system can be applied. Certainly such energy additions are not absolutely irreversible; the energy can be retrieved.

Two relevant points should be noted. First, consider the energy-addition mechanisms shown in Figure 1.13: bulk compression, stirring and electrical heating. The latter two are irreversible in the sense that this passive system is incapable, once it comes to equilibrium, of using them to do work on the surroundings. Secondly, by considering the system shown in Figure 1.12 it was shown that no clear distinction can be made between heat addition and 'irreversible' work mechanisms. It appears then that all of the energy addition, except reversible work of bulk compression, must be considered as a single unit. Henceforth this component of energy addition will be referred to, somewhat loosely, as the element of energy addition associated with irreversible processes, or simply as the *irreversible-energy addition*. This name should not be interpreted literally; the energy can in fact be withdrawn from the system. The name is merely a short-hand label for the arguments presented above.

Since the irreversible-energy addition is the only clearly defined quantity which we have been able to associate with the natural tendency to irreversibility, our attempt to give a precise mathematical expression to this tendency must make use of it. We begin by calculating this element of energy transfer. The net change in internal energy during a small change of state can be reckoned as the sum of three elements: reversible work, $dW_r = -p\,dV$; work associated with irreversible processes, dW_i; and heat flow, dQ. Then

$$\begin{aligned} dE &= dW_r + dW_i + dQ \\ &= -p\,dV + dW_i + dQ \end{aligned} \tag{1.5.1}$$

Since it is not possible to distinguish clearly between irreversible work and heat, the net energy change can be expressed as the sum of only two elements, reversible work and other energy inputs. The energy input associated with irreversibility is given by

$$dW_i + dQ = dE + p\,dV \tag{1.5.2}$$

For a finite process which is nearly in equilibrium at every stage, the irreversible-energy addition can be found by integration:

$$W_i + Q = \Delta E + \int_1^2 p\,dV \tag{1.5.3}$$

Since the integral depends on the process path, the sum $W_i + Q$ will do so as well. This is another way of saying that the sum W_i+Q is not a property of the system.

Now consider a process during which the system departs widely from equilibrium. The integral in equation (1.5.3) cannot now be defined, and the energy addition associated with irreversibility cannot be calculated in this way. Nor, since it is not a property, is it determined by the terminal states of the process. An example will make this limitation clearer. Energy can be added to the system shown in Figure 1.20 in two ways: heat transfer and work done by the piston. If the piston is moved in slowly and heat is supplied slowly, it is possible to determine the integral $\int p\ \mathrm{d}V$ and, knowing the net change in internal energy, to calculate the sum W_i+Q. If, on the other hand, the piston is rammed in, so that significant pressure differences are set up within the cylinder during the process, it is not possible to specify a unique pressure variation for the system, and the integral cannot be evaluated.

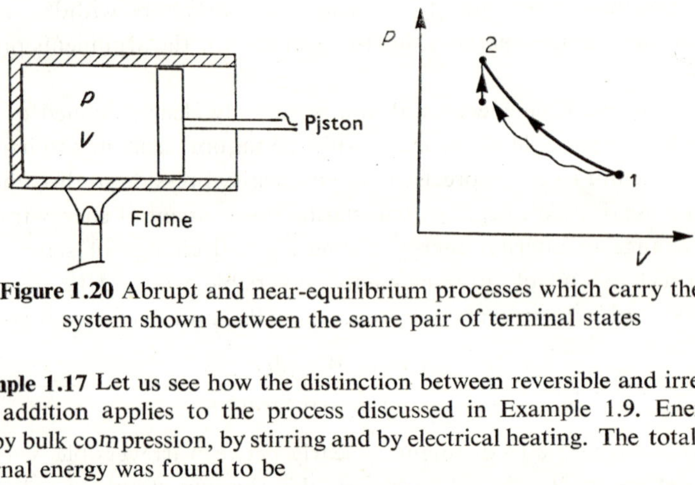

Figure 1.20 Abrupt and near-equilibrium processes which carry the system shown between the same pair of terminal states

Example 1.17 Let us see how the distinction between reversible and irreversible energy addition applies to the process discussed in Example 1.9. Energy was added by bulk compression, by stirring and by electrical heating. The total change in internal energy was found to be

$$\Delta E = W_c + W_s + W_e = 5270 \text{ ft lbf} = 7150 \text{ J}$$

So far as the internal energy is concerned, the three contributions are interchangeable. But when one considers the processes necessary to return the system to its original state, the energy inputs are seen to be essentially different.

To restore the original state, it is sufficient to set up the initial values of volume and internal energy. First allow the system to return to its original volume, expanding by 1/10 ft^3. Since the pressure has not increased significantly (as shown in Example 1.10), the work done *on* the system is very nearly

$$W = -p_1\,\Delta V = -1440 \text{ ft lbf} = -1950 \text{ J}$$

This has the magnitude of the original work of bulk compression W_c, but is of opposite sign. Thus the internal-energy change arising from reversible work is cancelled. The system is capable of restoring this amount of work to the surroundings while returning to its original state.

The residual energy change is associated with irreversible processes:

$$W_c + W_s + W_e + W = W_s + W_e = 3830 \text{ ft lbf} = 5200 \text{ J}$$

Once the system has come to equilibrium it is not capable of turning the stirring device to do work on the surroundings. Nor will an electric current flow spontaneously from the system. The only way in which the residual energy can be removed to restore the internal energy to its initial value is by a flow of heat. After the heat flow

$$Q = -3830 \text{ ft lbf} = -5200 \text{ J}$$

the system is restored to its original condition.

In this example the energy associated with irreversible work modes was removed from the system in the form of heat. Such a conversion of work to heat is sometimes described as energy degradation, since in practice only a portion of the energy thus transformed can be converted back into mechanical energy.

1.5.2 Entropy

Let us sum up our present understanding of irreversibility for systems exchanging energy with their surroundings. From experience it is known that there are preferred directions in energy flows; it is possible to predict that direction in a number of cases. The mechanisms of irreversibility have been identified; the total energy transfer associated with them can be calculated, but only for near-equilibrium processes. This state of understanding is unsatisfactory in at least two respects. First, we require a precise expression of the preferred direction in energy transfer; it should be applicable to the varied processes experienced by a flowing fluid and should be related to the existing structure of thermodynamic principles. Secondly, a method is required to describe irreversibility in processes during which the state departs widely from equilibrium.

The macroscopic properties introduced earlier represent various aspects of microscopic behaviour. Since the preferred direction of natural processes is also a manifestation of microscopic activity, it should be possible to find a macroscopic property to represent this effect as well. Moreover, since the change in a thermodynamic property during a process between two equilibrium states depends only on the terminal states, this property will provide a way of measuring irreversibility, no matter what the nature of the process. Hence the introduction of a property associated with irreversible-energy transfer has several potential advantages. This property, which we do not yet know how to define, will be called entropy, and will be denoted by the symbol S.

A strategy has now been established, namely, to introduce a thermodynamic property which can be used to express the direction of natural energy flows. It is not immediately obvious how this can be done. Certainly the definition of entropy must be consistent with the principles already established—the state postulate and the zeroth and first laws. Moreover, it must be related to the energy transfer associated with irreversibility, since this is the only measure of irreversibility we have been able to isolate. Finally, the behaviour of entropy must be consistent with the concept of irreversibility for an isolated system. Only when each of these constraints is taken into account does the appropriate definition of entropy emerge.

Certain of the characteristics of entropy are arbitrary and may be chosen to suit our ideas of irreversibility. Since entropy is to describe the essential irreversibility of natural processes, it will be required that

$$\mathrm{d}S > 0 \tag{1.5.4}$$

for any real process within an isolated system. For an ideal process, in which only reversible work is done within the system, the entropy change would be zero. Actually, the change in entropy might have been required to be negative: the choice (1.5.4) is always made, since this seems a more natural convention to most people.

Next the implications of the state postulate, the first law and finally the zeroth law will be considered in turn. If entropy is a thermodynamic property, there must exist, according to the state hypothesis, an equation of state

$$E = E(V, S) \tag{1.5.5}$$

The choice of this one of many possible relationships between properties is suggested by equation (1.5.1). For a small change of state

$$\mathrm{d}E = \left(\frac{\partial E}{\partial V}\right)_s \mathrm{d}V + \left(\frac{\partial E}{\partial S}\right)_v \mathrm{d}S$$

the two terms being directly comparable with those of equation (1.5.1). Equivalence of the two statements of internal-energy change requires that

$$\left(\frac{\partial E}{\partial V}\right)_s = -p \tag{1.5.6}$$

and

$$\left(\frac{\partial E}{\partial S}\right)_v \mathrm{d}S = \mathrm{d}W_\mathrm{i} + \mathrm{d}Q = \mathrm{d}E + p\,\mathrm{d}V \tag{1.5.7}$$

The first of the derivatives is identified with pressure. No reference to external forces is made in equation (1.5.6); the pressure defined in terms of the intrinsic properties of a system is called the *thermodynamic pressure.* Under equilibrium conditions it will be equal to the externally applied pressure. The properties of a substance are related by $p = -(\partial e/\partial v)_s$.

Equation (1.5.7) is of greater immediate importance. It indicates that a small change in entropy is proportional to the energy input associated with irreversibility, the one quantity associated with irreversibility which we know how to calculate. The factor of proportionality is the derivative $(\partial E/\partial S)_v$. Its characteristics are not yet clear, but will be investigated using the first law of thermodynamics and the state postulate.

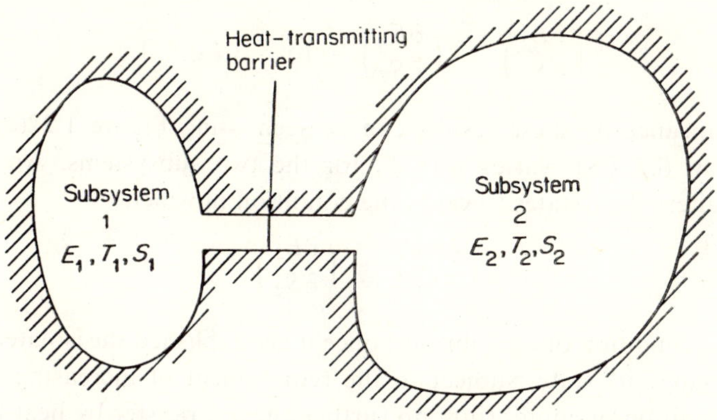

Figure 1.21 An isolated system containing two subsystems

Consider the isolated system of Figure 1.21. It contains two subsystems, separated by a barrier which does not move, but can transmit heat slowly so that the subsystems remain nearly in equilibrium throughout the energy addition. The total internal energy of this isolated system remains constant even though heat flows between the subsystems. Thus

$$E_1 + E_2 = \text{constant}$$

Hence in a small change between equilibrium states

$$dE_1 = -dE_2 \tag{1.5.8}$$

The state postulate requires that the subsystems will have equations of state of the form (1.5.5)

$$E_1 = E_1(V_1, S_1) \quad \text{and} \quad E_2 = E_2(V_2, S_2)$$

For energy transfer in which the volumes of the compartments do not change

$$dE_1 = \frac{\partial E_1}{\partial S_1} dS_1 \quad \text{and} \quad dE_2 = \frac{\partial E_2}{\partial S_2} dS_2$$

When the relation (1.5.8) is introduced, the net entropy change in the system is found to be

$$dS = dS_1 + dS_2 = \left[\frac{1}{\left(\dfrac{\partial E_1}{\partial S_1}\right)_v} - \frac{1}{\left(\dfrac{\partial E_2}{\partial S_2}\right)_v} \right] dE_1$$

The requirement (1.5.4) now comes into play. The entropy is to increase, irrespective of the direction of heat flow. This will be the case only if

$$\begin{aligned} \left(\frac{dE_1}{\partial S_1}\right)_v &< \left(\frac{\partial E_2}{\partial S_2}\right)_v \quad \text{for} \quad dE_1 > 0 \\ \left(\frac{\partial E_1}{\partial S_1}\right)_v &> \left(\frac{\partial E_2}{\partial S_2}\right)_v \quad \text{for} \quad dE_1 < 0 \end{aligned} \tag{1.5.9}$$

The significance of these results can be seen using Figure 1.22(a), which shows how $(\partial E/\partial S)_v$ varies with E_1 for the two subsystems. Any energy transfer alters their states towards the condition in which

$$\left(\frac{\partial E_1}{\partial S_1}\right)_v = \left(\frac{\partial E_2}{\partial S_2}\right)_v$$

This is the condition of equilibrium; once it is established the isolated system cannot change its state, subject to the requirement of increasing entropy. In this condition of equilibrium no further energy transfer by heat flow will occur between the two subsystems; in other words, they are in thermal equilibrium. Gathering together these two lines of thought, we conclude that the property $(\partial E/\partial S)_v$ is the same for the two subsystems when they are in thermal equilibrium.

This last conclusion has a rather similar structure to the zeroth law of thermodynamics, which implies that the temperatures of two systems are equal when they are in thermal equilibrium. Thus it appears that the property $(\partial E/\partial S)_v$ has a value defined by the temperature of the system it describes. In other words,

$$\left(\frac{\partial E}{\partial S}\right)_v = f(T) \quad \text{only}$$

Our final task is to decide upon the form of the function $f(T)$. In our work on thermodynamics it has proved convenient to measure temperatures using the absolute scale defined by the perfect-gas law $pv = RT$, and it is therefore appropriate to seek to express $(\partial E/\partial S)_v = f(T)$ in terms of the temperature defined by the gas laws. To do this we shall use the perfect-gas laws and our new results to examine the same kind of process, and then discover the conditions under which they are consistent.

The processes to be examined are depicted in Figure 1.22(b). The gas can move between states 1 and 2 via a number of different combinations of isen-

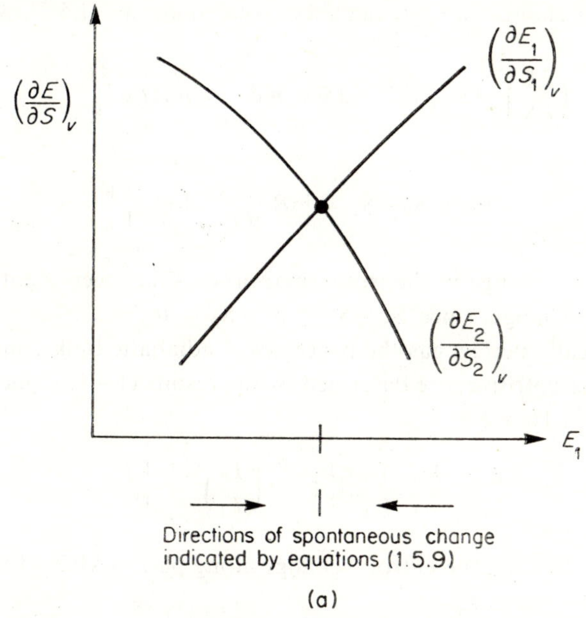

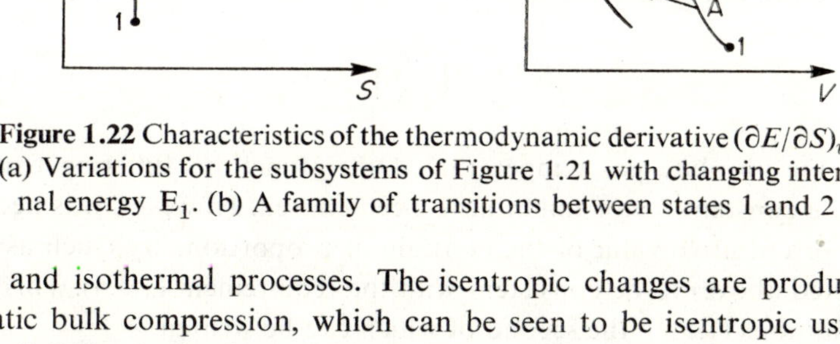

Figure 1.22 Characteristics of the thermodynamic derivative $(\partial E/\partial S)_v$. (a) Variations for the subsystems of Figure 1.21 with changing internal energy E_1. (b) A family of transitions between states 1 and 2

tropic and isothermal processes. The isentropic changes are produced by adiabatic bulk compression, which can be seen to be isentropic using the results (1.5.7); the isothermal changes are produced by compression with sufficient outwards heat flow to maintain uniform temperature. For a calorically perfect gas, the isothermal process AB is governed by

$$\mathrm{d}E = 0 \quad \text{and} \quad p\,\mathrm{d}V = mRT_{\mathrm{A}}\frac{\mathrm{d}V}{V}$$

and the entropy change can be calculated from equations (1.5.7), which give first

$$\left(\frac{\partial E}{\partial S}\right)_v \mathrm{d}S = f(T_A)\,\mathrm{d}S = p\,\mathrm{d}V = mRT_A\frac{\mathrm{d}V}{V}$$

and then

$$S_B - S_A = S_2 - S_1 = mR\frac{T_A}{f(T_A)}\log_e\frac{V_B}{V_A}$$

Here the entropy change in the isothermal process has been identified with the net entropy change, since $S_A - S_1 = S_2 - S_B = 0$.

For a calorically perfect gas the processes of adiabatic bulk compression, now seen to be isentropic, are described by the results (1.4.13), one of which is $V \propto T^{-1/(\gamma-1)}$. Hence

$$\frac{V_B}{V_A} = \frac{V_B}{V_2}\frac{V_1}{V_A}\frac{V_2}{V_1} = \left(\frac{T_2}{T_1}\right)^{\frac{1}{\gamma-1}}\frac{V_2}{V_1}$$

since $T_B = T_A$. Finally, the entropy change along path 1AB2 is found to be

$$S_2 - S_1 = mR\frac{T_A}{f(T_A)}\log_e\left[\left(\frac{T_2}{T_1}\right)^{\frac{1}{\gamma-1}}\frac{V_2}{V_1}\right]$$

Since S, T and V are all properties of the system, their values will depend only on the terminal states. The last result must then be independent of the intermediate temperature T_A. This will be so only if

$$f(T) \propto T$$

The simplest, and conventional choice which is consistent with this result is

$$f(T) = \left(\frac{\partial E}{\partial S}\right)_v = T \tag{1.5.10}$$

with T the absolute gas temperature. Other functions differing from this only by a constant factor might have been chosen. However, it is necessary to choose a positive value of the constant of proportionality (such as unity, as selected above) to be consistent with the requirement (1.5.4). This led to the conditions (1.5.9), the second of which can be written

$$f(T_1) > f(T_2) \quad \text{for} \quad \mathrm{d}E_1 < 0$$

corresponding to heat flow from subsystem 1 to subsystem 2, which will occur when $T_1 > T_2$.

Having established a suitable form for the factor relating entropy change to the energy input associated with irreversibility, we can now write equation (1.5.7) as

$$T\,dS = dW_i + dQ = dE + p\,dV \tag{1.5.11}$$

Although equations (1.5.10) define entropy, equation (1.5.11) relates it to the other thermodynamic properties in a more convenient way:

$$dS = \frac{dE + p\,dV}{T} \tag{1.5.12}$$

In terms of intensive properties, $e = E/m$, $v = V/m$, and $s = S/m$

$$ds = \frac{de + p\,dv}{T} \tag{1.5.13}$$

The entropy can also be related to enthalpy changes. Introducing $dh = de + d(pv)$, one has

$$ds = \frac{dh - v\,dp}{T} \tag{1.5.14}$$

The units of entropy are seen to be energy/temperature, and those of specific entropy to be energy/(mass × temperature).

The calculation of the entropy change for a *constant-density substance* is particularly simple. Here

$$T\,ds = de \tag{1.5.15}$$

Since no reversible work of bulk compression can be done on an incompressible substance, all of the energy input is associated with irreversibility. Note that entropy is, like e and T, independent of pressure. Hence

$$s = s(T) \tag{1.5.15}$$

for a constant-density substance. In this singular case there is no real advantage in using entropy to represent the irreversible element of energy addition. Equation (1.5.3) gives for this case

$$W_i + Q = \Delta E \tag{1.5.15}$$

The internal energy itself acts as a direct measure of the irreversible-energy addition. This fact will be found to be very useful when discussing the losses which occur in the flow of a liquid.

Example 1.18 Although the introduction of entropy required a rather lengthy discussion, the property, once defined, can be manipulated in much the same manner as other thermodynamic properties. To illustrate methods of determining entropy changes, look again at the processes in a steam-generating nuclear reactor which were discussed in Example 1.16.

For both cases the final condition of the steam is given by $T_2 = 600\ °C$, $p_2 = 20$ bar, for which the steam tables give $s_2 = 7{\cdot}701$ kJ/kg °K.

(a) Wet Steam

The average specific entropy at inlet to the reactor can be calculated in the same way as other extensive properties:

$$s = ys_f + (1-y)s_g$$

where y is the wetness fraction. For $p_s = 20$ bar and $y = 50$ per cent,

$$s_1 = 0\cdot50 \times 2\cdot447 + 0\cdot50 \times 6\cdot340$$
$$= 4\cdot394 \text{ kJ/kg °K}$$

Hence the net rise in specific entropy is

$$\Delta s = 7\cdot701 - 4\cdot394 = 3\cdot307 \text{ kJ/kg °K}$$

(b) Water at 150 °C

The constant-density model gives a reasonable picture of the properties of a liquid near the saturation line. Equations (1.5.15) give

$$s_1 = s_f(T_1)$$

for strictly constant density. Adopting this approximation, we have, by interpolation

$$s_1 = 1\cdot820 + 0\cdot040 \times \frac{2\cdot1}{3\cdot9} = 1\cdot842 \text{ kJ/kg °K}$$

and

$$\Delta s = 7\cdot701 - 1\cdot842 = 5\cdot859 \text{ kJ/kg °K}$$

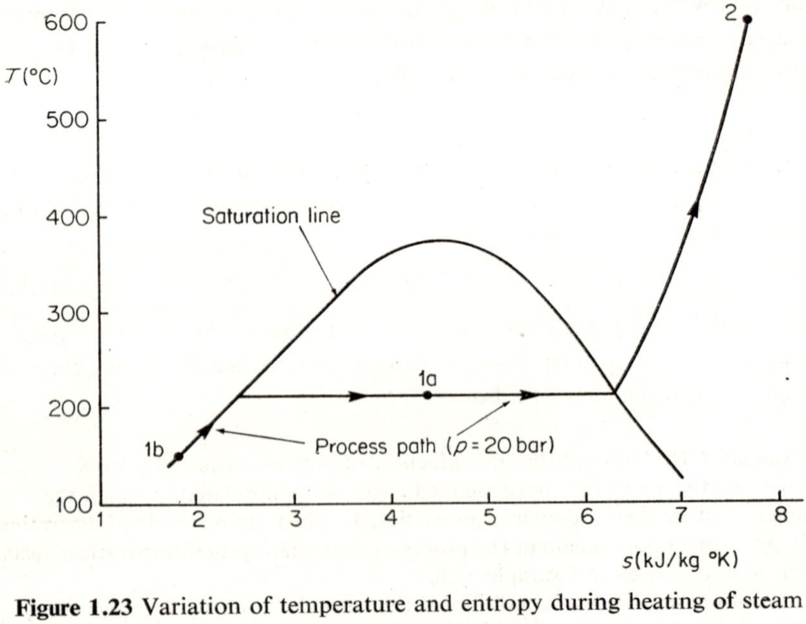

Figure 1.23 Variation of temperature and entropy during heating of steam at constant pressure

In each case there is a large entropy increase, corresponding to heat addition in the reactor core. Figure 1.23 shows these changes in a T, s-diagram. Some particularly simple results are possible for the flat part of the process path which represents boiling. Both temperature and pressure are uniform through this part of the process, and equation (1.5.14) gives

$$ds = \frac{dh}{T_s}$$

Then
$$T_s \, \Delta s = \Delta h$$

across the mixed-phase region. The consistency of the steam tables can be checked by introducing tabulated values into this equation.

1.5.3 The Entropy of a Perfect Gas

The differential expressions (1.5.13, 14) relate entropy to the other thermodynamic properties. One would, however, feel more confident in using entropy if it were related to the more familiar properties by an explicit algebraic equation. This is easy to do for the case of a calorically perfect gas. For this substance,

$$de = c_v \, dT \quad \text{and} \quad pv = RT$$

and equation (1.5.13) can be written

$$ds = c_v \frac{dT}{T} + R \frac{dv}{v} \tag{1.5.16}$$

This form can be integrated, giving

$$s - s_1 = c_v \log_e \frac{T}{T_1} + R \log_e \frac{v}{v_1}$$

where s_1, T_1 and v_1 are the properties at some reference state used to evaluate the constant of integration. This result can be written, using the relationship $R/c_v = \gamma - 1$, valid for a perfect gas:

$$\frac{s - s_1}{R} = \log_e \left[\left(\frac{T}{T_1} \right)^{\frac{1}{\gamma - 1}} \frac{v}{v_1} \right] \tag{1.5.17}$$

The entropy change given by equation (1.5.17) is, as expected, dependent only on the terminal equilibrium states of the process. It is the same for any process linking these states, whether a slow, near-equilibrium process, or an abrupt transition. This fact, which follows directly from the definition of entropy as a thermodynamic property, provides the justification for the method of deriving the entropy change (1.5.17). In integrating the result (1.5.16) we tacitly restricted consideration to smooth transitions in which the state of the mass of gas was at each point defined by a single temperature and specific volume. These are possible ways of passing between the two terminal states, and the entropy change between the states is therefore correctly calculated.

The result (1.5.17) relates the entropy change to changes in the product $T^{1/(\gamma-1)}v$. If the entropy is constant

$$v \propto T^{-\frac{1}{\gamma-1}}$$

Reference to equations (1.4.13) shows that this is one of the relationships between the properties of a perfect gas undergoing adiabatic bulk compression. This is to be expected, for an adiabatic process involving only bulk compression is easily seen to be isentropic:

$$T\,\mathrm{d}s = \mathrm{d}e + p\,\mathrm{d}v = -p\,\mathrm{d}v + p\,\mathrm{d}v = 0$$

The perfect-gas equation written in logarithmic-derivative form

$$\frac{\mathrm{d}p}{p} + \frac{\mathrm{d}v}{v} = \frac{\mathrm{d}T}{T}$$

can be combined with equation (1.5.16) to relate the entropy to the other properties in a number of ways. Thus

$$\begin{aligned} \mathrm{d}s &= c_v \frac{\mathrm{d}p}{p} + c_p \frac{\mathrm{d}v}{v} \\ \mathrm{d}s &= c_v \frac{\mathrm{d}p}{p} - c_p \frac{\mathrm{d}\varrho}{\varrho} \\ \mathrm{d}s &= c_p \frac{\mathrm{d}T}{T} - R \frac{\mathrm{d}p}{p} \end{aligned} \tag{1.5.16}$$

Each of these equations can be integrated to give a result similar to equation (1.5.17), and when restricted to isentropic processes gives another of the isentropic gas relations (1.4.13).

Since $e = c_v T$ for a perfect gas, equation (1.5.17) can be written

$$e = e_1 \left(\frac{v_1}{v}\right)^{\frac{R}{c_v}} \exp\left[\frac{s - s_1}{c_v}\right] \tag{1.5.18}$$

This result is termed the *canonical equation of state* of a perfect gas, the term canonical implying that this is in some sense a standard form. It has in fact the form (1.5.5) adopted for the equation of state during the discussion leading to the definition of entropy. Although pressure and temperature do not appear in the equation, they can be obtained from equations (1.5.6, 10):

$$p = -\left(\frac{\partial e}{\partial v}\right)_s \quad \text{and} \quad T = \left(\frac{\partial e}{\partial s}\right)_v$$

On carrying out the latter differentiation, one obtains $T = e/c_v$. Other special characteristics of a calorically perfect gas also follow easily from the equation of state written in the canonical form.

Example 1.19 One of the most important applications of perfect-gas results is in the analysis of the high-velocity flows around aircraft. Consider an aeroplane flying at a high altitude, the pressure and temperature being 0·18 bar and 218 °K in the still air well away from the aeroplane. The relationship between pressure, temperature and entropy is to be examined for the three points shown in Figure 1.24:

(a) the 'stagnation point' on the leading edge of the wing, where the pressure is 0·32 bar,

(b) a point in the low-pressure region above the wing, a few inches above the surface, where the pressure is 0·12 bar, and

(c) a point on the wing, beneath point (b), where the pressure is 0·12 bar and the temperature is 210 °K.

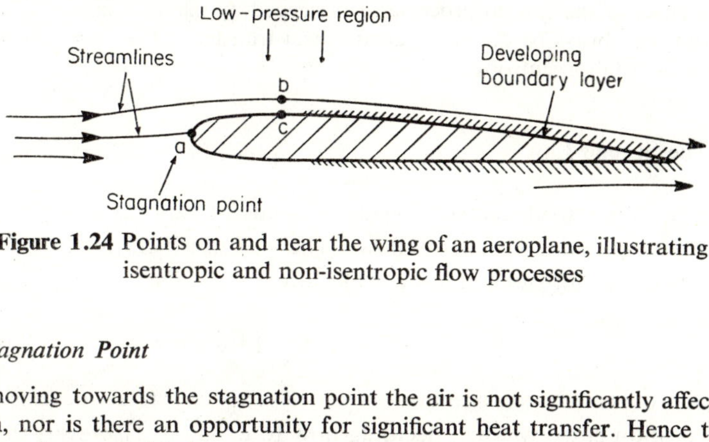

Figure 1.24 Points on and near the wing of an aeroplane, illustrating isentropic and non-isentropic flow processes

(a) Stagnation Point

In moving towards the stagnation point the air is not significantly affected by friction, nor is there an opportunity for significant heat transfer. Hence the entropy may be assumed to be constant, and the temperature and pressure are related by the isentropic relation

$$\frac{T_2}{T_1} = \left(\frac{p_2}{p_1}\right)^{\frac{\gamma-1}{\gamma}}$$

For air, with $\gamma = 1{\cdot}4$ and $(\gamma-1)/\gamma = 0{\cdot}286$

$$T_2 = 218 \times \left(\frac{0{\cdot}32}{0{\cdot}18}\right)^{0{\cdot}286} = 256\ {}^\circ\text{K}$$

(b) Above the Wing

The processes undergone by the air in reaching this point may also be taken to be isentropic, since the boundary layer, where friction and heat transfer are important, extends only a small distance from the surface. Then, proceeding as before

$$T_2 = 218 \times \left(\frac{0{\cdot}12}{0{\cdot}18}\right)^{0{\cdot}286} = 194\ °\text{K}$$

(c) On the Wing

The air reaching this point will have experienced various irreversible processes, including heat transfer to and from the wing and fluid friction. The entropy change is obtained from the last of equations (1.5.16) as

$$s_2 - s_1 = R \log_e \left[\frac{p_1}{p_2}\left(\frac{T_2}{T_1}\right)^{\frac{\gamma}{\gamma-1}}\right]$$

For air, with $\gamma/(\gamma-1) = 3{\cdot}5$ and $R = 287$ J/kg °K

$$s_2 - s_1 = 287 \log_e \left[\frac{0{\cdot}18}{0{\cdot}12}\left(\frac{210}{218}\right)^{3{\cdot}5}\right]$$

$$= 82\ \text{J/kg °K}$$

The net effect of the several processes experienced by the air is a rise in entropy. This need not always be the case; greater heat transfer to the wing might have given rise to a fall in entropy.

1.5.4 The Second Law

Entropy was introduced to provide a means of expressing the preferred direction of spontaneous energy flows. The way in which this principle will be expressed for an isolated system has already been set down in equation (1.5.4)

$$\mathrm{d}S > 0 \quad \text{in an isolated system}$$

for any process between equilibrium states. This characteristic was in fact imposed upon entropy when deciding how to define it. However, the more general statements of the second law which follow were not initially apparent. They incorporate the whole skein of ideas which led to the definition of entropy.

For a system whose internal energy is changed by work and heat addition, the result (1.5.11) suggests the appropriate expression of the second law. In any real energy addition, some of the work will be irreversible. Moreover, this element of work will always be positive, since the mechanisms with which it has been identified (for instance, the stirring device and electrical resistor of Figure 1.13) are incapable of doing work on the surroundings, but can only absorb work from them. Hence the actual entropy change will always be greater than that produced by heat addition. This idea is expressed by

$$T\,\mathrm{d}S > \mathrm{d}Q \quad \text{for any energy addition} \tag{1.5.19}$$

The limiting case, $T\,\mathrm{d}S = \mathrm{d}Q$, is that in which the work, if any, is purely reversible.

The general statement of the second law of thermodynamics (1.5.19) gives

$$\mathrm{d}S > 0 \quad \text{for any adiabatic process} \tag{1.5.20}$$

Although this has the same form as the condition (1.5.4) which was used in defining entropy, equation (1.5.20) applies more generally, including processes in which work is done on the system. Moreover, it can be applied not only by restricting consideration to adiabatic processes, but also by selecting a system which encompasses any heat flows which do take place. In this global sense, it is true to say that entropy can be produced but never destroyed.

The principle of increasing entropy has another interesting interpretation. If it is not invoked, any process consistent with the laws of mechanics and thermodynamics can proceed in either direction. In other words, any possible pair of initial and final states could be interchanged. Hence entropy has been termed 'the arrow of time'; it identifies that one of the terminal states which must occur at a later time.

The application of the second law will be demonstrated by seeing how it relates to some simple processes involving heat flow and work within a system, and heat flow and work at the boundaries of a system. In fact the situations to be considered are so simple that the nature of the results can be predicted without the use of the second law; thus they provide checks on its predictions. The processes examined will also demonstrate the various ways in which entropy changes can be produced in a system.

Internal and External Heat Flows

The flow of heat between two systems was considered in deciding how to define entropy. Nevertheless, it is instructive to examine this situation again making use of the second law.

Imagine a small amount of heat $\mathrm{d}Q$ to flow from a body at temperature T_1 to another at temperature T_2, the flow being so slow that the temperatures remain essentially uniform within each body, and so small that the temperatures are not changed significantly by the process. The system comprising the two bodies is isolated from work and heat flow. These conditions rule out work of all kinds, including irreversible elements, so that the limiting form of equation (1.5.19) may be applied to determine the entropy changes within the two bodies:

$$T_1\,\mathrm{d}S_1 = -\mathrm{d}Q \quad \text{and} \quad T_2\,\mathrm{d}S_2 = \mathrm{d}Q \tag{1.5.21}$$

Now consider the isolated system encompassing the two bodies. For this system

$$\mathrm{d}S = \mathrm{d}S_1+\mathrm{d}S_2 = \left(\frac{1}{T_2}-\frac{1}{T_1}\right)\mathrm{d}Q \tag{1.5.22}$$

The entropy of the isolated system must rise during this process. Hence

$$(T_1-T_2)\,\mathrm{d}Q > 0$$

or, in words, heat will flow from the hotter to the cooler body. As required, the second law incorporates this directional feature of natural energy flows.

The results above also illustrate the ways in which heat flows can change entropy. Equations (1.5.21) show that the entropy of a system can be increased by an inwards flow of heat and decreased by an outflow of heat. Note that entropy changes can take place even if the temperatures of the two systems exchanging heat are equal. This is not the case for the overall entropy change (1.5.22), which is attributable to an internal heat flow. This entropy change is necessarily positive, but can be made arbitrarily small if the temperatures T_1 and T_2 are nearly equal.

External Work

Next the second law will be applied to an adiabatic process involving both reversible and irreversible work. Figure 1.25 shows a quantity of a compressible substance confined within a cylinder by a piston. The system to be examined consists of the compressible substance together with the cylinder and piston. The entire system is at temperature T.

A friction force f (reckoned as positive when outwards) acts on the piston. Hence the force applied to the system by the piston rod is

$$F = (p-p_1)A+f$$

with A the piston area, p the pressure inside the cylinder and p_1 the pressure acting on the outer side of the piston.

The net work done on the system during displacement $\mathrm{d}x$ (reckoned as positive when inwards) is

$$\mathrm{d}W = (F+p_1A)\,\mathrm{d}x = pA\,\mathrm{d}x+f\,\mathrm{d}x$$

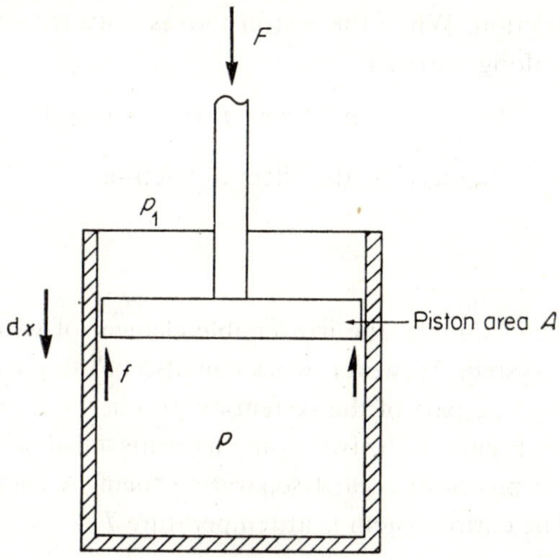

Figure 1.25 A compression process involving external irreversible work

This may be divided into reversible and irreversible elements

$$dW_r = pA\,dx = -p\,dV \quad \text{and} \quad dW_i = f\,dx$$

The reversible component is positive or negative, depending on whether the piston moves inwards or outwards. However, for this adiabatic case, the entropy change is given by equation (1.5.11) as

$$dS = \frac{dW_i}{T} \tag{1.5.23}$$

Then the second law requires that

$$dW_i = f\,dx > 0$$

This states that the friction force must act to resist the motion of the piston, whether it moves inwards or outwards. Note also that the entropy of the system is always increased by external irreversible work, whether the net work (including reversible work) is positive or negative.

So far only the work done on the system comprising cylinder and contents has been considered. In practice, the work transmitted along the piston rod is likely to be of greater interest. When the piston moves inwards, the work transmitted inwards along the rod is

$$F\,dx = (p - p_1)A\,dx + f\,dx > -(p - p_1)\,dV$$

The work required to compress the substance is greater than it would be in the absence of friction. When the piston moves outwards, the work transmitted outwards along the rod is

$$-F\,dx = -(p-p_1)A\,dx - f\,dx < (p-p_1)\,dV$$

The useful output is reduced by the effect of friction.

Internal Work

In the preceding example, the irreversible element of work was supplied from outside the system. However, work can also be done within an isolated system, that is, by one part of the system on another. Consider the isolated system shown in Figure 1.26, two compartments filled with compressible substances, and a piston of area A separating them. A friction force f acts on the piston. The entire system is at temperature T.

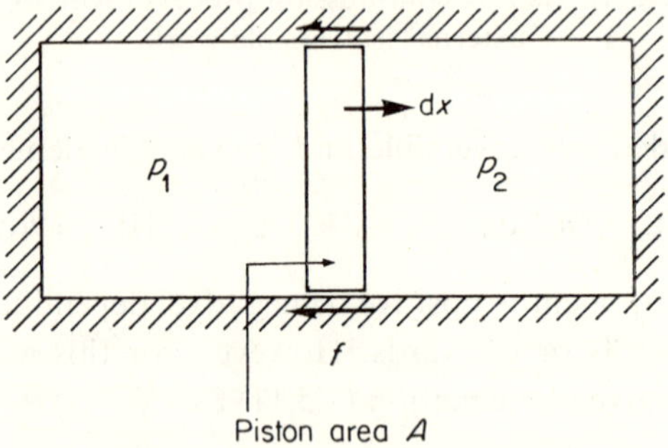

Figure 1.26 A system within which internal irreversible work can be done

The condition of static equilibrium gives

$$p_2A + f = p_1A$$

Imagine that the piston is displaced a small distance dx to the right and the system is then allowed to settle down. Evidently the only irreversible work is that done on the piston, the substances in the two compartments undergoing bulk compression and expansion. Hence for the three elements of the system the changes in internal energy, heat additions and work are related by

$$\begin{aligned} dE_1 &= dQ_1 - p_1A\,dx \\ dE_2 &= dQ_2 + p_2A\,dx \\ dE_p &= dQ_p + dW_i \end{aligned}$$

The individual changes in internal energy and heat additions are unknown.

However, since the system as a whole is isolated

$$\mathrm{d}E_1+\mathrm{d}E_2+\mathrm{d}E_p = 0$$

and

$$\mathrm{d}Q_1+\mathrm{d}Q_2+\mathrm{d}Q_p = 0$$

Hence, summing the three sets of contributions, one obtains

$$\mathrm{d}W_\mathrm{i} = (p_1-p_2)A\ \mathrm{d}x = f\,\mathrm{d}x$$

This element of work provides the net contribution to the entropy change of the system, which is thus found to be

$$\mathrm{d}S = \frac{\mathrm{d}W_\mathrm{i}}{T} \tag{1.5.24}$$

The second law, as applicable to an isolated system, gives

$$\mathrm{d}W_\mathrm{i} = f\,\mathrm{d}x > 0$$

This indicates that the only force which can arise naturally within the system is one retarding the motion—a friction force.

Example 1.11 dealt with a mechanical system on which reversible work was done (by compressing a spring) and within which irreversible work was done by friction. It was seen clearly that this internal irreversible work had the effect of reducing the mechanical energy and increasing the internal energy of the system. This energy transfer is called *dissipation*, and the energy transferred is the dissipated energy. Application of the second law shows that dissipation is always positive, that is, consists of a reduction in the mechanical energy of the system and an increase in its internal energy. This is another case in which energy may be said to be degraded, that is, converted from mechanical to thermal form.

The examples discussed above have revealed four ways of increasing the entropy of a system:

(1) external heat flow into the system,
(2) internal heat flow (across a finite temperature difference),
(3) external irreversible work, and
(4) internal irreversible work.

Only one way of reducing the entropy has been found—external heat flow out of the system.

Another way of expressing these results is to say that internal heat flow and external and internal irreversible work all give rise to *production of entropy*. For an infinitesimal heat addition, the entropy production within the system is

$$\mathrm{d}P = \mathrm{d}S - \frac{\mathrm{d}Q}{T}$$

where $\mathrm{d}S$ is the total entropy change within the system, including contributions from the three mechanisms of production. Hence the second law can be interpreted as

$$\mathrm{d}P > 0$$

simply the requirement that the entropy production be positive in any real process.

Using the idea of entropy production, we can express the entropy change in a more general way. Equation (1.5.11)

$$T\,\mathrm{d}S = \mathrm{d}W_{\mathrm{i}}+\mathrm{d}Q = \mathrm{d}E+p\,\mathrm{d}V$$

gave the entropy change for a process in which the only means of entropy production was external irreversible work. This can be generalized to

$$T\,\mathrm{d}S = T\,\mathrm{d}P+\mathrm{d}Q = \mathrm{d}E+p\,\mathrm{d}V \tag{1.5.25}$$

where $\mathrm{d}P$ is the total production of entropy in the process. The production $\mathrm{d}P$ includes not only external irreversible work, like that of equation (1.5.23), but also internal irreversible work, like that of equation (1.5.24), and the effect of internal heat flow, as represented in equation (1.5.22). This broadened interpretation of entropy change is particularly useful in studying fluid flows, in which internal friction and mixing are major causes of entropy change. These dissipative processes will be studied in detail in a later chapter.

In the past the principle of irreversibility has been expressed in a number of ways. No matter which form is chosen for the fundamental statement, the other possible forms can be derived from it. The statement adopted here (1.5.19) has an advantage over possible alternatives in leading rapidly and easily to these corollary results. Some of the *alternative expressions of the second law* are:

(1) Heat cannot of itself pass from a colder to a hotter body. This has already been shown to be a consequence of the form of the second law adopted here.

(2) It is impossible to construct an engine which would operate through a cycle and convert all the heat supplied to work. In the following section, this will be shown to follow from the present statement of the second law.

(3)

$$\Delta S = \int_{1\,\mathrm{rev}}^{2} \frac{\mathrm{d}Q}{T} \geqslant \int_{1}^{2} \frac{\mathrm{d}Q}{T}$$

the two integrations being carried out between the same terminal states, but the first following a reversible path, that is, one involving no entropy production. Our present formulation leads to these results as follows. In general, $T\,\mathrm{d}S = \mathrm{d}Q + \mathrm{d}P$ but for a reversible process, $\mathrm{d}P = 0$, and

$$\Delta S = \int_{1\atop \mathrm{rev}}^{2} \frac{\mathrm{d}Q}{T}$$

Integration of the elemental statement (1.5.19) gives

$$\Delta S \geqslant \int_{1}^{2} \frac{\mathrm{d}Q}{T}$$

(4) All the king's horses and all the king's men couldn't put Humpty Dumpty together again. This can be demonstrated by dropping an egg on the floor.

1.5.5 Thermodynamic Cycles

The discussion leading to the definition of entropy and the statement of the second law made no direct reference to thermodynamic cycles. However, by applying the second law to each of the processes making up a cycle, one can investigate the role of irreversibility in the cycle as a whole. Cyclic processes, in which a system passes repeatedly through the same series of states, are the basis of several much-used energy-converting devices. Two important examples are the refrigerator, which uses work to produce a flow of heat, and the steam-power cycle, which uses a heat flow to produce work. The discussion which follows will relate particularly to the latter case, but the general principles are more widely applicable.

Figure 1.27(a) shows schematically the hardware of the steam plant of an electricity generating station. The steam passes in turn through boiler, turbine, condenser, pump, and back into the boiler. Figure 1.27(b) indicates the variations of pressure and specific volume through the cycle. The pressure is raised by the pump and falls through the turbine, but is fairly constant in the boiler and condenser, where the major heat flows take place.

The essential external energy flows of the steam cycle are those shown in Figure 1.28. In this simplified cycle, the working fluid accepts heat Q_1 from a high-temperature source, rejects heat Q_2 to some low-temperature source, and does work W_0 ($= -W$, the work done on the fluid in the cycle). In a steam plant the boiler is the heat source, the condenser the sink, and the turbine absorbs work. Actually, the steam cycle involves other energy flows; a small amount of work is done on the condensed water by the pump, and there is some heat loss through the walls of the pipes through which the steam passes. Nevertheless, the simple cycle is close enough to a real cycle to provide a useful model of the actual operation.

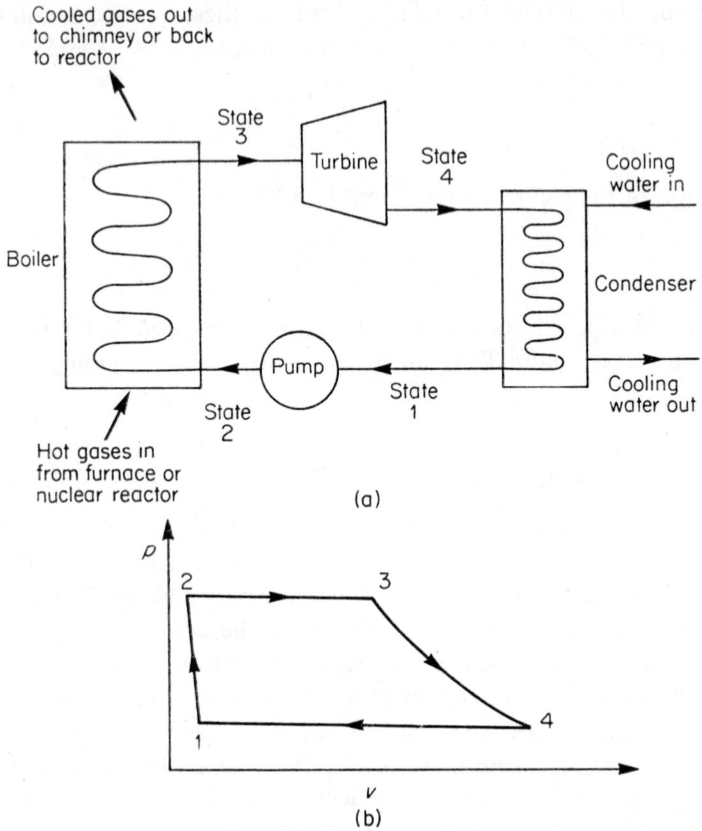

Figure 1.27 General features of a liquid/vapour power cycle. (a) Schematic layout of a plant driving a steam turbine. (b) Variations of pressure and specific volume during the cycle undergone by the steam

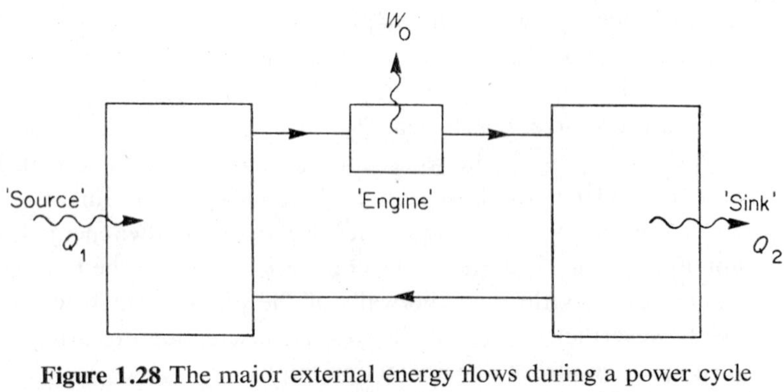

Figure 1.28 The major external energy flows during a power cycle

Let us see what the first law has to say about the simple cycle of Figure 1.28. The properties of the working fluid all recover their original values at the end of the cycle; this is the defining characteristic of a cycle. Then, applied over the cycle, the first law gives

$$\Delta E = Q_1 - Q_2 - W_0 = 0$$

Hence the heat flows and work output are related by

$$W_0 = Q_1 - Q_2$$

The *energy-conversion efficiency* of the simple cycle may be calculated as the ratio of work output to heat input; thus

$$\eta = \frac{W_0}{Q_1} = 1 - \frac{Q_2}{Q_1} \tag{1.5.26}$$

Obviously the greatest work output and highest efficiency are achieved when the rejected heat Q_2 has the least possible value. But how is it possible to reduce the ratio Q_2/Q_1? Since the second law relates heat flows to entropy changes, it may be expected to cast some light on this question.

We shall look first at a very special cycle—the *Carnot cycle.* For the moment the special feature which is of interest is simplicity; in this idealized cycle, the entropy changes can be calculated very easily. However, these idealizations make the Carnot cycle special in another way; it will be found that it provides the most efficient possible conversion of heat to work. In this ideal cycle no entropy is produced within the system; the only entropy changes result from heat acceptance and rejection while the temperature of the working substance is maintained constant at T_1 and T_2, respectively. These entropy changes are

$$\Delta S_1 = \frac{Q_1}{T_1} \quad \text{and} \quad \Delta S_2 = -\frac{Q_2}{T_2}$$

The net change in entropy over the cycle must be zero. Then

$$\Delta S = \Delta S_1 + \Delta S_2 = 0$$

and

$$\Delta S_2 = -\Delta S_1$$

or

$$\frac{Q_2}{T_2} = \frac{Q_1}{T_1} \tag{1.5.27}$$

For a Carnot cycle, the ratio Q_2/Q_1 depends only on the temperatures of acceptance and rejection of heat. Hence the efficiency (1.5.26) can be written for the Carnot cycle

$$\eta_c = 1 - \frac{T_2}{T_1} \tag{1.5.28}$$

The special features of the Carnot cycle are shown by the T, s-diagram in Figure 1.29(a). The three energy flows can be interpreted as areas on this diagram. The area below the cycle path is $-T_2\,\Delta S_2$, while this area plus that enclosed by the cycle path is $T_1\,\Delta S_1$. Since

$$W_0 = Q_1 - Q_2 = T_1\,\Delta S_1 - (-T_2\,\Delta S_2)$$

the area enclosed by the cycle path represents work output during the cycle. Figure 1.29(b) describes a cycle operating between the same temperature limits, but differing from the Carnot cycle in two ways: heat acceptance takes place with changing temperature, and internal production gives rise to additional entropy changes. The area below the cycle path is still a measure of the heat rejected. A comparison of this area with that enclosed by the cycle path indicates that the cycle efficiency is less than the Carnot efficiency for the same temperature limits.

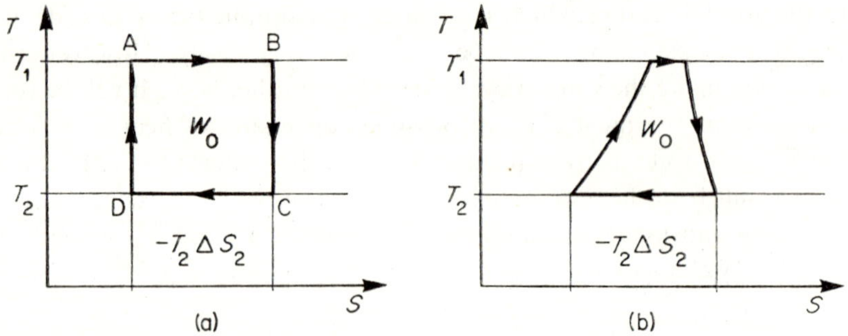

Figure 1.29 Temperature and entropy changes around idealized cycles. (a) A Carnot cycle; (b) A somewhat more realistic cycle

The reduction of the efficiency of a real cycle below the Carnot value can also be shown analytically. Consider a cycle during which the internal entropy production is S_i, and in which the average temperatures of heat input and rejection are f_1T_1 and f_2T_2, where T_1 and T_2 are the maximum and minimum temperatures of the cycle, so that $f_1 < 1$ and $f_2 > 1$. The net entropy change around this cycle is then

$$\Delta S = \frac{Q_1}{f_1T_1} - \frac{Q_2}{f_2T_2} + S_i = 0$$

Then

$$\frac{Q_2}{Q_1} = \frac{T_2}{T_1} + \left(\frac{f_2}{f_1} - 1\right)\frac{T_2}{T_1} + \frac{f_2 T_2 S_i}{Q_1}$$

and the general result (1.5.26) gives

$$\eta = \eta_c - \left(\frac{f_2}{f_1} - 1\right)\frac{T_2}{T_1} - \frac{f_2 T_2 S_i}{Q_1} \tag{1.5.29}$$

As expected, the terms representing temperature variation during heat flow and internal irreversibility are both negative. Hence the Carnot cycle has the highest efficiency of any cycle operating between specified temperature limits.

The predictions of the Carnot model cycle can be misleading if interpreted too literally. However, the significance of the temperature limits of real cycles is correctly indicated. It is generally desirable to make the temperature of heat addition as high as possible, and to lower the temperature of the working fluid as far as possible before rejecting heat. The upper temperature limit is usually determined by the mechanical properties of the materials enclosing the working fluid. For example, in order that the steel tubes in the hottest part of a boiler do not lose too much of their strength, the steam temperature is currently limited to about 1100 °F or 600 °C. In a real cycle the heat sink is usually the atmosphere or a body of water. Hence the lowest temperature of the working fluid must be somewhat above the environmental level. The minimum temperature achieved in an actual steam cycle might be expected to be around 90 °F or 30 °C. Hence the efficiency of a steam plant must be expected to be below the limiting Carnot value of

$$\eta_c = \frac{T_1 - T_2}{T_1} = \frac{1010}{1560} = 65 \text{ per cent}$$

Now let us see how a real cycle differs from the Carnot model. Compare Figure 1.30, giving the T, s-variation during the steam cycle of Figure 1.27, with the Carnot pattern of Figure 1.29(a). Only during evaporation (a–b) is the temperature constant during real heat addition; during pre-boiling (2–a) and superheating (b–3) stages the temperature of the fluid rises as heat is added at constant pressure. Note too that the flow through the turbine (3–4) involves a small increase in entropy. These features are characteristic of realistic power cycles: real heat addition usually involves considerable variations in temperature; on the other hand, nearly isothermal heat rejection can be achieved, because large heat sinks are available in the atmosphere, and in lakes, rivers and seas.

In order to get some idea of the effect of these departures from the ideal

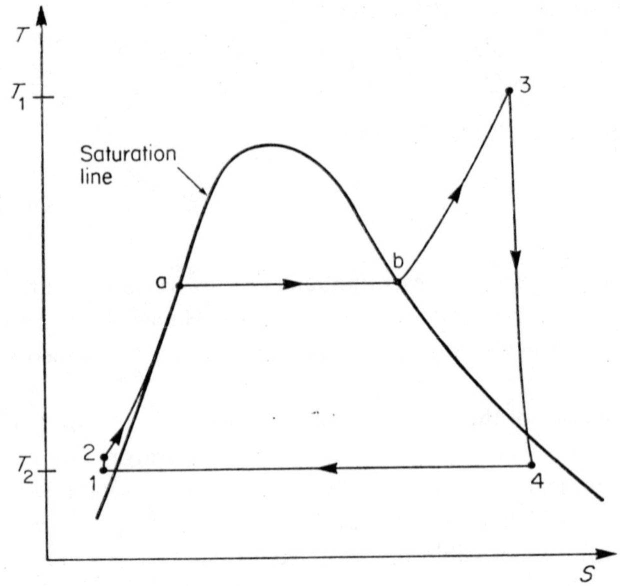

Figure 1.30 Variations of temperature and entropy during the steam cycle of Figure 1.27

pattern, take $f_1 = 2/3$ and $f_2 = 1$, and assume that internal entropy production gives rise to a 10 per cent increase in heat rejection. Then

$$f_2 T_2 S_i = \frac{1}{10}\left(\frac{f_2 T_2}{f_1 T_1}\right) Q_1$$

The result (1.5.29) now gives

$$\eta = \eta_c - \frac{1}{2} \cdot \frac{T_2}{T_1} - \frac{3}{20} \cdot \frac{T_2}{T_1}$$

For the temperature limits used above, 1100 °F and 90 °F, the efficiency is

$$\eta = 0{\cdot}65 - 0{\cdot}18 - 0{\cdot}05 = 42 \text{ per cent}$$

Efficiencies of this order are achieved in steam plants, little more than half the ideal Carnot value for the same temperature limits.

One more idea will be developed from the Carnot cycle before we leave it. The result (1.5.27) can be written

$$\frac{T_2}{T_1} = \frac{Q_2}{Q_1} \tag{1.5.30}$$

This relationship between temperatures and heat flows arises from the choice made in equation (1.5.10) in relating entropy and the absolute gas temperature. It is possible, however, to reverse the line of argument. The relationship (1.5.30) is then interpreted as a definition of a *thermodynamic temperature scale*. By calculating the heat flows Q_1 and Q_2 for a perfect gas undergoing a Carnot cycle, one can next show that the absolute gas scale is identical (except perhaps for a constant factor of proportionality) to the thermodynamic scale. Finally, it is concluded that the definition (1.5.10) is the appropriate relationship between entropy and the absolute temperature.

The study of thermodynamic cycles will not be taken further in this book. The general principles of thermodynamics will be used instead to investigate a variety of steady-flow processes. However, many of these flow processes form parts of cycles; for example, flows through pumps, turbines and heat exchangers are parts of the cycle undergone by steam in its power-generating circuit. Thus the subject matter of this book does ultimately relate to cyclic processes. It provides the basis for the detailed analysis of a cycle, in which the operation of each component must be given careful scrutiny.

Example 1.20 Although the Carnot cycle is not in fact used to generate power, it is a useful source of simple examples of thermodynamic calculations. Consider a cylinder of diameter 1 ft, which contains 1/10 lb of air, initially (point A in Figure 1.29(a)) at $p_A = 200$ lbf/in² and $T_1 = 1100$ °F. The air may be taken to be a perfect gas with $R = 53{\cdot}3$ ft lbf/lb °R and $\gamma = 1{\cdot}4$. The heat addition (AB) is sufficient to expand the gas by a factor of four. It is followed by isentropic expansion (BC) to 212 °F, the cylinder then being cooled by water boiling at atmospheric pressure on its outside surface during the heat rejection process, CD.

(a) Heat Addition

For this isothermal process undergone by a perfect gas

$$pV = p_A V_A = mRT_1$$

Hence the final pressure is given by

$$p_B = p_A \frac{V_A}{V_B} = 50 \text{ lbf/in}^2$$

Also

$$W_{AB} = -\int_B^A p\,dV = -mRT_1 \log_e \left(\frac{V_B}{V_A}\right)$$

$$= -\frac{53{\cdot}3}{10} \times 1560 \times \log_e 4 = -11{,}500 \text{ ft lbf}$$

Since $T = T_1$ throughout this process, $\Delta E = 0$ for this perfect gas, and

$$Q_1 = -W_{AB} = \frac{11{,}500}{778} = 14{\cdot}8 \text{ Btu}$$

The entropy change is given by

$$\Delta S = \frac{Q_1}{T_1} = 7{\cdot}38 \text{ ft lbf/ °R}$$

(b) Isentropic Expansion

Here $Q = 0$, and

$$W_{BC} = E_C - E_B = mc_v(T_2 - T_1)$$

$$= \frac{53{\cdot}3}{10} \times \frac{212-1100}{0{\cdot}4} = -11{,}800 \text{ ft lbf}$$

The pressure and volume changes are given by the isentropic gas relations (1.4.13):

$$p_C = p_B\left(\frac{T_2}{T_1}\right)^{\frac{\gamma}{\gamma-1}} = 50\times(0{\cdot}43)^{3{\cdot}5} = 2{\cdot}61 \text{ lbf/in}^2$$

The initial volume is $V_A = mRT_1/p_A = 0{\cdot}289$ ft^3. Hence

$$V_C = V_A \frac{V_B}{V_A}\left(\frac{T_2}{T_1}\right)^{\frac{-1}{\gamma-1}} = 0{\cdot}289\times4\times(0{\cdot}43)^{-2{\cdot}5} = 9{\cdot}51 \text{ ft}^3$$

The total volume change is $V_C - V_A = 9{\cdot}22$ ft^3, and the stroke of the piston is $9{\cdot}22/(\pi/4) = 11{\cdot}75$ ft. This very large stroke is obviously impractical, and goes some way to explaining why the Carnot cycle is not utilized in real machines. It also shows why compound steam engines proved attractive, despite their greater mechanical complexity. In a compound engine the steam is first expanded in a small-diameter high-pressure cylinder, and then in a low-pressure cylinder of larger diameter. Without this artifice, the large volume changes required to extract the available work from highly energized steam could not be realized within a piston engine.

(c) Net Effects

One could continue with calculations of the kind used above for the other two processes of the cycle. However, the results already found, taken with the general results for a Carnot cycle, are sufficient to determine the basic features of the operation. From equation (1.5.27)

$$Q_2 = Q_1\frac{T_2}{T_1} = 14{\cdot}8\times0{\cdot}43 = 6{\cdot}4 \text{ Btu}$$

The work done per cycle is

$$W_0 = Q_1 - Q_2 = 14{\cdot}8 - 6{\cdot}4 = 8{\cdot}4 \text{ Btu}$$

and the efficiency is

$$\eta_c = \frac{8{\cdot}4}{14{\cdot}0} = 57 \text{ per cent}$$

PROBLEMS FOR CHAPTER 1

Group A

1.1 Show that a water tap can be treated as an open or closed system, depending on the precise way in which the system boundaries are selected.

1.2 Consider an automobile as a thermodynamic system during the course of a day of its working life, in which it spends some time standing in a garage, some time running at high speed along the road, some time starting and stopping in traffic, some time parked in the sun and some time parked in the rain.

(a) Describe the system at various times during the day using the terms closed, open, equilibrium, homogeneous and isolated.

(b) Give five leading properties of the system for each of the states considered in (a).

(c) Describe the processes undergone by the system using the terms isothermal, isobaric, adiabatic, steady-flow, flow and non-flow.

(d) What cyclic processes are involved in the life of the vehicle?

1.3 (a) Estimate the mass of air in the room in which you are sitting.

(b) What mass of helium at the same temperature and pressure would the room contain?

1.4 During the compression stroke within the cylinder of an internal combustion engine, the mixture of air and vaporized fuel goes from a pressure of 0·95 atm to a pressure (after ignition) of 50 atm.

(a) Express the difference between these pressures and the atmospheric pressure (this difference is called the gauge pressure) in lbf/in^2, lbf/ft^2, in of water, in of mercury, bar, mbar, N/m^2, N/mm^2.

(b) Which units are of convenient size for pressure differences of these magnitudes?

1.5 An aeroplane flying horizontally at 600 m/sec near the equator at an altitude of 10,000 m experiences a local acceleration of gravity of 9·75 m/sec^2. The local atmospheric pressure and temperature are 0·25 bar and −40 °C. Embedded in the wing of the aeroplane is a fuel tank which extends 4 m in the fore-and-aft direction and 0·5 m in the vertical direction. The volume of the tank is 2·4 m^3; it is completely filled with fuel of specific gravity 0·8.

(a) What is the altitude in feet and miles?

(b) What is the speed of flight in km/hr, ft/sec, miles/hr, and knots?

(c) What is the local value of g in ft/sec^2?

(d) What is the ambient temperature in °K, °R and °F?

(e) What is the ambient pressure in N/m^2, kN/m^2, lbf/in^2 and lbf/ft^2?

(f) What is the specific weight of the fuel (that is, the weight per unit volume) in kgf/m^3 and lbf/ft^3?

(g) What are the density and specific volume of the fuel in kg/m^3, lb/ft^3, $slugs/ft^3$, and their reciprocals?

(h) What is the volume of the fuel tank in ft^3, litres and gallons?

(i) What is the weight of fuel in the tank in N, kgf, lbf, tons and metric tons?

(j) What is the mass of fuel in the tank in kg, lb and slugs?

(k) What is the difference in pressure between the uppermost and lowest points in the tank in bar, lbf/in^2, lbf/ft^2, N/m^2 and N/mm^2?

1.6 (a) What is the phase of water at

(1) 5 bar, 600 °C,
(2) 10 bar, 0·3 m^3/kg,
(3) 10 bar, 0·18 m^3/kg, and
(4) 5 bar, 100 °C?

(b) What is the density of the water for cases (1) and (4)?
(c) What is the temperature of the water for cases (2) and (3)?

1.7 Consider steam leaving a turbine in two conditions:

(1) 5 lbf/in^2 and 100 ft^3/lb, and
(2) 5 lbf/in^2 and 70 ft^3/lb.

(a) What is the temperature for these two conditions?
(b) What are the wetness and dryness fractions?

1.8 Imagine that the room in which you sit is filled with saturated steam.

(a) What must be its temperature, if the pressure is atmospheric? If the pressure is 10 bar?
(b) What mass of steam would there be in the room under these conditions?
(c) If the steam were condensed, what would be its volume?

1.9 A frictionless piston is maintained in equilibrium by a spring (spring constant 200 lbf/in) on one side and high-pressure steam on the other. Initially, the pressure in the steam is 100 lbf/in^2 (abs), and the spring force is 1000 lbf. The steam pressure is then increased to 200 lbf/in^2 (abs), the pressure on the spring side retaining the atmospheric level.

(a) How far does the spring deflect?
(b) How much energy does the spring absorb?
(c) How much work does the steam do?

1.10 One means of providing a variable-speed power supply from a shaft turning at constant speed is a hydraulic coupling, within which the energy is transferred between two sets of rotating vanes. There is necessarily some dissipation within the coupling, so that the output of mechanical power does not equal the input. If the input to such a coupling is 20 kW and the output shaft rotates at 70 rev/min against a torque of 200 kgf m,

(a) at what rate is energy dissipated within the coupling, and
(b) what is the efficiency of the transmission?

1.11 Consider the energy transfers which would occur in one of the processes postulated in (c) of Problem 1.8, condensation of steam at 1 atm.

(a) If the walls move in slowly as condensation takes place, so that the pressure remains constant, what work is done on the steam as it condenses?
(b) What is the net change in its internal energy as a result of this work?
(c) What are the changes in specific internal energy and enthalpy during condensation? How are they related?

1.12 In high-speed gas flows the kinetic energy of the fluid can be a significant proportion of the fluid's total energy. This is illustrated by the following calculations. The air approaching the rotor of a turbine has a velocity of 1000 m/sec; this is termed the absolute velocity. The rotor is in motion, and the air velocity relative to it is 500 m/sec. The temperature of the air is 800 °C.

(a) Express the specific kinetic energy of the absolute and relative flows in mechanical and thermal units: kJ/kg, ft lbf/sl, kcal/kg and Btu/lb.

(b) Find the ratio of the kinetic energy of the gas to its specific enthalpy, c_pT, for the absolute and relative flows.

1.13 Argon has specific heat $c_p = 0{\cdot}52$ kJ/kg °C, constant over a wide range of conditions as a result of the simple molecular structure of the gas.

(a) Express this specific heat in cal/g °C, kcal/kg °C, Btu/lb °F, and ft lbf/slug °F.

(b) In a cooler, 20 lb/min of argon are reduced in temperature by 400 °C, the pressure being sensibly constant. At what rate is heat removed from the flowing gas?

1.14 For the states defined in Problem 1.7, what are the specific enthalpies?

1.15 Use the methods developed in Section 1.4.2 to show that the approximation $\mathrm{d}v/v = -\varkappa\, \mathrm{d}p$ is appropriate when metals are compressed adiabatically. Consider the case of copper, whose properties are given in Appendix III.

1.16 Callendar's empirical equation of state for steam was introduced in Example 1.10.

(a) Write it in one of the consistent sets of units of Table 1.1.

(b) Write the perfect-gas law in the same form.

(c) For what value of γ do the two equations correspond?

1.17 For the states defined in Problem 1.7, what are the specific entropies?

1.18 An automobile radiator transfers energy to the air passing through it at the rate of 10 hp. The water temperature drops from 200 °F to 150 °F across the radiator and the air temperature rises from 70 °F to 90 °F.

(a) Find the air and water flow rates.

(b) Find the rate of entropy increase in the gas (assuming constant pressure and perfect-gas behaviour) and in the water (assuming constant density and specific heat).

(c) What is the net rate of entropy production?

1.19 A steam plant produces 500 MW of electrical power, rejecting heat to a nearby river whose flow is 5000 ft^3/sec. Assuming a realistic efficiency for the plant, estimate the rise in the temperature of the river water. Estimate also the rate at which the entropy of the river water rises.

1.20 (a) What is the change of entropy in the process discussed in Problem 1.11?

(b) How are the heat flow and entropy change related?

1.21 A manufacturer of steam turbines claims that one of his machines operates between the steam conditions 100 bar, 500 °C and 2 bar, 9·5 per cent wet.

(a) Is this a reasonable claim?

(b) How would you expect the outlet state to differ from that claimed, if the inlet condition and outlet pressure were those quoted?

1.22 Using the labels isentropic, adiabatic, isobaric and isothermal, classify the processes making up

(a) a Carnot cycle (Figure 1.29(a)), and

(b) a more realistic steam cycle (Figures 1.27, 30).

Group B

1.23 (a) What is the net uplift on a hydrogen-filled balloon of volume 1000 ft^3 at sea level? (Take the internal pressure and temperature to be those of the environment).

(b) What is the uplift on the same balloon at an altitude of 5000 m? (Appendix VI gives appropriate values of the atmospheric pressure, temperature and density.)

1.24 The airbrakes of the aeroplane considered in Problem 1.5 are extended, giving a deceleration of 1·7 g in the direction of flight. Here g has its standard value of 9·81 m/sec^2.

(a) What is the effective specific weight of the fuel?

(b) What force is applied to the tank by the fuel in it?

(c) What is the pressure difference between the furthest-forward, lowest point of the tank and the furthest-aft, highest point?

1.25 For air the parameters which appear in van der Waals equation are $a = 344$ atm$(ft^3/lb\ mol)^2$ and $b = 0{\cdot}585\ ft^3/lb\ mol$.

(a) What values should be used if absolute engineering units are to be introduced into the equation?

(b) What values are appropriate to SI units?

1.26 (a) Using binomial expansions, show that van der Waals equation can be written approximately as

$$v = \frac{RT}{p} - \frac{a}{RT} + b$$

(b) Calculate β and $\varkappa$ for this kind of gas, retaining only terms of the first order in a and b.

(c) Calculate β and $\varkappa$ using the full van der Waals equation.

(d) Under what circumstances are the results obtained in (b) a correct representation of the full results?

1.27 (a) Apply the results of Example 1.8 to find the work done in compressing water from atmospheric pressure to 1000 lbf/in^2 (abs); the water is initially at 100 °F. If you use the simple formula, justify that action.

(b) By what percentage does the volume change during this compression?

(c) By how much would the temperature rise if this amount of energy were added in the form of heat? By how much would the volume change in this case?

1.28 Consider again the compressive process discussed in the preceding problem, but now imagine that a valve at the back of the cylinder opens when the compression is completed, the compressed water then being expelled into a reservoir maintained at 1000 lbf/in^2 (abs).

(a) Show that the work done in the combined process is given approximately by $w \simeq v_1 p'(1+\frac{1}{2}\varkappa p')$.

(b) Hence show that the power expended when the fluid is compressed at rate $\dot{V}$ (volume per unit time) is $\dot{V}p'(1+\frac{1}{2}\varkappa p')$.

(c) Find the power required to compress water at the rate of 10 lb/sec.

(d) Can you suggest a simpler formula, appropriate in cases in which the volume does not change appreciably?

1.29 A tank containing 1 m^3 of water is heated by an electrical heater. The rate of heat transfer to the water is given by $(200 - T)$ kW with T the water temperature in °C. Heat is conducted away from the outer surface of the tank at rate $0{\cdot}1\,(T-20)$ kW.

(a) How long does it take to raise the temperature from 20 °C to 100 °C?

(b) What proportion of the heat supplied in this process remains in the water?

(c) The heater is controlled by a thermostat which maintains the temperature within the range 80 °C to 100 °C. What is the length of its operating cycle?

1.30 In an electricity generating station the boiler pressure is 100 bar, the steam temperature rising from 200 °C to 700 °C in the constant-pressure heating process. The steaming rate is 40,000 kg/hr.

(a) Find the enthalpy rise in the boiler in kJ/kg and Btu/lb.

(b) Find the rate at which heat is absorbed in the boiler in kcal.

(c) Forty percent of the energy of (b) is absorbed by the turbine driving the generator. Express this power in hp, kW and MW.

(d) Fifty-five per cent of the energy of (a) and (b) is rejected in the condenser. Express this in kcal/kg, Btu/lb and Btu/hr.

(e) Two per cent of the energy of (a) and (b) is absorbed in pumping the fluid around the circuit. Express this in kW, hp and metric hp.

1.31 Consider a compression stroke in an air compressor during which the volume of the air changes from 1 ft^3 to $0{\cdot}2$ ft^3; the initial pressure and temperature are the normal atmospheric values.

(a) Taking the pressure and velocity to be related by $pV^{1{\cdot}3} =$ constant, find the final pressure and temperature.

(b) What is the work done on the air during the compression?

(c) What is the change in internal energy during the process?

(d) What is the heat flow into the gas in the process?

(e) If the process were adiabatic, would more or less work be required during the compression stroke?

1.32 For isentropic processes in steam it is proposed to adopt a law of the form $pv^{\gamma} =$ constant. The appropriate value of γ is to be found using steam-table data.

(a) Show that

$$\gamma = -\frac{\Delta p/p}{\Delta v/v}$$

relates changes between nearby states whose entropy is the same.

(b) Find a pair of states in the steam tables and hence calculate γ. (There is a reasonably satisfactory pair near 120 bar and 425 °C; interpolation will provide others.)

(c) Is the value obtained about that which would be expected for a triatomic gas? (A simple model suggests $\frac{4}{3}$.)

(d) Can you think of any other way of determining γ from steam-table data?

1.33 Given the canonical equation of state (1.5.18) and the relations (1.5.6, 10), show that $pv = RT$. This is in effect working backwards through the derivation of the canonical equation of state.

1.34 The following simple experiment illustrates the diffusive effects of random molecular activity. Place eleven coins in a row, six copper coins at the right end, and five silver coins to the left. The locations of the coins can be identified by the digits 0, 1, ... 9, 10.

It is proposed to study the mixing of the copper and silver coins as a result of a series of interchanges between a coin and its right-hand neighbour. A random pattern of interchanges could be obtained by specifying the locations of the coins to be moved using one of the tables of random numbers which have been prepared for exercises of this kind. However, for this simple case the final digits in almost any table will do, for example, those in a table of logarithms, or in the telephone directory.

After carrying out this exercise you should have some feeling for the slowness of diffusion by small-scale processes.

1.35 (a) In the paint-drying process discussed in Example 1.12 air will be carried through the drying tunnel with the body shells. Decide whether this will have a significant effect on the power requirements for the heaters. To do this you will have to estimate some data, for example, the quantity of air travelling with each shell.

(b) From each of the shells evaporates 1 kg of a solvent with latent heat 1200 kJ/kg. Does this process absorb a significant amount of energy from the heaters?

1.36 In a steam turbine the velocity changes from 2000 ft/sec to 400 ft/sec, and the pressure and temperature from 1000 lbf/in^2 and 1000 °F to the values appropriate to saturation at atmospheric pressure.

(a) Find the changes in specific enthalpy, internal energy and entropy for this process.

(b) Find the ratio of the change in kinetic energy to the changes in enthalpy and internal energy.

(c) After it is condensed, the fluid passing through the turbine is returned to the boiler by a feed-water pump in which its pressure rises from atmospheric to 1000 lbf/in^2. The work done per unit mass may be estimated as $v_c\,\Delta p$, where v_c is the nearly constant specific volume of the water. Find the ratio of this energy addition to the enthalpy drop across the turbine.

(d) Comment on the relative importance of the various energy changes found above.

1.37 What is the entropy change in the process considered in Problem 1.31?

1.38 What are the changes in specific entropy in the processes discussed in (a) and (c) of Problem 1.27?

1.39 (a) Repeat the Carnot cycle calculations of Example 1.20, taking steam as the working fluid, but changing none of the other data. Without the simple formulae governing the perfect gas, a rather different method of attack is required. Note that

$$Q = \int T\,dS$$

for wholly reversible work, and that

$$W = \Delta E - Q = \Delta H - \Delta(pV) - Q$$

expresses the work and heat in terms of properties appearing in the steam tables. Then

$$Q = T\,\Delta S$$

gives the heat flow in an isothermal process, and

$$W = \Delta H - \Delta(pV)$$

gives the work for an adiabatic process.

(b) Why does the steam not condense when its temperature drops to 212 °F, at which boiling occurs on the outside of the cylinder?

(c) To what temperature would the steam have to be lowered in order that it reached saturation at one point of the cycle?

Group C

1.40 For many purposes an explicit formula for a property is more convenient than a table of its values at discrete points. For example, a designer of a gas liquefaction plant will have frequent use for properties of metals in the temperature range 100 °K to 250 °K. Suppose he needs to know the variation of the internal energy of copper over this range, and has available the data of Appendix III. Since the specific heat is known, he may proceed by obtaining a formula for its variation and then integrating it.

Formulae for temperature dependence of specific heats are usually given in the form

$$c_v = c_{v_1} + c_1\Delta + c_2\Delta^2 + c_3\Delta^3$$

where $\Delta = T - T_1$. The designer decides to develop such a formula, and is faced with the task of choosing the coefficients c_{v_1}, c_1, c_2 and c_3 to fit the data for $T_1 =$ 100 °K and for $T =$ 150°, 200°, and 250 °K. One way of doing this would be to substitute the three sets of values into the proposed formula, obtaining three equations for c_1, c_2 and c_3. The designer shrinks from the task of solving these equations, and decides to use a differencing technique, which involves merely subtracting successive tabled values, then subtracting these differences, and so on. Let us see how this procedure works.

(a) For equal increments Δ between successive values of T, show that the symbolic differences take the form:

T	c_v	First Differences	Second Differences	Third Differences
T_1	c_{v_1}			
		$c_1\Delta + c_2\Delta^2 + c_3\Delta^3$		
T_2	c_{v_2}		$2c_2\Delta^2 + 6c_3\Delta^3$	
		$c_1\Delta + 3c_2\Delta^2 + 7c_3\Delta^3$		$6c_3\Delta^3$
T_3	c_{v_3}		—	
		—		
T_4	c_{v_4}			

(b) Prepare the same kind of array using the numbers from the table which gives c_v as a function of T, and show that

$$c_v = 3{\cdot}86 + 1{\cdot}633\Delta_1 - 0{\cdot}635\Delta_1^2 + 0{\cdot}092\Delta_1^3$$

where

$$\Delta_1 = \frac{T - 100}{50}$$

(c) Integrate to find a formula for Δe.

1.41 Although the van der Waals equation cannot be expected to be accurate near the critical state, it can be used to obtain correlations between the parameters a and b and the properties of the fluid at the critical point, shown as point c in Figure 1.6(a). Below this point an isotherm, a constant-temperature contour, on the surface of state, is parallel to the v-axis over the portion of the surface corresponding to a mixture of liquid and vapour phases. Since the surface is continuous, the isotherm fcg will be 'horizontal' at the critical point. Then

$$\left(\frac{\partial p}{\partial v}\right)_T = 0 \quad \text{at c}$$

However, this isotherm is horizontal only at that point, so that, in addition

$$\left(\frac{\partial^2 p}{\partial v^2}\right)_T = 0 \quad \text{at c}$$

(a) Assuming that van der Waals equation applies at the critical point, use the conditions obtained above to show that

$$b = \frac{v_c}{3} \qquad a = \frac{9RT_c v_c}{8}$$

and hence that

$$\left(\frac{pv}{RT}\right)_c = \frac{3}{8}$$

(b) Show that the equation can then be written

$$\left(p^* + \frac{3}{v^{*2}}\right)\left(v^* - \frac{1}{3}\right) = \frac{8T^*}{3}$$

where $p^* = p/p_c$, etc., are non-dimensionalized properties.

Note that the gas constant R has been eliminated, as well as a and b, so that this equation applies for all gases. In fact, if one introduces three independent parameters to represent the characteristics of a particular gas, dimensional analysis suggests that a relationship of this kind must exist.

(c) Given that there exists a relationship of the form

$$f(p^*, v^*, T^*) = 0$$

show that this information can be presented in the form

$$\frac{pv}{RT} = f(p^*, T^*)$$

As would be anticipated from the result of (b), the quantity pv/RT, the compressibility factor, is found to be nearly the same function of p^* and T^* for many gases, within the limits of experimental accuracy.

1.42 It is often convenient to represent the real compression of a gas by a law of the form

$$pV^n = \text{constant}$$

This is called the polytropic gas law, since the index n can take on a variety of values; it includes the isentropic and isothermal laws when $n = \gamma$ and 1. The index n must be determined empirically, its value being uncalculable in realistic situations. However, the physical effects which determine the value of the polytropic index are known: internal dissipation and heat transfer to or from the compressor (usually from, since the gas is commonly raised above the ambient temperature by compression).

Here the first and second laws of thermodynamics will be applied to a real compressive process and the effects of irreversibility and cooling will be determined. Only the compression process itself will be considered here. However, if the delivery of the gas is included, as was done in Problem 1.28, a rather similar result is obtained.

(a) Making reference to Example 1.15, show that the reversible work done in compressing a perfect gas following the polytropic law is given by

$$W_r = \frac{mR}{n-1}(T_2-T_1) = \left[\frac{\gamma-1}{n-1}\right]\Delta E$$

(b) For this case the most easily applied form of the second law is the requirement that the remaining work, the irreversible element, be positive: $W_i > 0$. Show that

$$W_i = \Delta E - W_r + Q'$$

where Q' is the heat *removed from* the gas during compression.

(c) Show that

$$n = \gamma + (\gamma-1)\frac{W-Q'}{W_r}$$

and hence that the least value attainable is

$$n_{\min} = \gamma - \frac{(\gamma-1)Q'}{W_r}$$

Describe qualitatively how the polytropic index depends on internal dissipation and cylinder cooling.

(d) What values of n would you expect for compression in an insulated cylinder? How is it possible to achieve values of $n < \gamma$? What are the special features of the situation for $n = 1$? Is it possible to achieve values of $n < 1$?

1.43 The active core of a nuclear reactor occupies an annular region 6 ft high with inner and outer diameters of 5 ft and 2 ft. The fissile material is contained in a large number of parallel fuel pins running parallel to the axis of the core. These pins are 6 ft long and have diameter 0·25 in; they occupy two-thirds of the volume of the core.

The core is cooled by pumping liquid sodium through it in the axial direction, parallel to the pins. The liquid enters at 400 °C and emerges at 600 °C. The mean velocity of the flow amongst the pins is 15 ft/sec just after entry to the core, that is, when the temperature is 400 °C.

The properties of sodium at the low pressures maintained in the reactor core are as follows:

T (°K)	c_p (kJ/kg °C)	ϱ (kg/m³)
600	1·30	873
800	1·26	826
1000	1·26	777

(a) Find the flow area through the core, the number of pins required, and the total surface area of the pins.

(b) What is the power output of the core?

(c) What is the heat flux per unit surface area of the pins?

(d) What is the coefficient of thermal expansion of the liquid?

(e) If the pressure at the top is equivalent to a column of sodium 30 ft high, what are the changes in specific enthalpy and internal energy undergone by the liquid in the core?

1.44 The methods used to study the performance of a Carnot cycle can be applied to engines operating in a more realistic manner. The details of the processes making up the cycle will, of course, be altered. For example, in a conventional spark-ignition engine, the burning of the fuel and heating of the gas occur when the gas has been compressed by the piston nearly to its least volume. The rejection of heat in such an engine takes the form of expelling air in which combustion has taken place; this occurs when the cylinder volume is greatest, after the heated gas has done its work. Some of the essential features of such an engine can be studied by examining the following idealized cycle—the Otto cycle—

AB – addition of heat Q_1 at constant volume V_1
BC – isentropic expansion
CD – rejection of heat Q_2 at constant volume V_2
DA – isentropic compression

(a) Sketch the cycle in p,v and T,s-plots and mark in the states A, B, C, and D.

(b) If the fluid undergoing the cycle is a perfect gas, show that

$$\frac{T_B}{T_C} = \frac{T_A}{T_D} = r_c^{\gamma-1} = r_p^{\frac{\gamma-1}{\gamma}}$$

where the ratio $r_c = V_2/V_1$ is the compression ratio, the factor by which the volume is reduced before heat is added, and $r_p = p_A/p_D$ is the corresponding pressure ratio.

(c) Compute the heat and work flows in the four processes making up the cycle and show that $W_0 = Q_1 - Q_2$ around the whole cycle. Use the result (1.5.17) to calculate the entropy changes in the working fluid around the cycle.

(d) Using the results of (b), show that the energy-conversion efficiency is given by

$$\eta = 1 - \frac{T_D}{T_A} = 1 - r_c^{1-\gamma} = 1 - r_p^{\frac{1-\gamma}{\gamma}}$$

Evaluate for compression ratios typical of modern automotive practice.

(e) Show that the efficiency found in (d) is less than that of a Carnot engine operating between T_B and T_D, but can approach the Carnot efficiency if r_c is large.

(f) Show that the net entropy change per cycle is

$$\Delta S = \frac{Q_2}{T_D} - \frac{Q_1}{T_B} = \frac{Q_1}{T_B}\left(\frac{T_B}{T_A} - 1\right) = \frac{Q_1}{T_B}\frac{\eta_c - \eta}{1 - \eta_c}$$

if all the heat added comes from a source maintained at T_B, and all the heat rejected goes to a sink at T_D.

1.45 The situation to be studied here provides examples of systems of fixed contents and varying contents, and illustrates the kind of assumptions that have to be made in analysing realistic situations.

Twenty men attend a meeting in a room measuring 30 ft by 15 ft by 10 ft. The initial temperature of the room is 70 °F: an estimate is required for the rise in temperature during the meeting. Two energy sources can be distinguished:

(1) the men themselves, generating on average 1/10 hp each through metabolic processes; and

(2) radiation through the windows of intensity 30 langleys/hr; the window area is 60 ft^2, the sun's rays passing through at 45°. (The langley—one cal/cm^2—is a unit in which radiation intensity is often measured.)

In calculating the rise in internal energy of the contents of the room it is necessary to take into account the air in it, but not the bodies of the men, whose temperature is internally regulated. The increase in temperature of parts of the walls and furniture must also be accounted for. The materials of which they are made are not good conductors of heat, and the mass whose temperature does rise during the period of interest may be taken to be 100 kg, with specific heat 0·4 Btu/lb °F, a value near that for wood.

(a) Consider first the case in which the room is completely sealed, so that no air flows in or out. Find the rate of temperature rise. How long could the meeting last under these conditions?

(b) Now consider the case in which air is passed steadily through the room by a fan. Find the equilibrium temperature as a function of airflow, taking the air to enter at 70 °F, and assuming immediate, complete mixing. Plot the equilibrium temperature against the 'time for a change of air.'

(c) What airflow (in ft^3/min, the conventional unit) will be required for the meeting to be carried on in comfort?

(d) There is an intermediate case in which there is no forced ventilation, but air is allowed to leak out of the room as the temperature rises, so that the pressure remains constant. State in general terms how you would approach this problem.

1.46 Consider the system of Example 1.11 with one modification, the mass M now sliding between a pair of vertical guides which apply to it a constant friction force F, positive upwards. Careful consideration must be given to the energy transfer at the boundary of the two-mass system. It is clear that energy in the amount $F(z_0-Z)$ is transformed when the mass M moves through distance (z_0-Z). But where does this energy go? And what is the nature of the transfer—heat or work?

The answer to the first question depends on the details of the surfaces and the materials in contact. A definite answer can be avoided by denoting the proportion of the energy which ends up outside the two-mass system as r. The second question has no definite answer either. It is not fruitful to seek one, since only the sum of the work and heat transfers is of importance. Hence the overall energy addition to the system at the guides may be taken to be

$$Q+W_g=-rF(z_0-Z)$$

This quantity is in fact negative, since energy is imparted to the guides.

(a) Apply Newton's second law to the two masses, and hence find the mechanical energy equation for the two-mass system.

(b) Write the overall energy equation and obtain the thermodynamic energy equation by subtraction.

(c) Find the work terms distinguished in equations (1.3.9, 10, 11).

(d) What is the total dissipation arising in the motion?

1.47 In equation (1.5.6) a definition of thermodynamic pressure is given. Show that this quantity has the characteristic that its values in two systems are equal when the systems are in mechanical equilibrium, by considering systems like that of Figure 1.21, the isolating barrier being a frictionless piston.

2

Isenergic Flow

The first chapter dealt with the thermodynamics of a variety of processes occurring in engineering equipment. This chapter considers also the mechanics of flow processes, that is, changes in pressure and velocity and forces applied to surfaces over which the fluid passes. To simplify this introduction to fluid mechanics, the thermodynamic behaviour of the fluid will be restricted; most of the discussion relates to isenergic flows, that is, flows which do not exchange work or heat with the surroundings. This restriction rules out a complete analysis of devices such as pumps, turbines and heat exchangers, whose function is to extract energy from the flow or to supply energy to it. However, we can examine the flow through passive sections of energy-converting devices, for example, the intake of a pump or jet engine, or a row of fixed blades in a steam turbine. Even when a flowing fluid does not exchange heat or work with its surroundings, it is capable of a surprising range of behaviour. The examination of various isenergic flows will occupy not only this chapter, but the three which follow.

To the fundamental concepts of thermodynamics—state, energy conservation and irreversibility—must be added two basic principles of fluid mechanics. First the *geometry* of the flow must be taken into account. Many of the flows to be considered are confined by ducts or other solid boundaries, for example, a system of pipes or the blading of a compressor. Other flows are influenced to a smaller extent by solid boundaries, for example, those over the wing of an aeroplane and over the bed of a stream. In every case the presence of the flow boundaries must be taken into account. This can be done by requiring, first, that fluid not pass through the solid boundaries and, secondly, that mass be conserved in every part of the flow. These requirements are expressed in a *continuity equation*, linking the geometry of the flow to the dynamic and thermodynamic behaviour of the fluid.

The second important concept of fluid mechanics is the *momentum principle*. This simply states that the momentum change in a flow is equal to the force applied to the fluid. The application of the principle is not so simple. It is sometimes convenient to study a system of fixed contents, sometimes one with fixed boundaries, and sometimes one with neither characteristic. Hence the momentum principle must be expressed in forms appropriate to each kind of system.

The relationship between momentum and force provides a way of calculating the forces applied by the flow to its boundaries. This information is required for the design of the structures supporting the boundaries, for example, the propeller of a ship, the intake of a hydraulic turbine, the wing of an aeroplane or the roof of a building. It should not be inferred that the momentum principle alone determines the forces exerted by the fluid. In general, they are determined by combining the laws of mechanics with the thermodynamic characteristics of the fluid. For flow which is isentropic as well as isenergic, changes in velocity and thermodynamic properties are connected by simple results called *Bernoulli equations*. The relationship between pressure and velocity in isentropic flow has a wide range of application in calculating forces and in measuring flow rates.

Even when a fluid is stationary, it can apply forces on its boundaries. The force may result from a uniform pressure, as does that applied to an automobile tyre by the air inside. However, in many cases large pressures are generated simply by the weight of the fluid and therefore vary with height. The force on a dam blocking a valley to form a reservoir is of this kind. The atmospheric pressure at the earth's surface is simply the weight of the air in the atmosphere above. Evidently it is worthwhile giving special attention to *hydrostatic pressure*, that generated in still fluid by its own weight.

In the first chapter it was seen that the common fluids of engineering display a wide range of thermodynamic behaviour. As a result the techniques used to analyse the flows of gases, vapours and liquids are strikingly different, although the same fundamental principles apply to each species. In this book these principles will be applied to three *kinds of fluid:*

The perfect gas, which typifies a rarefied and highly compressible fluid;

The constant-density liquid, which often provides a sufficiently accurate representation of a real liquid; and

Steam, which is representative of substances with completely general thermodynamic response.

At certain points two intermediate substances will be considered:

The van der Waals gas, which represents vapours better than does the perfect gas; and

The ideal liquid, which gives an idea of the deviations of real liquids from the constant-density model.

By applying the same basic principles to these five kinds of fluid, one gest a balanced view of the mechanics and thermodynamics of fluid motion.

2.1 FLOW GEOMETRY

2.1.1 Integral Models of a Flow

It was seen earlier that thermodynamics can be developed in two ways, either by considering microscopic processes, or by looking at the gross properties of matter. The latter course—macroscopic thermodynamics—has been followed. There are also two ways of looking at a fluid flow, although both adopt the macroscopic point of view and discuss the macroscopic properties of the flowing material. One approach to fluid mechanics may be termed the *differential approach.* It uses differential equations expressing the conservation of mass, momentum and energy for each element of the fluid, and attempts to solve them to determine the whole pattern of the flow. The alternative is the *integral approach.* Here one applies the conservation principles to a simplified model which disregards the fine detail of the flow.

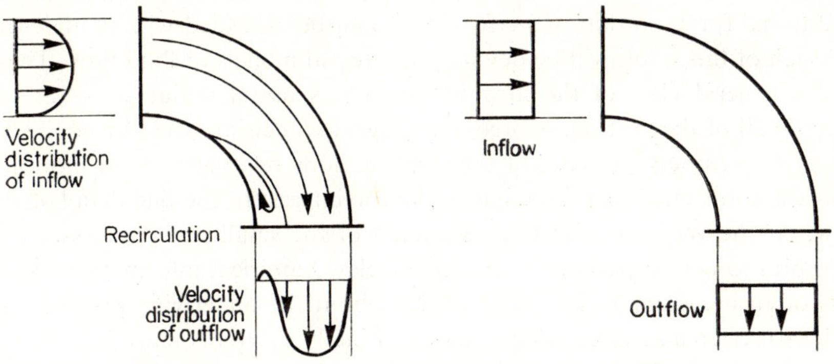

Figure 2.1 Flow through an elbow in a pipe line. (a) Detailed description of the flow; (b) Simplest integral model of the flow

Figure 2.1 illustrates the two ways of looking at the flow around an elbow in a pipe line. The differential representation takes account of the variation in velocity across each section of the elbow, of the fluctuations in velocity in time which are called turbulence, of the separation of the flow from the inside of the elbow, and of the formation of an unsteady recirculation downstream of the bend. The simplest integral model merely considers a uniform velocity across the pipe both upstream and downstream of the elbow. This example points to both the strength and the weakness of the integral approach. Since a detailed description of the flow within the elbow is not needed, the integral model can rapidly give some idea of the overall effect of the elbow on the flow. Equally, since it neglects the real complexity of the flow, the integral model cannot provide a fundamental understanding of the processes undergone by the fluid.

The integral approach is often dependent upon experimental information, which makes up for its inability to describe complex processes. Suppose that one needs to predict the drop in pressure when water flows around an elbow in a pipe. The integral approach is dependent upon having available an empirical loss coefficient, obtained by tests on an elbow of rather similar shape. However, even this weak position is better than that of the differential approach. A completely analytical discussion of the flow within the elbow can do no more than indicate the order of magnitude of the pressure drop. Thus the integral approach, supplemented by a body of empirical data, provides the best hope of answering many of the questions an engineer asks about flowing fluids. Special tests are sometimes carried out during the design of unusual equipment. However, this is often unnecessary, since experimental studies have been made of the flows which arise most frequently in engineering practice. Indeed, designers sometimes specify equipment of traditional form, so that its performance can be predicted with confidence.

Much of this book will be devoted to integral models of fluid flow. These give a general view of the flow which proves useful if one looks at the finer detail of the motion. Moreover, integral models are used by engineers for perhaps ninety per cent of their calculations relating to fluid motions. This does not mean that an engineer has no interest in the fine detail of the flow. At the very least, an understanding of the small-scale processes will help him to plan experiments and to correlate empirical information. Also, it is often necessary to know something about the underlying processes in order to set up a suitable integral model. Finally, entirely theoretical predictions are required for unfamiliar situations in which experiments are too costly or are impractical for some other reason. Aircraft development and space research provide many examples of this kind. It is often too expensive, or too dangerous, to carry out preliminary experiments, and every effort must be made to predict performance analytically. This final, ambitious goal is beyond the scope of this introductory treatment of fluid motions. Nevertheless, if integral flow models are to be intelligently used, one must have in mind a fairly realistic, though necessarily simplified picture of the small-scale processes within a flowing fluid.

For the most part we shall deal with confined flows whose direction and flow area are roughly defined by the solid boundaries surrounding the stream. Figure 2.1 shows a flow of this kind. As indicated there, the simplest integral model takes the velocity to be uniform across the whole flow passage, and takes the fluid velocity to be normal to the cross-section. It is important to understand clearly the relationship between the hypothetical uniform velocity of the integral model and the actual velocity field.

Let us begin by considering the velocity variation at a fixed point in the flow of Figure 2.1. This variation could be obtained by introducing a rapidly responding measuring instrument (such as a hot-wire anemometer) through the pipe wall. The signal from such an instrument can be displayed on an oscilloscope to present the history of the velocity variation. Figure 2.2 shows the kinds of velocity variations which might be measured within a pipe.

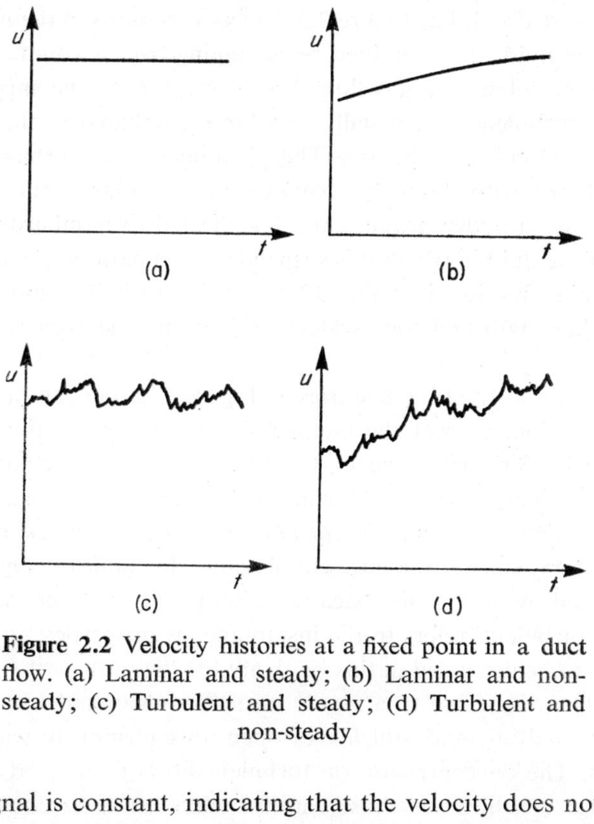

Figure 2.2 Velocity histories at a fixed point in a duct flow. (a) Laminar and steady; (b) Laminar and non-steady; (c) Turbulent and steady; (d) Turbulent and non-steady

The first signal is constant, indicating that the velocity does not change in time. The second signal changes slowly, according to a simple law. The other two flows exhibit frequent, apparently random fluctuations in velocity, one varying about a stable mean value, the other displaying a slow upwards trend. The second pair of flows are described as *turbulent*. In them the fluid is continually being distorted and mixed in a complex manner. The more regular flows are termed *laminar*. In them the fluid does not undergo rapid mixing, and the distortion is similar to that of a pile of flat plates (or laminae)

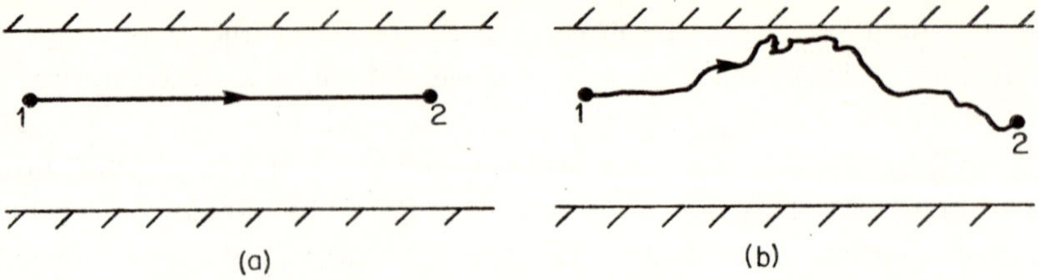

Figure 2.3 Particle motions in duct flow. (a) Laminar flow; (b) Turbulent flow

sliding over one another. Typical paths of particles within the two kinds of flow are shown in Figure 2.3.

It should be realized that Figure 2.2 depicts variations in the macroscopic velocity of the fluid, that obtained by averaging over a volume containing very many molecules. In a gas flow, for example, even the smallest length scales of the turbulence are usually very large compared to the mean-free-path of the molecules of the gas. The phenomenon of turbulence should not be confused with *Brownian motion*, the irregular motion of a tiny particle, perhaps a pollen grain, which is subjected to bombardment by the molecules of the fluid in which it is suspended. The path of a particle undergoing Brownian motion is rather like that for turbulent motion, Figure 2.3(b), but the length and time scales are very much shorter than those of turbulence.

Looking again at the time histories of Figure 2.2, one can make another kind of distinction. One of the laminar flows is steady in time, the other changes slowly. Similarly, one of the turbulent flows fluctuates about a steady mean value, while the other has an overall upwards trend. The former will be called a *steady turbulent flow*. The two adjectives may seem to be contradictory, since turbulence entails rapid fluctuations. The contradiction is resolved by noting that 'steady' refers to the mean properties of the flow, while 'turbulent' refers to the instantaneous properties.

The flows to be discussed in this book are for the most part steady turbulent flows. The restriction to a steady mean flow is dictated by a desire to simplify the analysis, and still leaves open an enormously wide range of applications. The concentration on turbulent flows is dictated by our aim of describing such flows as those within boilers, turbines and jet engines. It is found that the flows within large-scale engineering equipment are almost always turbulent. There are some notable exceptions: the flow of lubricants in bearings is usually laminar, and the flows in high-pressure hydraulic systems often are. In general, slow flows in narrow passages are laminar, while faster flows in larger passages are turbulent. The reason for the dependence of turbulence on the scale of the flow will be investigated later. For the moment it will be accepted as an experimental fact.

2.1.2 One-dimensional Flow

Let us see how the integral approach can be applied to steady turbulent flow through a duct. Figure 2.4(a) illustrates a particular case, flow through a slowly contracting passage. The fluid velocity varies through the duct and also in time, since the flow is turbulent. Then $u = u(x, y, z, t)$ gives the velocity distribution in the contraction. Figure 2.4(b) shows the instantaneous local velocities at points in a cross-section. A short time later the arrows representing the velocities will be somewhat different. However, the

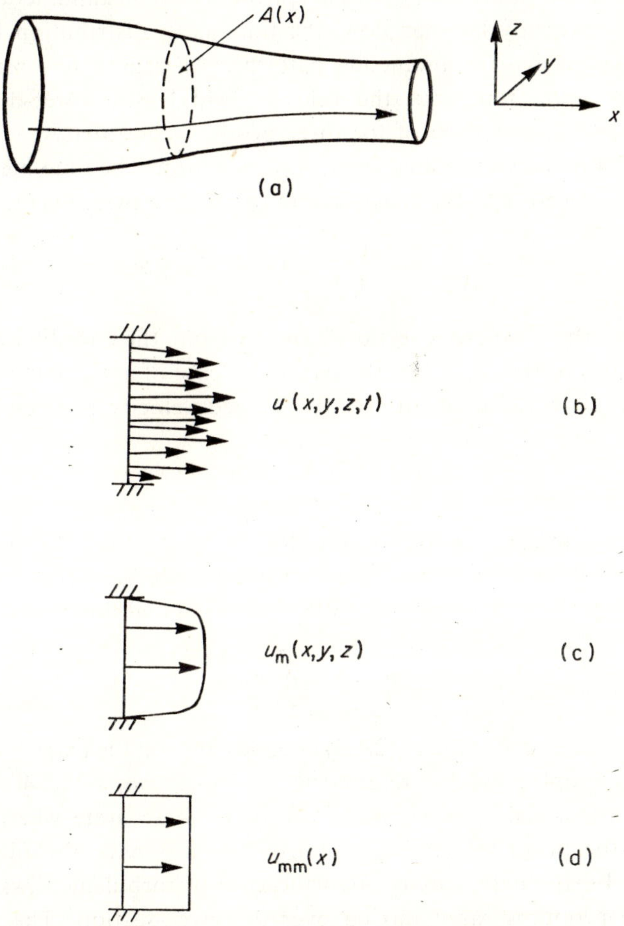

Figure 2.4 The velocity distribution within a flow through a duct. (a) Form of the duct. (b) Velocity distribution over a cross-section at an instant in time. (c) The time-mean velocity distribution over the cross-section. (d) The twice-averaged velocity of the one-dimentional model

essential features of the velocity distribution do not change; the velocities are generally larger towards the centre of the duct, fall to zero at the wall, and are mostly nearly parallel to the duct axis.

Figure 2.4(c) indicates the variation of the time-mean velocity, defined by

$$u_m(x, y, z) = \frac{1}{t_a}\int_0^{t_a} u(x, y, z, t)\, dt \tag{2.1.1}$$

Here t_a is the time over which the average is taken; it must be large compared to the time-scale of the turbulent fluctuations. The time-mean velocity variation retains some of the essential features of the instantaneous patterns. Indeed, for a steady laminar flow, the time-mean distribution is identical with the successive instantaneous patterns. Averaging has removed the dependence upon time, but the velocity field is still three-dimensional, that is, dependent on each of the three position coordinates.

Figure 2.4(d) shows the final stage of idealization. The time-mean velocity has been averaged over the cross-section to give the twice-averaged velocity

$$u_{mm}(x) = \frac{1}{A}\int_A u_m(x, y, z)\, dy\, dz \tag{2.1.2}$$

Here $A(x)$ is the local cross-sectional area, x being the coordinate measured along the axis of the duct. For the particular case of axisymmetric flow in a round duct with radius a, the twice-averaged velocity is given by

$$u_{mm}(x) = \frac{2}{a^2}\int_0^a r\, u_m(x, r)\, dr \tag{2.1.3}$$

The simplest integral model of the flow assumes that the fluid has this twice-averaged velocity across the entire cross-section. Thus the volume flux of the idealized flow is equal to that of the actual flow:

$$\dot{V} = u_{mm}A = \int_A u_m\, dy\, dz \tag{2.1.4}$$

The twice-averaged velocity (2.1.2) depends only on the axial coordinate x. Hence this simple model is referred to as *one-dimensional flow*. Obviously the one-dimensional approximation will be most accurate when the distribution of the mean velocity u_m is fairly flat, dropping abruptly near the duct wall. Fortunately, this is often the case in turbulent flows, since the turbulence produces rapid mixing over the cross-section. The result is a large central core with nearly uniform velocity, surrounded by thin boundary layers across which the velocity drops rapidly to zero at the wall. Since the velocity distribution in laminar pipe flow is not so flat, the one-dimensional model cannot be expected to be so successful with laminar flows.

So far only the variations in velocity have been considered. However, the discussion can be applied almost intact to the various thermodynamic properties of the fluid. Imagine that the walls of the duct of Figure 2.4 are cooled, so that their temperature is lower than that of the fluid in the duct. The instantaneous temperature distribution in the turbulent fluid will be very like the velocity variation shown in Figure 2.4(b). The time-mean temperature variation will be generally similar to the time-mean velocity variation of Figure 2.4(c). This is in fact the temperature distribution which would be found using a thermometer with a slow response which was not able to follow the rapid variations associated with the turbulence. Finally, it is possible to introduce a twice-averaged temperature, analogous to the one-dimensional velocity shown in Figure 2.4(d). This gives the one-dimensional temperature distribution, $T_{mm}(x)$.

In a similar manner it is possible to introduce a one-dimensional description of the variation of each thermodynamic property along the duct. Most of the calculations of this book will involve only these twice-averaged properties, and for convenience the subscripts 'mm' will be dropped, the one-dimensional variations being denoted simply by $u(x)$, $p(x)$, $T(x)$, $s(x)$, and so forth. Of course, in adopting the one-dimensional representation we are throwing away any chance of realistically representing certain processes of vital importance, for instance, friction and heat-transfer mechanisms of the boundary layer, the intensely sheared fluid adjacent to the duct wall. To account for these in a one-dimensional model it is necessary to introduce empirical friction and heat-transfer coefficients.

A consideration of Figure 2.1 makes it clear that the one-dimensional approximation will not be equally appropriate at every section of a flow. It gives an adequate description upstream and well downstream of the elbow, but would be quite unsuccessful if applied within the elbow or just downstream, where the flow is highly non-uniform. Figure 2.5 indicates some situations in which the one-dimensional approximation will be appropriate and inappropriate. Slow changes in flow direction and area will not invalidate the approximation. However, rapid changes are likely to do so, although the approximation will once again become valid well downstream. It should be noted that the influence of a distortion does not extend far upstream of the distorting section, but usually does extend downstream for many duct diameters. Similarly, gradual heating will not destroy the applicability of the one-dimensional model, but a sudden injection of heat by combustion would do so. Finally, it should be noted that the velocity and other properties may be reasonably uniform near the entry to a duct, where the boundary layers have not spread across the duct from the walls.

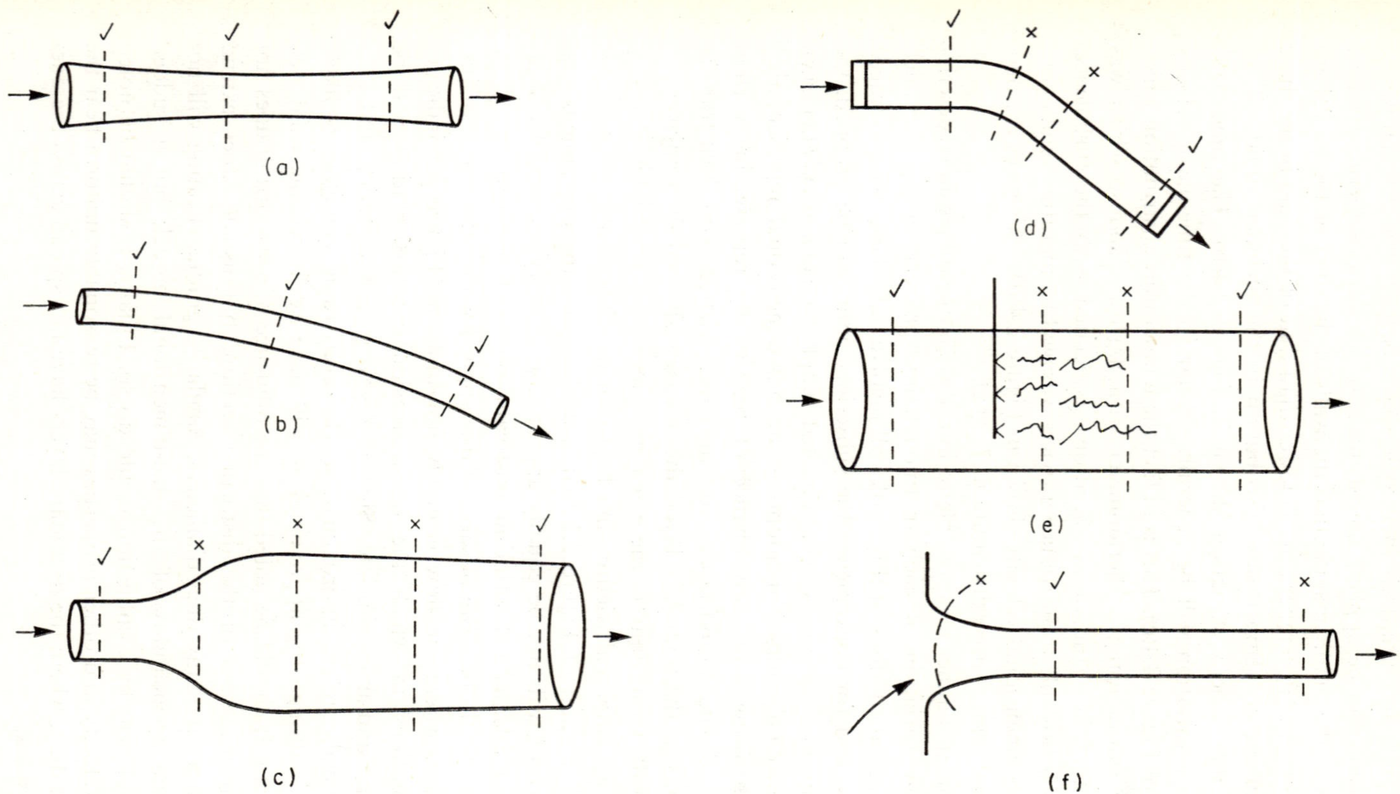

Figure 2.5 Sections of ducts at which the one-dimensional model is appropriate and inappropriate. The ticks denote acceptability of the model. (a) Slowly changing cross-section; (b) Slowly changing flow direction; (c) Rapidly changing cross-section; (d) Rapidly changing flow direction; (e) Sudden heating by combustion; (f) Developing laminar flow

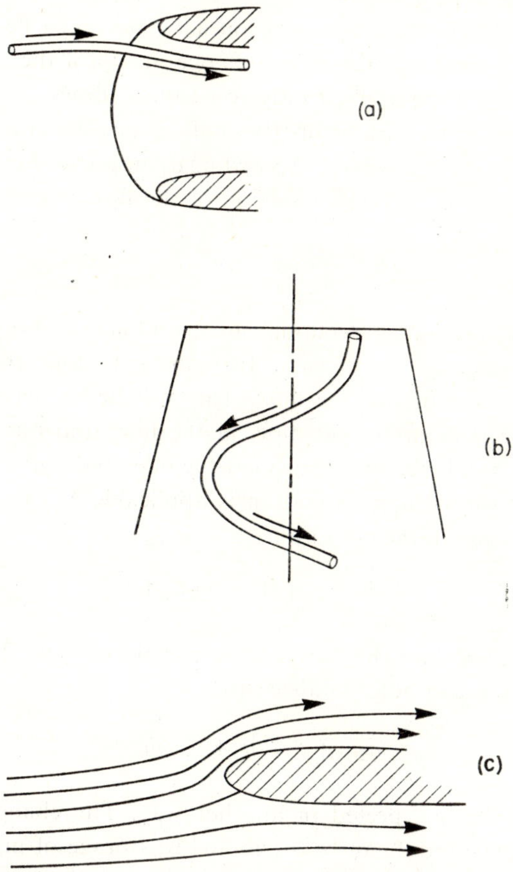

Figure 2.6 Streamtubes and streamlines. (a) Streamtube in the inlet of a jet engine; (b) Streamtube in the swirling flow leaving a turbine; (c) Streamlines near the leading edge of a wing or rotor blade

One-dimensional flow has been introduced with reference only to confined flows. However, the concept applies more widely, since any steady three-dimensional flow can be divided into *streamtubes*, surfaces parallel to the local velocity vector. Two-dimensional flows are usually described using *streamlines*, also parallel at each point to the velocity vector, so that no fluid passes across them. Figure 2.6 shows some examples of streamtubes and streamlines. Evidently the flow in any streamtube or between any pair of streamlines can be treated as a one-dimensional flow. However, many of the explicit results obtained for one-dimensional flow require a knowledge of the variation of the flow area along the flow passage. For a confined flow this information can often be supplied, but the positions of streamlines in an unconfined flow are not known initially, but must be determined in the course of the analysis. Hence, in practice, many of the results for one-dimensional flow can be applied only to confined flows.

The flow area and other properties of a one-dimensional flow are linked by the *continuity equation.* According to the one-dimensional approximation, the mass flux through a section of the flow is given by

$$\dot{m} = \varrho u A$$

where $A(x)$ is the local cross-sectional area, and $u(x)$ and $\varrho(x)$ are the local twice-averaged velocity and density. In many situations the mass flow is the same at two sections of the flow. This will be the case, for instance, at various sections of a straight pipe, at the inlet and outlet of a row of blades in a flow machine, and (nearly) at the inlet and outlet of a jet engine. If the one-dimensional approximation is applicable at two sections where the mass flow is equal, the condition

$$\dot{m} = \varrho_1 u_1 A_1 = \varrho_2 u_2 A_2 \tag{2.1.5}$$

must be satisfied.

If mass is injected into the flow between sections 1 and 2 at rate $\dot{m}_s$, the continuity relation can be generalized to

$$\varrho_2 u_2 A_2 - \varrho_1 u_1 A_1 = \dot{m}_s \tag{2.1.6}$$

This result can be interpreted in another way. The change in mass flux between sections 1 and 2 can be attributed to storage of mass between the sections at the rate $-\dot{m}_s$. For a single, steadily flowing stream, this possibility can be ruled out, since the mass between the two sections must remain constant.

Example 2.1 The flow through a steam turbine and condenser which was discussed in Example 1.5 involves some very large density changes; the cross-sections of the passages carrying the fluid must be varied accordingly. It was found that $v_1 = 1{\cdot}700$ ft^3/lb at entry to the turbine, that $v_2 = 38{\cdot}85$ ft^3/lb for steam leaving the turbine at 200 °F, and that $v_3 = 0{\cdot}01659$ ft^3/lb for the water leaving the condenser.

Take the design flow for the unit to be 20,000 lb/hr. If the steam velocity is to be limited to 100 ft/sec, and the water velocity to 5 ft/sec, what must be the diameters of the ducts at sections 1, 2 and 3?

The continuity relationship linking the three sections is

$$\frac{u_1 A_1}{v_1} = \frac{u_2 A_2}{v_2} = \frac{u_3 A_3}{v_3} = \dot{m}$$

For the prescribed maximum velocities and flow rate

$$\frac{100\,A_1}{1{\cdot}700} = \frac{100\,A_2}{38{\cdot}85} = \frac{5\,A_3}{0{\cdot}01659} = \frac{20{,}000}{3600}\ \text{lb/sec}$$

Then

$$A_1 = 0{\cdot}095\ \text{ft}^2, \quad A_2 = 2{\cdot}16\ \text{ft}^2, \quad \text{and} \quad A_3 = 0{\cdot}0184\ \text{ft}^2$$

and the minimum duct diameters are

$$d_1 = 4{\cdot}2\ \text{in}, \quad d_2 = 20\ \text{in} \quad \text{and} \quad d_3 = 1{\cdot}8\ \text{in}$$

Example 2.2 A common source of mass addition to a flow is fuel injection. Consider the effect of fuel injection in a ram jet; Figure 2.7 shows the parallel-sided duct in which burning occurs. The air/fuel ratio (air mass flow to fuel mass flow) is 30, and the air density before burning is 0·80 kg/m³. The velocities before and after burning are found to be 300 m/sec and 500 m/sec. An estimate of the density after burning is required, taking into account not only the acceleration of the fluid, but also the change in mass flow.

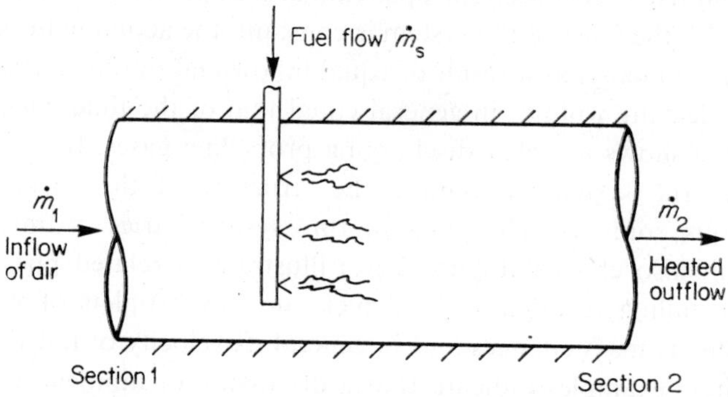

Figure 2.7 The combustion duct of a ram jet

The mass flows before and after fuel injection are related by the air/fuel ratio

$$r = \frac{\dot{m}_s}{\dot{m}_1} = \frac{\dot{m}_2 - \dot{m}_1}{\dot{m}_1}$$

For this parallel-sided duct

$$\frac{\dot{m}_2}{\dot{m}_1} = \frac{\varrho_2 u_2}{\varrho_1 u_1} = 1 + r$$

so that

$$\varrho_2 = \varrho_1 \frac{u_1}{u_2}(1+r)$$

For the flow specified

$$\varrho_2 = 0{\cdot}80 \times \frac{300}{500} \times \left(1 + \frac{1}{30}\right) = 0{\cdot}496\ \text{kg/m}^3$$

2.2 THE MOMENTUM PRINCIPLE

2.2.1 Newton's Second Law

In Section 1.3.1 Newton's law was used to study vertical motion under gravity, and hence led to the introduction of kinetic and potential energies. However, the law was then stated in a form applicable only to the motion of a body whose mass remained constant and whose every part had the same velocity. This does not include many of the systems which have to be discussed when fluid motion is studied.

The range of situations which should be dealt with by the momentum principle is indicated by the three situations illustrated in Figure 2.8. The first is a steady flow through a contracting duct. This situation is usually analysed by considering a system with fixed boundaries comprised of the duct wall and fixed cross-sections upstream and downstream of the contraction. Although the mass of the system is constant, the actual material within it changes continually, as a result of equal inflow and outflow. The velocity of the fluid leaving will not in general equal that of the fluid which enters. Figure 2.8(b) shows a rocket discharging propelling gases. It is convenient to consider the system containing the structure of the rocket and its instantaneous contents. The mass of this system varies in time, and its velocity will also change. Figure 2.8(c) illustrates a related phenomenon. A raindrop falling through a cloud picks up tiny droplets of water as it moves. Thus its mass changes and in general its velocity of fall will change as well. These examples indicate that a discussion of momentum changes in fluid motion must account for flows of momentum through system boundaries as well as accelerations within the system.

Newton's second law states that momentum changes take place when a force acts. The principle can be written in general terms as

$$\Sigma\frac{\mathrm{d}(mu)}{\mathrm{d}t} = \Sigma F$$

the symbol Σ indicating a summation over the several momentum changes and forces. Force and momentum are vector quantities, but for simplicity we shall consider unidirectional systems in the first instance. The results will later be generalized to include three-dimensional flows and force systems. The statement of the momentum principle can also be broken down into

$$\Sigma m\frac{\mathrm{d}u}{\mathrm{d}t}+\Sigma u\frac{\mathrm{d}m}{\mathrm{d}t} = \Sigma F \qquad (2.2.1)$$

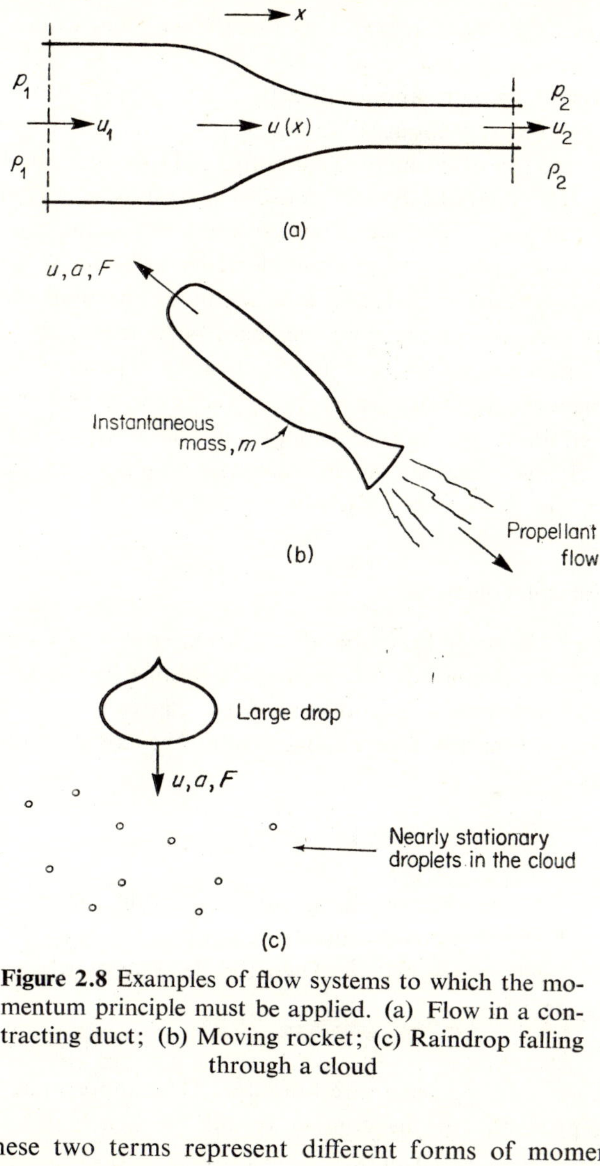

Figure 2.8 Examples of flow systems to which the momentum principle must be applied. (a) Flow in a contracting duct; (b) Moving rocket; (c) Raindrop falling through a cloud

Evidently these two terms represent different forms of momentum production. The action of a force may produce changes in velocity within a system, or may produce a net flow of momentum out of the system. Equation (2.2.1) can also be written

$$\Sigma m \frac{du}{dt} = F - \Sigma u \frac{dm}{dt} \tag{2.2.1}$$

This states that the momentum within a system changes for two reasons, because forces act on the system and because there is a net flow of momentum out of the system.

The results (2.2.1) can be applied directly to a system comprising a number of solid bodies moving about in space. The summations extend over the accelerations and forces acting on each body, and over those bodies entering and leaving the system chosen for study. The results (2.2.1) are also applicable to the flow of a continuous mass of fluid. But in this case the moving material is not separated into discrete elements; in order to carry out the summations, one must first decide how to divide the fluid into elements. There are two ways of doing this: the space occupied by the fluid can be divided into elemental volumes, or the fluid can be separated into elemental masses. Whichever approach is adopted, the same overall result will be obtained when the summations are completed. These two ways of calculating momentum changes, that based on elemental volumes and that based on elemental masses, will be examined in turn.

2.2.2 The Control Volume

Consider a volume of fixed shape which material may enter and leave. This system will be referred to as a control volume. It is divided into elements $\mathrm{d}V$, also of fixed shape, and hence fixed relative to the system boundaries. The rate of momentum change within the control volume can be calculated as

$$\int_V \frac{\partial(\varrho u)}{\partial t}\,\mathrm{d}V$$

where $\partial(\varrho u)/\partial t$ is the rate of change of momentum for the element $\mathrm{d}V$. The rate at which mass passes out of the system through some portion of the boundary will be denoted $\dot{m}$. Then the net rate at which momentum leaves the system is

$$\Sigma \dot{m} u$$

the summation covering the entire boundary. The appropriate form of the momentum principle for the control volume is then

$$\int_V \frac{\partial(\varrho u)}{\partial t}\,\mathrm{d}V = \Sigma F - \Sigma \dot{m} u \qquad (2.2.2)$$

Let us see how this result applies to the three situations depicted in Figure 2.8.

Although the velocity of each bit of fluid changes as it moves through the contraction shown in Figure 2.8(a), the velocity at each point in the steady flow is the same from one instant to the next. This is true also of the density. Hence at every point $\partial(\varrho u)/\partial t = 0$. This can be shown formally, by noting that, in the one-dimensional model, the velocity and density variations are given by $u = u(x)$ and $\varrho = \varrho(x)$. These arguments lead to the conclusion that the momentum within the control volume is constant, even though the fluid possessing that momentum changes continually. Indeed, this conclusion follows directly from the steadiness of the flow; one need not in this case break up the system into elements to demonstrate the constancy of momentum. Whichever line of argument is used, one finds that

$$\Sigma F = \Sigma \dot{m} u$$

Here the summation need include only the equal mass fluxes at sections 1 and 2. Hence

$$\Sigma F = \dot{m}(u_2 - u_1) \tag{2.2.3}$$

We shall return to this result later when making a detailed study of the forces developed by one-dimensional flows.

The rocket shown in Figure 2.8(b) provides a more complicated application of the general result (2.2.2). The system to be considered consists of the structure of the rocket and its remaining fuel. The velocity of the contents of the system changes, but the acceleration may be assumed with good accuracy to be uniform throughout the system. Forces, acceleration and velocities will all be taken to be positive forward along the axis of the rocket.

For a frame of reference fixed in the rocket

$$\begin{aligned}\frac{\partial(\varrho u)}{\partial t} &= \varrho\frac{\partial u}{\partial t} + u\frac{\partial \varrho}{\partial t}\\ &= \varrho a(t)\end{aligned}$$

since u is zero in this frame. Then

$$\int_V \frac{\partial(\varrho u)}{\partial t}\,\mathrm{d}V = a(t)\int_V \varrho\,\mathrm{d}V = a(t)\,m(t)$$

where $m(t)$ is the instantaneous mass of the system. The only mass flux which need be taken into account is the propelling jet, which leaves the rocket with velocity u_e. This is the velocity measured in the frame of reference moving with the rocket, and like u is positive in the forward direction. Equation (2.2.2) now gives

$$ma = F - \dot{m}_p u_e \tag{2.2.4}$$

The term F may represent aerodynamic drag or a component of the weight of the rocket. If the rocket is to overcome these, it is necessary that the momentum flux produce a thrust, that is, be equivalent to a force in the direction of flight. It must then be the case that

$$-\dot{m}_p u_e > 0$$

Since $\dot{m}_p > 0$ for fluid leaving the system, this condition will be satisfied only if fluid is ejected backwards, that is, if $u_e < 0$.

The motion of a rocket can also be analysed using absolute axes. Then the rocket has velocity u, and the velocity of fuel ejection is $u+u_e$, velocities again being taken to be positive in the direction of flight. In this case

$$\int_V \frac{\partial(\varrho u)}{\partial t}\,\mathrm{d}V = a(t)\,m(t)+u\int_V \frac{\partial \varrho}{\partial t}\,\mathrm{d}V$$

$$= a(t)\,m(t)-u\,\dot{m}_p$$

Equation (2.2.2) gives

$$ma-u\dot{m}_p = F-\dot{m}_p(u+u_e)$$

which is equivalent to the result (2.2.4).

The analysis of the motion of a raindrop, depicted in Figure 2.8(c), is very like that for the rocket. Again forces and velocities will be measured as positive in the direction of motion, in this case, downwards. Taking axes moving with the drop, one has

$$ma = F+\dot{m}_c u_c$$

Here $\dot{m}_c$ is the rate at which mass is captured by the drop, and u_c is the velocity of the captured mist droplets, as seen in the frame of reference moving with the larger drop. Since $\dot{m}_c > 0$ and $u_c < 0$, the momentum term $\dot{m}_c u_c < 0$, corresponding to an upwards force. Hence, as would be expected, the raindrop experiences an effective drag force which retards its downward motion. As before, F represents aerodynamic drag and the weight of the drop.

These examples demonstrate the importance of a clear statement of sign conventions for velocities and forces. To guard against mistakes in applying these conventions, always check to see whether the results make sense.

2.2.3 The Control Mass

The momentum principle (2.2.1) will now be applied to a system of fixed contents, which by analogy may be called a control mass. The basic statement is simpler than that for a control volume, as there are no momentum-flux terms. Hence the general result (2.2.1) reduces to

$$\int_m \frac{\mathrm{d}u}{\mathrm{d}t}\,\mathrm{d}m = \Sigma F \qquad (2.2.5)$$

where $\mathrm{d}u/\mathrm{d}t$ is the acceleration of the elemental mass $\mathrm{d}m$.

Let us first apply this result to the steady, one-dimensional flow shown in Figure 2.8(a). As the fluid moves along the duct its velocity changes according to the spatial variation $u(x)$. In moving distance dx, the particle undergoes a velocity change of

$$\frac{du}{dx}\,dx = \frac{du}{dx}\,\frac{dx}{dt}\,dt = u\frac{du}{dx}\,dt$$

Thus its acceleration is seen to be

$$\frac{du}{dt} = u\frac{du}{dx} \tag{2.2.6}$$

This is the acceleration of a particle moving through a velocity field which varies in space while remaining constant in time. It is called the *convective acceleration*, and must be distinguished from the *local acceleration* denoted by $\partial u/\partial t$. The latter is the rate of change of velocity at a fixed point in the flow; for a steady flow, $\partial u/\partial t = 0$.

For a one-dimensional flow a convenient element of mass is that contained in length dx of the flow; $dm = \varrho A\,dx$. Then, for the mass between sections x_1 and x_2 through the flow, the general result (2.2.5) gives

$$\int_{x_1}^{x_2} \varrho A u \frac{du}{dx}\,dx = \int_{x_1}^{x_2} \varrho A u\,du = \Sigma F$$

However, $\varrho A u = \dot{m}$ = constant in this steady, continuous flow. Hence

$$\Sigma F = \dot{m}\int_{x_1}^{x_2} du = \dot{m}(u_2 - u_1)$$

This is, of course, the result (2.2.3) obtained by considering a control volume.

Figure 2.9 illustrates the process considered here. The control mass between two cross-sections is shown at two instants separated by time dt.

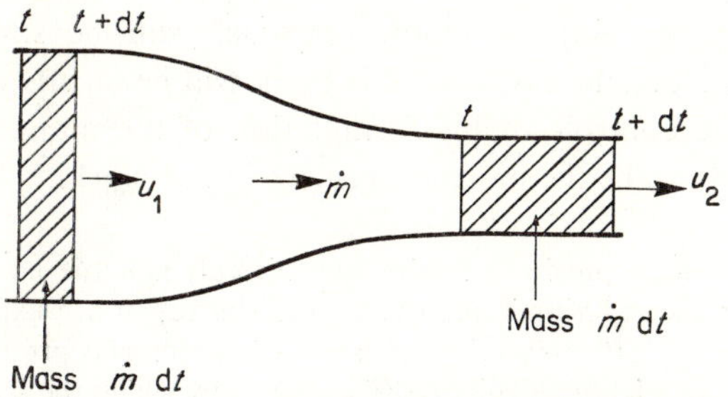

Figure 2.9 The displacement of a control mass in time dt for flow through a contracting duct

Its momentum at the later time has been increased by $u_2(\dot{m}\,dt)$, the momentum of the shaded volume to the right, and has been reduced by $u_1(\dot{m}\,dt)$, the momentum of the shaded volume to the left. The inner, unshaded volume is common to the two instants in time. Hence the net change in momentum over time dt is $\dot{m}(u_2 - u_1)\,dt$, and the rate of momentum change is that found above.

The motion of a rocket will be considered again, now using the control-mass approach. The system comprises the structure of the rocket and all its fuel, whether remaining on board or ejected. In evaluating the integral in equation (2.2.5) every part of the system must be considered. The exhausted fuel may be assumed to move at constant velocity, and hence does not contribute to the acceleration integral. This assumption might be literally correct if the rocket were travelling through empty space. Even if the jet mixes with the atmosphere, its momentum is conserved, and the effect is the same as if its velocity were constant. Now for the structure and remaining fuel. As before these will be taken to have mass $m(t)$ and uniform acceleration $a(t)$. Finally, consider the part of the fuel just leaving the rocket. In time dt, mass $\dot{m}_p\,dt$ changes velocity by u_e. The mean acceleration is then u_e/dt, and the product of mass and acceleration is $(\dot{m}_p\,dt)(u_e/dt) = \dot{m}_p u_e$. This accounts for all of the mass of the system, and the momentum equation (2.2.5) becomes

$$\Sigma F = \int_m \frac{du}{dt}\,dm = a(t)\,m(t) + \dot{m}_p u_e$$

Once again the result is equivalent to that (2.2.4) found using a control volume.

On comparing the application of the control-volume and control-mass formulations to identical situations, one must conclude that the former is the more direct. Although the fundamental statement of the momentum principle is simpler for a control mass, when one comes to particular cases the control-volume statement (2.2.2) is easier to use. Of course, we have considered only steady, or nearly steady flows, and have no grounds for concluding that the control-volume approach will be simpler for truly unsteady flows. Nevertheless, since this book will be mostly concerned with steady flows, the control-volume formulation of the momentum principle will usually prove to be the more convenient in our work.

Example 2.3 The propulsion of ships and aircraft is achieved by producing a momentum flux; the reaction force overcomes the resistance experienced by the moving craft. Figure 2.10 shows a propulsive unit which accepts a fluid at velocity U and sends it off in a jet of velocity U_j. The means by which the fluid is accelerated is not important; it might be a propeller, a jet engine or even a paddle wheel. However, we shall consider only propulsive units whose rates of mass acceptance

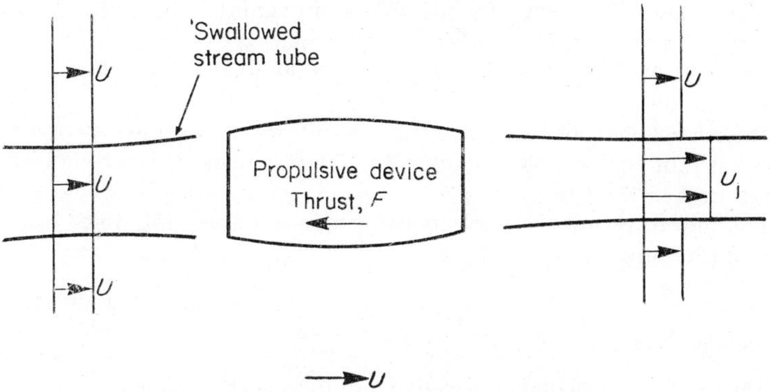

Figure 2.10 A generalized propulsive unit moving through the air or water with velocity U

and ejection are very nearly equal. This restriction rules out the rocket, which carries all its fuel on board, and is not 'air-breathing'.

The propulsive unit shown is surrounded on all sides by fluid. Well away from the disturbance caused by the unit, the pressure is uniform, and the velocity has (except in the jet) the uniform value U. Hence the only momentum change which has to be taken into account is that experienced by the jet flow, $\dot{m}(U_j - U)$. The only force acting to produce this change in momentum is that applied to the fluid by the propulsive unit. Hence the reaction force is given by the momentum principle as

$$F = \dot{m}(U_j - U)$$

This thrust on the propulsive unit is measured positive in the upstream direction, that is, in the direction of motion of the unit. The useful propulsive power of the device can be calculated as

$$\dot{W} = FU = \dot{m}U(U_j - U)$$

The situation sketched in Figure 2.10 is that seen by an observer moving with the propulsive unit. By adopting this frame of reference we have made the flow appear to be steady. If we adopt a frame of reference in which the surrounding fluid is stationary, the unit appears to move with velocity U through the fluid, and the situation is no longer steady. In this frame the fluid ahead of the unit is at rest, and that left behind it has velocity $U_j - U$.

This model propulsive process is not, of course, entirely representative of actual propulsive units. A real unit is mounted near the wing or fuselage of an aeroplane, or near the hull of a ship. Hence the pressure and velocity do not become uniform well away from the unit, and the velocity of the fluid approaching the unit does not have the uniform free-stream value. Moreover, the jet of a real propulsive unit is not the uniform axial flow assumed above. The axial component of velocity is non-uniform, the flow often swirls strongly and the jet mixes with the ambient fluid. Nevertheless, the one-dimensional integral model should represent the gross features of real propulsive processes. In practice, correction factors can be introduced to make the model more realistic. In any case, its performance may be thought of as an ideal towards which a designer may aspire.

In order to get some feeling for the orders of magnitude involved in propulsion of aircraft and ships, let us consider two particular cases, a large ship and a large aeroplane, both with a propulsive effort of 50,000 hp. The details of their operation are as follows:

(a) The aeroplane travels at 1500 ft/sec at 50,000 ft, swallows air from stream tubes (it has four engines) of total area $A = 20$ ft^2, and ejects it through jets whose total area is $A_j = 33{\cdot}3$ ft^2.

(b) The ship travels at 40 ft/sec, its two propellers accepting water from stream tubes with total area $A = 200$ ft^2.

(a) Aeroplane Propulsion

The thrust can be calculated directly from the data supplied:

$$F = \frac{\dot{W}}{U} = \frac{50{,}000 \times 550}{1500} = 18{,}300 \text{ lbf}$$

or 4600 lbf per engine.

The standard density at 50,000 ft (15,000 m) is $\varrho_a = 0{\cdot}00038$ sl/ft^3. The mass flow is then

$$\dot{m} = 0{\cdot}00038 \times 1500 \times 20 = 11{\cdot}3 \text{ sl/sec}$$

The velocity change through the engines is

$$U_j - U = \frac{\dot{W}}{\dot{m}U} = \frac{50{,}000 \times 550}{11{\cdot}3 \times 1500} = 1620 \text{ ft/sec}$$

and the densities are in the ratio

$$\frac{\varrho_j}{\varrho_a} = \frac{A}{A_j}\,\frac{U}{U_j} = 0{\cdot}6 \times \frac{1500}{3120} = 0{\cdot}288$$

If the fluid behaves like a perfect gas, despite the presence of combustion products, the temperature ratio will be

$$\frac{T_j}{T_a} = \frac{p_j}{p_a}\,\frac{\varrho_a}{\varrho_j} = \frac{\varrho_a}{\varrho_j} = \frac{1}{0{\cdot}288}$$

The standard atmospheric temperature is $T_a = 390$ °R, and the jet temperature may be estimated as

$$T_j = 1354 \text{ °R} = 894 \text{ °F}$$

Of course, mixing with cooler ambient air will rapidly reduce the temperature within the jet.

(b) Ship Propulsion

Here the total thrust is

$$F = \frac{\dot{W}}{U} = \frac{50{,}000 \times 550}{40} = 687{,}000 \text{ lbf}$$

or 344,000 lbf per propeller.

The density of sea water is nearly 2 sl/ft^3, and the mass flow is

$$\dot{m} = 2\times 40\times 200 = 16{,}000 \text{ sl/sec}$$

Then the velocity change is

$$U_j - U = \frac{\dot{W}}{\dot{m}U} = \frac{50{,}000\times 550}{16{,}000\times 40} = 43 \text{ ft/sec}$$

In this case the density is nearly constant, and the areas of the stream tubes in the jet and upstream are in the ratio

$$\frac{A_j}{A} = \frac{U}{U_j} = \frac{40}{83} = 0{\cdot}483$$

so that $A_j = 96{\cdot}5$ ft^2. The area of the flow at the propellers will be intermediate to A and A_j, so that the area swept out by each propeller will lie between 100 ft^2 and 48 ft^2. The propeller diameter must then be around 10 ft.

2.2.4 Change of Flow Direction

The examples which have been discussed so far have involved forces and momentum fluxes with a common line of action. The generalization to include two and three-dimensional force systems is not difficult in principle. The general statement of Newton's law (2.2.1) can be applied to a generar system by interpreting forces and velocities as the components in a particulal direction. Thus for the x-direction

$$\Sigma\, m\frac{du_x}{dt} + \Sigma\, u_x\frac{dm}{dt} = \Sigma F_x$$

Results derived from this general statement can be treated in the same way. Thus, for example, equation (2.2.3) gives

$$\Sigma F_x = \dot{m}(u_{x_2} - u_{x_1})$$

relating force components and velocities in the x-direction.

Let us apply this technique to the steady flow through the duct shown in Figure 2.11(a). The x-axis will be aligned with the initial direction of the flow. Then the fluid emerging from the duct has velocity components $u_2 \cos\theta$ and $u_2 \sin\theta$ in the x and z-directions. When the result (2.2.3) is applied it gives

$$\Sigma F_x = \dot{m}(u_2 \cos\theta - u_1)$$

and

$$\Sigma F_z = \dot{m}u_2 \sin\theta$$

It is instructive to examine the geometry of the force system. Figure 2.11(b) shows how the components add to give the resultant force on the

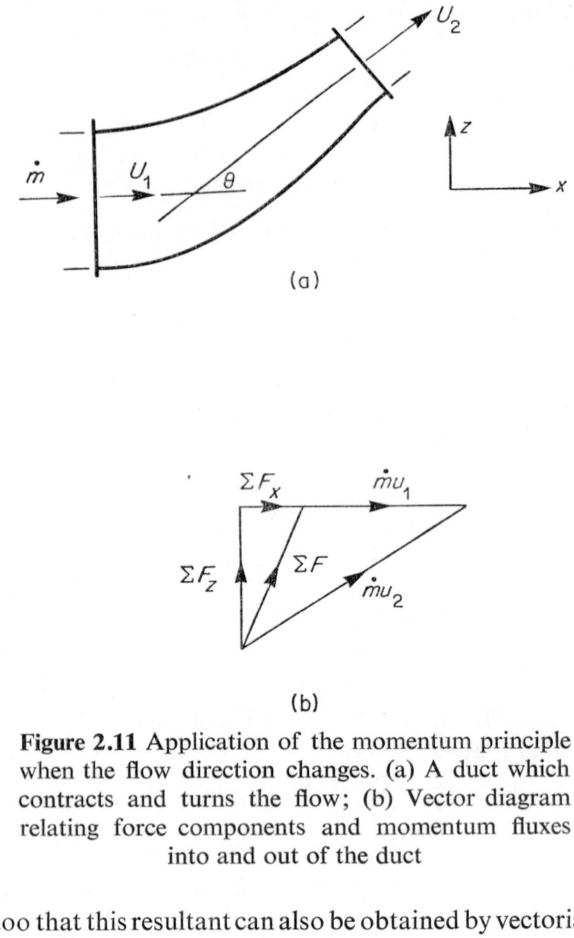

Figure 2.11 Application of the momentum principle when the flow direction changes. (a) A duct which contracts and turns the flow; (b) Vector diagram relating force components and momentum fluxes into and out of the duct

fluid. Note too that this resultant can also be obtained by vectorial subtraction of the initial and final momentum fluxes. This relationship is expressed in the vector equation

$$\Sigma \mathbf{F} = \dot{m}\mathbf{u}_2 - \dot{m}\mathbf{u}_1 \tag{2.2.7}$$

This is a generalization of equation (2.2.3), and reduces to it when the forces and momentum fluxes are collinear.

2.2.5 The Impulse Function

In the preceding discussion of the momentum principle, attention has been concentrated on the momentum side of the equation. Now the forces will be scrutinized. A particular case will be examined in detail, but the method of analysis is applicable to any one-dimensional flow. In order to indicate the true nature of the force system, it is necessary to consider a

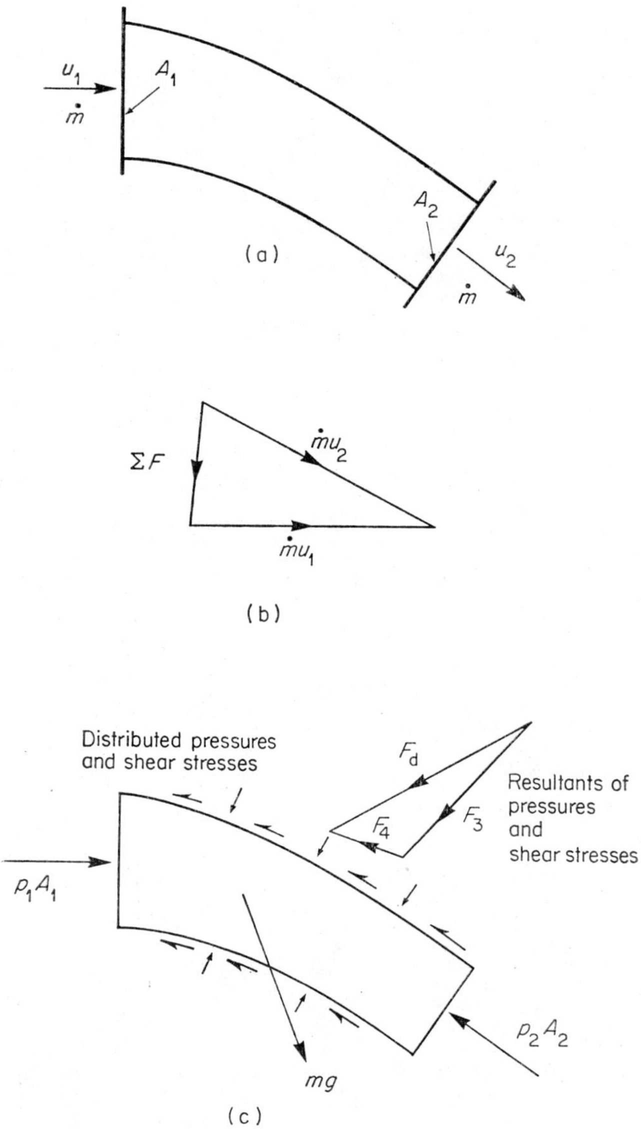

Figure 2.12 Forces on the fluid in a section of a duct. (a) A duct which deflects the flow in the vertical plane. (b) Relationship between net force on the fluid and the momentum fluxes into and out of the section of duct. (c) Elements of force contributing to the net force on the fluid.

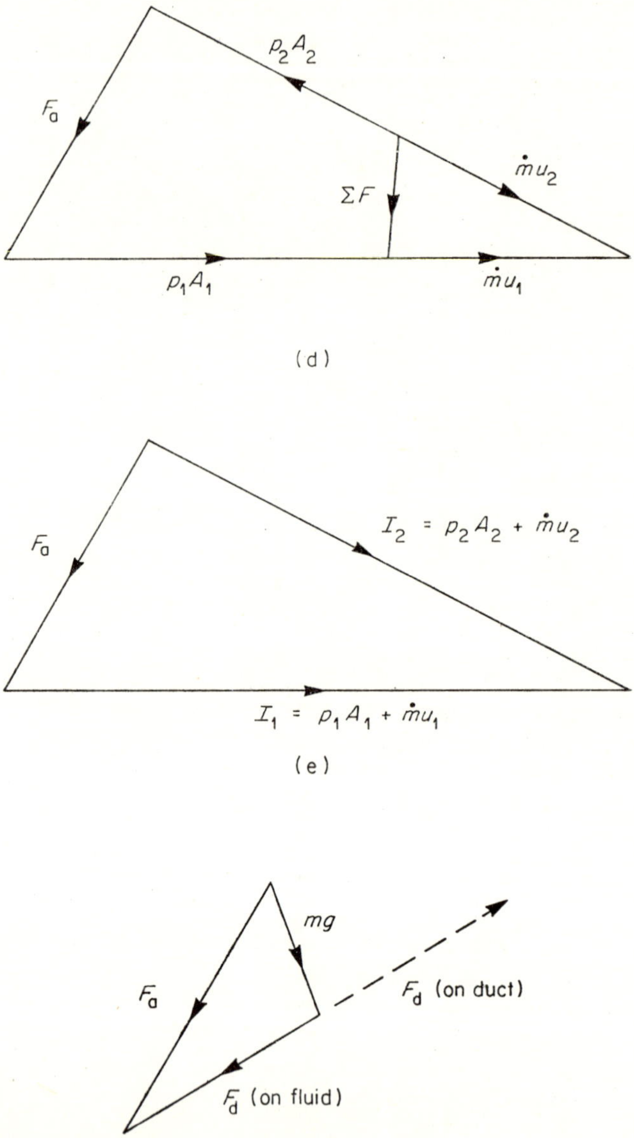

Figure 2.12 (d) Relationship between momentum fluxes, net force, terminal pressure forces and additional force on the fluid. (e) Relationship between impulse vectors and additional force on the fluid. (f) Relationship between additional force, fluid weight and duct force.

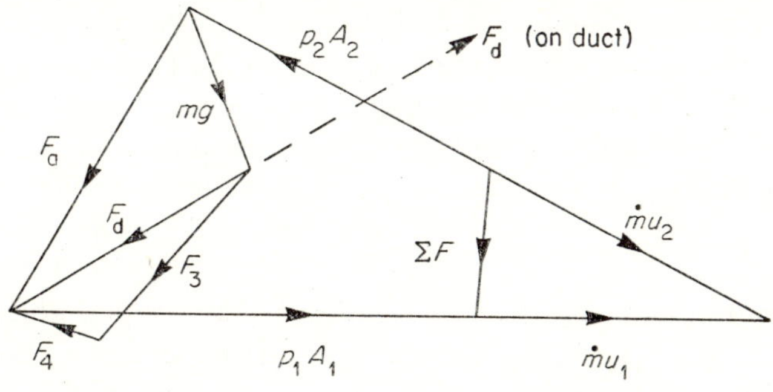

Figure 2.12 (g) Superposition of forces on the fluid

flow whose direction changes. The vector nature of the forces and momentum fluxes then becomes apparent.

Figure 2.12(a) shows a duct which turns the flow in the vertical plane. The initial and final velocity vectors and the vector representing the fluid weight all lie in the plane shown. The flow will be taken to be one-dimensional at sections 1 and 2. Its nature between these sections is not relevant; although a simple duct is shown for convenience, it might be arbitrarily complex and might even contain a pump or turbine exchanging energy with the flow. The momentum fluxes into and out of this section of duct are linked to the net forces on the fluid by the vector relationship shown in Figure 2.12(b); this is equivalent to equation (2.2.7) applied to this particular flow.

The forces which add together to give the net force $\Sigma\mathbf{F}$ are shown in Figure 2.12(c). They are of four kinds:

(1) the pressure forces acting inwards on the terminal sections, p_1A_1 and p_2A_2 in this one-dimensional model,

(2) the weight of the fluid, $m\mathbf{g}$, with m the mass of fluid between sections 1 and 2, and $\mathbf{g}$ directed downwards,

(3) a pressure distribution applied by the duct walls between sections 1 and 2, with resultant $\mathbf{F}_3$, and

(4) a shear-stress distribution applied by the duct walls between sections 1 and 2, with resultant $\mathbf{F}_4$.

The forces $\mathbf{F}_3$ and $\mathbf{F}_4$ add to give $\mathbf{F}_d$, the net force applied to the fluid by the duct.

Of the forces distinguished above, only the pressure forces (1) can be found directly without considering the details of the flow between sections 1 and 2. However, the momentum principle can be used to determine the unknown forces in terms of the directly calculable pressure forces and momentum fluxes at the terminal sections. Ultimately, it is $\mathbf{F}_d$ which is of greatest interest, for the reaction to it $-\mathbf{F}_d$ is the force applied to the duct by the fluid, and the structure supporting the duct walls must be designed to withstand this force. We shall not, however, seek $\mathbf{F}_d$ directly, but first determine the sum of the elements (2), (3) and (4) identified above. This will be called the additional force $\mathbf{F}_a$, since it comprises all the forces acting on the fluid between the terminal sections, except the pressure forces on those sections.

Figure 2.12(d) shows how the three forces F_a, p_2A_2 and p_1A_1 add vectorially to give the resultant $\Sigma\mathbf{F}$. The momentum flux vectors, which subtract vectorially to give $\Sigma\mathbf{F}$, are also shown. They are parallel to the corresponding pressure forces, since both pressure force and momentum flux are normal to the flow cross-section.

Figure 2.12(d) may be interpreted as giving $\mathbf{F}_a$ as the vector difference between two vectors with magnitudes

$$p_2A_2+\dot{m}u_2 \quad \text{and} \quad p_1A_1+\dot{m}u_1$$

The flows at sections 1 and 2 are related by

$$\dot{m} = \varrho_1u_1A_1 = \varrho_2u_2A_2$$

and these magnitudes may be written as

$$(p_2+\varrho_2u_2^2)A_2 \quad \text{and} \quad (p_1+\varrho_1u_1^2)A_1$$

It is apparent that the additional force $\mathbf{F}_a$ depends on the change in the vector quantity

$$\mathbf{I} = (p+\varrho u^2)\mathbf{A} \tag{2.2.8}$$

between the terminal sections. The vector $\mathbf{A}$ has the magnitude of the cross-sectional area and is parallel to the local velocity vector. The vector $\mathbf{I}$, which will be called the *impulse vector*, is like $\mathbf{A}$ normal to the cross-section. Its magnitude is the impulse function for that section

$$I = (p+\varrho u^2)A \tag{2.2.9}$$

This represents the net effect of pressure force and momentum flux at a section of a one-dimensional flow.

Figure 2.12(e) shows how the impulse vectors for sections 1 and 2 subtract vectorially to give the net additional force $\mathbf{F}_a$. Symbolically

$$\mathbf{F}_a = \mathbf{I}_2-\mathbf{I}_1 \tag{2.2.10}$$

We now have a convenient way of calculating the additional force. Since this force is the vector sum of the fluid weight and the force applied by the duct, a further vector subtraction is required to find $\mathbf{F}_d$. Figure 2.12(f) shows this step. Symbolically

$$\begin{aligned}\mathbf{F}_d &= \mathbf{F}_a-m\mathbf{g}\\ &= \mathbf{I}_2-\mathbf{I}_1-m\mathbf{g}\end{aligned} \tag{2.2.11}$$

The force applied by the fluid to the duct $-\mathbf{F}_d$ is also shown in Figure 2.12(f). It should be noted that this is the force applied by the fluid inside the duct. The stresses to be borne by the duct and its supports depend also on the loading on the outside of the duct. If the external loads arise from a uniform pressure, they can be accounted for simply by measuring internal pressures relative to the external pressure level. Each section of a duct must be able to withstand, first, the hoop stresses generated by the local pressure differential and, secondly, the bending moment, shear force and axial force resulting from the internal and external loading.

The forces which have been discussed above are shown together in Figure 2.12(g). Obviously, it is in general a complicated task to determine the force applied to a duct by the fluid moving through it. Note that only a two-

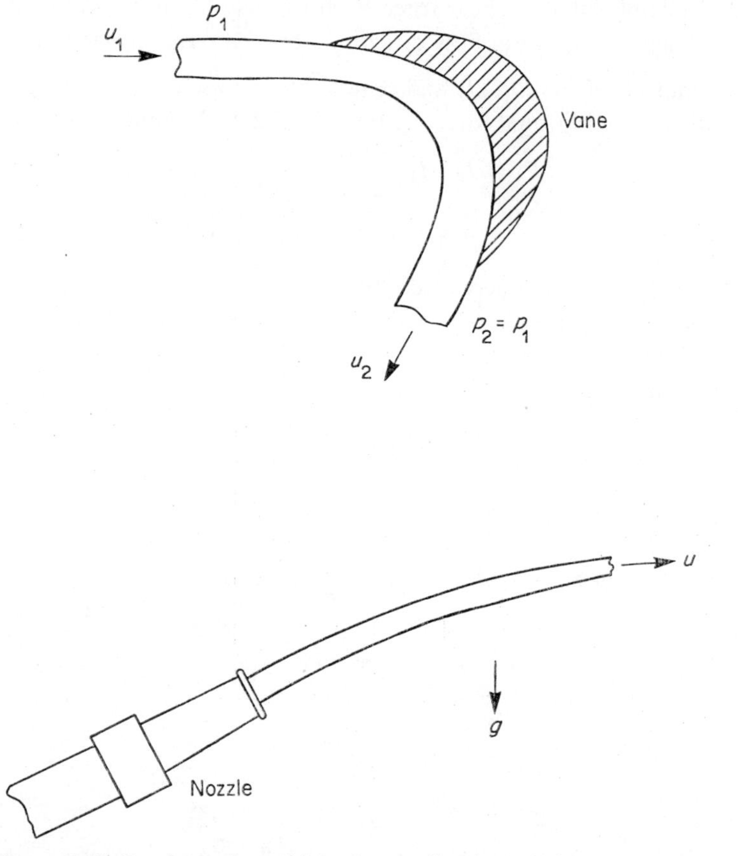

Figure 2.13 Jet flows for which the application of the momentum principle is somewhat simplified. (a) A jet striking a vane, with initial and final pressures equal. (b) A water jet moving through the air, with negligible friction and constant pressure

dimensional system of forces has been considered. The situation is simplified when, as is usually the case, some of the forces are of negligible magnitude. Some possible idealizations are:

(1) negligible momentum change, in particular, for a still fluid;

(2) negligible weight, usually the case for flows of gases and vapours;

(3) negligible pressure forces at the terminal sections, as when the fluid enters and leaves the control volume as a jet; a flow of this kind is shown in Figure 2.13(a);

(4) negligible duct force $\mathbf{F}_d$, as when the flow is a free jet; a flow of this kind is shown in Figure 2.13(b);

(5) negligible internal pressure force $\mathbf{F}_3$, for flow through a straight duct of uniform cross-sectional area;

(6) negligible internal shear force $\mathbf{F}_4$, for nearly frictionless flow; and

(7) negligible change in flow direction between terminal sections.

The final simplification (7) makes possible the use of a scalar equation in calculating the additional force F_a. Equation (2.2.10) reduces to

$$\begin{aligned} F_a &= I_2 - I_1 \\ &= (p_2 + \varrho_2 u_2^2)A_2 - (p_1 + \varrho_1 u_1^2)A_1 \\ &= p_2 A_2 - p_1 A_1 + \dot{m}(u_2 - u_1) \end{aligned} \tag{2.2.12}$$

The last form expresses separately the contributions of pressure forces and momentum change. Figure 2.14 shows the situation to which these results apply. Note that no restriction need be placed on the process occurring between the terminal sections, save that the mass inflow and outflow be

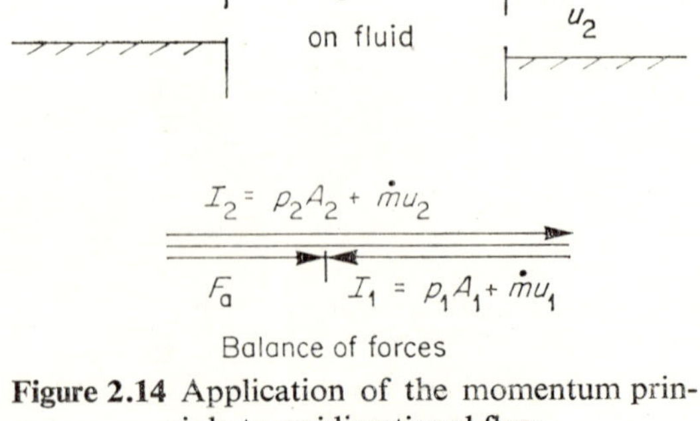

Figure 2.14 Application of the momentum principle to unidirectional flow

equal. It is still necessary to account for the weight of the fluid in finding the duct force $\mathbf{F}_d$. This is usually done most conveniently by a separate calculation after $\mathbf{F}_a$ has been found.

Example 2.4 The contracting pipe elbow shown in Figure 2.15 carries a water flow of $\dot{V} = 0{\cdot}11$ m^3/sec. The inlet and outlet diameters are 15 and 10 cm, and the inlet and outlet pressures are 700 and 560 kN/m^2. It is required to find the force which must be applied to hold the elbow in position against the forces imposed by the fluid within. The flow will be assumed to be one-dimensional at entry and outlet, and the density will be taken to be 1000 kg/m^3.

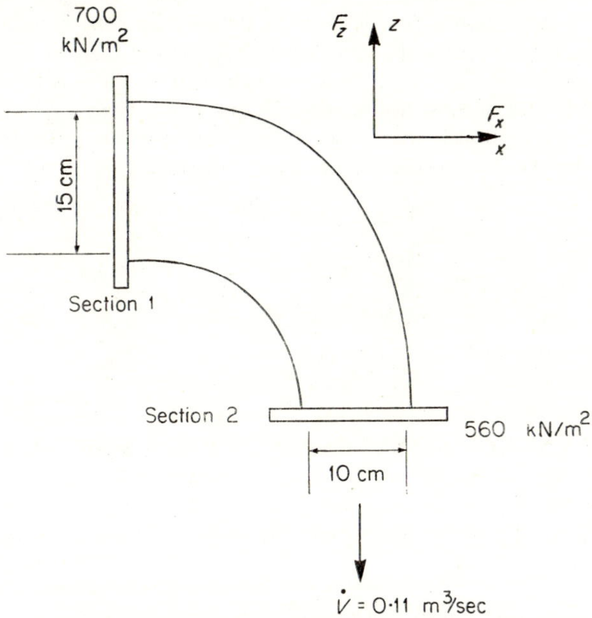

Figure 2.15 A reducing elbow in a water line

It is convenient to take axes aligned with the directions of inflow and outflow. For the axes of Figure 2.15, one finds the components of the force on the fluid to be

$$F_{a_x} = -I_1 \quad \text{and} \quad F_{a_z} = -I_2$$

For the flow specified

$$u_1 = \frac{\dot{V}}{A_1} = \frac{0{\cdot}11}{0{\cdot}01768} = 6{\cdot}23 \text{ m/sec}$$

and

$$u_2 = \frac{u_1 A_1}{A_2} = 6{\cdot}23 \times 2{\cdot}25 = 14{\cdot}0 \text{ m/sec}$$

Then the force components are

$$F_{a_x} = -p_1A_1 - \dot{m}u_1$$
$$= -12{\cdot}37 - 0{\cdot}68 = -13{\cdot}05 \text{ kN} = -1330 \text{ kgf}$$

and

$$F_{a_z} = -p_2A_2 - \dot{m}u_2$$
$$= -4{\cdot}40 - 1{\cdot}54 = -5{\cdot}94 \text{ kN} = -605 \text{ kgf}$$

Hence the net force on the fluid is

$$F_a = (1330^2 + 605^2)^{\frac{1}{2}} = 1460 \text{ kgf}$$

The weight of water within the elbow has now to be considered. For the duct shown, the average cross-section is perhaps 0·01 m^2 and the length of the flow path is perhaps 0·5 m. Hence the volume of water within the elbow is about 0·005 m^3, with a weight of about 5 kgf. Thus the weight of the fluid is almost negligible in this case. However, it must not be thought that this is typical of liquid flows.

If the vertical direction is that in which the fluid emerges from the elbow, the force applied by the duct to the fluid has components

$$F_{d_x} = F_{a_x} = -1330 \text{ kgf}$$

and

$$F_{d_z} = F_{a_z} - mg = -605 + 5 = -600 \text{ kgf}$$

In this case the weight acts to help deflect the flow, and in this way reduces the force applied to the duct by the fluid.

2.3 ENERGY CONVERSION IN ISENERGIC FLOW

2.3.1 The Energy Equation for Steady Flow

An isenergic flow is one which does not exchange energy with its surroundings, the total energy merely being redistributed amongst its several components. Nevertheless, we shall begin by developing a more general energy principle for a flowing fluid, one which does include energy addition by work and heat flow. In this way the special features of isenergic flows are revealed more clearly.

Figure 2.16 shows two sections of one-dimensional flow between which the fluid undergoes an unspecified process. The inflow at section 1 and outflow at section 2 are steady (though perhaps turbulent) and equal. Then

$$\dot{m} = \varrho_1 u_1 A_1 = \varrho_2 u_2 A_2$$

No further restrictions will be imposed on the process between sections 1 and 2: heat might be added or removed in a heat exchanger; work might be done on the fluid by a pump or compressor; energy might be extracted

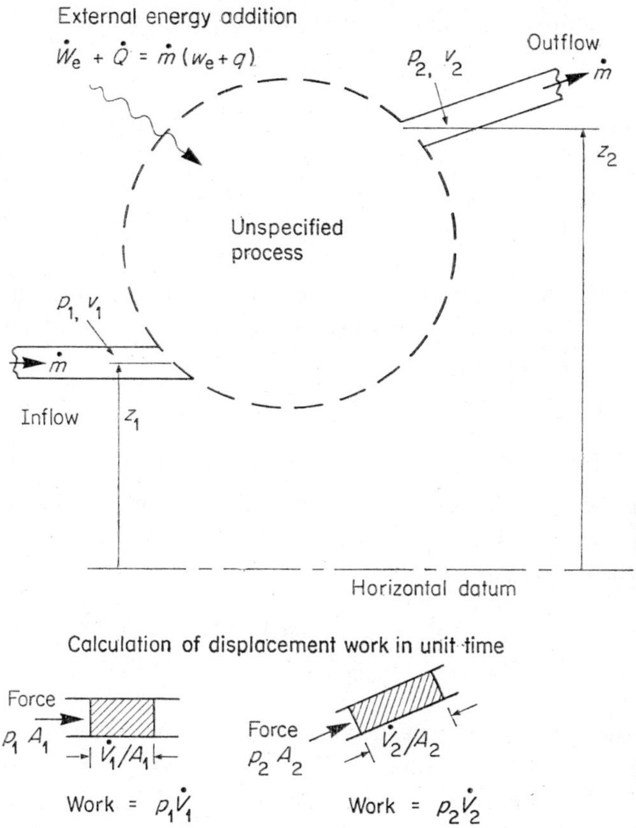

Figure 2.16 A steady-flow process with change of elevation, external energy addition and displacement work

from the fluid by a turbine or motor. Only the net energy addition to the fluid is relevant to the overall energy balance. The rate at which energy is added to the fluid between sections 1 and 2 will be denoted by $\dot{W}_e + \dot{Q}$, where $\dot{Q}$ is the rate of heat flow into the fluid, and $\dot{W}_e$ is the rate at which work is done on the flow by the surroundings. The subscript e labels this as *external work*, that is, work from outside the flow. It will be seen shortly that another work term has to be taken into account, that representing work done on one part of the flowing fluid by adjacent parts.

The energy balance for a steady-flow process will be derived from the overall energy equation developed in Section 1.3.5:

$$\Delta(TE) = W + Q + CE \tag{1.3.9}$$

This gives the change in the total energy of a system in any process. The convected energy CE accounts for the convection of energy with material passing through the system boundaries. The total energy

$$TE = E+KE+PE$$

was introduced as the sum of internal, kinetic and potential components. Two methods of applying this general result suggest themselves: the consideration of a control volume and of a control mass. Since the resulting energy equation is the basis of much of the work of this book, it is worth deriving in both ways.

A suitable *control mass*, or system of fixed contents, is the fluid between sections 1 and 2 which move along with the fluid. For this system, the convected energy $CE = 0$, and

$$\Delta(TE) = W+Q \tag{2.3.1}$$

Let us apply this result for unit time, in which mass $\dot{m}$ undergoes the process, and in which the energy added between sections 1 and 2 is $\dot{W}_e+\dot{Q}$. Work is also done at the moving boundaries; in unit time the net work done on the system at these boundaries is

$$\begin{aligned}\dot{W}_d &= p_1\dot{V}_1-p_2\dot{V}_2\\ &= \dot{m}(p_1v_1-p_2v_2)\end{aligned} \tag{2.3.2}$$

This element of work will be called *displacement work*. The total work done on the system during unit time is then $\dot{W} = \dot{W}_e+\dot{W}_d$, and the net energy addition is $\dot{W}+\dot{Q} = \dot{W}_e+\dot{W}_d+\dot{Q}$. The change in the total energy of the system during unit time can be calculated as

$$\Delta(TE) = \dot{m}(te_2-te_1) \tag{2.3.3}$$

with $te = e+\frac{1}{2}u^2+gz$, the total energy per unit mass for a uniform gravitational field. Equation (2.3.1) now gives

$$\dot{m}(te_2-te_1) = \dot{W}_e+\dot{Q}+\dot{m}(p_1v_1-p_2v_2) \tag{2.3.4}$$

This is the steady-flow energy equation.

Now look at the *control volume* defined by fixed sections 1 and 2. Since the flow is steady, the energy of the system does not change in time. Hence $\Delta(TE) = 0$, and equation (1.3.9) gives

$$W+Q+CE = 0 \tag{2.3.5}$$

Again consider a process occupying unit time. The external energy addition is still $\dot{W}_e+\dot{Q}$. Moreover, the fluid entering and leaving the system exchanges work with the adjacent fluid; in a sense it is pushed in by the fluid behind and has to push out the fluid ahead. This work done by the moving fluid is called *flow work;* it is precisely equivalent to the displacement work calculated in equation (2.3.2). The net convection of energy into the system in unit time is

$$CE = \dot{m}(te_1 - te_2)$$

This is the complement of the energy change (2.3.3) in the control mass in unit time. Introducing these results into equation (2.3.5), we get

$$\dot{W}_e + \dot{Q} + \dot{m}(p_1v_1 - p_2v_2) + \dot{m}(te_1 - te_2) = 0$$

This is, as expected, equivalent to equation (2.3.4) found for a system of fixed mass.

The energy equation can also be written

$$te_2 - te_1 = w_e + q + (p_1v_1 - p_2v_2) \tag{2.3.6}$$

where w_e and q are the external work and heat added per unit mass passing through the system. Alternatively

$$(te_2 + p_2v_2) - (te_1 + p_1v_1) = w_e + q \tag{2.3.6}$$

If the term pv is considered to be a form of energy, the *displacement energy*, this last result states simply that the change in the energy of the fluid is equal to the external energy addition. In taking pv to be a form of energy, we are acting as we did in Section 1.3.1, where gravitational potential energy was introduced in place of work done against gravity. In each case, one must be careful not to take account of the same mechanism twice by including both the work term and the corresponding energy term. Displacement energy replaces displacement work in the energy equation; the two are numerically equal, but have opposite signs and appear on opposite sides of the equation.

When the various components of energy are introduced explicitly, equation (2.3.6) becomes

$$\Delta(e + pv + \tfrac{1}{2}u^2 + gz) = w_e + q \tag{2.3.7}$$

Once again it is convenient to introduce the specific enthalpy

$$h = e + pv$$

which may now be interpreted as the sum of internal and displacement energies. Thus

$$\Delta(h + \tfrac{1}{2}u^2 + gz) = w_e + q \tag{2.3.8}$$

Enthalpy was first introduced as a measure of the energy addition to an expanding fluid. It is not surprising that enthalpy emerges again during an analysis of energy addition to a flowing fluid, whose boundaries are continually changing.

Finally, the breadth of application of the steady-flow energy equation should be noted. An obvious application is to flow through pipes and ducts leading to and from a flow machine or heat exchanger. The energy equation may also be applied to the flow through individual streamtubes. Since the flow area does not enter explicitly into the energy equation, its use does not depend upon a prior knowledge of the flow geometry. Figure 2.17 depicts

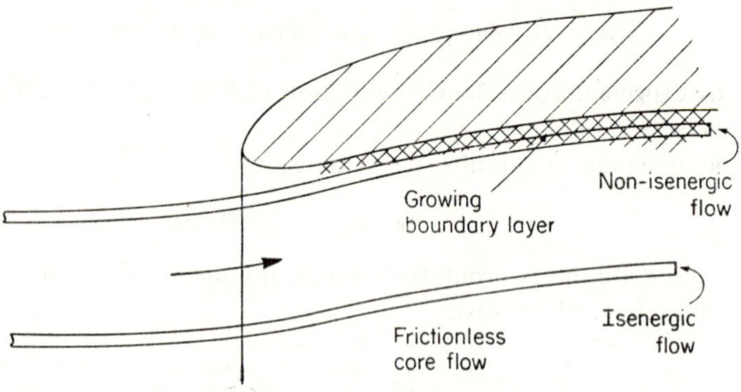

Figure 2.17 Streamtubes in the intake of a jet engine

two streamlines in the intake of a jet engine. The energy equation applies to both that in the frictionless core of the flow and that which passes into the boundary layer formed on the wall of the duct. For the former streamtube, heat addition and external work are negligible. For this isenergic process the energy equation (2.3.8) provides some fairly specific information about the variations in properties along the streamtube. The fluid in the streamtube which enters the boundary layer will exchange heat with the wall and will be worked on by frictional stresses. These energy inputs cannot be predicted unless a detailed study of the flow in the boundary layer is undertaken. In this second case the integral energy equation does not give immediate specific results.

Example 2.5 Although the result (2.3.7) applies to any fluid, the relative importance of the several contributions to the energy change depends very much on the thermodynamic behaviour of the fluid. Energy addition will not be considered in detail until a later chapter, but two simple energy additions will be discussed here to reveal the relationship of internal energy and displacement energy for liquids and gases. Consider adiabatic, isentropic compression in

(a) a water pump, and

(b) an air compressor drawing air from the atmosphere.

The two machines are shown in Figure 2.18. In each case changes in kinetic energy and potential energy may be neglected, and in each case the pressure rise is 100 ft of water, that is, a pressure difference capable of supporting the weight of a column of water 100 ft in height.

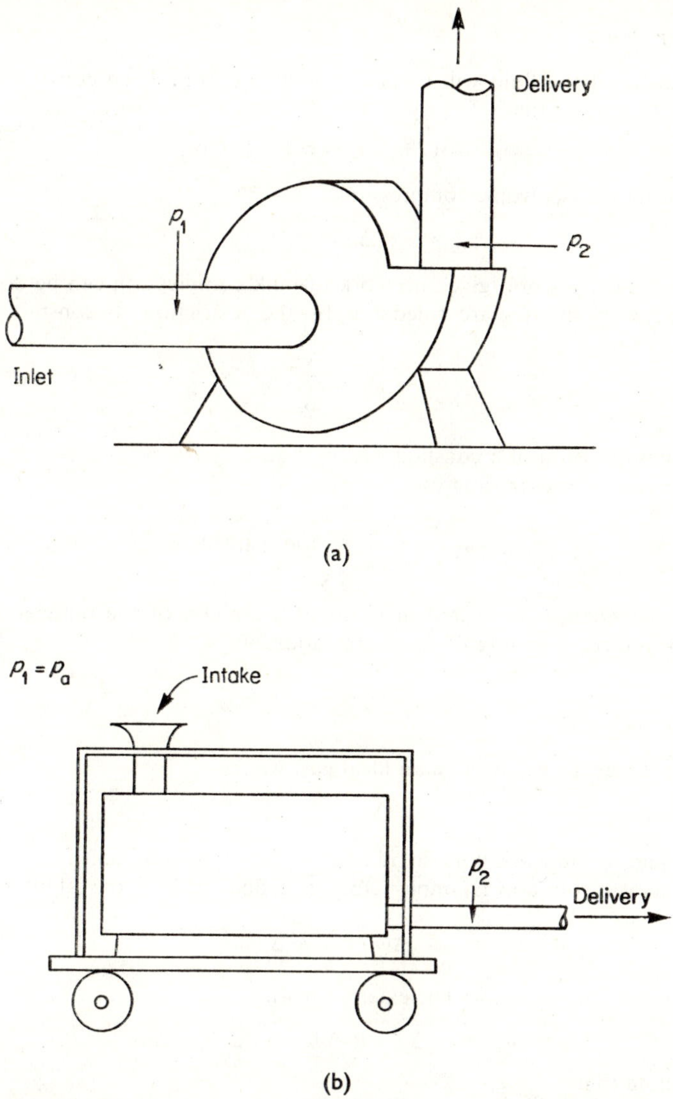

Figure 2.18 Two fluid pumping machines. (a) A water pump; (b) An air compressor

The external work required for compression is given by

$$w_e = \Delta e + \Delta(pv)$$

per unit mass, in each case. Also, for each process the pressure change is

$$\Delta p = 100 \text{ ft of water} = 6240 \text{ lbf/ft}^2$$

the specific weight of water being taken as 62·4 lbf/ft^3. There the similarity ends.

(a) Water Pump

The density of the fluid will be taken to be constant. Then equations (1.5.15) and (1.4.14) indicate that

$$s = s(T), \quad e = e(T) \quad \text{only}$$

Hence during the isentropic compression

$$\Delta e = 0$$

Since the water does not 'give', no work of bulk compression can be done on it; irreversible work modes are ruled out by the restriction to constant entropy. Hence

$$w_e = \frac{\Delta p}{\varrho_c}$$

for ideal compression of a constant-density liquid.

For the case of present interest

$$w_e = \frac{6240}{62 \cdot 4} = 100 \text{ ft lbf/lb}$$

The pressure change, measured in terms of a column of the pumped liquid, is seen to be a direct measure of the energy addition.

(b) Air Compressor

Taking the air to behave like an ideal gas, we have

$$\Delta e = c_v \, \Delta T$$

This is equal to the work done in adiabatically compressing unit mass of gas in a cylinder, as was found in Example 1.15. For a flow process, the additional energy change

$$\Delta(pv) = R \, \Delta T$$

must be taken into account. The external work is then

$$w_e = c_v \, \Delta T + R \, \Delta T = c_p \, \Delta T = \Delta h$$

For air, note that

$$\frac{\Delta e}{w_e} = \frac{c_v}{c_p} = \frac{5}{7} \quad \text{and} \quad \frac{\Delta(pv)}{w_e} = \frac{R}{c_p} = \frac{2}{7}$$

Thus for a perfect gas, the change in internal energy accounts for the larger part of the external work. This is, of course, contrary to our findings for a constant-density liquid.

The compression process to be considered is isentropic; hence

$$\frac{T_2}{T_1} = \left(\frac{p_2}{p_1}\right)^{\frac{\gamma-1}{\gamma}}$$

and

$$w_e = c_pT_1\left[\left(\frac{p_2}{p_1}\right)^{\frac{\gamma-1}{\gamma}} - 1\right]$$

For the case specified above, with $T_1 = 530$ °R and $p_1 = 2120$ lbf/ft^2, we have

$$w_e = 3{\cdot}5\times53{\cdot}3\times530\times\left[\left(\frac{8360}{2120}\right)^{0{\cdot}286} - 1\right] = 47{,}400 \text{ ft lbf/lb}$$

The energy stored in unit mass of a compressed gas is very much greater than that stored in a compressed liquid. This is in part the result of the greater volume of the gas, and in part the result of its compressibility, which allows the internal energy to change. Of course, the temperature of the gas rises during ideal compression, while that of the liquid is constant. Hence, if the gas is allowed to cool to its original temperature, much of the stored energy is 'lost'.

A comparison of the energy stored in the liquid and gas on a volumetric basis is also of interest. For the water

$$\varrho_c w_e = 6240 \text{ ft lbf/ft}^3$$

For the air it is not so obvious how to choose a representative density. Taking the mean between $\varrho_1 = 0{\cdot}075$ lbf/ft^3 and $\varrho_2 = 0{\cdot}200$ lbf/ft^3, we have

$$\bar{\varrho}w_e = 0{\cdot}137\times47{,}400 = 6500 \text{ lbf/ft}^3$$

Looked at in this way, the energy-storing capacities of the two fluids are seen to be more nearly the same. Of course, if the gas is cooled to its original temperature, its effective capacity will drop.

These considerations are relevant to the design of pneumatic and hydraulic systems, in which compressed fluids can be used to transfer energy in much the same way as an electric current.

2.3.2 Classification of Isenergic Flows

Isenergic flow may be defined precisely using the steady-flow energy equation. It is flow without external energy addition:

$$\dot{W}_e + \dot{Q} = 0$$

or

$$w_e + q = 0 \qquad (2.3.9)$$

This condition might be satisfied by cancellation of heat and work terms. More frequently, it is satisfied with heat flow and work separately equal to zero. The net displacement work cannot be arbitrarily specified, for it is a consequence of the fluid's entering and leaving the system.

For isenergic flow, the energy equation (2.3.6) gives

$$te_2 - te_1 = p_1v_1 - p_2v_2$$

showing that the change in total energy equals the net displacement work. Alternatively

$$te_2+p_2v_2 = te_1+p_1v_1$$

may be interpreted as an energy conservation principle for isenergic steady flow. Introducing the energy components, we get

$$e+pv+\tfrac{1}{2}u^2+gz = \text{constant}$$

or

$$h+\tfrac{1}{2}u^2+gz = \text{constant} \tag{2.3.10}$$

The balance between the energy components can be achieved in many ways. To ensure an orderly discussion, we shall consider a series of restricted cases.

Restrictions can be imposed on the conservation principle (2.3.10) by neglecting changes in kinetic energy, potential energy or density. The last two restrictions are virtually mutually exclusive, since heavy fluids (liquids) undergo only slight density changes, and light fluids (vapours and gases) are little influenced by gravity. An important exception is large-scale vertical motion in the atmosphere; this will be examined later. With this exception, nearly all isenergic flows fall into one of four categories:

(a) constant density with negligible kinetic-energy change;
(b) varying density with negligible kinetic-energy change;
(c) constant density with significant kinetic-energy change; and
(d) varying density with significant kinetic-energy change.

Each of the four cases defined above will be examined separately. All are governed by the principles of mass and energy conservation

$$\Delta(\varrho uA) = 0$$

and

$$\Delta(h+\tfrac{1}{2}u^2+gz) = 0 \tag{2.3.11}$$

Moreover, the force applied to the fluid is given, for unidirectional flows, by

$$F_a = \Delta[(p+\varrho u^2)A] \tag{2.3.12}$$

Despite this underlying unity, markedly different techniques are used to describe and analyse the four classes of isenergic flow.

Isenergic flows can be classified in yet another way, as isentropic or non-isentropic. In the discussion which follows, particular attention is given to entropy changes and the results for isentropic flow emerge as limiting cases. They will form the basis for the subsequent, more detailed discussion of flows that are isentropic as well as isenergic. There too it will prove convenient to distinguish between flows with constant density and varying density.

2.3.3 Low-velocity Isenergic Flow

The term low-velocity is chosen for its brevity, but is rather too restrictive. Here we require only that the change in kinetic energy is small; this is sometimes the case even though the velocities are large. Figure 2.19 shows three situations in which small changes in kinetic energy might be expected. The first is flow past a valve; a simple gate valve is shown, but the discussion applies to any kind of constriction. The valve might control the flow of air into an automobile engine, of water to a hydroelectric station, or of steam through a heating system. Figure 2.19(b) shows another very common flow,

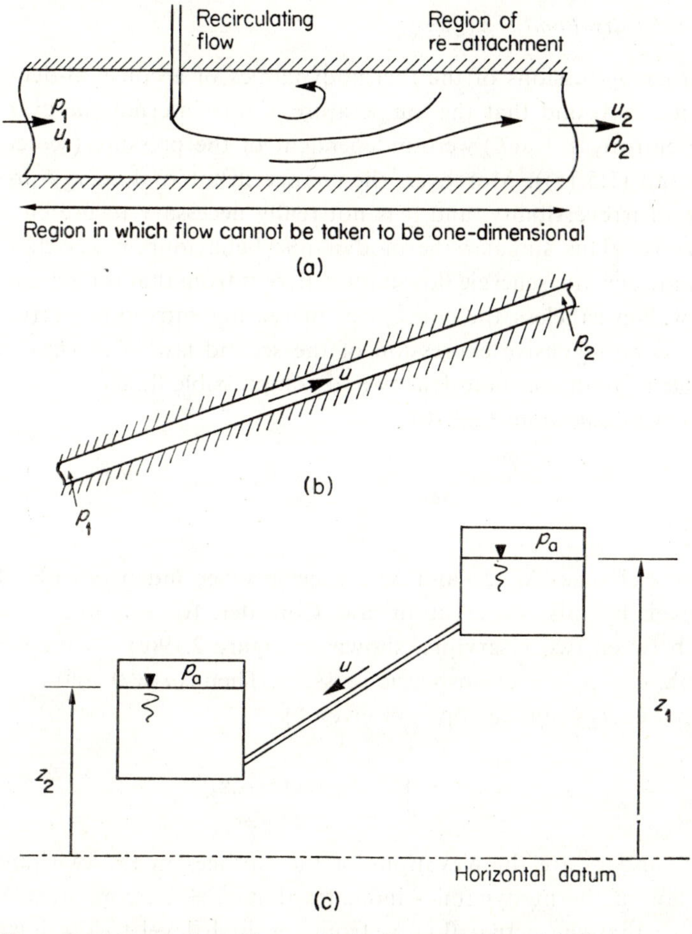

Figure 2.19 Examples of low-velocity isenergic flow. (a) Flow past a valve; (b) Flow through a straight pipe; (c) Flow between two reservoirs

that through a long uniform pipe. The restrictions imposed in this discussion require that the heat flow through the pipe wall be negligible, and that the kinetic-energy change along the pipe be small. The latter condition will certainly be satisfied in constant-density flow, no matter what the velocity. Finally, Figure 2.19(c) shows the flow between two tanks in which the fluid is nearly at rest. The kinetic energy in the pipe between them may be large, but the change in state between the two reservoirs is determined by the low-velocity conservation principle. In the sketch, the tanks contain a liquid with a free surface. However, the same general principles apply to a flow of steam or air through an insulated pipe between two tanks.

Constant-density Fluid

In earlier discussions of the thermodynamics of a constant-density substance, it was found that the temperature T, the internal energy $e = e(T)$ and the entropy $s = s(T)$ were independent of the pressure (see equations (1.4.14) and (1.5.15)). Moreover, the change in internal energy provides a measure of irreversibility, and it is not really necessary to use entropy for this purpose. This singular thermodynamic behaviour makes the analysis of constant-density isenergic flow quite different from that for a compressible fluid flow. Nevertheless, the concept of increasing entropy is useful in providing a comprehensive expression of the second law of thermodynamics, applicable to both constant-density and compressible fluids.

The first of equations (2.3.10) gives

$$\Delta e + \frac{\Delta p}{\varrho_c} + g\,\Delta z = 0 \tag{2.3.13}$$

Two of the changes Δe, Δp and Δz can be specified independently, the third being given by this conservation law. Consider, for example, flow under gravity between two reservoirs, shown in Figure 2.19(c). At the surface in each tank, $p = p_a$, the atmospheric pressure. Then $\Delta p = 0$, and the changes in internal energy and entropy are given by

$$\Delta e = \int T\,\mathrm{d}s = g(z_1 - z_2)$$

where z_1 and z_2 are the elevations of the surfaces in the two tanks. The second law of thermodynamics indicates that $z_1 > z_2$, consistent with our experience that the actual flow is from the high level to the lower tank. The potential energy $g(z_1 - z_2)$ is said to be dissipated in this process, that is, converted into internal energy.

Next consider a flow in which gravity is unimportant, pressure change providing the driving force; for example, the constant-area flow past a valve, Figure 2.19(a), or along a straight pipe, Figure 2.19(b). Equation (2.3.13) gives the dissipation per unit mass as

$$\Delta e = -\frac{\Delta p}{\varrho_c}$$

The dissipation per unit volume is simply the pressure drop:

$$\varrho_c \, \Delta e = -\Delta p$$

It is common practice to determine the dissipation in a low-velocity flow by measuring the changes in pressure along the flow. Since the pressure always falls as a result of dissipation (neglecting gravity, as here), its change is often referred to as a pressure loss, and energy dissipations are referred to as 'losses'.

The *rate of dissipation* in a one-dimensional flow process is given by $\dot{m} \, \Delta e$. If gravity is negligible

$$\dot{m} \, \Delta e = -\dot{V} \, \Delta p$$

If the flow area is constant, $u_1 = u_2 = u$ for constant density, and the momentum equation (2.3.12) reduces to $F = A \, \Delta p$. Hence

$$\dot{m} \, \Delta e = -\dot{V} \, \Delta p = -Fu \tag{2.3.14}$$

Here we have two simple interpretations of the dissipation rate, one as the product of volume flow rate and pressure drop, and the other as the product of flow velocity and the force applied to the flow.

Both interpretations of equation (2.3.14) apply to each of the flows shown in Figure 2.19(a) and (b). However, the *mechanisms of dissipation* are quite different in these two flows. In the flow at a valve, dissipation takes place in the locally unsteady, eddying flow just behind the gate. The flow separates from the lip of the gate and re-attaches some distance downstream, the separated stream surrounding a region of recirculation. Dissipation is the result of very large internal friction within these interacting flows. In pipe flow, dissipation is distributed uniformly along the pipe, being the result of uniform friction on the pipe wall. The dissipation per unit length of pipe is

$$\dot{m} \frac{\mathrm{d}e}{\mathrm{d}x} = -\dot{V} \frac{\mathrm{d}p}{\mathrm{d}x} \tag{2.3.15}$$

the product of the volume flow rate and the pressure gradient.

Returning now to the general case of low-velocity isenergic flow with constant density, we note that equation (2.3.13) gives the dissipation per unit volume as

$$\varrho_c \, \Delta e = -\Delta(p+\varrho_c g z) \tag{2.3.16}$$

Not all of the pressure change is the result of dissipation; some would occur even if the fluid were at rest. The gravity-dependent term accounts for this hydrostatic component of the pressure change. With this adjustment, the preceding discussion of dissipation may be carried over to flows with significant changes in elevation.

Compressible Fluid

Low-velocity isenergic flow with varying density has often been referred to as *throttled flow*. This term relates to flow past a valve, for example, the throttle valve controlling the flow of air to an engine, which has the effect of constricting the flow passage. Neglecting fluid weight, we find that

$$\Delta h = 0 \tag{2.3.17}$$

in low-velocity throttling. Hence the process may also be described as *isenthalpic flow*.

The condition of constant enthalpy implies that

$$h(p_1, T_1) = h(p_2, T_2) \tag{2.3.17}$$

together with similar relationships amongst other properties. If any three independent properties are specified, a fourth can be found. If, for example, the upstream state (p_1, T_1) and the downstream pressure p_2 are known, then the downstream temperature T_2 and all the other downstream properties can be determined. Note that the condition (2.3.17) is symmetrical with respect to the suffices 1 and 2. It does not distinguish between the transition from state 1 to state 2 and that from 2 to 1. One of these transitions will involve an increase in entropy and the other a decrease. Only that involving the increase in entropy can actually occur, according to the second law of thermodynamics.

The preceding discussion applies to any simple substance, even one of constant density. For a *perfect gas* more definite results are available. It has been seen in equations (1.4.11) that $h = h(T)$ and $e = e(T)$ for a semi-perfect gas. Hence $\Delta h = 0$ implies that

$$\Delta T = 0 \quad \text{and} \quad \Delta e = 0 \tag{2.3.18}$$

when a perfect gas is throttled at low velocity. Now consider the entropy change. Equation (1.5.14) gives, for h = constant

$$ds = -\frac{v}{T}\,dp = -R\frac{dp}{p}$$

for a semi-perfect gas. Integrating

$$s_2 - s_1 = R \log_e \frac{p_1}{p_2} \tag{2.3.18}$$

As would be expected, in a real throttling process the pressure always falls.

In constant-density flow the internal energy is a direct measure of energy degradation. For varying density, the less easily understood property entropy must be used as a measure of irreversibility. However, for particular processes, irreversibility can be measured more directly by determining its effect on the ability of the fluid to do work. Consider the air turbine shown in Figure 2.20. The flow of air is controlled by a throttle valve upstream of the turbine. The air exhausts directly to the atmosphere, where the pressure is p_a. We shall see that the effect of throttling is to reduce the energy available to the turbine as work.

The flow will be taken to be adiabatic throughout. Then application of the steady-flow energy equation (2.3.8) gives the work output for unit mass passing through the turbine as

$$w_0 = -w_e = h_2 - h_a = h_1\left(1 - \frac{h_a}{h_2}\right)$$

since $h_2 = h_1$, and taking h_a as the outlet enthalpy. For ideal expansion in the turbine

$$w_0 = h_1\left[1 - \left(\frac{p_a}{p_2}\right)^{\frac{\gamma-1}{\gamma}}\right]$$

This is the greatest amount of work which can be done by the air flowing adiabatically between states (h_1, p_2) and p_a.

Now consider the ideal expansion in the turbine for two operating conditions:

(1) throttle fully open, p_1 to p_a across turbine, and

(2) throttle partly closed, p_2 to p_a across turbine.

These two processes are shown in Figure 2.20(b). The work done in them is given by

$$w_{0_1} = h_1\left[1 - \left(\frac{p_a}{p_1}\right)^{\frac{\gamma-1}{\gamma}}\right] \quad \text{and} \quad w_{0_2} = h_1\left[1 - \left(\frac{p_a}{p_2}\right)^{\frac{\gamma-1}{\gamma}}\right]$$

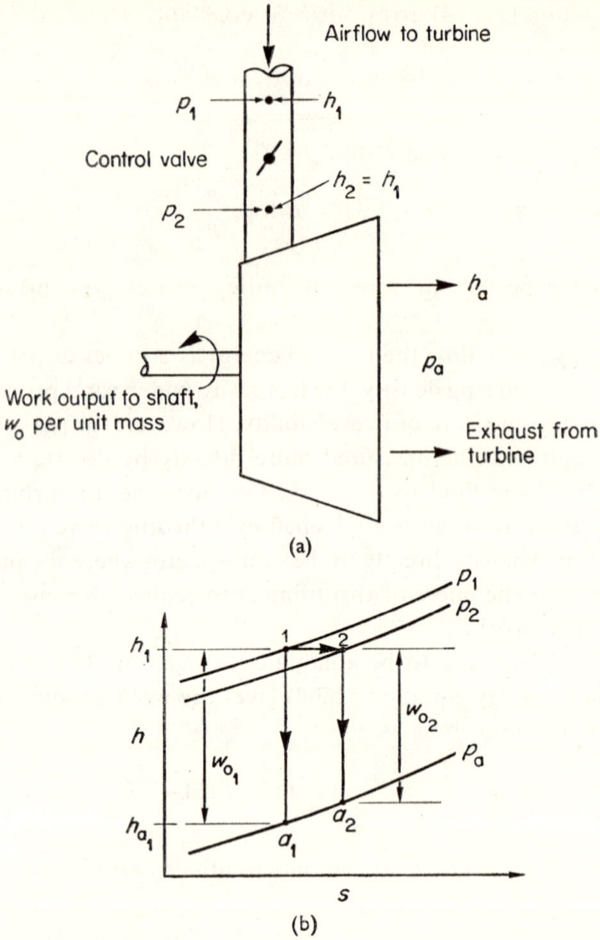

Figure 2.20 Operation of an air turbine with a throttle valve in the supply line. (a) General view of the installation; (b) Process paths with and without throttling

Hence

$$w_{0_1} - w_{0_2} = h_1\left(\frac{p_a}{p_1}\right)^{\frac{\gamma-1}{\gamma}}\left[\left(\frac{p_1}{p_2}\right)^{\frac{\gamma-1}{\gamma}} - 1\right]$$

$$= h_{a_1}\left[\left(\frac{p_1}{p_2}\right)^{\frac{\gamma-1}{\gamma}} - 1\right]$$

Since $p_1 > p_2$, it is always the case that $w_{0_1} > w_{0_2}$. The work done by unit mass in the turbine is reduced when throttling occurs upstream. Although the enthalpy of the fluid is not changed, the energy available as work is reduced. Note that the throttle valve acts to reduce the output of the turbine in two ways: by reducing the mass flow and by reducing the available energy per unit mass.

We have considered only the effect of a valve on the availability of energy, but other sources of entropy production will act in the same way. For example, friction in the duct will have a throttling effect even when the valve is fully open. Equation (1.5.29) indicates the consequences of entropy production for the efficiency of a thermodynamic cycle. The role of entropy production in the process shown in Figure 2.20 can also be displayed explicitly. The second of equations (2.3.18) gives $p_1/p_2 = \exp(\Delta s/R)$. Then

$$w_{0_1} - w_{0_2} = h_{a_1}\left[\exp\left(\frac{\Delta s}{c_p}\right) - 1\right]$$

Let us next consider how real gases and vapours behave when throttled. They will not, of course, have the simple response (2.3.18). The deviation from the perfect-gas pattern is called the *Joule–Thomson effect* and the quantity $(\partial T/\partial p)_h$, the Joule–Thomson coefficient, is a measure of the imperfection of a real gas.

The Joule–Thomson effect is the basis of a method of liquefying gases. The gas is subjected to the following processes:

(1) adiabatic compression with a large pressure rise,
(2) cooling with pressure maintained constant, and
(3) isenergic flow through a porous plug with large pressure drop.

By tracing these processes on a Mollier chart (that for steam will do), one can see how they move the state across the saturation line and so produce partial liquefaction. Figure 2.21 shows the general form of the process path.

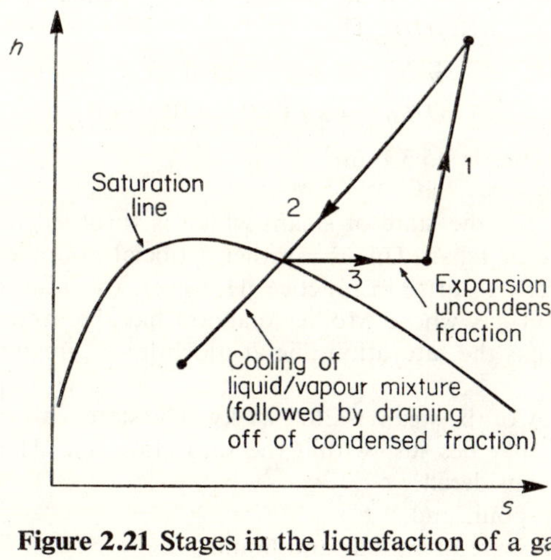

Figure 2.21 Stages in the liquefaction of a gas

A consideration of the last two stages of the liquefaction process leads to an important distinction. In each stage the temperature of the fluid is reduced, but only in the former stage (2) does heat flow from the fluid. This indicates that the verb 'to cool' must be used with care. Henceforth it will be used to mean 'to cause an outflow of heat'. The temperature will not necessarily fall during cooling defined in this way; for example, work may be done on the substance so that there is a net increase in its internal energy. The verb 'to heat' must be used with similar caution. Other situations will be encountered in which these distinctions are important.

Example 2.6 Water leaving the elbow discussed in Example 2.4 (with $p = 560$ kN/m^2, $u = 14{\cdot}0$ m/sec, and $\dot{V} = 0{\cdot}11$ m^3/sec) flows vertically downwards through a pipe of diameter 10 cm and length 10 m. At the bottom its pressure is 640 kN/m^2. Find

(a) the dissipation in the flow, and
(b) the friction stress on the pipe wall.

(a) Dissipation

Equation (2.3.16) gives the dissipation rate as

$$\dot{m}\,\Delta e = -\dot{V}\,\Delta(p+\varrho_c gz)$$
$$= -0{\cdot}11\times(80-9{\cdot}81\times10) = 2{\cdot}0 \text{ kW}$$

(b) Friction

The retarding force and dissipation are related by

$$\dot{m}\,\Delta e = -Fu$$

Here the retardation results from the friction stress τ acting on the inner surface of the pipe, which has area πDL. Then

$$\tau = \frac{\dot{m}\,\Delta e}{\pi DLu} = \frac{2000}{3{\cdot}14\times0{\cdot}1\times10\times14{\cdot}0}$$
$$= 45{\cdot}3 \text{ N/m}^2$$

Example 2.7 What is the state of steam which is throttled from 350 bar and 450 °C to a pressure of 1 bar. This is a rather artificial example, since the steam conditions are unlikely to arise in practice. However, this kind of transition can occur in a refrigerant gas whose Mollier diagram has the same general form as that for steam, but has the saturation line at much lower pressures and temperatures.

The initial enthalpy of the steam is 2673 kJ/kg. The state with the same enthalpy and a pressure of 1 bar lies just within the saturation line. Hence two possible final states may be considered:

(a) supercooled vapour, and
(b) equilibrium mixture of vapour and liquid.

(a) Supercooled Vapour

In the absence of condensation, the fluid behaves essentially like a superheated vapour, and the final properties can be found by extrapolating from the superheated region. Thus, using values for 1 bar, one has

$$T_2 = 100-50\times\frac{2676-2673}{2777-2676}$$

$$= 100-50\times\frac{3}{101} = 98{\cdot}5\ ^\circ\mathrm{C}$$

Similarly

$$v_2 = 1{\cdot}696-(1{\cdot}937-1{\cdot}696)\frac{3}{101} = 1{\cdot}689\ \mathrm{m^3/kg}$$

and

$$s_2 = 7{\cdot}360-(7{\cdot}614-7{\cdot}360)\frac{3}{101} = 7{\cdot}352\ \mathrm{kJ/kg\ ^\circ K}$$

(b) Vapour and Liquid

The temperature is now the saturation temperature for 1 bar:

$$T_2 = 99{\cdot}6\ ^\circ\mathrm{C}$$

The fraction of the steam which condenses is determined by

$$h_1 = yh_f+(1-y)h_g$$

Introducing values for 1 bar, one gets

$$y = \frac{h_g-h_1}{h_{fg}} = \frac{2}{2258} = 0{\cdot}09\ \text{per cent}$$

indicating that only a small part of the vapour will condense. The other properties follow:

$$v_2 = 0{\cdot}0009\times0{\cdot}001+0{\cdot}999\times1{\cdot}694 = 1{\cdot}692\ \mathrm{m^3/kg}$$

and

$$s_2 = 0{\cdot}0009\times1{\cdot}303+0{\cdot}999\times7{\cdot}351 = 7{\cdot}353\ \mathrm{kJ/kg\ ^\circ K}$$

Whether or not condensation occurs, there is a large rise in entropy. However, that just found is slightly higher than the entropy change for the supercooled vapour. Condensation involves a further irreversible process as the fluid moves towards its true equilibrium state.

Note the very marked Joule–Thomson effect in this process. The temperature drops by 350 °C without a change in enthalpy. It will be apparent why throttling proves useful in refrigeration cycles.

Example 2.8 The Joule–Thomson coefficient is given by the relationship

$$\left(\frac{\partial T}{\partial p}\right)_h = \frac{T(\partial v/\partial T)_p-v}{c_p}$$

which can be derived by lengthy manipulations of results obtained earlier. This result will be used to investigate the Joule–Thomson effect in a van der Waals gas.

Differentiation of equation (1.2.7) gives

$$T\left(\frac{\partial v}{\partial T}\right)_p = \frac{RT}{p-(a/v^2)(1-2b/v)} \simeq \frac{RT}{p}\left(1+\frac{a}{pv^2}\right)$$

when only terms of the first degree in the corrections a and b are retained. To the same order, the van der Waals equation gives

$$v \simeq \frac{RT}{p}\left(1-\frac{a}{pv^2}+\frac{b}{v}\right)$$

Then the Joule–Thomson coefficient is

$$\left(\frac{\partial T}{\partial p}\right)_h \simeq \frac{RT}{pc_p}\left(\frac{2a}{pv^2}-\frac{b}{v}\right) \simeq \frac{(2a/RT-b)}{c_p}$$

to the order of approximation used above.

This result indicates that the temperature will drop when throttling occurs at low temperatures, and will rise when the temperature is high. The temperature of inversion, at which there is no Joule–Thomson effect, is

$$T_i \simeq \frac{2a}{bR}$$

Consider the example of steam. The values from Appendix II give

$$T_i = \frac{2\times 1400}{0{\cdot}488\times 0{\cdot}73} = 7850\ ^\circ\mathrm{R}$$

In these units, the appropriate gas constant is

$$R = \frac{1545}{2117} = 0{\cdot}73 \text{ atm ft}^3/\text{lb mol }^\circ\mathrm{R}$$

This calculation is not strictly meaningful, since dissociation will have rendered the van der Waals equation inapplicable long before this temperature can be attained. However, the result does serve to show that the Joule–Thomson coefficient is positive for steam under all realistic conditions. Inspection of the steam tables confirms this prediction.

The preceding analysis demonstrates the utility of an analytical model, even a crude one. The value given for the temperature of inversion by the van der Waals equation will probably not be very accurate, but the calculation does indicate the existence of the phenomenon. Once known, it can be studied using more precise techniques.

2.3.4 High-velocity Isenergic Flow

The flows of Figure 2.19 may be included in this category if the changes in kinetic energy are large. On the other hand, those flows can occur without significant changes in velocity. There are, however, some isenergic flows which necessarily involve significant changes in kinetic energy. The inlet of a pump or turbine has the tasks of collecting fluid, accelerating it, and guiding it into the machine. Figure 2.22(a) depicts the penstock which carries water to a hydroelectric station from a reservoir. It is also necessary to

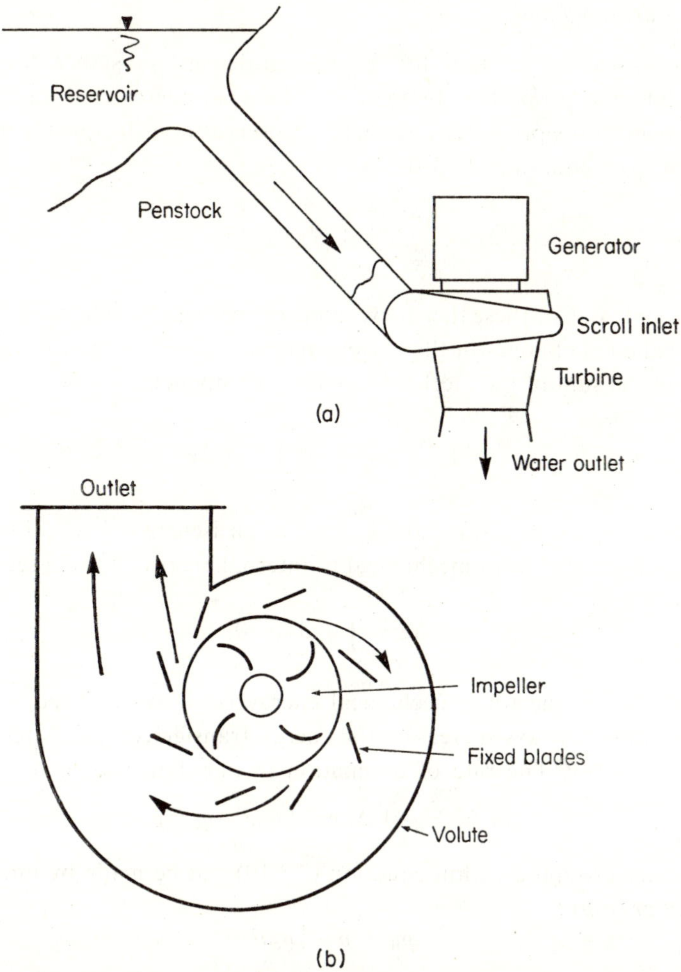

Figure 2.22 Examples of high-velocity isenergic flow. (a) Flow to and from the turbine of a hydroelectric station (the processes within the turbine are not isenergic). (b) Flow in the diffuser of a centrifugal compressor

decelerate the fast-moving fluid emerging from flow machines. Figure 2.22(b) shows the diffuser of a centrifugal compressor which performs this task. The gas emerging from the impeller slows down on passing through a row of stationary blades, and slows down further on moving through a volute spiral. The processes shown in Figure 2.22 are nearly isenergic, and both involve significant velocity changes.

Constant-density Fluid

It was noted earlier that for a constant-density substance the strictly thermodynamic properties, temperature, internal energy and entropy, are independent of the pressure. The energy conservation principle for this case, obtained from equations (2.3.10) as

$$\Delta e + \frac{\Delta p}{\varrho_c} + \Delta(\tfrac{1}{2}u^2) + g\,\Delta z = 0$$

contains only one of these thermodynamic properties. The change in internal energy balances changes in the various mechanical properties. It equals the change in the sum of the mechanical energy components:

$$\Delta\left(\frac{p}{\varrho_c} + \tfrac{1}{2}u^2 + gz\right) = -\Delta e \tag{2.3.19}$$

Thus the energy conversion taking place in an isenergic flow can be interpreted as a transfer from mechanical to internal energy. As was seen earlier

$$\Delta e = \int T\,\mathrm{d}s > 0$$

This shows that the total mechanical energy is always reduced, while the internal energy always increases; the energy transferred, Δe, is termed the *dissipated energy*. The rate of dissipation in one-dimensional flow is then

$$\dot{m}\,\Delta e = -\dot{V}\Delta(p + \tfrac{1}{2}\varrho_c u^2 + \varrho_c gz)$$

A further grouping within equation (2.3.19) can be made by introducing the *total pressure:*

$$p_0 = p + \tfrac{1}{2}\varrho_c u^2 \tag{2.3.20}$$

the sum of the pressure and the *dynamic pressure*, $\tfrac{1}{2}\varrho_c u^2$. For isentropic flow, the condition

$$\frac{p}{\varrho_c} + \frac{u^2}{2} + gz = \frac{p_0}{\varrho_c} + gz = \text{constant} \tag{2.3.21}$$

must be obeyed. This is *Bernoulli's equation* for a constant-density flow; the flow must also be steady, isenergic and isentropic. In this kind of flow, the temperature, internal energy, entropy and of course the density are all constant. However, the pressure, velocity and elevation of fluid can vary, subject to the Bernoulli condition (2.3.21). If changes in elevation are not important, Bernoulli's equation reduces to

$$p_0 = p + \tfrac{1}{2}\varrho_c u^2 = \text{constant}$$

The total pressure is seen to be that achieved when the fluid comes to rest isentropically and without change of elevation.

Equation (2.3.19) can be written

$$\frac{\Delta(p_0 + \varrho_c g z)}{\varrho_c} = -\Delta e = -\int_1^2 T\,ds$$

Temperature changes in adiabatic liquid flows are usually very small, as was found in discussing isentropic compression in Section 1.4.2. Hence the preceding result may be given with good accuracy as

$$\frac{-\Delta(p_0 + \varrho_c g z)}{\varrho_c} = \Delta e = c_v\,\Delta T = T_c\,\Delta s \qquad (2.3.22)$$

This provides a direct link between the fall in total mechanical energy, the dissipated energy, the rise in temperature, and the rise in entropy. Any of these quantities provides a measure of irreversibility in an isenergic, constant-density flow.

For hydraulic calculations, the equation (2.3.19) is often written in the form

$$\Delta\left(\frac{p - p_a}{\varrho_c g} + \frac{u^2}{2g} + z\right) = \Delta H_0 = -H_f \qquad (2.3.23)$$

Each term in this equation has units of length. The quantity H_0 is termed the *total head*. It is a measure of the total mechanical energy, being the sum of

(a) the local elevation z (the potential energy per unit weight),

(b) the velocity head $u^2/2g$ (the kinetic energy per unit weight), and

(c) the pressure head, $(p - p_a)/\varrho_c g$ (the displacement energy per unit weight).

The term

$$H_f = \frac{\Delta e}{g} > 0 \qquad (2.3.24)$$

gives the fall in total head, or *head loss*. This name is somewhat misleading, since energy is not lost, but is merely converted from mechanical to internal form. It follows that

$$\Delta H_0 < 0 \qquad (2.3.24)$$

in any isenergic process. This is the form of the second law of thermodynamics which is convenient for hydraulic applications.

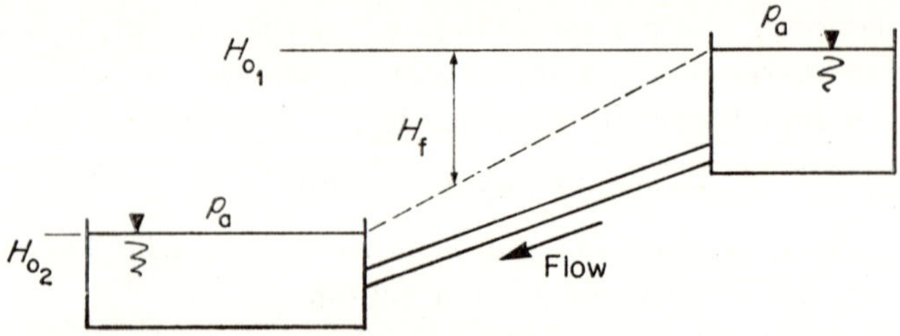

Figure 2.23 Fall in total head in the flow through a uniform pipe line between two reservoirs

Figure 2.23 shows how the total head changes in the flow between two reservoirs. In the reservoirs the total head is simply the surface elevation, and it falls linearly along the uniform pipe between them. Actually, the variation of total head near exit and entry to the pipe will be somewhat more complex, but these details are not of immediate importance. Chapter 5 will look at the mechanisms of dissipation or head loss in a flowing fluid.

Compressible Fluid

For this case, equations (2.3.10) give

$$h_0 = h+\tfrac{1}{2}u^2 = \text{constant} \tag{2.3.25}$$

where h_0 is the *total or stagnation enthalpy*. This is the enthalpy achieved if the fluid comes to rest isenergically. It is akin to the total pressure, introduced in equation (2.3.20), although to get the total pressure one must decelerate the fluid isentropically, as well as isenergically.

One other stagnation property can be introduced immediately, the *stagnation temperature*. For a perfect gas with $h = c_pT$, the temperature achieved when the gas comes to rest isenergically is $T_0 = h_0/c_p$. Equation (2.3.25) may be written, for a calorically perfect gas

$$T_0 = T+\frac{u^2}{2c_p} = \text{constant} \tag{2.3.26}$$

This serves to define the total temperature of a perfect gas, and also to show that the total temperature is constant in isenergic flow of a perfect gas. The term $u^2/2c_p$ represents the increase in temperature as the kinetic energy of the gas is converted into enthalpy. This contribution to the stagnation temperature is of great importance in high-speed flight. The air rushing towards an aeroplane must come to rest on the surface of the aeroplane. Thus the temperature of the air in contact with the skin is higher than the ambient temperature in the atmosphere. This temperature rise has already been considered in Example 1.19, where deceleration to the stagnation point was considered with the additional assumption of constant entropy. However, the temperature over most of the surface does not attain the stagnation value, since the flow in the boundary layer on the surface is neither isentropic nor isenergic.

While considering a perfect gas, we may note that the relations (2.3.18) are not followed in high-velocity throttling. Suppose, for example, that the cock on a cylinder of compressed air is opened. The air in the resulting jet would be cooler than that in the bottle for two reasons: the Joule–Thomson effect and the conversion of part of its energy into kinetic form.

For isenergic, constant-density flow it was found possible to describe irreversibility very conveniently in terms of the internal-energy change. It is tempting to seek a similar description for flow with varying density. The search must fail. In general, the internal energy and displacement energy are closely linked, and the strict separation of mechanical energy and internal energy achieved in equation (2.3.19) is not possible. Nor is it possible to measure irreversibility using the internal energy. As was seen in equation (1.5.3), when the density changes another property, entropy, must be introduced to represent irreversibility.

For isentropic flow, the constant-density treatment can be followed part way. In an infinitesimal throttling process, equations (1.5.14) and (2.3.25) give

$$\mathrm{d}h = T\,\mathrm{d}s + \frac{\mathrm{d}p}{\varrho} = -u\,\mathrm{d}u$$

Then

$$\frac{\mathrm{d}p}{\varrho} + u\,\mathrm{d}u = -T\,\mathrm{d}s < 0 \tag{2.3.27}$$

For an isentropic process

$$\frac{\mathrm{d}p}{\varrho} + u\,\mathrm{d}u = 0 \tag{2.3.28}$$

and integration gives Bernoulli's equation for varying density:

$$\int_1 \frac{\mathrm{d}p}{\varrho} + \frac{u^2}{2} = \text{constant} \tag{2.3.29}$$

In order to obtain an explicit relationship between velocity and pressure, like that for constant-density (2.3.21), one must specify the equation of state in the form $\varrho = \varrho(p, s_1)$, where s_1 is the constant entropy.

Although flows with constant density and varying density have been discussed separately, we have not indicated where, for practical purposes, the demarcation should be made. Seemingly, liquids can be described as having constant density in many circumstances. What about gases and vapours? The preceding results give

$$\varrho T\,\mathrm{d}s = -\left[\mathrm{d}p + \varrho\,\mathrm{d}\left(\frac{u^2}{2}\right)\right] = -\mathrm{d}\left(p + \frac{\varrho u^2}{2}\right) + \frac{u^2}{2}\,\mathrm{d}\varrho$$

for any small isenergic flow process. No matter what the equation of state, the dissipation per unit volume is

$$\varrho_1 T_1\,\mathrm{d}s = -\left[\mathrm{d}p + \varrho_1\,\mathrm{d}\left(\frac{u^2}{2}\right)\right] \qquad (2.3.30)$$

If the velocity is not too large, the result

$$\varrho_1 T_1\,\mathrm{d}s = -\mathrm{d}\left(p + \varrho\frac{u^2}{2}\right) = -\mathrm{d}p_{0_c} \qquad (2.3.30)$$

may also be obtained, where p_{0_c} is the total pressure found using the constant-density formula (2.3.20). For isentropic flow

$$p + \varrho\frac{u^2}{2} = \text{constant} \qquad (2.3.31)$$

Thus for any fluid flowing isenergically at low velocity, small isentropic changes are determined by the constant-density Bernoulli result, and the dissipation equals the drop in the sum of pressure and dynamic pressure. However, note that

$$\mathrm{d}e = T\,\mathrm{d}s - p\,\mathrm{d}v$$

does not give the dissipation in a gas flow, no matter how small the velocity.

Later it will be found that the constant-density results (2.3.30, 31) apply when the flow velocity is a small fraction of the speed of sound in the fluid. In sea-level conditions, the speed of sound in air is about 340 m/sec or 1100 ft/sec. Hence these results accurately relate pressure and velocity changes in many air flows; for example, that of the wind around a building, and that over the wing of an aeroplane travelling below, say, 100 m/sec. We shall also find that certain aspects of the flow of gases through fans and blowers can be described using the constant-density approximation.

Example 2.9 What are the losses in the flow around a pipe elbow considered in Example 2.4? The water starts with $p_1 = 700$ kN/m², $u_1 = 6{\cdot}23$ m/sec, and ends up with $p_2 = 560$ kN/m², $u_2 = 14{\cdot}0$ m/sec.

It has been seen earlier that the effect of gravity on this flow is small. Hence it is the change in total pressure which gives the dissipation:

$$-\Delta p_0 = p_1 - p_2 + \frac{\varrho_c}{2}(u_1^2 - u_2^2)$$

$$= 140 + \frac{1}{2}(6{\cdot}23^2 - 14{\cdot}0^2) = 61{\cdot}4 \text{ kN/m}^2$$

This is the dissipation per unit volume of the flow. The rate of dissipation in the entire flow is

$$-\dot{V}\,\Delta p_0 = 0{\cdot}11 \times 61{\cdot}4 = 6{\cdot}73 \text{ kW}$$

the flow rate being 0·11 m^3/sec. If these losses could be avoided, the same flow could be achieved with a reduction of at least 9 hp in pumping power.

Since $c_v \simeq 1$ kcal/kg °C = 4·19 kJ/kg °C, the rise in the temperature of the water is

$$\Delta T = \frac{\Delta e}{c_v} = -\frac{\Delta p_0}{\varrho_c c_v} = \frac{61{\cdot}4}{1000 \times 4{\cdot}19} = 0{\cdot}015\ ^\circ\text{C}$$

The specific heat of the water is so great that even this considerable energy dissipation produces only a slight rise in temperature. The entropy change is found to be

$$\Delta s = \frac{\Delta e}{T_1} = \frac{61{\cdot}4}{293} = 0{\cdot}21 \text{ J/kg °K}$$

on taking the temperature to be 20 °C.

Using Bernoulli's equation (2.3.21) one can also answer the question: For the same inlet conditions, what would have been the outlet conditions in the absence of losses? The velocity would not change, since it is determined by the condition of incompressibility. However, the pressure would be

$$p_2 = p_1 + \frac{\varrho_c}{2}(u_1^2 - u_2^2) = 700 - 78{\cdot}6 = 621{\cdot}4 \text{ kN/m}^2$$

This is the actual final pressure plus the losses found above.

Example 2.10 Consider again the ram jet of Example 2.2. Figure 2.24 shows the flow into the combustion chamber from the atmosphere. The flow may be taken to be isenergic prior to fuel injection. If the velocity of flight is 1200 m/sec

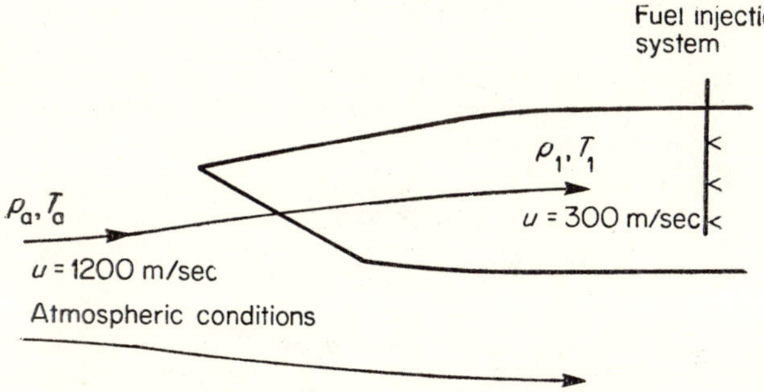

Figure 2.24 Flow into the combustion duct of a ram jet

at 20,000 m, what are the temperature and pressure in the combustion duct before the fuel is burned? The density and velocity there are $\varrho_1 = 0{\cdot}8$ kg/m³ and $u_1 = 300$ m/sec.

Assuming perfect-gas behaviour, the total temperature is constant; then

$$T_0 = T_a + \frac{u^2}{2c_p} = T_1 + \frac{u_1^2}{2c_p}$$

where T_a is the atmospheric temperature (217 °K for 20,000 m) and u is the velocity of flight. Then

$$T_0 = 217 + \frac{1200^2}{2010} = 933 \text{ °K}$$

and

$$T_1 = T_a + \frac{u^2 - u_1^2}{2c_p} = 888 \text{ °K}$$

The pressure follows as

$$p_1 = \varrho_1 R T_1 = 0{\cdot}8 \times 287 \times 888 \text{ N/m}^2 = 2{\cdot}04 \text{ bar}$$

As the flow decelerates, the temperature, density and pressure all rise far above their free-stream or atmospheric values. This compression by deceleration, or ram compression, allows the ram jet to operate without the compressor of a conventional turbojet. If efficient combustion is to be achieved, a considerable rise in pressure is necessary. Hence ram compression is effective only at high velocities.

Example 2.11 Steam is supplied to a whistle at 100 lbf/in² and 400 °F. It exhausts through a constriction to atmospheric pressure. With what velocity does it leave the whistle, a constant-diameter tube on either side of the constriction, if the velocity of approach is 100 ft/sec?

Steam tables give the enthalpy of the inlet steam as $h_1 = 1227{\cdot}6$ Btu/lb. The total enthalpy is

$$h_0 = h_1 + \frac{u_1^2}{2}$$

$$= 1227{\cdot}6 + \frac{100^2}{2 \times 778 \times 32{\cdot}2} = 1227{\cdot}8 \text{ Btu/lb}$$

The outlet state must be found by trial-and-error or some equivalent technique. It is determined by the two requirements of energy and mass conservation, which are to be satisfied simultaneously:

$$h_2 + \frac{u_2^2}{2} = h_0$$

and

$$\varrho_2 u_2 = \varrho_1 u_1 \quad \text{or} \quad u_2 = \frac{v_2}{v_1} u_1$$

None of v_2, h_2 and u_2 is known to start with.

Stage 1

Looking at steam-table values of specific volume for 1 atm, for enthalpies somewhat less than h_0, one sees that the specific volume will be of the order of $v_2 = 30$ ft³/lb. Hence the downstream velocity can be estimated as

$$u_2 = \frac{30}{4{\cdot}9} \times 100 = 600 \text{ ft/sec}$$

and the outlet enthalpy as

$$h_2 = h_0 - \frac{u_2^2}{2} = 1227{\cdot}8 - 7{\cdot}2 = 1220{\cdot}6 \text{ Btu/lb}$$

Stage 2

A value for v_2 can now be calculated and, using it, the values of h_2 and u_2 can be refined. Interpolation gives

$$v_2 = 30{\cdot}53 + (34{\cdot}68 - 30{\cdot}53) \times \frac{1220{\cdot}6 - 1192{\cdot}8}{1239{\cdot}0 - 1192{\cdot}8}$$

$$= 30{\cdot}53 + 2{\cdot}45 = 33{\cdot}0 \text{ ft}^3\text{/lb}$$

Then

$$u_2 = \frac{33{\cdot}0}{4{\cdot}94} \times 100 = 670 \text{ ft/sec}$$

and

$$h_2 = 1227{\cdot}8 - 8{\cdot}9 = 1218{\cdot}9 \text{ Btu/lb}$$

Stage 3

Proceeding as for Stage 2, we find

$$v_2 = 32{\cdot}8 \text{ ft}^3\text{/lb}, \quad u_2 = 665 \text{ ft/sec}, \quad h_2 = 1219{\cdot}0 \text{ Btu/lb}$$

There is no point in carrying the iteration beyond the point of practical utility, dictated either by the accuracy required in the results or by the accuracy of the initial data. Once sufficiently accurate values of the iterated properties are available, the others, T_2 and s_2, can be found by interpolation.

2.4 BERNOULLI'S THEOREM

2.4.1 Bernoulli Integrals

While considering energy conversion and dissipation in isenergic flow, we noted in passing that, with the additional restriction to constant entropy, the thermodynamic properties and velocity are related by

$$\int_1 \frac{dp}{\varrho} + \frac{u^2}{2} = \text{constant} \qquad (2.3.29)$$

We shall now discuss the consequences of this constraint, the Bernoulli condition, for three fluids whose equations of state are sufficiently simple to allow explicit integration: the constant-density fluid, the ideal liquid with constant compressibility, and the perfect gas. The Bernoulli results have a very wide range of applications and the whole of a later chapter is given over to isentropic flows. Here a few points which have arisen in the preceding discussions will be considered. We shall look at the circumstances in which flows of compressible fluids can be treated as if they were flows with constant density. Also, we shall take a more careful look at the stagnation properties which appear as constants of integration in particular forms of Bernoulli's equation.

Constant-density Fluid

For this case the explicit result

$$\frac{p}{\varrho_c}+\frac{u^2}{2}+gz = \frac{p_0}{\varrho_c}+gz = \text{constant} \tag{2.3.21}$$

has already been noted. One way of interpreting the constant is to introduce the total head and atmospheric pressure:

$$\frac{p}{\varrho_c}+\frac{u^2}{2}+gz = \frac{p_a}{\varrho_c}+gH_0 \tag{2.4.1}$$

Comparison with equation (2.3.13) shows that in this case the head loss H_f is zero. The total head is the elevation of the fluid when at rest at atmospheric pressure. Thus this is the highest elevation which can be reached in an isenergic flow with a free surface exposed to a uniform atmospheric pressure. If the flow is not isentropic, so that the total head falls during the motion, the maximum attainable height is less than the initial total head.

Consider, for example, a fire hose discharging upwards. The water cannot go higher than

$$z = H_0 = \frac{u_1^2}{2g}$$

elevation being measured from the level of the nozzle exit, where the velocity is u_1. In practice, losses will occur from mixing with the surrounding air and not even this height will be attained. Figure 2.25(a) shows the opposite kind of energy conversion. Water starting from rest in a reservoir flows over a spillway. To a reasonable approximation, it may be assumed that the total head remains constant. Hence, at any elevation z, the velocity is given by

$$u = [2g(H_0 - z)]^{\frac{1}{2}} \tag{2.4.2}$$

The effect of losses would be a reduction in the velocity below this value.

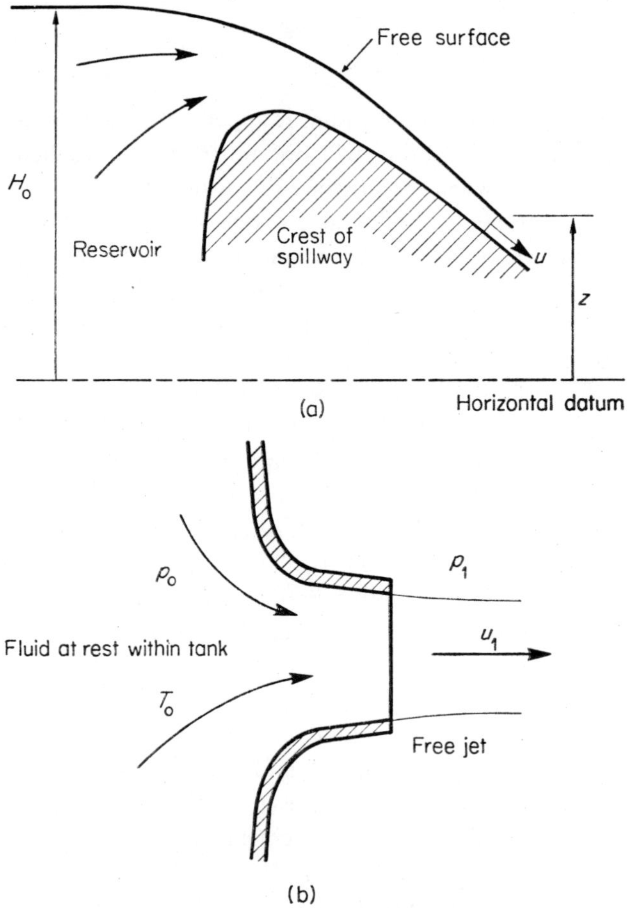

Figure 2.25 Examples of lossless, constant-density flow. (a) Flow down a spillway over a dam; (b) Flow from a pressurized tank through a nozzle

Equation (2.3.21) shows that the total pressure p_0 need not remain constant in isentropic flow under gravity. However, in the important special case in which only small changes in elevation occur, the total pressure does remain constant along each streamline. Figure 2.25(b) shows a flow of this kind; a jet is formed by liquid flowing from a pressurized tank through a rounded nozzle. The Bernoulli condition is here

$$p_0 = p + \frac{\varrho_c u^2}{2} = \text{constant}$$

and the jet velocity is given by

$$u_1 = \left[\frac{2(p_0 - p_1)}{\varrho_c}\right]^{\frac{1}{2}} \tag{2.4.3}$$

If the jet were discharged into a vacuum, it appears that the velocity would be $u_1 = (2p_0/\varrho_c)^{1/2}$. This condition will not normally be reached, for the liquid will begin to boil when the pressure drops to the saturation line. The phenomenon of spontaneous boiling as a result of pressure drop is termed *cavitation.* Since bubbles of vapour appear in the liquid, the constant-density assumption is invalid once cavitation commences. The phenomenon of cavitation will be encountered at several points in this book, and is therefore worth considering a little further here.

In isentropic flow of a liquid the temperature is nearly constant; hence the saturation pressure varies little from the value $p_s(T_0)$. This is the *vapour pressure*, that is, the pressure of vapour in equilibrium with the liquid at that temperature. Just as a vapour can exist in a metastable state beyond the equilibrium saturation line, so the liquid phase can be maintained at pressures somewhat below the saturation value. In fact, a very pure liquid can even sustain a tension, that is, a negative pressure. However, at the levels of purity usual in engineering equipment, there is a plentiful supply of nuclei at which bubble growth can start and the pressure does not fall much below the vapour pressure.

Consider a particular flow of the kind shown in Figure 2.25(b). Take the reservoir pressure to be $p_0 = 10$ lbf/in² and the water temparature to be $T_0 = 70$ °F, for which the steam tables give $p_s = 0{\cdot}36$ lbf/in². The maximum velocity which can be achieved without incipient cavitation is then

$$u = \left[\frac{2(p_0 - p_s)}{\varrho_c}\right]^{\frac{1}{2}} = 38 \text{ ft/sec}$$

This modest velocity can easily be achieved in water pumps and turbines; it follows that cavitation is a very real possibility in flow machines. In practice, a specious form of cavitation known as *gaseous cavitation* will often occur well before the vapour pressure is attained. This occurs when gases dissolved in the liquid begin to come out of solution to form bubbles. The two stages of bubble formation—gas and vapour—are easily distinguished when a pan of tap water is boiled.

Engineers are interested in cavitation not so much because cavities form, but because they collapse when the pressure rises again. When water flows through the impeller of a pump, for instance, its pressure at first drops as it accelerates into the impeller and then rises as it moves through the pump. The pressure changes very rapidly in a high-speed flow, and the sudden collapse of bubbles can produce very intense local shock loads on the solid boundaries around the fluid. Thus the blades of a turbine or pump can be seriously eroded by the repeated impact of collapsing cavitation bubbles. Gaseous cavitation does not give this problem, since the gas does not immediately go back into solution when the pressure rises. In fact, vapour cavities usually have some gas within them, and this provides a cushion when collapse occurs. Nevertheless, cavitation is potentially very damaging to engineering equipment, and considerable efforts are made to avoid it.

Ideal Liquid

The Bernoulli integral is so useful for constant-density flows that one naturally seeks to apply it to compressible fluids as well. For an ideal liquid, whose adiabatic behaviour is described by

$$\frac{d\varrho}{\varrho} = -\frac{dv}{v} = \varkappa\, dp \qquad (1.4.10)$$

where the compressibility $\varkappa$ = constant, the integral in equation (2.3.29) is

$$\int \frac{dp}{\varrho} = \int \frac{d\varrho}{\varkappa\varrho^2} = -\frac{1}{\varkappa\varrho}$$

Hence the Bernoulli integral is

$$-\frac{1}{\varkappa\varrho} + \frac{u^2}{2} = -\frac{1}{\varkappa\varrho_0}$$

where ϱ_0 is the stagnation density, that achieved by isentropic deceleration. Equation (1.4.10) can be integrated to give

$$\frac{\varrho}{\varrho_0} = \exp\left[\varkappa(p - p_0)\right]$$

where p_0 is the stagnation pressure corresponding to ϱ_0. Combining these results, we have

$$\frac{\varrho_0}{\varrho} = \exp\left[\varkappa(p_0 - p)\right] = 1 + \tfrac{1}{2}\varkappa\varrho_0 u^2 \qquad (2.4.4)$$

linking variations in pressure, density and velocity. These results differ fundamentally from the constant-density result (2.4.1), in that pressure is not the only property which changes with velocity. The same Bernoulli equation will be used for a relationship derived from the integral (2.3.29) between velocity and any property of the fluid. In practice the most frequently encountered Bernoulli results are those linking velocity and pressure.

When the pressure change is small, so that $\varkappa(p_0-p) \ll 1$,

$$\exp\,[\varkappa(p_0-p)] \simeq 1+\varkappa(p_0-p)$$

and equation (2.4.4) reduces to

$$p_0-p = \tfrac{1}{2}\varrho_0 u^2$$

This is, as would be anticipated, the constant-density result. For most liquids the isothermal compressibility $\varkappa$ is small; for example

$$\varkappa \simeq 5\times 10^{-5}/\text{bar} \simeq 3\times 10^{-6}\ \text{in}^2/\text{lbf}$$

for water under normal atmospheric conditions. Hence the constant-density result will apply to the vast majority of liquid flows, provided of course that they are isentropic.

Perfect Gas

For a calorically perfect gas the obvious way of carrying out the integration in equation (2.3.29) is to use the isentropic relation $p \propto \varrho^{\gamma}$. However, the integration can be avoided, by starting from the result

$$T_0 = T+\frac{u^2}{2c_p} \tag{2.3.26}$$

which is valid for any isenergic flow of a perfect gas. The other properties are linked to this by the isentropic relations

$$\left(\frac{p}{p_0}\right)^{\frac{\gamma-1}{\gamma}} = \left(\frac{\varrho}{\varrho_0}\right)^{\gamma-1} = \frac{T}{T_0} = 1-\frac{u^2}{2c_pT_0} \tag{2.4.5}$$

These results satisfy the requirements of constant entropy and constant total enthalpy as well as the perfect-gas laws. Hence they are the Bernoulli equations for a perfect gas.

As was noted earlier, the pressure/velocity relationship in low-velocity gas flow is commonly approximated by the constant-density result (2.3.31). Let us investigate the error introduced by this procedure. This can be done by examining the binomial expansion of the pressure/velocity relationship (2.4.5):

$$\frac{p}{p_0} = \left(1 - \frac{u^2}{2c_p T_0}\right)^{\frac{\gamma}{\gamma-1}}$$

$$= 1 - \frac{\gamma}{\gamma-1} \cdot \frac{u^2}{2c_p T_0} + \frac{1}{2}\frac{\gamma}{\gamma-1} \cdot \frac{1}{\gamma-1} \cdot \left(\frac{u^2}{2c_p T_0}\right)^2 - \dots$$

$$= 1 - \frac{u^2}{2RT_0} + \frac{1}{8\gamma} \cdot \left(\frac{u^2}{RT_0}\right)^2 - \dots$$

Finally

$$\frac{p}{p_0} = \frac{p_0 - \frac{1}{2}\varrho_0 u^2}{p_0} + \frac{1}{8\gamma}\left(\frac{u^2}{RT_0}\right)^2 - \dots \tag{2.4.6}$$

If only the first term on the right is kept, the constant-density result is recovered. The proportional error introduced by using the constant-density approximation is

$$er = \frac{u^2}{4\gamma RT_0} \tag{2.4.7}$$

Obviously, if the velocity is small enough and the temperature high enough, the constant-density approximation is entirely justified.

To see what 'small enough' means, consider the situation shown in Figure 2.26(a), air flowing from a reservoir into a pipe. The total properties, pressure, density and temperature, are known from reservoir conditions. Suppose also that the velocity of the fluid in the pipe is known; it might be found from continuity. The pressure in the pipe is to be calculated. The constant-density result will be in error by the fraction *er* when the velocity is

$$u = (4\gamma RT_0)^{\frac{1}{2}} er^{\frac{1}{2}} \tag{2.4.7}$$

For air at normal temperature, say $T_0 = 300$ °K, this gives

$$u \simeq 700 er^{\frac{1}{2}} \text{ m/sec} \quad \text{or} \quad 2300 er^{\frac{1}{2}} \text{ ft/sec}$$

If one is willing to accept an error of 1 per cent

$$u \simeq 70 \text{ m/sec} \quad \text{or} \quad 230 \text{ ft/sec}$$

is the highest allowable velocity in the pipe. If an error of 5 per cent is acceptable

$$u \simeq 150 \text{ m/sec} \quad \text{or} \quad 500 \text{ ft/sec}$$

is allowable.

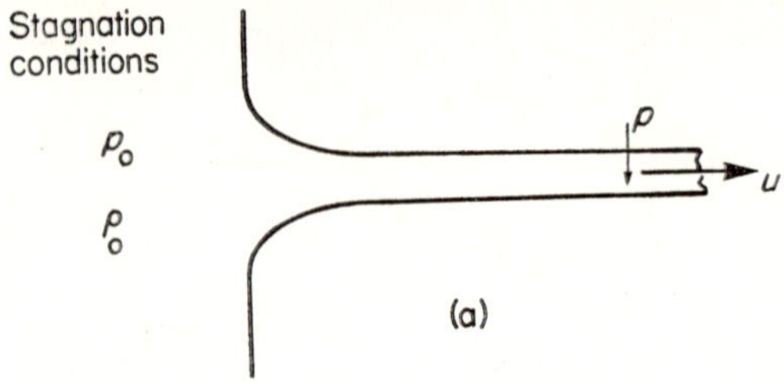

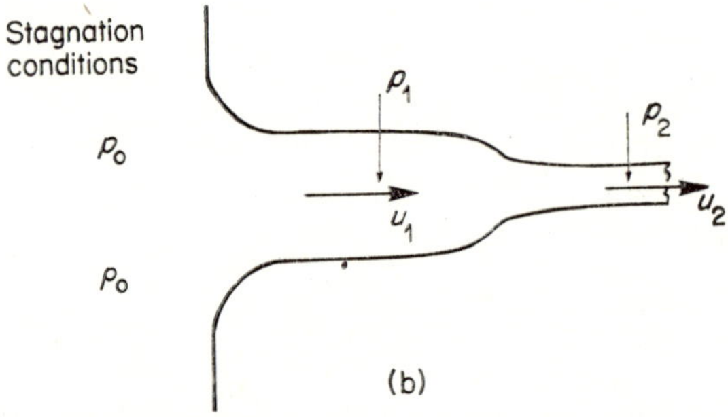

Figure 2.26 Flow of air from a reservoir. (a) Flow into a constant-diameter pipe; (b) Flow into a contracting pipe

The expansion (2.4.6) also allows us to see when the constant-density assumption gives a useful prediction of changes in pressure in a flowing gas. Figure 2.26(b) shows the same gas flow, now passing through a contraction, so that the velocity increases to $u_2 = ru_1$. To the constant-density approximation

$$r = \frac{u_2}{u_1} = \frac{A_1}{A_2}$$

is the contraction ratio of the duct. Applying equation (2.4.6) at the two sections to be related, we find

$$\frac{p_1 - p_2}{p_0} = \frac{\varrho_0(u_2^2 - u_1^2)}{2p_0} - \frac{(1/8\gamma)(u_2^4 - u_1^4)}{(RT_0)^2} + \dots$$

$$= \frac{(p_1 - p_2)_c}{p_0} - \frac{(1/8\gamma)(u_2^4 - u_1^4)}{(RT_0)^2} + \dots$$

where $(p_1 - p_2)_c$ is the pressure change which would be predicted on the basis of constant-density flow. As would be expected, the change in pressure depends on the change in the kinetic energy. However, the proportional error in the prediction of $(p_1-p_2)/p_0$ is

$$er = \frac{u_2^2+u_1^2}{4\gamma RT_0} \tag{2.4.8}$$

which depends on the mean kinetic energy of the contracting flow. The velocity at which this error is achieved is

$$u_1 = \left[\frac{4\gamma RT_0}{1+r^2}\right]^{\frac{1}{2}} er^{\frac{1}{2}} \tag{2.4.8}$$

The results of the last paragraph indicate that, although only a small pressure change occurs when the velocity change is small ($u_2 \simeq u_1$ or $r \simeq 1$), that pressure change cannot be accurately predicted using constant-density results unless the velocities (u_1, u_2) are themselves low enough. The constant-density result may be of use even when it is considerably in error, by providing a simple way of estimating the order of the pressure change.

For analytical convenience the approximation

$$p_0 = p+\frac{\varrho_0 u^2}{2} \tag{2.4.9}$$

has been considered in the preceding discussion. However, the alternative approximation

$$p_0 = p+\frac{\varrho u^2}{2} \tag{2.4.9}$$

introduced in equations (2.3.30, 31), is often easier to use. The errors introduced by the two approximations are generally of the same order. When they differ significantly in their predictions of p_0-p, neither is reliable. However, changes in pressure and velocity are more accurately related by

$$p_2-p_1 = \frac{\varrho_1}{2}(u_1^2-u_2^2)$$

which is independent of the total properties.

Example 2.12 An experimental rig for the study of cavitation is to be designed; the proposed arrangement is shown in Figure 2.27. Water flows from a large reservoir with a free surface through a constant-diameter pipe line (A = 100 cm^2), through a contraction where cavitation is to be studied, past a control valve, and finally emerges as a free jet into the atmosphere. The flow may be taken to be lossless, save at the valve, and to be steady, since the water level does not drop rapidly.

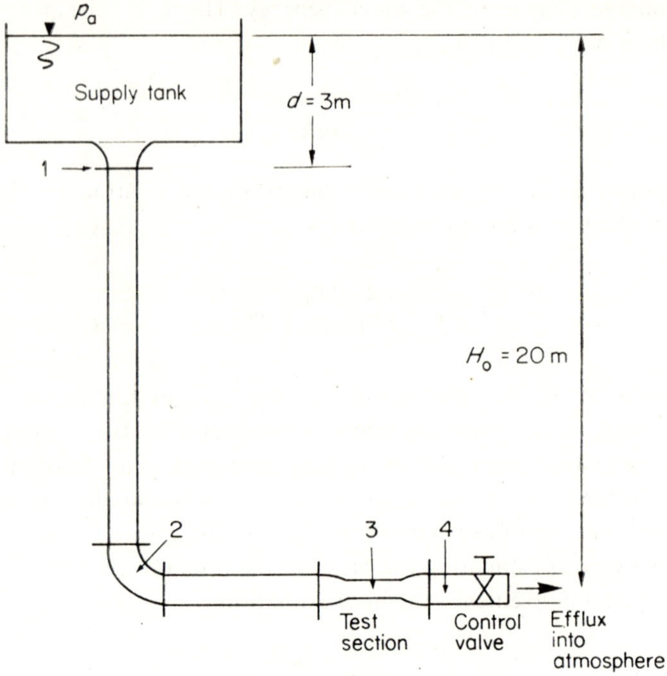

Figure 2.27 A cavitation test rig

It is required to find:

(a) the maximum possible flow through the system;

(b) the minimum contraction ratio required if cavitation is to occur with flow (a);

(c) the minimum flow and corresponding pressure drop across the valve for cavitation with a contraction ratio of two, and

(d) whether cavitation occurs at section 1 for flow condition (c).

Bernoulli's equation can be used to obtain all these results.

(a) The maximum flow occurs with the valve fully open. If there are no losses in the system

$$\frac{p_a}{\varrho_c}+\frac{u_4^2}{2}=\frac{p_a}{\varrho_c}+gH_0$$

Then

$$u_4=(2gH_0)^{\frac{1}{2}}=19{\cdot}8 \text{ m/sec}$$

and

$$\dot{V}_{max}=0{\cdot}198 \text{ m}^3\text{/sec}$$

(b) With incipient cavitation at section 3, and no losses at the valve

$$\frac{p_s}{\varrho_c}+\frac{u_3^2}{2}=\frac{p_a}{\varrho_c}+\frac{u_4^2}{2}$$

Then

$$\left(\frac{u_3}{u_4}\right)^2 = 1+\frac{p_a-p_s}{\varrho_c g H_0} = 1{\cdot}505$$

for a vapour pressure of $p_s = 2{\cdot}3$ kN/m^2. The contraction ratio is then

$$\frac{u_3}{u_4} = 1{\cdot}23$$

(c) For a contraction ratio of two, and lossless flow in the line

$$\frac{p_s}{\varrho_c}+\frac{(2u_4)^2}{2} = \frac{p_a}{\varrho_c}+gH_0 = \frac{p_4}{\varrho_c}+\frac{u_4^2}{2}$$

Then

$$u_4 = \left[\frac{p_a-p_s}{2\varrho_c}+\frac{gH_0}{2}\right]^{\frac{1}{2}} = 12{\cdot}1 \text{ m/sec}$$

and

$$\dot{V} = 0{\cdot}121 \text{ m}^3\text{/sec}$$

The pressure drop across the valve is

$$p_4-p_a = \varrho_c\left(gH_0-\frac{u_4^2}{2}\right) = 112{\cdot}5 \text{ kN/m}^2$$

(d) For section 1, with velocity equal to the outlet velocity

$$\frac{p_1}{\varrho_c}+\frac{u_4^2}{2} = \frac{p_a}{\varrho_c}+gd$$

Then

$$p_1 = p_a+\varrho_c\left(gd-\frac{u_4^2}{2}\right)$$

$$= 101{\cdot}3+29{\cdot}4-73{\cdot}7 = 57{\cdot}0 \text{ kN/m}^2$$

Cavitation need not occur at this section, if the inlet is rounded to avoid less local high velocities. However, doubt is cast on the realism of the calculations (a) and (b). For the flow rate predicted in (a), cavitation would undoubtedly occur at the inlet section 1. A considerable dissipation would result when the flow expanded again to fill the pipe. Hence the neglect of losses at the maximum flow condition is unrealistic and the flow rate predicted on that basis will not be achieved. Pipe friction, which has been neglected throughout this analysis, will also reduce the maximum flow through the system.

Example 2.13 In multi-stage flow machines, for example, the compressor of a jet engine or a steam turbine, large changes in the density of a compressible fluid occur. Nevertheless, the change in density across any one blade row of the machine is small, and an acceptable approximation to the change of state can often be obtained using the constant-density approximation. Consider the fixed blade row

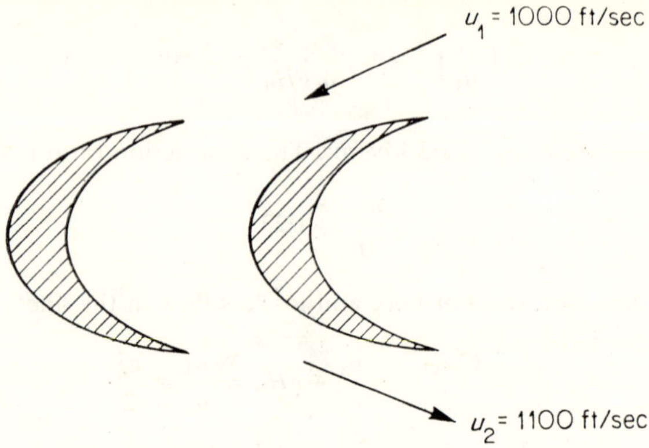

Figure 2.28 Flow through a stationary blade row in a steam turbine

of a steam turbine shown in Figure 2.28. The steam enters the row at $u_1 = 1000$ ft/sec, $p_1 = 400$ lbf/in^2, and $T_1 = 800$ °F, and emerges with $u_2 = 1100$ ft/sec.

Our first task is to discover whether the constant-density approximation is applicable in this situation. In estimating the error introduced by neglect of the density variation, we may assume that the steam behaves like a perfect gas. Then

$$\gamma = 1{\cdot}3, \qquad R = \frac{1545 \times 32{\cdot}2}{18{\cdot}0} = 2760 \text{ ft lbf/sl °R}$$

and

$$c_p = \frac{1{\cdot}3}{0{\cdot}3} \times 2760 = 11{,}960 \text{ ft lbf/sl °R}$$

The total temperature may be estimated as

$$T_0 = T_1 + \frac{u_1^2}{2c_p} = 1260 + 41{\cdot}8 = 1302 \text{ °R}$$

The proportional error (2.4.8) is then $er = 11$ per cent. In fact, we should be able to do somewhat better than this by using the local density, instead of the total density, to calculate the pressure change.

The steam tables give the inlet specific volume as $v_1 = 1{\cdot}816$ ft^3/lb. Then

$$p_1 - p_2 = \frac{1100^2 - 1000^2}{2 \times 1{\cdot}816 \times 32{\cdot}2} = 1800 \text{ lbf/ft}^2$$

$$= 12{\cdot}5 \text{ lbf/in}^2$$

As anticipated, the change in pressure is a small fraction of the absolute pressures p_1 and p_2. This pressure change can be estimated in two other ways, and it is of interest to compare them with the calculations above.

First, assuming the steam to behave like a perfect gas, one may use the isentropic relations (2.4.5). Thus

$$\frac{p_1}{p_2} = \left[\frac{1-u_1^2/2c_pT_0}{1-u_2^2/2c_pT_0}\right]^{\frac{\gamma}{\gamma-1}} = \left[\frac{1-0\cdot0321}{1-0\cdot0388}\right]^{4\cdot33}$$

$$= \frac{0\cdot8682}{0\cdot8423} = 1\cdot031$$

Then $p_1-p_2 \simeq 0\cdot031\times400 = 12\cdot4$ lbf/in^2. Note that the values of c_p and T_0 calculated earlier have been used. Note also the care which must be taken in handling these small changes in properties. Even so, the result could contain an error of a few per cent.

Next, let us use the Mollier diagram. We have

$$h_1-h_2 = \tfrac{1}{2}(u_2^2-u_1^2) = 4\cdot2 \text{ Btu/lb}$$

Locating the state 2 on the diagram, one may estimate

$$p_1-p_2 \simeq 10 \text{ lbf/in}^2$$

The chart is too coarse to allow a more precise estimate.

On comparing these three methods of calculation, we see that the constant-density approximation is the simplest. Moreover, it is little less accurate than the perfect-gas approximation. The use of the Mollier chart, though in principle the most realistic method, in practice turns out to be least accurate.

2.4.2 Total and Static Properties

The integration of the Bernoulli equation

$$\int_1 \frac{dp}{\varrho} + \frac{u^2}{2} = \text{constant} \tag{2.3.29}$$

produces a series of constants, the total pressure, total density, and so forth. These properties describe a state which can be reached by bringing the fluid to rest, isenergically and isentropically. Henceforth the properties of the moving fluid will be called static properties, since they are measured by instruments moving with the flow and therefore stationary with respect to the fluid. The total or stagnation properties (the two names are interchangeable) are measured by fixed instruments which bring the fluid to rest before determining its properties. The label 'total' is perhaps preferable, since it is less likely to be confused with 'static'. Of course, the total and static properties coincide when the flow velocity is small.

The total and static properties are related by

$$h_0 = h + \frac{u^2}{2}$$
$$s_0 = s \tag{2.4.10}$$

They also obey the same equations of state; for example

$$h = f(p, s) \quad \text{and} \quad h_0 = f(p_0, s_0)$$

Hence a knowledge of any two total properties determines the entire total state. The links (2.4.10) imply that both static and total states are determined if one state and the velocity are known. Conversely, if one knows the two states, the velocity by which they differ can be found.

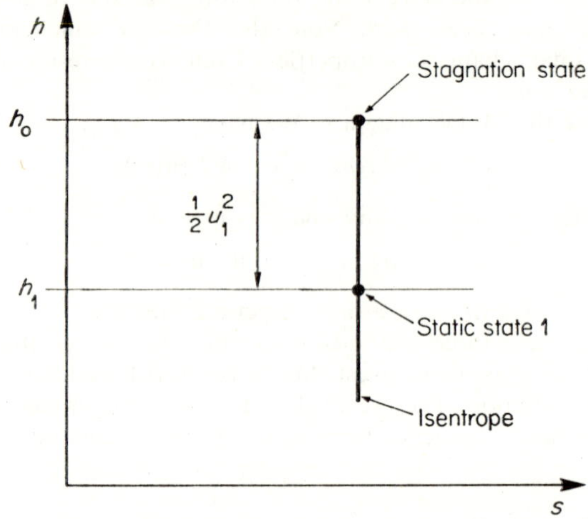

Figure 2.29 The Bernoulli line comprising states attainable by isentropic acceleration and deceleration

No Bernoulli equations have been developed for steam, since that fluid does not have a simple equation of state. However, the results (2.4.10), together with the Mollier chart, are essentially equivalent to the Bernoulli integrals. Figure 2.29 shows a 'Bernoulli line' on the Mollier diagram. States on the isentrope satisfy the Bernoulli condition (2.3.29).

In isenergic, isentropic flow,

$$\begin{aligned} h_0 &= \text{constant} \\ s = s_0 &= \text{constant} \end{aligned} \tag{2.4.11}$$

Hence all the total properties of a simple substance are constant; for example

$$p_0 = f(h_0, s_0) = \text{constant} \tag{2.4.11}$$

In less restricted flows these conditions will not be obeyed. For instance, in isenergic flow through an insulated pipe,

$$p_0 = f(s) \quad \text{with} \quad h_0 = \text{constant} \tag{2.4.12}$$

Again, in flow through an ideal compressor

$$p_0 = f(h_0) \quad \text{with} \quad s = \text{constant} \tag{2.4.13}$$

In general, the total state changes as a result of both entropy and enthalpy variations. In a boiler tube, for example, the entropy is changed by heat addition and friction, and the total enthalpy by heat addition. In a real pump or compressor, there will be internal entropy production, while the total enthalpy is increased by work.

Whatever the nature of the flow, the conditions (2.4.10) serve to relate *local* static and total properties. If the Bernoulli condition (2.3.29) can be integrated, it will provide equivalent relationships between local static and total properties.

The preceding results must be modified for flows in which changes in potential energy are important. The total properties are then defined as those attained with constant entropy and elevation. Then

$$h_0 + gz = \text{constant}$$

and for constant-density, isentropic flow

$$\frac{p_0}{\varrho_c} + gz = \text{constant}$$

The discussion can be carried over to include constant-density flow under gravity simply by replacing the total enthalpy by $h_0 + gz$ and the total pressure by $p_0 + \varrho_c gz$. However, the case of varying density is not so straightforward; fortunately, it is not as important in practice.

At present our interest is in isenergic flow, and we shall therefore concentrate on the relationship (2.4.12), linking entropy to changes in the total properties, in particular, the total pressure. For a constant-density isenergic flow we have already obtained an explicit statement of this relationship:

$$-\Delta(p_0 + \varrho_c gz)/\varrho_c = \Delta e = c_v \, \Delta T = T_c \, \Delta s \tag{2.3.22}$$

where T_c is the nearly constant temperature. Another important special case is the calorically perfect gas. Equations (1.5.16) give

$$\mathrm{d}s = c_p \frac{\mathrm{d}T}{T} - R \frac{\mathrm{d}p}{p}$$

Integrated between static and total states this gives

$$\begin{aligned} \Delta s &= R \log_e \left[\frac{p_1}{p_2} \left(\frac{T_2}{T_1} \right)^{\frac{\gamma}{\gamma-1}} \right] \\ &= \Delta s_0 = R \log_e \left[\frac{p_{0_1}}{p_{0_2}} \left(\frac{T_{0_2}}{T_{0_1}} \right)^{\frac{\gamma}{\gamma-1}} \right] \end{aligned} \tag{2.4.14}$$

Thus the entropy change can be related to changes in either static or total properties. These results apply to any process undergone by the gas. For the isenergic process of present interest, the relationship for total properties is particularly simple. Since $T_{0_1} = T_{0_2}$, as shown in equation (2.3.26)

$$\Delta s = R \log_e \frac{p_{0_1}}{p_{0_2}} \tag{2.4.15}$$

The second law of thermodynamics indicates that the total pressure will always fall in an isenergic flow.

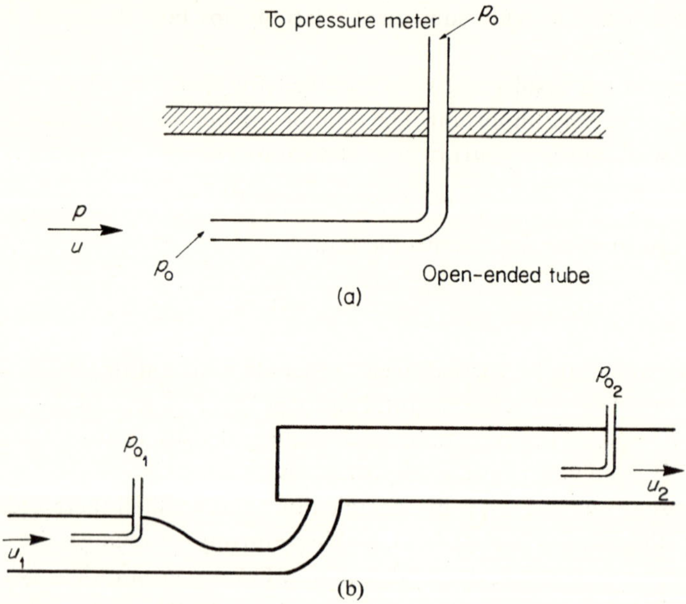

Figure 2.30 The Pitot tube and its use. (a) Measurement of total pressure; (b) Change in total pressure in a complicated, but isenergic flow

The connexion between entropy and total pressure would be only of academic interest, if total pressure were not in fact easy to measure. It can be found using a *Pitot tube*, shown in Figure 2.30(a). The open end of the tube points directly into the flow, and the stagnation pressure is generated there as the fluid is brought to rest, isenergically and isentropically. This pressure is communicated through the still fluid within the tube to a pressure gauge. Figure 2.30(b) shows Pitot tubes located at two sections of an isenergic flow. The fluid undergoes a complex series of processes between the two sections of nearly one-dimensional flow. However, the pressures p_{0_1} and p_{0_2} determine the net effect of irreversibility, no matter how it arises.

For a liquid, equation (2.3.22) gives the dissipation; for a gas, equation (2.4.15) gives the entropy change.

Up to this point entropy has appeared to differ from other properties in being difficult to visualize in physical terms and impossible to measure directly. Now it is seen that entropy changes in an isenergic flow can be measured easily. A Pitot tube may be thought of as an entropy meter.

Example 2.14 Example 2.10 dealt with the flow from the atmosphere into the duct of a ram jet flying at 20,000 m. The total temperature was found to be $T_0 =$ 933 °K both inside and outside the duct, the atmospheric temperature to be $T_a = 217$ °K, and the temperature in the duct (prior to combustion) to be $T_1 =$ 888 °K, with pressure $p_1 = 2{\cdot}04$ bar. What are the total pressures outside the duct and at section 1 within it?

For the undisturbed flow approaching the duct

$$p_{0_a} = p_a\left(\frac{T_0}{T_a}\right)^{\frac{\gamma}{\gamma-1}} = 0{\cdot}0553\left(\frac{933}{217}\right)^{3{\cdot}5} = 9{\cdot}12 \text{ bar}$$

Within the duct

$$p_{0_1} = p_1\left(\frac{T_0}{T_1}\right)^{\frac{\gamma}{\gamma-1}} = 2{\cdot}04\left(\frac{933}{888}\right)^{3{\cdot}5} = 2{\cdot}43 \text{ bar}$$

Then

$$\Delta s = R \log_e \frac{p_{0_a}}{p_{0_1}} = 380 \text{ J/kg °K}$$

It is natural to wonder how this large departure from isentropic flow arises, since there is no apparent way of getting heat into the flow before fuel is injected, and since the flow through this short duct should not be much affected by friction at the duct walls. The answer to this question lies in the necessity of suddenly decelerating the very rapid flow approaching the inlet. This is done through a series of shock waves at the mouth of the duct. These waves are very thin regions of intense dissipation, across which the velocity falls abruptly while the entropy rises abruptly.

Example 2.15 In Example 2.11 an isenergic, high-velocity flow of steam through a whistle was discussed. The following values may be noted:

Total enthalpy:	$h_0 = 1227{\cdot}8$ Btu/lb
Initial state:	$p_1 = 100$ lbf/in^2, $h_1 = 1227{\cdot}6$ Btu/lb
	$s_1 = 1{\cdot}652$ Btu/lb °R, $T_1 = 400$ °F
Final state:	$p_2 = 14{\cdot}7$ lbf/in^2, $h_2 = 1219{\cdot}0$ Btu/lb
	$s_2 = 1{\cdot}848$ Btu/lb

What are the corresponding total properties?

Figure 2.31 shows the relationship of the four states to be considered. The initial static and total states are virtually the same, since the velocity is small. The proper-

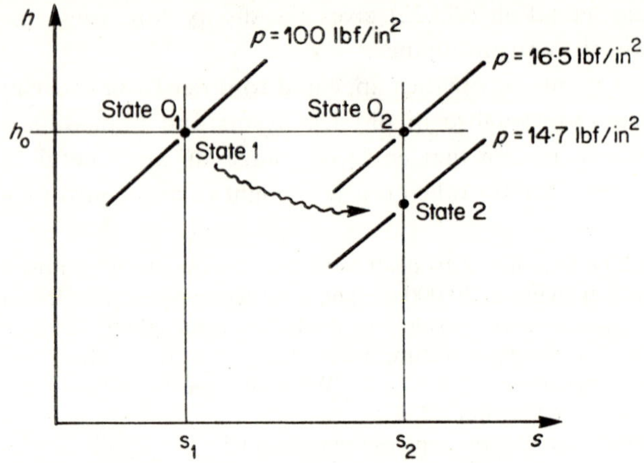

Figure 2.31 A high-velocity isenergic transition in steam

ties of the final total state can be estimated from the Mollier diagram: $p_{0_2} = 16{\cdot}5$ lbf/in^2 and $T_{0_2} = 377$ °F.

Note that the total temperature is not constant for this isenergic process in steam, even though the total enthalpy is unchanged.

2.5 PRESSURE VARIATIONS DUE TO GRAVITY

2.5.1 Hydrostatic Pressure

The pressure changes considered so far have been linked to dissipation and to changes in velocity. Now a still fluid will be considered; in it pressure changes are entirely the result of fluid weight.

Pressure is a stress applied to a fluid element by the neighbouring fluid or by a solid surface. Let us look more generally at the stresses which can be applied within a fluid or other continuous material. Figure 2.32 shows an

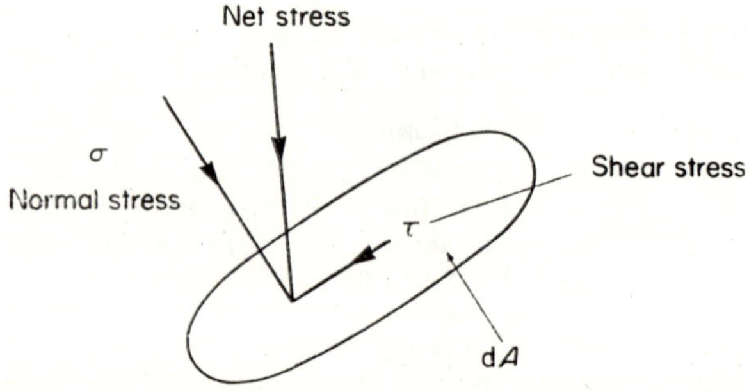

Figure 2.32 Stresses on an elemental area within a mass of fluid

elemental plane area within the material. The most general stress which can be applied to it is inclined at an angle to the plane, and can be broken down into a *normal stress* σ and a tangential component τ, the *shear stress*. Both components are measured in units of force/area, and the normal and shear forces on the elemental area are $\sigma\,\mathrm{d}A$ and $\tau\,\mathrm{d}A$.

A third kind of force which can be applied to the material is a distributed force, dependent on some local property of the substance, such as mass density, or density of electric charge. Such forces are called body forces. The body force to be considered here is due to a uniform gravitational field, the force depending on the fluid density. For an elemental volume $\mathrm{d}V$, the body force is the weight $\varrho g\,\mathrm{d}V$. A fluid acceleration gives rise to an effective body force, which is also dependent on the density of the fluid and can be reckoned in a similar manner.

The distinction between normal and shear stresses can be used to define a fluid, something we have not yet done. A *fluid* is a substance which, when at rest, can sustain no shear stresses. The words 'at rest' are important. An oil-filled buffer, used at the end of a railway siding to arrest the motion of a train, can exert a very large force while it is being driven in. A *solid* material, on the other hand, can resist applied shear stresses without continuously deforming. However, many substances deform slowly under shear stresses and cannot be strictly classified as either liquid or solid. Asphalt paving is an example. Indeed, the slow deformation of a road surface made of asphalt is one of its desirable features.

Figure 2.33 shows the forces acting on a triangular element of a fluid at rest. The geometry of the element gives

$$\frac{\mathrm{d}z}{\mathrm{d}l} = \sin\theta \quad \text{and} \quad \frac{\mathrm{d}x}{\mathrm{d}l} = \cos\theta$$

The conditions for static equilibrium in horizontal and vertical directions give

$$\sigma_x\,\mathrm{d}z \simeq (\sigma_\theta\,\mathrm{d}l\sin\theta) = \sigma_\theta\,\mathrm{d}z$$

and

$$\sigma_z\,\mathrm{d}x \simeq (\sigma_\theta\,\mathrm{d}l\cos\theta)+\tfrac{1}{2}\varrho g\,\mathrm{d}x\,\mathrm{d}z = \sigma_\theta\,\mathrm{d}x+\tfrac{1}{2}\varrho g\,\mathrm{d}x\,\mathrm{d}z$$

The symbols $\simeq$ denote errors of order $\mathrm{d}x^2$, $\mathrm{d}z^2$ and $\mathrm{d}x\,\mathrm{d}z$, introduced by neglect of variations in ϱ_x, σ_z and σ_θ over the sides of the element. Note that the weight itself is of this form. Hence in the limit, as $\mathrm{d}x$, $\mathrm{d}z \to 0$

$$\sigma_x = \sigma_\theta \quad \text{and} \quad \sigma_z = \sigma_\theta$$

the weight becoming negligible along with the other second-order terms. The normal stress within a fluid at rest is seen to be independent of direction, despite the presence of a directed body force. One can now introduce a single property, the pressure, to represent the normal stresses:

$$p = \sigma_x = \sigma_z = \sigma_\theta \tag{2.5.1}$$

This uniformity of pressure applies at each point in the still fluid, but of course the pressure can still vary from point to point, according to some law of the form $p(x, z)$.

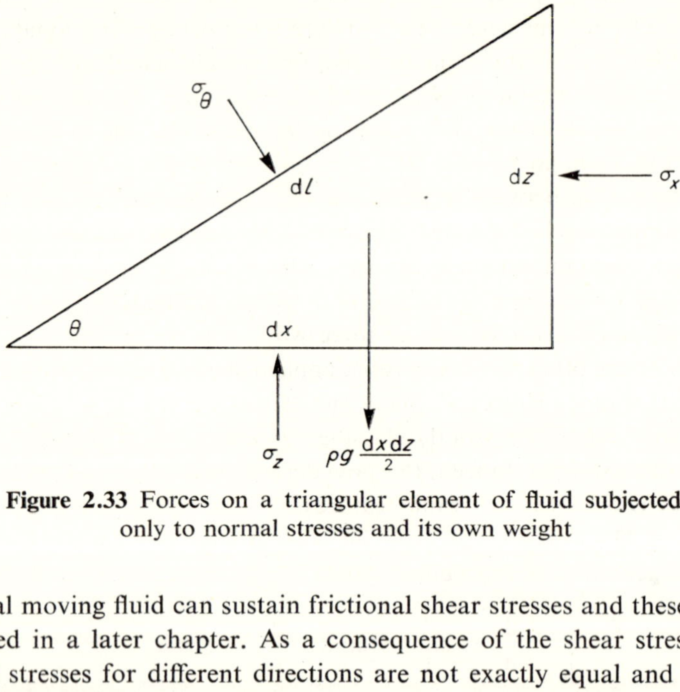

Figure 2.33 Forces on a triangular element of fluid subjected only to normal stresses and its own weight

A real moving fluid can sustain frictional shear stresses and these will be discussed in a later chapter. As a consequence of the shear stresses, the normal stresses for different directions are not exactly equal and a single pressure is not enough to describe them. However, the differences between the normal stresses are usually small and can often be neglected without significant error. The direct effects of the shear stresses are often of great importance, especially in thin regions near solid surfaces. In these boundary layers, the velocity changes rapidly from the zero value imposed by the stationary wall to a value determined by the outer flow. In the outer flow beyond the boundary layers, the velocity variations and shear stresses are smaller and a useful approximation to the fluid motion can often be obtained by neglecting the shear stresses. In these nearly frictionless regions, the force system within the fluid is that shown in Figure 2.33, with the addition of an effective force representing fluid acceleration. The latter force can be dealt with in the same way as fluid weight, and the result (2.5.1) is found to apply in the moving but still shear-stress-free fluid.

Having found how the pressure at a point is related to the other forces within the fluid, let us next consider pressure variations through a mass of stationary fluid. As before, we consider a uniform gravitational field, with constant acceleration of gravity g, and take the coordinate z to be measured upwards with respect to this field. Figure 2.34 shows a cubical element of fluid acted on by this gravitational field. Since there are no shear stresses, only weight and pressure forces need be considered. In general, the pressure in the fluid might be a function of position $p(x, z)$, and the forces on the opposed faces of the element will differ in the manner shown in the figure.

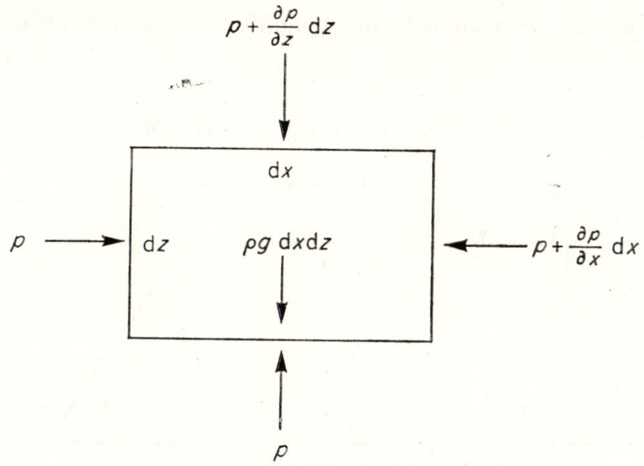

Figure 2.34 Forces on a cubical element subjected only to pressure forces and its own weight

However, for a still fluid the condition of horizontal equilibrium gives $\partial p/\partial x = 0$, so that $p = f(z)$ only. The condition of vertical equilibrium is

$$\frac{\partial p}{\partial z}\,\mathrm{d}z\,\mathrm{d}x + \varrho g\,\mathrm{d}x\,\mathrm{d}z = 0$$

Since the preceding result indicates that $p = f(z)$, this gives

$$\frac{\mathrm{d}p}{\mathrm{d}z} = -\varrho g \tag{2.5.2}$$

implying that $\varrho = f(z)$ as well. Hence both the pressure and density are constant along horizontal surfaces. The pressure variation with height given by equation (2.5.2) is termed the hydrostatic pressure distribution.

Since the other thermodynamic properties can be expressed as functions of pressure and density, they too depend only on elevation. For example, $T = T(p, \varrho) = f(z)$ only. The fluid is said to be horizontally stratified.

Constant Density

The simplest case is that in which there is no stratification, density being uniform. Integration of equation (2.5.2) gives

$$p + \varrho_c g z = \text{constant} \tag{2.5.3}$$

The method of determining the constant of integration depends on the application. One common use of the result is in determining pressure forces on tanks containing liquids with a free surface. Consider, for example, the simple geometry shown in Figure 2.35, a tank with vertical sides and flat bottom. The constant may be evaluated using conditions at the free surface:

$$p + \varrho_c g z = p_a + \varrho_c g d \tag{2.5.4}$$

elevation being measured from the bottom of the tank.

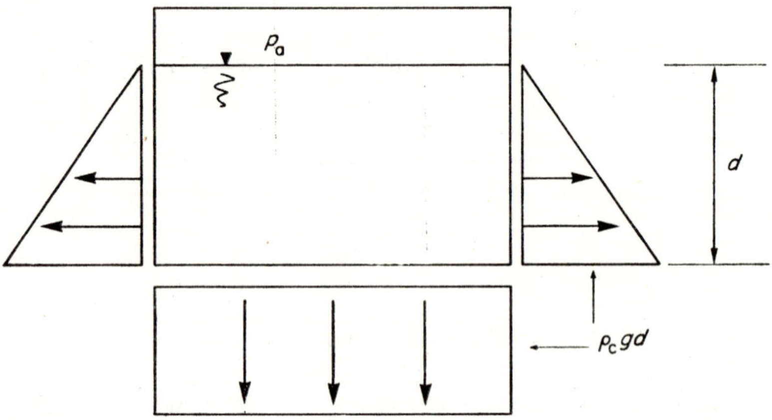

Figure 2.35 The variation of the pressure differential across the walls of a tank containing a liquid at rest

The pressure differential across the walls is given by

$$p - p_a = \varrho_c g(d - z)$$

as shown in Figure 2.35. Hence the force on a strip of wall of unit width, extending from surface to bottom, is

$$\int_0^d (p - p_a)\,\mathrm{d}z = \tfrac{1}{2}\varrho_c g d^2$$

The force on the bottom of the tank, whose area is A, is

$$(\varrho_c g d)A = \varrho_c g V$$

where V is the volume of liquid in the tank. This is, of course, simply the weight of the liquid in the tank.

Another very common application of the result (2.5.3) is in interpreting the readings of a *manometer*, a pressure gauge consisting of a tube open at

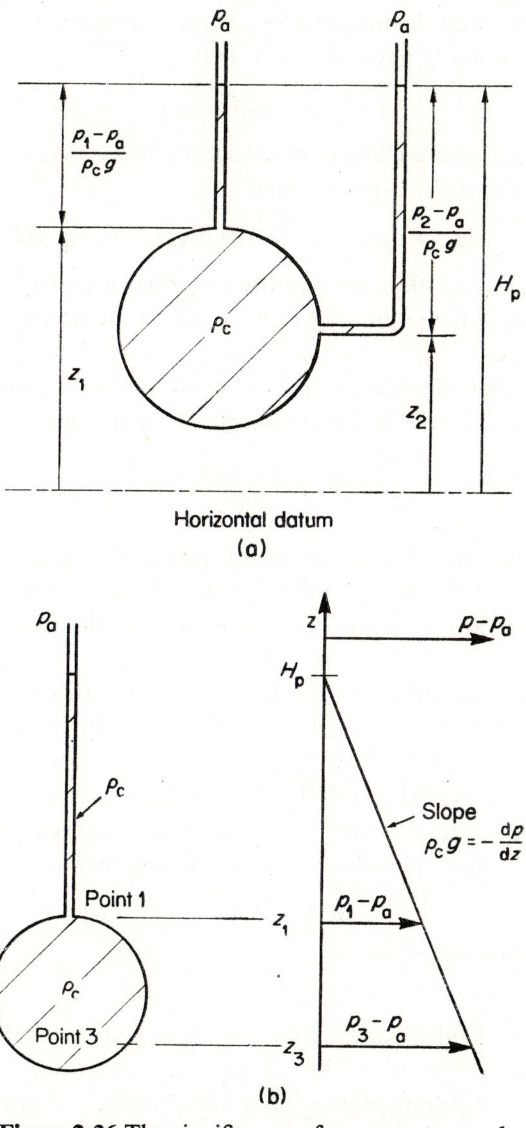

Figure 2.36 The significance of manometer readings. (a) Comparison of manometers connected to the same section of a pipe. (b) Determination of pressure from a manometer reading

one end to some fluid pressure, and at the other to the atmosphere.* Figure 2.36(a) shows a cross-section of a pipe at which two manometer tubes are connected. The liquid in the tubes is the same as that in the pipe. It is not important whether the liquid in the pipe is in motion or not, since motion parallel to the walls has only a small effect on the pressure sensed through a small hole in the wall. Moreover, the pressure variation normal to the flow within the pipe is nearly hydrostatic, so that

$$p_2 = p_1 + \varrho_c g(z_1 - z_2)$$

This is the same pressure change which occurs between z_1 and z_2 in the still fluid in the manometers. It follows that

$$p + \varrho_c g z = p_1 + \varrho_c g z_1 = p_2 + \varrho_c g z_2 = p_a + \varrho_c g H_p$$

gives the pressure variation both within the tubes and over the pipe cross-section. Here H_p is the common height to which the liquid rises in the two manometer tubes.

Although the pressure varies over the cross-section of the pipe shown in Figure 2.36, and also within the manometers, the quantity

$$H_p = \frac{p - p_a}{\varrho_c g} + z \tag{2.5.5}$$

is constant throughout. This is termed the *piezometric head*, that is, the elevation of the liquid in a pressure meter or piezometer. Figure 2.36(b) shows how the pressure at any point can be found once the piezometric head is known.

Referring back to equation (2.5.4) we see that it too can be interpreted as a statement that the piezometric head is constant through the tank. It follows that, if a manometer were connected to any point on the wall of the tank, the fluid would simply rise to the level of the surface within the tank. The results found earlier for low-velocity isenergic flow can also be interpreted in terms of piezometric head. The dissipation

$$\varrho_c \, \Delta e = -\Delta(p + \varrho_c g z) \tag{2.3.16}$$

can be represented as

$$\Delta e = -g \, \Delta H_p$$

Thus a manometer filled with the flowing liquid automatically compensates for pressure changes from gravity; changes in manometer readings along a low-velocity or constant-velocity flow are therefore a direct measure of dissipation in the flow.

* Other forms of manometer are often used in making flow measurements; here for simplicity we consider only one simple type.

At this point the distinction between gauge pressures and absolute pressures should be noted. A manometer open to the atmosphere indicates the pressure relative to the atmospheric level. Certain other kinds of pressure gauge do the same, for example, the Bourdon gauge, which contains a curved, flexible metal tube, whose deflexion depends upon the pressure difference between the inside and outside of the tube. The pressure indicated by such gauges is termed the *gauge pressure* p_g, and is related to the absolute pressure by

$$p_g = p - p_a$$

where p_a is usually the atmospheric pressure, but might in some circumstances be another reference pressure.

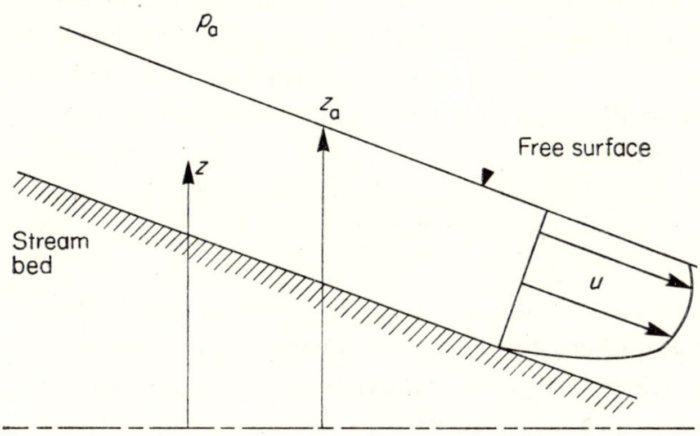

Figure 2.37 Free-surface flow down an incline

As a final example of the application of equation (2.5.3), consider parallel, free-surface flow down a slope, shown in Figure 2.37. Since there is no acceleration in the z-direction, equation (2.5.3) gives

$$p + \varrho_c g z = p_a + \varrho_c g z_a$$

The local surface elevation is in fact the local piezometric head. The total head at any point in the flow is given by

$$\frac{p}{\varrho_c g} + z + \frac{u^2}{2g} = \frac{p_a}{\varrho_c g} + H_0$$

Hence,

$$H_0 = z_a + \frac{u^2}{2g} \tag{2.5.6}$$

the sum of piezometric and velocity heads. The total head will, in general, vary across the flow, since u varies with depth. However, in the one-dimensional model, with u = constant

$$H_0 = z_a + \frac{u^2}{2g} = \text{constant}$$

Thus, in one-dimensional flows, the total head at any point beneath the surface has the surface value. This result provides a very convenient way of calculating changes in total head in flows with a free surface.

Stratified Liquid

We next consider a stratified liquid comprising distinct layers with differing densities or a liquid whose properties vary gradually with height. An example of the first class is oil floating on water, and an example of the second is the ocean, in which density varies with salinity and temperature. The vertical equilibrium condition

$$\frac{dp}{dz} = -\varrho g \tag{2.5.2}$$

gives the pressure distribution with depth when the appropriate density distribution is prescribed. There is, however, a limitation on the density distributions which can persist—the requirement of *static stability*. A stable system is one which will return to its original configuration if slightly disturbed. A stratification in which the density decreases with elevation has this characteristic. On the other hand, a lighter liquid underneath will tend to move upwards to replace the heavier above. Hence only density variations with

$$\frac{d\varrho}{dz} < 0 \tag{2.5.7}$$

so that $d^2p/dz^2 > 0$ can persist. They are described as stably stratified.

When a stable stratification is upset, commonly by heating at the bottom, convection currents are established which tend to produce a new stable system. Such currents are clearly visible in a pan of heating soup.

Stratified Gas

The pressure of a gas is not much changed by gravity over the differences in elevation which occur in most engineering equipment. In the *atmosphere*, however, large changes in properties with altitude occur, and we shall concentrate on this case. The criterion (2.5.7) would seem to indicate that the atmosphere is usually stably stratified; the weight of the gas dictates that the density always falls as the altitude increases. However, this criterion assumes that the displaced fluid retains its density on moving to a new level. This is very nearly so for a stratified liquid, but is far from being the case in a stratified gas. In the latter case the density will be modified by the pressure change during vertical motion. This process has a destabilizing effect, since the density will fall as an element of gas moves upwards into a region of lower pressure.

We shall for the moment set aside the question of stability and investigate the properties of the atmosphere in general terms. For a gas, equation (2.5.2) gives

$$\frac{\mathrm{d}p}{p} = -\frac{g}{R}\frac{\mathrm{d}z}{T} \tag{2.5.8}$$

This can be integrated once the temperature variation $T(z)$ is specified.* For an isothermal atmosphere, typical of the stratosphere, this has already been done in Example 1.3. In lower levels of the atmosphere the temperature falls with altitude. It is convenient to adopt a linear variation, and this is, moreover, a fair representation of the temperature changes over considerable ranges of altitude. Take

$$T = T_r - \alpha z \tag{2.5.8}$$

where T_r is the reference temperature at the level from which z is measured, and α is the *temperature lapse rate*. The standard atmosphere of Appendix VI is determined on this basis. Examination of the temperature variation shows that $\alpha = 6{\cdot}5$ °C/km in this model atmosphere.

The pressure variation is found by integrating equations (2.5.8):

$$\frac{p}{p_r} = \left(\frac{T_r - \alpha z}{T_r}\right)^{\frac{g}{\alpha R}} = \left(\frac{T}{T_r}\right)^{\frac{g}{\alpha R}} \tag{2.5.9}$$

This integral does not hold for $\alpha = 0$, the isothermal case which has been dealt with in Example 1.3.

Example 2.16 The hydrostatic pressure distribution can be used in several ways to determine the forces applied by a still or slowly moving liquid to the structures containing it. An obvious method is integration over the surface of contact between the fluid and the structure. However, a detailed integration can sometimes be avoided by a judicious selection of control volume or, as it is often called in mechanics, free body. This approach will be exemplified by finding the forces on the roller gate shown in Figure 2.38(a). Such gates are often used to control the

* Here it is assumed that g does not vary with altitude.

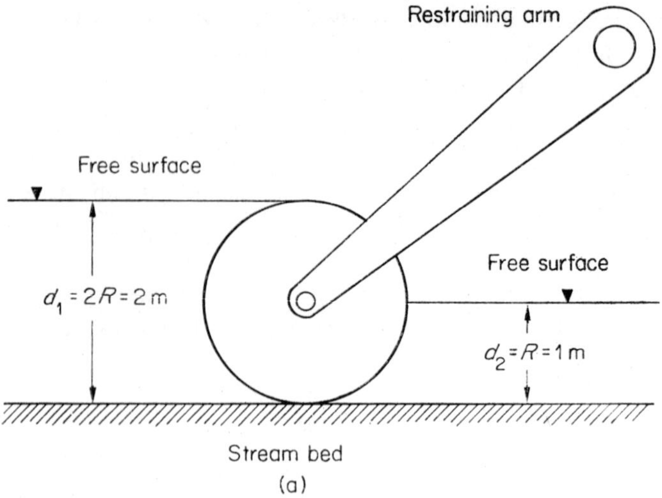

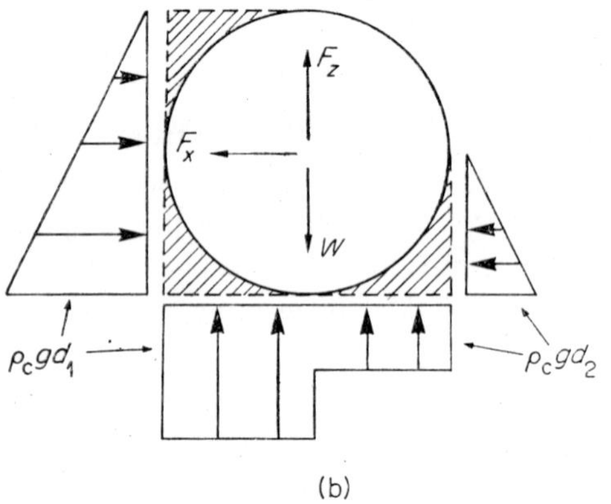

Figure 2.38 The forces on a roller gate. (a) Geometry; (b) Forces

level of water in canals and rivers. The water upstream will be taken to reach the top of the gate, while that downstream comes half way up it. The gate diameter is 2 m, and its weight is 1000 kgf/m of length.

One could determine the components F_x and F_z of the force required to hold the gate in position by integrating suitable components of the pressure force over the whole surface. Thus

$$F_x = \int_{0^\circ}^{360^\circ} (p - p_a) \cos\theta \, R \, d\theta$$

with p given appropriate values in the four quadrants. We shall, however, consider the forces acting on a plane-sided surface which encloses not only the circular gate, but also some of the surrounding water. Figure 2.38(b) shows this surface and the distribution of gauge pressures on it, as given by equation (2.5.4).

The only horizontal forces are the component of the restraining force F_x and the pressure forces upstream and downstream. Then

$$F_x = \tfrac{1}{2}\varrho_c g(2R)^2 - \tfrac{1}{2}\varrho_c gR^2 = \tfrac{3}{2}\varrho_c gR^2$$

The vertical forces acting on the material within the control volume are the component F_z of the restraining force, the weight of the gate W, the weight of the water

$$-\varrho_c gV = -\tfrac{3}{4}\varrho_c g(4R^2 - \pi R^2)$$

and the net pressure force at the bottom

$$R(\varrho_c gR + 2\varrho_c gR)$$

Then

$$F_z = W - \frac{3\pi}{4}\varrho_c gR^2$$

The second term may be interpreted as the net weight of the water 'displaced' by the gate.

For the case specified, the restraining force has components

$$F_x = \frac{3}{2}\times 1000 = 1500 \text{ kgf/m} \quad \text{(upstream)}$$

and

$$F_z = 1000 - \left(\frac{3\pi}{4}\times 1000\right) = -1360 \text{ kgf/m}$$

The resultant of these forces must act through the centre of the roller, since every element of the pressure force acts normal to the cylindrical surface, and the weight may be presumed to act through the centre as well.

Example 2.17 Figure 2.39 shows a phenomenon which occurs when a fast-moving flow with a free surface is forced to decelerate, in this case by a weir. In order that the fluid pass over the weir its depth must increase; the increase occurs quite suddenly across a hydraulic jump. This phenomenon can also be seen in a flat-bottomed sink. If a tap is allowed to discharge onto the bottom, a circular hydraulic jump is formed, a disordered ring across which the depth rises abruptly.

The hydraulic jump is of considerable technical importance. Here however its analysis is of interest chiefly because it gathers together several of the ideas developed in this chapter. In analysing this flow one must make use of the continuity and momentum principles and must also be able to calculate the dissipation in flow under gravity.

At sections upstream and downstream of the jump, the flow will be taken to be steady, with uniform velocities from surface to stream bed. The continuity condition is then

$$u_1 d_1 = u_2 d_2$$

If the bed is level and bed friction is negligible over the short length of the transition, the only horizontal forces are the pressure forces upstream and down-

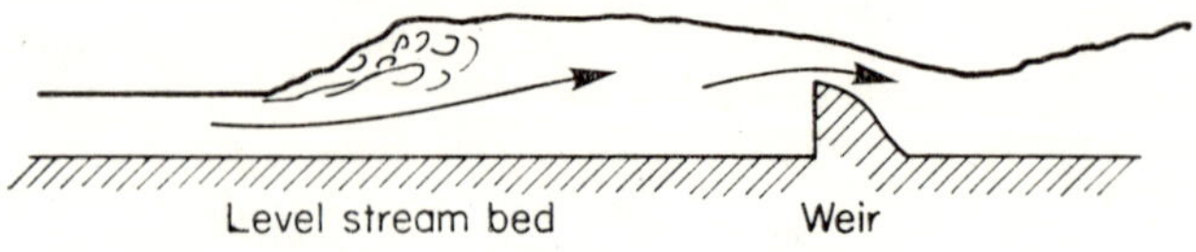

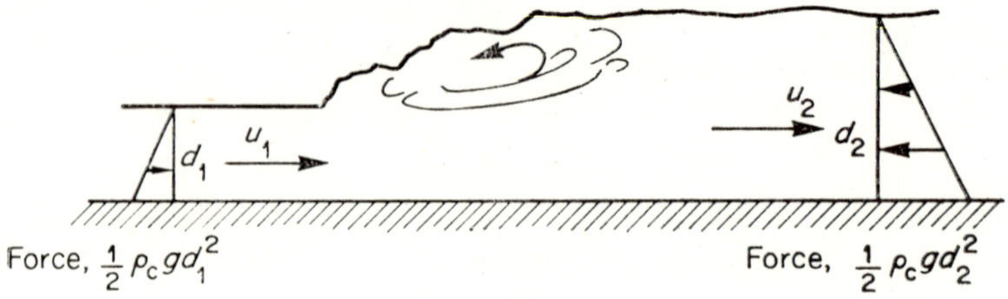

Figure 2.39 The hydraulic jump. (a) Formation of the jump at a weir; (b) Details of the flow through the jump

stream. The change in momentum flux across the jump balances these forces. Then

$$\dot{m}(u_2-u_1) = \varrho_c u_1 d_1(u_2-u_1) = \tfrac{1}{2}\varrho_c g(d_1^2-d_2^2)$$

This result may be rewritten

$$\frac{u_1^2}{gd_1}\left(\frac{u_2}{u_1}-1\right) = \frac{1}{2}\left(1-\frac{d_2^2}{d_1^2}\right)$$

The continuity condition may be introduced through the depth ratio

$$r = \frac{d_2}{d_1} = \frac{u_1}{u_2}$$

Then

$$\frac{u_1^2}{gd_1} = \frac{1}{2}r(1+r)$$

As may be seen from equation (2.5.6), the change in total head is the same for every element of fluid passing through the jump, no matter what its elevation on entering and leaving. This change is most conveniently evaluated using the surface conditions

$$\begin{aligned}\Delta H_0 &= \left(d_2+\frac{u_2^2}{2g}\right)-\left(d_1+\frac{u_1^2}{2g}\right)\\ &= d_1\left[\left(\frac{d_2}{d_1}-1\right)+\frac{u_1^2}{2gd_1}\left(\frac{u_2^2}{u_1^2}-1\right)\right]\\ &= d_1(r-1)\left[1-\frac{u_1^2}{gd_1}\,\frac{1+r}{2r^2}\right]\end{aligned}$$

Since the momentum and continuity conditions require that $u_1^2/gd_1 = \frac{1}{2}r(1+r)$, the dissipation is given by

$$\frac{\Delta H_0}{d_1} = -\frac{\Delta e}{gd_1} = (r-1)\left[1-\frac{(1+r)^2}{4r}\right] = -\frac{(r-1)^3}{4r}$$

The only physically meaningful case is that for which $\Delta H_0 < 0$, in accordance with the second law of thermodynamics. Hence $r > 1$; the depth must always increase across the jump, while the kinetic energy falls. In fact, the fall in kinetic energy is greater than the increase in potential energy, and the result is a net decrease in mechanical energy.

2.5.2 Low-velocity Flow under Gravity

The preceding discussion relates to the effect of gravity on a stationary fluid. It is reasonable, however, to expect pressure variations within a slowly moving fluid to be approximately hydrostatic, so that

$$\frac{\mathrm{d}p}{\mathrm{d}z} = -\varrho g \tag{2.5.2}$$

with fair accuracy. If the flow is also steady and isenergic

$$\mathrm{d}e+\mathrm{d}(pv)+g\,\mathrm{d}z = 0 \tag{2.5.10}$$

is obtained from equations (2.3.10). Combined, these restrictions lead to

$$\mathrm{d}e+p\,\mathrm{d}v = 0$$

Referring to equation (1.5.13) we see that $\mathrm{d}s = 0$ in this process. It is not really surprising that this should be so, for the net work done on the moving fluid in this slow motion is simply reversible work associated with volume change. The assumption of the hydrostatic pressure variation (2.5.2) ensures that work done by the pressure gradient balances that done by gravity.

For a constant-density fluid, the requirement $\mathrm{d}s = 0$ implies that $\mathrm{d}T = 0$ as well. This result may be interpreted in two ways. If a fluid element moves through a mass of fluid, its temperature remains constant unless it mixes with its surroundings. However, if one does mix a liquid thoroughly, its temperature will tend to become uniform despite the effect of gravity.

These conclusions do not apply to a compressible fluid. If an element of gas is moved vertically, its temperature will change, since $T = T(p, s) = T(p)$, in this case. In the absence of rapid horizontal mixing and heat transfer, the temperature of the moving element will not equal that of its surroundings and, more importantly, neither will its density. Looking at these results in another way, one might expect a thoroughly mixed atmosphere to have $s =$ constant.

This brings us to the point at which the *stability* of a stratified gas can be discussed. The mechanism by which air moves from one altitude to another is easily observed in nature. The fluffy cumulus clouds mark the tops of 'thermals', ascending columns of moist air, while the dramatic, billowing cumulo-nimbus clouds denote even more vigorous vertical convection. We shall therefore examine a column of air ascending through a still atmosphere, as shown in Figure 2.40. The pressure variation of the surrounding atmosphere is imposed on the air moving in the column, and its other properties change accordingly.

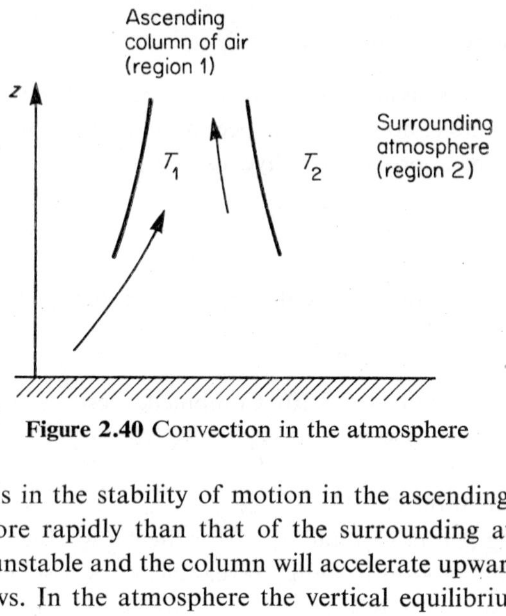

Figure 2.40 Convection in the atmosphere

Our interest is in the stability of motion in the ascending column. If its density falls more rapidly than that of the surrounding atmosphere, the motion will be unstable and the column will accelerate upwards. This can be shown as follows. In the atmosphere the vertical equilibrium condition is satisfied:

$$\frac{\mathrm{d}p}{\mathrm{d}z}+\varrho_2 g=0$$

Inside the column the pressure gradient is the same, but the density is not:

$$\frac{\mathrm{d}p}{\mathrm{d}z}+\varrho_1 g=-\varrho_1 a$$

where a is the vertical acceleration of the fluid in the column. Hence

$$\frac{a}{g}=\frac{\varrho_2-\varrho_1}{\varrho_1}$$

The fluid accelerates upwards if $\varrho_2 > \varrho_1$, the outer fluid being the heavier. The column is then buoyant in the atmosphere.

For the moment we shall consider the case of neutral stability, in which the motion of the column is steady. It may also be taken to be isenergic, since 'thermals' are in fact many metres across and diffusion of heat to the surrounding air is relatively slow. Finally, it will be assumed that changes in kinetic energy are small compared to those in potential energy and enthalpy. Hence equation (2.5.10) applies to the neutrally stable motion, and for a perfect gas gives

$$c_p T_1 + gz = \text{constant}$$

This result will apply so long as water does not condense out from the air; if moisture condenses to form clouds, the effective specific heat will change. With this reservation

$$\frac{\mathrm{d}T_1}{\mathrm{d}z} = -\frac{g}{c_p}$$

the temperature falling linearly with height within the thermal.

For neutral stability, the changes of state in the column must match those in the atmosphere around. Then

$$\frac{\mathrm{d}T_2}{\mathrm{d}z} = \frac{\mathrm{d}T_1}{\mathrm{d}z} \quad \text{or} \quad \alpha_2 = \frac{g}{c_p}$$

This is the lapse rate in a neutrally stable atmosphere; $g/c_p = 10\ ^\circ\mathrm{C/km}$. It is termed the dry adiabatic temperature lapse rate since it holds only when condensation does not occur. When introduced into equation (2.5.9), the adiabatic lapse rate gives

$$\frac{p}{p_r} = \left(\frac{T}{T_r}\right)^{\frac{c_p}{R}}$$

the isentropic relation, showing that the neutrally stable atmosphere is isentropic.

The conditions for instability can be determined by comparing the density in the ascending column with that outside. For the air outside

$$\frac{\mathrm{d}\varrho_2}{\varrho} = \frac{\mathrm{d}p}{p} - \frac{\mathrm{d}T_2}{T}$$

For the air inside

$$\frac{\mathrm{d}\varrho_1}{\varrho} = \frac{\mathrm{d}p}{p} - \frac{\mathrm{d}T_1}{T}$$

Then

$$\frac{\mathrm{d}\varrho_2 - \mathrm{d}\varrho_1}{\varrho} = \frac{(\mathrm{d}T_1 - \mathrm{d}T_2)}{T}$$

The temperature change outside is $-\alpha_2\,dz$, and that inside is approximately that for steady flow, $-(g/c_p)\,dz$. Hence

$$d\varrho_2 - d\varrho_1 = \frac{\varrho}{T}\left(-\frac{g}{c_p}+\alpha_2\right)dz$$

Evidently the density change outside will be the greater when

$$\alpha_2 > \frac{g}{c_p}$$

that is, when the atmospheric temperature falls more rapidly with altitude than the adiabatic lapse rate. This unstable situation will occur when the sun's rays heat the earth's surface, which in turn heats the air near it. The atmosphere is stable when

$$\alpha_2 < \frac{g}{c_p} \tag{2.5.11}$$

This is the case at night, when the earth's surface radiates heat, cools down, and then cools the lower layers of air. Convection currents are then slowed down by negative buoyancy forces and the atmosphere becomes much quieter. The phenomena discussed here are experienced in a very direct way by glider pilots, who often depend upon thermals to maintain their flight. Only by flying most of the time in updrafts can the glider maintain or gain altitude. When night approaches, the pilot is unable to find thermals and is forced to land.

In a gas the condition $\alpha_2 < g/c_p$ replaces (2.5.7) as the requirement for stability. It is of interest to express it in terms of the density gradient, so that the two may be compared more directly. As before

$$\frac{d\varrho_2}{\varrho} = \frac{dp}{p} - \frac{dT_2}{T}$$

Equations (2.5.8) give

$$\frac{d\varrho_2}{dz} = \frac{\varrho}{T}\left(\alpha_2 - \frac{g}{R}\right)$$

For stability, $\alpha_2 < g/c_p$, and

$$\frac{d\varrho_2}{dz} < \frac{\varrho g}{T}\left(\frac{1}{c_p}-\frac{1}{R}\right) = -\frac{\varrho g}{\gamma RT} \tag{2.5.11}$$

As expected, it is not enough that the atmospheric density merely decreases with altitude. It must decrease rapidly enough to compensate for the expansion of the rising gas.

The standard atmosphere is stable, with $\alpha = 6{\cdot}5$ °C/km in the lower atmosphere and $\alpha = 0$ in the stratosphere. It is not surprising that the lapse rate typical of the atmosphere lies between $\alpha = 0$, the value which would be produced by molecular conduction in the absence of vertical mixing, and $\alpha = g/c_p$, the value which would be produced by thorough mixing. In the lower atmosphere, these two processes continually battle with one another. However, the convective activity does not extend into the stratosphere; hence the temperature is nearly constant there.

The consideration of the thermodynamics and fluid mechanics of the atmosphere will not be taken further in this book. It is hoped that enough has been said about them to show that meteorology povides many interesting applications of the principles developed in this book.

PROBLEMS FOR CHAPTER 2

Group A

2.1 The laminar flow in a straight pipe of uniform radius R has the velocity profile $U(1-r^2/R^2)$.

(a) Find the mean velocity of the flow.

(b) Does this seem a likely application for the one-dimensional flow model?

2.2 A measure of the efficiency of a propulsive process is

$$\frac{\text{Energy expended}}{\text{Distance travelled} \times \text{Weight moved}}$$

Since energy costs money, this leads quickly to the relative cost, often expressed as cost/(ton-mile).

(a) Considering only the energy directly expended in propulsion, show that this ratio is simply F/W, where F is the thrust of the propulsive unit, and W is the weight moved.

(b) Find this ratio for the aeroplane and ship considered in Example 2.3, taking their weights to be 100 tons and 50,000 tons. Are the results in accord with the use to which aircraft and ships are usually put?

2.3 For the flow discussed in Example 2.6, find

(a) the pressure forces at the ends of the section of duct,

(b) the additional force F_a,

(c) the fluid weight, and

(d) the force F_d applied by the duct.

2.4 Consider again the elbow discussed in Example 2.4.

(a) What force does the water apply if there is no flow, the internal pressure being uniform at 700 kN/m^2?

(b) What is the net fluid force on the elbow, taking into account normal atmospheric pressure on the outer surface of the elbow?

2.5 Consider again the ram jet of Example 2.2. Find the pressure change required to accelerate the flows of air and fuel. Assume that the fuel is injected with zero velocity along the duct axis and that no forces are applied to the flow save the upstream and downstream pressure forces.

2.6 The continuity requirements for the pump and compressor of Example 2.5 are to be considered.

(a) If the supply and delivery pipes of the water pump have diameters 3 cm and 2 cm, and the inlet velocity is 10 ft/sec, what is the outlet velocity?

(b) If the inlet and delivery ducts for the air compressor are of equal cross-sectional area, 1 ft^2, and the inlet velocity is 50 ft/sec, what is the outlet velocity?

2.7 (a) What powers are absorbed by the pump and compressor considered in the preceding problem? (Additional power will have to be supplied in real machines to provide for internal losses.)

(b) Are the changes in kinetic energy in fact negligible in these two flows?

2.8 (a) Use data from Appendix II to find the temperature of inversion for the Joule–Thomson effect for (1) air, (2) hydrogen and (3) helium.

(b) If these gases are throttled at temperatures near the normal atmospheric level, will their temperature rise or fall?

2.9 What does Callendar's equation

$$pv = 1{\cdot}247(h-835)$$

in English steam-table units, imply about the change in internal energy when steam is throttled at low velocity? Are these predictions correct?

2.10 (a) Trace the process of Exercise 2.7 on the Mollier chart.

(b) Does it seem likely that the final state will be free from condensation? Would the answer be modified if the transition occurred via an isentropic expansion to high-velocity states, followed by isenergic deceleration?

2.11 As water flows past a partially closed valve in a uniform pipe, its pressure drops by 7 lbf/in^2. The temperature is near 40 °F, and the density may be assumed constant.

(a) What is the change in total pressure? In the mechanical energy per unit mass? In the specific internal energy of the fluid? In its temperature?

(b) How much energy is dissipated in this process? What is the entropy change?

2.12 Saturated steam at 20 bar escapes through a blow-off valve into the atmosphere.

(a) Show that the maximum velocity with which it can escape without incurring the danger of condensation within the valve is about 500 m/sec.

(b) What fraction of the steam would condense if the velocity of the uncondensed vapour were 700 m/sec at atmospheric pressure?

2.13 The result (2.4.3) gives the velocity of a jet emerging from a pressurized tank. If the driving pressure arises from hydrostatic pressure in the liquid within the tank, show that the result (2.4.2) gives the velocity. Is this merely a coincidence?

2.14 Example 1.19 deals with the flow of air over the wing of an aeroplane.

(a) What is the speed of the aeroplane?

(b) What is the velocity at point (b) above the wing?

(c) What are the total pressure, temperature and density at the three points considered in the example?

2.15 Figure 2.41 shows a leaf gate in a barrage across a river; the sluice-way is of square section. Find the force on the gate.

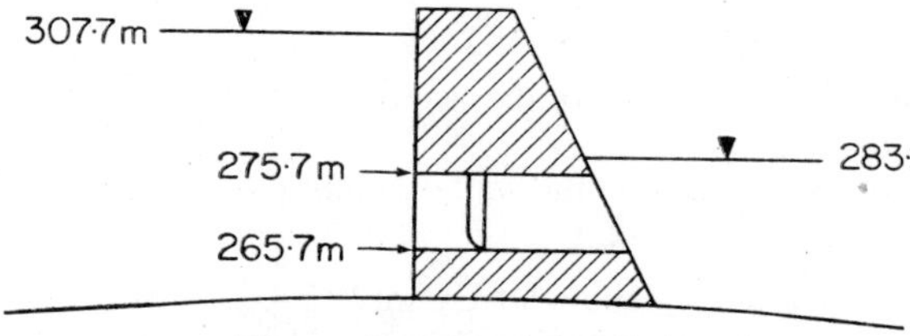

Figure 2.41 See Problem 2.15

2.16 Water reaches the foot of a spillway over a dam with a velocity of 20 m/sec and a depth of 80 cm. In order that this energetic flow does not erode the soil at the base of the dam, it is arranged that a hydraulic jump decelerates the fluid on a concrete apron at the bottom of the spillway.

(a) What is the velocity of the water after the jump?

(b) Find the rate of energy dissipation if the spillway breadth is 7·0 m.

(c) The efficiency of the hydraulic jump as an energy dissipator may be defined as

$$\frac{\text{energy dissipated}}{\text{initial mechanical energy}}$$

What is this efficiency for the case described above?

(d) By what amount is the temperature of the water raised?

Group B

2.17 The velocity distribution in a tube whose radius $R(x)$ changes along the tube is given by

$$u(x, r, t) = U\left(1-\frac{r^2}{R^2}\right)[1-\varepsilon \sin (kx-\omega t)]$$

with $U = U(x)$ and $\varepsilon = \varepsilon(r, x)$. This represents a wave travelling along the tube with constant speed ω/k; the velocity fluctuates with amplitude $U\varepsilon$.

(a) Find the time-mean velocity distribution $u_m(x, r)$.

(b) Find the twice-averaged velocity $u_{mm}(x)$.

(c) If the mean flow is at constant density, what relationship exists between $R(x)$ and $U(x)$?

2.18 Consider ideal flow from a water line through a nozzle into the atmosphere; a fire hose is an example of this kind of flow.

(a) Show that $F = p_{g_1}A_1-\dot{m}(u_2-u_1)$ gives the force on the nozzle, p_{g_1} being the gauge pressure at entry to the nozzle, and u_2, u_1 being the exit and entry velocities.

(b) Using Bernoulli's equation and the continuity condition, show that

$$F = p_{g_1}A_1\left[1-\frac{2}{(1+r)}\right] = \frac{1}{2}\,\varrho_c u_1^2 A_1(r-1)^2$$

where $r = A_1/A_2$ is the ratio by which the flow area is reduced in the nozzle.

(c) Explain why the force on the nozzle is always (for $r \gtrless 1$) in the same direction.

(d) Is the limiting behaviour ($r \to 0, 1, \infty$) of the results (b) sensible?

(e) Water enters the nozzle of a fire hose of diameter 3 in with a pressure of 40 lbf/in^2 (abs), and leaves in a jet of diameter $1\frac{1}{2}$ in. What force is required to maintain the nozzle in position?

(f) If the jet impinges directly on the wall of a building, what force does it apply to the wall?

2.19 Suppose that the reducing elbow of Example 2.4 consists of two sections, as sketched in Figure 2.42, an area-reducing section, followed by a constant-diameter turning section.

(a) Find the force on the reducing section, on the assumption that the flow in it is one-dimensional and lossless.

(b) Find the force on the turning section, taking the losses to be those found in Example 2.9, namely, 61·4 kJ/m^3.

(c) What is the total force on the two-piece elbow?

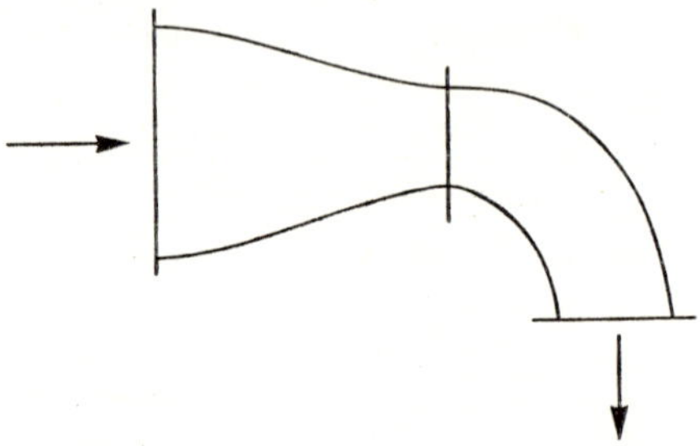

Figure 2.42 See Problem 2.19

2.20 The calculations of Example 2.4 give the force applied by the fluid within the elbow. This would be the net force applied to the elbow if the pipeline were located in a vacuum. More usually, it will be surrounded by air at atmospheric pressure.

(a) Show that the net force applied by the fluid inside and atmosphere outside can be obtained by a calculation like that of Example 2.4, but with gauge pressures rather than absolute pressures used at sections 1 and 2.

(b) What is the net force on the elbow, if the external pressure has the standard atmospheric value?

(c) What would be the net force on the elbow if it turned the flow through 45°, rather than 90°, all the other defining parameters being unchanged?

(d) What would be the net fluid force on a 90°-elbow located inside a tank within which the pressure is 630 kN/m^2?

2.21 A jet of water, as shown in Figure 2.43, strikes a curved vane and is deflected through an angle β.

(a) On the assumption that the flow is one-dimensional and lossless, show that the speed with which the fluid leaves the vane is identical with the incident speed.

(b) What is the ratio of flow areas before and after deflection?

(c) Show that the net force on the blade acts along a line bisecting the directions

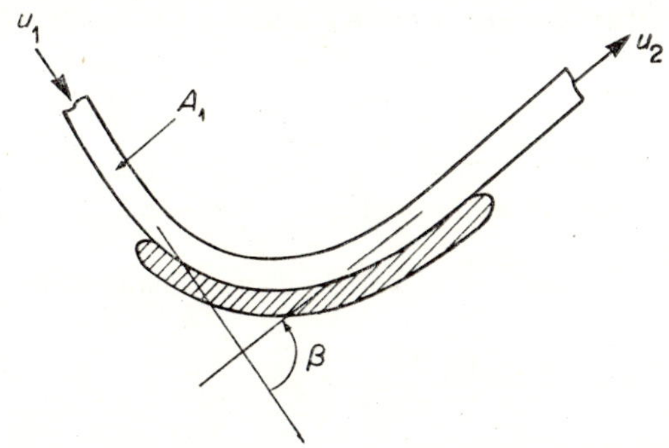

Figure 2.43 See Problem 2.21

of approach and departure. This is most easily done by adopting axes one of which bisects the deflexion angle.

(d) Show that the force is $F = 2\varrho_c A_1 u_1^2 \sin\left(\frac{1}{2}\beta\right)$. Find the special results appropriate when β is small and when it is near 180°.

2.22 Show that an appropriate impulse function for horizontal flow with a free surface is $\varrho_c\left(u^2+\frac{1}{2}gd\right)d$. Use this formula to obtain the momentum condition for a hydraulic jump. Compare with the result found in Example 2.17.

2.23 A rocket on a test stand exhausts fuel at the rate of 100 lb/sec through a nozzle whose exit diameter is 3 ft. The velocity in the exit plane is 10,000 ft/sec, and the pressure there is 5 lbf/in^2 (abs).

(a) What is the net thrust on the rocket, on the assumption that sea-level atmospheric pressure acts over the entire outer surface?

(b) The jet behind the rocket adjusts to atmospheric pressure. What is the velocity then, on the assumption that no air is carried along with the jet, and that the pressure on the surface of the jet is atmospheric?

2.24 When a fluid slowly leaves a rigid, insulated tank, conditions within the tank change according to the isentropic law. This can be shown most easily by considering a fixed amount of material, that remaining within the tank while a small quantity leaves.

(a) For the mass remaining, show that $\mathrm{d}E = -p\,\mathrm{d}V$, and hence that $\mathrm{d}s = 0$.

(b) In a tank of air initially at 1000 lbf/in^2 and normal atmospheric temperature, what is the temperature when the pressure has dropped to atmospheric? Assume perfect gas behaviour to get the answer. How much confidence do you have in the result?

2.25 Use the result of Example 2.8 to predict the temperature change in steam which undergoes the expansion specified in Example 2.7. Compare with the change found using steam tables.

2.26 Steam at 100 lbf/in^2 and 800 °F expands isentropically and isenergically (that is, with total enthalpy constant) until it is saturated.

(a) Use the Mollier chart to find the change in kinetic energy of the steam.

(b) If it starts from rest, what is the ultimate velocity?

(c) If it starts with a velocity of 30 m/sec, what is the final speed?

2.27 The throttling of a calorically perfect gas can be analysed quite simply if the initial conditions are known and the pressure drop is specified.

(a) Using the perfect gas law $p = \varrho RT$ and the continuity statement $\varrho uA =$ constant, show that

$$T_2 = \frac{u_2}{u_1}\frac{p_2A_2}{p_1A_1}T_1$$

relates the temperatures at sections of one-dimensional flow, no matter what process intervenes.

(b) For isenergic flow in which kinetic energy changes may be significant, show that

$$\frac{p_2A_2}{p_1A_1}\frac{T_1}{u_1}u_2+\frac{u_2^2}{2c_p} = T_0$$

(c) Air passes through a valve in a uniform pipe of diameter 10 cm. The upstream conditions are 500 kN/m^2 and 300 °C, with velocity 60 m/sec; the downstream pressure is atmospheric. Find the downstream temperature and velocity, and the force applied to the valve.

2.28 It is required to find the final state of steam which undergoes the process defined in (c) of the preceding problem. Use the methods of that problem to estimate the final state, then refine the results using steam tables.

2.29 What is the change in entropy in the transition specified in Problem 2.27(c)?

2.30 For a perfect gas, equation (2.4.15) relates changes in total pressure and entropy in isenergic flow. Using the expansion $\log_e(1+x) = x - x^2/2 + x^3/3 - \dots$ obtain the approximate result (2.3.30). Discuss the error introduced by using the approximate result.

2.31 Moving hydraulic jumps occur in several engineering applications. The translating waves in canals and locks are usually called surges; the waves sent into tidal rivers by the rising tide are called bores.

(a) Show that the speed of a surge relative to the water ahead of it is given by $c^2 = \frac{1}{2}gd_2(1+d_2/d_1)$, where d_1 and d_2 are depths ahead of and behind the wave.

(b) Consider a bore of height 2 ft advancing up a river of depth 6 ft, which flows at velocity 10 ft/sec (both values relate to conditions before the bore arrives).

(1) With what speed will the bore advance upstream?

(2) What will be the absolute velocity of the water downstream of the bore; that is, where the depth is 8 ft?

Group C

2.32 For the throttling process of Example 2.7, take the initial velocity of the steam to be 1 ft/sec. Assume that the enthalpy is conserved in the transition.

(a) Find the mass flow through the valve if the pipe diameter is 4 in.

(b) Taking the flow area to be the same downstream, find the downstream velocity achieved if condensation does not take place. (See Problem 2.10).

(c) Find the velocity downstream when condensation does occur. Take the flow area for the vapour to be equal to the upstream area; for a constant-area pipe this could be achieved by draining away the condensate. Show first that

$$\frac{u_2}{u_1} = (1-y)\frac{v_g}{v_1}$$

(d) What will be the answer to (b) if the flow area is reduced by a factor of two?
(e) Is the assumption of constant enthalpy justified in these cases?

2.33 Problem 1.45(d) raised the question of the rise in temperature in a room from which air is allowed to leak, the pressure remaining constant as the temperature rises. This process involves displacement work done by the air of the room on the surroundings. However, this is not a steady-flow system, and the control mass approach proves most convenient.

(a) Show that the mass of air displaced when the temperature rises by $\mathrm{d}T$ is $(p_a V_r/RT^2)\,\mathrm{d}T$, where p_a is the uniform pressure and V_r is the volume of the room.

(b) Show that the volume of air displaced when the temperature rises by $\mathrm{d}T$ is $V_r\,\mathrm{d}T/T$.

(c) Show that the energy equation for the process is

$$\left[\frac{c_p}{R}\frac{p_a V_r}{T} + m_c c\right]\frac{\mathrm{d}T}{\mathrm{d}t} = \dot{Q}$$

where $\dot{Q}$ is the constant rate of heat addition, and $m_c c$ is the 'heat capacity' of the contents of the room, exclusive of the air.

(d) Solve this equation, and plot T against t for the data given in Problem 1.45. How long could the meeting last under these conditions?

2.34 An experimenter draws air from an insulated reservoir. Knowing that conditions within the tank will vary as a result of isentropic cooling during the period in which measurements are made, he decides to determine the varying supply conditions using perfect-gas laws. He has two worries:

(1) Will the initial state be so dense that the perfect-gas laws are invalid?

(2) Will the cooling which occurs move the state towards the saturation condition, and hence ultimately invalidate these laws, even if they are initially valid?

The experimenter decides to use van der Waals equation to answer these questions, and decides that the quantities a/pv^2 and b/v must be less than 1 per cent to ensure the accuracy required.

(a) Answer question (2).

(b) If the reservoir temperature is initially at the normal atmospheric level, what initial pressure can be allowed without violating the criteria?

2.35 Three processes have been identified which lead to a reduction in the temperature of a gas emerging from a pressurized tank:

(1) isentropic cooling as the pressure within the tank falls (discussed in Problem 2.24);

(2) the Joule–Thomson effect; and

(3) adiabatic conversion of enthalpy into kinetic energy in the jet from the tank.

Consider the flow of air from an insulated tank within which the initial pressure and temperature are 12 bar (gauge) and 20 °C. Estimate the temperature of the emerging gas when the gauge pressure is 6 bar and the jet velocity is 300 m/sec. Note that there will be Joule–Thomson cooling in both expansion processes, that within the tank, and that in the outlet nozzle.

2.36 Use the results of Problem 1.41 to relate the temperature of inversion for the Joule–Thomson effect to the critical temperature.

2.37 A syphon spillway is installed at a reservoir to maintain it at a safe level. It consists of a tube 3-ft in diameter passing through the dam, as shown in Figure 2.44.

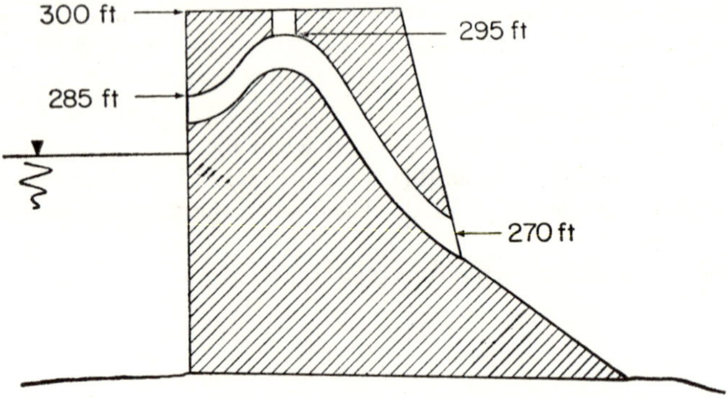

Figure 2.44 See Problem 2.37

(a) Explain the principle of operation, specifying the reservoir levels at which the syphon begins to function upon filling, at which it ceases to function on emptying, and at which the flow is a maximum.

(b) On the basis of lossless flow, prepare a plot of flow rate against reservoir level. Mark in the limits found in (a).

(c) For what range of pressures should an inspection hatch at the top of the syphon be designed?

2.38 Example 2.16 dealt with the hydrostatic pressure forces on a roller gate. Now the forces applied to the gate by a flowing fluid are to be determined. When the gate is raised from the bottom, it may be assumed that (1) the downstream depth d_2 is equal to the distance between the bottom of the gate and the stream bed, and (2) the flow is lossless.

(a) Find the variation of the horizontal component of the force with d_2, taking the total head H_0 to be maintained constant.

(b) Repeat (a), now taking the upstream depth d_1 to remain constant.

(c) How might the conditions H_0 = constant and d_1 = constant be achieved in practice?

(d) At the surface on the upstream face of the gate there is a stagnation point. Find its elevation for conditions (a) and (b).

(e) How would you go about finding the vertical component of force on the gate?

(f) How realistic are the assumptions (1) and (2) above? When are they most nearly correct?

2.39 Equations (2.5.11) give the criterion for the stability of a stratified gas. Find a similar criterion for a stratified ideal liquid.

2.40 (a) Show that the maximum speed which can be attained by a perfect gas flowing isenergically is $(2c_pT_0)^{1/2}$. What are the temperature and pressure when this velocity is achieved? How could this effect be achieved in practice?

(b) The maximum speed for a general fluid is $(2h_0 - 2h_L)^{1/2}$, where h_L is a limiting enthalpy determined by the datum level from which enthalpy is measured. Callendar's equation is $pv \times 10^2 = 0{\cdot}231(h - 1943)$ for SI steam-table units. What does this tell you about the enthalpy of steam at its maximum velocity?

(c) Calculate the maximum speed for steam initially at rest with (1) 500 °C and 0·1 bar, and (2) 500 °C and 350 bar.

Do this in two ways: using the results of (b), and using the result of (a) with $c_p = [\gamma/(\gamma - 1)]\mathbf{R}/MW$ and $\gamma = 1{\cdot}3$. Comment on the results.

3

Shock Waves and Sound Waves

A compression wave passing through a fluid increases its density, thus doing work on it and setting it in motion. Hence the thermodynamic and mechanical response to the wave must be closely related. It is their interaction which determines the speed of propagation of the wave, the pressure rise across it, and the velocity it induces in the fluid. There are several branches of engineering in which strong compression waves, or shock waves, are of great importance. The shock waves generated in high-speed flight are the most familiar example; they impose serious limitations on the operation of supersonic aircraft, first, by providing a large drag to be overcome by the propulsive plant and, secondly, by producing a sonic boom which even at ground level can be annoying and perhaps dangerous.

Shock waves can also form in compressors and in steam and gas turbines. When they do, the efficiency of the machine can be significantly reduced. The dissipative effect of shock waves is not always harmful, however. It is put to use when a space vehicle re-enters the earth's atmosphere. The enormous kinetic energy is converted into thermal energy in the enveloping shock wave, and in this way the vehicle is slowed down and returned safely to the earth's surface.

Important though strong compressive waves are, weak waves, or sound waves, are of more far-reaching significance. Architects, musicians and air-conditioning engineers, among others, are interested in sound waves because they are heard by human beings. Our interest in them is less subjective. Here it is their implications for the fluid motion rather than for the human listener which are of chief interest. The *speed of sound* in a compressible material depends upon the thermodynamic response to compression; hence it is a thermodynamic property. The sound speed also has a geometric significance, and can be used to distinguish between supersonic and subsonic flows. This distinction is most conveniently made using the *Mach number*, a dimensionless speed obtained by dividing the flow velocity by the speed of sound. The Mach number is a most useful measure of the state of a flowing fluid, since it relates the thermodynamic state and the state of motion, that is, the velocity.

In this chapter we shall for the first time make use of *flow tables* giving variations of the properties of a perfect gas as functions of Mach number. Here the tables of particular interest are those for a shock transition, but in later chapters somewhat similar tables will be used to analyse isentropic flows and flows subject to heat addition and friction.

The results of the preceding chapter relate for the most part to steady flows. One can apply them to many shock and sound waves simply by choosing axes in which the wave appears to be stationary. For example, if one wishes to discuss the shock waves formed by an aeroplane it is convenient to adopt axes fixed in the aeroplane rather than axes fixed in the surrounding air. The process to be considered is then a steady, isenergic flow. However, since the process is fundamentally the same, no matter what the frame of reference, the results can easily be adapted to waves propagating at constant speed. We shall also see how the ideas of wave propagation can be applied to truly non-steady flows.

Since compressibility plays a vital role in the propagation of waves through a continuous medium, a constant-density fluid is of no interest in this discussion. Even if one wishes to discuss waves in a liquid, compressibility must be taken into account; here the ideal-liquid model will be used to describe the response of a liquid to compression waves. However, waves in highly compressible fluids are of greater interest. Three such fluids will be considered: the perfect gas, steam and the van der Waals gas. The perfect gas will be given most attention since it is of greatest practical importance.

3.1 SHOCK WAVES

3.1.1 Classification of Waves

Before starting to analyse compression waves, we shall look at some of the ways in which they can be generated. Stationary waves will be considered first. Figure 3.1(a) shows the wave formed by a body travelling at high speed through the atmosphere. The rounded body might be the nose of an aeroplane or bullet, or the leading edge of a wing. A similar wave can be formed by a model held fixed in a wind tunnel, or by the leading edge of a blade in a turbomachine. In each case the motion is steady when viewed from axes fixed in the body.

The waves discussed in the following pages are *normal shock waves*, so named because the fluid moves along a line normal to the wave front and is not deflected laterally on passing through it. It is apparent from Figure 3.1(a) that these conditions are satisfied only by the part of the wave directly

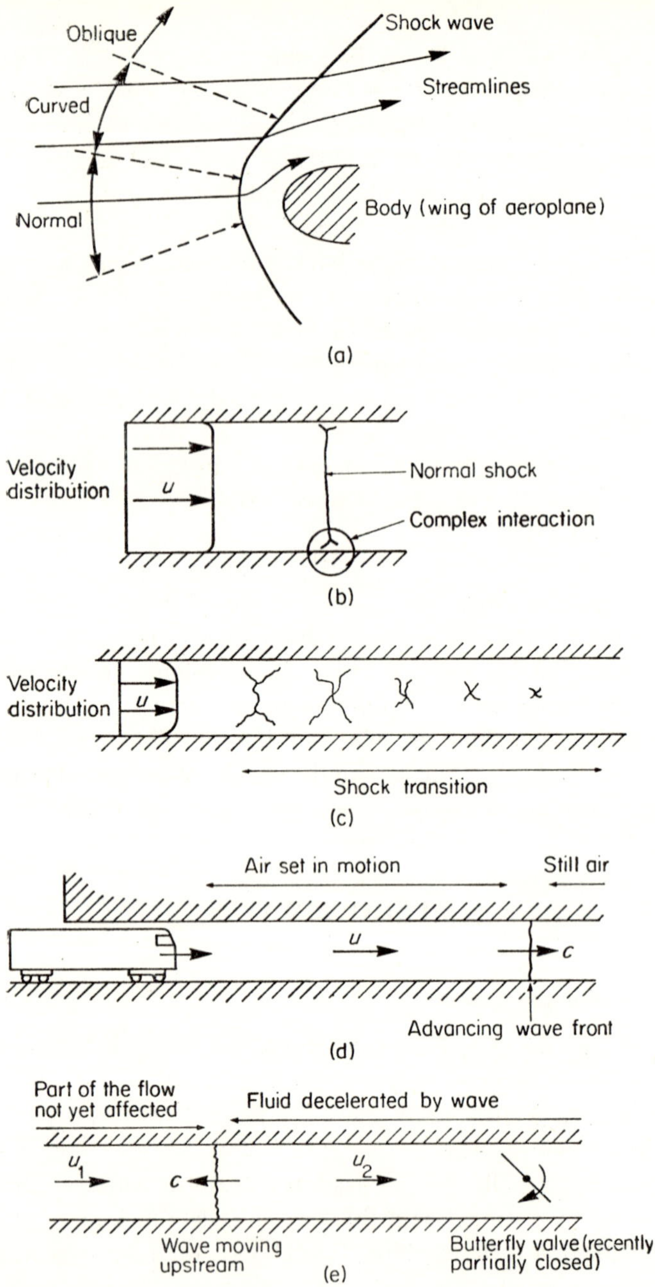

Figure 3.1 Examples of shock waves. (a) Shock formed ahead of a body in a uniform high-speed flow. (b) Shock in a duct showing boundary-layer interactions. (c) Extended shock in a fully developed turbulent pipe flow. (d) Compression wave generated by a train entering a tunnel. (e) Wave generated by the closing of a valve in a pipe line

ahead of the blunt body. The straight part of the wave well away from the body is called an oblique shock wave, and the section joining the oblique and normal waves is termed a curved shock wave. In general, the state of the fluid behind the wave will vary along it. However, all the fluid passing through the normal-shock section will experience the same change of state. Note that the flow is uniform ahead of the shock wave; the disturbance set up by the body does not penetrate into this region.

Figure 3.1(a) shows the shock wave simply as a line across the flow. This is in fact a fairly realistic representation, for the shock waves ahead of fast-moving bodies are very thin, say, 10^{-5} cm in a typical case. Hence within the shock the velocity and temperature gradients are very large, and viscous stresses and heat transfer can produce large changes in state, even though the fluid passes through the wave in an instant. Within strong shock waves there are significant departures from thermodynamic equilibrium. However, the methods of classical thermodynamics can be used to relate conditions upstream and downstream of the wave.

Stationary waves can also form in non-uniform flows, but their structure is more complex. Figures 3.1(b) and (c) show waves formed in a duct; such patterns might appear in the intake of a jet engine, in a high-speed wind tunnel, or in the passage between the blades of a flow machine. In Figure 3.1(b) the boundary layer is thin and a normal shock extends over much of the duct cross-section, with a small region of interaction between shock and boundary layer near the wall. Figure 3.1(c) shows a shock transition in a fully-developed turbulent pipe flow, where turbulent mixing extends right across the channel. Since the velocity is not uniform across the duct, the shock cannot be either. The transition consists of a chain of oblique and normal shocks extending over several duct diameters. The flow may be assumed to be essentially one-dimensional upstream and downstream of this transition region, but not necessarily within it. Hence the extended shock can be discussed in much the same way as the abrupt normal shock shown in Figure 3.1(a).

Next consider moving shock waves. When energy is introduced at some point in a compressible medium, it spreads out through pressure waves transmitted through the medium. To form a strong wave, a large concentration of energy is required, for example, that from a violent impact or explosion. These are often man-made, but sometimes occur naturally, for example, by volcanic action, earthquakes, lightning or the impact of meteorites. The pressure waves from explosions are called *blast waves*, and those produced by geological processes are called *seismic waves*. Moving waves also occur in ducts, sometimes as a result of explosions, for example, in mine tunnels and experimental shock tubes. As indicated in Figures 3.1(d) and (e), they can also be formed by the motion of trains in tunnels and by the opening and closing of valves in pipelines. Many readers will have noticed the hammering sound sometimes emitted by household plumbing systems when a tap is turned; this is the result of waves reflecting back and forth within the pipes. Compression waves in water are of great importance in hydro-electric stations. Whenever the flow through the turbine is altered, waves pass back through the intake system, where they may produce very large changes in pressure.

Sound waves are weak shock waves. A spreading blast wave grows progressively weaker as the wave front expands. Initially, the wave may be strong enough to demolish buildings. Ultimately, it is only a bang and can be perceived only by a very sensitive pressure gauge—such as the ear. More complex patterns of sound, such as speech or music, can be thought of as a succession of pressure waves augmenting or cancelling one another to produce a meaningful signal. It is significant that the pattern of sound does not change much over long distances of propagation. This indicates that sound waves of differing frequencies travel at the same speed. Waves with this characteristic are said to be non-dispersive.

Some of the waves mentioned above, particularly blast and seismic waves, can be formed in a variety of media. The explosion of a ship, for example, will generate compression waves in the air, in the water and in the sea bed beneath. Compression waves in solids and composite materials, such as a bed of sand grains permeated with water, can be analysed using the methods which we shall apply to waves in fluids. However, the explosion of a ship will generate other kinds of waves which cannot be discussed in this way. These are waves at an interface between dissimilar materials, for example, those at the surface of the sea. These waves depend upon properties other than compressibility. Thus waves on the surface of deep water depend chiefly on the difference in the weight of the water and the air above. It is of interest to note that gravity waves on deep water are dispersive, unlike weak compression waves. Interface waves will not be discussed in detail in this book. They have been mentioned here only to round out the picture of wave propagation and to indicate the special features of compression waves in a uniform material.

3.1.2 General Results for a Normal Shock Wave

Figure 3.2 shows a steady flow through a normal shock transition; with a suitable choice of axes this might be any of the flows shown in Figure 3.1. During the transition no energy is supplied to the fluid, and no external force acts on it. Hence the flow may be taken to be isenergic and the impulse function may be taken to be unchanged across the shock. It is true that each of the shocks shown in Figure 3.1 is established either by applying a force to the flow or by introducing energy into the fluid. However, the propagation of the wave away from the source of the disturbance, and the choice of axes in which the shock is at rest makes it possible to discuss the process using results for steady, isenergic, constant-impulse flow.

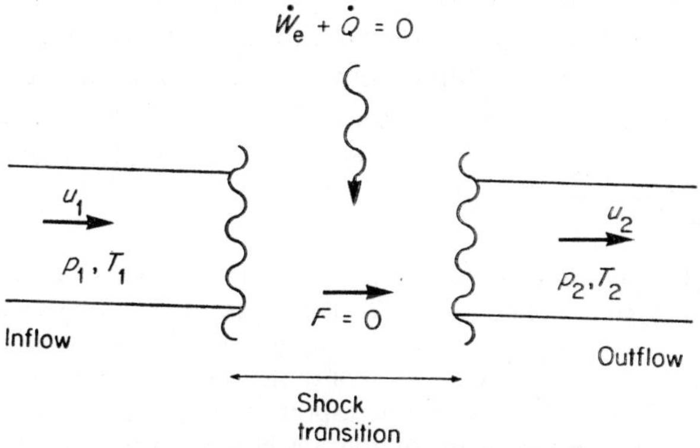

Figure 3.2 Generalized shock transition in the frame of reference in which it appears stationary and the flow appears steady

Across a normal shock the flow area is unchanged, and

$$\varrho_2 u_2 = \varrho_1 u_1 \tag{3.1.1}$$

For isenergic flow

$$h_2 + \tfrac{1}{2}u_2^2 = h_1 + \tfrac{1}{2}u_1^2 \tag{3.1.2}$$

The potential energy terms have been omitted since they are not important in realistic cases. Finally, the momentum requirement is

$$p_2 + \varrho_2 u_2^2 = p_1 + \varrho_1 u_1^2 \tag{3.1.3}$$

The principles of continuity, energy and momentum can be summarized compactly by the three conservation statements:

$$\begin{aligned} \dot{m} &= \text{constant} \\ h_0 &= \text{constant} \\ I &= \text{constant} \end{aligned} \tag{3.1.4}$$

across a normal shock.

The three equations (3.1.1, 2, 3) involve eight parameters, the upstream and downstream values of the velocity, pressure, density and enthalpy.

These quantities are also related by an equation of state, giving, for example

$$h_1 = f(p_1, \varrho_1) \quad \text{and} \quad h_2 = f(p_2, \varrho_2) \tag{3.1.5}$$

In all there are five independent relationships amongst the eight quantities. Hence if three independent quantities are specified, the other five can be determined. The required information can be supplied in many ways, one of the most common being the prescription of the upstream conditions, the downstream conditions having to be found. If the substance undergoing the shock has a simple equation of state, the basic constraints can be combined to give explicit relationships between the known quantities and those which have to be found. In general, only a tabular equation of state is available, and one must work directly from the fundamental conservation statements (3.1.1, 2, 3).

Example 3.1 Equation (1.4.10) describing the compression of an ideal liquid is simple enough to give explicit results for the shock transition. These results provide a useful approximation for shocks in liquids and also for those in some solid materials. This shock wave is not typical of those in highly compressible materials, since the dynamic and thermodynamic aspects of the response are virtually decoupled. However, the simplicity of the algebra compensates for this lack of generality.

The equation of state $\varkappa\, \mathrm{d}p = \mathrm{d}\varrho/\varrho$ can be integrated to give

$$\varkappa(p_2 - p_1) = \log_e \frac{\varrho_2}{\varrho_1}$$

This indicates, not unexpectedly, that the pressure rises across a compression wave. Moreover, the continuity equation

$$\frac{\dot{m}}{A} = \varrho_1 u_1 = \varrho_2 u_2$$

shows that the velocity falls across a compression wave. Combined with the momentum condition (3.1.3), this gives

$$p_2 - p_1 = \left(\frac{\dot{m}}{A}\right)^2 \left(\frac{1}{\varrho_1} - \frac{1}{\varrho_2}\right)$$

Hence the upstream and downstream densities are related by

$$\log_e \left(\frac{\varrho_2}{\varrho_1}\right) = \varkappa \left(\frac{\dot{m}}{A}\right)^2 \left(\frac{1}{\varrho_1} - \frac{1}{\varrho_2}\right)$$

Given $\varkappa$ and the upstream conditions ϱ_1 and u_1, the downstream density ϱ_2 can be found. The pressure differential follows and, with p_1 specified, the downstream pressure can be calculated.

The relationships used here are, like the general conservation principles (3.1.1, 2, 3), symmetrical with respect to the suffices 1 and 2. Thus, according to them, for any shock transition there is an equally satisfactory reverse transition. However, only one of these possible transitions is consistent with the second law of thermodynamics. Equation (2.3.27) relates the dissipation in isenergic flow to the changes in pressure and velocity:

$$\int_1^2 T\,\mathrm{d}s = -\int_1^2 \frac{\mathrm{d}p}{\varrho} - \Delta\left(\frac{u^2}{2}\right)$$

The evaluation of the right-hand side for the special case of constant-area flow of an ideal liquid requires a rather lengthy calculation. The steps are set out in one of the problems at the end of this chapter and can be followed through by those interested. Here only the final result is required:

$$\int_1^2 T\,\mathrm{d}s \simeq -\frac{(v_2-v_1)^3}{12\varkappa v_1^2} \simeq -\frac{(u_2-u_1)^3}{12u_1}$$

It is apparent that only shocks with $v_2 < v_1$ and $u_2 < u_1$, that is, compressive waves, will occur in nature. Expansion waves do exist, of course, but they spread out as they propagate and hence are not predicted by this analysis, which assumes a constant wave profile and constant speed of propagation.

Example 3.2 When a tabular equation of state is used to determine a shock transition, a trial-and-error solution is necessary. A procedure rather like that of Example 2.11 can be used, but is somewhat more difficult to start since none of the downstream properties is known initially.

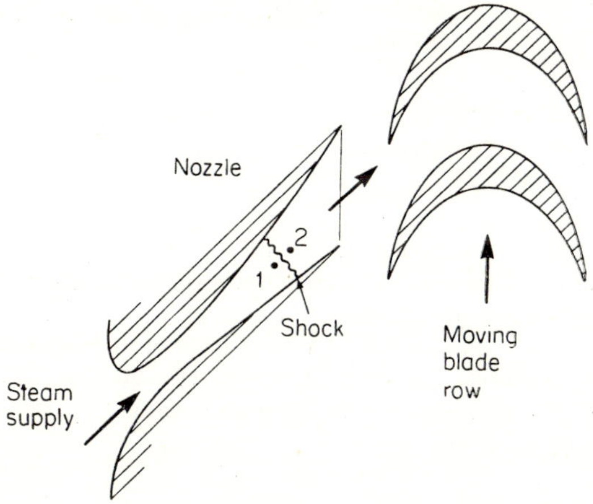

Figure 3.3 A shock in the inlet nozzle of **a** steam turbine

Consider a shock which might occur in the flow passages of a steam turbine, as shown in Figure 3.3. This shock would not appear in normal operation near the design point, but might occur if the downstream pressure had an unusually high value. The steam is initially at $p_1 = 10$ lbf/in^2, $T_1 = 400$ °F, and moves at $u_1 = 3000$ ft/sec. The other properties are found from steam tables as $h_1 = 1240{\cdot}6$ Btu/lb, $v_1 = 51{\cdot}04$ ft^3/lb and $s_1 = 1{\cdot}917$ Btu/lb °R. The conservation principles

(3.1.4) give

$$\frac{u_2}{v_2} = \frac{3000}{51\cdot 04} = 58\cdot 75 \text{ lb/sec ft}^2$$

$$h_2 + \frac{u_2^2}{2} = 1240\cdot 6 + \frac{3000^2}{2 \times 32\cdot 2 \times 778}$$

$$= 1420\cdot 7 \text{ Btu/lb}$$

$$p_2 + \frac{u_2^2}{v_2} = 10\cdot 0 + \frac{3000^2}{51\cdot 04 \times 32\cdot 2 \times 144}$$

$$= 48\cdot 0 \text{ lbf/in}^2$$

Note that the units have been kept consistent with those of the tables available.

Stage 1

Something must be guessed to start the calculation. Velocity is best, since it appears in each of the conservation statements. The examples of Figure 3.1 and the preceding discussion of a shock in an ideal liquid suggest that the flow decelerates across a shock, so let us start by assuming that $u_2 = 1000$ ft/sec. This gives immediately

$$v_2 = \frac{1000}{58\cdot 75} = \frac{51\cdot 04}{3} = 17\cdot 0 \text{ ft}^3/\text{lb}$$

and

$$h_2 = 1420\cdot 7 - \frac{180\cdot 1}{9} = 1400\cdot 7 \text{ Btu/lb}$$

It is now possible to see whether this is the correct end state, that is, the state satisfying the third conservation statement as well. The steam tables available are too coarse for the present calculation, although they could be used to prepare a large-scale chart of the part of the surface of state of immediate interest. On this chart accurate interpolation would be possible. However, for steam Callendar's equation is available, and is sufficiently accurate for at least a preliminary estimate of the final state. It gives

$$p_2 = 1\cdot 247 \frac{h_2 - 835}{v_2} = 1\cdot 247 \times \frac{565\cdot 7}{17\cdot 0} = 41\cdot 4 \text{ lbf/in}^2$$

Then, for the velocity assumed

$$p_2 + \frac{u_2^2}{v_2} = 41\cdot 4 + 38\cdot 0 \times \frac{3}{9} = 54\cdot 1 \text{ lbf/in}^2$$

This impulse function is too high.

Stage 2

It is not immediately obvious whether the assumed velocity is too high or too low. To see, try $u_1 = 1200$ ft/sec. Then

$$v_2 = \frac{51{\cdot}04 \times 1200}{3000} = 20{\cdot}4 \text{ ft}^3/\text{lb}$$

$$h_2 = 1420{\cdot}7 - 180{\cdot}1 \times \left(\frac{1{\cdot}2}{3}\right)^2 = 1391{\cdot}9 \text{ Btu/lb}$$

$$p_2 = 1{\cdot}247 \times \frac{556{\cdot}9}{20{\cdot}4} = 34{\cdot}1 \text{ lbf/in}^2$$

and

$$p_2 + \frac{u_2^2}{v_2} = 34{\cdot}1 + 38{\cdot}0 \times \frac{1{\cdot}2}{3} = 49{\cdot}3 \text{ lbf/in}^2$$

This is closer to the required result.

Stage 3

The process can be repeated to improve upon the values already obtained. Another approach is extrapolation from the two trial states considered above. The impulse functions for Stages 1 and 2 (I_1 and I_2) are related to the correct impulse function (I_c) by

$$\frac{I_2 - I_c}{I_1 - I_2} = \frac{1{\cdot}3}{4{\cdot}8} = 0{\cdot}27$$

Hence the correct values can be estimated as

$$\begin{aligned} u_2 &= 1200 + 0{\cdot}27 \times 200 = 1254 \text{ ft/sec} \\ v_2 &= 20{\cdot}4 + 0{\cdot}27 \times (20{\cdot}4 - 17{\cdot}0) = 21{\cdot}3 \text{ ft}^3/\text{lb} \\ h_2 &= 1391{\cdot}9 - 0{\cdot}27 \times (1400{\cdot}7 - 1391{\cdot}9) = 1389{\cdot}5 \text{ Btu/lb} \\ p_2 &= 34{\cdot}1 - 0{\cdot}27 \times (41{\cdot}4 - 34{\cdot}1) = 32{\cdot}1 \text{ lbf/in}^2 \end{aligned}$$

The Mollier chart, or steam tables, show that $s_2 \simeq 1{\cdot}93$ Btu/lb °R. Since an entropy increase has occurred, the assumed deceleration is realistic. The reversed process, which would involve an acceleration of the flow, will not occur spontaneously, since it corresponds to a decrease in entropy.

3.2 SOUND WAVES

3.2.1 The Speed of Sound

It may appear to be impossible to draw any conclusions from the conservation statements for a normal shock

$$\begin{aligned} \varrho u &= \text{constant} \\ h + \tfrac{1}{2}u^2 &= \text{constant} \\ p + \varrho u^2 &= \text{constant} \end{aligned} \qquad (3.2.1)$$

without specifying an equation of state linking the variables. This is not so for very weak waves; indeed, the constraints then give information about the equation of state itself.

A vanishingly weak compression wave is a sound wave, and the speed with which it propagates through a still fluid is termed the speed of sound. The small changes across a weak wave can be obtained most easily by differentiating the equations (3.2.1). Then

$$\begin{aligned} \varrho \,\mathrm{d}u + u\,\mathrm{d}\varrho &= 0 \\ \mathrm{d}h + u\,\mathrm{d}u &= 0 \\ \mathrm{d}p + \varrho u\,\mathrm{d}u &= 0 \end{aligned} \tag{3.2.2}$$

The simple form of the last result depends on the continuity requirement, ϱu = constant.

Consider first the thermodynamic implications of the relationships (3.2.2). Eliminating the velocity, one gets

$$\mathrm{d}h + u\,\mathrm{d}u = \mathrm{d}h - \frac{\mathrm{d}p}{\varrho} = 0$$

However, quite generally $T\,\mathrm{d}s = \mathrm{d}h - \mathrm{d}p/\varrho$

Then
$$\mathrm{d}s = 0 \tag{3.2.3}$$

across this weak wave. Sound waves travel with very little dissipation of energy; in other words, they are only weakly attenuated. Of course, when sound radiates out into the atmosphere, its intensity drops off rapidly, not through dissipation, but through the spreading of the propagating energy over a progressively larger wave front. When sound is channelled through a tunnel, its intensity drops off less rapidly.

Another way of combining the results (3.2.2) is

$$\mathrm{d}p + \varrho u\,\mathrm{d}u = \mathrm{d}p - u^2\,\mathrm{d}\varrho = 0$$

But the process has been shown to be isentropic. Then

$$\frac{\mathrm{d}p}{\mathrm{d}\varrho} = \left(\frac{\partial p}{\partial \varrho}\right)_s = u^2$$

specifies the flow velocity in terms of a thermodynamic property, the derivative $(\partial p/\partial \varrho)_s$. For a wave propagating through a still fluid, this is the speed of propagation. Hence the speed of sound is given by

$$a^2 = \left(\frac{\partial p}{\partial \varrho}\right)_s \tag{3.2.4}$$

Given a simple equation of state, a more explicit formula for the sound speed can be found. For a perfect gas, obeying the isentropic law, $p \propto \varrho^{\gamma}$, logarithmic differentiation gives

$$\frac{\mathrm{d}p}{p} - \gamma\frac{\mathrm{d}\varrho}{\varrho} = 0$$

whence

$$a = \left(\frac{\gamma p}{\varrho}\right)^{\frac{1}{2}} = (\gamma RT)^{\frac{1}{2}} \tag{3.2.5}$$

For an ideal liquid, with $\varkappa\,\mathrm{d}p = \mathrm{d}\varrho/\varrho$, one gets

$$a = (\varkappa\varrho)^{-\frac{1}{2}} \tag{3.2.6}$$

The determination of these wave speeds did not require the specification of the wave form. Hence they may be expected to apply to any pattern of waves which does not produce significant changes in velocity and properties of state. Another way of stating this conclusion is to say that small compression waves (and negative compression or rarefaction waves) can be superposed to give any pattern of weak waves. The whole pattern will propagate at the sound speed, without dispersion. It follows, in particular, that the speed of propagation of a periodic train of waves will be independent of the wavelength and frequency of the train. As was noted earlier, not all periodic waves have this characteristic. The crests of small-amplitude gravity waves on deep water, for example, propagate at speeds given by

$$c = \left(\frac{g\lambda}{2\pi}\right)^{\frac{1}{2}} = \frac{g}{2\pi f} \tag{3.2.7}$$

where λ and f are the wavelength and frequency of the wave train.

Example 3.3 A bell-buoy is set ringing by water waves of period 1 sec, and radiates sound through air and water, as shown in Figure 3.4. Let us find the velocities of propagation of the three waves involved.

Take the air temperature to be 20 °C. Then

$$a_a = (\gamma RT)^{\frac{1}{2}} = (1{\cdot}4\times 287\times 293)^{\frac{1}{2}} = 343 \text{ m/sec}$$

For water at about the same temperature, $\varkappa = 2\times 10^{-8}$ ft²/lbf and for sea water $\varrho = 2$ sl/ft³. Then

$$a_w = (\varkappa\varrho)^{-\frac{1}{2}} = \left(\frac{10^8}{4}\right)^{\frac{1}{2}} = 5000 \text{ ft/sec} = 1500 \text{ m/sec}$$

Note that the sound speed for water is some $4\frac{1}{2}$ times that for air. Typically, the

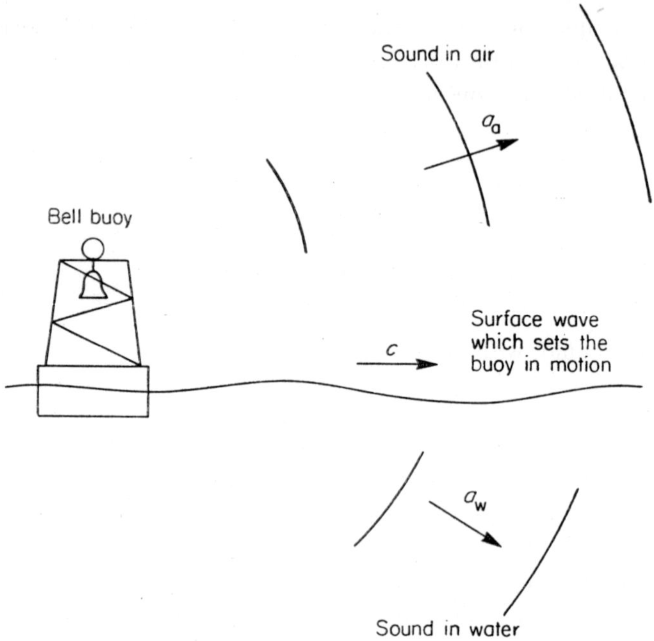

Figure 3.4 Sound transmission in air and water

speeds of sound in liquids are considerably greater than those in highly compressible materials.

Now consider the surface wave:

$$c = \frac{g}{2\pi f} = \frac{9{\cdot}81}{6{\cdot}28} = 1{\cdot}56 \text{ m/sec}$$

This is several orders-of-magnitude slower than the compression waves in water and air. Of course, a longer wave will propagate more quickly. For example, the gravity wave whose crest speed is a tenth of the speed of sound in air has period

$$\frac{1}{f} = \frac{2\pi c}{g} = \frac{6{\cdot}28 \times 34{\cdot}3}{9{\cdot}81} = 22{\cdot}0 \text{ sec}$$

and wavelength

$$\lambda = \frac{c}{f} = 34{\cdot}3 \times 22{\cdot}0 = 750 \text{ m}$$

It appears that even for very long water waves the crest speed is much less than the sound speeds.

Example 3.4 The general relationship

$$\left(\frac{\partial p}{\partial \varrho}\right)_s = \gamma \left(\frac{\partial p}{\partial \varrho}\right)_T$$

can be used to derive an expression for the speed of sound in a van der Waals gas. This result is easily proved for a perfect gas (do this), and can be shown to apply to any simple compressible substance.

The van der Waals equation can be written

$$p = \frac{\varrho RT}{1-b\varrho} - a\varrho^2$$

(Here a is the van der Waals constant; the speed of sound will be denoted a_s.) Then to the first order in the correction terms

$$\left(\frac{\partial p}{\partial \varrho}\right)_T \simeq RT(1+2b\varrho) - 2a\varrho$$

To the same order

$$a_s = \left[\gamma\left(\frac{\partial p}{\partial \varrho}\right)_T\right]^{\frac{1}{2}} \simeq (\gamma RT)^{\frac{1}{2}}\left[1-\varrho\left(\frac{a}{RT}-b\right)\right]$$

In general, when the gas is rarefied and hot, the correction terms are small. However, there is a particular temperature, $T = a/Rb$, at which the simple formula is correct. This is just half the temperature of inversion for the Joule–Thomson effect which was found in Example 2.8.

Let us find the sound speed in the steam considered in Example 3.2, for which $T_1 = 860$ °R, and $\varrho_1 = 0{\cdot}00109$ lb mol/ft³. The appropriate values of a, b and R are given in Example 2.8. Using them, we find

$$\varrho\left(\frac{a}{RT} - b\right) = 0{\cdot}00109 \times (2{\cdot}23 - 0{\cdot}49) = 0{\cdot}0019$$

It appears that the correction term is negligible in this situation.

Now to calculate the first approximation to the sound speed, $(\gamma RT)^{1/2}$. For an accurate determination of the speed, $\gamma = c_p/c_v$ must be determined with care. Appropriate values of the specific heats can be found from the steam tables by considering small temperature changes. Thus, using values for $p_1 = 10$ lbf/in²

$$c_p = \frac{\Delta h}{\Delta T} = \frac{1287{\cdot}5 - 1193{\cdot}9}{500-300} = 0{\cdot}468 \text{ Btu/lb °R}$$

$$c_v = \frac{\Delta h}{\Delta T} - \frac{p\,\Delta v}{\Delta T}$$

$$= 0{\cdot}468 - 1440 \times \frac{57{\cdot}05 - 45{\cdot}00}{200 \times 778}$$

$$= 0{\cdot}356 \text{ Btu/lb °R}$$

Here the dependence of e upon v has been neglected.

Then

$$\gamma = \frac{c_p}{c_v} = 1{\cdot}313 \text{ for steam in this condition.}$$

The sound speed is given by

$$a_s = (\gamma RT)^{\frac{1}{2}}$$

$$= \left(\frac{1{\cdot}313 \times 32{\cdot}2 \times 860 \times 1545}{18{\cdot}0}\right)^{\frac{1}{2}} = 1765 \text{ ft/sec}$$

The application of the van der Waals correction changes this value only a little, to about 1760 ft/sec.

3.2.2 Mach Number

In preparing tables and graphs, and in calculations generally, it is often convenient to use dimensionless or unit-free quantities. These make the presentation of data independent of any particular unit system, and also reduce the number of inter-related variables. This technique has been introduced at a number of points in the preceding work, and will be used extensively in the rest of the book. Indeed, it will be found to provide a powerful tool of analysis and not merely a convenient way of expressing results already available.

An obvious way of rendering the fluid velocity unit-free is to divide it by the sound speed. The resulting dimensionless quantity is called the Mach number:

$$M = \frac{u}{a} \qquad (3.2.8)$$

In general, both u and a will vary through a flow. For $M < 1$, the flow is locally *subsonic*, that is, slower than the speed of sound. For $M > 1$, the flow is described as *supersonic*. The case $M = 1$ is termed sonic; and near-sonic flows are sometimes called transonic.

The *geometric significance* of Mach number is indicated in Figure 3.5 which shows the relationship between the absolute speed of a wave and the local fluid velocity. A disturbance in a uniform duct produces waves which travel upstream and downstream. When the flow is sonic, the sound wave propagating upstream appears to be stationary ($u-a = 0$), and that travelling downstream moves at twice the sound speed ($u+a = 2a$). In a subsonic stream the upstream wave can make some advance against the flow ($u-a < 0$, $u+a > a$). In a supersonic stream, even the upstream-propagating wave is in fact carried downstream ($u-a > 0$, $u+a > 2a$).

The upstream region into which the sound does not penetrate in sonic and supersonic flows is called the region of forbidden signals or the zone of silence. Although we have discussed only wave propagation in one dimension, the same principles apply to two and three-dimensional propagation. Thus

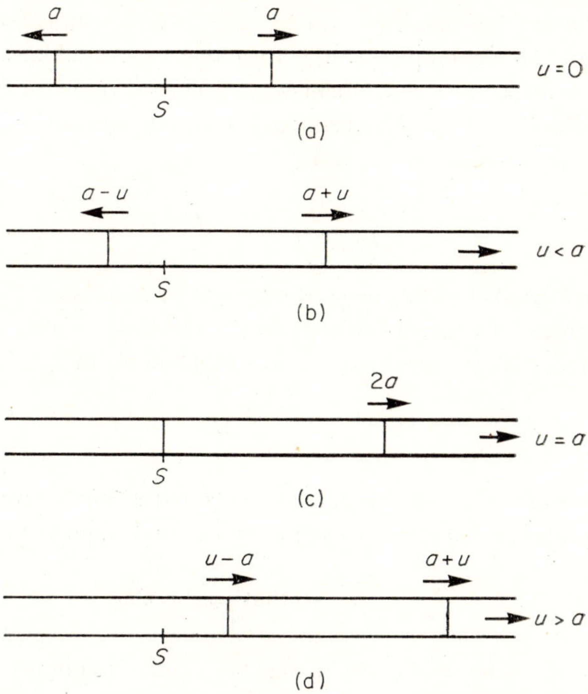

Figure 3.5 Sound transmission in a flowing fluid. In each case, *S* indicates the position of the stationary source in the duct. (a) Negligible motion; (b) Subsonic flow ($M < 1$); (c) Sonic flow ($M = 1$); (d) Supersonic flow ($M > 1$)

the sound generated by a supersonic aeroplane is heard only within a cone extending backwards around its flight path. The region ahead and to the side of the cone is the zone of silence. A good deal of the energy radiated by the aeroplane is concentrated on the boundary cone. This gives rise to the sonic boom where the cone passes over the ground.

The Mach number also has a *thermodynamic significance*, which is most clearly seen by considering the special case of a perfect gas. Note that for a perfect gas

$$M^2 = \frac{u^2}{\gamma RT} = \frac{2}{\gamma - 1}\cdot\frac{u^2}{2c_pT} = \frac{2}{\gamma - 1}\cdot\frac{u^2}{2h} \tag{3.2.9}$$

This shows that the Mach number is a measure of the division of the total energy ($h_0 = h + \frac{1}{2}u^2$) between the ordered mean motion ($\frac{1}{2}u^2$) and the random microscopic motion (h). In a flow at high Mach number, much of the energy is kinetic and only a small part is thermal.

Although it has been defined quite generally, the Mach number proves most useful when the sound speed is given by a fairly simple formula. For a perfect gas and an ideal liquid, simple formulae are available, and the Mach number provides a most convenient way of characterizing a flow. For example, the relationship between total and static temperatures (2.3.26) can be written in terms of Mach number as

$$\frac{T_0}{T} = 1 + \frac{\gamma - 1}{2} M^2 \tag{3.2.10}$$

Many other formulae of this general type will be used later.

Some significant features of flow at very high Mach number are suggested by equation (3.2.10). It appears that the simplified result

$$\frac{T_0}{T} = \frac{\gamma - 1}{2} M^2$$

will then provide a useful estimate of the relationship between total and static temperatures. Note that this approximation is equivalent to

$$\frac{h_0}{h} = \frac{u^2}{2h} \quad \text{so that} \quad u^2 = 2h_0$$

Thus nearly all of the total enthalpy has been converted into kinetic energy. Accordingly the flow velocity is nearly constant, and changes in Mach number result primarily from changes in sound speed. Unfortunately, no sooner do these simple results become applicable than the perfect-gas model ceases to be realistic. For air, with $T = 250\ ^\circ\text{K}$ and $M = 5$, equation (3.2.10) gives the total temperature as

$$T_0 = 250 \times 6 = 1500\ ^\circ\text{K}$$

Hence the assumption of constant specific heats is no longer valid if the gas is decelerated, for example, by a shock wave. For a somewhat higher Mach number, say $M = 9$, equation (3.2.10) gives, with $\gamma = 1{\cdot}3$, appropriate to air at high temperatures

$$T_0 = 250 \times 13{\cdot}15 = 3300\ ^\circ\text{K}$$

Now dissociation provides another reason for the failure of the simple gas laws.

It is apparent that the perfect-gas model is an inadequate tool for the study of gas flows at high Mach number, say $M > 5$, where real-gas effects become important as the flow is decelerated. Flows of this class are termed *hypersonic*. The most important class of hypersonic flows are those established when space vehicles re-enter the atmosphere. In that case the temperature rise across the shock is so large that the gas becomes incandescent.

Example 3.5 To see how the introduction of dimensionless quantities can simplify the statement of physical laws, consider once again the results obtained in Example 3.1 for a shock in an ideal liquid.

The sound speed of this kind of substance is given in equation (3.2.6): $a^2 = 1/\varkappa\varrho$. Hence the continuity relation gives

$$\frac{u_2}{u_1} = \frac{v_2}{v_1} = \frac{\varrho_1}{\varrho_2} = \left(\frac{a_2}{a_1}\right)^2 = \left(\frac{M_2}{M_1}\right)^2$$

The pressure change is

$$p_2 - p_1 = \varrho_1 u_1^2 - \varrho_2 u_2^2 = \frac{M_1^2 - M_2^2}{\varkappa}$$

and the relationship connecting the upstream and downstream flows becomes

$$M_1^2 - M_2^2 = \log_e \left(\frac{M_1}{M_2}\right)^2$$

Since $\varkappa$ does not enter explicitly, this result applies to any substance whose behaviour can be described by taking $\varkappa$ = constant.

The choice $M_1 = M_2 = 1$ satisfies this equation; it corresponds to a very small change in state across the wave—a sound wave, in fact. To investigate slightly stronger shocks, take

$$\left(\frac{M_1}{M_2}\right)^2 = 1+\varepsilon$$

so that the condition relating the two Mach numbers becomes

$$M_2^2 = \frac{1}{\varepsilon} \log_e (1+\varepsilon)$$

The expansion of the logarithmic function for small ε is

$$\log_e (1+\varepsilon) = \varepsilon - \frac{\varepsilon^2}{2} + \frac{\varepsilon^3}{3} - \ldots$$

Hence, to the first order in ε

$$M_2^2 \simeq 1 - \frac{\varepsilon}{2}, \quad M_1^2 = (1+\varepsilon) M_2^2 \simeq 1 + \frac{\varepsilon}{2}$$

and

$$M_1^2 - M_2^2 \simeq \varepsilon$$

It is apparent that, for weak shocks, the upstream and downstream Mach numbers are equidistant on either side of the sonic condition.

These results can be used to show that pressure waves in liquids and solids can almost always be taken to be weak waves. The data for water in Appendix III show that $\varkappa \simeq 50 \times 10^{-6}$/atm. Then for a wave across which the pressure rises by 1000 atm

$$\varepsilon \simeq M_1^2 - M_2^2 = \varkappa(p_2 - p_1) \simeq 50 \times 10^{-6} \times 10^3 = 0{\cdot}050$$

Then

$$M_1^2 \simeq 1{\cdot}025 \quad \text{and} \quad M_2^2 \simeq 0{\cdot}975$$

Although the wave considered propagates with a velocity not very different from that of sound, the density ratio across the 'weak' wave is $\varrho_2/\varrho_1 = (M_1/M_2)^2 \simeq 1{\cdot}05$. A five per cent reduction in volume is a savage one for an only slightly compressible material. However, such waves can be generated by explosions and high-velocity impacts. Predictions from an idealized model under these extreme conditions should be treated with caution. They indicate the order-of-magnitude of the changes, but the precise values cannot be relied upon.

3.2.3 Subsonic and Supersonic Isenergic Flow

There may appear to be a contradiction between the conclusions concerning shock waves and those reached earlier when discussing throttling flow. For a shock, the requirement of increasing entropy implies that the velocity falls while the pressure rises. For constant-area throttling flow, an entropy rise corresponds to a fall in pressure and a rise in velocity. The flow discussed in Example 2.11 displays these characteristics.

The apparently contradictory conclusions noted above can be reconciled by considering an infinitesimal isenergic change in a steady, constant-area flow. For a small isenergic change in state

$$\frac{\mathrm{d}p}{\varrho}+u\,\mathrm{d}u = -T\,\mathrm{d}s < 0 \tag{2.3.27}$$

For this constant-area flow, the momentum principle (2.2.12) gives the additional force on the fluid as

$$\mathrm{d}F_{\mathrm{a}} = A(\mathrm{d}p+\varrho u\,\mathrm{d}u) = -\varrho TA\,\mathrm{d}s$$

Hence the force applied by the fluid always acts upstream. The only infinitesimal process which can occur spontaneously, without the application of a force, is an isentropic change.

The constant-area continuity condition gives

$$\frac{\mathrm{d}u}{u} = -\frac{\mathrm{d}\varrho}{\varrho}$$

When introduced into equation (2.3.27) this gives

$$\frac{\mathrm{d}p}{\varrho}\left[1-\frac{u^2}{\mathrm{d}p/\mathrm{d}\varrho}\right] = \frac{\mathrm{d}u}{u}\left(u^2-\frac{\mathrm{d}p}{\mathrm{d}\varrho}\right) = -T\,\mathrm{d}s < 0$$

Since the changes are small, we may approximate $\mathrm{d}p/\mathrm{d}\varrho \simeq (\partial p/\partial \varrho)_s = a^2$ in investigating the general nature of the results. Then

$$\frac{\mathrm{d}p}{\varrho}\left(1-\frac{u^2}{a^2}\right) \simeq \frac{\mathrm{d}u}{u}(u^2-a^2) \simeq -T\,\mathrm{d}s < 0$$

These results indicate that the signs of the changes in velocity and pressure are reversed for subsonic and supersonic throttling processes. A small transition in a sonic flow ($u = a$) will be isentropic. It was noted earlier that $dF_a = 0$ for this isentropic process. Hence the transonic transition has the characteristics of a sound wave.

The preceding discussion relates to an infinitesimal process, but indicates the general nature of finite transitions. The examples discussed in Chapter 2 involved subsonic flow past a blockage, and exhibited a fall in pressure and rise in velocity. Similar calculations could be performed for finite supersonic transitions, but reversed changes in pressure and velocity must be expected. In every case a force is applied to the flow, and to determine the transition one must specify that force or some other characteristic of the transition, perhaps the pressure change. When the force is taken to be zero, the transition is a shock wave. For a finite shock, there is an increase in entropy and the changes in pressure and velocity are like those in a supersonic transition. Thus the flow past a valve, the flow along a pipe, and the flow through a shock wave are fitted together into a family of constant-area, isenergic flows.

In the preceding discussion, the Mach number emerges naturally as an index of the state of the flowing fluid. This suggests that it will play a key role in the subsequent discussions of the response of a flowing fluid to other kinds of disturbance, for example, heat addition or a change in flow area.

3.3 SHOCKS IN A PERFECT GAS

3.3.1 Formulae and Tables

The shock waves which have attracted the greatest attention are those set up by aeroplanes and missiles, both externally and within internal ducts. Since air conforms to the perfect-gas model over a considerable range of temperatures, formulae derived for an ideal gas are often useful. The chief limitation on their applicability is high temperature, where the specific heats vary and where dissociation and ionization occur. This limitation is particularly important for shock flows. Even though the temperature is low initially, that behind a strong shock can be very high, and perfect-gas results may not describe the transition adequately.

One kind of shock has already been discussed in some detail, that in an ideal liquid. This lacks the close interaction of dynamic and thermodynamic response which is typical of shock waves in highly compressible substances. The perfect gas does display this coupling and, as a result, the algebra is somewhat complicated. Although the details depend on the equation of state, there is a basic similarity in the shock transitions in gases and vapours. Hence perfect-gas results indicate the general nature of shocks in a wider class of highly compressible substances.

When perfect-gas behaviour is assumed, the energy and state equations take on definite and related forms. Equations (3.1.2, 5) are now

$$c_p T_2 + \tfrac{1}{2}u_2^2 = c_p T_1 + \tfrac{1}{2}u_1^2 \tag{3.3.1}$$

$$p_1 = \varrho_1 R T_1 \quad \text{and} \quad p_2 = \varrho_2 R T_2 \tag{3.3.2}$$

The first result implies that

$$T_0 = \text{constant across the shock} \tag{3.3.3}$$

In equation (2.3.26) this was seen to be the case for any isenergic flow of a perfect gas.

The normal shock is also governed by the constant-area continuity and momentum relationships:

$$\varrho_2 u_2 = \varrho_1 u_1 \tag{3.1.1}$$

and

$$p_2 + \varrho_2 u_2^2 = p_1 + \varrho_1 u_1^2 \tag{3.1.3}$$

When faced with a set of equations involving many parameters, one must either follow a definite plan or run the risk of getting bogged down in algebra. Example 3.5 has shown how shock relations can be simplified by introducing the sound speed and Mach number. This procedure will be followed here. For a perfect gas, equations (3.2.5) give

$$a_1^2 = \gamma R T_1 = \frac{\gamma p_1}{\varrho_1}$$

and

$$a_2^2 = \gamma R T_2 = \frac{\gamma p_2}{\varrho_2}$$

With these results, pressure, density and temperature can be eliminated. The energy equation (3.3.1) becomes

$$a_2^2\left(1 + \frac{\gamma - 1}{2} M_2^2\right) = a_1^2\left(1 + \frac{\gamma - 1}{2} M_1^2\right) \tag{3.3.4}$$

With the use of the continuity equation (3.1.1), the momentum equation can be written

$$\frac{p_2}{\varrho_2 u_2} + u_2 = \frac{p_1}{\varrho_1 u_1} + u_1$$

or, when sound speed and Mach number are introduced

$$a_2\left(\frac{1}{\gamma M_2}+M_2\right) = a_1\left(\frac{1}{\gamma M_1}+M_1\right)$$

Now a_2/a_1 can be eliminated, and a little algebra gives

$$M_2^2 = \frac{1+\frac{\gamma-1}{2}M_1^2}{\gamma M_1^2-\frac{\gamma-1}{2}} \tag{3.3.5}$$

As would be anticipated $M_2 = 1$ when $M_1 = 1$. Moreover, for the realistic case of $\gamma > 1$, $M_2 < 1$ for $M_1 > 1$. Earlier discussions of entropy changes in shocks (Examples 3.1 and 3.2) suggest that $M_2 > 1$ will correspond to a decrease in entropy, and hence that this case will not arise in any real flow. This will later be shown to be correct.

Although the downstream Mach number M_2 has been expressed explicitly in terms of the upstream Mach number M_1, the formula is rather complicated. It proves convenient to use a tabular presentation of the relationships between M_1 and M_2 and other dimensionless parameters characterizing the shock. No single method of tabulating the results will be convenient for all possible problems, but that adopted should deal directly with the common task of finding the downstream conditions corresponding to a known upstream state. Hence the standard tabulation takes M_1 as argument, and presents the other properties as functions of M_1.

The formulae required to generate the tables follow easily from the results (3.3.4) and (3.3.5). For the perfect gas under consideration

$$\frac{T_2}{T_1} = \left(\frac{a_2}{a_1}\right)^2 = 1+\frac{2(\gamma-1)}{(\gamma+1)^2}\left(\gamma+\frac{1}{M_1^2}\right)(M_1^2-1) \tag{3.3.6}$$

The other ratios follow from continuity (3.1.1) and the equations of state (3.3.2):

$$\frac{\varrho_2}{\varrho_1} = \frac{u_1}{u_2} = \frac{M_1}{M_2}\frac{a_2}{a_1} = \frac{\gamma+1}{\gamma-1+2/M_1^2} \tag{3.3.6}$$

$$\frac{p_2}{p_1} = \frac{\varrho_2}{\varrho_1}\frac{T_2}{T_1} = 1+\frac{2\gamma}{\gamma+1}(M_1^2-1) \tag{3.3.6}$$

These results are given in tabular form in Appendix V for $\gamma = 1{\cdot}3$ and $1{\cdot}4$. The tables for $\gamma = 1{\cdot}4$ are most often used, since they apply closely to air. However, one should not forget their limitation to this single value of the ratio of specific heats.

Example 3.6 The shock tables for $\gamma = 1{\cdot}3$ are those most closely applicable to the shock in steam examined in Example 3.2. The sound speed for the flow upstream of the shock was found in Example 3.4 to be 1760 ft/sec. Hence

$$M_1 = \frac{3000}{1760} = 1{\cdot}704$$

The shock tables give $M_2 = 0{\cdot}628$, $p_2/p_1 = 3{\cdot}15$, $\varrho_2/\varrho_1 = 2{\cdot}33$ and $T_2/T_1 = 1{\cdot}36$. Then the velocity downstream of the shock can be calculated either as

$$u_2 = u_1 \frac{\varrho_1}{\varrho_2}$$

or as

$$u_2 = M_2 a_2 = M_2 a_1 \left(\frac{T_2}{T_1}\right)^{\frac{1}{2}} = 1290 \text{ ft/sec}$$

Also

$$v_2 = v_1 \frac{\varrho_1}{\varrho_2} = \frac{51{\cdot}04}{2{\cdot}33} = 21{\cdot}9 \text{ ft}^3/\text{lb}$$

and

$$p_2 = p_1 \frac{p_2}{p_1} = 10 \times 3{\cdot}15 = 31{\cdot}5 \text{ lbf/in}^2$$

These values are quite close to those found by the lengthy calculations of Example 3.2. This is so because steam conforms fairly well to the perfect-gas model under the conditions considered. Moreover, in Example 3.2 the Callendar equation was used to avoid laborious interpolation; this equation is in essence a perfect-gas model for steam. Nevertheless, the methods of Example 3.2 are applicable, in principle, to any shock transition, while those used here are appropriate only to a perfect gas. The perfect-gas methods may prove more widely useful in providing an estimate from which the iterative procedure of Example 3.2 can start.

Example 3.7 Figure 3.6 shows the flow into the intake of a ram jet flying at 1200 m/sec at an altitude of 20,000 m. This is the flight condition considered earlier in Example 2.10. Now, however, a normal shock is assumed to stand across the mouth of the inlet. The Mach number of flight is $M_1 = 1200/295 = 4{\cdot}07$;

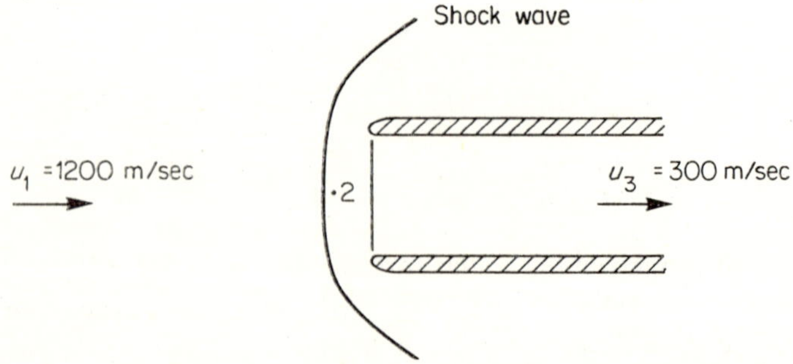

Figure 3.6 Flow into a ram jet through a normal shock wave

this is also the Mach number of the flow approaching the shock, as seen in the frame of reference in which the shock and duct are stationary. Note that conditions are uniform throughout the air ahead of the shock. Disturbances from the moving body cannot penetrate into this region, since they propagate more slowly than the shock wave, which is driven along by the moving body.

The state of the gas inside the duct (Section 3) will be determined using the normal-shock tables for a perfect gas with $\gamma = 1{\cdot}4$, assuming further that the flow behind the shock is isentropic. As in Example 2.10, the velocity of the air within the duct will be taken to be 300 m/sec. The total temperature throughout this isenergic flow is

$$T_0 = T_1\left(1+\frac{\gamma-1}{2}M_1^2\right) = 216{\cdot}7\times 4{\cdot}31 = 933\ ^\circ\mathrm{K}$$

as found in Example 2.10. The temperature just behind the shock is given by normal-shock tables as

$$T_2 = \frac{T_2}{T_1}T_{\mathrm{a}} = 4{\cdot}16\times 216{\cdot}7 = 901\ ^\circ\mathrm{K}$$

When the velocity is 300 m/sec, the temperature is

$$T_3 = T_0 - \frac{u_3^2}{2c_p} = 888\ ^\circ\mathrm{K}$$

This too is a value found in Example 2.10.

The Mach number at Section 3 is

$$M_3 = \frac{u_3}{(\gamma R T_3)^{\frac{1}{2}}} = 0{\cdot}502$$

while that just behind the shock is found from shock tables as

$$M_2 = 0{\cdot}432$$

Evidently the fluid accelerates slightly on entering the duct.

Now consider the pressure variation in the flow. That just behind the shock is found using shock tables:

$$p_2 = \frac{p_2}{p_1}p_{\mathrm{a}} = 19{\cdot}2\times 0{\cdot}0553 = 1{\cdot}06\ \mathrm{bar}$$

The pressure in the duct is found using the condition s = constant for the flow after the shock:

$$p_3 = \left(\frac{T_3}{T_2}\right)^{\frac{\gamma}{\gamma-1}} p_2 = 1{\cdot}01\ \mathrm{bar}$$

In Example 2.10 the pressure within the duct was found to be 2·04 bar. This higher value could be achieved if the fluid entered the duct, not through a normal shock, but through a series of oblique shocks, which decelerated the flow with a smaller increase in entropy and hence a smaller fall in pressure. The smaller the entropy increase, or in other words the smaller the losses in the intake, the more efficient is the ram compression of the ingested air. Hence the highest pressure which can be generated by ram compression to $u_3 = 300$ m/sec is that achieved in a wholly isentropic flow. In this case

$$p_{3_i} = \left(\frac{T_3}{T_1}\right)^{\frac{\gamma}{\gamma-1}} p_a = 7{\cdot}71 \text{ bar}$$

The temperatures calculated above deserve special attention. Deceleration raises the temperature to values at which many metals would be seriously weakened. Hence the specified flight condition could be achieved only if

(1) it were maintained for only a short time, or
(2) materials retaining strength at high temperature were used, or
(3) the surfaces in contact with the air were internally cooled.

3.3.2 Entropy and Total Pressure

The change in entropy across a shock wave can be found by introducing the results (3.3.6) into

$$\Delta s = R \log_e \left[\frac{p_1}{p_2}\left(\frac{T_2}{T_1}\right)^{\frac{\gamma-1}{\gamma}}\right] \tag{2.4.14}$$

$$= R \log_e \left\{\left[1+\frac{2\gamma\Delta}{\gamma+1}\right]\left[1+\frac{2(\gamma-1)}{(\gamma+1)^2}\left(\gamma+\frac{1}{1+\Delta}\right)\Delta\right]^{\frac{\gamma}{\gamma-1}}\right\}$$

where $\Delta = M_1^2 - 1$. For $\Delta = 0$, it is apparent that $\Delta s = 0$, as was earlier seen to be the case for a sound wave. However, the complexity of the relationship obscures other features of the dependence of Δs on M_1. A more directly interpretable result can be obtained by using binomial and logarithmic expansions, making use of the fact that Δ is small for weak shocks. The algebra is lengthy, although straightforward, and only the final result need be considered:

$$\frac{\Delta s}{R} = \frac{2\gamma}{3(\gamma+1)^2}(M_1^2-1)^3 + \text{higher-order terms in } (M_1^2-1) \tag{3.3.7}$$

As might be anticipated from the study of shocks in other fluids

$$\Delta s > 0 \quad \text{only when} \quad M_1 > 1$$

Coupled with equation (3.3.5) this indicates that

$$M_1 > 1, \qquad M_2 < 1 \tag{3.3.8}$$

for any shock in a perfect gas.

For any isenergic process, the entropy change in a perfect gas is related to the total pressures by

$$\Delta s = R \log_e \frac{p_{0_1}}{p_{0_2}} \tag{2.4.15}$$

Naturally this result applies to the particular case of a normal shock. Since the total pressure of a flowing fluid is more directly perceived than its entropy, the ratio p_{0_1}/p_{0_2} is commonly used as a measure of irreversibility in a shock wave, and is given in the shock tables as a function of M_1. The appropriate relationship can be found by comparing equations (2.4.14) and (2.4.15) given above:

$$\frac{p_{0_1}}{p_{0_2}} = \frac{p_1}{p_2}\left(\frac{T_2}{T_1}\right)^{\frac{\gamma}{\gamma-1}} \tag{3.3.9}$$

Both the ratios on the right are expressed in equations (3.3.6) as functions of M_1 and γ.

In Section 2.4.2 it was seen that total pressure could be measured using a Pitot tube. It is this fact which makes total pressure a convenient measure of energy degradation in an isenergic flow. In a subsonic stream, the deceleration to the stagnation state can be nearly isentropic, and the Bernoulli results can be used to relate velocity and total pressure. But a shock wave forms ahead of a body introduced into a supersonic stream. This is true, in particular, of a Pitot tube, and the fluid reaching the mouth of the tube will pass through the normal portion of the shock directly ahead of it. Thus the deceleration will take place in two stages; an abrupt, dissipative deceleration in the shock, followed by a nearly isentropic, subsonic deceleration to the mouth of the tube. The pressure measured in the tube will therefore be p_{0_2}, the total pressure just behind the shock.

Although the free-stream total pressure p_{0_1} cannot be measured directly, the free-stream static pressure can. It is the pressure measured through a small hole in a flat plate or pointed tube placed parallel to the stream. Only a weak oblique shock is formed by a pointed body, and the pressure on the parallel side is very little above the free-stream value. The measured values p_{0_2} and p_1 can be related to the free-stream Mach number using

$$\frac{p_{0_2}}{p_1} = \frac{p_{0_2}}{p_{0_1}}\frac{p_{0_1}}{p_1} \tag{3.3.10}$$

Equation (3.3.9) gives p_{0_2}/p_{0_1} as a function of M_1 and γ, while equation (3.2.10) gives

$$\frac{p_{0_1}}{p_1} = \left(\frac{T_0}{T_1}\right)^{\frac{\gamma}{\gamma-1}} = \left(1+\frac{\gamma-1}{2}M_1^2\right)^{\frac{\gamma}{\gamma-1}}$$

for the isentropic link between total and static states. The function p_{0_2}/p_1 derived in this way is given in the shock tables to facilitate the interpretation of Pitot-tube measurements in a supersonic stream.

Example 3.8 A Pitot tube on board a supersonic airliner registers 1 bar. It is known that the atmospheric conditions are 0·250 bar and 220 °K. What is the velocity of flight?

For $p_{0_2}/p_1 = 4$, shock tables for $\gamma = 1{\cdot}4$ give

$$M_1 = 1{\cdot}647 \quad \text{and} \quad M_2 = 0{\cdot}655$$

The speed of sound is $a_1 = (\gamma R T_1)^{1/2} = 297$ m/sec. Hence

$$u_1 = 490 \text{ m/sec}$$

It is also of interest to see how the pressure change is divided between the two stages of deceleration:

$$\frac{p_{0_2}}{p_1} = \frac{p_2}{p_1}\frac{p_{0_2}}{p_2}$$

From shock tables, $p_2/p_1 = 3{\cdot}00$ across the shock. The pressure change in the isentropic deceleration after the shock is

$$\frac{p_{0_2}}{p_2} = \left(1+\frac{\gamma-1}{2}M_2^2\right)^{\frac{\gamma}{\gamma-1}} = 1{\cdot}33$$

3.3.3 The Hydraulic Analogy

There is a marked similarity between a shock wave and the hydraulic jump studied in Example 2.17. Each is a dissipative transition with no external force applied to the flow. For the hydraulic jump it was found that the depth always increases:

$$r = \frac{d_2}{d_1} = \frac{u_1}{u_2} > 1$$

Hence the dimensionless quantity

$$\frac{u_1}{(gd_1)^{\frac{1}{2}}} = \left[\frac{1}{2}r(1+r)\right]^{\frac{1}{2}} > 1$$

for a real hydraulic jump. These characteristics too correspond generally to those of a shock.

For a weak wave, $u_1 \rightarrow (gd)^{1/2}$, as $r \rightarrow 1$, Hence $(gd)^{1/2}$ is the velocity of propagation of small waves in still, shallow water, analogous to the speed of sound in a compressible medium. This result differs from (3.2.7), which gives the speed of wave crests on deep water, that is, water so deep in comparison to the wavelength that the waves cannot 'feel bottom'. The present result $(gd)^{1/2}$ applies only for wavelengths considerably greater than the water depth. For such long waves the speed of propagation is independent of wavelength and they are therefore non-dispersive. Moreover, the long waves propagate more rapidly than any of the shorter waves whose speed is influenced by wavelength. Hence the long-wave speed $(gd)^{1/2}$ is the 'signal speed' for disturbances in water of uniform depth d. These remarks apply strictly only to waves of small amplitude. Waves of finite amplitude can propagate faster, as the example of the finite hydraulic jump indicates.

The quantity $u/(gd)^{1/2}$ is called the *Froude number*. It can be interpreted as the ratio of the local flow velocity to the local signal speed, that is, the speed of long, shallow-water waves. Hence the Froude number is analogous to the Mach number of a compressible fluid flow. Ahead of a hydraulic jump the Froude number is greater than unity, while behind the jump it is less than unity.

The analogy between the free-surface and compressible flows can be extended to two dimensions. Although the physical processes are quite different, the geometry of the wave patterns in the two species of flow is broadly similar. Hence shallow-water flows are often used to demonstrate the general nature of the waves found in gas flows. It is a very simple matter to obtain in the laboratory a thin uniform sheet of flowing water. The waves produced by introducing bodies or distorting walls into this flow have much the same shape as those set up in compressible fluid flows. In fact, the wave shapes are those which would be developed in a gas with a ratio of specific heats of $\gamma = 2$.

3.4 MOVING WAVES

3.4.1 General Results

The results obtained for stationary shocks can be applied to waves which propagate at a constant speed through a still fluid simply by adopting a frame of reference in which the shock appears to be stationary. The velocities for the two frames of reference are shown in Figure 3.7.

The static properties, those measured by instruments moving with the fluid, are not changed by this transformation. However, the total properties are altered. An example will make this point clearer. In the axes of Figure 3.7(a), the total temperature both before and after the shock is, for a perfect gas

$$T_0 = T_1 + \frac{u_1^2}{2c_p} = T_2 + \frac{u_2^2}{2c_p}$$

For the axes of Figure 3.7(b), it is

$$T_{0_1} = T_1 \quad \text{before the shock}$$

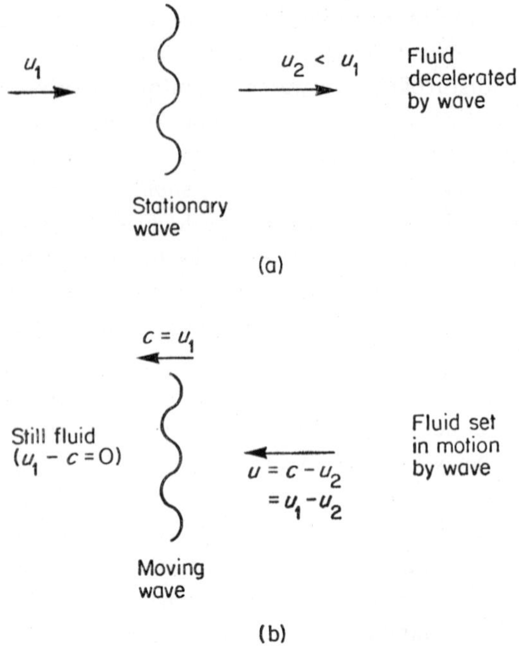

Figure 3.7 A shock wave in two frames of reference. (a) Frame in which the wave appears stationary; (b) Frame in which the air ahead of the wave appears stationary

and

$$T_{0_2} = T_2 + \frac{(u_1 - u_2)^2}{2c_p} \quad \text{after the shock.}$$

In this frame of reference the flow is not steady, and $T_{0_1} \neq T_{0_2}$. Nor are the other total properties specified by the steady-flow results. Hence the procedure for moving between the two frames of reference is to carry over the static properties unchanged, to transform the velocities suitably, and to leave the total properties behind.

It is also possible to analyse a moving shock directly, without using the frame of reference in which it appears stationary. The analysis is based on the continuity and momentum requirements for the non-steady flow. The system to be considered is shown in Figure 3.8—a control volume through which the shock is passing. The rate at which fluid is compressed as the wave advances is $(\varrho_2 - \varrho_1)c$; this is the rate at which the mass of fluid within the system is increasing. The rate of inflow at the rear is $\varrho_2 u$. Hence the conservation of mass requires that

$$(\varrho_2 - \varrho_1)c = \varrho_2 u$$

The rate at which the momentum of the system is increasing is $\varrho_2 uc$, an increase attributable in part to the pressure differential $(p_2 - p_1)$, and in part to the inflow of momentum at the rear at rate $\varrho_2 u^2$. Hence, the momentum equation is

$$\varrho_2 uc = p_2 - p_1 + \varrho_2 u^2$$

These two equations give

$$\varrho_2 - \varrho_1 = \varrho_2 \frac{u}{c} \tag{3.4.1}$$

and

$$p_2 - p_1 = \varrho_2 u(c-u) = \varrho_1 uc \tag{3.4.2}$$

Dividing corresponding members, one finds

$$c^2 = \frac{\varrho_2}{\varrho_1} \frac{p_2 - p_1}{\varrho_2 - \varrho_1} \tag{3.4.3}$$

Multiplication of the members gives

$$u^2 = (v_1 - v_2)(p_2 - p_1) \tag{3.4.4}$$

These two relationships express the wave speed and fluid velocity in any material in terms of the pressure rise and the densities on either side of the wave.

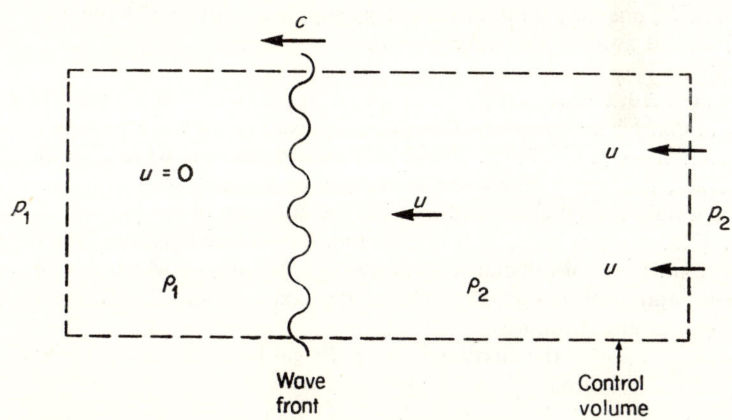

Figure 3.8 A fixed volume through which a shock wave is passing

Example 3.9 What is the absolute velocity of the air behind the normal shock formed ahead of the airliner considered in Example 3.8? The Mach number of flight is 1·647, and the speed of sound in the atmosphere is 297 m/sec.

Consider the frame of reference in which the wave is stationary.

$$u_1 = M_1 a_1 = 490 \text{ m/sec}$$

The velocity behind the shock is determined using shock tables:

$$u_2 = \frac{\varrho_1}{\varrho_2} u_1 = \frac{490}{2{\cdot}11} = 232 \text{ m/sec}$$

In the absolute frame of reference, for which the air ahead of the shock is stationary, the wave propagation speed is

$$c = 490 \text{ m/sec}$$

and the velocity behind the wave is

$$u_1 - u_2 = 258 \text{ m/sec}$$

in the direction of flight and of wave propagation.

It is also of interest to determine the rise in total temperature across the wave in the absolute frame:

$$\Delta T_0 = \left(\frac{T_2}{T_1} - 1\right) T_a + \frac{(u_1 - u_2)^2}{2c_p}$$

$$= 0{\cdot}421 \times 220 + \frac{258^2}{2010}$$

$$= 92{\cdot}6 + 33{\cdot}1 = 125{\cdot}7 \text{ °K}$$

In this frame of reference work is done on the air as the wave passes through it. This energy addition corresponds to a major component of the drag experienced by an aeroplane flying supersonically. This part of the resistance is termed 'wave drag' since it is associated with the formation of shock waves. A somewhat similar wave drag is experienced by ships generating waves on the surface of the sea. In both cases, one may think of the energy supplied to overcome the wave drag as being radiated away in the wave system.

Example 3.10 The results for a propagating wave can be applied easily to a variety of composite materials. Consider, for example, a wave passing through a liquid containing gas bubbles. It will be assumed that the wave compresses only the bubbles, that the temperature of the gas is constant, that absorption of the gas by the fluid is negligible, and that the pressure within the bubbles equals that in the liquid. The last restriction implies that the surface tension of the bubbles is unimportant. It is not difficult to conceive of situations in which one or more of these assumptions breaks down. Hence the result obtained will apply only in rather special circumstances.

Take the fraction of the mixture occupied by gas bubbles to be n. Then the initial density of the mixture is

$$\varrho_1 = n\varrho_{g_1} + (1-n)\varrho_L$$

Since the temperature and mass of the gas remain constant

$$p_1 V_{g_1} = p_2 V_{g_2} = mRT$$

After compression by the wave, the mass of the mixture which initially occupied unit volume occupies the volume

$$n\left(\frac{p_1}{p_2}\right) + (1-n) = 1 - n\left(1 - \frac{p_1}{p_2}\right) = \frac{v_2}{v_1}$$

Hence

$$\frac{\varrho_2-\varrho_1}{\varrho_2} = 1-\frac{v_2}{v_1}$$

$$= n\left(1-\frac{p_1}{p_2}\right)$$

and equation (3.4.3) gives

$$c^2 = \frac{p_2-p_1}{\varrho_1 n(1-p_1/p_2)} = \frac{p_2}{n\varrho_1}$$

If there are no bubbles, the medium as defined is incompressible. Thus, as $n \to 0$, the wave speed approaches infinity.

3.4.2 Weak Waves

If the intensity of the moving wave is small, the analysis can be simplified. To see how, take the small density and pressure changes to be $d\varrho$ and dp, and the corresponding wave speed to be $a+dc$. Then the continuity and momentum equations (3.4.1, 2) become

$$(a+dc)\, d\varrho = (\varrho+d\varrho)u$$

and

$$dp = \varrho u(a+dc)$$

Now neglect all terms but those of first order in products of the small perturbations u, dp, $d\varrho$, and dc. Then

$$a\, d\varrho = \varrho u \qquad (3.4.5)$$

and

$$\varrho u a = dp \qquad (3.4.6)$$

These are, not surprisingly, two of the equations (3.2.2) for a weak stationary wave, rewritten in the notation for a propagating wave. On eliminating u, one has as before $a^2 = dp/d\varrho$ for the sound speed.

This technique for dealing with small perturbations can be used to study waves moving through a channel of varying cross-section, as shown in Figure 3.9. Now an additional axial force must be taken into account, the axial component of the pressure force normal to the sides of the duct. The sum of the pressure forces acting on the elemental volume is

$$p\, dA+(p+dp)A-p(A+dA) = A\, dp \qquad (3.4.7)$$

and the rate of momentum change is, to the linear approximation, $(\varrho aA)u$. Then the momentum principle gives $A dp = \varrho aAu$, whence

$$dp = \varrho au \qquad (3.4.8)$$

This is identical with equation (3.4.6), but has a wider application since the change of area of the duct was taken into account in the derivation.

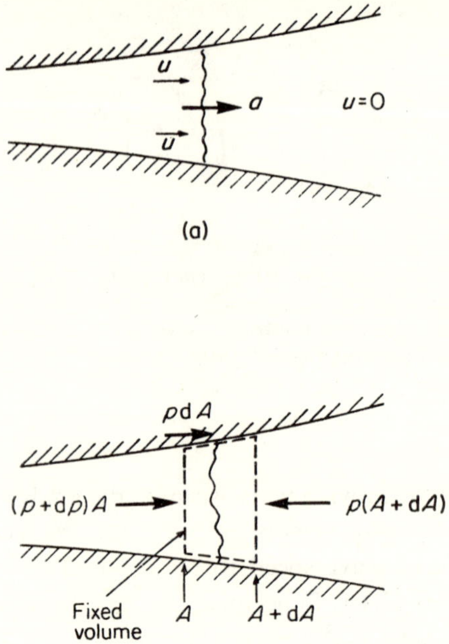

(b)

Figure 3.9 A weak wave in a duct of varying cross-sectional area. (a) Kinematics of the flow; (b) Forces on a control volume around the wave

The continuity condition across the wave can be obtained by imagining the elemental volume to extend only a small distance on either side of the wave. Then the change in area is negligible, and equation (3.4.5) applies again. The sound speed is still given by $a^2 = \mathrm{d}p/\mathrm{d}\varrho$ for this weak wave; it does not depend on the duct cross-section, but only on the equation of state of the medium through which the wave travels. Hence, if the properties of the undisturbed fluid are uniform, the wave will propagate with a constant speed.

There is another continuity condition to be satisfied. The steadily advancing waves that have been studied here, which involve a simple jump in pressure and other properties, are impelled forward by a steady flow of material from the rear. Then for a duct of varying section

$$\varrho u A = \dot{m}, \quad \text{a constant}$$

for this class of waves. Now consider the simple case in which the fluid properties (ϱ and a) are uniform before the wave passes. Since ϱ does not change much across the wave, nor in the slow flow behind it

$$u = \frac{\dot{m}}{\varrho A} = \frac{\text{constant}}{A} \tag{3.4.9}$$

and

$$\mathrm{d}p = \frac{\dot{m}a}{A} = \frac{\text{constant}}{A} \tag{3.4.9}$$

Hence the strength of the weak wave in a uniform material varies inversely as the area of the duct through which it moves.

Although the preceding analysis referred specifically to a wave moving through a duct, it applies rather more generally. Consider a spherical wave, which might be produced by an explosion in the air, or a hemispherical wave, which might be produced by a blast on the ground. Here $A \propto 1/r^2$, where r is the radius of the wave front. Equations (3.4.9) give

$$u,\ \mathrm{d}p \propto \frac{1}{r^2} \tag{3.4.10}$$

These results predict the rate at which the effects of a blast wave die away as it propagates away from its source They do not apply, of course, to wave fronts across which there is a large density change, although even there they give a general idea of the decay.

Example 3.11 It was found in Example 3.5 that waves in liquids are nearly always 'weak' in the sense that they travel at a velocity only a very little higher than the sound speed, with a small change in fluid velocity taking place across them. Despite the simplification this allows, the prediction of the wave speed in a pipe is not entirely straightforward. The pipe itself can expand radially as the pressure rises, and this expansion alters the continuity condition and so affects the wave speed.

Figure 3.10(a) shows a wave moving through a thin-walled pipe, with a resultant change in the cross-sectional area from A to $A+\mathrm{d}A$. The momentum equation offers no difficulty. It is

$$\mathrm{d}p = \varrho uc$$

the change in area playing no explicit role, as was seen above in equation (3.4.8).

The continuity condition is

$$(\varrho+\mathrm{d}\varrho)(A+\mathrm{d}A)u = c[(\varrho+\mathrm{d}\varrho)(A+\mathrm{d}A)-\varrho A]$$

or, linearized

$$\frac{\mathrm{d}\varrho}{\varrho}+\frac{\mathrm{d}A}{A}-\frac{u}{c} = 0$$

In order to relate the area change to the pressure change we must examine the deflexion of the pipe under the increased internal pressure. Figure 3.10(b) shows

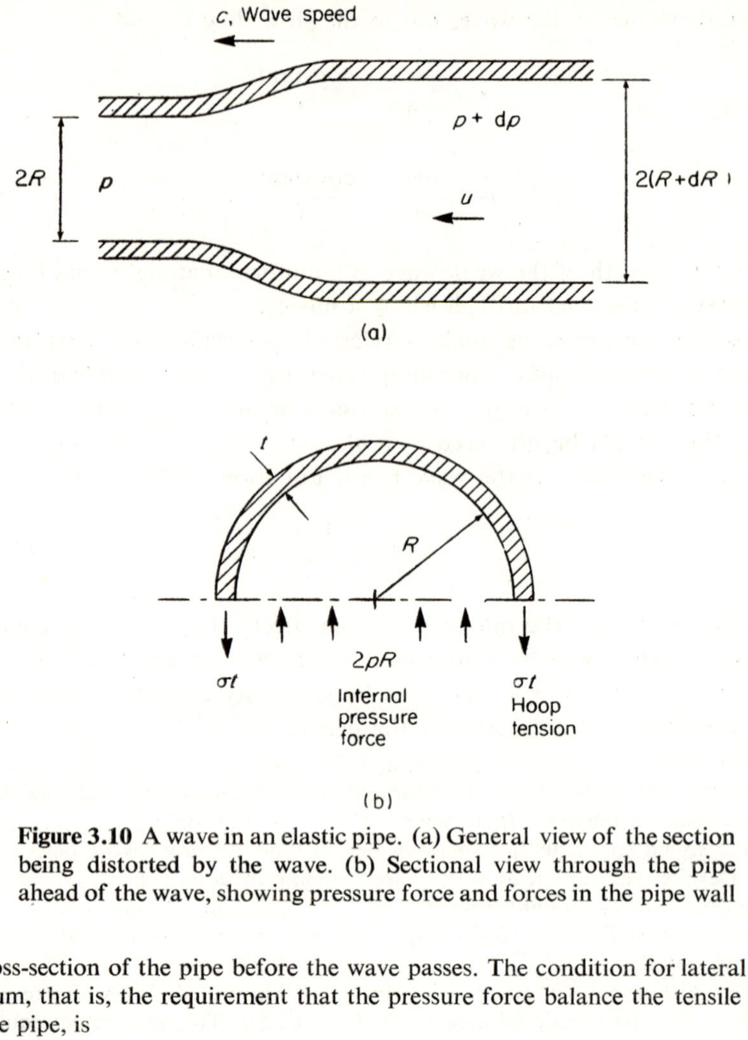

Figure 3.10 A wave in an elastic pipe. (a) General view of the section being distorted by the wave. (b) Sectional view through the pipe ahead of the wave, showing pressure force and forces in the pipe wall

a cross-section of the pipe before the wave passes. The condition for lateral equilibrium, that is, the requirement that the pressure force balance the tensile force in the pipe, is

$$\sigma t = pR$$

Here p is the pressure difference between the inside and outside of the pipe, and σ is the hoop stress in the pipe wall. The same kind of condition applies behind the wave:

$$(\sigma + \mathrm{d}\sigma)t = pR + \mathrm{d}(pR)$$

Then the changes are related by

$$t\,\mathrm{d}\sigma = R\,\mathrm{d}p + p\,\mathrm{d}R$$

or

$$\frac{t}{pR}\,\mathrm{d}\sigma = \frac{\mathrm{d}p}{p} + \frac{\mathrm{d}R}{R} = \frac{\mathrm{d}p}{p} + \frac{1}{2}\frac{\mathrm{d}A}{A}$$

Finally, two material equations must be introduced:

$$d\sigma = Y\frac{dR}{R} = \frac{Y}{2}\frac{dA}{A}$$

relates the changes in stress and strain in the pipe wall, with Y the elastic or Young's modulus of the material of the pipe; for the liquid

$$\frac{d\varrho}{\varrho} = \varkappa\, dp$$

The two equations governing the distortion of the pipe give

$$d\sigma = \frac{Rp}{t}\left(\frac{dp}{p}+\frac{1}{2}\frac{dA}{A}\right) = \frac{Y}{2}\frac{dA}{A}$$

so that

$$\frac{dA}{A} = \frac{2(dp/p)}{Yt/pR-1}$$

If this result, together with $u = dp/\varrho c$ and $d\varrho/\varrho = \varkappa\, dp$, is introduced into the continuity equation, this becomes

$$\varkappa\, dp+\frac{2(dp/p)}{Yt/pR-1}-\frac{dp}{\varrho c^2} = 0$$

or

$$\frac{1}{c^2} = \varrho\varkappa+\frac{2\varrho R/Yt}{1-pR/Yt}$$

For $Y \to \infty$, the case of a very rigid pipe wall, $c^2 \to a^2 = 1/\varrho\varkappa$, as would be expected.

In order to get some feeling for the orders of magnitude involved in the rather complicated equation obtained above, consider a particular case, a wave generated in the penstock of a hydroelectric station. Such a wave propagates up the penstock when the flow through the turbines is reduced; it acts to decelerate the water flowing towards the turbines. The penstock is a tube of diameter 6 ft, and wall thickness 1 in. Its outside is exposed to the atmosphere, and the pressure within is 10 atm (gauge). For water, take $\varkappa = 3\times 10^{-6}$ in^2/lbf; for steel, take $Y = 30\times 10^6$ lbf/in^2.

The dimensionless parameter

$$\frac{pR}{Yt} = \frac{147\times 36}{30\times 10^6\times 1} \simeq 2\times 10^{-4}$$

In this case, and others like it, the approximation

$$\frac{1}{c^2} = \varrho\varkappa+\frac{2\varrho R}{Yt}$$

is entirely adequate, the term accounting for the increase in pipe diameter being negligible. Note that a wave in a pipe containing an exactly incompressible material, such that $\varkappa = 0$, would have speed $c_i = (Yt/2\varrho R)^{1/2}$. In general, the wave speed is given by

$$\frac{1}{c^2} = \frac{1}{a^2} + \frac{1}{c_i^2}$$

and is less than either of the two limiting speeds, a and c_i.

The relative importance of the two contributions can be assessed as follows. Write

$$c^2 = \frac{a^2}{1+2R/t\varkappa Y}$$

For the particular case specified above, the parameter

$$\frac{2R}{t\varkappa Y} = \frac{72}{3\times10^{-6}\times30\times10^{6}} = 0{\cdot}8$$

Hence the contributions of pipe and liquid elasticity are of comparable magnitude. The wave speed is given by

$$c = \frac{a}{\sqrt{1{\cdot}8}} = 0{\cdot}75a$$

3.5 UNSTEADY FLOWS

3.5.1 The Formation of a Shock Wave

The only strong waves that have so far been considered are those propagating without change of form or speed; they can be reduced to steady flows by a change of axes. Now a number of truly non-steady flows will be examined.

The simplest way of imagining a shock wave to be formed is that indicated in Figure 3.11(a). A piston is started impulsively and moves with constant velocity along a pipe which contains a compressible material, initially at rest. If friction at the walls is neglected, the velocity ahead of the piston will be $u = U$, the piston speed, and a wave will propagate into the fluid with speed c. Equations (3.4.3, 4) relate these speeds to the changes in material properties.

But what would happen if, more realistically, the piston is imagined to accelerate gradually to an ultimate velocity U_0? The resulting situation is shown in Figure 3.11(b). Here the process of acceleration has been represented by a series of small impulsive velocity changes, each propagating into fluid which has been traversed by its predecessors. Each of the wavelets travels at the sound speed a relative to the fluid. But since that fluid has been set in motion by earlier wavelets, the absolute velocity of propagation is

$$c = a+u \tag{3.5.1}$$

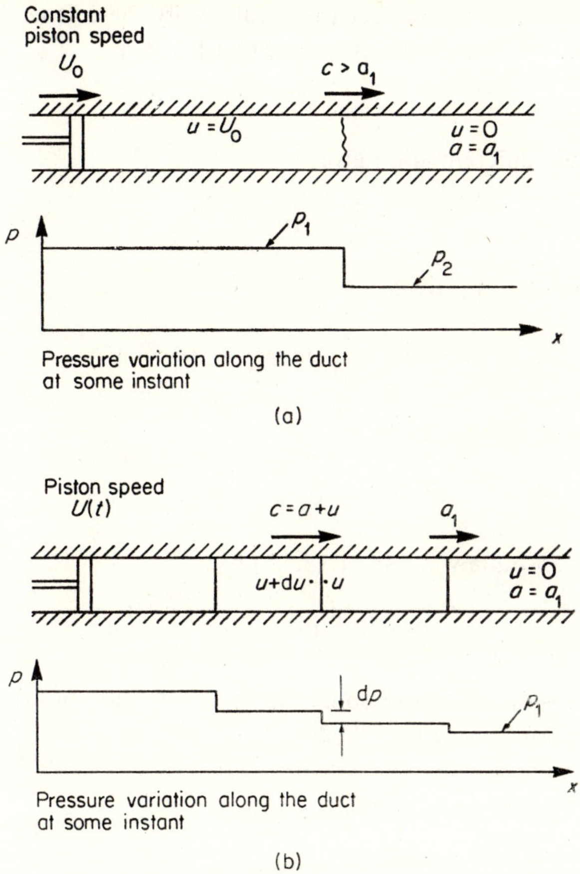

Figure 3.11 Generation of compression waves by a piston. (a) Wave produced when the piston starts impulsively and moves with constant speed along a duct. (b) Waves produced when the piston accelerates gradually

The small change in velocity across any wavelet is given by equation (3.4.5), rewritten as

$$du = a\frac{d\varrho}{\varrho}$$

Since each of the small steps is isentropic, one can in principle specify the function $a = a(\varrho, s_1)$, and calculate

$$u = \int_{\varrho_1}^{\varrho} a(\varrho, s_1)\frac{d\varrho}{\varrho} \tag{3.5.2}$$

The wave speed (3.5.1) follows.

A particular case—the perfect gas—will clarify these general arguments. For this case, the isentropic relations (1.4.13) give

$$T \propto a^2 \propto \varrho^{\gamma-1}$$

and logarithmic differentiation gives

$$2\frac{\mathrm{d}a}{a} = (\gamma-1)\frac{\mathrm{d}\varrho}{\varrho}$$

Then equation (3.5.2) provides

$$u = \int_{a_1}^{a} \frac{2}{\gamma-1}\,\mathrm{d}a = \frac{2}{\gamma-1}(a-a_1)$$

The local sound speed is given by

$$a = a_1 + \frac{\gamma-1}{2}u \tag{3.5.3}$$

and the local absolute wave speed (3.5.1) is

$$c = a_1 + \frac{\gamma-1}{2}u + u = a_1 + \frac{\gamma+1}{2}u \tag{3.5.4}$$

For $u > 0$, $c > a_1$ for two reasons: the fluid is in motion (u) and has had its sound speed increased $((\gamma-1)/2u)$.

The last results can also be written

$$c = \frac{\gamma+1}{\gamma-1}a - \frac{2}{\gamma-1}a_1 \tag{3.5.4}$$

The two forms (3.5.4) show that with c fixed, u and a are also fixed. Then the fluid properties remain the same all along the path of each wavelet. In particular, the wave path in the x, t-plane is a straight line, since

$$\frac{\mathrm{d}x}{\mathrm{d}t} = c = a+u = \text{constant}$$

Figure 3.12(a) shows the wave paths in the x, t-plane for the finite wave of Figure 3.11(b). The path of the piston is also shown. The fluid velocity on each of the wave paths is constant, with a value equal to the piston speed at the moment at which that wavelet was generated. As a result of their higher absolute propagation velocity, later wavelets catch up the first ones. In the x, t-plane this corresponds to a convergence of the straight wave paths. As the series of wavelets coalesces, a shock front forms, its speed increasing as further wavelets catch up. Two stages in the steepening of the wave front are shown in Figure 3.12(b).

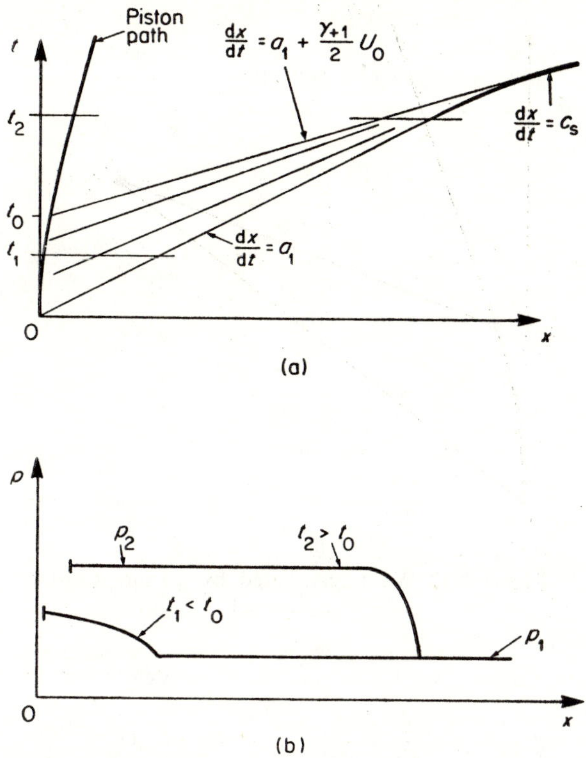

Figure 3.12 The development of a shock wave. (a) Wave paths in the x, t-plane; (b) Pressure variations along the duct at two instants during the generation and formation of the shock

This process explains the extreme thinness of shock waves in a uniform flow. The wavelets at the rear of the wave gain on those ahead until the gradients of velocity and temperature within the wave are so large that heat conduction and viscous stresses balance the steepening. Only then can an 'equilibrium' shock profile be established.

Example 3.12 An underground railway train accelerates out of a station at a constant rate of 1 m/sec^2, until it reaches a speed of 20 m/sec. How long a uniform tunnel must lie ahead, if a fully developed shock wave is to form ahead of the train?

Strictly speaking, the velocity ahead of the train will not equal its speed, since some air can pass between the train and tunnel wall. Also, there are, in actual tunnels, vents and connexions with other tunnels which spill air from the path of the train. These complications will be neglected. The waves to be studied are shown in Figure 3.13; only the leading and trailing waves sent out by the train will be considered.

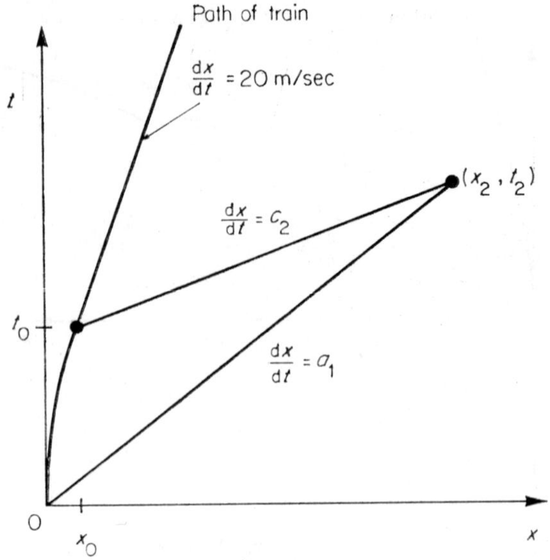

Figure 3.13 Waves generated by a train starting in a tunnel

Consider first the process of acceleration, which determines the point (x_0, t_0). For $U < U_0$

$$U = \dot{U}t \quad \text{and} \quad x = \frac{\dot{U}t^2}{2} = \frac{U^2}{2\dot{U}}$$

Then $t_0 = 20$ sec, and $x_0 = 200$ m.

Now to match the two waves at the point of intersection:

$$x_L = a_1 t_L = x_0 + c_L(t_L - t_0)$$

so that

$$x_L = a_1 t_L = \frac{a_1(c_L t_0 - x_0)}{c_L - a_1}$$

This result can be simplified considerably. Note that

$$t_0 = \frac{U_0}{\dot{U}} \quad \text{and} \quad x_0 = \frac{U_0^2}{2\dot{U}}, \quad \text{while} \quad c_L - a_1 = \frac{\gamma+1}{2} U_0$$

for a perfect gas. Then a little algebra gives

$$x_L = \frac{2}{\gamma+1} \frac{a_1}{\dot{U}} \left(a_1 + \frac{\gamma U_0}{2}\right)$$

$$\simeq \frac{2}{\gamma+1} \frac{a_1^2}{\dot{U}} \quad \text{in this case, since } U_0 \ll a_1.$$

A fairly high value must be chosen for the sound speed, since the air in much used train tunnels is warmer than that of the atmosphere. Then

$$x_L \simeq \frac{2}{2 \cdot 4} \frac{350^2}{1} = 10^5 \text{ m} = 100 \text{ km}$$

It appears that there will not be an opportunity for fully developed shock waves to develop in an actual tunnel. The waves generated by trains will be reflected at the open ends of the tunnel, and in this way can obtain a longer 'run' in which to steepen. Nevertheless, dissipation, through wall friction, leakage through vents, and partial reflexion where the tunnel area changes, can be expected to attenuate the wave before it heads up into a shock.

3.5.2 Thermodynamic and Mechanical Pressures

The process shown in Figure 3.11(a), impulsive shock generation, provides an insight into the meaning of pressure. Consider the situation sketched in Figure 3.14; a piston is driven impulsively into a rigid cylinder with constant velocity, so that the volume of the fluid within is reduced by the small amount $-\mathrm{d}V$. Afterwards the shock will move back and forth within the cylinder, reflecting at the ends, and the substance will ultimately come to rest with properties little changed from their initial values.

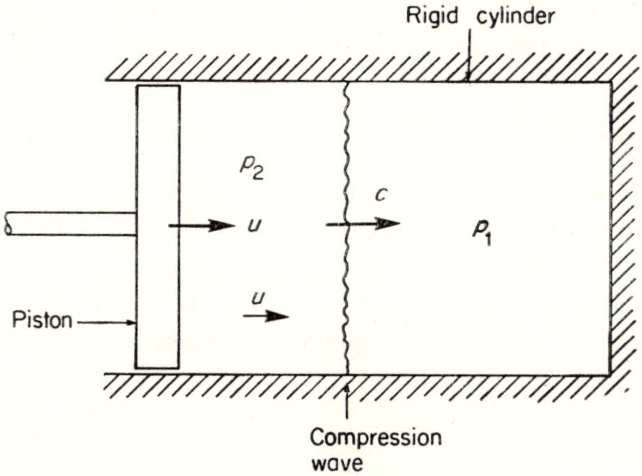

Figure 3.14 The transient wave produced in a cylinder during rapid compression

The work done on the fluid during the displacement is $-p_2\,\mathrm{d}V$. The reversible component is $-p_1\,\mathrm{d}V$, that which is done if the substance expands again without losses. Then

$$(p_2 - p_1)(-\mathrm{d}V)$$

is the irreversible element of the work, and

$$p_2 - p_1$$

is the energy dissipated per volume displaced in the piston's motion.

In earlier discussions the pressure has been tacitly used in two senses. It has been taken to be a thermodynamic property, defined in terms of other properties in equation (1.5.6). The pressure has also been related to an externally applied force. In the case considered above, this dual interpretation breaks down; during the process of compression more than one pressure can be distinguished. The thermodynamic pressure is p_1, that existing before compression and (very nearly) after compression as well. The applied mechanical pressure is p_2. Using equation (3.4.2) the difference between the two pressures can be written

$$p_2 - p_1 = \varrho_1 uc \geqslant \varrho_1 a_1 u \tag{3.5.5}$$

This shows that the difference between the two pressures rises steadily as u is increased, but can be made arbitrarily small by moving the piston slowly enough. Hence the mechanical and thermodynamic pressures will be identical if the process is performed slowly, so that the pressure remains essentially uniform throughout the cylinder. In this case the process of compression is reversible, and the external work associated with it is $-p_1\,\mathrm{d}V$.

3.5.3 Expansion Waves

Figure 3.11 shows waves generated when a piston is driven into a fluid at rest. It is natural to ask what would happen if the piston were withdrawn rapidly. Figure 3.15 shows a series of wavelets generated in this way.

The analytical results obtained earlier still apply, although the signs of the quantities involved in the equations must be reconsidered. Now the fluid velocity u is negative. Hence the absolute wave speed for a perfect gas, given by

$$c = a + u = a_1 + \frac{\gamma+1}{2}u \tag{3.5.4}$$

is less than a_1. The wavelets are convected backwards and propagate through a gas whose sound speed has been reduced. The waves generated later fall further and further behind those generated first, and the wave paths in the x, t-plane fan out instead of converging.

This behaviour explains why there is no such thing as a steadily propagating negative shock. Finite expansion waves do exist, but they change their form continually as they propagate. Hence the preceding analysis, which assumed steady flow and unchanging wave form, gave no indication that such waves existed.

Since the elemental wavelets of an expansion wave do not coalesce to form a concentrated shock, their propagation can in principle be isentropic.

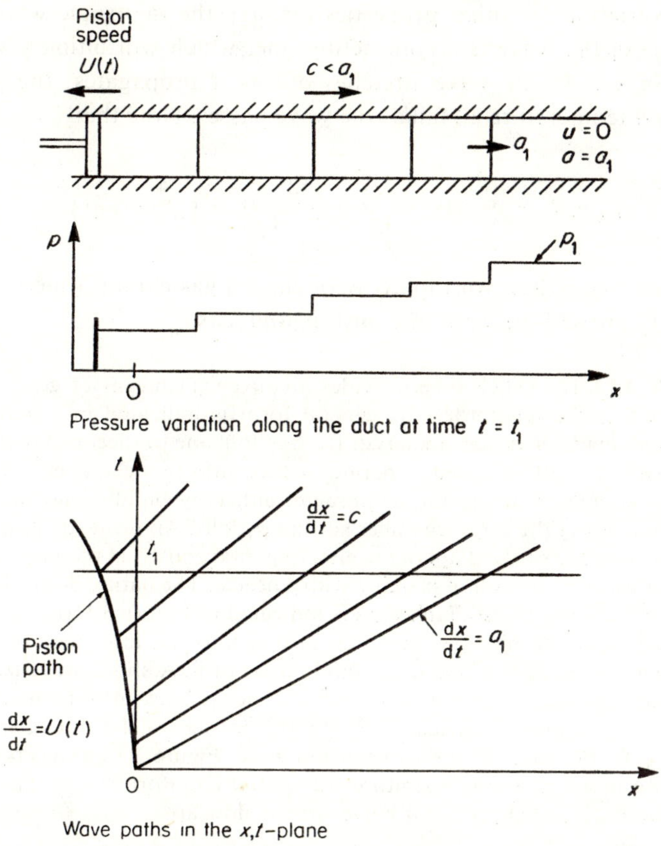

Figure 3.15 Generation of an expansion wave by a piston

In practice other sources of dissipation will act to modify them, for instance, friction at the walls of the duct in which they move. However, the assumption of isentropic behaviour is certainly more appropriate than it would be for a shock wave. The connecting link between conditions in the still gas ahead of the wave and the state of the moving gas is

$$a = a_1 + \frac{\gamma - 1}{2} u \tag{3.5.3}$$

which incorporates the constancy of entropy. Since $a^2 = \gamma RT$ for a perfect gas

$$\left(\frac{T}{T_1}\right)^{\frac{1}{2}} = \left(\frac{\varrho}{\varrho_1}\right)^{\frac{\gamma-1}{2}} = \left(\frac{p}{p_1}\right)^{\frac{\gamma-1}{2\gamma}} = 1+\frac{\gamma-1}{2}\frac{u}{a_1} \qquad (3.5.6)$$

give the variations in other properties through the isentropic wave.

For $u > 0$, the wave is a contracting one, which will ultimately form a shock. For $u < 0$, the wave stretches out as it propagates; the pressure, density and temperature all fall as the wave passes. Note that

$$p, \varrho, T \to 0 \quad \text{as} \quad -u \to \frac{2}{\gamma-1}a_1 = \left(\frac{4}{\gamma-1}c_pT_1\right)^{\frac{1}{2}}$$

If the piston is withdrawn rapidly enough, the gas cannot follow it, and a vacuum is created between the piston and gas.

Example 3.13 The shock tube provides a convenient means of generating unsteady flows in the laboratory. Its general form is indicated in Figure 3.16(a). A diaphragm made of various materials from cellophane to sheet metal, depending on the shock strength required, separates a tube into two chambers. A pressure differential is created across the diaphragm, either by building up the pressure p_4 or by evacuating the other chamber so that p_1 falls. An unsteady flow is generated by the bursting of the diaphragm, either spontaneously as the pressure difference increases, or as a result of piercing with a needle. The nature of the flow which follows the breaking of the diaphragm is indicated in Figure 3.16(b).

The fluid from the high-pressure side pushes into the low-pressure chamber, much like a piston, driving the other fluid ahead of it. A shock wave travels into the low-pressure side, and an expansion wave moves back into the high-pressure vessel. The temperature is raised (to T_2) by the passage of the shock, and is lowered (to T_3) by the passage of the expansion wave. Figure 3.16(b) has been drawn for the case in which the temperature was initially uniform, $T_1 = T_4$.

The conditions linking the two halves of the flow are

$$p_2 = p_3 \quad \text{and} \quad U_2 = U_3$$

where U_2 and U_3 are the absolute velocities of the fluid traversed by shock and expansion waves. Using these conditions it is possible to obtain explicit formulae describing certain aspects of shock-tube performance. However, for the purposes of this example where a single case is to be studied, a direct attack will be made on the problem.

It is required to make an experimental study of shocks in steam, using helium (molecular weight, $MW_4 = 4{\cdot}00$; $\gamma_4 = 1{\cdot}66$) as the driver gas. The relationship between shock Mach number $M_1 = c/a_1$ and driving pressure ratio p_4/p_1 is needed, in order that the experimental programme can be planned. Both fluids may be assumed to be perfect gases for this purpose.

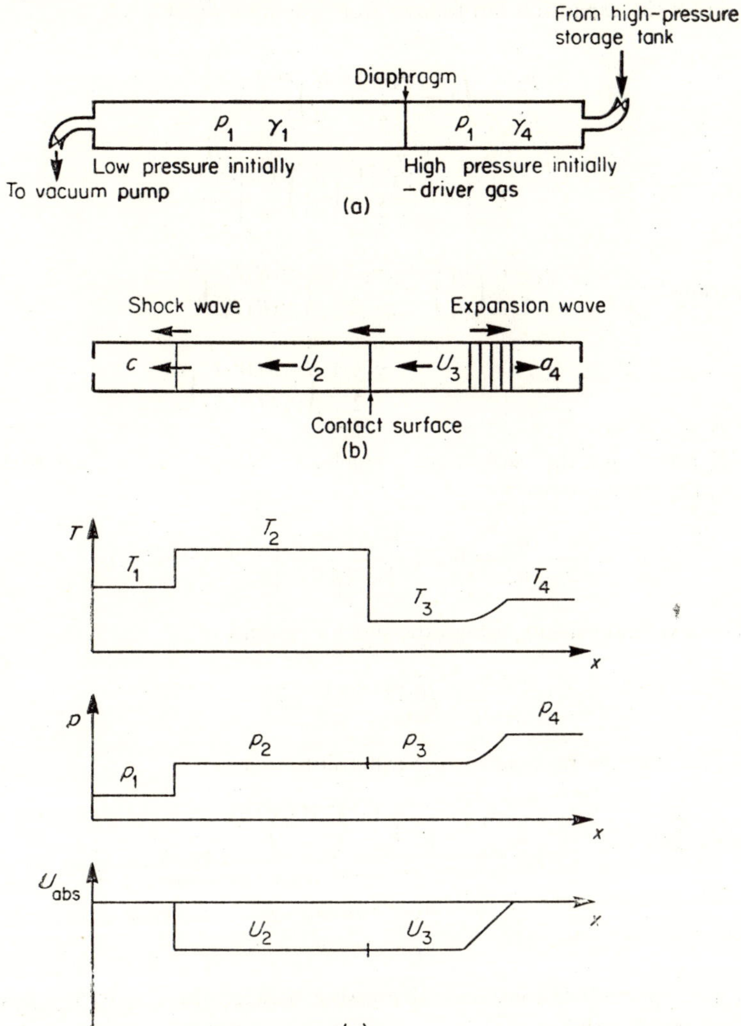

Figure 3.16 The operation of a shock tube. (a) Initial configuration; (b) Wave propagation during operation; (c) Variation of properties during operation

The calculation proceeds as follows. For a given M_1, shock tables for $\gamma_1 = 1{\cdot}3$ give p_2/p_1. Also

$$\frac{u}{a_1} = \frac{u_1 - u_2}{a_1} = M_1 - \frac{a_2}{a_1} M_2 = -\frac{U_2}{a_1} = -\frac{U_3}{a_1}$$

gives the non-dimensional flow velocity between the two waves. The pressure ratio across the expansion fan follows from equation (3.5.6):

$$\frac{p_3}{p_4} = \left(1 - \frac{\gamma_4 - 1}{2}\frac{u}{a_4}\right)^{\frac{2\gamma_4}{\gamma_4 - 1}}$$

$$= \left[1 + \frac{\gamma_4 - 1}{2}\left(\frac{a_1}{a_4}\right)\frac{U_3}{a_1}\right]^{\frac{2\gamma_4}{\gamma_4 - 1}}$$

The parameter
$$\frac{\gamma_4 - 1}{2}\left(\frac{a_1}{a_4}\right) = \frac{\gamma_4 - 1}{2}\left(\frac{\gamma_1 R_1 T_1}{\gamma_4 R_4 T_4}\right)^{\frac{1}{2}}$$

$$= \frac{\gamma_4 - 1}{2}\left(\frac{\gamma_1 MW_4 T_1}{\gamma_4 MW_1 T_4}\right)^{\frac{1}{2}}$$

provides a limit for the performance which can be achieved using a particular pair of gases. In this case

$$\frac{\gamma_4 - 1}{2}\frac{a_1}{a_4} = 0{\cdot}33\left[\frac{1{\cdot}66 \times 18{\cdot}0}{1{\cdot}3 \times 4{\cdot}00}\right]^{-\frac{1}{2}} = 0{\cdot}1375$$

and the maximum velocity behind the shock is given by

$$\frac{u}{a_1} = (0{\cdot}1375)^{-1} = 7{\cdot}27$$

Finally, the pressure ratio across the expansion wave is

$$\frac{p_3}{p_4} = \left(1 + \frac{0{\cdot}1375 U_3}{a_1}\right)^{5{\cdot}03}$$

and

$$\frac{p_4}{p_1} = \frac{p_4}{p_3}\frac{p_2}{p_1}$$

which incorporates the constancy of pressure between the two waves, gives the driving pressure ratio.

It is usually most convenient to arrange calculations of this kind in a table.

Table 3.1. Performance of a shock tube

M_1	p_2/p_1	a_2/a_1	M_2	u/a_1	$(p_3/p_4)^{1/5{\cdot}03}$	p_4/p_3	p_4/p_1
2	4·39	1·235	0·563	1·305	0·820	2·72	11·9
4	17·96	1·822	0·406	3·26	0·552	19·8	35·6
8	72·2	3·22	0·357	6·85	0·058	$1{\cdot}68 \times 10^6$	$1{\cdot}22 \times 10^8$

It is clear that an enormous pressure ratio is required if the limiting Mach number is to be approached. There is no way of exceeding this Mach number for the particular combination of gases and initial temperatures considered here.

3.5.4 The Equations of Unsteady Flow

Up to this point the unsteady motion of a compressible fluid has been studied by superposing the effects of a series of small wavelets, each locally equivalent to a steadily propagating wave. Now an alternative approach through explicit equations governing the unsteady flow will be developed.

The general continuity and momentum equations set down in the preceding chapter provide the starting point. These are

$$\varrho_2 u_2 A_2 - \varrho_1 u_1 A_1 = \dot{m}_s \tag{2.1.6}$$

where $-\dot{m}_s$ can be interpreted as the rate at which mass is retained between sections 1 and 2, and

$$\int_V \frac{\partial \varrho u}{\partial t}\, \mathrm{d}V + \Sigma \dot{m} u = \Sigma F \tag{2.2.2}$$

where the two momentum-change terms give the effects of velocity changes within the system and of momentum flux through its boundaries.

Equations giving progressive changes in $u(x,t)$ and $\varrho(x,t)$ are sought; they must be expected to contain derivatives of u and ϱ with respect to x and t. Such equations can be obtained by applying the general integral relations given above to a short section of duct of length δx. This system is shown in Figure 3.17; the duct shape is taken to be invariant in time, and the sections across the duct are also fixed. For brevity, the changes in properties across the system have been denoted by δ; since they occur

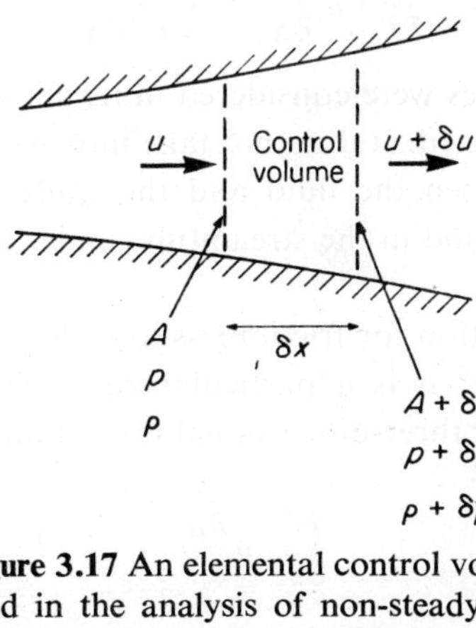

Figure 3.17 An elemental control volume used in the analysis of non-steady flow in a duct of varying cross-sectional area

over the small length δx, they may be interpreted in terms of derivatives with respect to x using

$$\delta = \frac{\partial}{\partial x}\,\delta x \tag{3.5.7}$$

The rate at which material is retained within the system is

$$-\dot{m}_s = \frac{\partial \varrho}{\partial t} A\,\delta x$$

and the continuity equation (2.3.3) gives

$$\delta(\varrho u A) = -\frac{\partial \varrho}{\partial t} A\,\delta x$$

whence, with the use of the relation (3.5.7)

$$\frac{\partial \varrho}{\partial t} + \frac{1}{A}\frac{\partial}{\partial x}(\varrho u A) = 0 \tag{3.5.8}$$

This is the *differential continuity equation* for non-steady, one-dimensional flow.

The rate of momentum change within the elemental system is $(\varrho A \delta x)$ $\partial u/\partial t$. The net flux outwards is $(\varrho u A)\,\delta u$ and the net axial pressure force on all the faces of the element is $-A\,\delta p$. The last result follows from equation (3.4.7), the sign being changed in accordance with the different convention adopted here for the pressure rise. Hence the appropriate form of equation (2.2.2) is

$$(\varrho A\,\delta x)\frac{\partial u}{\partial t} + (\varrho u A)\,\delta u = -A\,\delta p$$

When (3.5.7) is introduced, this becomes

$$\frac{\partial u}{\partial t} + u\frac{\partial u}{\partial x} = -\frac{1}{\varrho}\frac{\partial p}{\partial x} \tag{3.5.9}$$

Since only pressure forces were considered in its derivation, the application of this equation is restricted. It does not take into account the weight of the fluid, nor friction between the fluid and the walls of the duct or, more generally, between the fluid in the streamtube under consideration and that in adjacent streamtubes.

The momentum equation for frictionless, one-dimensional flow (3.5.9) is termed *Euler's equation;* it is a particular case of the Euler's equations which can be written for three-dimensional flow of an ideal, that is, frictionless fluid. The derivative

$$\frac{\partial}{\partial t} + u\frac{\partial u}{\partial x} \tag{3.5.10}$$

is termed the *Eulerian* or *convective derivative.* It was obtained above by considering the flux through a fixed volume and the rate of change within that volume. It can also be obtained by considering a system of fixed contents whose boundaries move with the fluid. The total rate of change in time then consists of two elements, that specifying the variation of the field in time ($\partial/\partial t$), and that giving the rate of change as a result of convection at velocity u through the spatially varying field ($u\partial/\partial x$). This way of looking at the process gives rise to the label 'convective'. The convective acceleration was encountered earlier in equation (2.2.6).

Although it was obtained above by considering velocity variations, the Eulerian derivative gives the change in any property of a moving substance. For example, the net rate of change of temperature of a moving element is

$$\frac{\partial T}{\partial t}+u\frac{\partial T}{\partial x}$$

The Eulerian derivative can easily be generalized to describe the effects of convection by two and three-dimensional flows. The combination

$$\frac{\partial}{\partial t}+u\frac{\partial}{\partial x}+v\frac{\partial}{\partial y}+w\frac{\partial}{\partial z} \tag{3.5.10}$$

gives the total rate of change for an element moving in a three-dimensional flow field; the velocity components u, v and w are parallel to the x, y and z-axes.

For a specified area distribution $A(x)$, the two equations (3.5.8, 9) relate the three functions u, ϱ and p. One way of rendering the problem determinate is to specify $\mathrm{d}s = 0$. This is most conveniently done by noting that $\mathrm{d}p = (\partial p/\partial \varrho)_s \, \mathrm{d}\varrho$, so that

$$\frac{\partial p}{\partial x} = \left(\frac{\partial p}{\partial \varrho}\right)_s \frac{\partial \varrho}{\partial x} = a^2 \frac{\partial \varrho}{\partial x} \tag{3.5.11}$$

With the pressure eliminated in this way, the momentum equation becomes

$$\frac{\partial u}{\partial t}+u\frac{\partial u}{\partial x} = -\frac{a^2}{\varrho}\frac{\partial \varrho}{\partial x} \tag{3.5.12}$$

The results obtained to this point apply to any compressible substance. To go further one must introduce one of two kinds of restriction:

(1) small disturbances, that is, discussion limited to sound waves, or
(2) a particular substance with a simple equation of state.

These two courses will be taken in turn. To simplify the discussion, only the case of motion in a constant-area streamtube will be considered. Then A = constant for the subsequent discussion.

For small disturbances in a uniform still fluid, the equations (3.5.8, 12) can be linearized:

$$\frac{\partial \varrho}{\partial t}+\varrho_1\frac{\partial u}{\partial x}=0 \quad \text{and} \quad \frac{\partial u}{\partial t}=-\frac{a_1^2}{\varrho_1}\frac{\partial \varrho}{\partial x}$$

Then

$$-\frac{\partial^2 \varrho}{\partial x\,\partial t}=\varrho_1\frac{\partial^2 u}{\partial x^2}=\frac{\varrho_1}{a_1^2}\frac{\partial^2 u}{\partial t^2}$$

and

$$\frac{\partial^2 u}{\partial t^2}=a_1^2\frac{\partial^2 u}{\partial x^2} \tag{3.5.13}$$

A similar manipulation gives

$$\frac{\partial^2 \varrho}{\partial t^2}=a_1^2\frac{\partial^2 \varrho}{\partial x^2} \tag{3.5.13}$$

Equations of the same form can be derived for any of the properties of the fluid. This equation—the simple *wave equation*—is satisfied by any solution of the form

$$f_1(x+a_1t)+f_2(x-a_1t)$$

as may be seen by substitution in the equation. The two terms represent waves propagating without change of form to the left and right. This is the kind of non-dispersive propagation which, as was argued earlier, characterizes sound waves in a uniform medium.

Consider next the particular case of a perfect gas, for which the results (3.5.3, 4) have already been obtained. Since these results involve the sound speed we must, to obtain comparable results from the present approach, express equations (3.5.8, 12) in terms of the sound speed. For isentropic processes

$$\frac{\mathrm{d}a}{a}=\frac{\gamma-1}{2}\frac{\mathrm{d}\varrho}{\varrho}$$

and the equations for constant area become

$$\frac{\partial a}{\partial t}+\frac{\gamma-1}{2}a\frac{\partial u}{\partial x}+u\frac{\partial a}{\partial x}=0$$

and

$$\frac{\gamma-1}{2}\frac{\partial u}{\partial t}+\frac{\gamma-1}{2}u\frac{\partial u}{\partial x}=-a\frac{\partial a}{\partial x}$$

The nature of the solutions can be inferred from equation (3.5.3), which shows that

$$a-\frac{\gamma-1}{2}u=a_1, \quad \text{a constant}$$

for a wave moving with speed

$$\frac{dx}{dt} = c = a + u$$

Arranging the equations as

$$\frac{\partial}{\partial t}(a) + a\frac{\partial}{\partial x}\left(\frac{\gamma-1}{2}u\right) + u\frac{\partial}{\partial x}(a) = 0$$

and

$$\frac{\partial}{\partial t}\left(\frac{\gamma-1}{2}u\right) + a\frac{\partial}{\partial x}(a) + u\frac{\partial}{\partial x}\left(\frac{\gamma-1}{2}u\right) = 0$$

we can see groups of this kind emerging. Addition of corresponding members of these equations gives

$$\left[\frac{\partial}{\partial t} - (a-u)\frac{\partial}{\partial x}\right]\left(a - \frac{\gamma-1}{2}u\right) = 0 \qquad (3.5.14)$$

while subtraction gives

$$\left[\frac{\partial}{\partial t} + (a+u)\frac{\partial}{\partial x}\right]\left(a + \frac{\gamma-1}{2}u\right) = 0 \qquad (3.5.14)$$

This pair of equations is equivalent to the pair from which they were derived. They state that

$$\begin{aligned} a + \frac{\gamma-1}{2}u &= \text{constant on each curve given by } \frac{dx}{dt} = a + u \\ a - \frac{\gamma-1}{2}u &= \text{constant on each curve given by } \frac{dx}{dt} = -a + u \end{aligned} \qquad (3.5.15)$$

The curves $dx/dt = \pm a + u$ are termed *characteristic curves;* from the earlier discussion it is obvious that they correspond to the paths of wavelets propagating to the right and left in a moving fluid. For expansion waves generated by the withdrawal of a piston, shown in Figure 3.15, the two sets of characteristics are shown in Figure 3.18. Since all the negative characteristics ($dx/dt = -a + u$) run through the quiescent region, where $a = a_1$ and $u = 0$, it must be the case that

$$a - \frac{\gamma-1}{2}u = a_1, \quad \text{a constant through the entire flow} \qquad (3.5.16)$$

This result was obtained earlier by another line of argument. The name *simple wave* is applied to a flow in which one of the sets of constants of the equations (3.5.15) is the same for the whole family of characteristic curves.

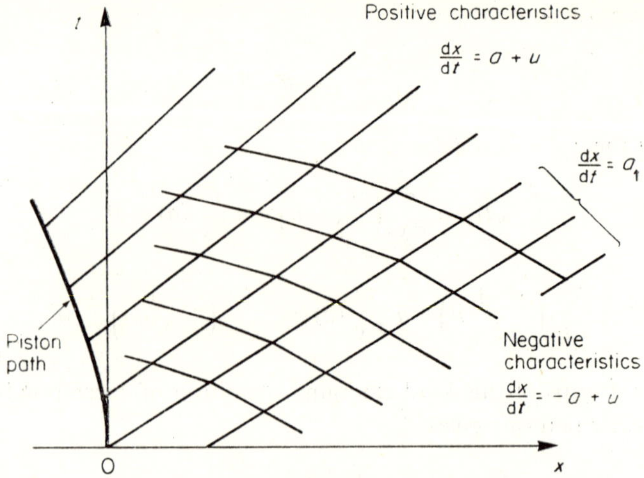

Figure 3.18 Characteristic curves for an expansion wave generated by a piston. This is the kind of flow depicted in Figure 3.15

For the set of positive characteristics ($\mathrm{d}x/\mathrm{d}t = a+u$)

$$a+\frac{\gamma-1}{2}u = \text{a different constant for each} \qquad (3.5.17)$$

Addition of corresponding members of equations (3.5.16, 17) gives

$$a = \text{constant for each positive characteristic}$$

and subtraction gives

$$u = \text{constant} = U$$

where U is the piston speed at the moment of generation of the wavelet. Then the constants for the several characteristics are given by

$$a+\frac{\gamma-1}{2}u = a_1+\frac{\gamma-1}{2}u+\frac{\gamma-1}{2}u$$

$$= a_1+(\gamma-1)U$$

The slopes of the positive characteristics are given by

$$\frac{\mathrm{d}x}{\mathrm{d}t} = a+u = a_1+\frac{\gamma-1}{2}U$$

so that each is a straight line.

This approach through the differential equations governing unsteady flow has yielded no results which were not obtained earlier by simpler techniques. However, suppose that the still gas into which the waves move did not have a constant temperature and sound speed. Then the structure evolved above would break down, and a complete rethinking of the problem would be required. The equations (3.5.8, 12) provide a basis for the study of this more complicated situation. In particular, the results (3.5.15) make possible a step-by-step numerical solution which proceeds by small increments along the net of characteristic lines. However, a discussion of these techniques is beyond the scope of this introductory survey.

PROBLEMS FOR CHAPTER 3

Group A

3.1 For air, show that the speed of sound is given approximately by 33 $T^{1/2}$ miles/hr, with T in °R.

3.2 (a) Show that the formula for the speed of sound in a perfect gas is $a = (p/\varrho)^{1/2}$ on the assumption that the small compressions and rarefactions occur isothermally.

(b) Newton thought that the result obtained above was correct, and attempted to predict the time elapsed between seeing the flash of a gun on the other side of a lake and hearing the bang. For a lake two miles across, by how much would his prediction of the propagation time be in error?

3.3 At what Mach number is the aeroplane considered in Example 1.19 flying? Note that T_2 was used there to denote the total temperature.

3.4 Show that sound speed, wavelength and frequency are related by $a = \lambda f$.

(a) A loudspeaker emits sound of frequency 256 c/sec (or Hz) in both directions along a duct filled with still air. Estimate the wavelength.

(b) If the same loudspeaker, operating at the same frequency, emits sound when the air is moving through the duct at 100 m/sec, what are the frequency and wavelength perceived at stations upstream and downstream of the loudspeaker?

3.5 An underwater explosion generates a shock wave which travels outwards at 1700 m/sec. The speed of sound is 1500 m/sec in the fluid ahead of the wave. Starting from the ideal-liquid result $M_1^2 - M_2^2 = \log_e (M_1/M_2)^2$, use results of Examples 3.1 and 3.5 to find

(a) the Mach number M_2 behind the wave,
(b) the density of the fluid behind it,
(c) the pressure rise across it, and
(d) the fluid velocity behind it.

3.6 For the shock wave specified in Example 3.8, find

(a) the temperature just behind the wave, and
(b) the temperature at the nose of the plane, where the air is brought to rest.

3.7 (a) For the shock wave specified in Example 3.8, find the following pressures

(1) static pressures ahead of and just behind the shock,

(2) total pressures for the frame of reference moving with the missile and shock, and

(3) total pressures for the frame of reference in which the air ahead of the shock is still.

(b) Find the entropy change across the shock.

(c) Why do the total pressures of (3) above not give the entropy change if introduced into equation (2.4.15)?

3.8 The wing of an aeroplane is made of a steel whose strength depends upon temperature in the following way:

Temperature (°F)	500	1000	1200	1400	1600
Ultimate strength (lbf/in²)	74,400	70,600	64,500	50,200	30,000

(a) Make an estimate of the speed at which the aeroplane should be able to fly in the stratosphere (i.e., above 11,500 m) and near the ground. Use standard atmospheric data from Appendix VI.

(b) Why is it not necessary to know whether the air passes through a shock before reaching the wing?

(c) How would an aluminium alloy do under these conditions?

3.9 (a) Show that the speed of shock waves in a material for which $p \propto \varrho$ is given by $(p_2/\varrho_1)^{1/2}$.

(b) Use this result to give a simple interpretation to the formula for the wave speed in a bubbly liquid found in Example 3.10.

3.10 In laboratory experiments illustrating waves in liquids contained within elastic tubes it is common practice to examine waves in rubber hoses.

(a) Use the result of Example 3.11 to find the wave speed in a rubber hose of the kind used for watering lawns. Take the Young's modulus to be $Y = 500$ lbf/in².

(b) Why is rubber more suitable than steel for a laboratory study? Does the experiment really reproduce the phenomenon analysed in Example 3.11?

3.11 The spherical wave described in Problem 3.5 is to be considered when it has travelled outwards three times farther from the point at which the explosion took place. Estimate

(a) the pressure rise across the wave,

(b) the velocity behind it, and

(c) the density behind it.

The results of Section 3.4.2 will prove useful.

Group B

3.12 (a) Show that, so long as the accuracy of Callendar's equation is acceptable, the calculation of the state behind a shock wave in steam is equivalent to the solution of the quadratic equation

$$\frac{I}{\dot{m}}u - \frac{u^2}{144g} = 1{\cdot}247\left(h_0 - 835 - \frac{u^2}{2gJ}\right)$$

with all quantities in English steam-table units, and $g = 32{\cdot}2$, $J = 778$.

(b) What is the corresponding result in SI units?

(c) Solve for the state resulting from the shock of Example 3.2.

3.13 By manipulating the results of Example 3.4, show that the sound speed in a van der Waals gas can be given as

$$\left(\frac{\gamma p}{\varrho}\right)^{\frac{1}{2}}\left[1-\frac{\varrho}{2}\left(\frac{a}{RT}-b\right)\right]$$

3.14 Consider again the situation discussed in Problem 3.4. If the loudspeaker, while oscillating at 256 c/sec, were in some way to move along the duct with velocity 100 m/sec, what frequency and wavelength would be perceived at fixed stations upstream and downstream if

(a) the air is still,

(b) the air is moving at 100 m/sec with the speaker, and

(c) the air is moving at 100 m/sec in the direction opposite to the speaker?

3.15 Without restricting consideration to constant-area flow, apply the momentum principle to the infinitesimal throttling process considered in Section 3.2.3, and combine the result with $T\,\mathrm{d}s = -\mathrm{d}p/\varrho - u\,\mathrm{d}u$ to show that

$$\frac{\mathrm{d}F_a}{pA} = \frac{\mathrm{d}A}{A} - \frac{\varrho T}{p}\,\mathrm{d}s$$

(a) In an isentropic flow, what is the direction of the force applied by the fluid to an expanding duct?

(b) What is the direction of the force applied to a blockage in a constant-area duct for supersonic and subsonic flows?

3.16 (a) Find the limiting forms of the ratios of temperature, pressure, sound speed, density, Mach number and total pressure as the strength of a shock in a perfect gas becomes very large. For example,

$$\frac{p_2}{p_1} \to \frac{2\gamma}{\gamma+1}M_1^2$$

(b) Use theseresults to estimate changes in properties across a shock in front of a satellite moving through the atmosphere at $M_1 = 6$.

(c) To what sources of error are the results of (b) subject?

3.17 A quick-acting valve suddenly closes across a pipe carrying air at 20 °C, 450 kN/m², and 150 m/sec.

(a) Estimate the pressure change using the weak-wave result

$$\Delta p = \varrho a\,\Delta u$$

(b) Use shock tables to find the properties behind the wave which travels back from the valve. A trial-and-error solution will be required.

3.18 (a) Reconsider the last problem, now with the valve reducing the flow to 50 m/sec rather than bringing the fluid to rest on its upstream side.

(b) What additional non-dimensional parameter could be included in the shock tables in order that this kind of problem might be solved directly?

3.19 Find the speed of wave propagation in a gas containing small drops of an incompressible liquid. Assume the gas to be perfect and to be isentropically compressed. An approach generally similar to that of Example 3.10 may be used.

3.20 Pulsating sound waves pass through bubbly water with no transfer of heat or mass between bubbles and water. On the basis of this assumption, show that the speed of sound is given by $a^2 = \gamma p/\varrho n$, where n is the fraction of the volume occupied by gas bubbles.

3.21 (a) Use equations (3.2.6) and (3.5.2) to show that the fluid velocity induced in an ideal liquid by an isentropic wave is related to the local sound speed by $u = 2(a_1 - a)$.

(b) In what fundamental way does this result differ from that for a perfect gas?

(c) Do waves in this material steepen as they propagate?

3.22 The results of Example 3.12 may be applied to a wider range of problems than that studied in detail there. In particular, they can be related to the waves set up by shells and bullets in gun barrels.

A shell leaves the barrel of a gun at 200 m/sec. The acceleration may be taken to be constant over a distance of 2 m. Does a shock form within the barrel? Discuss the validity of the analysis of Example 3.12 for this situation.

3.23 (a) For a perfect gas, show that the difference between mechanical and thermodynamic pressures given in equation (3.5.5) may be written

$$\frac{p_2 - p_1}{p_1} = \gamma \frac{c}{a_1} \frac{u}{a_1}$$

where

$$\frac{u}{a_1} = M_1\left(1 - \frac{\varrho_1}{\varrho_2}\right)$$

(b) Use shock tables to prepare a plot of $(p_2 - p_1)/p_1$ against u/a_1 for air.

3.24 (a) Examine the performance of a shock tube generating shocks in steam using air as the driver gas. Compare with the performance of the helium driver considered in Example 3.13.

(b) What is the advantage of helium? Suggest other gases which might be suitable for high-performance shock tubes.

(c) The performance of a shock tube can be improved by juggling with the initial temperatures of the two gases. How?

Group C

3.25 The dissipation in an isenergic flow of an ideal liquid is given by

$$\int_1^2 T\,\mathrm{d}s = -\int_1^2 \frac{\mathrm{d}p}{\varrho} - \Delta\left(\frac{u^2}{2}\right) = \Delta\left(\frac{v}{\varkappa} - \frac{u^2}{2}\right)$$

For the common case of constant-area flow, which includes the normal-shock transition, the changes Δu and Δv are related, and the dissipation can be expressed in terms of one or the other. The following calculations indicate how this can be done.

(a) For constant-area flow, show that the dissipation per unit mass is

$$\frac{\Delta v}{\varkappa}\left[1 - \frac{\varkappa}{2}\left(\frac{\dot{m}}{A}\right)^2 (2v_1 + \Delta v)\right]$$

(b) Use the results of Example 3.1 to write this as

$$\frac{\Delta v}{\varkappa}\left[1-\frac{v_1}{\Delta v}\left(1+\frac{1}{2}\frac{\Delta v}{v_1}\right)\log_e\left(1+\frac{\Delta v}{v_1}\right)\right]$$

(c) Using the logarithmic expansion introduced in Example 3.5, show that the dissipation is approximately

$$-\frac{1}{12}\frac{(\Delta v)^3}{\varkappa v_1^2} \quad \text{per unit mass}$$

(d) Show that the preceding result is equivalent to

$$-\frac{1}{12}\frac{(\Delta u)^3}{u_1}$$

3.26 When waves travelling in a uniform tube reach a section where the area changes abruptly, a reflexion takes place. Two limiting cases can be distinguished:

(1) reflexion at a closed end, which results in a reflected wave which brings the fluid to rest at that end, and

(2) reflexion at an open end, which results in a reflected wave such that the pressure at the end of the tube retains its original value.

For weak waves these processes can be discussed quite simply; only this case will be considered here.

(a) Show that the condition for reflexion (1) can be satisfied if the reflected weak wave is exactly like the incident wave; and that the pressure rise produced by the passage of the two waves is twice that of the incident wave.

(b) Show that the condition for reflexion (2) can be satisfied by an 'inverted' reflexion, that is, by a reflected wave which produces the opposite change in pressure; and that the fluid velocity is doubled by the passage of the second wave.

(c) Apply these results to estimate the pressure on the front of a railway train which runs at 200 km/hr into a tunnel of length 1 km.

3.27 (a) Show that for weak waves the continuity equation (3.5.8) for non-steady, one-dimensional flow reduces to

$$\frac{\partial \varrho}{\partial t}+\varrho_1\frac{\partial u}{\partial x}+\frac{\varrho_1 u}{A}\frac{dA}{dx}=0$$

the corresponding momentum equation being

$$\frac{\partial u}{\partial t}=-\frac{a_1^2}{\varrho_1}\frac{\partial \varrho}{\partial x}$$

(b) Hence show that the motion is governed by

$$\frac{\partial^2 u}{\partial t^2}=a_1^2\left[\frac{\partial^2 u}{\partial x^2}+\frac{\partial}{\partial x}\left(\frac{u}{A}\frac{dA}{dx}\right)\right]$$

(c) An explicit solution of an equation of this complexity can be expected only under rather special circumstances. One such case will be discussed here.

For disturbances that are periodic in time, $u = U(x)\sin\sigma t$, in a duct whose area is specified by $A = A_0 e^{kx}$, show that the equation reduces to

$$\frac{d^2U}{dx^2} + \frac{dU}{kdx} + \frac{\sigma^2}{a_1^2} U = 0$$

The exponential form of duct is not as artificial as it might seem; loudspeakers, musical instruments and hearing trumpets are not far from this shape.

(d) Find a solution of the equation of (c) by assuming it to have the form $U = \exp(c_1 x) \sin c_2 x$.

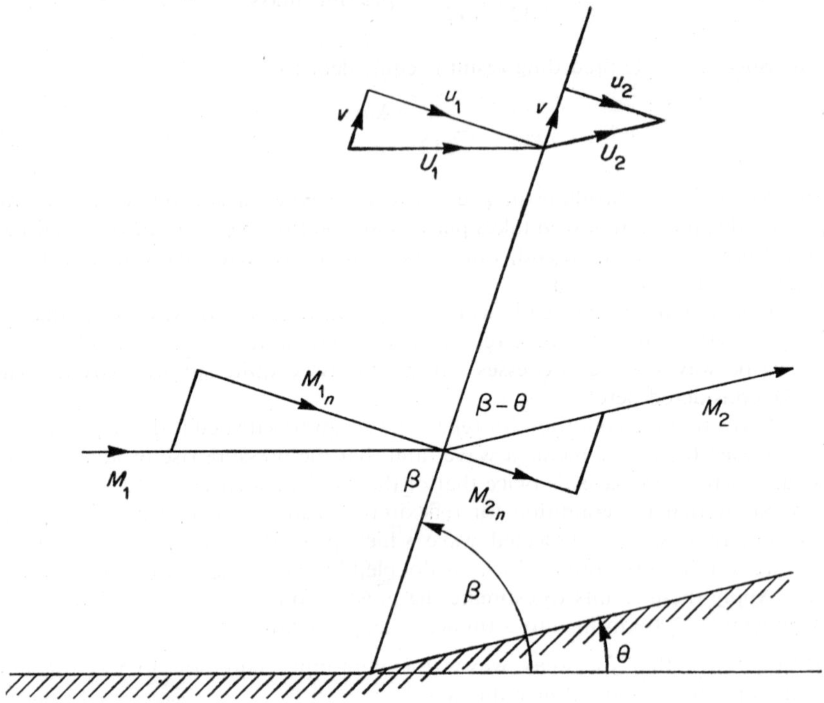

Figure 3.19 See Problem 3.28

3.28 The oblique shock shown in Figure 3.1(a) can be analysed in a simple way by adopting axes translating along the line of the shock. The resulting geometry is shown in Figure 3.19. Any streamline of a frictionless flow can be taken as the contour of a body bounding the flow. Hence the flow of Figure 3.19 can be interpreted as that past a sharp inwards turning of a wall, or as that past the sharp leading edge of a wing. The angle by which the wall deflects is θ, and the angle of the wave to the initial flow direction is β. Since the static properties are unchanged by the adoption of translating axes, the ratios in the shock tables still apply, if the Mach number in terms of which they are given is interpreted as the normal Mach number M_{1_n}.

(a) Show that the Mach number M_1 based on the upstream velocity is related to that based on the velocity normal to the flow by $M_1 = M_{1_n}/\sin\beta$.

(b) For an upstream Mach number of $M_1 = 4$, and a wave angle of 30°, show that the pressure ratio across an oblique shock in air is 4·5.

(c) Show that

$$\frac{\tan \beta}{\tan (\beta-\theta)} = \frac{u_1}{u_2} = \frac{\varrho_2}{\varrho_1}$$

What is the deflexion angle for the situation specified in (b)?

(d) What is the Mach number $M_2 = U_2/a_2$ for the case specified in (b)? Is M_{2n} equal to the Mach number M_2 opposite M_{1n} in the shock tables?

(e) Use the results of (a) and (c) to show that

$$\frac{\tan (\beta-\theta)}{\tan \beta} = \frac{(\gamma-1)M_1^2 \sin^2 \beta+2}{(\gamma+1)M_1^2 \sin^2 \beta}$$

Show that $\sin \beta \rightarrow 1/M_1$ as $\theta \rightarrow 0$.

This weak oblique wave is called a Mach wave, and the angle $\mu = \arcsin (1/M_1)$ is the Mach angle. What is the Mach angle for the particular case studied above?

(f) What is M_{1n} for a Mach wave?

3.29 The results of the last problem can be used to predict the lift and drag on a supersonic aerofoil, of which a simple example—a flat plate—is shown in Figure 3.20(a). This supersonic flow pattern is very different from the subsonic flow around an aerofoil; since small disturbances cannot propagate against a supersonic flow, the influence of one part of the flow does not extend to all other parts as it does in a subsonic flow.

In the supersonic case, expansion fans and oblique shocks at the leading and trailing edges give rise to uniform pressures on the upper and lower surfaces; these

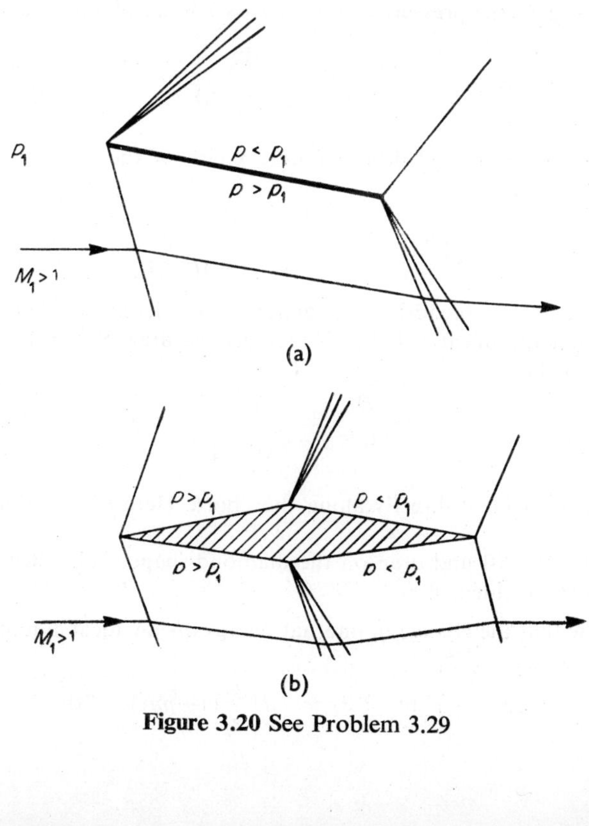

Figure 3.20 See Problem 3.29

are, respectively, below and above the free-stream level. Hence there is a net force normal to the surface, which can be resolved into lift and drag components. The drag thus found is termed wave drag, since it arises from the production of shock waves. Other components of drag will arise in a real, viscous fluid.

In the simplest, linear theory to be used here, the expansion fans and shock waves are both taken to be weak, essentially isentropic waves.

(a) Show that

$$\frac{U_1}{U_2} = \frac{\cos(\beta-\theta)}{\cos\beta}$$

and that

$$-\frac{\Delta U}{U_2} = \cos\theta - 1 + \tan\beta\sin\theta$$

for any oblique wave. Hence obtain

$$\frac{\mathrm{d}U}{U} = -\tan\mu\,\mathrm{d}\theta = -\frac{\mathrm{d}\theta}{(M^2-1)^{\frac{1}{2}}}$$

giving the change in velocity across the weak wave set up by a flow deflexion $\mathrm{d}\theta$.

(b) Since a weak wave is virtually isentropic

$$\mathrm{d}p + \varrho U\,\mathrm{d}U = 0$$

Show that this gives the pressure change across a weak oblique wave as

$$\mathrm{d}p = \varrho U^2 \frac{\mathrm{d}\theta}{(M^2-1)^{\frac{1}{2}}}$$

(c) Show that the lift and drag per unit length of the flat-plate aerofoil are given by

$$\frac{L}{\varrho U^2 c\alpha} = \frac{D}{\varrho U^2 c\alpha^2} = \frac{2}{(M^2-1)^{\frac{1}{2}}}$$

where c is the chord or breadth of the plate and α is its angle of attack.

(d) Show that the linearized theory predicts the drag of the diamond-shaped foil of Figure 3.20(b) to be

$$\frac{D}{\varrho U^2 c} = \frac{2(t/c)^2}{(M^2-1)^{\frac{1}{2}}}$$

when the flow is incident directly along the chord. Here t is the thickness of the aerofoil.

(e) What are the lift and drag on the diamond-shaped foil when it is inclined at angle α to the incident flow?

3.30 (a) Show that the change in internal energy in an ideal liquid can be expressed as

$$\mathrm{d}e = -p\,\mathrm{d}v + T\,\mathrm{d}s = -d\left[\frac{v}{\varkappa}(1+p\varkappa)\right] + T\,\mathrm{d}s$$

(b) Under what conditions does the simple result

$$de = -\frac{dv}{\varkappa} = \frac{dp}{\varrho}$$

give the internal energy change in a shock in an ideal liquid? Discuss the sign of the change.

(c) Why, for a weak shock in an ideal liquid, is the internal energy change not equal to the dissipation?

3.31 In Section 3.3.3 a geometric interpretation is given for the Froude number. Devise a dynamic interpretation, analogous to that given for Mach number in equation (3.2.9).

3.32 (a) What is the hoop stress in the penstock considered in Example 3.11?

(b) A compression wave which decelerates the flow by 20 ft/sec passes up the penstock as a result of a reduction in the flow to the turbines. Estimate the hoop stress in the penstock just after the wave has passed. (The change in diameter may be neglected.)

3.33 The pressure difference between the inside and outside of a bubble is given by $2S/R$, where S and R are the surface tension of the interface and the bubble radius. Repeat the calculation of Example 3.10, now taking into account bubble size and surface tension.

3.34 In Section 3.5.2 the difference between thermodynamic and mechanical pressures was discussed by considering the rapid motion of a piston into a cylinder. Determine the entropy change in this process, taking the compressed substance to be a perfect gas.

4

Isentropic Flow

In Chapter 2, while considering the general class of steady, isenergic flows, we obtained a number of simple results for flows that are isentropic as well. These results, called Bernoulli equations, link changes in velocity with changes in other fluid properties and thus lead to more explicit predictions of velocities, pressure variations and forces applied by the flowing fluid. These predictions would be of academic interest only, were it not for the fact that many flows of engineering interest are nearly isentropic. Moreover, the isentropic model can sometimes be modified, using empirical data, to provide a more realistic picture of the development of flows in which dissipation is significant.

Nearly isentropic flows are of interest not only because they can be analysed easily, but also because the dissipation of mechanical energy in them is small. A designer usually aims to extract as much work as possible from a flowing fluid, or to minimize the work he supplies to it; hence he tries to reduce the dissipation within the flow so that it follows closely the isentropic ideal. Of course, a design conducive to isentropic flow may entail certain disadvantages, for example, increased bulk and cost, and the optimum design is that with the greatest overall effectiveness.

There are some situations in which the aim of the designer is reversed, and maximum dissipation is sought. Consider, for example, water flowing over a dam, down a spillway and into a river below. Unless the mechanical energy of the fluid can be markedly reduced, the velocity of the water entering the river will be high; this could lead to erosion of the river bed and undermining of the dam. In this situation, a conversion of mechanical energy to internal energy is welcome, and the designer makes sure that a hydraulic jump decelerates the flow before it enters the river.

Given that the engineer wishes in most instances to minimize dissipation, how does he achieve this end? In the preceding chapters we have distinguished three mechanisms of dissipation: friction at a solid boundary, rapid spreading and mixing at an abrupt change in flow area, and the shock wave. Evidently, an isentropic flow will occur if these mechanisms are avoided.

Hence the designer will seek to reduce the boundary friction to which his flow is exposed, will avoid abrupt changes in cross-section, and will avoid supersonic flows in which shocks can form. Since the losses in friction and mixing are strongly dependent on the local velocity, an overall prescription for small dissipation is simply to keep the flow velocity down. However, this can usually be achieved only by increasing the size of the flow passage, with a consequent increase in capital cost. Hence a balance must be struck between running costs, which generally increase when the dissipation increases, and capital costs, which can usually be reduced by accepting a higher dissipation.

This chapter will examine four aspects of isentropic flow: the application of Bernoulli's equation to an entire flow field, not merely to a single streamtube; the flow of a compressible fluid through a duct of varying section; the use of isentropic results for flow measurements; and the adaptation of isentropic results to represent dissipative flows in nozzles and diffusers. Each of these developments leads directly to a family of engineering applications. The Bernoulli equation for a flow field provides a way of calculating the forces applied to a body immersed in a flowing fluid, and is a basic tool of designers of aeroplanes and turbomachines. The analysis of duct flows with varying density can be applied to the design of wind tunnels for high-speed testing and of air intakes for the engines of high-speed aircraft. Flow measurements are vital for the control of flow processes as varied as those in the reactors of a chemical plant and the ventilating system of a coal mine. Finally, the realistic prediction of nozzle and diffuser performance is required for the design of the passages within flow machines of all kinds.

4.1 BERNOULLI EQUATIONS

4.1.1 Bernoulli's Theorem for a Streamtube

The properties of an element of fluid which undergoes an isentropic process as part of an isenergic steady flow are related by

$$\int_1 \frac{\mathrm{d}p}{\varrho} + \frac{u^2}{2} + gz = \text{constant} \tag{4.1.1}$$

This is a generalization of the result (2.3.29) to include potential energy. In a steady flow each of the particles moving in succession along a streamtube undergoes the same process. Hence the relationship (4.1.1) applies not only to each particle, but also to conditions at all points along a streamtube. Once it has been decided that the Bernoulli equations relate conditions at certain points in a flow, it is natural to ask whether they can relate conditions at all points in the flow. This question will be answered later; for the moment the consequences for a single streamtube will be examined.

In Chapter 2 the Bernoulli theorem emerged from a discussion of energy conversions in isenergic flow; it entails the additional requirement that the conversions occur via isentropic processes. The results obtained in Chapter

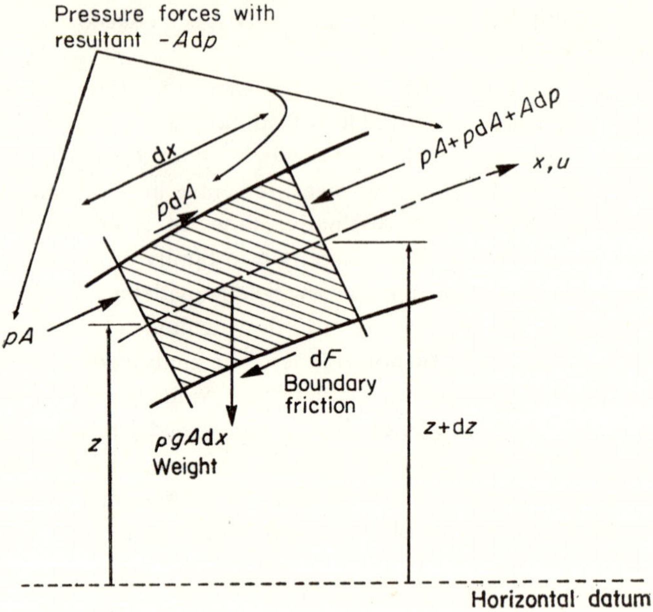

Figure 4.1 Forces on an element of one-dimensional flow

3 make it possible to derive the Bernoulli theorem in another way, and this second derivation gives added insight into the significance of the Bernoulli result. The momentum equation for non-steady, frictionless, one-dimensional flow was found to be

$$\frac{\partial u}{\partial t}+u\frac{\partial u}{\partial x}=-\frac{1}{\varrho}\frac{\partial p}{\partial x} \tag{3.5.9}$$

If x is measured along the axis of a streamtube, this relates the fluid acceleration to pressure changes along the tube. In the present discussion, the effects of fluid weight and friction are also of interest. Figure 4.1 shows these components of force for an element of a streamtube. The streamwise component of the fluid weight is $-g\,dz/dx$ per unit mass, and the component of the friction force is $-dF/(\varrho A)$ per unit mass, where A is the local cross-sectional area of the streamtube. Hence equation (3.5.9) can be generalized to

$$\frac{\partial u}{\partial t}+u\frac{\partial u}{\partial x} = -\frac{1}{\varrho}\frac{\partial p}{\partial x}-g\frac{\mathrm{d}z}{\mathrm{d}x}-\frac{\mathrm{d}F}{\varrho A} \tag{4.1.2}$$

Some of the terms of this momentum equation are related to those appearing in the Bernoulli equation (4.1.1). Indeed, if the flow is steady and frictionless, the two results correspond, the momentum equation (4.1.2) becoming

$$\frac{\mathrm{d}p}{\varrho}+u\,\mathrm{d}u+g\,\mathrm{d}z = 0$$

which can be integrated to give the Bernoulli result. This derivation leads directly to the constancy of the Bernoulli sum along a streamtube, that is, for varying x. Moreover, the necessity of steady conditions is more explicit. A final point of interest in this derivation is its failure to use the condition of constant entropy. The requirements that the flow be continuous, one-dimensional and frictionless replace the restriction to isentropic flow. Let us see why this is so. A continuous flow, that is, one in which abrupt changes in properties do not occur, was assumed in setting down equation (4.1.2), for the derivatives of that equation would not be defined at discontinuities. Hence the formation of shock waves is ruled out. Moreover, in a strictly one-dimensional flow internal mixing cannot occur. Thus the only mechanism of entropy production is friction at the boundaries of the streamtube. When this too is eliminated, the flow is automatically isentropic.

For three simple equations of state the relationships between velocity and the other fluid properties have been derived:

$$\frac{p}{\varrho_c}+\frac{u^2}{2}+gz = \frac{p_a}{\varrho_c}+gH_0 \tag{2.4.1}$$

for a constant-density fluid;

$$\frac{\varrho_0}{\varrho} = \exp\,[\varkappa(p_0-p)] = 1+\frac{1}{2}\varkappa\varrho_0 u^2 \tag{2.4.4}$$

for an ideal liquid with compressibility $\varkappa$ = constant; and

$$\left(\frac{p}{p_0}\right)^{\frac{\gamma-1}{\gamma}} = \left(\frac{\varrho}{\varrho_0}\right)^{\gamma-1} = \frac{T}{T_0} = 1-\frac{u^2}{2c_pT_0} \tag{2.4.5}$$

for a perfect gas. In each case the total properties, denoted by subscript 0, are constant in isentropic, isenergic flow.

The results for compressible fluids can be given in alternative forms by introducing the speed of sound and Mach number. For an ideal liquid, equations (2.4.4) can be rewritten using equations (3.2.6, 8) as

$$\frac{\varrho}{\varrho_0} = \exp\,[(\varkappa(p-p_0)] = \left(\frac{a_0}{a}\right)^2 = 1-\frac{M^2}{2} \tag{4.1.3}$$

For a perfect gas, equations (2.4.5) become, using equations (3.2.5, 8)

$$\left(\frac{p_0}{p}\right)^{\frac{\gamma-1}{\gamma}} = \left(\frac{\varrho_0}{\varrho}\right)^{\gamma-1} = \frac{T_0}{T} = \left(\frac{a_0}{a}\right)^2 = 1+\frac{\gamma-1}{2}M^2 \tag{4.1.4}$$

Since the total properties are constant, a knowledge of the changes in Mach number in an isentropic flow serves to determine the variations in all of the fluid properties. The results for a perfect gas are frequently used, and tables giving the ratios p/p_0, etc., as functions of Mach number are widely available. Their use will be discussed later in this chapter.

Example 4.1 In Section 2.4.1, we discussed the applicability of the constant-density approximation to the pressure/velocity relationship in isentropic flow. It is worth looking at this matter again using the results just obtained, for the Mach number provides a convenient index of the role of compressibility.

For an ideal liquid, the logarithmic expansion of one of the results (4.1.3) gives

$$\begin{aligned} p_0-p &= -\frac{1}{\varkappa}\log_e\left(1-\frac{M^2}{2}\right) \\ &= \frac{\varrho u^2}{2}\left(1+\frac{M^2}{4}+\frac{M^4}{12}+\dots\right) \end{aligned}$$

since $M^2/2\varkappa = \varrho u^2/2$ for this kind of fluid.

For a perfect gas, binomial expansion of the pressure relationship from equations (4.1.4) gives

$$\frac{p_0}{p} = \left(1+\frac{\gamma-1}{2}M^2\right)^{\frac{\gamma}{\gamma-1}} = 1+\gamma\frac{M^2}{2}\left(1+\frac{M^2}{4}+(2-\gamma)\frac{M^4}{24}-\dots\right)$$

and

$$p_0-p = \frac{\varrho u^2}{2}\left(1+\frac{M^2}{4}+(2-\gamma)\frac{M^4}{24}-\dots\right)$$

since $\gamma p M^2/2 = \varrho u^2/2$ in this case.

For each of the fluids considered above the constant-density result is obtained for $M \ll 1$, while the first error term is $\frac{1}{4}M^2$. Hence the proportional error introduced by using the constant-density approximation is approximately

$$er = \frac{M^2}{4}$$

This result is equivalent to equations (2.4.7). If an accuracy of 1 per cent is required, $M < \frac{1}{4}$. If an accuracy of 5 per cent is adequate, the simple result may be used so long as $M < \frac{1}{2}$.

Example 4.2 The most frequently used Bernoulli results are those linking changes in pressure and velocity. For a constant-density fluid this is the only Bernoulli relationship; the other fluid properties are constant in isentropic flow. Even when the density and temperature do change, it is the pressure variation which is most often of interest to engineers, for the structures over which the fluid passes must be capable of resisting the forces associated with these pressures. One important application of isentropic-flow results is in calculating the forces acting on wings of aeroplanes and blades of turbomachines. Another is in predicting the forces applied by the wind to buildings, chimneys, cooling towers and other structures. Calculations of this kind, which involve a knowledge of the flow pattern over the entire surface, are necessarily complicated, and we shall consider here a simple and rather artificial example—flow over a building with a semicircular roof. We shall calculate the net force on a building with roof diameter 20 m for a wind speed of 150 km/hr.

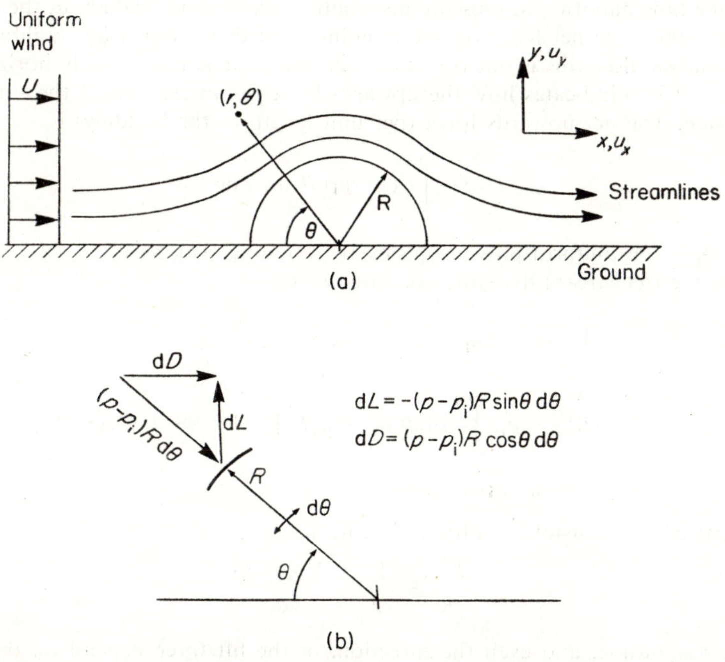

Figure 4.2 Flow over a semi-circular building. (a) Ideal flow pattern and coordinate system; (b) Calculation of elements of lift and drag on the surface of the building

The calculation is artificial not only in relating to a building of unusual shape, but also in assuming the flow to be steady, two-dimensional and wholly isentropic. A classical theory gives the velocity components for isentropic, constant-density flow around a circular cylinder:

$$u_x = U\left[1 - \left(\frac{R}{r}\right)^2 \cos 2\theta\right] \quad \text{and} \quad u_y = U\left(\frac{R}{r}\right)^2 \sin 2\theta$$

where U is the wind speed, that is, the velocity achieved everywhere in the flow well away from the cylinder whose radius is R. Figure 4.2 shows the situation and indicates how the polar coordinates r, θ are defined. This flow pattern is unrealistic in being symmetrical about the plane $\theta = 90°$, and also in being uniform well upstream and downstream of the building. In reality, the flow would separate from the top of the round roof and a wake would form downstream. Moreover, the wind fluctuates in time and varies with height above the ground. Despite these limitations, this flow serves to illustrate the methods which can be used to calculate the forces associated with more realistic pressure distributions.

Since the pressures on the surface of the building are required, our attention is concentrated on the streamtube running along the ground and over the roof. The flow in this streamtube will be taken to be isentropic. Then, for $r = R$

$$p = p_0 - \frac{\varrho_c}{2}(u_x^2 + u_y^2) = p_0 - \frac{\varrho_c U^2}{2}(2 - 2\cos 2\theta)$$

Since the flow pattern and pressure distribution are symmetrical about the centreline ($\theta = 90°$), the net force on the building is vertical. (For a tall building in a more realistic flow this is not the case; then the force is more nearly horizontal.)

Figure 4.2(b) indicates how the upwards force on an element of the surface is calculated. The net upwards force (per unit length of the building) is

$$L = -\int_0^\pi (p - p_i)\, R \sin\theta \, d\theta$$

where p_i is the uniform pressure inside the building. Taking $q_1 = \varrho_c U^2/2$ to represent the free-stream dynamic pressure, we have

$$L = -\int_0^\pi R[p_0 - p_i - 2q_1(1 - \cos 2\theta)] \sin\theta \, d\theta$$

$$= -R(p_0 - p_i)\int_0^\pi \sin\theta \, d\theta - 4q_1 R \int_0^\pi (1 - \cos^2\theta)\, d(\cos\theta)$$

$$= -2R(p_0 - p_i) + \tfrac{16}{3} q_1 R$$

In terms of a dimensionless lift coefficient, this is

$$\frac{L}{q_1 R} = \frac{16}{3} - \frac{2(p_0 - p_i)}{q_1}$$

The magnitude, and even the direction, of the lift force depend on the value of the internal pressure, which is determined by the way in which the interior is vented to the atmosphere. For instance, if the walls were airtight, except for a hole at the top of the roof where $\theta = 90°$, the internal pressure would be

$$p_i = p_0 - q_1(2+2) = p_0 - 4q_1$$

and

$$\frac{L}{q_1 R} = \frac{16}{3} - 8 = -\frac{8}{3}$$

In this case the wind would push the building onto the ground. However, if the only vents were at ground level where the air is at rest, $p_i = p_0$ and $L/q_1 R = 16/3$, giving a strong uplift. These two cases provide limits for the vertical force applied by the wind.

Consider the particular case in which ventilation is through windows extending from 1 m to 3 m from the ground. At the middle of the windows, $\sin\theta = 1/5$, and

$$p = p_0 - 2q_1(1-\cos 2\theta) = p_0 - \frac{4}{25} q_1$$

Then

$$\frac{L}{q_1 R} = \frac{16}{3} - \frac{8}{25} \simeq 5$$

For a wind of 150 km/hr, the dynamic pressure is

$$q_1 = \frac{\varrho_c U^2}{2} = 1{\cdot}225 \times \frac{(150/3{\cdot}6)^2}{2} = 1060 \text{ N/m}^2$$

and the uplift is

$$L \simeq 5q_1 R = 5 \times 1060 \times \frac{10}{9{\cdot}8} = 5400 \text{ kgf/m}$$

Buildings are usually vented at a number of points, at windows, doors and various chinks. Hence the pressure is not uniform through the interior but varies over a range somewhat smaller than that of the pressure distribution outside. These variations within a building and between the interior and the outside are of importance in the design of ventilating systems.

4.1.2 Bernoulli's Theorem for a Flow Field

The conditions under which

$$\int_1 \frac{dp}{\varrho} + \frac{u^2}{2} + gz = \text{constant} \tag{4.1.1}$$

applies throughout a steady flow, and not merely along a single streamtube, can be found by examining the changes in fluid properties between adjacent streamlines. Figure 4.3 shows a pair of streamlines separated by distance δn, measured normal to the flow direction x, and away from the local centre of curvature of the streamlines, which is distance n from the point considered. In general, the total enthalpy, pressure, elevation and velocity will differ from streamline to streamline; the change over the distance δn will be denoted by $\delta = (\partial/\partial n)\delta n$. It will be assumed that equation (4.1.1) applies along each streamline. However, it is possible that

$$\int \frac{dp}{\varrho} + \frac{u^2}{2} + gz = f(n)$$

The Bernoulli constant will be the same from streamline to streamline only if

$$\int^{p+\delta p} \frac{dp}{\varrho} + \frac{1}{2}(u+\delta u)^2 + g\,\delta z - \int^{p} \frac{dp}{\varrho} - \frac{u^2}{2} = 0$$

In the limit as $\delta n \rightarrow 0$, this gives

$$\frac{\delta p}{\varrho} + u\,\delta u + g\,\delta z = 0 \tag{4.1.5}$$

The definitions of entropy (1.5.14) and total enthalpy (2.3.25) relate the changes in the properties of a simple compressible substance:

$$T\,ds = dh - v\,dp = dh_0 - u\,du - \frac{dp}{\varrho}$$

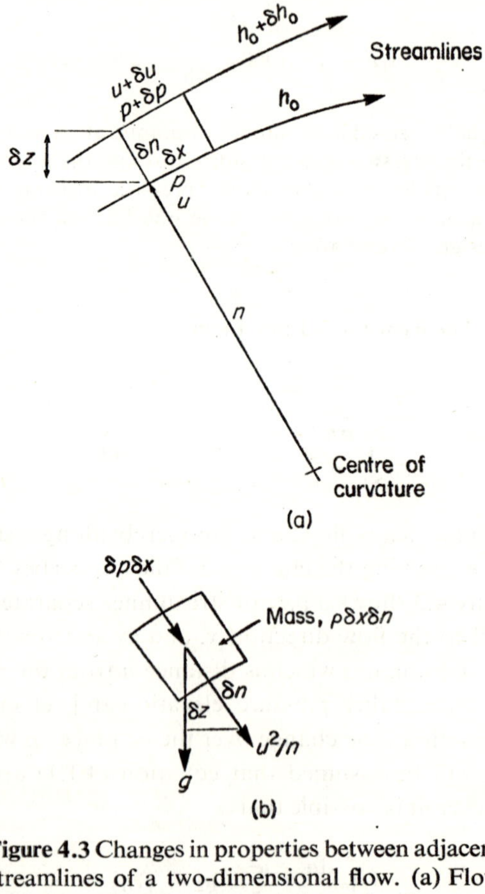

Figure 4.3 Changes in properties between adjacent streamlines of a two-dimensional flow. (a) Flow geometry; (b) Momentum balance for the transverse direction

In particular, they relate the changes between adjacent streamlines in a flow of such a substance. Hence the change in the Bernoulli sum between adjacent streamlines is given by

$$\frac{\delta p}{\varrho}+u\ \delta u+g\ \delta z = \delta(h_0+gz)-T\ \delta s \tag{4.1.6}$$

Over a region in which

$$h_0+gz = \text{constant} \quad \text{and} \quad s = \text{constant} \tag{4.1.7}$$

the condition (4.1.5) will be satisfied, and the Bernoulli sum will be the same from streamline to streamline. Such a flow is termed *homenergic* and *homentropic.*

Something further can be extracted from this analysis, for no use has been made of the momentum equation for the n-direction. For steady flow, the changes in pressure and elevation must provide the acceleration u^2/n towards the centre of curvature. For the element shown in Figure 4.3(b)

$$\delta p\ \delta x+g(\varrho\ \delta n\ \delta x)\frac{\delta z}{\delta n} = \frac{u^2}{n}(\varrho\ \delta n\ \delta x)$$

whence

$$\frac{\partial p}{\partial n} = \varrho\left(\frac{u^2}{n}-g\frac{dz}{dn}\right) \tag{4.1.8}$$

Equation (4.1.6) can be written

$$\frac{u^2}{n}\ \delta n+u\ \delta u = \delta(h_0+gz)-T\ \delta s$$

or, with $\delta = (\partial/\partial n)\delta n$

$$u\left(\frac{\partial u}{\partial n}+\frac{u}{n}\right) = \frac{\partial(h_0+gz)}{\partial n}-T\frac{\partial s}{\partial n} \tag{4.1.9}$$

This statement is called *Crocco's theorem.*

Comparison of equations (4.1.5, 6, 9) leads to a startling conclusion: the satisfaction of the purely kinematic condition

$$\frac{\partial u}{\partial n}+\frac{u}{n} = 0 \tag{4.1.10}$$

implies, first, that the Bernoulli theorem holds throughout the flow field and, secondly, that the flow is homentropic and homenergic, unless the enthalpy and entropy are related in a rather artificial manner. We shall now investigate the kinematic quantity $\partial u/\partial n+u/n$ to see why it has such far-reaching implications for the energy conversions within the flow.

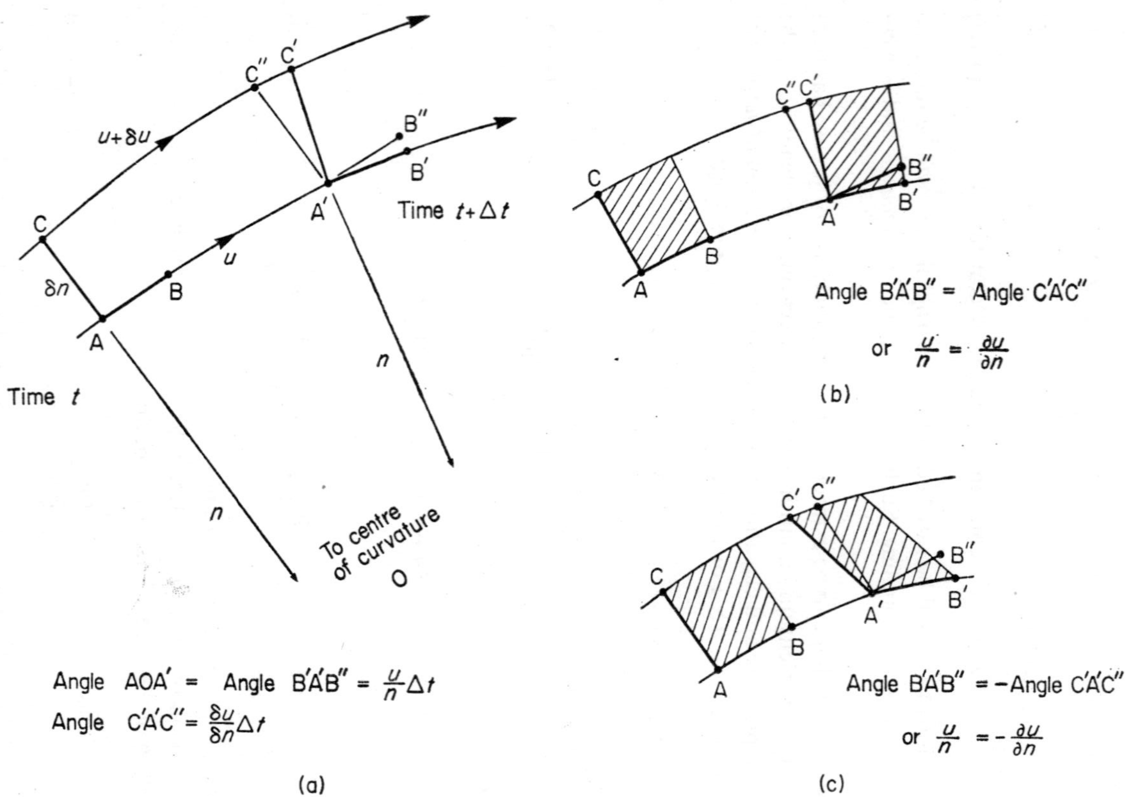

Figure 4.4 Fluid rotation illustrated by following two initially perpendicular line elements which move with the fluid. (a) General case; (b) Case of equal rotation; (c) Case of zero net rotation

In Figure 4.4 the terms $\partial u/\partial n$ and u/n are interpreted as angles of rotation of lines moving with the fluid. Two initially perpendicular lines of fluid particles are shown, AC along the radius from the centre of curvature, and AB along the streamline. After the short time Δt, they have moved to A′C′ and A′B′. The lines A′C″ and A′B″ mark the initial directions of AC and AB, the angles C′A′C″ and B′A′B″ give the rotations of the two lines in the time Δt. The line AC rotates at rate $\delta u/\delta n = \partial u/\partial n$, and the line AB rotates at rate u/n, equal to that of the radius from the centre of curvature. The total rate of rotation of the fluid element containing AB and AC can be reckoned as

$$\zeta = \frac{\partial u}{\partial n} + \frac{u}{n} \tag{4.1.11}$$

This quantity is termed the *rotation* or *vorticity* of the fluid. It can be introduced in more rigorous ways which show that it is a unique property of the flow, independent of the system of axes and pair of lines used to define it here.

Two particular cases will help in understanding the significance of vorticity. First take $\partial u/\partial n = u/n$. This situation is sketched in Figure 4.4(b); the line A′C′ is normal to A′B′, having performed the same rotation. Hence

$$\zeta = \frac{\partial u}{\partial n} + \frac{u}{n} = 2\omega$$

where ω is the local angular velocity of the fluid. This kind of motion, with every element experiencing the same rotation, is also found in a moving solid body. Hence the fluid flow with ω = constant is termed *solid-body rotation*. This pattern of rotation is established in fluid in a closed tank which rotates with a uniform angular velocity for a long time. Fluid friction ultimately imposes the motion of the tank on the whole of the fluid. Hence this pattern of motion is also called a forced vortex.

A second simple pattern of motion is sketched in Figure 4.4(c). Here

$$\frac{\partial u}{\partial n} = -\frac{u}{n}$$

and

$$\zeta = \frac{\partial u}{\partial n} + \frac{u}{n} = 0$$

Although each of the lines rotates, the fluid element as a whole does not; this kind of flow is described as *irrotational*. However, in this case the fluid element is distorted as it moves. The *rate of shear*

$$\varepsilon = \frac{\partial u}{\partial n} - \frac{u}{n} \tag{4.1.12}$$

describes this distortion; this is the rate at which two mutually perpendicular lines rotate towards one another. Note that the shear is zero in solid-body rotation, in which each fluid element merely rotates without change of shape.

It is now apparent that the condition of zero vorticity, which must be satisfied if the Bernoulli condition is to apply throughout the flow, defines a flow in which the angular velocity of every fluid element is zero. This is reasonable, since the Bernoulli condition is an energy conservation principle which includes the kinetic energy of translation, but not the kinetic energy of rotation of the fluid.

At this point it is appropriate to summarize the conditions under which Bernoulli's equation is applicable. The flow must be steady, isenergic, isentropic and frictionless along any streamtube to which the equation is applied. If it is applied to a flow field, the fluid must in addition be homentropic, homenergic and irrotational. The requirements that the flow be isentropic, irrotational and frictionless are inter-related, since friction is the mechanism by which entropy variations and rotation are introduced into an initially uniform fluid.

For the special case of a constant-density fluid, the results obtained above take on simplified forms. In this special case

$$\varrho = \varrho_c, \quad s = s_0 \quad \text{and} \quad T = T_0.$$

Then

$$T\,\mathrm{d}s = T_0\,\mathrm{d}s_0 = \mathrm{d}h_0 - \frac{\mathrm{d}p_0}{\varrho_c}$$

Hence Crocco's theorem becomes

$$\frac{\delta p}{\varrho_c} + u\,\delta u + g\,\delta z = u\zeta\delta n = \delta\left(\frac{p_0}{\varrho_c} + gz\right)$$

$$= \delta(gH_0) \tag{4.1.13}$$

where H_0 is the total head introduced in equation (2.3.23). These results show that the Bernoulli theorem applies through a constant-density flow if the flow is irrotational, if the total head is constant, and (when gravity is not important) if the total pressure is constant.

The preceding discussion has been tacitly limited to two-dimensional flow fields, that is, to flows that are the same in a series of parallel planes. Such flows can be completely represented by a two-dimensional drawing, and by the use of two coordinates. The essential features of three-dimensional flows are revealed by the two-dimensional treatment, although the formal mathematical discussion of the three-dimensional case differs considerably from that given above.

Example 4.3 Let us see how the conditions derived above apply to some of the situations discussed earlier in the book.

(1) In the idealized flow of wind past a building of Example 4.2 and Figure 4.2, the entropy does not change; this is demonstrated by the reversibility of the flow pattern, that is, by its symmetry about the mid-plane $\theta = 90°$. In this flow gravity has been neglected, and the upstream conditions are uniform, $u = U$, $p = p_a$. Hence the flow is homenergic as well as homentropic. It follows that the flow is irrotational throughout, that friction is everywhere zero, that the total pressure is uniform, and that a single Bernoulli equation links pressure and velocity throughout the field.

(2) Consider next the flow shown in Figure 2.25, lossless, constant-density flow from a reservoir. Within the reservoir the total head is constant, and thereafter the flow is isenergic and isentropic. Hence the flow field is homentropic and homenergic, that is, has a uniform total head throughout. Crocco's theorem indicates that it is also irrotational, with changes in pressure, velocity and elevation related by a single Bernoulli equation.

(3) Next some situations will be considered in which the rather stringent conditions of Crocco's theorem are not satisfied. Behind the curved shock shown in Figure 3.1(a) the total enthalpy is constant, since the shock process is isenergic and gravity is negligible. Then $\partial h_0/\partial n = 0$. However, $\partial s/\partial n \neq 0$ and $\zeta \neq 0$, since the entropy change and deceleration across the shock depend on the angle at which the fluid approaches it. The entropy gradient and vorticity are related by

$$u\zeta = -T\frac{\partial s}{\partial n}$$

(4) Figure 2.17, which shows the flow into the intake of a jet engine, provides another example in which the conditions of Crocco's theorem are not satisfied everywhere. The flow in the boundary layer is obviously rotational; this implies in turn that the flow is subject to variations in entropy and total enthalpy. There are further complications in this case, since the boundary layer is not isenergic either. However, the frictionless core of the flow into the duct is nearly isentropic and isenergic, and since the properties are initially uniform it must also be homenergic, homentropic and irrotational.

Example 4.4 Some simple irrotational flows can be discovered by asking whether there are any motions of a restricted class that are irrotational. Are there, for example, any flows with parallel streamlines, that is, with $u = u(y)$, that are free of rotation? The radius of curvature of the streamlines is infinite in this case, and the condition for zero vorticity is simply

$$\frac{\partial u}{\partial n} = \frac{\mathrm{d}u}{\mathrm{d}y} = 0$$

Hence the only parallel flow satisfying the requirement of irrotationality is $u =$ constant.

A more interesting case is that of concentric circular streamlines, for which $u = u(r)$. Here $n = r$, the radial coordinate, and

$$\zeta = \frac{\mathrm{d}u}{\mathrm{d}r} + \frac{u}{r} = \frac{1}{r}\frac{\mathrm{d}(ru)}{\mathrm{d}r} = 0$$

Hence the rotation-free flow of this class has ru = constant. This is called a free vortex, since it is formed in the absence of friction and blades which apply torque to the fluid. The spiral flows of water around the drain of a bath and of a gas in the vaneless diffuser of a compressor have variations of swirl velocity which correspond approximately to this law.

In a forced vortex, or solid-body rotation, the velocity variation is given by $u = \omega r$, with ω the angular velocity. A variety of swirling flows can be manufactured by superposing forced and free vortex motions.

Example 4.5 The pressure difference between the inside and outside walls of a bend in a pipe provides a means of metering the flow through the pipe. In order to investigate the essential features of this kind of meter, an idealized model of the flow may be used, the fluid being assumed irrotational throughout. Figure 4.5(a) shows the external geometry, while Figure 4.5(b) shows the velocity profiles across three sections, upstream and downstream of the bend, and through the middle. The flow upstream is taken to be uniform and parallel to the pipe walls. Hence when the flow becomes parallel downstream of the bend, it will again be uniform. In the middle of the bend, the streamlines are assumed to be circular, in conformity with the walls of the elbow, with velocity distribution $ru = k$, a constant which is a measure of the angular momentum of the fluid.

For simplicity consider a constant-density fluid in a duct of rectangular section, as sketched in Figure 4.5(c). The constant k can be obtained from continuity considerations:

$$u_1 ab = b\int_{R_1}^{R_2} u\,\mathrm{d}r = bk\log_e\frac{R_2}{R_1}$$

Then

$$u = \frac{k}{r} = u_1\frac{a/r}{(\log_e R_2/R_1)}$$

The pressure distribution in Section 2 can now be found using the fact that the Bernoulli sum is constant. Neglecting gravity

$$p_0 = p + q = p_1 + q_1$$

throughout the frictionless flow, where $q = \varrho_c u^2/2$ is the local dynamic pressure, and q_1 is that upstream of the bend. The pressure at the inner wall is given by

$$p_{2_i} = p_0 - q_1\left(\frac{a}{R_1}\right)^2\left[\log_e\frac{R_2}{R_1}\right]^{-2} < p_1$$

and that at the outer wall by

$$p_{2_o} = p_0 - q_1\left(\frac{a}{R_2}\right)^2\left[\log_e\frac{R_2}{R_1}\right]^{-2} > p_1$$

The rise in pressure on the outside and fall on the inside of the bend are associated with the deceleration of the outer fluid and the acceleration of that on the inner wall.

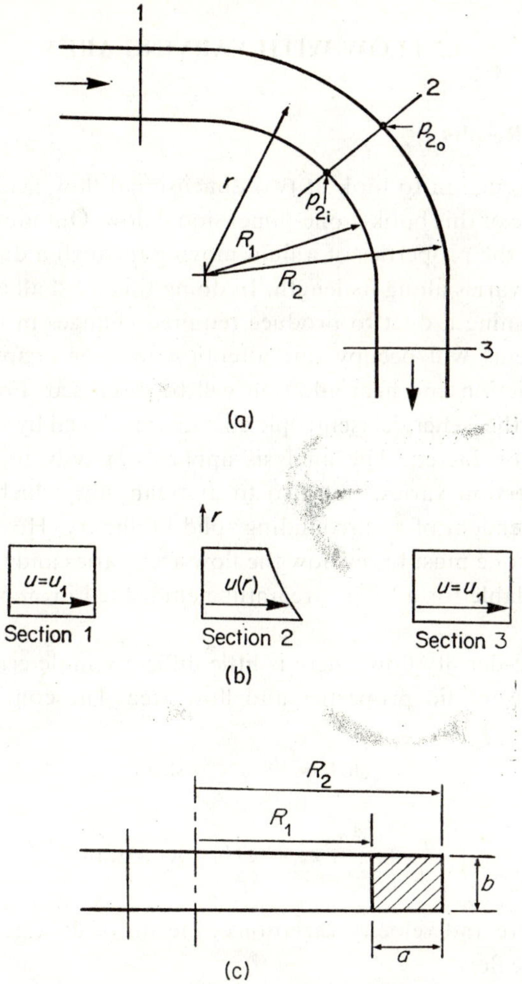

Figure 4.5 Ideal flow through a bend meter. (a) View of the bend normal to its plane; (b) Velocity distributions; (c) View of the bend in its plane

For a particular bend geometry

$$p_{2_o} - p_{2_i} \propto q_1 \propto u_1^2$$

The flow equation for an actual meter cannot be determined on the assumption of irrotational flow. However, the processes which the fluid undergoes are generally similar to those of the idealized model, and the same form of flow law

$$\Delta p \propto \dot{V}^2$$

can be expected to apply. The constant of proportionality must be determined empirically.

4.2 FLOW WITH VARYING AREA

4.2.1 General Results

After our excursion to look at two-dimensional flow fields, we return to the main theme of this book—one-dimensional flow. Our aim is to determine the changes in the properties of a fluid moving through a duct whose cross-sectional area varies along its length. In doing this we shall also develop the means of designing a duct to produce required changes in fluid properties. Similar problems will occupy our attention in later chapters, where the response to friction and heat addition will be discussed. For the present, however, only the isenergic, isentropic changes produced by variation in flow area will be considered. The analysis applies not only to flow in a duct whose cross-section varies, but also to a streamtube which contracts and expands independent of a surrounding solid boundary. However, to obtain explicit results one must know how the flow area varies and this information is seldom available for a 'free' streamtube embedded in a mass of flowing fluid.

For constant-density flow, there is little difficulty in determining the relationships between fluid properties and flow area. The continuity and Bernoulli equations

$$\varrho_c u A = \dot{m} = \text{constant}$$

and

$$\frac{p}{\varrho_c}+\frac{u^2}{2}+gz = gH_0 = \text{constant}$$

give the pressure and velocity variations in terms of duct geometry and the constants of the flow:

$$u = \frac{\dot{m}}{\varrho_c A}$$

and

$$p = \varrho_c g\left[H_0 - z - \frac{1}{2}\left(\frac{\dot{m}}{\varrho_c A}\right)^2\right] \qquad (4.2.1)$$

The other fluid properties, s, $T(s)$ and $e(s)$, remain constant in an isentropic flow.

Since the constant-density case is fully dealt with in equations (4.2.1), our attention will be directed to the study of flows of compressible fluid. Accordingly the effect of gravity will be assumed to be negligible. Our ultimate aim is to obtain results generally similar to those for constant density, giving explicitly the dependence of the fluid properties on the flow area, but also taking into account density variations in the flow. However, to avoid the restriction to a particular fluid in this initial discussion, we shall consider for the time being an infinitesimal process. For such a process, the Bernoulli condition is

$$\frac{dp}{\varrho}+u\,du = 0 \tag{4.2.2}$$

for any fluid. The process is also governed by the basic energy conservation statement for isenergic flow:

$$dh+u\,du = 0 \tag{4.2.2}$$

The duct geometry and flow properties are linked by the differential continuity equation

$$\frac{d\varrho}{\varrho}+\frac{du}{u}+\frac{dA}{A} = 0 \tag{4.2.2}$$

Since the process under discussion is isentropic, the changes in pressure and density are linked quite generally by

$$\frac{dp}{d\varrho} = a^2$$

where a is the speed of sound.

When the conditions (4.2.2) are combined, a few lines of algebra give

$$\frac{dh}{a^2} = \frac{dp}{\varrho a^2} = \frac{d\varrho}{\varrho} = -M^2\frac{du}{u} \tag{4.2.3}$$

This indicates that the change in velocity is always opposite in sign to those in enthalpy, pressure and density. However, the relative size of the proportional changes depends on the local Mach number of the flow. The changes in properties are linked to those in flow area by

$$\frac{dh}{a^2} = \frac{dp}{\varrho a^2} = \frac{d\varrho}{\varrho} = \frac{M^2}{1-M^2}\frac{dA}{A}$$

and

$$\frac{du}{u} = -\frac{1}{1-M^2}\frac{dA}{A} \tag{4.2.4}$$

This last result is termed the *area–velocity relation*. In these results the role of Mach number is more significant. The sign of the changes in each property in response to a change in flow area is seen to depend on whether the flow is locally subsonic or supersonic.

The reason for this rather surprising behaviour can be found by examining the way in which the continuity equation is converted into the area–velocity relation. With the use of one of equations (4.2.3), the continuity relationship becomes

$$-M^2\frac{du}{u}+\frac{du}{u}+\frac{dA}{A}=0$$

The first term represents the change in density, $d\varrho/\varrho$. At very low Mach numbers ($M \ll 1$) this change is negligible, and the constant-density result

$$uA = \text{constant}$$

emerges. For subsonic flow ($M < 1$) this result is valid qualitatively, although the effects of area change are magnified by the factor $1/(1-M^2)$. However, for supersonic flow ($M > 1$) the density changes so rapidly that the direct effect of velocity change is overbalanced, and the qualitative relationship between velocity and area is reversed. Indeed, for very high Mach numbers ($M \gg 1$), the continuity requirement reduces to the remarkable result

$$\varrho A = \text{constant}$$

Table 4.1 indicates the direction of the changes in properties in contracting and expanding flow passages, as indicated by equations (4.2.3, 4). A duct which accelerates the fluid and reduces its pressure and enthalpy is called a *nozzle*, while a duct which produces the opposite effects is called a *diffuser*. A subsonic nozzle, for example that placed at the end of a fire hose, contracts the flow in order to accelerate the fluid. A supersonic nozzle, on the other hand, expands in the direction of the flow. An example is the nozzle used to exhaust the propellant gases of a rocket; see Figure 1.1. The nozzle opens outwards in order that the supersonic flow is accelerated, and thus produces a larger thrust reaction on the rocket.

Table 4.1. Changes in properties in isentropic flow

	Nozzle	Diffuser
Subsonic or Supersonic Flow ($M \lesseqgtr 1$)	$dh < 0$ $dp < 0$ $d\varrho < 0$ $du > 0$	$dh > 0$ $dp > 0$ $d\varrho > 0$ $du < 0$
Subsonic Flow Only ($M < 1$)	$dA/A < 0 \rightarrow$	$dA/A > 0 \rightarrow$
Supersonic Flow Only ($M > 1$)	$dA/A > 0 \rightarrow$	$dA/A < 0 \rightarrow$

A subsonic diffuser expands the flow, thus reducing its kinetic energy and increasing enthalpy and pressure. A typical application of a diffuser is shown in Figure 2.22(b); high-velocity (but subsonic) flow emerging from the impeller of a compressor is decelerated in an expanding flow passage. A diffuser is said to 'recover' pressure from the kinetic energy of the fluid. This energy conversion is desirable because a high-velocity flow is subject to large frictional dissipation. Moreover, in many applications a high velocity jet is not required, but rather a slow-moving fluid at high pressure. In this latter state the fluid can store mechanical energy for later use.

The word 'diffusion' is used in another sense, quite unrelated to the way in which it is used above. In this alternative use, diffusion denotes the spreading of a substance or property by molecular interactions. If, for example, coal gas is released from a cock in one corner of a room, it can soon be smelt throughout the room; the gas diffuses through the air even if there are no marked convection currents across the room. It is usually obvious from the context of the word 'diffusion' whether it denotes molecular diffusion or the bulk flow process discussed above.

Finally, it is necessary to consider the implications of the results (4.2.4) for locally parallel flow at a section of maximum or minimum flow area. Two subcases must be examined:

$$\mathrm{d}A = 0 \quad \text{with} \quad M = 1 \quad \text{and} \quad \mathrm{d}u \neq 0 \tag{4.2.5}$$

$$\mathrm{d}A = 0 \quad \text{with} \quad M \neq 1 \quad \text{and} \quad \mathrm{d}u = 0 \tag{4.2.6}$$

Figure 4.6 indicates four conceivable ways in which the conditions (4.2.5) might be satisfied. The transitions at a section of maximum flow area are in fact inconsistent with the changes shown in Table 4.1. In Figure 4.6(c), for example, the expanding subsonic flow approaching the proposed transition will in fact decelerate, instead of accelerating towards the postulated sonic condition. This inconsistency does not arise at a *throat*, or section of minimum flow area. However, as will be seen later, the continuous supersonic-to-subsonic transition of Figure 4.6(b) is unstable, and will break down into an unsteady flow. Hence the only continuous, steady-flow, isentropic transition across the sonic condition is that shown in Figure 4.6(a), the subsonic-to-supersonic transition at a throat. In practice, the reverse transition, supersonic-to-subsonic, is usually accomplished through a non-isentropic shock wave.

The second subcase (4.2.6) does not involve a sonic transition. The conditions (4.2.6) can be satisfied in four ways: $M \lesseqgtr 1$ for both maximum and minimum flow area. These cases are illustrated in Figure 4.7, which also shows which properties take on maximum and minimum values when $\mathrm{d}A = 0$.

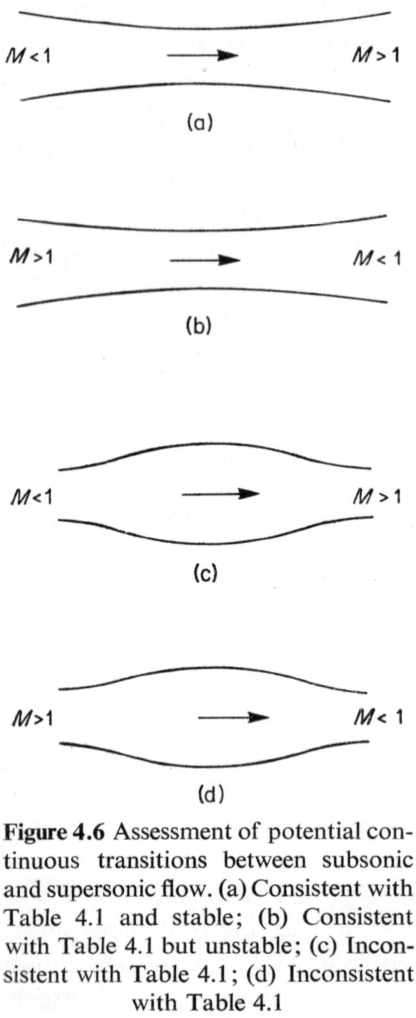

Figure 4.6 Assessment of potential continuous transitions between subsonic and supersonic flow. (a) Consistent with Table 4.1 and stable; (b) Consistent with Table 4.1 but unstable; (c) Inconsistent with Table 4.1; (d) Inconsistent with Table 4.1

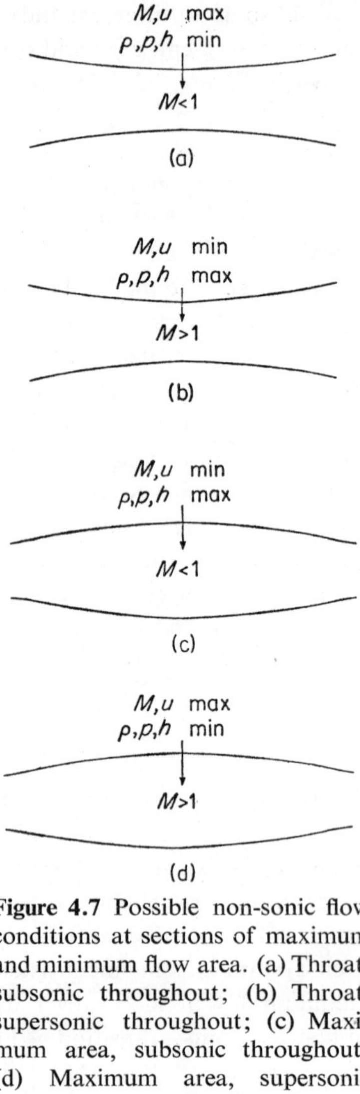

Figure 4.7 Possible non-sonic flow conditions at sections of maximum and minimum flow area. (a) Throat, subsonic throughout; (b) Throat, supersonic throughout; (c) Maximum area, subsonic throughout; (d) Maximum area, supersonic throughout

Example 4.6 Supersonic wind tunnels provide a fine variety of nearly isentropic flows. It is necessary that the flow through the tunnel be nearly isentropic in order that it represent the uniform conditions which an aeroplane meets in the atmosphere, and in order that the power requirements of the tunnel be as low as possible. One type of supersonic tunnel is shown in Figure 4.8, an intermittent-flow tunnel. With the valve closed, the reservoir is pumped slowly down to a pressure well below atmospheric. When the valve is opened, air rushes through the duct and a steady supersonic flow is established in the test section. The reservoir gradually fills and, when the pressure there reaches a high enough level, the supersonic flow in the test section breaks down. The reservoir must then be pumped down once again before further supersonic running is possible.

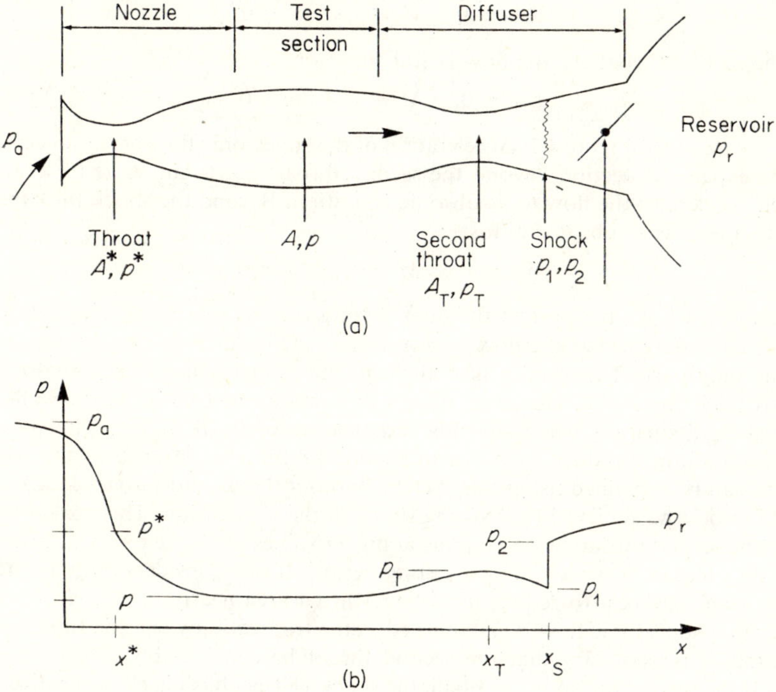

Figure 4.8 A supersonic wind tunnel of the intermittent-flow, blow-down type

During the period of supersonic operation, the flow accelerates through the bell-mouth inlet from rest outside:

$$dA < 0, \quad M < 1, \quad du > 0$$

At the throat A^*, sonic flow is achieved:

$$dA = 0, \quad M = 1, \quad \text{and} \quad du > 0$$

This is case (a) of Figure 4.6.

Following the throat the flow continues to accelerate, now supersonically, through the expanding part of the nozzle:

$$dA > 0, \quad M > 1, \quad du > 0$$

Thus the whole of the converging–diverging inlet acts as a nozzle, the transition from subsonic to supersonic flow being achieved at the throat. A converging–diverging nozzle is often referred to as a Laval nozzle, since it was used by de Laval to generate a high-velocity jet of steam to drive the turbine which is also named after him.

In the test section, parallel flow is established:

$$dA = 0, \quad M > 1, \quad du = 0$$

This is case (d) of Figure 4.7. It is in this uniform flow that an aeroplane model would be located.

The flow decelerates from the test section in the supersonic diffuser:

$$dA < 0, \quad M > 1, \quad du < 0$$

At the second throat A_T the flow is still supersonic:

$$dA = 0, \quad M > 1, \quad du = 0$$

This is case (b) of Figure 4.7. Acceleration of the supersonic flow begins once again in the expanding section beyond the second throat, but is cut short by a shock, which decelerates the flow to a subsonic condition. Beyond the shock the expanding duct acts as a subsonic diffuser:

$$dA > 0, \quad M < 1, \quad du < 0$$

Taken as a whole, the part of the duct beyond the test section acts as a diffuser, which ultimately brings the flow to rest in the evacuated reservoir.

The only part of the entire flow through the tunnel which is essentially non-isentropic is the shock. However, there will in reality be friction at the walls and a further dissipation when the flow exhausts into the reservoir. The pressure variation along the duct is shown in Figure 4.8(b). The purpose of the second throat can be explained using this sketch. Without the second throat, deceleration could at best be achieved by a strong shock in the test section. The weaker shock after the second throat allows running at higher values of p_r, the reservoir pressure, and thus makes possible a longer testing period. If one considers only the steady flow, it appears best to reduce the second-throat area nearly to A^*. However, in order that supersonic flow be established in the test section when the tunnel starts running, it is necessary that the second throat be considerably larger than the first. If the second throat is too small, the shock will establish itself in the diverging part of the inlet and the flow in the test section will be subsonic.

The flow upstream of the shock is the same over a range of reservoir conditions. Within this range, the influence of the downstream conditions cannot make itself felt through the region of supersonic flow. However, as the pressure p_r rises the shock moves towards the second throat, in order that the pressure and velocity behind it be sufficient for the flow to force its way into the reservoir against the rising pressure. When the shock reaches the second throat the situation changes; an advance past the throat does not give a stable flow, that is, a flow with a high enough impulse function to enter the reservoir. Hence the shock moves abruptly through the second throat and establishes itself in the diverging part of the inlet nozzle. The steady supersonic flow in the test section is destroyed, although the tunnel will continue to run unless the valve is closed, first with sonic flow at the first throat, and finally with subsonic flow throughout.

Although weak waves — sound waves — cannot make their way upstream against a supersonic flow, the passage of the shock through the test section shows that a strong enough shock wave can do so. Its speed of propagation is greater than the local flow velocity. This discussion makes it clear why the transition of Figure 4.6(b) is unstable.

In this description of the operation of an intermittent wind tunnel no use has been made of the special properties of air. When results for a perfect gas have been derived, some of the details of this flow will be worked out.

4.2.2 The Perfect Gas

For this important special case, equations (4.1.4) relate the fluid properties to the Mach number of the flow. Introducing

$$\alpha = 1+\frac{\gamma-1}{2}M^2 \tag{4.2.7}$$

for compactness, we have

$$\begin{aligned} \frac{T_0}{T} &= \alpha & \frac{a_0}{a} &= \alpha^{\frac{1}{2}} \\ \frac{\varrho_0}{\varrho} &= \alpha^{\frac{1}{\gamma-1}} & \frac{p_0}{p} &= \alpha^{\frac{\gamma}{\gamma-1}} \end{aligned} \tag{4.2.8}$$

Tables of these functions are given in Appendix V for $\gamma = 1.2$, 1.3 and 1.4. The results for T_0/T and a_0/a are more widely applicable; they hold for any steady, isenergic flow of a perfect gas, not merely for the more restricted isentropic case.

In order to use these results to find the variations in properties along a duct, it is necessary to relate the Mach number to the duct geometry. This can be done using the continuity relation

$$\varrho u A = \varrho^* a^* A^*$$

where the superscript * denotes sonic conditions. It is convenient to express the results in dimensionless form, but the stagnation properties, although useful in equations (4.2.8), do not provide a suitable reference area. Hence the sonic properties will be used as reference values in the continuity relation. They are related to the stagnation values by

$$\left(\frac{p_0}{p^*}\right)^{\frac{\gamma-1}{\gamma}} = \left(\frac{\varrho_0}{\varrho^*}\right)^{\gamma-1} = \left(\frac{a_0}{a^*}\right)^2 = \frac{T_0}{T^*} = \alpha^* = \frac{\gamma+1}{2} \tag{4.2.9}$$

The ratios p^*/p_0, T^*/T_0, etc., are the values appearing opposite $M = 1$ in the isentropic flow tables. Incidentally, the sonic and stagnation states can be used as reference states even if they do not exist in the flow under consideration. These are the states which would be achieved if the flow were to move isentropically and isenergically to $M = 1$ and to $M = 0$, respectively.

When the results (4.2.8) are introduced, the continuity requirement becomes

$$\frac{A}{A^*} = \frac{\varrho^*}{\varrho}\frac{a^*}{u} = \frac{\varrho^*}{\varrho}\frac{(a^*/a)}{M} \tag{4.2.10}$$

$$= \frac{1}{M}\left(\frac{\alpha}{\alpha^*}\right)^{\frac{\gamma+1}{2(\gamma-1)}} = \frac{1}{M}\left[\frac{2}{\gamma+1}\left(1+\frac{\gamma-1}{2}M^2\right)\right]^{\frac{\gamma+1}{2(\gamma-1)}} \tag{4.2.11}$$

The first results (4.2.10) hold for any substance and for any one-dimensional flow process. For the particular case of interest here, a perfect gas flowing isentropically, equations (4.2.7, 8, 9) give the more restricted results (4.2.11). The function A/A^* is also given in the isentropic tables. In a sense it provides the key to them, since it relates the other columns to the duct geometry.

Example 4.7 The wind-tunnel flow discussed in general terms in Example 4.6 can be analysed in detail with the results now available, on the assumption that the air flowing through the tunnel behaves like a perfect gas with $\gamma = 1{\cdot}4$. Consider a tunnel with a test Mach number of 4·00, and with a ratio of test-section area to second-throat area of $A/A_T = 1{\cdot}5$. The test section is 1 m wide and 2 m high.

The ratio of the flow area at $M = 4$ to that for sonic flow is given by the isentropic flow tables for $\gamma = 1{\cdot}4$ as $A/A^* = 10{\cdot}72$; this is the value given by equation (4.2.11). Since the test-section area is $A = 2\ \text{m}^2$, that of the first throat, where sonic flow is achieved, is $A^* = 2/10{\cdot}72 = 0{\cdot}1864\ \text{m}^2$. To simplify manufacture, the inlet nozzles of supersonic tunnels are usually curved in only one plane. Thus the area variation results entirely from changes in the height of the flow passage. In the case considered here, the height of the throat would be 18·64 cm.

The area of the second throat is specified by $A/A_T = 1{\cdot}5$, so that

$$\frac{A_T}{A^*} = \frac{A_T}{A}\frac{A}{A^*} = 7{\cdot}15$$

The corresponding Mach number is found by interpolation in the flow tables: $M_T = 3{\cdot}55$.

The tunnel can run supersonically as long as a shock can be maintained downstream of the second throat. It moves towards the throat as the reservoir fills. In the limiting position when the shock stands in the second throat, $M_1 = 3{\cdot}55$ upstream, and the shock tables give $p_{0_2}/p_{0_1} = 0{\cdot}206$. If the deceleration following the shock is isentropic, $p_{0_2} = p_r$. If the flow up to the shock is isentropic, $p_{0_1} = p_a$. In this ideal case the shock will pass through the second throat when the reservoir pressure is 0·206 atm. In reality, the disgorging of the shock occurs at a somewhat lower pressure, since the subsonic diffusion involves considerable dissipation.

In the absence of a second throat, the limiting case would be that in which the shock stood in the test section. Then $M_1 = 4{\cdot}00$ upstream, and $p_{0_2}/p_{0_1} = 0{\cdot}139$. Supersonic operation would cease at a lower value of $p_{0_2} = p_r$, and the running time of the tunnel would be reduced.

It is instructive to determine the values of the fluid properties in the test section. The stagnation state is that in the atmosphere. Then for $M = 4{\cdot}00$

$$\frac{p}{p_a} = 0{\cdot}00658 \qquad \frac{\varrho}{\varrho_a} = 0{\cdot}0277$$

$$\frac{T}{T_a} = 0{\cdot}238 \qquad \frac{a}{a_a} = 0{\cdot}488$$

The pressure drops to a very low value indeed; this has important implications for measuring technique. With standard sea-level temperature outside the tunnel, the test-section temperature is $T = 0{\cdot}238 \times 288 = 68{\cdot}5\ ^\circ\text{K}$. This very low value points to the problem of freezing of water vapour in the air supplied to the tunnel. The air cannot in fact be drawn directly from the atmosphere, as indicated in Figure 4.8, but must be dried before entering the supersonic tunnel. The velocity in the test section is

$$u = \frac{u}{a}\,\frac{a}{a_a}\,a_a = 4\times 0{\cdot}488\times 340 = 664 \text{ m/sec}$$

This is less than might have been anticipated for $M = 4$, since the sound speed drops markedly as the gas accelerates into the tunnel.

4.2.3 Thrust Calculations for a Perfect Gas

In this discussion of isentropic flow in a duct no use has been made of the momentum principle. The force acting on the fluid between two sections of one-dimensional flow is given by

$$F = I_2 - I_1 \tag{2.2.12}$$

the difference between the values of the impulse function

$$I = (p + \varrho u^2)A \tag{2.2.9}$$

at the two sections. In keeping with the earlier treatment, this result will be written in dimensionless form:

$$\frac{F}{I^*} = \frac{I_2}{I^*} - \frac{I_1}{I^*}$$

where I^* is the impulse function for sonic flow.

For a perfect gas, with $a^2 = \gamma p/\varrho$

$$I = pA\left[1 + \frac{\varrho}{p}u^2\right] = pA(1 + \gamma M^2) \tag{4.2.12}$$

Hence

$$I^* = p^* A^*(1+\gamma) \tag{4.2.13}$$

and

$$\frac{I}{I^*} = \frac{p}{p^*}\,\frac{A}{A^*}\,\frac{1+\gamma M^2}{1+\gamma} \tag{4.2.14}$$

These results apply to any one-dimensional flow of a perfect gas.

For the case of immediate interest, isentropic flow, the ratios of equations (4.2.8, 9, 11) can be introduced to give

$$\begin{aligned}\frac{I^*}{I} &= \frac{a}{a^*}\,\frac{1}{M}\,\frac{1+\gamma M^2}{1+\gamma}\\ &= \frac{1}{M}(1+\gamma M^2)\left[2(1+\gamma)\left(1+\frac{\gamma-1}{2}M^2\right)\right]^{-\frac{1}{2}}\end{aligned} \tag{4.2.15}$$

This ratio too is given in the isentropic flow tables.

Example 4.8 A rocket motor discharges from a supersonic nozzle with an area ratio $A_e/A^* = 5$, and a throat area of $A^* = 5$ in^2. The chamber diameter is equal to that of the nozzle exit, and the pressure at the entrance to the nozzle is 300 atm. The combustion products may be taken to behave like a perfect gas with $\gamma = 1{\cdot}2$. The net thrust of the motor and the thrust components acting on the converging and diverging parts of the nozzle are to be found, on the assumption that the pressure over the whole of the exterior of the rocket is 5 lbf/in^2.

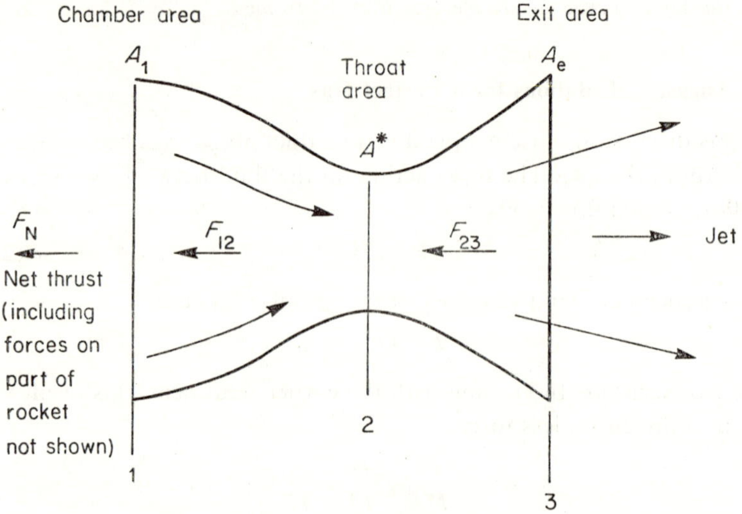

Figure 4.9 Forces generated by the flow through a rocket nozzle

The rocket is sketched in Figure 4.9. The three forces to be determined are F_N, the net thrust on the rocket, F_{12} on the converging part of the nozzle, and F_{23} on the diverging part. They are given in general terms by

$$F_N = I_3 - p_a A_e$$
$$F_{12} = I_2 - I_1$$

and

$$F_{23} = I_3 - I_2$$

Note that

$$F_{12} + F_{23} = I_3 - I_1 \neq F_N$$

The difference is the thrust applied on the forward face of the chamber.

The calculations must be directed towards the determination of the conditions at the throat, since sonic values have been used to render the results dimensionless. For $A_1/A^* = 5{\cdot}00$, isentropic flow tables for $\gamma = 1{\cdot}2$ give $M_1 = 0{\cdot}12$ and $p_1/p_0 = 0{\cdot}990$. Then

$$\frac{p^*}{p_1} = \frac{p^*}{p_0}\frac{p_0}{p_1} = \frac{0{\cdot}564}{0{\cdot}990} = 0{\cdot}570$$

For a chamber pressure of $p_1 = 300$ atm, the sonic impulse function is

$$I^* = p^* A^* (1+\gamma) = (0{\cdot}570 \times 14{\cdot}7 \times 300) \times 5 \times 2{\cdot}2$$
$$= 27{,}600 \text{ lbf}$$

Then

$$\frac{p_a A_e}{I^*} = \frac{5 \times 25}{27{,}600} = 0{\cdot}005$$

The flow tables give the values of I/I^* for the three sections of interest as 4·04, 1·00, and 1·324. Hence the thrust components are

$$F_N = (1{\cdot}324 - 0{\cdot}005)I^* = 36{,}400 \text{ lbf}$$
$$F_{12} = (1{\cdot}00 - 4{\cdot}04)I^* = -83{,}900 \text{ lbf}$$
$$F_{23} = (1{\cdot}324 - 1{\cdot}00)I^* = 8{,}900 \text{ lbf}$$

As would be expected, the internal pressure gives rise to a large outwards force on the converging part of the nozzle. Note that the force on the diverging part is only a quarter of the total thrust; most of the thrust has already been achieved in the acceleration to the throat.

4.3 FLOW MEASUREMENT

4.3.1 Pressure and Temperature

An important class of applications of isentropic-flow theory is in the measurement of fluid velocity and flow rate. Whenever a fluid flow is distorted, for example, by a change in flow area, by a change of flow direction or by deceleration, there occur pressure changes related to the velocity of the flow. Since pressure is relatively easy to measure, these changes provide a convenient way of determining the fluid velocity and flow rate. The other easily measured fluid property is temperature. Taken together, pressure and temperature measurements determine the thermodynamic state of the flowing fluid and, suitably interpreted, its velocity as well.

Here only flow metering devices which depend upon fluid-mechanical processes will be considered. However, a wide range of transducers utilizing electrical phenomena are used to measure flow rate and pressure. Some of the *electrical transducers* in common use are hot-wire anemometers, thermistors, piezo-electric crystals and strain gauges. The chief advantages of electrical devices are their ability to register very rapid changes in velocity and pressure, and their ability to provide an electrically generated record of the process. Such devices are necessary, for example, in studying turbulence or propagating shock waves. The chief disadvantages of electrical measuring techniques are expense, the necessity of calibration, and the necessity of using complex electrical filters, amplifiers and recorders. Hence, unless rapid response and continuous recording are required, purely fluid-mechanical meters are generally preferred.

Techniques of measurement have been discussed at several points in the preceding chapters: the use of manometers for pressure measurement in

Section 2.5.1 and the use of a Pitot tube to find total pressure in Sections 2.4.2 and 3.3.2. Here we shall look at the problems which arise in measuring static pressure and temperature.

The determination of the *static pressure* in a moving fluid is simple in principle, but requires some care in practice. The pressure in a fluid flowing through a pipe can, in principle, be determined simply by connecting a manometer or other gauge to a hole through the pipe wall. However, if an accurate measurement is required, the hole must be small, so that an eddy will not form in it, and free from burrs extending into the flow. Moreover, if the pressure varies in time, the measured value will lag behind that in the fluid, chiefly because a considerable time is required for the fluid to flow through the small hole and along the manometer tubing. As was suggested above, this problem can be avoided by using an electrical pressure transducer.

The pressure within a flow field can be determined using a *static-pressure probe*, shown in Figure 4.10(a). The precautions noted above are also required here. Moreover, care must be taken in locating the measuring holes relative to the tip and stem of the probe, in order that the effects of acceleration near the tip and deceleration near the stem cancel one another. Probes of standard design, for the most part developed in government laboratories, have this characteristic. The static-pressure probe, like the Pitot tube, must be precisely aligned with the local flow direction if an accurate measure of the free-stream pressure is to be obtained.

The *Pitot-static probe* shown in Figure 4.10(b) combines in one instrument the means of finding total and static pressures, and is therefore particularly suited to the determination of fluid velocity. For constant-density flow, the velocity is related to the pressures by

$$u = \left[\frac{2(p_0 - p)}{\varrho_c}\right]^{\frac{1}{2}} \tag{4.3.1}$$

In Example 4.1 it has been seen that this result will apply with good accuracy for flow at low Mach numbers, no matter what the fluid, the error in $p_0 - p$ being only 1 per cent for $M = 1/4$.

At higher Mach numbers the nature of the equation of state affects the relationship between the fluid velocity and the total and static pressures. For the important case of a perfect gas, the Mach number is related to the pressures by one of equations (4.1.4):

$$\left(\frac{p_0}{p}\right)^{\frac{\gamma-1}{\gamma}} = 1 + \frac{\gamma-1}{2} M^2$$

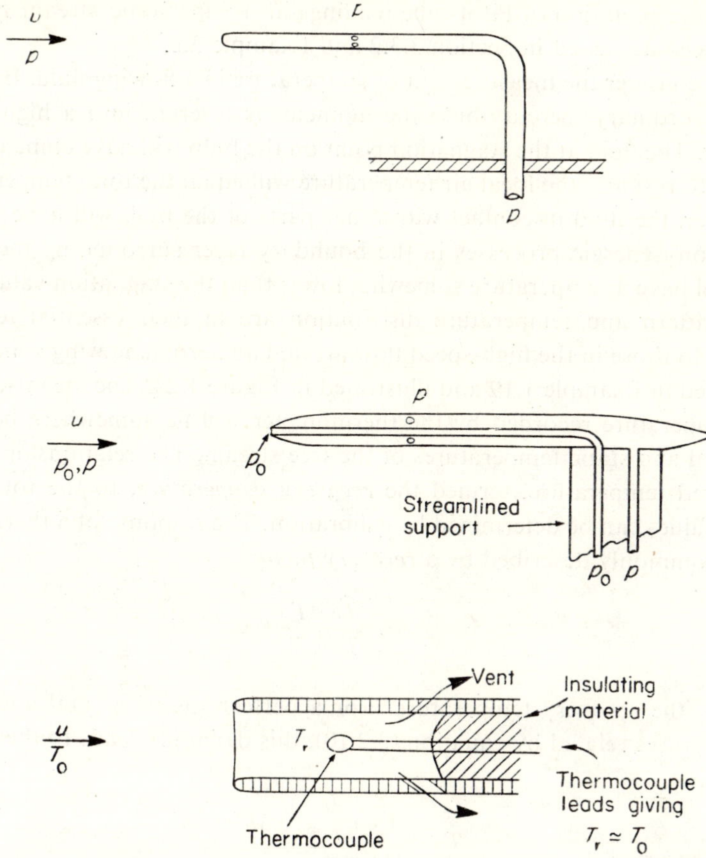

Figure 4.10 Measurement of pressure and temperature in a flowing fluid. (a) The static-pressure probe; (b) The Pitot-static probe (c) The total-temperature probe

This relationship is given in isentropic flow tables. The determination of the velocity itself requires a knowledge of some temperature associated with the flow. For a perfect gas, the velocity is given by

$$
\begin{aligned}
u &= [2c_p(T_0 - T)]^{\frac{1}{2}} \\
&= \left\{2c_pT\left[\left(\frac{p_0}{p}\right)^{\frac{\gamma-1}{\gamma}} - 1\right]\right\}^{\frac{1}{2}} \qquad (4.3.2) \\
&= \left\{2c_pT_0\left[1 - \left(\frac{p}{p_0}\right)^{\frac{\gamma-1}{\gamma}}\right]\right\}^{\frac{1}{2}}
\end{aligned}
$$

The interpretation of Pitot-tube readings in a supersonic stream has already been discussed in Section 3.3.2 and Example 3.8.

Next consider the measurement of temperature in a flowing fluid. Imagine that an ordinary mercury-bulb thermometer is inserted into a high-speed air flow. The fluid at the stagnation point on the bulb will have come isenergically to rest and the local air temperature will equal the total temperature. However, the fluid in contact with other parts of the bulb will have undergone non-isenergic processes in the boundary layer surrounding the bulb, and will have a temperature somewhat lower than the stagnation value. The flow pattern and temperature distribution are in their essential features similar to those in the high-speed flow around an aeroplane wing which was discussed in Example 1.19 and illustrated in Figure 1.24. The mean value of the temperature recorded by the thermometer will lie somewhere between the total and static temperatures of the free stream. The relationship of the measured temperature, termed the *recovery temperature*, to the total and static values can be determined by calibration. The response of a thermometer is commonly described by a *recovery factor*

$$r = \frac{T_r - T}{T_0 - T} \tag{4.3.3}$$

with T_r the recovery temperature. For a perfect gas with total and static temperatures related by equation (3.2.10), this definition leads to the simple result

$$\frac{T_r}{T} = 1 + r\frac{\gamma - 1}{2}M^2 \tag{4.3.4}$$

The recovery factor depends on the geometry of the temperature probe and on the heat-transfer characteristics of the fluid. For a particular thermometer and fluid it remains nearly constant over considerable ranges of flow conditions. However, the recovery factor must be expected to be significantly different in subsonic and supersonic flows.

From what has been said above, it will be apparent that the static temperature of a fast-moving fluid is very difficult to measure directly. However, a temperature near the stagnation value can be measured using a device rather like a Pitot tube. A suitable arrangement is sketched in Figure 4.10(c). Within the measuring chamber the flow velocity is small. Hence $r \simeq 1$ for this thermometer.

Another problem encountered in measuring temperature is the effect of *radiation*. A thermometer exposed to strong sunlight, or held near a piece of very hot or very cold metal, will not register precisely the temperature of the air around it. The radiation of energy to or from the bulb will raise or lower its temperature from the ambient level. This problem arises in still as well as moving fluids. It can be avoided by shielding the thermometer, that is, by placing an opaque barrier between it and the source (or sink) of radiant energy.

Example 4.9 It is required to find the ranges of total pressure and recovery temperature which have to be measured by the instruments of a supersonic airliner. The temperature probe has a recovery factor of 0·8 in subsonic flow and 0·9 in supersonic flow. Typical flight conditions to be considered are

(1) take-off at sea level at 200 miles per hour
(2) subsonic climb at $M = 0{\cdot}6$ at 10,000 ft, and
(3) supersonic cruise at $M = 2$ at 40,000 ft.

Standard atmospheric conditions are to be adopted throughout.

Table 4.2 presents the data and derived information. Most of the results come directly from tables for $\gamma = 1{\cdot}4$.

Table 4.2. Calculation of Pitot pressures and recovery temperatures

Case	Altitude (m)	p_a(bar)	T_a (°K)	M_1	p_1/p_{0_1}	p_{0_1}(bar)
1	0	1·013	288	0·263	0·953	1·063
2	3040	0·697	268	0·600	0·784	0·890
3	12,200	0·187	217	2·00	0·1278	1·462

Case	p_{0_2}/p_{0_1}	p_{0_2}(bar)	T_1/T_0	T_0 (°K)	T_r (°K)
1	—	—	0·986	292	291
2	—	—	0·933	287	282
3	0·721	1·056	0·556	390	373

The take-off Mach number is calculated as

$$M = \frac{200 \times 88}{60 \times 3{\cdot}28 \times 340} = 0{\cdot}263$$

Isentropic flow tables for $\gamma = 1{\cdot}4$ then give p_1/p_{0_1}.
Alternatively, since the Mach number is low, the constant-density result can be used:

$$\begin{aligned} p_0 &= p + \tfrac{1}{2}\varrho_c u^2 \\ &= 101{,}300 + \tfrac{1}{2} \times 1{\cdot}225 \times (89{\cdot}5)^2 \text{ N/m}^2 \\ &= 1{\cdot}062 \text{ bar} \end{aligned}$$

4.3.2 The Venturi Meter

We now turn to the problem of measuring the rate at which fluid flows through a duct. The mass flow through a section at which the flow is approximately one-dimensional is given by

$$\dot{m} = \varrho u A \tag{4.3.5}$$

This relationship is conceptually simple, but does not in practice provide a convenient way of finding the flow rate, especially when the density varies along the flow. Since temperature and pressure are the easily measured properties, it is expedient to express the flow rate directly in terms of these properties.

As was pointed out earlier, almost any deflexion or distortion of the flow in a duct will give rise to a change in pressure which can be related, either theoretically or by empirical calibration, to the flow rate. The bend meter of Example 4.5 illustrates this principle. There are two advantages in using a nearly isentropic distortion to generate a pressure change: first, simple theoretical results are available to relate the measured values and flow rate and, secondly, the losses introduced in making the measurement are small. The Venturi meter shown in Figure 4.11 is a device possessing these advantages. In the short nozzle the flow accelerates nearly isentropically; in the slowly expanding diffuser the flow returns to its original area with relatively small dissipation.

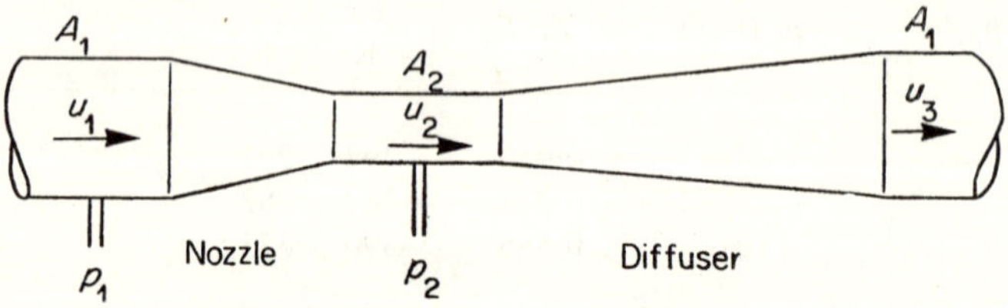

Figure 4.11 A Venturi meter

If the change in elevation and dissipation are negligible, and if the flow is taken to be one-dimensional at the throat (section 2) and just upstream of the meter (section 1), Bernoulli's equation (4.1.1) gives

$$u_2^2 - u_1^2 = 2\int_2^1 \frac{dp}{\varrho} \tag{4.3.6}$$

The continuity requirement is $u_2/u_1 = \varrho_1 A_1/\varrho_2 A_2$, giving

$$u_1^2 = \frac{2}{(\varrho_1 A_1/\varrho_2 A_2)^2 - 1}\int_2^1 \frac{dp}{\varrho} \tag{4.3.7}$$

The flow rate follows from equation (4.3.5):

$$\dot{m}^2 = (\varrho_1 \dot{V}_1)^2 = \frac{2}{(\varrho_2 A_2)^{-2} - (\varrho_1 A_1)^{-2}} \int_2^1 \frac{dp}{\varrho} \tag{4.3.8}$$

These results apply for any fluid and for any density change.

For flow at low Mach number it may be assumed that $\varrho = \varrho_c$ between sections 1 and 2. Hence the velocity and flow rate are given by

$$u_1 = \frac{\dot{m}}{\varrho_c A_1} = \frac{\dot{V}}{A_1} = \left[\left(\frac{A_1}{A_2}\right)^2 - 1\right]^{-\frac{1}{2}} \left[\frac{2(p_1 - p_2)}{\varrho_c}\right]^{\frac{1}{2}} \tag{4.3.9}$$

These results too apply for any fluid, so long as the density change within the meter is small.

An explicit result can also be obtained for a perfect gas. With

$$\frac{p}{p_1} = \left(\frac{\varrho}{\varrho_1}\right)^{\gamma} \quad \text{and} \quad c_p T = \left[\frac{\gamma}{\gamma - 1}\right] \frac{p_1}{\varrho_1}$$

equation (4.3.7) gives

$$u_1^2 = 2c_p T_1 \left[\left(\frac{p_1}{p_2}\right)^{\frac{2}{\gamma}} \left(\frac{A_1}{A_2}\right)^2 - 1\right]^{-1} \left[1 - \left(\frac{p_2}{p_1}\right)^{\frac{\gamma-1}{\gamma}}\right] \tag{4.3.10}$$

in terms of the easily measured properties, pressure and temperature. The relationships

$$\frac{u_1^2}{2c_p T_1} = \frac{\gamma - 1}{2} M_1^2 = \frac{\gamma - 1}{2\gamma} R \left(\frac{\dot{m} T_1^{\frac{1}{2}}}{A_1 p_1}\right)^2$$

provide the means of calculating the Mach number of the flow in the pipe or the mass flow through it.

Example 4.10 The flow of air through a pipe is to be measured using a Venturi meter whose throat area is one-half the pipe area. How does the velocity parameter $u_1^2/(2c_p T_1)$ vary with p_2/p_1?

Equation (4.3.10) gives, with $\gamma = 1{\cdot}4$ and $A_1/A_2 = 2$

$$\frac{u_1^2}{2c_p T_1} = \frac{M_1^2}{5} = \frac{1 - (p_2/p_1)^{0{\cdot}286}}{4(p_1/p_2)^{1{\cdot}428} - 1}$$

The values given by this formula are shown in Figure 4.12. Only ratios $p_2/p_1 < 1$ have been considered; hence in all the flows represented the pressure falls and the velocity rises as the gas passes through the contraction. These are initially subsonic flows. For an initially supersonic flow, $p_2/p_1 > 1$ would result from this contraction of the flow passage.

Despite the restriction to initially subsonic flows, the dependence on pressure ratio is somewhat puzzling at first sight. For $p_2/p_1 \simeq 1$, there is no difficulty; as the flow speed increases, the pressure at the throat falls. But for low values of p_2/p_1 this trend is reversed; this indicates that the velocity at Section 2 is then supersonic. The small sketches in Figure 4.12 show the kinds of ducts to which the plotted results apply. The left-hand part of the velocity versus pressure curve corresponds to ducts which contract to $A_T/A_1 < \frac{1}{2}$, and then expand again to $A_2/A_1 = \frac{1}{2}$. This part of the curve does not represent actual flows in a duct of fixed, strictly converging shape.

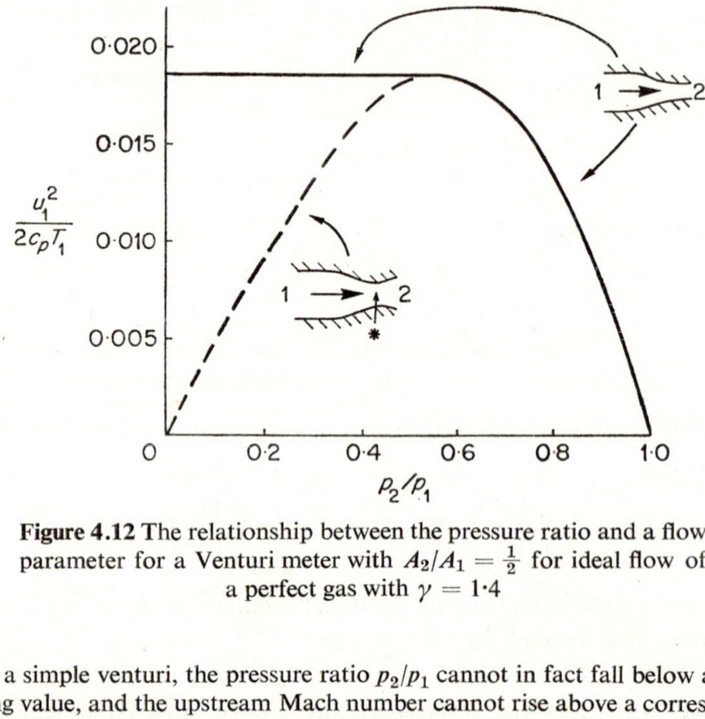

Figure 4.12 The relationship between the pressure ratio and a flow parameter for a Venturi meter with $A_2/A_1 = \frac{1}{2}$ for ideal flow of a perfect gas with $\gamma = 1{\cdot}4$

For a simple venturi, the pressure ratio p_2/p_1 cannot in fact fall below a certain limiting value, and the upstream Mach number cannot rise above a corresponding limit. This condition corresponds to sonic flow at the throat: $M_2 = 1$. The limiting conditions can be determined using isentropic tables. For

$$\frac{A}{A^*} = \frac{A_1}{A_2} = 2$$

the tables give

$$M_1 = 0{\cdot}306 \quad \text{and} \quad \frac{p_1}{p_0} = 0{\cdot}937$$

For $M_2 = 1{\cdot}00$, the tables give $p_2/p_0 = 0{\cdot}528$.

Then the limiting values are

$$\frac{u_1^2}{2c_pT_1} = \frac{M_1^2}{5} = 0{\cdot}0187$$

and

$$\frac{p_2}{p_1} = \frac{0{\cdot}528}{0{\cdot}937} = 0{\cdot}563$$

4.3.3 The Sonic Nozzle

The preceding example indicates that the acceleration in a converging nozzle is limited by the requirement that the velocity in the throat not exceed the local speed of sound. When the flow at the throat is sonic or critical the nozzle is said to be choked. Changes in downstream conditions which leave the nozzle choked have no influence on the flow upstream of the throat. This characteristic of a choked nozzle can be put to use in measuring the flow through a system which is choked at some section.

Figure 4.13 shows a nozzle flow supplied from a reservoir in which the fluid is sensibly at rest with pressure p_0. The nozzle is smoothly rounded and has parallel walls at the exit, so that the flow may be taken to be one-dimensional with $\mathrm{d}A = 0$ on leaving the nozzle. The pressure in the region into which the nozzle exhausts is often referred to as the back pressure on the nozzle, and will be denoted by p_b. For flow out of the reservoir, it is always the case that $p_\mathrm{b} < p_0$. If $p_\mathrm{b} > p^*$, the pressure for sonic flow, the flow remains subsonic at the throat and is not choked. Then

$$p_0 > p_\mathrm{e} = p_\mathrm{b} > p^*$$

For $p_\mathrm{b} < p^*$ the flow is choked, and a further expansion takes place outside the nozzle. Then

$$p_0 > p_\mathrm{e} = p^* > p_\mathrm{b}$$

These relationships amongst the pressures of the system are shown in Figure 4.13(b).

Although the mass flow through the nozzle does not depend on the back pressure, so long as $p_\mathrm{b} < p^*$, it does depend on the properties of the fluid entering the nozzle. This dependence can conveniently be taken into account by expressing the mass-flow equation in dimensionless form. Quite generally

$$\frac{\dot{m}}{\varrho_0 a_0 A} = \frac{\varrho}{\varrho_0}\frac{u}{a}\frac{a}{a_0} \tag{4.3.11}$$

For a section at which sonic flow exists, $A = A^*$, and

$$\frac{\dot{m}}{\varrho_0 a_0 A^*} = \frac{\varrho^*}{\varrho_0}\frac{a^*}{a_0} \tag{4.3.12}$$

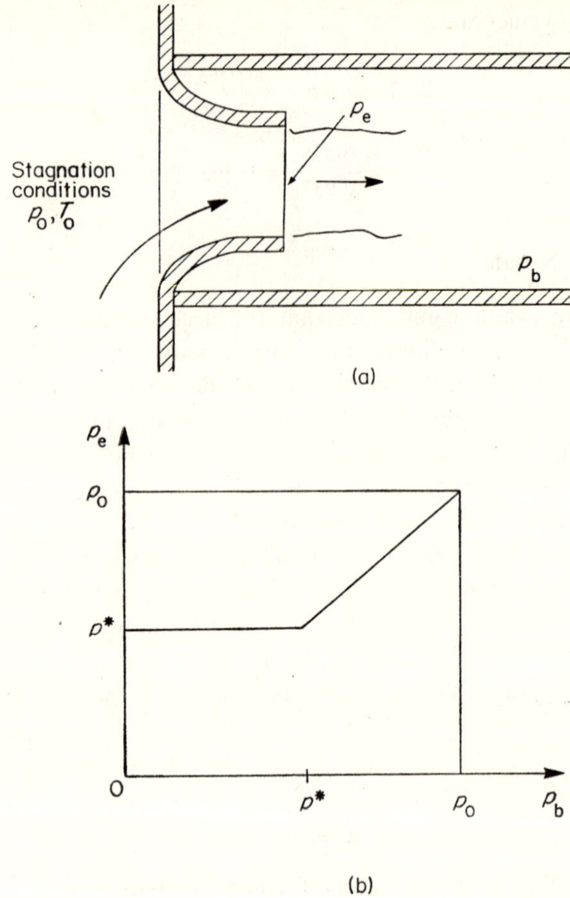

Figure 4.13 Flow through a contracting nozzle. (a) Geometry and limiting properties; (b) Relationships between pressures

To express these results in terms of the easily measured properties, pressure and temperature, one must consider a particular equation of state. The kind of fluid most commonly treated in this way is the perfect gas, for which

$$\frac{\dot{m}}{\varrho_0 a_0 A} = \left(\frac{R}{\gamma}\right)^{\frac{1}{2}} \frac{\dot{m}T_0^{\frac{1}{2}}}{p_0 A} \tag{4.3.13}$$

The subsequent discussion will relate to this special case, and the mass-flow parameter used will be $\dot{m}T_0^{1/2}/p_0A$. This is not in fact dimensionless, although it can easily be made dimensionless as indicated in equation (4.3.13).

Up to this point no restriction has been put on the process linking the reservoir and critical states. In a later chapter other processes linking reservoir and critical states will be considered, but our interest is for the moment in an isentropic transition, since the flow in a short, rounded nozzle like that shown in Figure 4.13 should be very nearly isentropic. For a perfect gas, the relations (4.2.8) give

$$\frac{\dot{m}T_0^{\frac{1}{2}}}{p_0 A} = \left(\frac{\gamma}{R}\right)^{\frac{1}{2}} M\alpha^{-\frac{\gamma+1}{2(\gamma-1)}} \tag{4.3.14}$$

This relates the mass-flow parameter to the local Mach number. Indeed, at low Mach numbers

$$\frac{\dot{m}T_0^{\frac{1}{2}}}{p_0 A} \simeq \left(\frac{\gamma}{R}\right)^{\frac{1}{2}} M$$

For sonic flow, the relations (4.2.9) give

$$\frac{\dot{m}T_0^{\frac{1}{2}}}{p_0 A_T} = \left(\frac{\gamma}{R}\right)^{\frac{1}{2}} \left[\frac{2}{\gamma+1}\right]^{\frac{\gamma+1}{2(\gamma-1)}} \tag{4.3.15}$$

For particular values of γ and R, the mass flow through a choked nozzle is given by

$$\dot{m} = \text{constant} \times \frac{p_0 A_T}{T_0^{\frac{1}{2}}} \tag{4.3.16}$$

The constant is not dimensionless. For air, it is 0·532 for the English engineering units given in the first column of Table 1.1, that is, mass flow in lb/sec, pressure in lbf/ft^2, area in ft^2, and temperature in °R.

A law of the form (4.3.16) gives the flow of a perfect gas through a choked nozzle or orifice of any shape, provided that the flow from stagnation conditions is steady and isentropic. However, the geometric area of the outlet A_e will equal the throat area A_T only if the flow is parallel in the exit plane. At a sharp-edged orifice, for example, the effective throat is located in the jet outside the orifice, and the throat area A_T is considerably smaller than the outlet area A_e. In such a case, it is convenient to use the known area A_e in a law of the form (4.3.16), the constant being determined by experiment. Note that the mass flow can still vary if p_0 and T_0 change. It is the mass-flow parameter $\dot{m}T_0^{1/2}/p_0$ which is independent of reservoir conditions.

A nozzle can also be used to measure the flow rate when the flow is not choked at the exit. Equation (4.3.14) gives for the exit conditions

$$\frac{\dot{m}T_0^{\frac{1}{2}}}{p_0 A_e} = \left(\frac{\gamma}{R}\right)^{\frac{1}{2}} M_e \alpha_e^{-\frac{\gamma+1}{2(\gamma-1)}}$$

When $p_e = p_b$ and p_0 are known, M_e and α_e can be found from the known value p_e/p_0. Hence a law of the form

$$\dot{m} = f\left(\frac{p_e}{p_0}\right) \times \frac{p_0 A_e}{T_0^{\frac{1}{2}}} \tag{4.3.17}$$

gives the flow rate when the nozzle is unchoked. For $p_e = p^* \geqslant p_b$, the result (4.3.16) is recovered. In general, three quantities, p_e, p_0 and T_0, must be measured in order to calculate the mass flow. For the special case of choked flow, only p_0 and T_0 need be measured; the condition $M_e = 1$ is equivalent to a third piece of measured data.

Example 4.11 The mass-flow parameter introduced to describe the performance of a sonic nozzle has a wider application. It can be used to relate flows in differing environmental conditions and to relate the results of small-scale experiments to the performance of full-scale equipment.

An air intake system for a supersonic aeroplane is to be tested in the wind tunnel discussed in Examples 4.6 and 4.7. The geometry is sketched in Figure 4.14. The diameter of the duct of the scale model is 10 cm, while that in the aeroplane is 100 cm. For comparable conditions, what is the ratio of the mass flows through the model and through the actual intake in flight at 12,000 m?

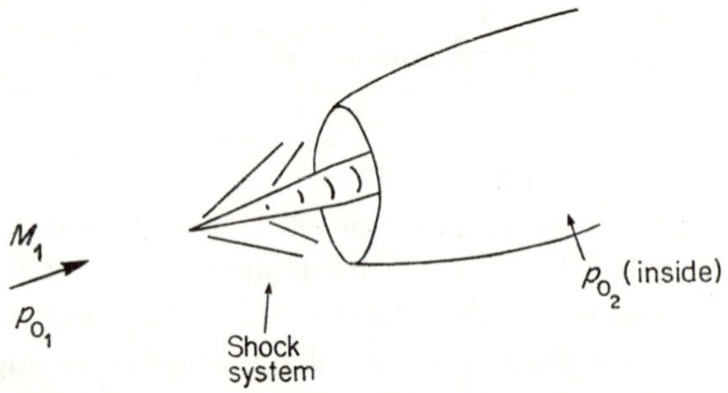

Figure 4.14 The air intake of a supersonic aeroplane

The phrase 'comparable conditions' implies that the distribution of Mach number through the two flows, model and prototype, is similar. Thus the Mach number of the flow approaching each duct is $M = 4$. Moreover, equation (4.3.14) indicates that there is a unique relationship between Mach number and the parameter $\dot{m}T_0^{1/2}/p_0A$ (unique, that is, for a particular perfect gas). Hence the two duc flows are related by

$$\left(\frac{\dot{m}T_0^{\frac{1}{2}}}{p_{0_2}A}\right)_{\text{model}} = \left(\frac{\dot{m}T_0^{\frac{1}{2}}}{p_{0_2}A}\right)_{\text{prototype}}$$

for comparable flow conditions. Since the flow geometry, including the shock system, is geometrically similar under these conditions, the ratio p_{0_2}/p_{0_1} will be the same for the test and in-flight conditions. Thus we conclude that the model and prototype mass flows are in the ratio

$$\frac{\dot{m}_{\rm m}}{\dot{m}_{\rm p}} = \left(\frac{p_{0_1}A}{T_0^{\frac{1}{2}}}\right)_{\rm m}\left(\frac{T_0^{\frac{1}{2}}}{p_{0_1}A}\right)_{\rm p}$$

For the tests in the blow-down tunnel, the total properties are those of the atmosphere outside:

$$p_{0_1} = 1{\cdot}013 \text{ bar} \quad \text{and} \quad T_0 = 288\ ^\circ\text{K}$$

For flight at 12,000 m, the local atmospheric properties are modified using the isentropic compression ratios:

$$p_{0_1} = p_{\rm a}\frac{p_{0_1}}{p_1} = \frac{0{\cdot}194}{0{\cdot}00658} = 29{\cdot}5 \text{ bar}$$

and

$$T_0 = T_{\rm a}\frac{T_0}{T_1} = \frac{217}{0{\cdot}238} = 910\ ^\circ\text{K}$$

Then

$$\frac{\dot{m}_{\rm m}}{\dot{m}_{\rm p}} = \frac{1{\cdot}013}{29{\cdot}5}\cdot\left(\frac{910}{288}\right)^{\frac{1}{2}}\cdot\left(\frac{1}{10}\right)^2 = \frac{1}{1640}$$

gives the ratio of the mass flows.

4.3.4 Critical Flow at a Weir

In Section 3.3.3 it was noted that there is a general similarity between compressible flow in a duct and open-channel flow. In particular, the shock wave and hydraulic jump were found to act in a similar manner in decelerating the fluid from a velocity above the local signal speed to a velocity less than that speed. For a gas, the signal speed is the speed of sound; for a free-surface flow, it is the speed of long waves of small amplitude $(gd)^{1/2}$ where d is the water depth.

This analogy extends to isentropic flow. Consider two-dimensional flow over the smooth-crested weir shown in Figure 4.15. If the flow is lossless and one-dimensional at each section, the Bernoulli equation evaluated at the surface gives

$$\frac{u^2}{2}+g(d+z) = gH_0, \quad \text{a constant}$$

and the continuity condition is

$$ud = \dot{v}$$

the constant flow per unit breadth of the channel. Then

$$\frac{u^2}{2}+g\left(\frac{\dot{v}}{u}+z\right) = gH_0$$

Differentiation gives

$$u\,\mathrm{d}u - \frac{g\dot{v}}{u^2}\,\mathrm{d}u + g\,\mathrm{d}z = 0$$

or

$$(u^2 - gd)\frac{\mathrm{d}u}{u} = -g\,\mathrm{d}z$$

This result is equivalent to the area–velocity relation (4.2.4). At the weir crest, where $\mathrm{d}z = 0$, one might have $\mathrm{d}u = 0$; this is the case shown in Figure 4.15(a). Alternatively, one may have $u_c^2 = gd_c$, the crest velocity being equal to the local speed of long waves. This situation analogous to sonic flow is

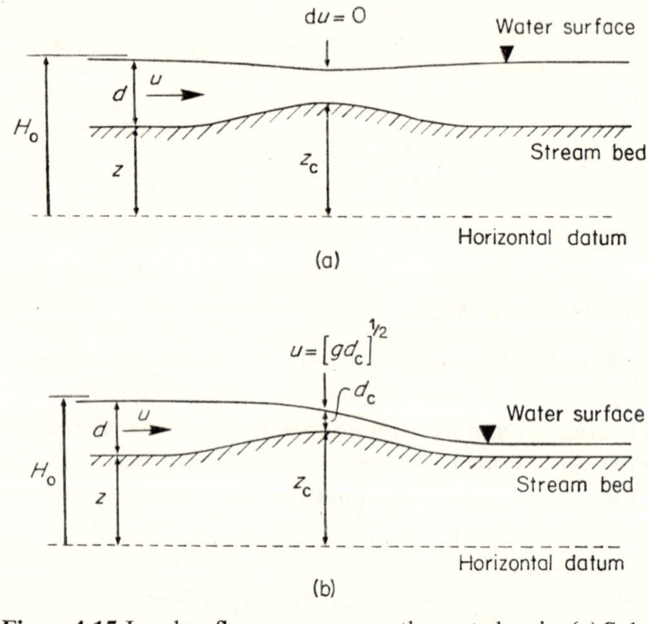

Figure 4.15 Lossless flow over a smooth-crested weir. (a) Subcritical throughout; (b) Transition from subcritical to supercritical flow

termed critical flow. When the flow becomes critical, it can pass from the subcritical to the supercritical condition at the weir crest; this process is shown in Figure 4.15(b).

The critical depth is related to the total head by

$$H_0 = z_c + d_c + \frac{u_c^2}{2g} = z_c + \frac{3d_c}{2}$$

Hence, when the flow is critical at the crest, the flow rate is found to be

$$\dot{v} = u_c d_c = (g d_c^3)^{\frac{1}{2}} = \left(\frac{8g}{27}\right)^{\frac{1}{2}} (H_0 - z_c)^{\frac{3}{2}}$$

In this case a single measurement, that of the difference in elevation $H_0 - z_c$, is sufficient to determine the flow rate. Somewhat similar formulae can be used to relate depth to flow rate in other open-channel meters.

As the preceding discussion illustrates, there is a broad similarity between the techniques of measurement appropriate to compressible flows in ducts and to constant-density flows with a free surface. Since open-channel flows are not of much interest from the point of view of thermodynamics, they will not be considered further in this book. It is hoped that the reader will have other opportunities to study the fascinating behaviour of flows with a free surface. Such flows are of great importance in hydraulic and marine engineering, and are also of interest to geologists and oceanographers.

4.4 NEARLY ISENTROPIC FLOWS

4.4.1 Nozzle and Diffuser Flows

Real flows are never exactly isentropic or one-dimensional. Nevertheless, as has been assumed in the preceding discussions, a sufficiently accurate representation can sometimes be obtained using results for one-dimensional isentropic flow. These results can prove useful, even when the deviations from the ideal are significant, by providing a pattern with which real flows can be compared and on which a more realistic representation can be based.

Although the physical processes which cause departures from the isentropic, one-dimensional model can be identified, their effects cannot easily be calculated. Hence the ideal results are usually modified by introducing empirical factors which combine the effects of irreversibility and of deviations from one-dimensional flow. Frequently these coefficients also depend on the way in which the measurements are made and interpreted.

Before considering the way in which the empirical coefficients are defined and used, we shall look briefly at the processes which they represent. Separate consideration must be given to nozzle and diffuser flows, for the response of the boundary layer on the duct wall is strongly dependent on the sign of the pressure change which the fluid encounters. The intermediate case of flow at constant area cannot usefully be discussed on the basis of the isentropic model, which gives no change in properties along the flow, and hence fails to provide a meaningful pattern for the actual development.

Figures 4.16(a) and (b) show the distributions of time-mean velocity at entry and exit from ducts which contract and expand slightly. Only the more common subsonic case will be considered; hence these ducts act, respectively, as a nozzle and as a diffuser. The effect of the contraction is to reduce the thickness of the boundary layer, while that of the expansion is to increase its thickness. This difference can be ascribed to the retardation of the boundary layer advancing against an adverse pressure gradient, that is, against a rising pressure. The deceleration involves the conversion of momentum into pressure. Since the boundary layer has already been slowed down by friction, it does not have as large a stock of momentum with which to overcome the rising pressure and hence undergoes a larger deceleration.

Figures 4.16(c) and (d) indicate the nature of the flows in ducts which impose a larger change in flow area. Consider the diffuser flow (d) first. Here the retardation of the boundary layer is so marked that the fluid near the wall comes to a stop, and the through-flow separates from the duct wall. This flow re-attaches at the end of the expansion, enclosing a region of unsteady, recirculating flow next to the wall. This flow system, which is generally similar to the eddying flow downstream of a valve (see Figure 2.19(a)) and to the hydraulic jump (see Figure 2.39), involves very large local dissipation. In the rapidly contracting duct of Figure 4.16(c), the fluid separates from the duct walls in a region where there is a local adverse pressure gradient, but re-attaches soon after, enclosing a small separation bubble. Some dissipation will be associated with this local separation and re-attachment.

It is apparent that departures from the isentropic, one-dimensional pattern are most marked in diffuser flows. Indeed, the augmented losses in the rapid contraction of Figure 4.16(c) can be attributed to inefficient diffusion; the fluid is over-accelerated as a result of separation, and then diffuses to fill the whole duct once more. Evidently, fairly gradual changes in duct area are necessary for efficient, that is, nearly isentropic, diffusion. The restriction on an accelerating flow is not so stringent, but care must be taken to avoid large local adverse pressure gradients.

In a very long diffuser, the direct effect of wall friction provides an important source of dissipation. Hence, for a specified change in flow area, there is a limit beyond which further lengthening of the diffuser serves to increase, rather than reduce, the net dissipation. For the large-scale, turbulent flows which are usual in engineering equipment, optimum diffusion is achieved in a cone having an included angle between 5° and 8°. This criterion is not universally valid. If, for example, the flow passage bends as well as expands,

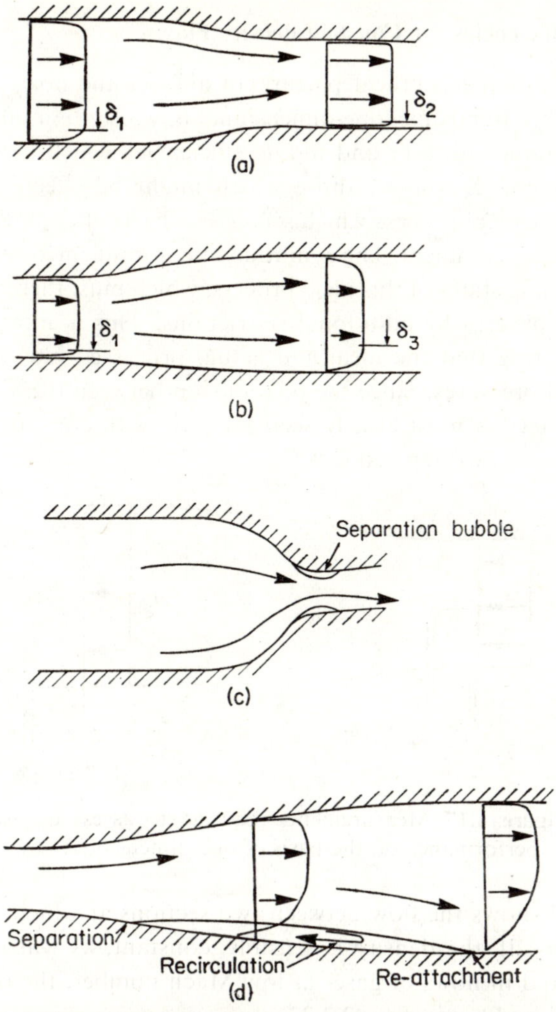

Figure 4.16 Subsonic flows in nozzles and diffusers. (a) Slight contraction with reduction in boundary-layer thickness. (b) Slight expansion with increase in boundary-layer thickness. (c) Abrupt contraction with local separation. (d) Rapid expansion with major separation

or if the flow entering the diffuser is highly non-uniform or unsteady, the rate of expansion for best performance will be reduced. It is not always expedient to achieve optimum diffusion. Sometimes geometric constraints make it impossible to fit in a long, gradually tapering diffuser. Here one must accept inefficient diffusion as the price of some other benefit, perhaps compactness or low cost.

4.4.2 Duct Efficiencies for Constant-density Flow

Our task is to represent the departures of diffuser and nozzle performance from the ideal of isentropic, one-dimensional flow. A common way of doing this is to introduce diffuser and nozzle efficiencies which are the ratios of actual energy conversions to those which might be effected ideally. The definition of the ideal process which serves as a basis of comparison requires some preliminary comment. The condition of constant entropy is not enough to fix the terminal states of this ideal process, which must therefore be related to the actual process by additional restrictions. This is most conveniently done by requiring that the ideal and actual processes extend between the same terminal pressures. Since the relationship between the actual flow and isentropic pattern is most clearly seen for flow with constant density, this restricted case will be discussed first.

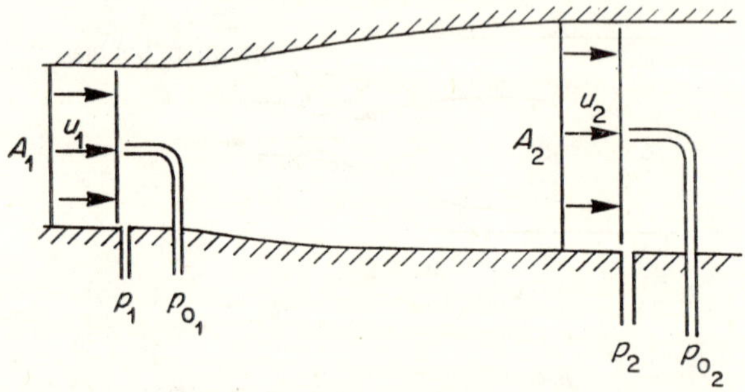

Figure 4.17 Measurements required to assess diffuser performance on the basis of one-dimensional flow

Figure 4.17 shows the flow between two sections at which the flow areas are A_1 and A_2. If the density is sensibly constant, as will be the case in liquid flows and in flows of gases at low Mach number, the dissipation per unit mass is given by equation (2.3.22) as

$$-\frac{\Delta(p_0+\varrho_c gz)}{\varrho_c}>0$$

for steady, isenergic flow. For simplicity, we shall henceforth omit the gravity term and take the dissipation to be $-\Delta p_0/\varrho_c$ per unit mass. The effect of changes in elevation can be introduced into the results simply by replacing p_0 by $p_0+\varrho_c gz$.

With this restriction, the changes in pressure and kinetic energy are related by

$$p_2-p_1=\tfrac{1}{2}\varrho_c(u_1^2-u_2^2)+\Delta p_0 \qquad (4.4.1)$$

with $\Delta p_0<0$. Two ideal pattern processes can be derived from this result:

(1) an isentropic pressure change from velocity u_1 to velocity u_2, giving

$$p_i - p_1 = \tfrac{1}{2}\varrho_c(u_1^2 - u_2^2) \tag{4.4.2}$$

and

(2) an isentropic velocity change from pressure p_1 to pressure p_2, giving

$$p_2 - p_1 = \tfrac{1}{2}\varrho_c(u_1^2 - u_i^2) \tag{4.4.3}$$

The first seems more sensible for a diffuser, whose job it is to convert kinetic energy into pressure; the second is more meaningful for a nozzle, whose job it is to accelerate the fluid as efficiently as possible. The second ideal pattern has the advantage of being readily extended to flow with varying density. However, it has the less desirable characteristic of being inconsistent with the prescribed duct geometry. Note that

$$A_2 = \frac{u_1}{u_2} A_1 \neq A_i = \frac{u_1}{u_i} A_1 \tag{4.4.4}$$

For constant-density flow, it does not matter which of the pattern processes (4.4.2, 3) is adopted; the resulting definitions are the same. For a *diffuser* equations (4.4.1, 2, 3) give

$$p_i - p_1 = \tfrac{1}{2}\varrho_c(u_1^2 - u_2^2) > p_2 - p_1 = \tfrac{1}{2}\varrho_c(u_1^2 - u_i^2)$$

Hence an efficiency can be defined as

$$\eta_D = \frac{p_2 - p_1}{p_i - p_1} = \frac{u_1^2 - u_i^2}{u_1^2 - u_2^2} \tag{4.4.5}$$

The actual pressure change is less than that which might be achieved by isentropic deceleration in the duct, and the actual velocity change is greater than that which would be achieved by isentropic deceleration between the terminal pressures.

For a *nozzle* equations (4.4.1, 2, 3) give

$$\tfrac{1}{2}\varrho_c(u_i^2 - u_1^2) = p_1 - p_2 > \tfrac{1}{2}\varrho_c(u_2^2 - u_1^2) = p_1 - p_i$$

and

$$\eta_N = \frac{p_1 - p_i}{p_1 - p_2} = \frac{u_2^2 - u_1^2}{u_i^2 - u_1^2} \tag{4.4.6}$$

Here the actual pressure change required for acceleration is greater than would be required for isentropic acceleration, and the actual velocity change is smaller than would be achieved by isentropic acceleration between the same terminal pressures. Suppose, for example, that a nozzle is used to direct a jet of water onto the rotor of a turbine. The departure from the isentropic pattern implies either that water must be supplied at a higher pressure, or that the outlet velocity will be smaller with a consequent reduction in the energy available to the rotor. In either case, the effect of irreversibility is to reduce the overall efficiency of energy extraction in the system comprising nozzle and rotor.

The efficiencies defined above can be expressed in alternative forms which are more useful in practice. The relationship

$$p_i - p_1 = p_2 - p_1 - \Delta p_0 = \Delta p - \Delta p_0$$

allows equations (4.4.5, 6) to be written

$$\eta_N = \frac{1}{\eta_D} = 1 - \frac{\Delta p_0}{\Delta p} \tag{4.4.7}$$

These forms are particularly suitable for the experimental determination of duct efficiencies. The measurements indicated in Figure 4.17, total and static pressures at entry and exit sections, determine the efficiency of a section of duct. In practice, there will be considerable variations of total pressure over the cross-section, and suitable mean values must be found by traversing over the flow.

Other useful expressions for the efficiencies are obtained by combining the actual velocity and pressure changes of equations (4.4.5, 6):

$$\eta_D = \frac{1}{\eta_N} = \frac{p_2 - p_1}{\frac{1}{2}\varrho_c(u_1^2 - u_2^2)} \tag{4.4.8}$$

When the continuity requirement, $u_1A_1 = u_2A_2$, is introduced:

$$\eta_D = \frac{1}{\eta_N} = \frac{p_2 - p_1}{\frac{1}{2}\varrho_c u_1^2(1 - A_1^2/A_2^2)} \tag{4.4.9}$$

These forms are particularly useful for predicting pressure variations associated with a specified flow rate. Frequently a knowledge of the duct geometry makes possible an estimate of the efficiency using the available body of empirical data. The corresponding pressure change can then be calculated using the results obtained above.

As might be supposed from a consideration of the flow processes shown in Figure 4.16, experimentally determined values of the efficiencies of short, carefully shaped nozzles are very near unity, while the efficiencies of diffusers often indicate a marked departure from the ideal pattern. Figure 4.18 shows the dependence of diffuser efficiency upon angle of expansion for a particular case. As anticipated, very rapid expansion and very slow expansion are both inefficient, but proper design can achieve a reasonably close approximation to the isentropic model.

meter is satisfactory from two points of view: the ideal-flow model gives a fairly accurate prediction, easily corrected to give high precision; and the tapering diffuser minimizes the losses. The chief advantage of the orifice plate is low initial cost. Not only is the intrinsic cost of the meter low, but it is easy to install in a pipe line.

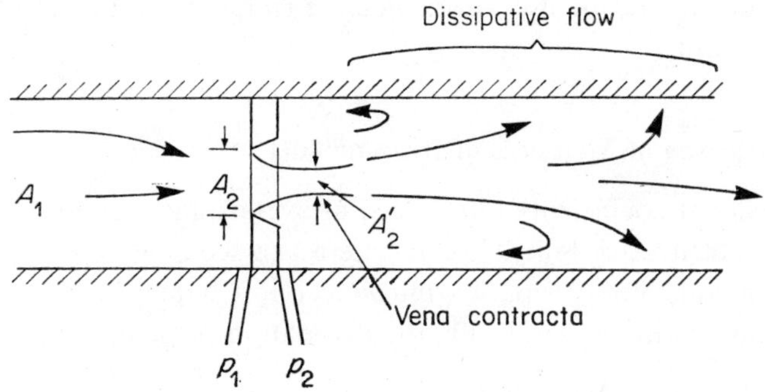

Figure 4.21 The flow through an orifice meter in a pipe

Despite the very different flow boundaries provided by the Venturi and orifice meters, the flows within them are rather similar. Accordingly, the Venturi result (4.4.14) is used to relate the measured pressure difference and the flow rate. The discharge coefficient C is now considerably less than unity; it depends on the position of the pressure taps, the details of orifice manufacture, and the ratio A_2/A_1, among other factors. Values between 0·6 and 0·7 are found by calibration.

The single coefficient C accounts not only for losses and non-uniformity of velocity across the flow, but also for the marked contraction of the flow beyond the orifice. By introducing a *contraction coefficient*,

$$C_c = \frac{\text{area of vena contracta}}{\text{orifice area}}$$

$$= \frac{A_2'}{A_2}$$

one can express the effective ratio of flow areas as

$$\frac{A_2'}{A_1} = \frac{C_c A_2}{A_1} = C_c m$$

Hence a somewhat more rational form for the flow equation is

$$\dot{V} = \frac{C_v C_c A_2}{(1 - C_c^2 m^2)^{\frac{1}{2}}} \left(\frac{2\,\Delta p}{\varrho_c} \right)^{\frac{1}{2}} \tag{4.4.15}$$

The *velocity coefficient* C_v accounts for a residual deviation from the ideal pattern, resulting from local friction and turbulence, and from non-uniformity of the velocity distribution across the two sections at which measurements are made.

This dissection of the discharge coefficient shows why it differs considerably from unity, but for use in practice the simple form (4.4.14) is usually more convenient.

4.4.5 Comparison of Measures of Irreversibility

The empirical coefficients introduced above are most simply thought of as measures of irreversibility. It is of interest to see how they are related to the other measures of dissipation introduced in Chapter 2. For a constant-density fluid flowing isenergically, the dissipation per unit mass is given by

$$-\frac{\Delta(p_0+\varrho_c gz)}{\varrho_c} = \Delta e = c_v \, \Delta T = T_c \, \Delta s \qquad (2.3.22)$$

where T_c is the nearly constant temperature. The expressions (4.4.7) for constant-density duct efficiencies give the additional relationships

$$\Delta(p_0+\varrho_c gz) = (1-\eta_N)\,\Delta p = \left(1-\frac{1}{\eta_D}\right)\Delta p \qquad (4.4.16)$$

Here the gravity term has been introduced to generalize the results. Finally, the flow coefficient can be expressed in the form

$$C^2 = \tfrac{1}{2}\varrho_c u_1^2\left(\frac{A_1^2/A_2^2-1}{p_1-p_2}\right)$$

by manipulating equation (4.4.14). Comparison with equations (4.4.9) indicates that

$$C^2 = \eta_N = \frac{1}{\eta_D} \qquad (4.4.17)$$

relates this coefficient to the other measures of dissipation.

In the preceding discussion we have introduced only a few of the many ways of describing irreversibility in a flowing fluid. It will be apparent that they are essentially equivalent, although in particular circumstances one or the other may prove more convenient. The results given above apply strictly to the flow of a constant-density fluid. However, most of these measures of dissipation apply to a compressible fluid flowing at low Mach number. The exceptions are Δe and $c_v \, \Delta T$, whose use depends on the possibility of distinguishing clearly between mechanical and thermal energy components.

PROBLEMS FOR CHAPTER 4

Group A

4.1 Check the prediction made at the end of Example 4.1, namely, that the constant-density Bernoulli equation (2.4.1) is accurate within 5 per cent for Mach numbers less than one-half.

To do this consider lossless flow of air from a reservoir through a smoothly rounded entry into a uniform pipe. Within the pipe the Mach number is 0·5, the pressure is 1 bar, and the temperature is 400 °C.

(a) What are the velocity and density in the pipe?
(b) Find the pressure in the tank on the assumption of constant-density flow.
(c) Find the pressure in the tank using the results (4.1.4).
(d) For the assumptions of (b) and (c), find the temperature in the tank.

4.2 Consider a flow field, like that of Example 4.2, in which the Bernoulli sum is constant and specified by conditions in a region of uniform, parallel flow: p_1 and $q_1 = \frac{1}{2}\varrho_c U^2$.

(a) Show that the pressure distribution can be specified by the pressure coefficient

$$c_p = \frac{p-p_1}{q_1} = 1-\frac{u^2}{U^2}$$

where u is the local velocity.

(b) Find the pressure variation $c_p(r, \theta)$ for the flow of Example 4.2. Where (for $r \geqslant R$) is the pressure greatest and least? What happens inside the closed streamline $r = R$?

(c) How far ahead and above must one go in order to find a flow velocity within 1 per cent of the free-stream value, that is, $u = (1 \pm 0{\cdot}01)U$?

4.3 Consider a mass of fluid which has a rotatory motion about a vertical axis. Equation (4.1.8) prescribes the pressure variation through $\mathrm{d}p/\mathrm{d}r = \varrho u^2/r$. In the following calculations, take the density to be uniform.

(a) Find the variation $p(r)$ for solid-body rotation with $u = \omega r$.
(b) Find the variation $p(r)$ for a free vortex with $u = k/r$.
(c) In the latter flow

$$H_0 = \frac{p}{\varrho_c g} + \frac{u^2}{2g} + z = \text{constant}$$

What is the shape of the free surface where $p = p_a$? Is this consistent with your observations of vortices in the bath tub?

4.4 (a) Equations (4.2.4) give the changes in enthalpy, density, pressure and velocity resulting from a change in flow area. Find the corresponding expression for the change in internal energy in an isentropic flow, starting from $\mathrm{d}e = -p\,\mathrm{d}v$.

(b) For a perfect gas more explicit results are available. Show that in this case

$$\frac{\mathrm{d}T}{(\gamma-1)T} = \frac{\mathrm{d}p}{\gamma p} = \frac{M^2}{1-M^2}\frac{\mathrm{d}A}{A}$$

4.5 The relationships (4.2.3, 4) apply to an isentropic, steady-flow process undergone by any fluid. Apply them to the nearly isentropic change in state in steam between 120 bar, 425 °C and 1000 m/sec, and 140 bar and 450 °C.

(a) Use the energy equation $\left(h+\frac{1}{2}u^2 = \text{constant}\right)$ to find the velocity when the pressure is 140 bar.

(b) Use the relations (4.2.3) to estimate the Mach number and sound speed in the stream.

(c) Use the continuity condition to find the ratio of flow areas.

(d) Use the relations (4.2.4) to estimate the Mach number.

4.6 Draw h,s and p,v-diagrams for the flow discussed in Example 4.6. Mark the points corresponding to

(a) stagnation conditions at either end of the process,

(b) the two throats,

(c) the test section, and

(d) points just upstream and downstream of the shock.

4.7 For the wind-tunnel flow discussed in Examples 4.6 and 4.7 find the properties of the fluid at the two throats.

4.8 For the wind-tunnel flow of Examples 4.6 and 4.7, what is the force on the nozzle between the first throat and test section?

4.9 For the flight condition discussed in Example 1.19, use isentropic flow tables to find

(a) the Mach number of flight,

(b) the Mach number of the flow at point (b) above the wing, and

(c) the temperature at points (a) and (b).

4.10 During a test flight the following values are measured on board an aeroplane: $p = 58$ lbf/ft^2, $p_0 = 4030$ lbf/ft^2, $T_r = 2090$ °C. The pressures may be taken as correct measures of the atmospheric pressure and the pressure achieved after deceleration through a normal shock. The recovery factor of the temperature probe is 0·90.

(a) Estimate the temperature and density of the atmospheric air, and the velocity of flight, assuming the air to behave like a perfect gas with $\gamma = 1{\cdot}4$.

(b) At about what altitude is the aeroplane flying?

4.11 A meteorite makes a hole of diameter $\frac{1}{2}$ cm in the side of a satellite in orbit around the earth. The pressure of the air inside is 0·6 bar and the temperature is 25 °C. Use the result (4.3.16) to estimate the rate at which air will leak out of the capsule.

4.12 During spring when the water level is unusually high, water runs to waste from the reservoir of a hydroelectric station. The flow onto the spillway is shown in Figure 2.25; it may be taken to be lossless up to the crest.

(a) If the surface level of the reservoir is 30 cm above the crest, and the crest is horizontal and 5 m across, estimate the flow down the spillway.

(b) If the reservoir level is 100 m above the tailrace of the turbines supplied from the reservoir, estimate the loss of generating capacity per day which is represented by the discharge calculated above.

4.13 Although the flow-meter formula (4.3.9) was derived with reference to the flow of a constant-density fluid, it applies to gases and vapours when the velocity changes within the meter are small. In such cases the density is often not known initially, but must be inferred from pressure and temperature readings. Show that, for a particular meter and gas, assumed perfect, the meter formula may be written

$$\dot{m} = \text{constant} \times \left(1 - \frac{p_2}{p_1}\right)^{\frac{1}{2}} \left(\frac{p_1}{T_1^{\frac{1}{2}}}\right)$$

4.14 A draft tube extends vertically for 4 m below the outlet from a hydraulic turbine, its diameter increasing by a factor of 1·5 between the top and bottom sections. The pressure of the fluid emerging from the turbine is −0·5 bar (gauge) and that at outlet from the draft tube is 0·1 bar (gauge). The swirl in the flow is negligible and a one-dimensional approximation to the flow may be adopted. If the diffuser efficiency is 70 per cent, what is the velocity of the water at the top and bottom of the draft tube?

4.15 Use the perfect-gas results (4.4.13) to estimate the efficiencies for the components of the Venturi meter of Example 4.13 (Use $\gamma = 1{\cdot}3$). Compare with the results found using steam tables, and comment on any differences.

4.16 The flow of water through a horizontal pipe of diameter 10 cm is measured using a Venturi meter whose throat has diameter 5 cm.

(a) Find the flow rate for a pressure difference of 20 cm of water between taps at the constriction and upstream of the meter. Take the discharge coefficient to be 0·95.

(b) If the same differential was obtained for taps whose elevation differed by 10 cm, what would be the flow rate? (There will be two answers.)

(c) If the same differential was obtained from taps just upstream and downstream of an orifice plate with orifice diameter 5 cm, what would be the flow rate? Take the discharge coefficient to be 0·6.

(d) If the same differential was registered between total and static pressures when a Pitot-static tube was introduced on the centre-line of the 10 cm pipe, what would the flow rate be?

4.17 In a high-pressure hydraulic system a hydraulic fluid (specific gravity 0·90, specific heat 0·50) passes through a constrictor into a line of smaller diameter. The temperature is found to rise by 0·007 °F, while the pressure drops by 2 lbf/in^2. The temperature is initially 120 °F. Find

(a) the energy dissipated per unit mass,
(b) the change in specific entropy,
(c) the change in total pressure,
(d) the nozzle efficiency, and
(e) the discharge coefficient.

Group B

4.18 (a) By expanding the appropriate expression in equations (4.1.4), show that

$$\frac{\varrho_0 - \varrho}{\varrho} \simeq \frac{M^2}{2}$$

for a perfect gas flowing at low Mach number.

(b) What is the corresponding result for pressure changes?

(c) Explain why density changes in gas flows are often considered negligible, even though pressure changes of the same order of magnitude are thought to be important.

4.19 The flow generated by the wind in the angle between the ground and the vertical wall of a hangar is described by

$$u_x = -2x \quad \text{m/sec}, \qquad u_y = 2y \quad \text{m/sec}$$

where y (in metres) is measured vertically from the ground, and x (in metres) is measured horizontally away from the wall. An estimate of the force on a door in the wall, 20 m wide and 7 m high, is required. There is a gap at the bottom of the door, so that the pressure generated by the wind at that point is communicated to the interior of the hangar. The flow may be taken to be lossless.

4.20 Consider the irrotational, swirling flow of a perfect gas with $ur = k$, a constant. This is a particular case of the flows with concentric streamlines considered in Example 4.4.

(a) If the total enthalpy is uniform throughout the flow, show that the entropy is as well.

(b) By integrating the radial momentum equation, show that the pressure distribution is given by

$$\left(\frac{p}{p_0}\right)^{\frac{\gamma-1}{\gamma}} + \frac{\gamma-1}{2\gamma}\cdot\left(\frac{\varrho_0}{p_0}\right)\frac{k^2}{r^2} = 1$$

(c) Obtain the same result from the energy equation

$$c_p T_0 = c_p T + \tfrac{1}{2}u^2$$

subject to the requirement of uniform entropy.

(d) What is the least radius at which this kind of flow can be maintained? What happens within this radius?

4.21 Consider the bend meter discussed in Example 4.5.

(a) By integrating the radial momentum equation across the bend, show that

$$\Delta p = q_1\left(\frac{a^2}{R_1^2} - \frac{a^2}{R_2^2}\right)\left[\log_e \frac{R_2}{R_1}\right]^{-2} = f q_1$$

where f represents the geometric factors. Show that this result is consistent with the pressures calculated for the inner and outer walls of the bend in Example 4.5.

(b) By expanding the logarithmic expression, show that the result found above reduces to $\Delta p = 2q_1 a/R_1$ when a/R_1 is small. Is this the result you would expect?

(c) For a bend meter in a pipe of diameter 6 in, with $R_1 = 2$ ft and $R_2 = 2{\cdot}5$ ft, it is found that a discharge coefficient of 0·7 must be introduced into the ideal result

$$\dot{V} = A\left(\frac{2\,\Delta p}{f\varrho_c}\right)^{\frac{1}{2}}$$

in order that the actual performance be described. Prepare a plot of $\dot{V}$ against Δp for this meter.

4.22 (a) For an ideal liquid, use equations (4.1.3) to show that the sonic properties achieved by isentropic acceleration from stagnation conditions are given by

$$\sqrt{2}a_0, \quad \frac{1}{2}\varrho_0, \quad p_0-\left(\frac{1}{\varkappa}\right)\log_e 2$$

and hence find the flow (per unit area) through a sonic nozzle.

(b) What reservoir pressure p_0 is required if a sonic water jet is to be produced in the atmosphere?

4.23 (a) Find the mass flow through the wind tunnel of Example 4.7, for supersonic operation.

(b) If the volume of the reservoir is 1000 m^3, find the running time for the tunnel, that is, the length of time it can operate with supersonic flow in the test section. Take the reservoir pressure to be 1/100 atm initially, and assume isentropic diffusion after the shock. The air may be taken to be a perfect gas, so that the reservoir temperature is T_a, the atmospheric value, throughout the run.

(c) What would be the running time in the absence of a second throat, the flow again being assumed isentropic except in the shock?

4.24 The area of a converging nozzle is given by

$$\frac{A}{A_t} = \frac{1}{2}\left(x+\frac{1}{x}\right) \quad \text{for} \quad 0 < x \leqslant 1$$

where x is distance measured along the duct axis, and A_t is the area of the throat. The flow of air through the nozzle is from stagnation to atmospheric conditions, and may be assumed isentropic.

(a) Plot the pressure variation along the nozzle for the case of sonic flow at the throat.

(b) Repeat for $M = 0{\cdot}3$ at the throat.

(c) Plot the exit Mach number as a function of p_b/p_0.

4.25 A fluid flows steadily at 1 kg/sec from a reservoir where its state is specified by 20 bar and 0·1995 m^3/kg. The fluid passes isenergically and isentropically through a duct of varying cross-section. It is required to find the dependence of the flow velocity u on the duct area A on the basis of three assumptions about the fluid:

(1) that its density remains constant,
(2) that it is a perfect gas with $\gamma = 1{\cdot}4$, and
(3) that it is dry steam.

Although the results of the calculations are generally similar for each kind of fluid, the technique by which they are obtained differs markedly for these three fluids.

(a) To discover the range of velocity which must be examined, find the limiting speeds appropriate to the three fluids. These are the speeds at which the pressure drops to zero and, for steam, the speed at which saturation is achieved.

(b) Plot A against u for the three fluids on the same graph. The following methods should be used in preparing the plots: (1) for constant density, the velocity is inversely proportional to the flow area; (2) for air, isentropic-flow tables give a/a_0 and A/A^* as functions of M; corresponding values of u and A can easily be calculated; and (3) for steam, a table with columns p, h, v, u and A should be prepared.

4.26 A nozzle like that considered in Example 4.10 is installed in a pipe of diameter 10 cm, and exhausts directly into the atmosphere. The dependence of the air flow on the pressure p_1 upstream of the nozzle is required over the range 1 to 3 atm, taking $T_1 = 250$ °C over the whole range. Use the values

p_2/p_1	0·9	0·8	0·7	0·6 (and below)
$u_1^2/(2c_pT_1)$	0·0081	0·0138	0·0171	0·0187

(a) How does the mass flow depend on these parameters?
(b) What is the source of the values given above?
(c) Prepare the required graph.

4.27 Find the function $f(p_e/p_0)$ which occurs in equation (4.3.17).

4.28 Water flows through a Venturi meter in which $A_1/A_2 = 3$ with $u_1 = u_3 =$ 20 ft/sec.

(a) For nozzle and diffuser efficiencies of 98 per cent and 80 per cent, find the pressure differences p_1-p_2 and p_1-p_3.

(b) Consider flow through the venturi in the reverse direction. What are the pressure changes, for efficiencies 95 per cent and 60 per cent?

(c) Why do the efficiencies change in this way?

4.29 Consider the flow of Example 4.13, now assuming the steam to flow at constant density, that within the reservoir. The calculations will be based on the following measured values: reservoir conditions, actual flow rate, pressures at end of nozzle and diffuser, and actual area ratio, $A_2/A_3 = 0{\cdot}46$.

(a) What is the actual flow rate per unit area of the throat?

(b) What are the velocities u_2 and u_3 for this flow rate, on the assumption of constant density?

(c) What are the diffuser efficiencies, on the assumption of constant density?

4.30 Consider the flow of steam through the venturi discussed in Example 4.13.

(a) Find the velocities at stations 2 and 3 for (1) the actual flow and (2) ideal flow between the stations.

(b) Use the continuity condition to make a rough calculation of the ratio A_2/A_3 for the values (1) and (2) obtained above, taking the density to be the same at stations 2 and 3 for both flows. (This assumption can be shown to be fairly realistic by comparing the quantities p/T for the several states.)

(c) Why is the area ratio not the same for the actual and ideal flows?

4.31 A wind tunnel for undergraduate instruction is to have a test-section area of 1/10 m² and test velocity of 30 m/sec. The losses in the channel up to the end of the test section may be specified as $1{\cdot}1q$, where $q = \frac{1}{2}\varrho u^2$ is the dynamic pressure in the test section. Air is drawn directly from the atmosphere into the tunnel and is exhausted again into the atmosphere; such a tunnel is referred to as an open-circuit tunnel.

It is required to estimate the power which will be absorbed by the electric motor driving the fan in this tunnel. Ideally, the power might be estimated as $\dot{V}\,\Delta p$; to account for losses in the motor and for inefficiency in the fan, the power may be taken to be $1{\cdot}5\dot{V}\,\Delta p$.

(a) What is the pressure at the end of the test section? Remember to take account of the dynamic pressure as well as the losses.

(b) Estimate the power required to drive the tunnel if the fan is placed at the end of the test section with no diffuser following it.

(c) Estimate the power required if there is a diffuser with $A_2/A_1 = 3$ and $\eta_D =$ 90 per cent.

(d) The fan would probably be located at the end of the diffuser. Why?

4.32 Using Figure 4.18, estimate the length of the diffuser consistent with the data specified in the preceding problem. The duct may be taken to be of circular section throughout. If you had the job of designing the wind tunnel, would you select this diffuser?

4.33 (a) Suppose that there are available a series of pressure tappings along the wall of a water pipe containing an orifice meter. Sketch the pressure variation to be expected, starting well upstream, and extending well downstream of the meter.

(b) Suppose also that it is possible to traverse a Pitot-static probe along the axis of the pipe, measuring the variations of total and static pressures. Sketch these variations with that of wall pressure, and comment on the relationships amongst the three.

(c) What are the advantages and disadvantages of locating pressure taps for metering purposes right against the orifice plate?

Group C

4.34 In Example 4.1 it was found that the first-order error terms in the constant-density Bernoulli equation are the same for a perfect gas and ideal liquid. This is not a coincidence, as the following generalized analysis shows.

(a) Show that

$$\varrho = \varrho_0 + \frac{p - p_0}{a_0^2}$$

to the first order, using a truncated Taylor expansion.

(b) Hence show that Bernoulli's equation is

$$\frac{p}{\varrho_0} + \frac{(p - p_0)^2}{2\varrho_0 a_0^2} + \frac{u^2}{2} = \text{constant}$$

(c) Hence show that

$$\varrho = \varrho_0\left(1 - \frac{u^2}{2a_0^2}\right)$$

and

$$\frac{p_0 - p}{\varrho} = \frac{u^2}{2}\left(1 + \frac{u^2}{4a_0^2}\right)$$

4.35 A more realistic representation of the pressure applied by the wind to the semicircular building considered in Example 4.2 can be obtained by taking the pressure to be uniform over some portion of the rear of the building.

(a) Find the lift force applied to the building on the assumption that

$$p = p\,(120°) \quad \text{for} \quad 180° < \theta < 120°$$

(b) Find the corresponding drag

$$D = \int_0^{\pi} (p - p_i) R \cos\theta \, d\theta$$

(c) Evaluate for the particular case specified in Example 4.2, taking p_i to have the value used there.

4.36 Derive the result (4.2.11) from the results (b) of Problem 4.4.

4.37 In Example 4.8 the thrust of a rocket engine was determined. For the motor specified this thrust will depend on the external pressure and the exit area of the nozzle.

(a) Investigate this dependence by plotting the thrust against exit area (for $A_e > A^*$) for operation with $p_a = 5$ lbf/in² and with $p_a = 15$ lbf/in².

(b) How do these results relate to the problem of choosing the exit area?

4.38 Consider the performance of the diffuser of the supersonic wind tunnel of Examples 4.6 and 4.7, assuming the flow to be isentropic save in the shock.

(a) Show that

$$\eta_D = \frac{\left(\dfrac{p_r}{p_1}\right)^{\frac{\gamma-1}{\gamma}} - 1}{\dfrac{T_a}{T_1} - 1}$$

gives the efficiency of diffusion. The subscripts r, a and 1 refer, respectively, to reservoir, atmospheric and before-shock conditions.

(b) Show that, for a specified M_1 upstream of the shock, T_a/T_1 and p_a/p_1 can be found from isentropic flow tables, and that p_r/p_a is obtainable from normal-shock tables.

(c) Plot η_D against p_r/p_a for $M_1 = 1$ to 5.

(d) For operation with test Mach number 4·0, as in Example 4.7, mark in the range of efficiency encountered during a run for

(1) second throat with $A/A_T = 1{\cdot}5$, and

(2) no second throat.

4.39 Consider again the situation examined in the preceding problem, but now take $\eta_D = 60$ per cent to characterize the subsonic diffusion after the shock. The flow up to the shock may still be assumed to be isentropic.

(a) Show that

$$\eta_D = \frac{\left(\dfrac{p_r}{p_2}\right)^{\frac{\gamma-1}{\gamma}} - 1}{\dfrac{T_a}{T_2} - 1}$$

gives the efficiency of subsonic diffusion, while the result (a) of the preceding problem applies to the whole process of diffusion.

(b) Show that, for a specified M_1, the use of isentropic flow tables makes possible the calculation of T_a/T_2 and hence of p_r/p_2, p_r/p_1 and p_r/p_a.

(c) Plot the overall η_D against M_1 on this basis. What effect does a reduction in diffuser efficiency have on supersonic running time?

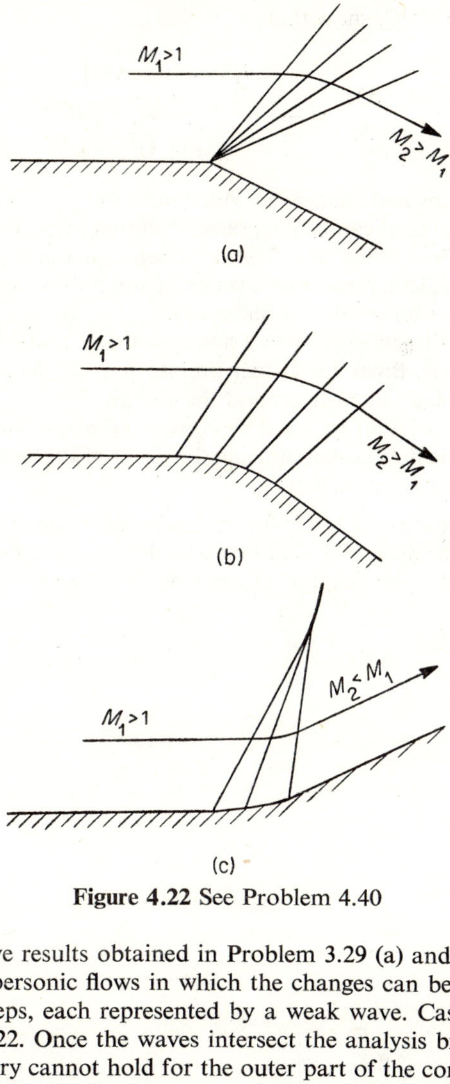

Figure 4.22 See Problem 4.40

4.40 The weak-wave results obtained in Problem 3.29 (a) and (b) can be used to describe certain supersonic flows in which the changes can be broken down into a series of small steps, each represented by a weak wave. Cases of this kind are shown in Figure 4.22. Once the waves intersect the analysis breaks down. Hence the weak-wave theory cannot hold for the outer part of the compressive wave (c); there the weak wavelets coalesce into an oblique shock. For a hypothetical infinite flow field, the expansion waves (a) and (b) will not intersect. However, they will do so on reflexion from another wall, or on meeting waves generated elsewhere, for example, those produced by a deflexion of the opposite wall of the duct.

(a) It is convenient to use the local Mach number as a parameter, particularly since isentropic flow tables are presented on this basis. Show that the result of Problem 3.29 (a) can be written

$$\frac{dM}{M}+\frac{da}{a}=-\frac{d\theta}{(M^2-1)^{\frac{1}{2}}}$$

Hence, using $a/a_0 = \alpha^{-1/2}$, show that

$$\frac{\mathrm{d}M}{M} = -\frac{\mathrm{d}\theta\left(1+\frac{\gamma-1}{2}M^2\right)}{(M^2-1)^{\frac{1}{2}}}$$

relates small deflexions and changes in Mach number.

(b) The preceding result can be integrated in closed form, thus providing a link between isentropic flow tables and flow deflexion. Alternatively, it can form the basis for numerical calculations. For a series of small deflexions, it gives the corresponding small changes in Mach number.

Consider a flow initially with $M = 2$ and $p = 0{\cdot}3$ bar which is deflected by 8° over a rounded corner. Break the corner up into four 2° elements and hence find the distribution of Mach number around the corner.

(c) The pressure variation in the flow considered above can be found in two ways: using isentropic flow tables, or using the result of Problem 3.29(b). Compare the results of these two approaches.

4.41 Following equation (4.4.7) it is stated that in calculating the diffuser efficiency 'suitable' mean values must be found by traversing over the flow. What would be a 'suitable' mean value for Δp_0 for constant-density flow?

5

Dissipative Processes

Most of the discussion of fluid mechanics in this book relates to a one-dimensional model which neglects variations in properties across the flow. When this idealization was introduced in Section 2.1 it was realized that some account would have to be taken of the actual flow pattern, which in general is three-dimensional and perhaps turbulent. In particular, the small-scale processes within the flow determine the friction and the rate of heat exchange at the boundaries. The generation of fluid friction will be considered in this chapter and heat transfer will be dealt with in the next. In each case the ultimate aim is to find ways of representing the effects of the small-scale processes using a one-dimensional flow model. Empirical data are usually required to specify friction and heat transfer within a fluid, particularly a turbulent fluid. Even so, it is helpful to have a general picture of the processes underlying the experimental facts.

The generation of friction within a fluid is closely related to the gradient of velocity across the flow. Velocity-profile variations along the flow are also related to dissipation, since they are achieved through mixing and spreading of the flow, processes linked to friction stresses within the fluid. These two mechanisms of dissipation have been identified in particular flows discussed earlier. Figure 2.19 shows an example of each, a flow subjected to uniform boundary friction in a long uniform pipe, and a flow undergoing rapid mixing and spreading behind a gate valve. In the first case the dissipation is concentrated in a thin boundary layer at the wall, while in the second it takes place in a region of intense turbulence within the body of the flow and is not related directly to friction at the wall.

Even though the two mechanisms of dissipation have the same root cause and the same ultimate effect, the methods used to analyse them are markedly different. Hence this chapter falls into three parts: a study of friction between a flowing fluid and a solid boundary, a discussion of the effects of changes in velocity profile, and finally a consideration of their combined effect on the flow through a system of pipes. This provides the basis for the most frequently performed calculations relating to a flowing fluid, namely, the prediction of the flow rate through a piping system for a specified pressure difference, and the converse problem of finding the pressure difference required to achieve a specified flow rate.

The mechanism of *boundary friction* and its magnitude depend critically upon whether the friction-generating flow is laminar or turbulent. The simpler case of viscous friction in laminar flow will be examined first and a fairly complete description of laminar pipe flow will be obtained. The treatment of turbulent pipe flow is by no means as satisfactory. The present understanding of turbulence does not provide an adequate theoretical description even for such an apparently simple flow as that in a uniform pipe. It is true that the essential features of turbulent motion are known and can be described mathematically. However, the description is too complicated for application to the flows of interest to engineers, and the understanding is not sufficiently penetrating to lead to useful simplifications.

In view of these rather pessimistic remarks, you may wonder how it is possible to say anything about a turbulent flow without falling back on experimental data. The answer lies in adopting an indirect approach, not trying to analyse the motion in detail, but only to describe its main features. In this way a skeletal description of some simple flows can be built up; this provides a pattern into which empirical information can be fitted. This indirect approach requires special analytical tools, and a family of techniques has been developed to examine the general features of a process without requiring a knowledge of its detailed development. These techniques are known as *similarity methods*, since they reveal the relationships between families of geometrically similar flows.

The similarity technique which will be used here is *dimensional analysis*. This is based on the idea that the laws governing a physical process must be independent of the system of units used to measure its characteristics. This concept leads to the determination of unit-free groups characterizing the process, in our case, a particular kind of flow. The relationships amongst these groups give, first, a skeleton of the detailed laws governing the flow and, secondly, the conditions which must be satisfied if two flows are to be similar not only in geometry but also in their dynamics and thermodynamics. These results are useful in the planning and interpretation of experiments using scaled models. Moreover, in certain simple cases they provide functional relationships which, with the introduction of a few empirical constants, become working formulae for engineering calculations. It will be apparent that dimensional analysis, like other similarity methods, is potentially applicable to a wide range of problems; in the final chapter of this book, for example, it is applied to families of turbomachines. Hence the technique will be developed in a fairly general way, even though its principal use in this chapter is in analysing turbulent friction.

As was suggested earlier, a quite different approach is required in studying dissipation in *mixing and spreading flows*. In general, the prediction of these losses depends on empirical information, and dimensional analysis is again useful in planning experiments and presenting data. However, theoretical predictions can be obtained for certain cases by using integral flow models that are somewhat more realistic than the simplest one-dimensional description. The jet pump, a device which utilizes mixing to achieve a pumping effect, can be analysed in this way. This discussion of dissipation in flows with changing velocity profiles leads naturally to a consideration of some other problems associated with velocity variations across the flow: the validity of the one-dimensional approximation, the losses in a diffuser, and the relationship of the drag on a body to the flow in its wake.

This chapter deals almost entirely with constant-density flow, but most of the results apply with reasonable accuracy to compressible fluids flowing subsonically. The application to supersonic flow requires more careful consideration, since in that case dissipation can also occur through shock-wave formation, and since the considerable changes in temperature will give rise to significant changes in the fluid viscosity. The flows discussed in this chapter will be assumed to be essentially isenergic; in other words, it will be assumed that the heat transfer (if any) within the fluid does not affect the friction. Obviously, this assumption will cease to be valid when the heat-transfer rate and temperature variations across the flow are very great. However, it should be valid for a range of moderate heat-transfer rates. Hence the results developed here will be applicable, for example, to the prediction of the friction experienced by water or steam flowing through a boiler tube, but could not be expected to apply in the evaporator section of the boiler, where the fluid is actually vaporizing.

5.1 VISCOUS FRICTION

5.1.1 Viscosity

The easiest way to start thinking about fluid friction is to imagine two very large plates separated by a thin, uniform layer of fluid. The layer is sheared by the steady motion of one plate parallel to the other. This situation is shown in Figure 5.1(a). The upper plate moves with velocity U relative to the lower, while the force F produces the relative motion and balances the friction force acting over the wetted area A. The flow will be non-uniform near the edges of the wetted area, but may be expected to be uniform over the central region. In this uniform region, the fluid is uniformly sheared, as indicated in Figure 5.1(b). This simple parallel flow is called Couette flow or, to distinguish it from the rather similar concentric flow between rotating circular cylinders, *plane Couette flow*.

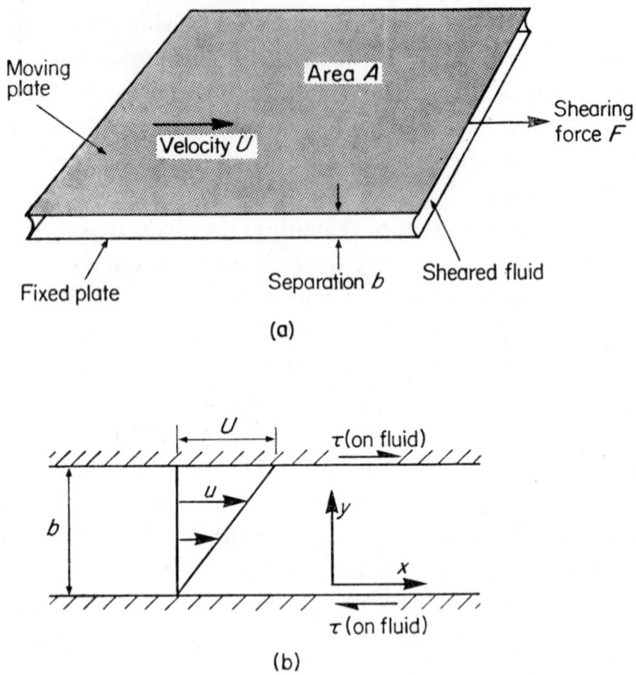

Figure 5.1 A simple shearing flow used to introduce viscosity. (a) Means of imposing a uniform shearing motion on a fluid. (b) Velocity distribution in the laminar plane Couette flow of a fluid of uniform viscosity

In the situation described above and in other laminar flows, it is usually found that the friction stress within the fluid, $\tau = F/A$, is proportional to the rate at which the fluid is sheared, U/b. Then

$$\tau = \text{constant} \times U/b$$

This statement is referred to as *Newton's law of friction*, and a fluid obeying it is termed Newtonian. The value of the constant depends on the fluid which is sheared and on its thermodynamic state, conveniently specified by its pressure and temperature. Accordingly, the constant is a property of the fluid; it is called the *dynamic viscosity*, coefficient of viscosity, or simply viscosity, and is denoted by the symbol μ. Thus

$$\tau = \mu \frac{U}{b} \tag{5.1.1}$$

defines the viscosity. The units of dynamic viscosity are (force×time)/area.

The viscosity coefficient is not very strongly dependent on pressure, but varies markedly with temperature, falling by a factor of ten between 0 °C and 100 °C for typical liquids, and rising by some 50 per cent over this range for typical gases. The viscosities of some common fluids are given in Appendix VII. This marked dependence of viscosity on temperature is to be expected, since temperature is a measure of the molecular activity of a substance, and it is this activity which is represented by macroscopic viscosity. In a gas, the momentum transfer accomplished by migrating molecules gives rise to the viscosity; accordingly, a rise in temperature, signifying an increase in molecular activity, is accompanied by a rise in viscosity. In a liquid, on the other hand, viscosity arises primarily from inter-molecular forces; these weaken as the temperature rises, and the viscosity of the liquid falls.

Another much used measure of viscosity is the *kinematic viscosity*

$$\nu = \frac{\mu}{\varrho}$$

The name kinematic is explained by the units, those of area/time. This is a measure of the relative magnitudes of viscous and inertial reactions to the deformation of a fluid. In this sense liquids are less viscous than gases, their higher dynamic viscosity being more than compensated by their much greater density.

Since viscous friction arises from molecular migrations, the molecules moving only very short distances before interacting with others, the viscous stress at a point in a flow is dependent on conditions very near that point. Hence it is reasonable to take

$$\tau = \mu \frac{\partial u}{\partial y} \tag{5.1.2}$$

to relate the local stress and shear rate in a more general parallel flow, where $u = u(y, t)$ so that the rate of shear varies. Although this result gives the stress in a time-dependent parallel flow, it does not apply to motions which vary through space in a more complex manner, that is, with $u = u(x, y, z)$. In general, the velocity variations in other directions must be taken into account in determining the friction stress.

Three limitations on the preceding discussion should be noted. First, the moving surfaces must not be so close that high spots on them come into contact. If they do, the friction will depend on the nature of the surfaces and not merely on the properties of the fluid between them.

Secondly, in certain fluids the shear stress generated is not proportional to the rate of shear; in other words, the viscosity is dependent on the rate of straining. Such fluids are called *non-Newtonian* to distinguish them from fluids which obey the laws (5.1.1, 2). The behaviour of Newtonian and non-Newtonian fluids can be related to their microscopic structure. To simplify the comparison, let us assume that fluid viscosity arises from momentum transfer by molecules moving through a region of varying mean momentum. Newtonian fluids have small molecules which migrate at high velocity and interact with other molecules at discrete intervals. Their random motion is little influenced by the mean shear, and the momentum transfer they accomplish increases linearly with the mean velocity gradient. Non-Newtonian fluids typically contain long-chain organic molecules, or perhaps larger elements such as the wood fibres of paper pulp, whose motion is influenced by the mean shear. Since the velocity gradient influences the momentum transfer in two ways, it is no longer simply proportional to the friction stress.

The third limitation on the discussion leading to equations (5.1.1, 2) is turbulence. Under circumstances to be looked at later, laminar Couette flow is unstable, and a turbulent flow will be generated between the plates of Figure 5.1. Once this occurs the law (5.1.1) is no longer obeyed. In a sense a turbulent fluid may be thought of as non-Newtonian; the mechanism of friction—turbulent mixing—is generated by the mean velocity gradient.

Example 5.1 The viscosity of a liquid can be measured by shearing it between concentric cylinders, as indicated in Figure 5.2. In the case to be considere da torque of 10 gf cm is applied to the shaft of the inner cylinder by the falling weight; that cylinder rotates at 1·7 rev/min, the other one being fixed. The radius of the inner cylinder is 3 cm, the gap between the cylinders is 5 mm, and the fluid fills the gap to a depth of 20 cm. The viscosity will be calculated on the assumption that non-uniformity at the top and bottom of the liquid mass is negligible, so that the flow consists of concentric Couette flow over the entire depth of 20 cm.

The result (5.1.2) does not apply to this non-parallel flow. For a flow with concentric streamlines, equation (4.1.12) gives the rate of shear, and the shear stress is

$$\tau = \mu\left(\frac{du}{dr} - \frac{u}{r}\right) = \mu r \frac{d(u/r)}{dr}$$

The torque applied to the inner cylinder is transmitted by viscous friction to the outer cylinder. Hence the torque on any cylindrical surface within the fluid will be the same. This torque is

$$(2\pi rb)\tau r = 2\pi b \mu r^3 \frac{d(u/r)}{dr} = L, \quad \text{a constant}$$

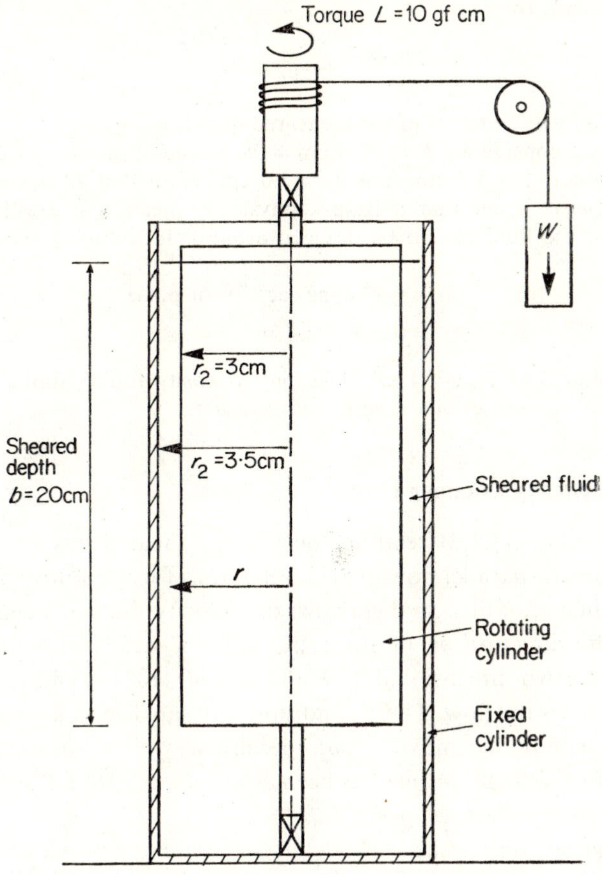

Figure 5.2 A viscometer generating axisymmetric Couette flow

Integration gives

$$u = -\frac{L/4\pi b\mu}{r} + Ar$$

The two boundary conditions to be satisfied by this solution are:

$$u = 0 \quad \text{for} \quad r = r_2 \text{ on the outer cylinder}$$

and

$$u = u_1 \quad \text{for} \quad r = r_1 \text{ on the inner cylinder}$$

The first condition gives

$$A = \frac{L}{4\pi b\mu r_2^2}$$

The second leads to

$$\mu = -\frac{L}{4\pi b u_1}\,\frac{1-r_1^2/r_2^2}{r_1}$$

giving the viscosity in terms of the measured quantities.

For the case considered, $L = 10$ gf cm $= 9810$ dyne cm, $u_1 = -3\times 2\pi\times 1{\cdot}7/60 = -0{\cdot}534$ cm/sec, $r_2 = 3{\cdot}5$ cm, and $r_1 = 3{\cdot}0$ cm. Note that L and u_1 must have opposite signs in order that μ have a physically meaningful positive value; in other words, the fluid has to be dragged around. Introducing these values, we have

$$\begin{aligned}\mu &= 6{\cdot}47 \text{ dyne sec/cm}^2 \text{ or poise}\\ &= 0{\cdot}647 \text{ N sec/m}^2\end{aligned}$$

For water at 20 °C, $\mu \simeq 1$ cP. The high viscosity found above is typical of heavy lubricating or fuel oils at room temperature.

5.1.2 Some Steady Laminar Flows

The expression (5.1.2) relating the viscous shear stress to the velocity gradient across a parallel flow provides the basis for the study of a family of simple laminar motions. In these flows the velocity does not change significantly in the direction of motion; hence for a steady flow of this kind, $u = u(y)$ for a two-dimensional flow, and $u = u(r)$ for pipe flow, with y and r measured across the flow. These conditions are satisfied in a long channel or pipe, once the fluid has moved a considerable way along the channel so that an equilibrium velocity profile has been established. Such a flow is described as established or *fully developed.*

The simplest flow of the class defined above is plane Couette flow, for which the shear stress is uniform. Then equation (5.1.2) can be integrated to give

$$u = \frac{\tau_0}{\mu}y = \frac{Uy}{b} \tag{5.1.3}$$

the velocity varying linearly between the boundaries, as was indicated in Figure 5.1(b).

A second simple case is *plane Poiseuille flow*, in which fluid flows between fixed parallel plates, driven by pressure forces or by its own weight. This situation is sketched in Figure 5.3(a). The velocity is shown as dropping to zero at the stationary walls. This behaviour was assumed earlier without comment and no doubt seemed plausible to most readers. A careful study of the interaction between the wall and the molecules of the fluid is necessary to show directly that this assumption is appropriate. However, this *no-slip hypothesis* can also be justified by comparing experimental results with predictions incorporating the hypothesis. The predictions made on this basis are usually found to be correct with a high degree of accuracy. In highly rarefied gases the no-slip condition breaks down; the rate of interaction of molecules and surface is so low that a mean relative motion—a slip velocity—can be established.

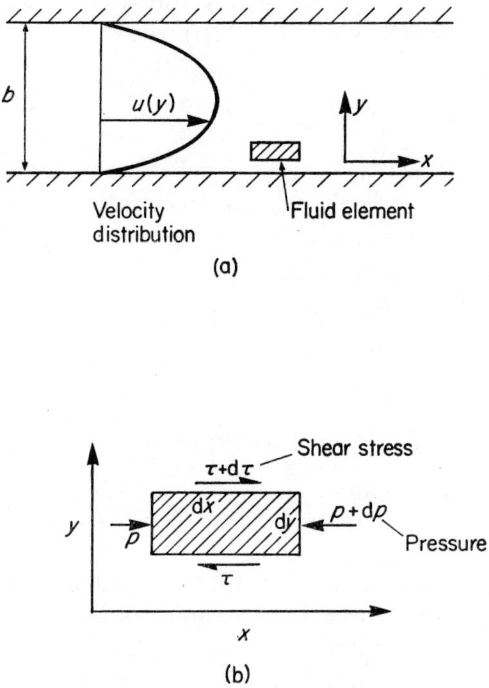

Figure 5.3 Poiseuille flow: parallel, laminar flow between fixed plates. (a) Flow geometry; (b) Forces on the element of fluid shown in (a)

For simplicity we shall consider the shearing motion established by a pressure gradient, neglecting the effect of gravity. There are then no pressure variations normal to the established flow $u(y)$ or $u(r)$, and the pressure distribution is given by $p = p(x)$, where x is measured in the direction of the flow. The relationship between the velocity distribution and the pressure is determined by the requirement that the friction stresses balance the pressure forces. Figure 5.3(b) shows the forces acting in the direction of motion on a cubical element which has unit depth perpendicular to the plane shown. In steady flow along a uniform channel, the fluid element will move at constant velocity. Then the net shear force $\mathrm{d}\tau\, \mathrm{d}x$ and net pressure force $\mathrm{d}p\, \mathrm{d}y$ will balance, and

$$\frac{\mathrm{d}\tau}{\mathrm{d}y} = \frac{\mathrm{d}p}{\mathrm{d}x} \tag{5.1.4}$$

If the shear stress depends solely on the velocity distribution $u(y)$, it follows that the two members of this equation are constants, for this is the only way in which functions of x and y can be equal. This will be the case, for example, in laminar flow with constant viscosity and, as will be seen later, in turbulent flow with constant density. When the pressure gradient and stress gradient are constants, equation (5.1.4) can be integrated:

$$\tau = \frac{\mathrm{d}p}{\mathrm{d}x}y + \tau_0 \tag{5.1.5}$$

The stress is found to vary linearly across the channel as a result of the uniform pressure gradient. Similarly, the pressure varies linearly along the channel.

For flow between fixed walls distance b apart, the requirement that the net pressure force balances the equal friction stresses at the two walls is

$$-b\frac{\mathrm{d}p}{\mathrm{d}x} = 2\tau_0$$

and the stress variation can be given in the forms

$$\tau = \frac{\mathrm{d}p}{\mathrm{d}x}\left(y - \frac{b}{2}\right) = \tau_0\left(1 - \frac{2y}{b}\right) \tag{5.1.6}$$

Note that the stress is zero at the centre of the channel. The results of this paragraph do not depend upon the assumption of Newtonian friction; hence they hold for non-Newtonian and even turbulent fluids. In the latter case the relationships apply to time-mean values of pressure gradient and stress.

Now consider the case of Newtonian friction and take the viscosity to be uniform in space as well. For the stress given by equation (5.1.2), equation (5.1.4) becomes

$$\frac{\mathrm{d}^2u}{\mathrm{d}y^2} = \frac{1}{\mu}\frac{\mathrm{d}p}{\mathrm{d}x} \quad \text{a constant} \tag{5.1.7}$$

Two integrations give

$$u = \frac{1}{2\mu}\frac{\mathrm{d}p}{\mathrm{d}x}y^2 + Ay + B \tag{5.1.8}$$

For $\mathrm{d}p/\mathrm{d}x = 0$, the linear velocity variation of Couette flow (5.1.3) is recovered. For Poiseuille flow between fixed plates, the constants of integration of equation (5.1.7) are determined by the boundary conditions

$$u = 0 \quad \text{at} \quad y = 0, b$$

Then

$$u = \frac{1}{2\mu}\frac{\mathrm{d}p}{\mathrm{d}x}y(y-b) \tag{5.1.9}$$

The rate at which fluid flows through a strip of unit width across the channel can now be found:

$$\dot{v} = \int_0^b u \, dy = \frac{1}{2\mu}\frac{dp}{dx}\int_0^b (y^2 - by)\, dy = -\frac{b^3}{12\mu}\frac{dp}{dx} \qquad (5.1.10)$$

The direction of flow shown in Figure 5.3(a) corresponds to $dp/dx < 0$, the pressure dropping in the direction of flow.

The results obtained above can be generalized to describe *combined Couette–Poiseuille flow*. This is not just a mathematical curiosity, but represents certain features of the flow of a lubricant in a hydrodynamic bearing film. The relative motion of the two bearing surfaces can be utilized to generate large pressures in the thin film separating them. Thus contact between the surfaces is prevented and friction and wear are greatly reduced. The assumption of two-dimensional flow is unrealistic for actual bearings, in which the lubricant is able to flow sideways as well as in the direction of the relative motion of the bearing surfaces, but the results suggest the essential features of hydrodynamic lubrication.

The velocity distribution for the combined flow could be determined from equation (5.1.8), using the conditions $u = 0$ for $y = 0$ and $u = U$ for $y = b$.

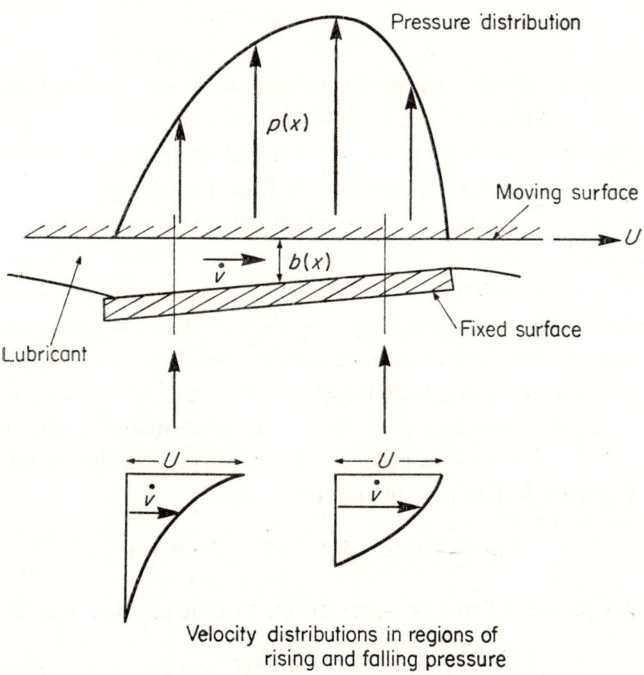

Figure 5.4 Laminar, two-dimensional flow in a slipper bearing

There is, however, an easier way. Since the governing equation (5.1.7) is linear, the Couette and Poiseuille solutions can be superposed to give a more general solution. Thus equations (5.1.3, 4) give the velocity distribution

$$u = \frac{Uy}{b} + \frac{1}{2\mu}\,\frac{\mathrm{d}p}{\mathrm{d}x}\,y(y-b)$$

The net flow rate can also be found by superposition:

$$\dot{v} = \frac{Ub}{2} - \frac{b^3}{12\mu}\,\frac{\mathrm{d}p}{\mathrm{d}x}$$

This result is the key to the analysis of two-dimensional bearing flows, such as that shown in Figure 5.4. The bearing film is represented as a succession of combined Couette–Poiseuille flows, with common values of the relative velocity U and flow rate $\dot{v}$. This assumption is realistic since a typical lubrication film has a thickness of the order of 0·10 mm, with a length of order 10 cm. Hence the flow at any section of the slightly tapering channel is very close to a combined Couette–Poiseuille flow in a channel of constant breadth. A rewriting of the last equation gives

$$\frac{1}{12\mu}\,\frac{\mathrm{d}p}{\mathrm{d}x} = -\frac{\dot{v}}{b^3} + \frac{U}{2b^2} \tag{5.1.11}$$

expressing the local pressure gradient as a function of the film thickness $b(x)$, the parameters μ, $\dot{v}$ and U being constants. For a specified $b(x)$ the pressure gradient is fixed by this equation and the pressure distribution, net separating force, and net friction can be found by integration.

Another laminar flow of great practical importance is *axisymmetric Poiseuille flow*. This kind of flow occurs in narrow-bore tubing; for instance, a fuel line, a manometer tube, or the lines of a hydraulic or pneumatic control system. The corresponding turbulent motion is usually called turbulent pipe flow. The preliminary discussion applies to both laminar and turbulent cases. For steady, fully-developed flow in a uniform tube, $\mathrm{d}p/\mathrm{d}x =$ constant, and $u = u(r)$ and $\tau = \tau(r)$. The relationship between pressure gradient and shear stress can be derived by considering the annular element shown in Figure 5.5. The net shear force is now

$$2\pi[(\tau+\mathrm{d}\tau)(r+\mathrm{d}r)-\tau r] = 2\pi\,\mathrm{d}(\tau r)$$

while the net pressure force is $-2\pi r\,\mathrm{d}r\,\mathrm{d}p$. For steady uniform flow

$$\frac{1}{r}\,\frac{\mathrm{d}(\tau r)}{\mathrm{d}r} = \frac{\mathrm{d}p}{\mathrm{d}x} \tag{5.1.12}$$

5.1.3 A Non-steady Laminar Flow

Viscous flows which vary in time are more difficult to analyse than the steady flows considered above. Indeed, only a small number of exact solutions are known. These do, however, reveal some of the basic features of this class of motions, in particular, the way in which viscous and inertial forces combine to determine the response of the fluid.

First an appropriate equation of motion must be found. Consider the element of a plane parallel flow depicted in Figure 5.3(b). The velocity is taken to be uniform along the flow; hence $\partial u/\partial x = 0$, and the acceleration of the element can be reckoned as $\partial u/\partial t$. Newton's law applied to the element of mass $\varrho \, dx \, dy$ gives

$$(\varrho \, dx \, dy)\frac{\partial u}{\partial t} = -dp \, dy + d\tau \, dx$$

which leads to the differential equation of motion:

$$\varrho \frac{\partial u}{\partial t} = -\frac{\partial p}{\partial x} + \frac{\partial \tau}{\partial y} \tag{5.1.17}$$

This involves $u(y, t)$, $p(x, t)$ and $\tau(y, t)$.

For simplicity, a situation in which the pressure is uniform and the viscosity is constant will be considered. Then equation (5.1.17) reduces to

$$\frac{\partial u}{\partial t} = \nu \frac{\partial^2 u}{\partial y^2} \tag{5.1.18}$$

an equation for the velocity alone. Consider a large volume of fluid adjacent to a plane wall, a situation shown in Figure 5.6. The wall will be taken to move back and forth in its own plane, according to the law $u = U \cos \omega t$, so that the displacement amplitude is $a = U/\omega$. The motion of the wall sets the fluid into periodic motion as well.

This form of excitation has been adopted primarily for mathematical convenience, but does have physical applications. For instance, when surface waves pass through shallow water they set up a cyclic velocity variation at the bottom; this situation is not very different from that to be examined here.

The task of determining the solution of equation (5.1.18) corresponding to the prescribed motion of the wall does not involve any very difficult techniques, but is rather lengthy. It will be left as an exercise for those who wish to follow through the procedures outlined in one of the problems at the end of this chapter. All that is of interest here is the solution:

$$u = U \exp\left[-\left(\frac{\omega}{2\nu}\right)^{\frac{1}{2}} y\right] \cos\left[\omega t - \left(\frac{\omega}{2\nu}\right)^{\frac{1}{2}} y\right] \tag{5.1.19}$$

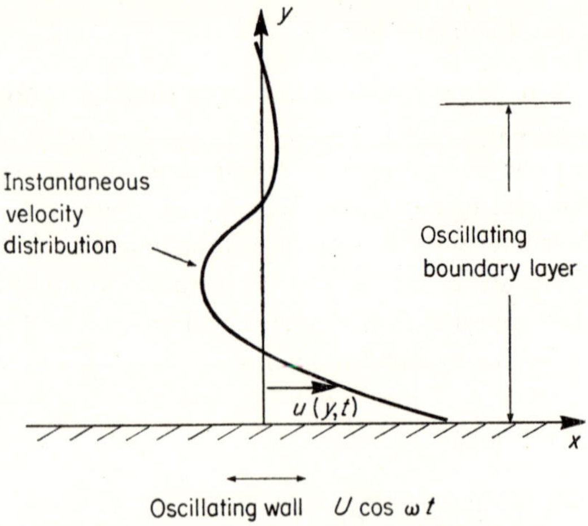

Figure 5.6 Laminar flow generated by oscillating a plane wall back and forth in its own plane

This can be shown to satisfy equation (5.1.18) by direct substitution. It has the required behaviour $u \to U\cos\omega t$ as $y \to 0$, and $u \to 0$ as $y \to \infty$. These are the boundary conditions which distinguish the solution given above from other solutions of the equation.

The solution (5.1.19) can be thought of as a damped wave propagating away from the wall. The damping factor

$$\exp\left[-\left(\frac{\omega}{2\nu}\right)^{\frac{1}{2}} y\right]$$

is of particular interest. The distance which the disturbance extends from the wall depends on the parameter $(\omega/2\nu)^{1/2}$; for example, the disturbance in the fluid has only one hundredth of the wall amplitude when $(\omega/2\nu)^{1/2}\, y = 4{\cdot}6$.

The disturbed region near the wall is a form of *boundary layer*. Its thickness is measured more significantly in non-dimensional terms as a fraction of the displacement amplitude at the plate, $a = U/\omega$. The layer has no edge in the strict sense, but for the purposes of this discussion the point at which the fluid motion is 1 per cent of that at the wall can be used.

Then

$$\frac{\delta}{a} = 4{\cdot}6\,\frac{(U/2\nu a)^{-\frac{1}{2}}}{a} = 6{\cdot}5\left(\frac{Ua}{\nu}\right)^{-\frac{1}{2}} \qquad (5.1.20)$$

gives the thickness as a fraction of the wall amplitude. The dimensionless parameter Ua/ν which determines this relative thickness is called the *Reynolds number* of the flow.

The significance of the combination of parameters called the Reynolds number is made clearer by an inspection of the governing equation (5.1.18). The inertia and viscous forces can be estimated as

$$\varrho \frac{\partial u}{\partial t} \sim \varrho U \omega = \frac{\varrho U^2}{a}$$

and

$$\mu \frac{\partial u^2}{\partial y^2} \sim \frac{\mu U}{\delta^2}$$

with δ a distance over which the velocity changes significantly. The symbol $\sim$ may be read 'behaves like', not quite as strong a connexion as 'is proportional to' or 'is of the order of'. The inertia and viscous forces must balance one another; hence $\varrho U^2/a \sim \mu U/\delta^2$, or

$$\frac{\delta}{a} \sim \left(\frac{Ua}{\nu}\right)^{-\frac{1}{2}} \tag{5.1.21}$$

as was found above. The Reynolds number is seen to prescribe the flow geometry required in order that the inertia and viscous forces balance one another. When it is high, the boundary layer is thin; but for low Reynolds number, the disturbance spreads a long way into the fluid.

Although the role of the Reynolds number has been demonstrated for only one flow, a parameter of the form

$$\frac{\text{typical velocity} \times \text{typical length}}{\text{kinematic viscosity}} \tag{5.1.22}$$

is invariably found to be an important index of the behaviour of flows in which inertia and viscous forces are significant.

The process by which the disturbance propagates normal to the direction of fluid motion is called *diffusion.* As was pointed out earlier, this process is quite different, except in name, from the deceleration occurring in a diffuser. The kind of diffusion considered here is akin to the conduction of heat; the transfer of heat, momentum, or some other quantity is the result of mixing. In a laminar flow the mixing is at the microscopic level, and in a turbulent flow it results from the transfer of finite lumps of fluid.

The equation (5.1.18) is an example of the simple diffusion equation which will later be seen to govern one-dimensional heat conduction as well. The constant appearing in the general diffusion equation is the diffusivity of the transferred quantity in the medium through which it passes. Thus the kinematic viscosity is the diffusivity of momentum in the fluid. In more general two-dimensional flows it is the vorticity of the fluid which is diffused, the velocity variation itself being influenced by the pressure gradient, as shown in equation (5.1.17). Hence the spreading of a disturbance by viscosity is usually referred to as the diffusion of vorticity.

5.1.4 Laminar Dissipation

Let us next see how dissipation takes place in some of the laminar flows considered above. This is most simply done by converting the steady-flow momentum equation for two-dimensional motion

$$\frac{d\tau}{dy} = \frac{dp}{dx} \tag{5.1.4}$$

into an energy equation. Multiplication by the local velocity u gives an energy balance for unit volume of the fluid:

$$u\frac{dp}{dx} = u\frac{d\tau}{dy} = \frac{d(\tau u)}{dy} - \tau\frac{du}{dy} \tag{5.1.23}$$

Further multiplication by the volume element $dx\ dy \times 1$ gives an energy balance for the elemental volume shown in Figure 5.7:

$$u\ dp\ dy = d(\tau u)\ dx - \tau\ du\ dx$$

The terms of this equation can be interpreted as follows:

$u\ dp\ dy$ is the rate of working of the pressure gradient;

$d(\tau u)\ dx$ is the net rate of working of the shear forces on the top and bottom of the element;

hence the balancing term $\tau\ du\ dx$ must represent the dissipation in the elemental volume. We conclude that $\tau\ du/dy$ gives the rate of dissipation per unit volume. This interpretation applies to both laminar and turbulent flows, it being understood that it is the time-mean properties of the turbulent flow that are relevant.

In Chapter 2 it was found that a steady flow through a duct gave rise to dissipation at the rate

$$\dot{m}\frac{de}{dx} = -\dot{V}\frac{dp}{dx} \tag{2.3.15}$$

per unit length of the duct. Is this result consistent with a local dissipation rate of $\tau\ du/dy$? Consider flow in a two-dimensional channel, for which $u = 0$ when $y = 0, b$, and

$$\frac{d\tau}{dy} = \frac{dp}{dx} = \text{constant}$$

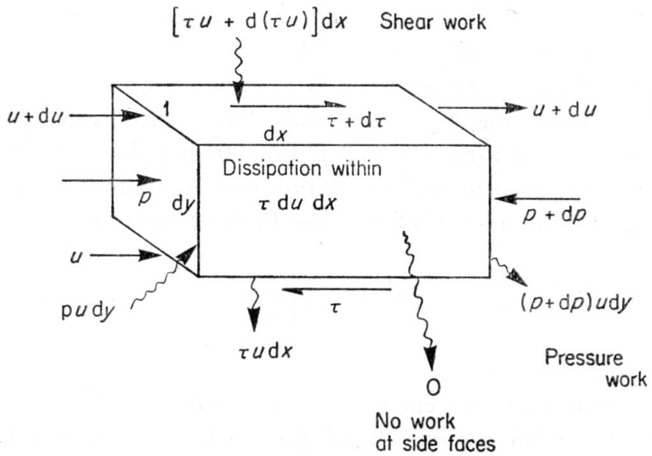

Figure 5.7 The energy balance for a fluid element in a parallel, two-dimensional flow

as found in equation (5.1.4). Using integration by parts and the information given above, the dissipation per unit length of the channel is found to be

$$\int_0^b \tau \frac{\mathrm{d}u}{\mathrm{d}y}\,\mathrm{d}y = \tau u\Big|_0^b - \int_0^b u \frac{\mathrm{d}\tau}{\mathrm{d}y}\,\mathrm{d}y$$
$$= -\frac{\mathrm{d}\tau}{\mathrm{d}y}\int_0^b u\,\mathrm{d}y = -\dot{V}\frac{\mathrm{d}p}{\mathrm{d}x}$$

As expected, this result is in agreement with equation (2.3.15).

To find the way in which dissipation is distributed within a flow, it is necessary to know $u(y)$ and $\tau(y)$; this in turn requires that τ and u be related in some way. For a Newtonian fluid the dissipation per unit volume is

$$\tau \frac{\mathrm{d}u}{\mathrm{d}y} = \mu\left(\frac{\mathrm{d}u}{\mathrm{d}y}\right)^2 = \frac{\tau^2}{\mu} \qquad (5.1.24)$$

In order that the dissipation be positive, the viscosity must be positive. Common sense suggested that this would be the case, but here it is shown to be a consequence of the second law of thermodynamics, not a separate hypothesis. Note that the rate of dissipation depends on the square of the velocity gradient, and will therefore be strongly concentrated in boundary layers where the velocity changes rapidly. For example, in plane Poiseuille flow equation (5.1.6) gives

$$\mu\left(\frac{\mathrm{d}u}{\mathrm{d}y}\right)^2 = \frac{\tau^2}{\mu} \propto \left(y - \frac{1}{2}b\right)^2$$

measurable in the same units as work. (In practice, distinct units of heat are often used—the thermal units, Btu and kcal—but they are not strictly necessary.) In the same way, Newton's second law links the units of mass and force, and the units of entropy are prescribed in terms of those of energy and temperature.

Since the same physical quantity can be measured in a variety of equivalent units (for example, work in lbf ft, N m, kgf m, or dyne cm), it is convenient to introduce the concept of a dimension as a generalization of a unit. Thus work is said to have dimensions of force×length. Since force has dimensions of mass×acceleration or mass×length/time², the dimensions of work can also be stated as mass×length²/time². These statements are more conveniently expressed symbolically:

$$W = [FL], \quad F = \left[\frac{ML}{T^2}\right] \quad \text{and} \quad W = \left[\frac{ML^2}{T^2}\right]$$

The symbols F, L, M and T indicate the dimensions of the quantities; in other words, they indicate the kinds of units which must be used to measure them. The square brackets indicate that the symbols within do not stand for physical quantities.

By virtue of the interlocking of units, all mechanical quantities can be measured in terms of three *fundamental units*, while thermodynamics uses an additional unit, that of temperature, whose dimension will be denoted θ. The units of length, time and temperature are almost universally taken to be fundamental. The fourth fundamental unit is sometimes taken to be that of mass, and sometimes to be that of force. In practice, no clear-cut decision is necessary or desirable; it is customary to express derived units in a combination of mass and force units. Thus specific energy is usually given in kJ/kg rather than the fundamental units m^2/sec^2. Nevertheless, it is important to realize that it is possible to express all of the quantities encountered in this book in terms of four dimensions, usually taken to be L, T, θ, and one of F and M.

In earlier chapters it has often proved convenient to express formulae in terms of dimensionless parameters. This has several advantages. Calculations can be made more simply; tables and graphs take on a unique form and can be used for calculations in various unit systems; and the number of parameters involved in each formula is reduced, so that the relationships amongst them are clearer. The formulae (3.3.6) describing a shock wave in a perfect gas illustrate these advantages.

Certain of the advantages of dimensionless parameters require that they be unit-free as well, in other words, that the quantities entering into them are measured in consistent units. A dimensionless quantity need not be unit-free. For example, in Callendar's equation

$$pv = 1{\cdot}247(h-835)$$

in English steam-table units, the number 1·247 is dimensionless, since both pv and h have dimensions of energy/mass, but nevertheless has units of

$$(\text{lbf ft/Btu})\,(\text{ft}^2/\text{in}^2) = \frac{144}{778}$$

Although the units of a dimensionless quantity are equivalent to a pure number, that number is unity only for consistent units. When pv and h are measured in any set of consistent units, the constant will be

$$1{\cdot}247 \times \frac{144}{778} = 0{\cdot}231$$

which is unit-free as well as dimensionless.

5.2.2 Dimensional Homogeneity

It has been noted that the number of dimensionless parameters appearing in an equation is less than the number of dimensioned quantities required to describe the process governed by the equation. Our next task is to see why this is so.

We shall consider a process which has been analysed in detail in an earlier chapter—the change in temperature undergone by a perfect gas which passes through a normal shock wave. Without analysing the problem in detail, one might postulate that this temperature change was given by a relationship amongst the properties

$$T_2,\ T_1,\ u_1,\ u_2,\ a_1,\ a_2,\ c_v,\ c_p \qquad (5.2.1)$$

This does indeed turn out to be the case, as was shown in Section 3.3.1. The relevant equation of the set (3.3.6) has the form

$$\frac{T_2}{T_1} = f\left(\frac{u_1}{a_1}, \frac{c_p}{c_v}\right) \qquad (5.2.2)$$

Three dimensionless parameters have replaced the eight of the list (5.2.1). This reduction can be explained in two ways:

(1) Some of the quantities (5.2.1) are specified by the others by virtue of the special features of a normal shock and perfect gas. Thus u_2 and a_2 are determined in terms of T_2, T_1, u_1, a_1, c_p and c_v by the conditions for isenergic flow of a perfect gas:

$$c_pT+\frac{u^2}{2} = \frac{a^2}{c_p/c_v-1}+\frac{u^2}{2} = \text{constant}$$

If one had realized this, the list of related properties might have been reduced to

$$T_2,\ T_1,\ u_1,\ a_1,\ c_p,\ c_v \tag{5.2.3}$$

Thus even an incomplete specification of the process suggests something about the structure of the solution. By applying further constraints, one can ultimately deduce the form (5.2.2); indeed, this was done in deriving equations (3.3.6).

(2) Part of the reduction from the parameters (5.2.1) to those of equation (5.2.2) can be explained without referring to the particular process. This reduction is implied by the principle of dimensional homogeneity: each physical law must be independent of the system of units used to describe the process, and can therefore be expressed entirely in terms of dimensionless groups. It is obvious that this principle applies to any law derived by combining fundamental statements that are themselves dimensionally homogeneous. We now take the further small step of extending the principle to relationships which we do not know how to obtain from basic laws and definitions, but which must nevertheless be governed by those principles.

Let us consider the implications of the principle of dimensional homogeneity for the relationship between the quantities (5.2.3). First note their dimensions:

$$\begin{aligned} T_2,\ T_1 &= [\theta] \\ u_1,\ a_1 &= \left[\frac{\mathrm{L}}{\mathrm{T}}\right] \\ c_p,\ c_v &= \left[\frac{\mathrm{FL}}{\mathrm{M}\theta}\right] = \left[\frac{\mathrm{L}^2}{\mathrm{T}^2\theta}\right] \end{aligned} \tag{5.2.4}$$

Next rewrite the relationship

$$T_2 = f(T_1,\ u_1,\ a_1,\ c_p,\ c_v) \tag{5.2.5}$$

in a series of equivalent forms, in order to build up dimensionless groups in succession:

$$\frac{T_2}{T_1} = f_1(T_1,\ u_1,\ a_1,\ c_p,\ c_v)$$

$$u_1 = f_2\left(\frac{T_2}{T_1},\ T_1,\ a_1,\ c_p,\ c_v\right)$$

$$\frac{u_1}{a_1} = f_3\left(\frac{T_2}{T_1},\ T_1,\ a_1,\ c_p,\ c_v\right)$$

$$c_p = f_4\left(\frac{T_2}{T_1},\ \frac{u_1}{a_1},\ T_1,\ a_1,\ c_v\right)$$

$$\frac{c_p}{c_v} = f_5\left(\frac{T_2}{T_1}, \frac{u_1}{a_1}, T_1, a_1, c_v\right)$$

$$a_1 = f_6\left(\frac{T_2}{T_1}, \frac{u_1}{a_1}, \frac{c_p}{c_v}, T_1, c_v\right)$$

$$\frac{a_1}{(c_v T_1)^{\frac{1}{2}}} = f_7\left(\frac{T_2}{T_1}, \frac{u_1}{a_1}, \frac{c_p}{c_v}, T_1, c_v\right)$$

At this point the process fails, since $T_1 = [\theta]$ and $c_v = [L^2/T^2\theta]$ cannot by themselves form a dimensionless group. If the result is to be dimensionally homogeneous, these quantities cannot enter into it. Hence

$$\frac{T_2}{T_1} = f_8\left(\frac{u_1}{a_1}, \frac{c_p}{c_v}, \frac{a_1}{(c_v T_1)^{\frac{1}{2}}}\right) \tag{5.2.6}$$

specifies the relationships of the form (5.2.5) which are dimensionally homogeneous. No physically significant statements have been eliminated in adopting the more restricted form (5.2.6).

The procedure used in passing from equation (5.2.5) to equation (5.2.6) can be applied to any set of related variables. However, this rather tedious series of manipulations can be avoided by deducing from this one example the features common to the whole class of problems. It is of particular interest to know how many dimensionless groups will enter into the dimensionless statement of the relationship between a set of dimensioned variables. The procedure illustrated above indicates that the reduction in the number of parameters results from the rejection of residual variables which cannot by themselves be combined into dimensionless groups. The number of residual variables will always equal the number of *independent dimensions* occurring in the list of dimensions for the original variables. Thus two dimensioned quantities are rejected in passing from (5.2.5) to (5.2.6); this is the number of independent dimensions (θ and L/T in this case) appearing in the list (5.2.4). Note that the number of 'independent dimensions' can be less than the number of 'fundamental dimensions' in the original list. It cannot exceed the number of relevant fundamental dimensions, and is usually equal to it.

These conclusions are summarized in a statement which is often called the *pi-theorem*, since the dimensionless groups were denoted by the letter Π in the original formulation: the number n of independent dimensionless groups required to express a physical relationship is equal to the number p of physical quantities entering into the relationship minus the number d of independent dimensions appearing in the p quantities. Symbolically

$$n = p - d \tag{5.2.7}$$

In the list (5.2.4), $p = 6$ and $d = 2$. Then $n = 4$, as in equation (5.2.6). Note that the number of dimensionless groups appearing in the governing equation may be less than the number suggested by the pi-theorem, by virtue of special features of the problem not taken into account in specifying the original p quantities.

In the statement of the pi-theorem the word 'independent' is used twice. We have seen what is meant by independent dimensions; we must also understand what is meant by *independent dimensionless groups.* Note first that dimensionless groups can be combined to form further dimensionless parameters. For example, some combinations of the groups of equation (5.2.6) are

$$\frac{a_1}{(c_v T_1)^{\frac{1}{2}}} \left(\frac{c_v}{c_p}\right)^{\frac{1}{2}} = \frac{a_1}{(c_p T_1)^{\frac{1}{2}}}$$

and

$$\frac{u_1}{a_1} \frac{a_1}{(c_v T_1)^{\frac{1}{2}}} = \frac{u_1}{(c_v T_1)^{\frac{1}{2}}}$$

However, the process leading to equation (5.2.6) produces a series of independent groups, that is, groups which cannot be generated by combining the others. These original groups can be replaced by alternative combinations without changing the essential nature of the relationship, so long as the number of independent groups remains the same. Hence

$$\frac{T_2}{T_1} = f\left[\frac{u_1}{(c_v T_1)^{\frac{1}{2}}}, \frac{c_p}{c_v}, \frac{a_1}{(c_p T_1)^{\frac{1}{2}}}\right]$$

is equivalent to equation (5.2.6), but

$$\frac{T_2}{T_1} = f\left[\frac{c_p}{c_v}, \frac{a_1}{(c_v T_1)^{\frac{1}{2}}}, \frac{a_1}{(c_p T_1)^{\frac{1}{2}}}\right]$$

is not.

A second important conclusion is suggested by the result (5.2.6). It differs from equation (5.2.2) by including the group $a_1/(c_v T_1)^{1/2}$. For a perfect gas, this parameter equals $(\gamma R/c_v)^{1/2} = [\gamma(\gamma - 1)]^{1/2}$, and the apparent discrepancy is resolved. However, the fact that the dimensional analysis generated this unwanted group provides a warning. The results of dimensional analysis are, to a great extent, determined by the selection of the original variables. Their pertinence depends on the insight of the selector into the process being analysed. Suppose, for example, that dimensional analysis were applied to the 'unpruned' quantities (5.2.1). The pi-theorem indicates that the relationship will involve six independent groups, and these can easily be constructed:

$$f\left(\frac{T_2}{T_1}, \frac{u_1}{a_1}, \frac{u_2}{a_2}, \frac{c_p}{c_v}, \frac{a_2}{a_1}, \frac{a_1}{(c_v T_1)^{\frac{1}{2}}}\right) = 0$$

It is true that the change in properties across a shock in a perfect gas is governed by a law encompassed by this general result. However, the actual relationship is so much more restricted that the general result is for practical purposes useless, and may even be harmful in giving an incorrect impression of complexity. If, on the other hand, one had sufficient insight to select only the quantities

$$T_2, T_1, u_1, c_p, c_v$$

the result could be given as a three-parameter relationship:

$$f\left(\frac{T_2}{T_1}, \frac{c_p}{c_v}, \frac{u_1}{(c_v T_1)^{\frac{1}{2}}}\right) = 0$$

which is easily seen to be equivalent to the result (5.2.2).

Example 5.3 Here the physical quantities defining a variety of processes will be picked out and the pi-theorem will be used to find the number of dimensionless groups required to describe each situation. Later (in Example 5.5) the actual groups will be determined.

(1) Load supported by a slipper bearing

On what quantities does the load $W = [F]$ supported by the slipper bearing shown in Figure 5.4 depend? Obviously the geometry of the bearing film is relevant; it is specified by the length $l = [L]$, the angle between the surfaces $\alpha = [-]$, and the minimum clearance between the surfaces $b' = [L]$. The relative velocity $U = [L/T]$ is significant, since it is the relative motion which gives rise to the pressure in the film. Certain properties of the lubricant are also relevant: the viscosity $\mu = [FT/L^2]$, since viscous stresses are the means by which the pressure is generated; and the surface tension $S = [F/L]$, since this helps to determine the area of the surface wetted by the film. Normally the fluid density is not important, since viscous forces are much greater than inertia forces in the fluid.

Hence we anticipate that the load will depend upon six quantities which define one of a family of essentially similar bearings. These quantities involve three independent dimensions. Hence the pi-theorem gives $n = 7 - 3 = 4$.

(2) Flow over a V-notch weir

This device for metering an open-channel flow is shown in Figure 5.8(a). The geometric parameters influencing the flow rate $\dot{V} = [L^3/T]$ are the angle of the notch $\theta = [-]$ and the elevation of the upstream surface above the tip of the notch $d = [L]$. The weight of the fluid, characterized by the specific weight $\varrho g = [F/L^3]$, provides the driving force for the flow. One might think at first sight that these quantities alone defined the flow rate in the absence of significant viscous and surface tension forces. However, since each of the dimensions F and T appears only once in the set of quantities mentioned above, it is not possible to form dimensionless groups with this set. When the fluid density $\varrho = [M/L^3] = [FT^2/L^4]$ is included, this difficulty is resolved. Physically the density is important since the fluid weight acts to accelerate the fluid as it approaches the notch.

The fluid weight and density must appear in the dimensionless groups in the combination $\varrho g/\varrho = g = [L/T^2]$. Hence we may take

$$\dot{V} = f(\theta, d, g)$$

as the general form of the flow law. These four related quantities involve two independent dimensions. Hence $n = 4-2 = 2$ gives the number of groups. In this case dimensional analysis has provided some guidance in the selection of relevant quantities.

(3) Heat transfer from a sphere

Figure 5.8(b) shows a sphere maintained at temperature T_s immersed in a fluid whose temperature is T well away from the sphere. The rate of heat transfer from the sphere to the fluid $\dot{Q} = [FL/T]$ is specified by an empirically determined coefficient $C_h = [F/LT\theta]$. The heat flow may also be supposed to depend on the temperature difference $(T_s - T) = [\theta]$ and on the surface area of the sphere $A = [L^2]$.

Even though four fundamental dimensions are involved, the four related quantities can be expressed in terms of three independent dimensions, say, FL/T, L^2, and θ. Hence the number of dimensionless groups to be found is $n = 4-3 = 1$.

(4) Drag of a sphere

Figure 5.8(c) shows a sphere immersed in a uniform gas flow. A law specifying the drag $D = [F]$ of the sphere is required. This will involve the diameter of the sphere $d = [L]$ and the velocity of the stream $U = [L/T]$. It will also depend on certain of the fluid properties, namely, the density $\varrho = [FT^2/L^4]$, since this will influence the pressure distribution over the surface, and the viscosity $\mu = [FT/L^2]$, since this will influence the shear-stress distribution over the surface. The response of the gas may be expected to be of some significance; it will be specified by the gas constant $R = [FL/M\theta] = [L^2/T^2\theta]$ and the ratio of specific heats $\gamma = [-]$. The flow pattern will also depend on the compressibility of the fluid, which is in turn dependent on the temperature. Hence the temperature of the stream $T = [\theta]$ is relevant as well.

The eight quantities involve four independent dimensions, so that $n = 8-4 = 4$.

Note that the properties used to specify the flow are those of the undisturbed stream, it being tacitly assumed that the entire flow pattern is specified once these have been fixed. For particular values of the defining parameters, the laws governing the fluid motion allow only one pattern of development for the flow around the sphere.

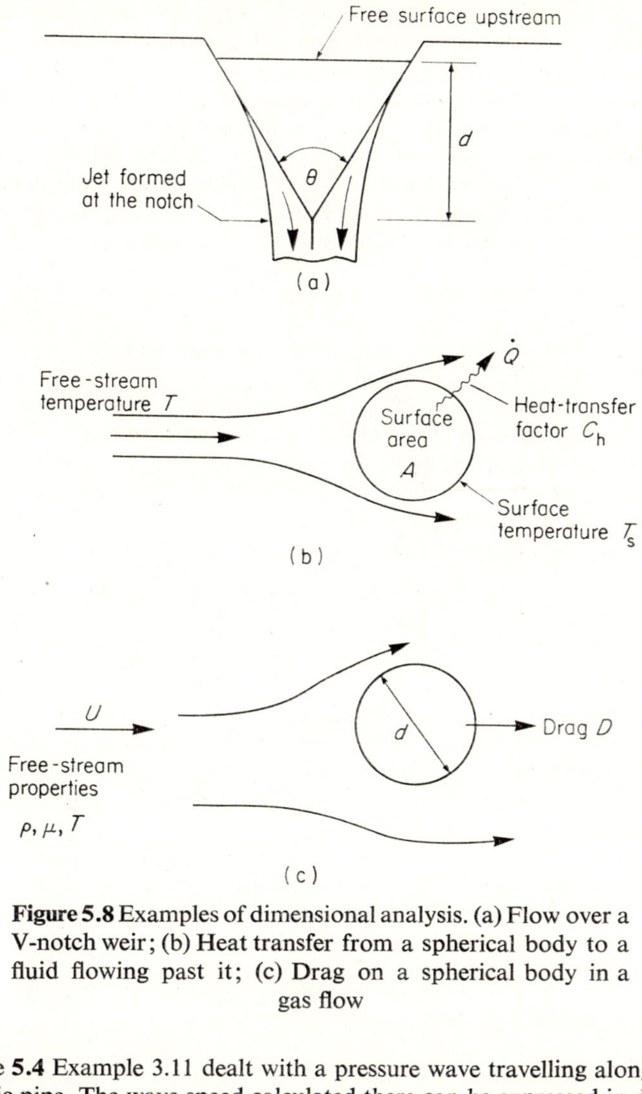

Figure 5.8 Examples of dimensional analysis. (a) Flow over a V-notch weir; (b) Heat transfer from a spherical body to a fluid flowing past it; (c) Drag on a spherical body in a gas flow

Example 5.4 Example 3.11 dealt with a pressure wave travelling along a liquid-filled elastic pipe. The wave speed calculated there can be expressed in dimensionless form as

$$c^2 \varrho \varkappa = \frac{1}{1+(2R/\varkappa Yt)/(1-pR/Yt)}$$

Let us compare this result with the indications of dimensional analysis for the same problem.

Without undertaking any calculations, one might anticipate that the wave speed $c = [\mathrm{L/T}]$ would depend on the geometry of the pipe, specified by the radius $R = [\mathrm{L}]$ and wall thickness $t = [\mathrm{L}]$, and on the density $\varrho = [\mathrm{FT^2/L^4}]$ and compressibility of the fluid $\varkappa = [\mathrm{L^2/F}]$. The modulus of rigidity of the pipe wall $Y =$

$[F/L^2]$ and the pressure within the pipe $p = [F/L^2]$ may also be expected to play a part by determining the proportional expansion of the pipe. Hence the number of independent dimensionless groups will be $n = 7-3 = 4$.

One can immediately write down the groups t/R, $\varkappa Y$ and $\varkappa p$. Another dimensionless combination is p/Y. However, it is not suitable for use in conjunction with $\varkappa Y$ and $\varkappa p$, since it is not independent of them. It could, however, replace one of those groups. Earlier discussions have made familiar the dimensions of $(\varrho\varkappa)^{-1/2} = [L/T]$. Hence a fourth independent group is easily found, $c^2\varrho\varkappa$. Thus dimensional analysis suggests the relationship

$$c^2\varrho\varkappa = f(\varkappa Y, \varkappa p, R/t)$$

The following points emerge on comparing this result with the explicit formula given above.

(1) Only three dimensionless groups appear in the explicit formula. Special features of the process dictate this more rigid grouping, which cannot be predicted by dimensional considerations alone. However, if one had argued that the modulus of rigidity and wall thickness must appear together in the combination tY, since they act only through the net hoop stress in the pipe wall, the number of related quantities would be reduced by one, and $n = 6-3 = 3$. Then

$$c^2\varrho\varkappa = f\left(\frac{\varkappa Yt}{R}, \varkappa p\right)$$

(2) The dimensionless groups selected are not those in which the formula is most conveniently expressed. There is nothing in the procedures of dimensional analysis to suggest the most appropriate combinations. The recasting of dimensionless laws in useful forms requires an understanding of the physical situation or a knowledge of conventional groupings.

(3) In Example 3.11 it was seen that the parameter pR/Yt is not relevant in certain important cases. This fact cannot be deduced from the general functional relationship. The mere fact that a parameter is very small (or very large) is not significant if one has no idea of its relationship to the other parameters.

5.2.3 Dimensionless Groups

When one looks over the sets of variables listed in Example 5.3, it is apparent that there are many ways of combining them into dimensionless groups and that some of these groups will have a rather complex structure. The more obvious groups will not necessarily be the most useful, nor will all the necessary groups be apparent by inspection. Hence we require methods of generating dimensionless parameters which can be applied with confidence to unfamiliar combinations of dimensions. It must be emphasized, however, that a formal technique need not be used to determine dimensionless groups. Nor is it necessary to determine all of the groups using the same method. The pi-theorem indicates the number of independent groups required, but places no restriction on the way in which they are found.

Four methods of finding dimensionless parameters will be illustrated with reference to the same physical law. Consider the aerodynamic drag on the projectile shown in Figure 5.9. The drag $D = [F]$ will depend on the length $l = [L]$ of the body, and on its diameter $d = [L]$ and nose angle $\beta = [-]$. The speed of flight $U = [L/T]$ is also important, as are the density $\varrho = [FT^2/L^4]$, viscosity $\mu = [FT/L^2]$, and sound speed $a = [L/T]$ of the air. The pi-theorem indicates that $n = 8-3 = 5$ groups must be sought.

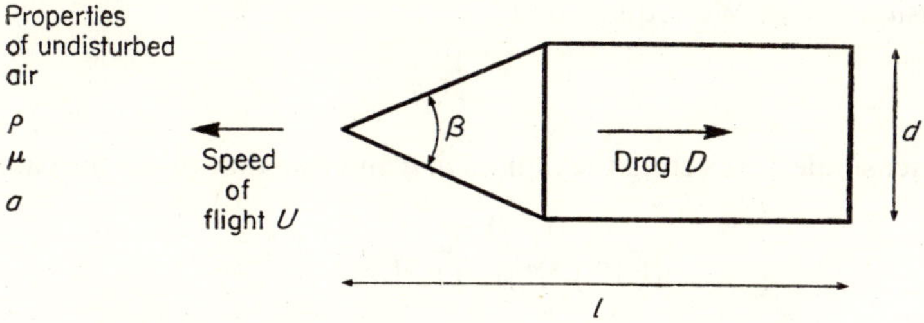

Figure 5.9 Factors influencing the drag of a projectile

Method I. Inspection and Familiarity

The dimensionless parameters l/d, β and U/a are immediately obvious. Remembering that ϱu^2 has dimensions of pressure, we can obtain a fourth, somewhat less obvious group: $D/\varrho U^2 d^2$. The fifth group is not easy to spot without experience, and must be sought using some more systematic technique.

Method II. Cancellation of Dimensions

Looking at the groups already found, we note that viscosity is absent. Hence the final group must involve this quantity, which has dimensions FT/L^2. Let us generate a dimensionless group by combining it with other quantities so as to cancel out these dimensions. The cancelling quantities must be capable of being combined to form each of the dimensions which has to be removed. Suitable combinations are ϱ, U, d; D, U, l; and μ, a, l.

Let us use the first set, which can be combined to give

$$d = [L], \quad \frac{d}{U} = [T] \quad \text{and} \quad \varrho U^2 d^2 = [F]$$

Such combinations are referred to as a characteristic length, time and force for the system. Using these characteristic quantities to cancel the dimensions of viscosity, we obtain

$$\frac{\mu}{\varrho U^2 d^2 (d/U) d^{-2}} = \frac{\mu}{\varrho U d} = [-]$$

for the fifth dimensionless group.

Method III. Simultaneous Elimination

The cancelling operation proposed in Method II can be carried out in another way. Consider once again the combining of ϱ, U and d to cancel the dimensions of μ. We require that

$$\frac{\mu}{\varrho^a U^b d^c}$$

be dimensionless, selecting the indices a, b and c so that this is the case. Then

$$\frac{[\mathrm{FT/L^2}]}{[\mathrm{FT^2/L^4}]^a\,[\mathrm{L/T}]^b\,[\mathrm{L}]^c} = [-]$$

requiring

$$1-a = 0 \quad \text{for the elimination of F}$$

$$-2+4a-b-c = 0 \quad \text{for the elimination of L}$$

and

$$1-2a+b = 0 \quad \text{for the elimination of T}$$

Solving, we find $a = 1$, $b = 1$, and $c = 1$, giving the grouping found earlier using Method II.

The cancelling quantities used in this procedure cannot be selected entirely at random. They must still satisfy the criterion of Method II, namely, that they can be combined to give each of the dimensions which has to be cancelled. If this condition is not obeyed, the set of equations generated cannot be solved for the indices a, b and c.

Although they have been used here to construct a single group, Methods II and III can be used to produce all of the dimensionless groups in turn. The quantities ϱ, U and d (for example) are simply used to render each of the remaining quantities dimensionless. This approach is particularly useful for unfamiliar situations where one cannot pick out many of the groups by inspection.

Method IV. Progressive Elimination

Once again we decide to form a group involving $\mu = [\mathrm{FT/L^2}]$. Now the dimensions will be eliminated one at a time, suitable cancelling quantities being selected from the other quantities, D, l, d, β, U, a and ϱ. Thus, selecting

ϱ to cancel the dimension F, we have

$$\frac{\mu}{\varrho} = \left[\frac{\mathrm{L}^2}{\mathrm{T}}\right]$$

Selecting U to cancel the dimension T without re-introducing F, we have

$$\frac{\mu}{\varrho U} = [\mathrm{L}]$$

Finally, we select d to remove the dimension L, and obtain

$$\frac{\mu}{\varrho U d} = [-]$$

This method involves fewer arithmetic operations than do Methods II and III, and also has the advantage of splitting up the task of selecting cancelling quantities into easy stages. It is particularly suited to situations like that considered here, in which the general nature of the missing group is fairly clear. However, it can be applied systematically to find all the independent groups, simply by using the same three quantities to eliminate the dimensions from the rest.

It is now apparent that the dimensional analysis of a particular process involves the following steps:

(1) select the quantities which define the aspect of immediate interest,

(2) note their dimensions and use the pi-theorem to determine the number of independent dimensionless groups to be found,

(3) use Method I to find as many independent groups as possible,

(4) if further groups are required, use one of Methods II, III and IV to find them (you need be familiar with only one of these methods, since they perform the same task), and

(5) arrange the groups in the most convenient form.

The final stage (5) requires further comment. We have noted earlier that the dimensionless groups can be combined to form other groups. Which of the infinite number of alternatives are to be selected?

A person familiar with the literature of fluid mechanics would be likely to express the results of the preceding dimensional analysis in the form

$$\frac{D}{\frac{1}{2}\varrho U^2(\frac{1}{4}\pi d^2)} = f\left(\frac{Ud}{\nu}, \frac{l}{d}, \beta, \frac{U}{a}\right) \tag{5.2.8}$$

In doing this, he would be guided by the desire to use groups that are

(1) physically significant,

(2) simple in form, and

(3) conventional.

It is not always possible to satisfy all of these requirements. For example, the drag coefficient used in equation (5.2.8) is not as simple as $D/(\varrho U^2 d^2)$. However, it can be interpreted as the ratio of the actual drag to the force equivalent to the dynamic pressure acting on the frontal area of the projectile. This ratio will be significant if the drag is mostly associated with an inertia-dependent pressure differential between the nose and base of the projectile. Since this is the case for large bodies moving through the air, the grouping suggested in (5.2.8) is in common use. Thus it satisfies the requirements (1) and (3) given above, being interpretable in physical terms and having conventional form.

Next consider the viscosity parameter of equation (5.2.8). Comparison with equations (5.1.21, 22) shows that this is a Reynolds number. The parameter $\mu/(\varrho U d)$ which was isolated in the preceding discussion has been replaced by its reciprocal which is in common use. As shown in Section 5.1.3, the Reynolds number can be interpreted as a measure of the relative roles of inertia forces and viscous forces in the flow. Viscous forces are important in the boundary layer on the body where the velocity falls rapidly to zero at the fixed surface. Hence the Reynolds number also characterizes the thickness of the region in which viscous forces are important. Thus it appears that the choice of the Reynolds number as the parameter characterizing the role of viscosity combines the advantages of physical significance, simplicity and conventionality.

The other parameters of equation (5.2.8) require little comment, since they can be given direct geometrical interpretations.

Before leaving the example of projectile drag, let us look further at the problem of choosing a drag coefficient. The adoption of the coefficient of equation (5.2.8) was justified on the grounds that the main source of drag is an inertia-dependent pressure differential between nose and base. Now consider the drag on a body of the same general shape moving slowly through a viscous liquid. Further, take it to be a slender, needle-like body with $l/d \gg 1$. No new quantities enter into the drag law, and the result (5.2.8) is still valid. However, the precise form is no longer appropriate. Now the drag arises primarily from viscous friction along the sides of the body, and its magnitude may be expected to be something like

$$\mu\left(\frac{U}{d}\right)(\pi d l) = \mu U(\pi l)$$

where $\pi d l$ is the wetted area, that exposed to the shear stress. Hence the viscous drag coefficient $D/[\mu U(\pi l)]$ should prove more appropriate in this instance. Note that this is a simple combination of the dimensionless groups of equation (5.2.8):

$$\frac{D}{\varrho U^2 d^2} \cdot \frac{\varrho U d}{\mu} \cdot \frac{d}{l} = \frac{D}{\mu U l}$$

In the situation postulated above, slow motion of a slender body through a very viscous fluid, the Mach number is not relevant, and we might suppose also that the nose angle β is relatively unimportant. Hence the form

$$\frac{D}{\mu U(\pi l)} = f\left(\frac{Ud}{\nu}, \frac{l}{d}\right) \tag{5.2.9}$$

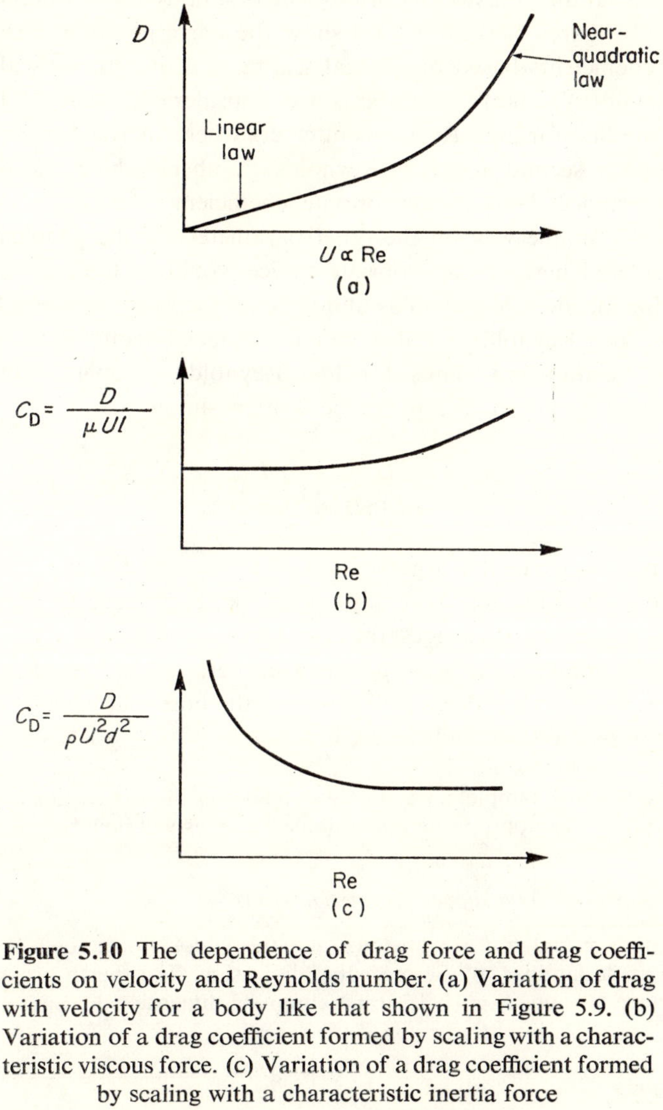

Figure 5.10 The dependence of drag force and drag coefficients on velocity and Reynolds number. (a) Variation of drag with velocity for a body like that shown in Figure 5.9. (b) Variation of a drag coefficient formed by scaling with a characteristic viscous force. (c) Variation of a drag coefficient formed by scaling with a characteristic inertia force

seems more appropriate for viscosity-dominated flow past a slender body. Two of the groups have been dropped as being unimportant in this case, while another has been recast in what seems to be a more appropriate form.

What is meant by the word 'appropriate' when applied to a dimensionless group? For the case discussed above, this question is equivalent to: What is the advantage of using $D/\mu Ul$ for highly viscous flow instead of $D/\varrho U^2d^2$? These questions can be answered by referring to Figure 5.10. This indicates in (a) the variation with velocity of the drag of a body such as that shown in Figure 5.9. Figures 5.10(b) and (c) show the corresponding variations of the drag coefficients based on inertial scaling and on viscous scaling. The coefficients of (b) and (c) are constants over considerable ranges of Reynolds number, the first for high Reynolds numbers where inertia forces predominate, and the second for low Reynolds numbers where viscosity is of prime importance. Thus an 'appropriate' coefficient is seen to be one which depends in a simple way on the other parameters of the problem. In the case considered here, an appropriate choice would be the drag coefficient which is independent of Reynolds number over the range of interest. Within this range the Reynolds number can be dropped from the appropriate functional relationship. Thus for low Reynolds numbers, Figure 5.10 indicates that equation (5.2.9) can be written simply as

$$\frac{D}{\mu U(\pi l)} = f\left(\frac{l}{d}\right)$$

Although the particular arguments used in treating the example of drag in a viscous fluid are not generally applicable, they serve to illustrate the kinds of argument which suggest the most useful formulation of the results of dimensional analysis. As was indicated earlier, an understanding of the essential features of a process is necessary if the potential benefits of dimensional analysis are to be fully realized.

Example 5.5 In Example 5.3 a start was made on the dimensional analysis of four situations. The appropriate dimensionless groups will now be found.

(1) Load supported by a slipper bearing ($n = 4$)

The groups α and b'/l are obvious, and the viscous-force coefficient $W/\mu Ul$ has been made familiar by the preceding discussion. The fourth group must involve surface tension, $S = [\text{F/L}]$. First eliminate dimension F using μ:

$$\frac{S}{\mu} = \left[\frac{\text{L}}{\text{T}}\right]$$

Next eliminate the dimensions L and T using U:

$$\frac{S}{\mu U} = [-]$$

This parameter can be interpreted as the ratio of a typical surface tension force to a typical viscous force. Then

$$\frac{W}{\mu U l} = f\left(\alpha, \frac{b'}{l}, \frac{S}{\mu U}\right)$$

expresses the relationship in dimensionally homogeneous form.

(2) Flow over a V-notch weir (n = 2)

The angle θ of the notch serves as one dimensionless group. Now eliminate the dimensions of $\dot{V} = [L^3/T]$. Using g

$$\frac{\dot{V}}{g^{\frac{1}{2}}} = \left[L^{\frac{5}{2}}\right]$$

and using d, we obtain the group $\dot{V}/g^{1/2}d^{5/2}$. Thus

$$\frac{\dot{V}}{(gd^5)^{\frac{1}{2}}} = f(\theta)$$

is the form of the flow law. For a given installation,

$$\dot{V} = \text{constant} \times d^{\frac{5}{2}}$$

relates flow rate and depth, and only one measurement is required to find the flow

(3) Heat transfer from a sphere (n = 1)

The form of the group sought is not familiar from earlier work. However, $\dot{Q}$ and C_h must occur in the combination

$$\frac{\dot{Q}}{C_h} = [L^2\theta]$$

Hence

$$\frac{\dot{Q}}{C_h A(T_s - T)}$$

is a suitable group, and the heat-transfer law has the form

$$\dot{Q} = \text{constant} \times C_h A(T_s - T)$$

In practice this relationship is used to define the heat-transfer factor C_h, the constant being taken to be unity.

(4) Drag of a sphere (n = 4)

Earlier discussions have made familiar dimensionless groups suitable for all of the significant quantities. Hence we can write immediately

$$\frac{D}{\frac{1}{2}\varrho U^2(\frac{1}{4}\pi d^2)} = f\left(\frac{\varrho Ud}{\mu}, \gamma, \frac{U}{(\gamma RT)^{\frac{1}{2}}}\right)$$

It is convenient to select the Mach number as one of the groups even though it is not the simplest dimensionless combination.

When, as in these examples, the groups are found by inspection or by combining inspection and a more formal method, it is necessary to check that they are in fact independent of one another. A sufficient, but not necessary condition is that each group contains a quantity which appears in no other group. This condition is satisfied by the combinations selected above.

5.2.4 Dynamic Similarity

When the idea of dimensional analysis was introduced at the beginning of this chapter, it was associated with the geometric similarity of flow fields. This relationship has great practical importance in providing the basis for the design and interpretation of experiments using scale models. In relating dimensional analysis to the ideas of similarity, it is necessary to distinguish several ways in which one flow system can be similar to another. There follow definitions of four kinds of similarity, involving progressively closer identification of the two flow systems.

Geometric Similarity

This term relates to the boundaries of the flow, for example, the solid bodies over which the fluid passes, the ducting through which it moves, or the interface of a jet with the ambient fluid. The boundaries of two flows are said to be geometrically similar when a uniform expansion of one set of boundaries makes it identical with the second set. Thus geometrically similar boundaries have the same basic shape, although they are in general of differing size. Points occupying similar positions in the two systems are termed corresponding points.

Kinematic Similarity

Two flows are said to be kinematically similar when their flow patterns are similar, that is, when their streamlines are geometrically similar. It follows that the velocities at each pair of corresponding points have the same direction, and that there is a common ratio between the velocities for all pairs of corresponding points.

Dynamic Similarity

Two flows are said to be dynamically similar when the forces on the fluid at each pair of corresponding points have the same direction, and when there is a common ratio between the forces at all pairs of corresponding points. Generally, the satisfaction of these conditions requires the similarity of each kind of force which contributes significantly to the net force on the fluid, that is, similarity in the distributions of inertia and pressure forces, viscous stresses, surface tensions, fluid weight and so forth.

Thermodynamic Similarity

Two flows are said to be thermodynamically similar when the distributions of each thermodynamic property have the same basic shape. This kind of similarity can be extended to include processes involving mass transport, by requiring that the distributions of concentration have the same shape.

Figure 5.11 illustrates the consequences of these four kinds of similarity. The relationships between these levels of similarity are investigated in the following paragraphs.

Geometric similarity of the flow boundaries is the prerequisite for each of the other forms of similarity, since it is necessary for the definition of corresponding points. Hence for most tests considerable care is taken to provide accurate scale models. It need hardly be remarked that the geometric similarity of two flows is not sufficient to establish their kinematic or dynamic similarity. Indeed, a better overall correspondence of model and prototype flows is sometimes achieved using a model whose geometry is distorted in some way. For example, the friction on a ship hull is best represented, not by a model with a smooth surface, but by one whose roughness is much larger than direct scaling would indicate.

Kinematic and dynamic similarity are very closely related. The former must certainly be achieved if dynamic similarity is required. Moreover, two flows that are kinematically similar will have similar acceleration patterns and will thus have similar distributions of the net force on the fluid, provided that the density is sensibly constant or is otherwise similar in distribution. Thus it is generally true that exact kinematic similarity can be achieved only in flows that are dynamically similar as well. However, an adequate degree of similarity can sometimes be achieved without full dynamic similarity. Consider, for example, the flow shown in Figure 5.11. If one is interested primarily in the main features of the flow, and not in the details of the boundary layer, it should not be necessary to achieve exact dynamic similarity in the boundary layers. The hydraulic analogy

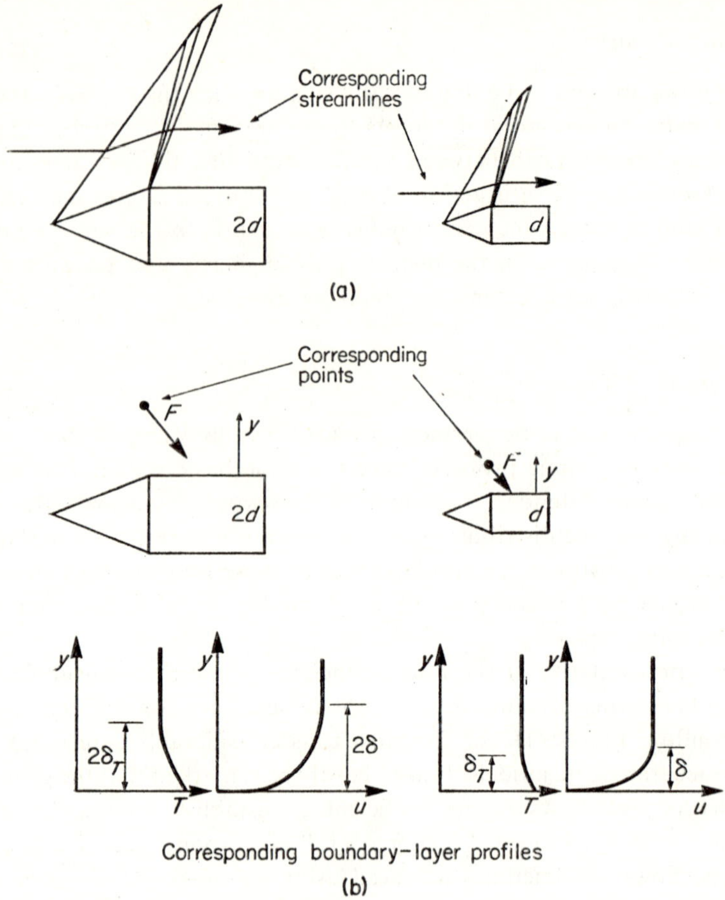

Figure 5.11 Geometric consequences of the similarity of two flows. (a) Geometric and kinematic similarity; (b) Dynamic and thermodynamic similarity

discussion in Section 3.3.3 provides a second example. Obviously, the flows of water in a thin sheet with a free surface and of gas in a duct are not dynamically similar. Nevertheless, their kinematic similarity is close enough to make the analogy useful in studying gas flows.

Finally, consider the relationship between dynamic and thermodynamic similarity. It is the main theme of this book that the dynamics and thermodynamics of a flowing fluid are closely linked. Strictly interpreted, this implies that dynamic similarity can never be achieved without thermodynamic similarity as well, and hence that kinematic similarity also requires thermodynamic similarity. However, for practical purposes the interaction between dynamic and thermodynamic response is often weak enough to be neglected. Thus one can sometimes achieve an adequate degree of dynamic similarity between air and water flows, or between air and steam flows, or between air flows at very different temperatures, even though the thermodynamic behaviour of the fluids is quite different. For example, in investigating the general features of the flow of combustion gases through a furnace, one might use a model supplied with water or with air at room temperature. Of course, such models will not represent the effects of density changes and heat transfer. This limitation can be justified either on the grounds that the effects are small or that their proper simulation would be too costly.

We have concluded that dynamic similarity does not always require thermodynamic similarity. Nor does a practical level of thermodynamic similarity always require dynamic similarity. For example, the general form of the temperature distribution in a combustion chamber may be only weakly dependent on the flow speed and boundary-layer thickness on the walls of the chamber. On the other hand, when the thermodynamic changes result primarily from dynamic processes, for example, when the temperature rises as a result of isenergic deceleration, the similarity of the relevant dynamic processes is the prerequisite of thermodynamic similarity.

The preceding discussion leads to the conclusion that exact kinematic, dynamic or thermodynamic similarity can be achieved only when all are achieved. However, the satisfaction of all of these requirements by two flows of differing scales is virtually impossible, and we must usually accept less than perfect similarity. Often this does not entail a serious loss, since the more important aspects of the process can be correctly scaled, leaving the less significant aspects to deviate from the similarity pattern.

The relationship between dimensional analysis and similarity is implicit in the definitions of similarity. The similarity of two flows requires that the ratios of certain forces, lengths and thermodynamic properties be the same for the two flows, while the dimensional analysis of a class of broadly similar flows leads to dimensionless groups that are ratios of forces, lengths and fluid properties. Hence the conditions for similarity of two flows reduce to the requirement that the dimensionless groups have the same values for the two flows. Accordingly, the dimensionless groups are often referred to as *similarity parameters*.

Let us see how these ideas apply to two flows of the kind illustrated in Figure 5.9, for which the dimensionless drag law is

$$\frac{D}{\frac{1}{2}\varrho U^2(\frac{1}{4}\pi d^2)} = f\left(\frac{Ud}{\nu}, \frac{l}{d}, \beta, \frac{U}{a}\right) \tag{5.2.8}$$

If the parameters l/d and β have the same values for the two flows, the flow boundaries are geometrically similar. If, in addition, the Mach number U/a is the same for the two flows, the shock wave configurations will be similar, so that a general kinematic similarity is achieved and the basic requirement for thermodynamic similarity is satisfied. Finally, if the Reynolds number Ud/ν also has the same value for the two members of the family, the viscous and inertia forces will have the same ratio and the boundary layers will be similar. When all of these conditions are satisfied, the drag coefficient $D/\varrho U^2 d^2$ is also the same for the two flows. In other words, the drag coefficient will be the same for model and prototype flows when the other similarity parameters are the same.

If, as is often the case, it proves impossible to model every aspect of the flow correctly, one must try to reproduce those parameters which are of most significance. For supersonic flow, for example, one might feel that establishing the proper Mach number was of greatest importance. On the other hand, for velocities well below the sonic speed, one could argue that Mach number is unimportant and attempt to reproduce the Reynolds number as closely as possible. Moreover, in certain flow regimes the drag coefficient does not depend strongly upon the Reynolds number (see Figure 5.10), and it may be sufficient to ensure that the Reynolds number falls somewhere within the range over which the coefficient is constant.

A final point concerning similarity parameters should be noted. The defining quantities appearing in equation (5.2.8) describe the body and the undisturbed flow; nothing is said about the complex flow pattern near the body. It is implicit in the preceding discussion that, when the defining quantities are properly balanced, the small-scale processes will automatically fall into the similarity pattern under the action of the universal laws of motion. Of course, if we have omitted some parameter relevant to these small-scale processes, for example, the turbulence level of the free stream, or heat transfer at the body surface, or surface roughness, these processes may not adopt the similarity pattern. Ultimately, the similarity analysis can only be as meaningful as one's choice of parameters to define the modelled process.

Example 5.6. Example 4.11 dealt with the testing of a one-tenth-scale model of an air intake for a supersonic aeroplane. The duct geometry and Mach number ($M = 4$) were modelled correctly, but no consideration was given to the Reynolds number of the flow. The relationship between model and prototype Reynolds numbers for this particular case gives some insight into the problems of dynamic modelling in a wider context.

The Reynolds number of greatest significance is that based on the characteristics of the flow within the intake duct where the effects of friction are most significant. However, we do not have the information required to calculate this parameter.

For the purpose of comparing test and prototype the less significant grouping

$$\mathrm{Re} = \frac{\varrho_0 U d}{\mu_0}$$

will be used, with U the velocity of the flow approaching the inlet, d the duct diameter and ϱ_0 and μ_0 the density and viscosity corresponding to stagnation conditions. The former values are used because they are readily available measures of the overall flow pattern, and the latter values are chosen as being representative of the fluid within the duct, which has undergone a considerable deceleration. It is particularly important to select a meaningful value of the viscosity, since this property varies markedly with temperature.

Table 5.1 gives the relevant properties for prototype operation and model test, based on results given in Example 4.11. The model Reynolds number is found to be smaller than the prototype value by a factor of 78.

Table 5.1. Characteristics of prototype and model flows

Property	Units	Prototype value	Model value
p_0	bar	29·5	1·013
T_0	°K	910	288
ϱ_0	kg/m^3	11·3	1·225
μ_0	N sec/m^2	$3{\cdot}9 \times 10^{-5}$	$1{\cdot}8 \times 10^{-5}$
d	m	1·00	0·100
a	m/sec	295	166
U	m/sec	1180	664
Re	—	$3{\cdot}5 \times 10^{8}$	$4{\cdot}5 \times 10^{6}$

A marked shortfall of Reynolds number is typical of model tests. It usually proves impractical to compensate for the reduction in geometric scale by increasing velocity and density. Indeed many tests, like that considered above, involve a reduction in the velocity as well as the length scale.

5.2.5 Inspectional Analysis

Dimensional analysis is one of a family of similarity techniques. It is the least specific, that is, that requiring the least detailed knowledge of the process under analysis. Now we shall consider briefly one of the similarity methods which make use of the equation governing the process. Even though this equation cannot be solved, an inspection of its structure reveals something about the nature of the solution. This information makes possible a more penetrating dimensional analysis of the problem.

The technique of inspectional analysis will be applied to a non-steady parallel laminar flow, like that depicted in Figure 5.6. Now a different motion of the wall will be considered, the wall being assumed to be set impulsively into motion at time $t = 0$ and to continue to move at constant velocity thereafter. Thus

$$u = U \quad \text{for} \quad y = 0 \quad \text{for} \quad t > 0 \tag{5.2.10}$$

If the fluid is initially at rest, its motion must be expected to be described by a relationship of the form

$$u = f(U, t, y, \varrho, \mu)$$

However, ϱ and μ are the only quantities involving the dimensions F and M. Hence an immediate simplification is possible:

$$u = f(U, t, y, \nu)$$

The pi-theorem suggests that three dimensionless parameters are involved in this relationship.

The equation of motion (5.1.18) governs this flow; it can be written

$$\frac{\partial u}{\partial(\nu t)} = \frac{\partial^2 u}{\partial y^2}$$

suggesting that ν and t will occur in the solution in the combination νt. This grouping is consistent with the boundary condition (5.2.10). Hence a solution of the form

$$u = f(U, \nu t, y)$$

is anticipated, and dimensional homogeneity requires that

$$\frac{u}{U} = f\left(\frac{y^2}{\nu t}\right) \tag{5.2.11}$$

An inspection of the role of the defining parameters in the governing equation and boundary conditions has restricted the relationship so that dimensional analysis is able to provide a more definite result, involving only two dimensionless parameters. A solution of the form (5.2.11) which satisfies the boundary condition (5.2.10) can be found, but it is not of immediate importance and will be left as a problem for those interested in the method of derivation.

We have seen that dimensional analysis is bound up with geometric similarity of a family of flows. This relationship leads to the use of the name similarity methods for a number of related methods of analysis. However, the families of flows linked by other similarity methods are not always geometrically and kinematically similar. In general, they are similar only when described using suitably distorted systems of coordinates. The velocity field (5.2.11) is an example of this kind of generalized similarity relation. The velocity profiles for all times appear the same when plotted as a function of $y/(\nu t)^{\frac{1}{2}}$. Further examples of coordinate scaling will be found in the discussion of turbulent friction which follows.

5.3 TURBULENT FRICTION

5.3.1 Mixing Stresses

Relatively modest goals will be set for this introduction to the analysis of turbulence. The general nature of the stresses and dissipative processes within a turbulent flow will be considered. Then the velocity distribution near a wall and the resistance offered to the flow by smooth and rough walls will be discussed. For these latter investigations, dimensional analysis, guided by the physical picture developed earlier, offers the only practicable approach.

The reader is advised to refer to the general introduction to turbulence in Section 2.1 before starting this more detailed discussion. In a turbulent flow near a wall or within a pipe, the time-mean flow is of necessity nearly parallel to the wall, and the velocity components can be represented by

$$u = u_m + u', \quad v = v', \quad w = w'$$

with u_m the time-mean velocity shown in Figure 2.4(c) and defined in equation (2.1.1). The following discussion relates to steady turbulent flows, that is, flows in which u_m is independent of time. The turbulent fluctuations u', v', w' are functions of time as well as the space coordinates; typical time histories are shown in Figure 2.2. For compactness, the operation of taking a mean value over time, as in equation (2.1.1), will be denoted by an overbar. Thus

$$\bar{u} = \bar{u}_m + \overline{u'} = u_m, \quad \overline{v'} = 0, \quad \overline{w'} = 0$$

since the mean of a non-varying quantity is that quantity itself, while the mean of the random perturbations is zero by definition.

Now consider the stresses within the turbulent flow. A unit area parallel to the wall and to the mean flow is shown in Figure 5.12. The time-mean of the viscous stress $\mu \, \partial u/\partial y$ is

$$\tau_v = \mu \frac{\partial \overline{(u_m + u')}}{\partial y} = \mu \frac{\partial u_m}{\partial y} \qquad (5.3.1)$$

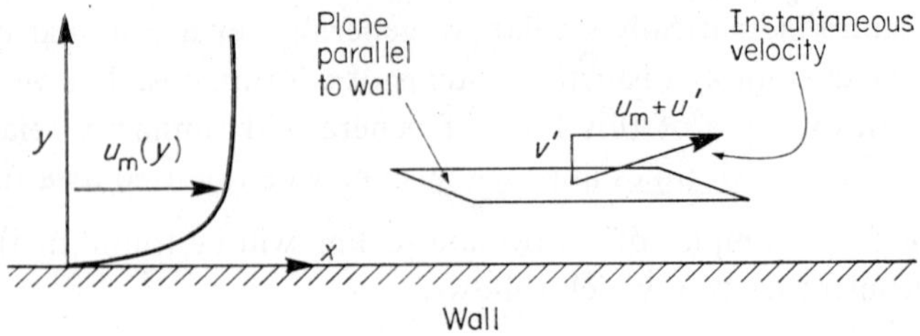

Figure 5.12 An element of area parallel to the mean flow near a wall, illustrating the generation of mixing stresses in turbulent flow

This corresponds to a force on the fluid below the unit area tending to drag that fluid along the wall. The turbulent mixing also contributes an effective stress, the counterpart of the net momentum flux across the unit area. The instantaneous upwards flux of x-momentum through the unit plane is

$$\varrho uv = \varrho(u_m + u')v' = \varrho(u_m v' + u'v')$$

Hence the mean effective force on the fluid below the unit area is

$$\tau_t = -\varrho(\overline{u_m v'} + \overline{u'v'}) = -\varrho\overline{u'v'} \tag{5.3.2}$$

The total time-mean stress is then

$$\tau = \tau_v + \tau_t = \mu \frac{\partial u_m}{\partial y} - \varrho\overline{u'v'} \tag{5.3.3}$$

the sum of terms representing momentum transfer by microscopic and macroscopic mixing.

Turbulence does not always give rise to a mixing stress; it is possible that the sum $\overline{u'v'}$ will average out to zero. This will happen, for example, at the centre-line of a channel, where the effects of the two walls cancel. Near a wall, however, the fluid coming from below the unit area has a quite different history from that approaching from above. That from below has been slowed down by the wall, while that from above has been speeded up in the fast-moving flow beyond the boundary layer. Hence it is more probable that the combinations

$$u' > 0, \quad v' < 0 \quad \text{and} \quad u' < 0, \quad v' > 0$$

will occur together. One may expect therefore that

$$\overline{u'v'} < 0 \quad \text{when} \quad \frac{\partial u_m}{\partial y} > 0$$

Thus, as would be expected

$$\tau_t = -\varrho\overline{u'v'} > 0$$

the effective mixing stress acting to drag the fluid along the wall.

The turbulent mixing which gives rise to the mixing stresses, also called *Reynolds stresses*, is the macroscopic counterpart of the microscopic mixing which is represented by the viscosity coefficient μ. The parallel cannot be carried very far, for molecular mixing usually takes place over length and time scales that are very short compared to those of the macroscopic flow. Hence the molecular viscosity can be taken to be independent of the velocity gradient; it is a property of the fluid in the thermodynamic sense. Turbulent mixing, on the other hand, extends over distances comparable to the channel width; hence its 'effective viscosity' is not uniform and this quantity only occasionally provides a useful way of representing turbulent friction.

The mean stress within a fully developed channel flow is specified by the pressure gradient: for two-dimensional flow in a channel of width b

$$\tau = \frac{\mathrm{d}p}{\mathrm{d}x}\left(y-\frac{b}{2}\right) \tag{5.1.6}$$

where y is measured from one wall; and for flow in a pipe of radius R, equation (5.1.13) gives

$$\tau = \frac{1}{2}\frac{\mathrm{d}p}{\mathrm{d}x}r$$

where r is the radial coordinate. Although the stress varies smoothly, in fact linearly, across the duct, the process by which it is generated changes markedly across a turbulent flow. Very near the wall, $v' \to 0$, and τ_t must drop to zero, giving $\tau \simeq \tau_v$. Away from the wall, where the turbulence is not inhibited, its activity may be expected to be far more significant than that of the microscopic mixing, so that $\tau \simeq \tau_t$.

The ratio of the turbulent and viscous contributions to the stress can be estimated as

$$\frac{\tau_t}{\tau_v} = -\frac{\varrho\overline{u'v'}}{\mu\,\partial u_m/\partial y} \sim \frac{\varrho u_{mm}^2}{\mu u_{mm}/b} = \frac{u_{mm}b}{\nu} \tag{5.3.4}$$

a Reynolds number characterizing the mean flow. Here u_{mm} is the spatial mean velocity in the channel, that shown in Figure 2.4(d), and defined in equation (2.1.2). The turbulent fluctuations have been taken to be roughly proportional to this mean velocity, and the length over which the mean velocity varies significantly has been taken to be comparable with the channel width. Thus

$$\frac{\tau_t}{\tau_v} \sim \mathrm{Re}$$

in the turbulent part of the flow, and one may expect that turbulence stresses will be dominant in a flow with a high Reynolds number, except very near the wall.

The process of *dissipation* within a turbulent flow is much more complicated than that in a laminar flow. In equation (5.1.23) the term $\tau\, du/dy$ was identified as the rate at which energy is extracted from the flow by the friction stresses. For a turbulent fluid the terms must be interpreted as time-mean values, and equation (5.3.3) gives

$$\tau \frac{du_m}{dy} = \mu\left(\frac{du_m}{dy}\right)^2 - \varrho\overline{u'v'}\frac{du_m}{dy} \tag{5.3.5}$$

The first term represents immediate viscous dissipation. The second term, which is far larger through much of the flow, gives the rate at which energy is extracted from the mean flow and converted into turbulence energy. Strictly speaking, this energy has not been dissipated, for the turbulence energy can still be described as mechanical energy, corresponding to macroscopic velocity and pressure fluctuations. However, there is no way of extracting this energy from the random fluctuations, and for all practical purposes it has been dissipated.

The mechanical energy extracted from the mean flow by turbulence is converted to internal energy only after it has been transferred through a series of progressively smaller scales of the turbulence. It is helpful to think of the turbulence as a system of superposed eddies, swirling masses of fluid of various scales. In the largest eddies, those which interact directly with the mean motion, direct viscous dissipation is negligible. They pass the extracted energy to smaller eddies, and the *cascade of energy* to ever smaller scales of motion ultimately reaches eddies so small that viscous dissipation operates directly. This final process is essentially like that represented by the first term of equation (5.3.5), but in a turbulent fluid the velocity components depend upon all the space coordinates, and the dissipation is given by a complicated function involving quantities such as $\mu\overline{(\partial u'/\partial x)^2}$.

Under some conditions steady flow in a channel remains laminar, while in others it breaks down into turbulence. Equation (5.3.5) casts some light on the reasons for this breakdown, often called *transition.* Consider a small disturbance developing in an initially laminar flow. The term $-\varrho\overline{u'v'}\, du_m/dy$ represents the flow of energy from the mean flow to the disturbance, while $\mu(du_m/dy)^2$ represents the dissipative processes tending to damp it out. We take the perturbation velocities to be proportional to the mean velocity, and consider the largest possible disturbance (that of scale b), for which the damping is least. Then the ratio of production to dissipation is

$$-\varrho\overline{u'v'}\frac{du_m}{dy}\bigg/\left[\mu\left(\frac{du_m}{dy}\right)^2\right] \sim \frac{u_{mm}b}{\nu} = \text{Re}$$

the ratio which was found in equation (5.3.4). Thus the Reynolds number is an index of the development of a disturbance in a laminar flow. In a flow with low Reynolds number, a disturbance will be eroded away by viscous dissipation. For a high Reynolds number the reverse is true; energy is gained from the mean flow more rapidly than it is lost through viscous friction. Hence disturbances grow, and the flow becomes turbulent.

This prediction is in accord with observation. When the Reynolds number of a pipe flow, $u_{mm}d/\nu$, exceeds a certain limiting value, the flow becomes turbulent. The critical value does not drop below about 2000, but can be raised to very much higher values by ensuring smooth entry into the pipe. A similar process is found in a boundary layer, for example, that on the wing of an aeroplane. Near the start of the layer the flow will be laminar; later it breaks down into turbulence. The length of the laminar layer depends on the roughness of the surface and the level of turbulence in the outer stream, among other factors. Usually transition occurs when $U\delta/\nu \simeq 1000$, with U the free-stream velocity, and δ the boundary-layer thickness. This criterion is not markedly different from that for pipe flow.

5.3.2 Flow near a Smooth Wall

With the general picture of turbulence now available, dimensional analysis can provide a surprising amount of information about turbulent flow in a pipe or channel. The relationship between flow rate and driving pressure gradient is of great practical utility, and the investigation will be directed towards finding it. A step towards the determination of the flow rate is the study of the variation in velocity within a turbulent flow near a wall.

For simplicity, two-dimensional flow in a channel with smooth walls will be considered in the first instance. This situation is shown in Figure 5.13(a). One may expect that the variation in the time-mean velocity across the channel will be of the form

$$u_m = f(y, \tau_0, b, \varrho, \mu) \tag{5.3.6}$$

The inclusion of the distance from the wall y, the channel width b, and the fluid viscosity μ requires no comment. The fluid density ϱ is included since the stresses in a turbulent flow depend upon it. Finally, the shear stress at the wall τ_0 provides a convenient and generally applicable means of specifying the interaction between the flow and the wall. For pressure-driven flow in a two-dimensional channel, for instance, it is related to the pressure gradient by

$$2\tau_0 = -b\left(\frac{dp}{dx}\right)$$

The wall stress can also be calculated for open-channel flows in which fluid

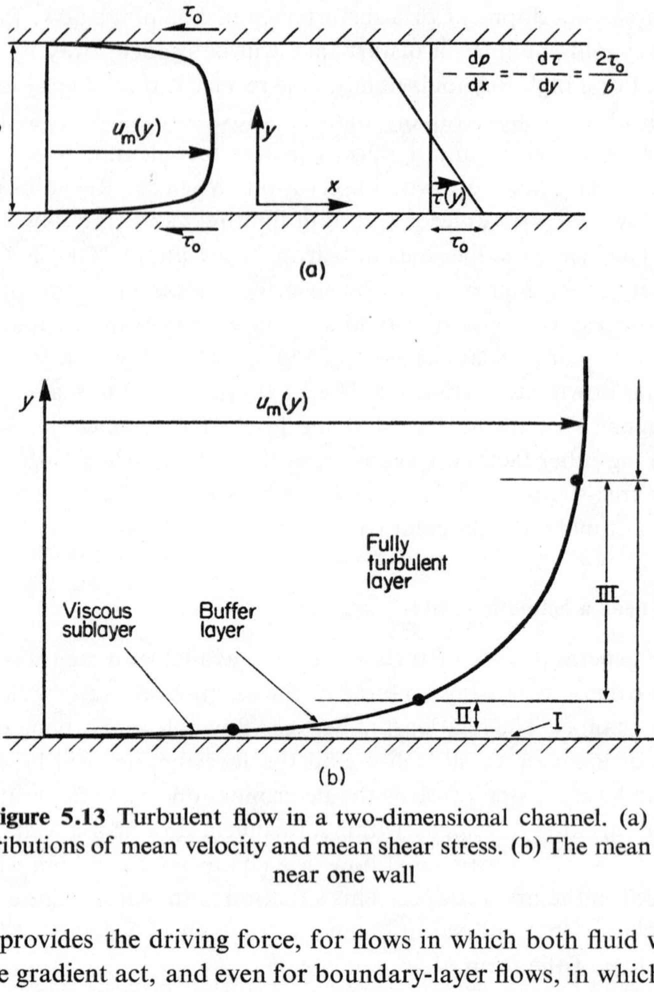

Figure 5.13 Turbulent flow in a two-dimensional channel. (a) Distributions of mean velocity and mean shear stress. (b) The mean flow near one wall

weight provides the driving force, for flows in which both fluid weight and pressure gradient act, and even for boundary-layer flows, in which the shear stress corresponds to a deceleration of the fast-moving fluid at the edge of the boundary layer.

The requirements of dimensional homogeneity are satisfied by expressing the relationship (5.3.6) as

$$\frac{u_m}{(\tau_0/\varrho)^{\frac{1}{2}}} = f\left[\frac{(\tau_0/\varrho)^{\frac{1}{2}} y}{\nu}, \frac{y}{b}\right] \tag{5.3.7}$$

The quantity $(\tau_0/\varrho)^{1/2}$, which has dimensions of velocity, is often called the *friction velocity*, and will be denoted henceforth by u_f. Thus

$$\frac{u_m}{u_f} = f\left(\frac{u_f y}{\nu}, \frac{y}{b}\right) \tag{5.3.7}$$

The symbol u_f is merely a convenient notation; a geometrical interpretation is not particularly useful. In the equations (5.3.7) the velocity has been rendered dimensionless using an inertial scale, rather than the alternative viscous scale ν/b, since inertia-dependent mixing stresses are expected to determine the velocity variation over much of the flow. The distance from the wall has been scaled using the length scale ν/u_f which incorporates the same velocity scale.

Although the form (5.3.7) applies across the whole channel, rather simpler relationships can be obtained for limited parts of the flow. Four layers can be distinguished in the turbulent flow near a wall; they are shown in Figure 5.13(b). The fundamental division is into a *wall layer*, in which most of the velocity variation takes place, and a *core flow*, where the velocity is relatively uniform. Let us consider the core flow first. It will depend on the geometry of the channel or, in the case of a boundary layer, on the geometry of the essentially inviscid outer flow field. The absolute value of the velocity in the core is related to the velocity change across the wall layer, and is thus dependent on the viscosity of the fluid. However, the shear stresses in the core result almost entirely from turbulent mixing, and the corresponding velocity variations are therefore independent of viscosity. Hence in the core flow within a two-dimensional channel the velocity gradient will be related to the other parameters by

$$\frac{\mathrm{d}u_m}{\mathrm{d}y} = f(y, \tau_0, b, \varrho) \tag{5.3.8}$$

Now consider the wall layer. In a fully developed turbulent flow with a high Reynolds number, its thickness is a small proportion of the channel width. Hence the stress will not vary much across it, and the flow in it will be virtually independent of the position of the other wall of the channel. Then

$$u_m = f(y, \tau_0, \varrho, \mu) \tag{5.3.9}$$

in the wall layer, and dimensional homogeneity requires that

$$\frac{u_m}{u_f} = f\left(\frac{u_f y}{\nu}\right) = f(\eta) \tag{5.3.10}$$

This is a particular form of equation (5.3.7). It implies that all velocity profiles near smooth walls will have the same form when u_m/u_f is presented in terms of the scaled coordinate $\eta = u_f y/\nu$. The assumption that this part of the flow depends only on the stress at the wall makes it independent of the core flow and of whatever happens beyond. Thus equation (5.3.10) may be expected to describe a variety of turbulent flows near smooth walls, for example, those in a two-dimensional channel, in a round pipe, or at the bottom of a boundary layer. The wall stress τ_0 is, of course, determined by the outer flow, and to this extent the scaling adopted represents the outer flow.

The wall layer can be further dissected, as indicated in Figure 5.13(b). Immediately next to the wall there is a region (I) called the *viscous sublayer*, in which turbulent mixing stresses are suppressed by the presence of the wall. Then

$$u_m = f(y, \tau_0, \mu)$$

and

$$\frac{\mu u_m}{\tau_0 y} = \text{constant} = 1 \tag{5.3.11}$$

by virtue of the definition of viscosity (5.1.1). This result can be written in terms of the variables of equations (5.3.10):

$$\frac{u_m}{u_f} = \frac{u_f y}{\nu} = \eta \tag{5.3.12}$$

for the viscous sublayer.

Beyond the sublayer there is a region (II), often called the *buffer layer*, in which viscous and turbulence stresses are of comparable magnitude. In this layer the general result (5.3.10) applies; no further simplification is possible. However, beyond the buffer layer there is a region (III), still close to the wall and with shear stress nearly equal to τ_0, in which viscous stresses are unimportant. The absolute velocity in this *fully turbulent layer* is determined by the change across the inner layers (I and II), but variations in velocity within region III are independent of viscosity. Thus

$$\frac{du_m}{dy} = f(y, \tau_0, \varrho)$$

and

$$\frac{y}{u_f}\frac{du_m}{dy} = \text{constant} = \frac{1}{K} \tag{5.3.13}$$

The dimensionless constant K is often referred to as *von Karman's constant*. Subject to the assumptions made above, it characterizes the whole family of constant-stress wall layers, whether over smooth or rough walls.

In terms of the variables of equation (5.3.10), equation (5.3.13) is

$$\eta\frac{d(u_m/u_f)}{d\eta} = \frac{1}{K} \tag{5.3.14}$$

Integration gives

$$\frac{u_m}{u_f} = \frac{1}{K}(\log_e \eta + B) \tag{5.3.15}$$

This has the required form (5.3.10). The constant B is independent of y, and hence of the other parameters specifying the flow, which can appear only in conjunction with y. Thus B is a universal constant for flows over a smooth wall. It specifies the net change of velocity across the viscous sublayer and buffer layer which lie between the smooth wall and the fully turbulent part of the wall layer.

The law (5.3.15) cannot apply right down to the wall. Experiments indicate that it holds for $\eta > 70$. The sublayer law (5.3.12) is found to hold for

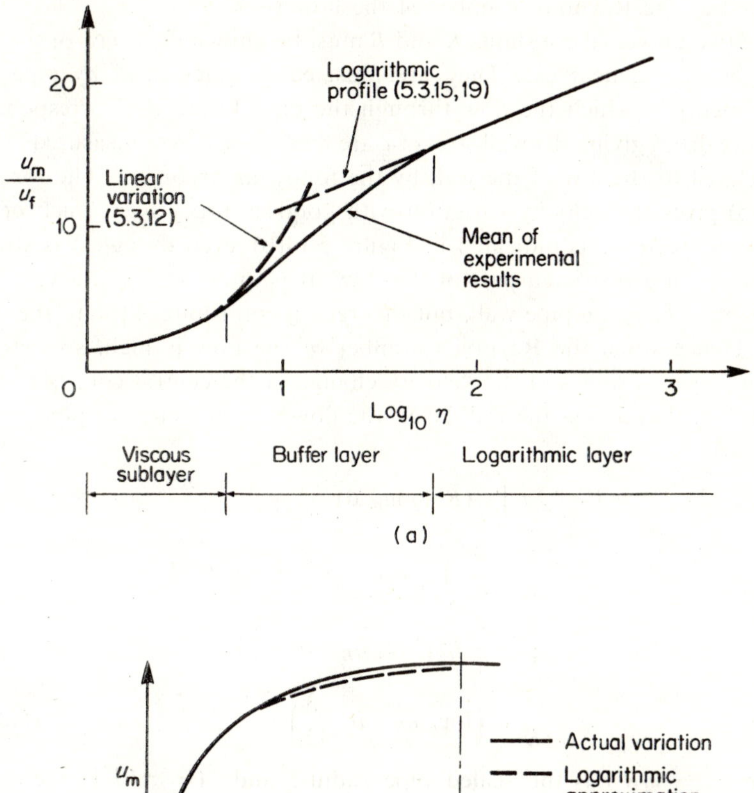

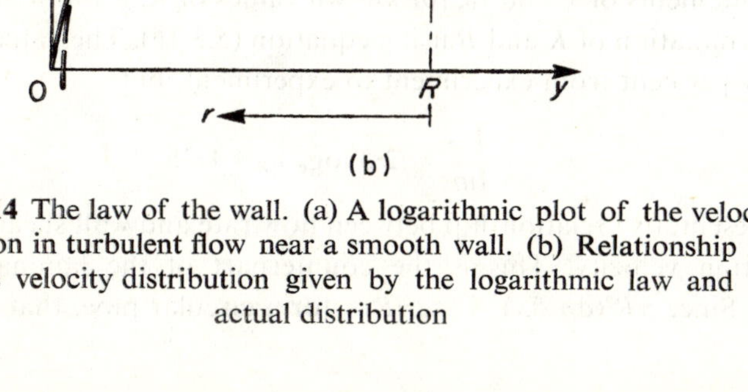

Figure 5.14 The law of the wall. (a) A logarithmic plot of the velocity distribution in turbulent flow near a smooth wall. (b) Relationship between the velocity distribution given by the logarithmic law and the actual distribution

$\eta < 5$, and the buffer layer, also of universal form, but less easily predicted, extends between the two. The three segments together make up the *universal law of the wall*. The form of the velocity variation is shown in Figure 5.14(a). Note that $\eta = c$, a constant, corresponds to

$$\frac{y}{b} = c\frac{\nu}{u_f b}$$

This indicates that the actual thickness of the sublayer and buffer layer is reduced as the Reynolds number of the flow rises.

The two universal constants K and B must be known if the law of the wall is to be of practical use. They are most easily determined by pipe-flow experiments in which the flow through the pipe $\dot{V}$ and the corresponding pressure drop, giving the wall stress τ_0, are measured. These measured values are related to the law of the wall by the following argument. The formula (5.3.15) gives the velocity variation with good accuracy over much of the pipe cross-section, as indicated in Figure 5.14(b), even though it is strictly applicable in a restricted part of the flow. It is quite wrong in a very thin region very near the pipe wall, but this region contributes little to the total flow. Hence when the Reynolds number of the flow is high, so that the viscous layer is thin and the velocity change in the central core is a small fraction of that across the wall layer, the flow rate in a circular pipe can be estimated as

$$\begin{aligned}\dot{V} &= 2\pi\int_0^R (R-y)u_m\,\mathrm{d}y \\ &= \frac{2\pi}{K}\frac{\nu^2}{u_f}\int_0^{\eta_R}(\eta_R-\eta)(\log_e\eta+B)\,\mathrm{d}\eta \\ &= \frac{\pi}{K}\frac{\nu^2}{u_f}\eta_R^2\left(\log_e\eta_R+B-\frac{3}{2}\right) \\ &= \frac{A}{K}u_f\left(\log_e\eta_R+B-\frac{3}{2}\right) \qquad (5.3.16)\end{aligned}$$

where $\eta_R = u_f R/\nu$ is the scaled pipe radius, and $A = \pi R^2$ is the cross-sectional area.

Measurements of $\dot{V}$ and τ_0, for known values of R, ϱ and μ, make possible the determination of K and B using equation (5.3.16). The values found vary by a few per cent from experiment to experiment, but

$$\frac{\dot{V}}{Au_f} = 2{\cdot}5\log_e\eta_R + 1{\cdot}75 \qquad (5.3.17)$$

is a representative relationship between flow rate and wall shear, specified by the friction velocity. This is the counterpart of the laminar-flow result (5.1.16). Since $\pi R^2(\mathrm{d}p/\mathrm{d}x) = -2\pi R\tau_0$ for a circular pipe, that result can be written

$$\frac{\dot{V}}{Au_f} = \frac{u_f R}{4\nu} = \frac{\eta_R}{4} \qquad (5.3.18)$$

It is now possible to give an explicit equation for the logarithmic portion of the law of the wall (5.3.15). By relating the constants of equations (5.3.17) and (5.3.16), one finds $K = 0{\cdot}40$ and $B = 2{\cdot}2$. Then

$$\frac{u_m}{u_f} = 2{\cdot}5 \log_e \eta + 5{\cdot}5 \qquad (5.3.19)$$

prescribes the *logarithmic law of the wall.* It must be remembered that this identification of constants is based on the integration of equation (5.3.16). Although that calculation reveals the general form of the flow law and provides a useful basis for the data correlation (5.3.17), the constants appearing in the flow law need not be precisely identical to those defining the velocity variation (5.3.15). Nevertheless, these results are adequate for many practical calculations.

Example 5.7. Although the constants in the logarithmic law of the wall are most conveniently found from measurements of pressure-driven flow through a round pipe, the wall layers generated in other turbulent flows are described by this law. Here it will be used to relate the flow of a liquid under gravity down a uniform slope and the angle the slope makes with the horizontal. On a very long slope the water depth will ultimately become uniform with a velocity distribution which does not vary along the flow. In this condition, described as 'normal' flow by hydraulic engineers, the shear stress at the bed balances the component of fluid weight which acts downstream:

$$\tau_0 = \varrho_w g d \sin \theta$$

in terms of the symbols defined in Figure 5.15.

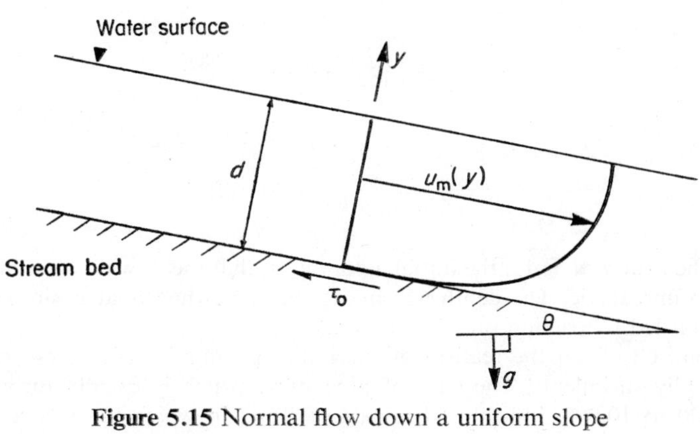

Figure 5.15 Normal flow down a uniform slope

A natural stream bed will not of course be smooth, but a laboratory channel or a spillway constructed as part of hydraulic works might be smooth enough for the result (5.3.19) to be applicable. In such a case, the flow rate in a wide channel of width b can be calculated as

$$\dot{V} = b\, u_f \int_0^d (2{\cdot}5 \log_e \eta + 5{\cdot}5)\, dy$$

$$= b\, \nu \eta_d (2{\cdot}5 \log_e \eta_d + 3)$$

$$= u_f\, d\, b\, (2{\cdot}5 \log_e \eta_d + 3)$$

Here $\eta_d = u_f d/\nu$ is a dimensionless measure of the water depth. The distribution of velocity used here does not satisfy the condition $\tau = 0$ at $y = d$ which is very nearly applicable for water flowing beneath air. However, the error introduced is no greater than others arising from the approximate representation of the velocity variation. The relationship between flow rate, water depth and bed slope is given explicitly by

$$\dot{V} = b(gd^3 \sin\theta)^{\frac{1}{2}} \left[2{\cdot}5 \log_e \left(\frac{gd^3 \sin\theta}{\nu^2} \right)^{\frac{1}{2}} + 3 \right]$$

Now consider a particular case, water flowing at 1 ft^3/sec down a laboratory channel with vertical sides. The channel width is 2 ft and its slope is 1/100. The viscosity may be taken to be $\nu = 10^{-5}$ ft^2/sec, the value appropriate to room temperature. What is the normal depth?

In this case

$$\frac{\dot{V}}{\nu b} = 50{,}000 = \eta_d (2{\cdot}5 \log_e \eta_d + 3)$$

This equation must be solved iteratively. Start with $\eta_d = 2000$. An improved value is

$$\eta_d = \frac{20{,}000}{1{\cdot}2 + \log_e 2000} = 2300$$

Using this estimate, a still better value is found:

$$\eta_d = \frac{20{,}000}{1{\cdot}2 + \log_e 2300} = 2240$$

With this value

$$d = \left(\frac{\nu^2 \eta_d^2}{g \sin\theta} \right)^{\frac{1}{3}} = 0{\cdot}116 \text{ ft}$$

Since the ratio $b/d \simeq 17$, the initial assumption that the flow is two-dimensional is not too unrealistic. The depth will in fact be somewhat greater, since the flow is also retarded by friction on the side walls.

A second check on the realism of the initial assumptions is necessary. Is the flow actually turbulent? The case of pipe flow, which is usually turbulent for Re $>$ 2000 to 10,000, indicates the scale of flows in which turbulence is to be expected. A suitable Reynolds number for open-channel flow is

$$\mathrm{Re} = \frac{ud}{\nu} = \frac{\dot{V}}{\nu b}$$

with u the velocity averaged over the depth. For the case considered above Re = 50,000, and it is reasonable to expect that the flow will be turbulent.

5.3.3 Flow near a Rough Wall

Since no real surface is a featureless plane, what meaning can be attached to the preceding discussion of flow past a smooth wall? The following criterion provides an approximate guide: so long as the irregularities on the surface (such as rivet heads, rust scale, pebbles or tool marks) do not project beyond the viscous sublayer, and hence do not greatly affect the turbulent part of the flow, they do not greatly alter the drag of the surface. A wall with roughness elements of uniform height k may be expected to be effectively smooth if $\eta_k < 5$ or $k < 5\nu/u_f$. This situation is often described by saying that the surface is *aerodynamically smooth*, though the distinction is universally applicable, and not to air flows alone. For $\eta_k > 5$, the wall is in some measure rough. For $\eta_k > 70$, when the roughness elements project into the fully turbulent region and interact directly with the mixing stresses, the wall is described as *fully rough*.

Note that the effective roughness is not dependent merely on the nature of the wall, but also on the thickness of the viscous layer. Hence the effective roughness of a wall is greater when the Reynolds number of the flow is high.

Since most surfaces have a variety of roughness elements, the distinction between smooth and rough conditions is not always clear cut. The height k can be identified with the average height of the projections on the wall, but the largest will project beyond the viscous sublayer even for $\eta_k < 5$, and the smallest will still be enclosed within the viscous layer when $\eta_k > 70$. Thus the transition between smooth and rough conditions is more gradual than for roughness of uniform size.

For the moment let us assume that it is possible to find a length scale k which adequately represents the roughness of a wall. The velocity variation in the wall layer may be expected to be of the form

$$u_m = f(y, \tau_0, k, \varrho, \mu)$$

The dimensionally homogeneous equivalent is

$$\frac{u_m}{u_f} = f\left(\eta, \frac{y}{k}\right) \qquad (5.3.20)$$

For the fully rough condition, it is plausible to expect that the shear stress will be transmitted directly between roughness elements and mixing stresses. Then in Region III of Figure 5.13(b), where the Reynolds stresses are dominant

$$u_m = f(y, \tau_0, k, \varrho)$$

and

$$\frac{u_m}{u_f} = f\left(\frac{y}{k}\right) \quad \text{only} \tag{5.3.21}$$

Equation (5.3.13) still applies in this fully turbulent part of the flow, and can be written in terms of the variables of (5.3.21):

$$\frac{y}{k}\,\frac{d(u_m/u_f)}{d(y/k)} = \frac{1}{K} \tag{5.3.22}$$

Integration gives

$$\frac{u_m}{u_f} = \frac{1}{k}\left[\log_e \frac{y}{k} + C\right] \tag{5.3.23}$$

Naturally this has the required form (5.3.21). Since C is independent of y, it is also independent of k. In other words, C is the same for all flows past fully rough walls, so long as the single parameter k provides an adequate characterization of the roughness.

Using the distribution (5.3.23) the flow rate in a pipe can be calculated as in equation (5.3.16):

$$\dot{V} = \frac{A}{K}\, u_f\left[\log_e \frac{R}{K} + C - \frac{3}{2}\right]$$

Note that the value of the constant C depends on the way in which the roughness height k is defined. In a classic series of experiments carried out by Nikuradse, C was determined by studying the flow through pipes whose inner surfaces were uniformly coated with uniform sand grains, the parameter k being defined as the sand-grain diameter. Such experiments indicate that

$$\frac{\dot{V}}{Au_f} = 2{\cdot}5 \log_e \frac{R}{k} + 4{\cdot}75 \tag{5.3.24}$$

where k is the *sand-grain roughness*. This result is directly applicable to the flow past a wall whose surface is rather like that made up of close-packed sand grains. Since most surfaces are not like this, it is necessary to assign to them equivalent sand-grain roughnesses, obtained by fitting their friction characteristics to the law (5.3.24). We shall see how this is done later (Section 5.5.1).

Example 5.8. What would have been the water depth for the situation described in Example 5.7, if the bottom of the channel were covered with gravel or coarse sand with diameter around 1/10 in?

It is necessary first to see whether the original calculation is still relevant; it will be if the surface is effectively smooth. For $d = 0{\cdot}116$ ft, as found in Example 5.7

$$u_f^2 = \frac{\tau_0}{\varrho} = gd\sin\theta = 0{\cdot}037 \text{ ft}^2/\text{sec}^2$$

so that

$$\eta_k = \frac{u_f k}{\nu} = 160$$

There can be little doubt that the surface is fully rough. The roughness height would have to be kept below 0·003 in for the wall to be considered smooth.

Taking the velocity distribution (5.3.23) for the entire flow, with $K = 0{\cdot}40$ and $C/K = 1{\cdot}5/K + 4{\cdot}75 = 8{\cdot}5$, we find the flow rate to be

$$\dot{V} = b\,u_f\,k \int_0^{\frac{d}{k}} \left[2{\cdot}5 \log_e \frac{y}{k} + 8{\cdot}5\right] \mathrm{d}\left(\frac{y}{k}\right)$$

$$= b\,u_f\,k\left(\frac{d}{k}\right)\left[2{\cdot}5 \log_e \frac{d}{k} + 6{\cdot}0\right]$$

Since $u_f = (gd\sin\theta)^{1/2}$, this result can be written

$$\dot{V} = b(gk^3\sin\theta)^{\frac{1}{2}}\left(\frac{d}{k}\right)^{\frac{3}{2}}\left[2{\cdot}5 \log_e \frac{d}{k} + 6{\cdot}0\right]$$

For the specific case to be considered, $\dot{V} = 1$ ft³/sec, $b = 2$ ft, $\sin\theta = 0{\cdot}01$, and $k = 1/120$ ft. Hence

$$\left(\frac{d}{k}\right)^{\frac{3}{2}}\left[\log_e \frac{d}{k} + 2{\cdot}4\right] = 464$$

determines the depth, though not explicitly. Beginning the iteration with $d/k = 10$, we find the improved values

$$\frac{d}{k} = \left[\frac{464}{(\log_e 10 + 2{\cdot}4)}\right]^{\frac{2}{3}} = 21{\cdot}3$$

and

$$\frac{d}{k} = \left[\frac{464}{(\log_e 21{\cdot}3 + 2{\cdot}4)}\right]^{\frac{2}{3}} = 19{\cdot}4$$

Then $d = 0{\cdot}16$ ft is a reasonable estimate of the water depth, with the sidewall friction neglected as before. As would be expected, the rough bottom gives rise to a slower, deeper flow.

5.4 MIXING AND SPREADING FLOWS

5.4.1 The Jet Pump

We shall now look at some situations in which dissipation is associated with changes in the way velocity varies across the flow. Here dissipation arises not from friction at solid boundaries, but from friction within the body of the fluid, associated with mixing and spreading of the flow. The first case to be examined is the flow in a jet pump or ejector, shown in Figure 5.16. This device is of some interest in itself, but the results obtained have greater importance in their application to more frequently encountered situations, for example, the flow following an abrupt expansion or local constriction in a duct. The analysis of a jet pump is also of interest in illustrating the treatment of a system through which more than one stream passes.

In the jet pump, fluid is injected with a high velocity along the axis of a uniform pipe which carries a second flow. It will be shown later that the mixing of the two flows is accompanied by a rise in pressure. Thus the high-velocity jet has the effect of pumping the low-velocity flow along the pipe. One of the early applications of the jet pump was in the steam locomotive. The exhausted steam was ejected up the funnel, inducing a flow through the firebox, and thus providing a measure of forced draught. In many modern applications the energetic pumping fluid is steam. However, in the following discussion it will be assumed, for simplicity, that the densities of the two fluids are the same and are uniform throughout the flow. Moreover, the cross-section of the tube within which mixing takes place will be taken to be uniform, although this is often not so in actual jet pumps.

Each of the flows into the jet pump, that in the jet and that in the surrounding pipe, will be taken to be one-dimensional in the plane of injection. Considered as a whole the flow at the inlet section is far from one-dimensional. However, after the two streams have mixed, the total flow may be assumed one-dimensional with fair accuracy. The assumed velocity distributions are shown in Figure 5.16(b). The distance required for thorough mixing depends on the relative sizes of the outer and injection pipes, and can be several times the diameter of the larger pipe. To achieve rapid mixing, it is usual to locate the high-velocity jet centrally, so that the mixing occurs all around it. However, the integral analysis which follows is in no way dependent on this inlet geometry, and is applicable to a variety of inflow distributions.

The continuity, momentum and energy principles can be generalized in a straightforward manner to represent the non-uniform flow at section 1.

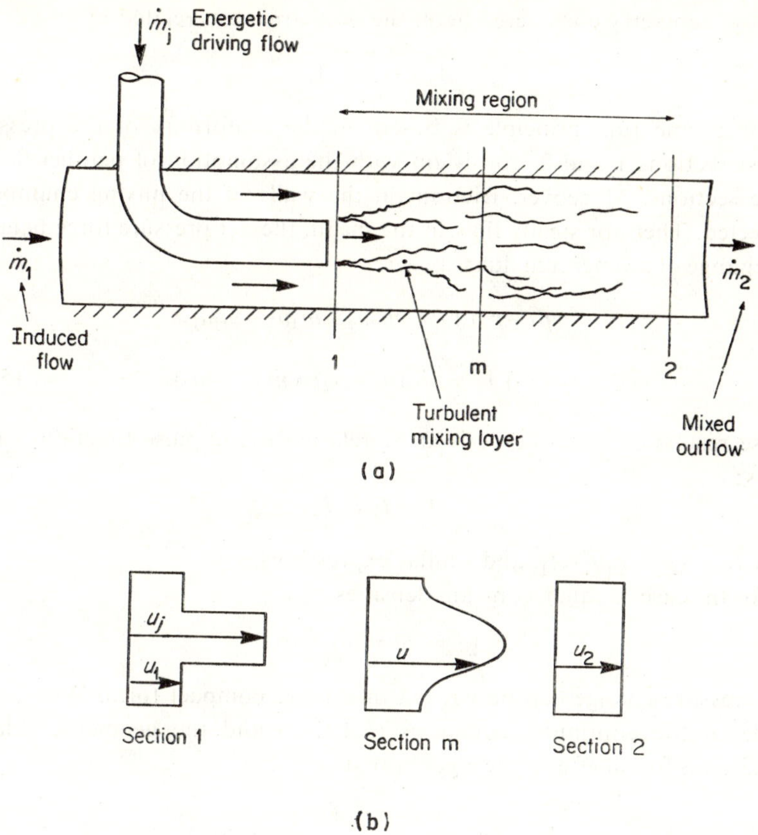

Figure 5.16 Flow in a jet pump. (a) General features; (b) Velocity distributions at inlet section, at a section through the mixing chamber, and at a section some distance downstream

The continuity requirement is simply

$$\dot{m}_1 + \dot{m}_j = \dot{m}_2 \tag{5.4.1}$$

or

$$\varrho_1 u_1 A_1 + \varrho_j u_j A_j = \varrho_2 u_2 A_2 \tag{5.4.1}$$

If the densities of the two fluids are taken to be constant, though not necessarily equal, and if the possibility of their reacting or absorbing one another is ruled out, there is a second continuity requirement; the volume flows in and out of the mixing region are also equal:

$$\dot{V}_1 + \dot{V}_j = \dot{V}_2 \tag{5.4.2}$$

or

$$u_1 A_1 + u_j A_j = u_2 A_2 \tag{5.4.2}$$

For the geometry considered here, the flow areas are related by

$$A_1 + A_j = A_2 \tag{5.4.3}$$

The momentum principle is based on the uniformity of the pressures across sections 1 and 2, consistent with the assumption of parallel flow at these sections. Moreover, friction on the walls of the mixing chamber is neglected. Then for steady flow in the mean, the net pressure force balances the change of momentum flux:

$$(p_1 - p_2)A_2 = \dot{m}_2 u_2 - \dot{m}_1 u_1 - \dot{m}_j u_j$$

or

$$(p_2 - p_1)A_2 = \dot{m}_1(u_1 - u_2) + \dot{m}_j(u_j - u_2) \tag{5.4.4}$$

These results can also be derived by relating the impulse functions of the flows:

$$I_1 + I_j = I_2$$

with $I_1 = (p_1 + \varrho_1 u_1^2)\, A_1$, and similar expressions.

For the case of equal, constant densities

$$\varrho_1 = \varrho_j = \varrho_2 = \varrho_c$$

the pressure change can be written in a more compact form. When use is made of the continuity equations (5.4.1, 2) and the geometric relation (5.4.3), it is found after some algebra that

$$\Delta p = p_2 - p_1 = \varrho_c \frac{A_1}{A_2} \frac{A_j}{A_2} (u_j - u_1)^2 \tag{5.4.5}$$

Note that the pressure always rises as a result of mixing. If other factors remain fixed, the maximum pressure rise is achieved when there is no primary flow, that is, when $u_1 = 0$.

Next consider the energy balance. For an unfamiliar situation, it is judicious to start from the most fundamental form of the energy conservation principle. For this steady, adiabatic flow, equation (1.3.9) gives

$$W + CE = 0$$

relating the net work done on the fluid between the fixed sections 1 and 2 to the net convection of energy into the system. The only work done is displacement work at the rate

$$p_1(u_1 A_1 + u_j A_j) - p_2 u_2 A_2$$

The convected energy is measured by $e + \frac{1}{2}u^2$ per unit mass, potential energies being neglected. Then the net rate at which energy is convected out of this system is

$$\dot{m}_2\left(e_2+\frac{u_2^2}{2}\right)-\dot{m}_1\left(e_1+\frac{u_1^2}{2}\right)-\dot{m}_j\left(e_j+\frac{u_j^2}{2}\right)$$

$$= p_1(u_1A_1+u_jA_j)-p_2u_2A_2 \tag{5.4.6}$$

As might have been anticipated, this result can be written

$$\dot{m}_1h_{0_1}+\dot{m}_jh_{0_j} = \dot{m}_2h_{0_2} \tag{5.4.6}$$

If the mixing streams consist of the same constant-density fluid, the dissipation can be discussed without difficulty, as there is no additional irreversibility associated with mixing of different species of fluid. In this case the entropy changes can be reckoned using

$$T\,\mathrm{d}s = \mathrm{d}e = -\frac{\mathrm{d}p_0}{\varrho_c}$$

and the rate of dissipation is equivalent to the rate at which internal energy is produced in the system:

$$\frac{\mathrm{d}E}{\mathrm{d}t} = \dot{m}_1(e_2-e_1)-\dot{m}_j(e_2-e_j)$$

$$= -\frac{\dot{m}_1(p_{0_2}-p_{0_1})}{\varrho_c}-\frac{\dot{m}_j(p_{0_2}-p_{0_j})}{\varrho_c} \tag{5.4.7}$$

$$= -\frac{(\dot{m}_1+\dot{m}_j)\,\Delta p}{\varrho_c}-\frac{\dot{m}_1}{2}(u_2^2-u_1^2)-\frac{\dot{m}_j}{2}(u_2^2-u_j^2)$$

With the use of equations (5.4.1, 4) and some algebra

$$\frac{\mathrm{d}E}{\mathrm{d}t} = \frac{\dot{m}_1}{2}(u_1-u_2)^2+\frac{\dot{m}_j}{2}(u_j-u_2)^2 \tag{5.4.8}$$

The dissipation is seen to be necessarily positive, consistent with the second law of thermodynamics. The reverse process of spontaneous separation into high and low-velocity streams is impossible.

The efficiency of energy transmission within a jet pump can be reckoned as

$$\eta = \frac{\text{mechanical energy out}}{\text{mechanical energy in}} \tag{5.4.9}$$

as for any other form of transmission, for example, a gear box. Then

$$\eta = 1-\frac{\varrho_c\,\mathrm{d}E/\mathrm{d}t}{\dot{m}_1p_{0_1}+\dot{m}_jp_{0_j}} \tag{5.4.10}$$

For a given rate of dissipation, the amount by which the efficiency is reduced below 100 per cent depends upon the ratio of the dynamic-pressure terms of equation (5.4.8) to the input of total pressure. In essence

$$\frac{\varrho_c \, \mathrm{d}E/\mathrm{d}t}{\dot{m}_1 p_{0_1} + \dot{m}_j p_{0_j}} \sim \frac{\frac{1}{2}\varrho_c u^2}{p_1 + \frac{1}{2}\varrho_c u^2}$$

where u is a typical velocity of the system. Hence the efficiency of energy transmission can be made arbitrarily close to 100 per cent by increasing the pressure level.

The preceding results have wider implications. They suggest that it is advantageous to raise the pressure level in hydraulic transmission systems, for example, those used to move the controls of an aeroplane or the tools of metal-cutting machines. The energy transmitted is of the order $\dot{m}\,p$, while the dissipation is of the order $\dot{m}(\frac{1}{2}u^2)$. As the pressure is raised, there is a reduction in the proportion of the energy which is dissipated.

Example 5.9. The jet-pump principle is applied in small filter pumps and aspirators, which use water flowing from a tap to create a low-grade vacuum. Figure 5.17 shows the arrangement. The water jet spreads out to fill the outlet tube, mixing with the gas as it does so, and carrying some of it from the chamber. The reduced pressure in the chamber induces a flow through the side tube.

The explicit results obtained above apply only when the fluid densities are the same. However, a prediction of the maximum suction which can be created is possible even if the densities differ, for in that case there is no induced flow, and

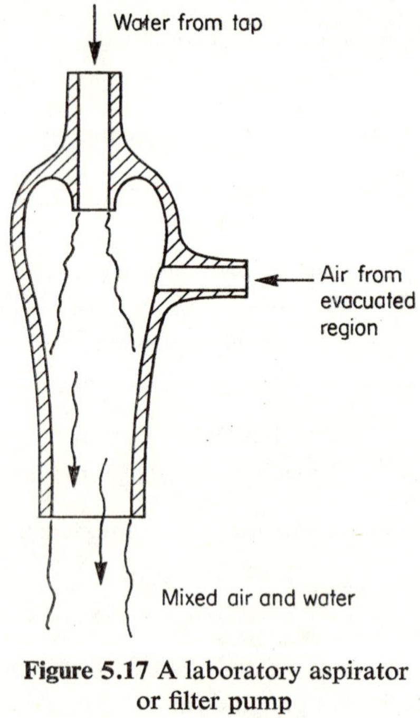

Figure 5.17 A laboratory aspirator or filter pump

the density of the pumped component is not relevant. The pressure change arising from the spreading of the jet is then

$$\Delta p = \varrho_c \frac{A_1}{A_2} \frac{A_j}{A_2} u_j^2$$

Consider the particular case of a water jet of diameter 1 cm directed into a tube of diameter 1·5 cm with a velocity of 18 m/sec. The jet may be taken to fill the cross-section of the tube at outlet. Then $A_j/A_2 = 4/9$ and $A_1/A_2 = 5/9$, and

$$\Delta p = 1000 \times \tfrac{4}{9} \times \tfrac{5}{9} \times 18^2 \text{ N/m}^2 = 0{\cdot}80 \text{ bar}$$

This degree of suction can be obtained only if the vapour pressure of the water is less than the predicted pressure in the chamber, 0·20 bar. Tables of the properties of saturated water and steam give $p_s = 0{\cdot}023$ bar for 20 °C, and $p_s = 0{\cdot}20$ bar for 60 °C. Hence the suction predicted appears to be possible if the water is at normal room temperature or at one of a range of higher temperatures.

5.4.2 The Validity of the One-dimensional Model

The pumping action of a jet pump can be ascribed to non-uniformity of velocity across the inflow section. Since non-uniformity of velocity has significant dynamic effects, we are led to question the one-dimensional model on which most of our work has been based. May not the velocity variation imposed by the no-slip condition at the walls of a duct introduce significant dynamic effects not accounted for in the one-dimensional model?

The expression of the principles of mass, momentum and energy conservation for a flowing fluid requires the calculation of the mass flow, the momentum flux and the flux of kinetic energy. For a one-dimensional flow model, these are given by

$$f_{1_0} = \varrho_c u_{mm} A$$

$$f_{2_0} = \varrho_c u_{mm}^2 A \qquad (5.4.11)$$

and

$$f_{3_0} = \tfrac{1}{2} \varrho_c u_{mm}^3 A$$

with u_{mm} and ϱ_c the mean velocity and density over the flow area A. For a flow whose velocity, and perhaps density, varies over the cross-section, these results can be generalized by the three flux integrals:

$$f_1 = \int_A \varrho u \, dA$$

$$f_2 = \int_A \varrho u^2 \, dA \qquad (5.4.12)$$

and

$$f_3 = \tfrac{1}{2} \int_A \varrho u^3 \, dA$$

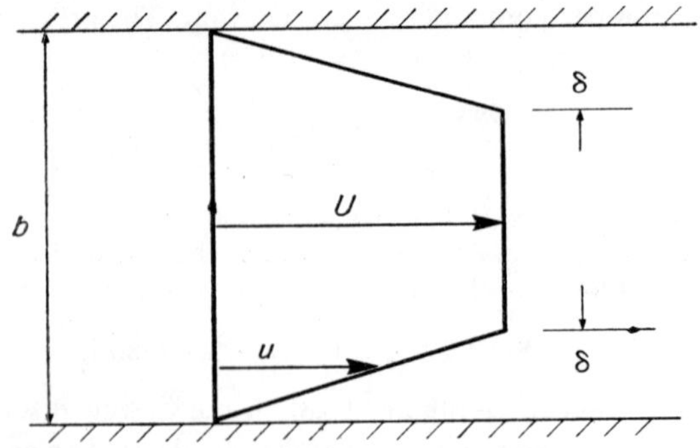

Figure 5.18 A simplified model of flow in a channel

While it is possible to introduce one of the velocity variations found earlier into these integrals, for instance, equation (5.1.15) for laminar pipe flow, or equation (5.3.15) for turbulent pipe flow, the algebra is tedious and may hide the point of the calculations. Instead, a simple, yet representative model will be considered: two-dimensional, constant-density flow in a channel of height b, with a boundary layer of thickness δ on each wall. This situation is shown in Figure 5.18. In the boundary layer

$$u = U\frac{y}{\delta} \quad \text{for} \quad y < \delta$$

Introducing $r = \delta/b$ to represent the proportional thickness of the boundary layer, we find the integrals (5.4.12) to be

$$f_1 = 2\varrho_c \int_0^\delta U\frac{y}{\delta}\,\mathrm{d}y + 2\varrho_c \int_\delta^{\frac{1}{2}b} U\,\mathrm{d}y = \varrho_c Ub(1-r) \qquad (5.4.13)$$

and similarly

$$f_2 = \varrho_c U^2 b\left(1-\frac{4r}{3}\right)$$
$$f_3 = \frac{1}{2}\varrho_c U^3 b\left(1-\frac{3r}{2}\right) \qquad (5.4.13)$$

Now consider the relevance of these results for the one-dimensional model. The obvious way of defining the mean velocity of the one-dimensional flow is to require that the mass flow be the same in the actual and model flows. Then

$$f_1 = \int_A \varrho_c u\,\mathrm{d}A = \varrho_c u_{mm} A \qquad (5.4.14)$$

as in equation (2.1.4). For the flow described in Figure 5.18 and equations

(5.4.13), b corresponds to A, and the mean velocity is given by

$$u_{mm} = U(1-r) \tag{5.4.14}$$

The other two flux integrals can now be written

$$\begin{aligned} f_2 &= \varrho_c u_{mm}^2 b \frac{1-4r/3}{(1-r)^2} \simeq \varrho_c u_{mm}^2 b\left(1+\frac{2r}{3}\right) \\ f_3 &= \frac{1}{2}\varrho_c u_{mm}^3 b \frac{1-3r/2}{(1-r)^3} \simeq \frac{1}{2}\varrho_c u_{mm}^3 b\left(1+\frac{3r}{2}\right) \end{aligned} \tag{5.4.14}$$

The approximate forms are obtained from a truncated binomial expansion. The terms in brackets give the deviations from the one-dimensional approximations (5.4.11). The actual values are larger than the one-dimensional estimates, since the higher central velocities make an augmented contribution to integrals in which they are squared and cubed.

To represent laminar flow in a channel, which has a parabolic velocity profile, the proportional thickness of the 'boundary layer' might be taken to be $r = 1/3$. Comparing the results (5.4.11, 14), we find that, although $f_1/f_{1_0} = 1$ by definition

$$\frac{f_2}{f_{2_0}} = 1{\cdot}25 \quad \text{and} \quad \frac{f_3}{f_{3_0}} = 1{\cdot}69$$

In a turbulent channel flow the velocity variation is confined to a region very near the wall, say, $r = 1/20$ at a moderately high Reynolds number. Then

$$\frac{f_2}{f_{2_0}} = 1{\cdot}03 \quad \text{and} \quad \frac{f_3}{f_{3_0}} = 1{\cdot}08$$

from the approximate forms (5.4.14). As anticipated in Section 2.1, the one-dimensional approximation provides an adequate description of the gross features of a turbulent channel flow, but is seriously in error for laminar flow.

5.4.3 The Flow in a Diffuser

As a second application of the flow integrals (5.4.14), let us consider the flow in a diffuser. When discussing diffuser performance in Section 4.4.1, we noted that the velocity distribution across the duct changed markedly along the diffusing section; this is indicated in Figure 4.16(b). The effects of these changes will now be considered in more detail.

Figure 5.19 gives a crude model of the flow within a two-dimensional diffuser. The boundary-layer thickens in the expanding section and becomes

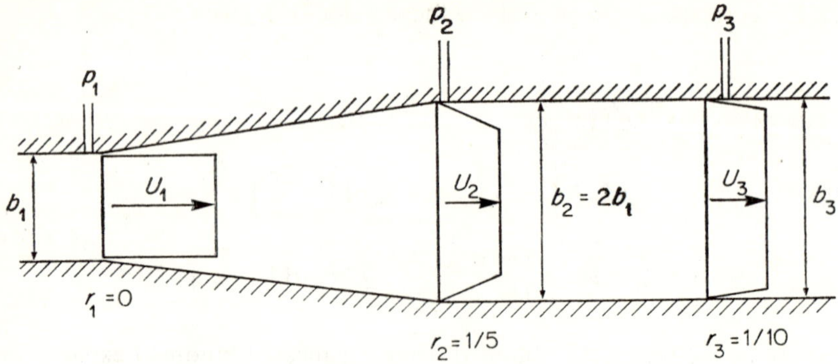

Figure 5.19 A simplified model of flow through a diffusing duct

thinner again in the uniform duct downstream. The continuity conditions governing this development are

$$u_1 b_1 = u_2 b_2 = u_3 b_3 \tag{5.4.15}$$

where u_1, u_2, u_3 are the mean velocities of the one-dimensional model. (The subscripts mm have been dropped at this point for simplicity). Note that $u_2 = u_3$, since $b_2 = b_3$. The mean and core velocities are related by

$$u = U(1-r) \tag{5.4.15}$$

at each section. Even though $u_2 = u_3$, it is not the case that U_2 and U_3 are equal.

In this simplified model of diffuser performance, we shall relate conditions at sections 1 and 2 by taking the total pressures in the core flow to be the same:

$$p_1 + \tfrac{1}{2}\varrho_c U_1^2 = p_2 + \tfrac{1}{2}\varrho_c U_2^2$$

Using equations (5.4.15), we rewrite this as

$$p_1 + \frac{\varrho_c u_1^2}{2(1-r_1)^2} = p_2 + \frac{\varrho_c u_2^2}{2(1-r_2)^2} \tag{5.4.16}$$

Conditions at sections 2 and 3 will be related by assuming friction on the walls to be negligible. Thus

$$p_2 b_2 + (f_2)_2 = p_3 b_3 + (f_2)_3$$

where f_2 is the momentum flux of equations (5.4.14). This condition may be written

$$p_2 + \varrho_c u_2^2 \frac{1-4r_2/3}{(1-r_2)^2} = p_3 + \varrho_c u_3^2 \frac{1-4r_3/3}{(1-r_3)^2} \tag{5.4.17}$$

Equation (4.4.9) gives the diffuser efficiency in a form suitable for this discussion:

$$\eta_D = \frac{p_2 - p_1}{\frac{1}{2}\varrho_c u_1^2(1 - A_1^2/A_2^2)}$$

where $p_2 - p_1$ is the actual pressure rise and u_1 is the mean velocity at entry. In the present case, $A_1/A_2 = b_1/b_2$. Introducing the pressure rise given by equation (5.4.16), we find the efficiency of the duct between sections 1 and 2 to be

$$\eta_D = \frac{\dfrac{1}{(1-r_1)^2} - \dfrac{(b_1/b_2)^2}{(1-r_2)^2}}{1 - \dfrac{b_1^2}{b_2^2}} \tag{5.4.18}$$

For the postulated values of Figure 5.19, $r_1 = 0$, $r_2 = 0{\cdot}2$, $b_2/b_1 = 2$, we find

$$\eta_D = 81{\cdot}3 \text{ per cent}$$

In practice the total pressure in the core would not remain constant as assumed in this calculation, and the duct efficiency would be less than that calculated above.

Suppose that we decide instead to measure the diffuser efficiency in terms of the pressure change between sections 1 and 3. Equations (5.4.15) give $u_2 = u_3 = (b_1/b_2)u_1$. When these results are introduced into equation (5.4.17), it gives

$$\frac{p_3 - p_2}{\frac{1}{2}\varrho_c u_1^2(1 - b_1^2/b_2^2)} = \frac{2\left[\dfrac{1-4r_2/3}{(1-r_2)^2} - \dfrac{1-4r_3/3}{(1-r_3)^2}\right]}{\dfrac{b_2^2}{b_1^2} - 1} = 5{\cdot}1 \text{ per cent}$$

for the values postulated in Figure 5.19. If duct friction is taken into account, the pressure recovery in the constant-area duct will not be as large. Indeed, the pressure may continue to fall throughout the duct.

The overall efficiency of the duct between sections 1 and 3 can now be calculated as

$$\eta_D = 81{\cdot}3 + 5{\cdot}1 = 86{\cdot}4 \text{ per cent}$$

For the reasons pointed out above, this will be an overestimate of the efficiency associated with the velocity distribution of Figure 5.19.

This simple example illustrates several of the points made in the general discussion of diffuser performance in Section 4.4:

(1) the thickening of the boundary layer between sections 1 and 2 is linked to the diffuser efficiency;

(2) not all of the pressure drop between sections 1 and 2 arises from immediate dissipation; some is associated with an 'acceleration' of the core flow (relative to the ideal pattern flow) produced by boundary-layer thickening;

(3) even though the dissipation associated with this 'acceleration' has not yet occurred, it will inevitably do so when the non-uniform velocity distribution is evened out by mixing and spreading of the flow; and

(4) different values of the diffuser efficiency will be obtained depending on the section at which measurements are made.

5.4.4 The Drag Integral

The flow integrals introduced in equations (5.4.12) can be used to investigate many other aspects of fluid motion, for example, the development of boundary layers, wakes and jets. One simple application will be developed here, an indirect method of determining the drag on a body in a uniform stream.

In Figure 5.20 are shown a body on which a drag force acts and a control volume surrounding the body. The drag is the streamwise component of the force on the body; the cross-stream component, if any, is called lift. The

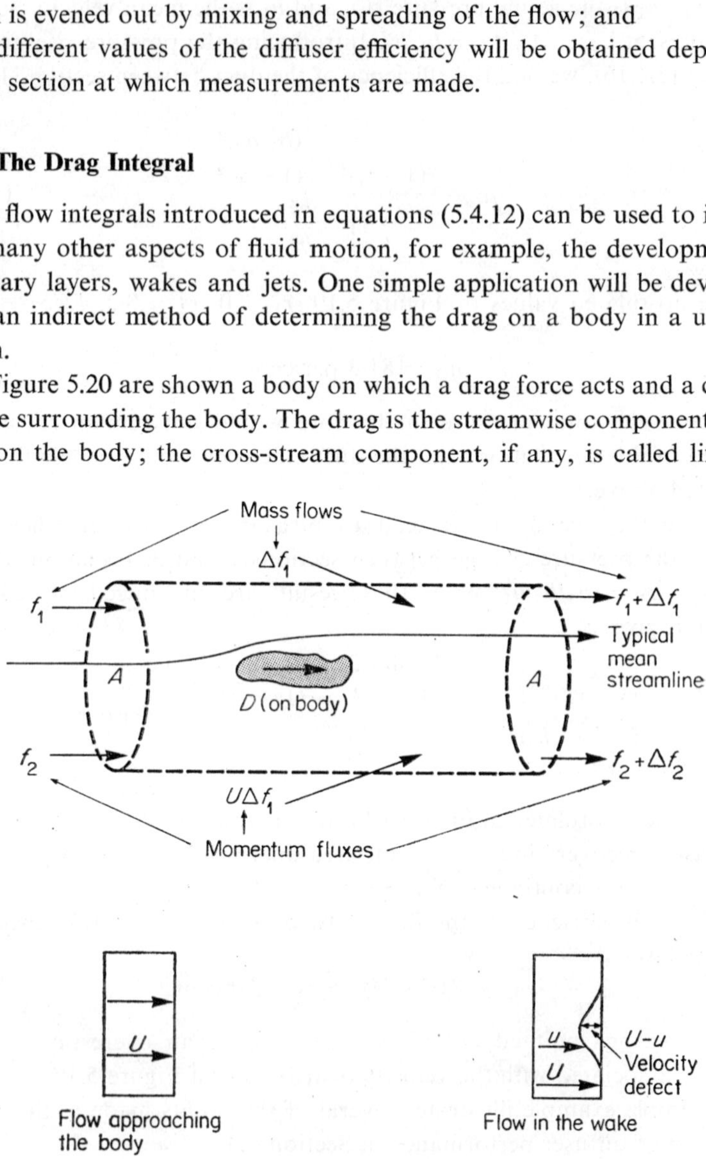

Figure 5.20 Mass and momentum fluxes through a control volume surrounding a body on which a drag force acts

control volume is defined by a surface fixed with respect to the body and having sides normal and parallel to the undisturbed flow. Since this surface is well away from the body, the pressure may be taken to be uniform over the whole of it. Moreover, the velocity is nearly uniform, except in the wake behind the body, although the displacement effect of the body and wake produces a slight outwards deflexion of the streamlines.

At the top of the control volume are shown the mass flows through its forward and rear faces, and through the whole of the side area joining them. At the bottom are shown the fluxes of momentum through these surfaces. Since the drag component of the force is to be found, the momentum considered is that in the direction of the undisturbed flow. Note particularly the momentum flux $U\,\Delta f_1$ associated with the net lateral displacement of the streamlines. The changes in mass and momentum fluxes are shown conventionally as positive, although they will in fact be negative, since the fluid is deflected outwards when the wake is formed.

The force and momentum balance for the control volume gives

$$D = U\,\Delta f_1 - \Delta f_2$$

for the drag force on the body. In terms of the velocity variations in the wake and ahead of the body

$$\Delta f_1 = \varrho_c \int_A u\,\mathrm{d}A - \varrho_c \int_A U\,\mathrm{d}A = \varrho_c \int_A (u-U)\,\mathrm{d}A$$

and

$$\Delta f_2 = \varrho_c \int_A u^2\,\mathrm{d}A - \varrho_c \int_A U^2\,\mathrm{d}A = \varrho_c \int_A (u^2-U^2)\,\mathrm{d}A$$

Then

$$D = \varrho_c \int_A u(U-u)\,\mathrm{d}A$$

where A is the rear face of the control volume. When the velocity defect $(U-u)$ is small, that is, when $u \simeq U$, the drag integral can be approximated by the simple result

$$D = \varrho_c U \int_A (U-u)\,\mathrm{d}A$$

This will apply far downstream along the wake, since wake spreading progressively reduces the velocity defect.

The drag on a body is frequently described using a dimensionless drag coefficient

$$C_D = \frac{D}{\frac{1}{2}\varrho_c U^2 A_b} = 2\int_A \frac{u}{U}\left(1-\frac{u}{U}\right)\mathrm{d}\left(\frac{A}{A_b}\right) \simeq 2\int_A \left(1-\frac{u}{U}\right)\mathrm{d}\left(\frac{A}{A_b}\right)$$

with A_b a characteristic area of the body, usually either the maximum cross-section or the wetted area, that on which wall friction acts. Which of these areas is most meaningful depends on whether the drag is primarily the result of streamwise shear stresses on the body surface, or primarily the result of pressure differentials between the front and rear of the body. The total drag on most bodies contains significant components of both friction and form drag. For both the inertial scaling used above is appropriate for the large-scale, high-Reynolds-number flows that are usually of interest to engineers.

Example 5.10. Consider the testing of a submarine model with hull diameter 20 cm in a wind tunnel with test velocity $U = 100$ m/sec. A Pitot traverse across the nearly axisymmetric wake gives the variation of mean velocity there

$$u = U\left[1-0{\cdot}05 \exp \frac{-r^2}{100}\right]$$

the radial coordinate r being measured in centimetres. An estimate of the power required to drive the actual submarine is required; its hull diameter is 7 m and its submerged operating speed is 20 m/sec. In earlier tests of bodies of this general shape, it has been found that the variation of drag coefficient with Reynolds number is of the form $C_D \propto \mathrm{Re}^{-1/7}$.

For the specified velocity distribution, u is within 5 per cent of U, and the approximation

$$D = \varrho_c U^2 \int_A \left(1-\frac{u}{U}\right) \mathrm{d}A$$

is adequate. Then

$$\frac{D}{\varrho_c U^2} = 2\pi \int_0^\infty 0{\cdot}05 \exp \frac{-r^2}{100}\, r\, \mathrm{d}r = \frac{\pi}{20}\int_0^\infty \exp \frac{-r^2}{100}\, \mathrm{d}r^2 = 5\pi \text{ cm}^2$$

The integration has been extended to $r \to \infty$ since the contribution of the outer tail of the distribution is insignificant. The drag coefficient of the model, based on frontal area $A_b = \frac{1}{4}\pi \times 20^2 = 78{\cdot}5 \text{ cm}^2$, is

$$C_{D_m} = \frac{1}{10}$$

If the drag coefficient were independent of Reynolds number, the drag of the full-scale submarine could be calculated directly. However, the drag of a streamlined body such as a submarine hull arises largely from skin friction. (This is the case only when the submarine is well below the water surface, so that wave-making is unimportant.) Hence a considerable Reynolds-number dependence or scale effect must be expected. The suggested empirical correlation gives

$$\frac{C_{D_p}}{C_{D_m}} = \left(\frac{\mathrm{Re}_m}{\mathrm{Re}_p}\right)^{\frac{1}{7}} = \left[\frac{1{\cdot}5}{17}\times\frac{20}{700}\times\frac{100}{20}\right]^{\frac{1}{7}} = 0{\cdot}535$$

where the viscosities have been taken to be $\nu_m = 17$ centistokes for air at 100 °F and 1 atm, and $\nu_p = 1{\cdot}5$ centistokes for water at 40 °F. Hence for the prototype submarine $C_{D_p} = 0{\cdot}054$ is a better estimate of the drag coefficient. Its drag is then of the order

$$D = C_D(\tfrac{1}{2}\varrho_c U^2)A_b = 4{\cdot}16\times10^5 \text{ N}$$
$$= 42{,}300 \text{ kgf}$$

The propulsive power is then approximately

$$P = DU = 8{\cdot}3\times10^6 \text{ J/sec} = 8300 \text{ kW}$$

The power output calculated above is ultimately dissipated by viscous processes in the wake. This is not the total power requirement for the propulsive process. Further energy will be expended as a result of inefficiencies of the propellers, in the driving motors, and in the process providing energy to drive the motors. Hence the total energy expenditure required to drive the submarine could well be three or four times that found above.

5.5 LOSSES IN PIPE FLOW

5.5.1 Friction Losses

The results obtained in this chapter provide the means of predicting the dissipation in fully developed pipe flows and in the mixing flows near abrupt changes in flow area. These are the dominant sources of dissipation in piping systems made up, as most are, of constant-diameter tubes joined together by compact fittings, such as elbows and tees. Hence we are in a position to answer the two questions that are most often asked about a flowing fluid: 'What will be the flow through a given pipe line for a specified pressure drop?' and 'What pressure difference is required to produce a specified flow through a given pipe line?' The techniques commonly used to answer these questions are explained in detail in engineering handbooks, which provide a variety of graphical aids to calculation and extensive tables of empirical coefficients. There is little point in a textbook attempting to compete with these compendious works, and here we shall aim only to discuss the ideas underlying this body of engineering lore.

The contribution of wall friction to the resistance to flow through a pipe line will be considered first; for most piping systems this is the larger source of dissipation. The equations (5.3.17, 18, 24) linking flow rate and wall stress make possible a direct calculation of the flow through a pipe if the wall stress is specified in terms of a pressure gradient. These laws apply, respectively, to laminar flow in a pipe, to turbulent flow in a smooth pipe, and to turbulent flow in a fully rough pipe. For a pipe of radius R the dimensionless parameters of the friction laws are linked with the pressure gradient and fluid properties by

$$u_f^2 = \frac{\tau_0}{\varrho} = -\frac{R}{2\varrho}\frac{dp}{dx}$$

and

$$\eta_R = \frac{\varrho u_f R}{\mu} \tag{5.5.1}$$

The friction laws obtained earlier are less well suited to the inverse problem of predicting the pressure gradient required to generate a specified flow. For this problem an iterative calculation, like those of Examples 5.7 and 5.8, can be used. It is also possible to recast the friction laws, which now give $\dot{V}/Au_f$ against $u_f R/\nu$ or R/k, into forms suitable for the direct solution of this second problem. To see how this can be done, consider the dependence of friction on the other parameters defining a pipe flow. It is to be expected that

$$\tau_0 = f(u, d, k, \varrho, \mu) \tag{5.5.2}$$

where u is the mean velocity of the one-dimensional model, and d and k are the pipe diameter and a length characterizing the wall roughness. The conventional dimensionless form of the relationship (5.5.2) is

$$c_f = f\left(\text{Re}, \frac{k}{d}\right) \tag{5.5.3}$$

Here

$c_f = \tau_0/\frac{1}{2}\varrho u^2$ is the *friction coefficient*, the local dynamic pressure being used to render the wall stress dimensionless;

Re $= ud/\nu$ is the Reynolds number based on the pipe diameter; and

k/d is a roughness parameter, often termed the *relative roughness*.

Using the first of equations (5.5.1) we see that the friction coefficient is related to the pressure gradient by

$$c_f = 2\left(\frac{u_f}{u}\right)^2 = -\frac{d}{2\varrho u^2}\frac{dp}{dx} \tag{5.5.4}$$

Another much used dimensionless measure of friction is the *friction factor* defined as

$$f = -\frac{d}{\frac{1}{2}\varrho u^2}\frac{dp}{dx} \tag{5.5.5}$$

The two friction parameters, sometimes called the Fanning and Darcy friction factors, are related by

$$f = 4c_f$$

Since the symbol f is commonly used for both parameters it is necessary to check the definition of friction factors whenever they are encountered.

The parameters appearing in equation (5.5.3) are related to those of equations (5.3.17, 18, 24) by

$$c_f = \frac{\tau_0}{\frac{1}{2}\varrho u^2} = 2\left(\frac{u_f}{u}\right)^2 = 2\left(\frac{Au_f}{\dot{V}}\right)^2$$

and (5.5.6)

$$\text{Re} = \frac{ud}{\nu} = 2\,\frac{u_f R}{\nu}\cdot\frac{\dot{V}}{Au_f} = 2\,\eta_R\left(\frac{2}{c_f}\right)^{\frac{1}{2}}$$

Hence the earlier equations can be converted to the form (5.5.3) by simple substitutions. They become:
for laminar flow

$$c_f = \frac{16}{\text{Re}} \tag{5.5.7}$$

for turbulent flow over a smooth wall

$$\left(\frac{2}{c_f}\right)^{\frac{1}{2}} = 2{\cdot}5\,\log_e\left[\frac{\text{Re}}{2}\left(\frac{c_f}{2}\right)^{\frac{1}{2}}\right] + 1{\cdot}75 \tag{5.5.8}$$

for turbulent flow over a fully rough wall

$$\left(\frac{2}{c_f}\right)^{\frac{1}{2}} = 2{\cdot}5\,\log_e\left(\frac{d}{2k}\right) + 4{\cdot}75 \tag{5.5.9}$$

The result (5.5.8) is less convenient than the others since the friction coefficient is not given explicitly in terms of the other parameters. Hence this result is commonly approximated by a power law, for example

$$c_f = 0{\cdot}079\,\text{Re}^{-\frac{1}{4}} \quad \text{for turbulent flow with} \quad \text{Re} < 10^5.$$

Note that $\tau_0 \propto u^{7/4}$ according to this result. Hence, even for a smooth wall, turbulent friction is very nearly proportional to the square of the velocity.

The variations of the friction coefficient given by equations (5.5.7, 8, 9) are shown in Figure 5.21; the form of variation typical of the pipes in common use is also shown. Figure 5.21 represents many of the phenomena discussed earlier, for instance, the transition from laminar to turbulent flow (Section 5.3.1), and the change from effectively smooth to fully rough wall (Section 5.3.3). For the non-uniform roughness of commercial piping, this latter change occurs gradually over a wide range of Reynolds number.

Note that the friction coefficient based on the dynamic pressure of the mean flow varies only slightly at high Reynolds numbers; this inertial scaling has accounted for most of the velocity dependence. Thus for the large-scale flows typical of engineering equipment, the friction coefficient is primarily dependent upon the roughness of the wall. Unfortunately, it is not possible to predict or to measure directly the effective roughness of a surface; consequently the friction of turbulent flows often cannot be predicted with great precision.

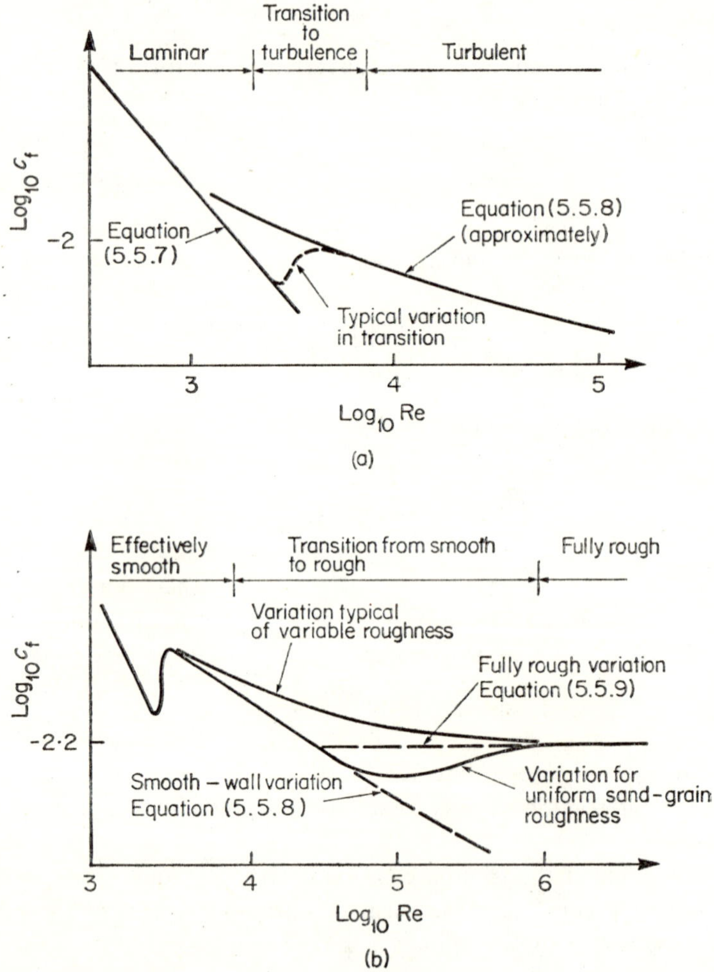

Figure 5.21 Variation of pipe friction coefficient with Reynolds number. (a) Smooth-walled pipe; (b) Rough-walled pipes: variations typical of commercial pipe and artificially roughened pipe, both with sand-grain roughness specified by $k/d = 0{\cdot}002$

In practice, friction calculations are performed with the aid of widely available charts, which give empirically determined curves of the general form indicated in Figure 5.21. The curves correspond to particular values of the sand-grain roughness, but are modified to give the more usual gradual transition from the smooth wall to the fully rough condition. Even when such a chart—a Moody diagram—is available, there still remains the problem of determining the sand-grain roughness which is equivalent at high Reynolds numbers to the actual roughness of a particular pipe. This can be done in two ways, by using data from tests on the particular pipe (such information is often supplied by manufacturers), or by using roughness values found to be typical of piping of its general class. Some 'standard' values of roughness are given in Table 5.2; these are not heights of actual roughness elements, but are the values of the sand-grain roughness which would give the same friction coefficient at high Reynolds numbers.

Table 5.2. Guide to the roughness of pipes

Type of pipe	Equivalent roughness k in	Equivalent roughness k mm
Drawn tubing	0·0001	0·0025
Commercial steel	0·002	0·05
Cast iron	0·01	0·25
Concrete	0·05	1·25
Riveted steel	0·2	5·0

Given the equivalent roughness of a pipe and the flow rate, the friction can be found directly with the aid of a Moody diagram, or by using formulae similar to equations (5.5.7, 8, 9). However, an iterative solution is necessary if the friction is known and the flow rate has to be found using a Moody diagram. The alternative presentation $\dot{V}/Au_f = f(u_f d/\nu, k/d)$, which would allow a direct solution of the latter problem, is not commonly available.

The preceding discussion of friction relates almost entirely to pipes of circular cross-section. Although this is the most important case in practice, one does sometimes have to discuss the flow in ducts of non-circular section, for example, heating or ventilating ducts. Unless the cross-section has a very complex shape, the round-pipe results can be used, as explained below, to provide a fairly accurate estimate of the friction.

For a round pipe the wall stress is uniform around the perimeter, and the pressure drop and friction are related by

$$\tau_0 P = -A \frac{dp}{dx}$$

where P is the perimeter length (πd for a round pipe) and A is the cross-sectional area ($\frac{1}{4}\pi d^2$ for a round pipe). Then

$$-\frac{\tau_0}{\mathrm{d}p} = \frac{4\,\mathrm{d}x}{d}$$

relates the stress and pressure drop. For a non-circular section the shear stress and boundary-layer structure are fairly uniform over considerable portions of the perimeter. This uniformity is the result of turbulence-induced secondary flows which carry energetic fluid into the corners of the cross-section. The average wall stress and pressure drop are related by

$$-\frac{\tau_a}{\mathrm{d}p} = \frac{P}{A}\,\mathrm{d}x$$

Comparing the last two results, we see that the quantity

$$d_e = \frac{4A}{P}$$

termed the effective or *hydraulic diameter*, replaces the diameter in the relationship between the average wall stress and pressure drop. To the extent that it is realistic to assume the equality of the average stresses τ_0 and τ_a, it is also reasonable to assume that the average wall stress depends on roughness and mean velocity $\dot{V}/A$ in the same way as for a round pipe. This suggests that a fair approximation to the turbulent friction in a non-circular section can be obtained using the results for a round pipe, with the hydraulic diameter replacing the actual diameter.

5.5.2 Loss Coefficients

Earlier it was suggested that we are in a position to calculate the losses arising in rapid mixing and spreading of a flow. Strictly speaking, we can do this only for a few special cases, and must use experimental results to find the losses in most such flows. However, the general nature of the dissipative processes is suggested by these special cases; this gives some insight into the significance of the experimental results and suggests ways in which they can conveniently be presented.

The jet-pump model discussed in Section 5.4.1 can be used to predict the losses occurring in a single stream under certain circumstances. Although that analysis represented the mixing of two streams, the results apply when one of the flows is absent, the other merely expanding from area A_1 to area A_2. Then equation (5.4.5) gives the pressure recovery as

$$p_2 - p_1 = \varrho_c\left(1 - \frac{A_1}{A_2}\right)\frac{A_1}{A_2}u_1^2 \tag{5.5.10}$$

Equation (5.4.8) gives the fall in total pressure and rise in internal energy in this constant-density flow as

$$\frac{p_{0_1} - p_{0_2}}{\varrho_c} = \frac{1}{\dot{m}}\frac{dE}{dt} = \frac{1}{2}(u_1 - u_2)^2 \tag{5.5.11}$$

Although the static pressure rises, as seen in equation (5.5.10), the fall in kinetic energy more than compensates for this, and the net effect is a reduction in mechanical energy.

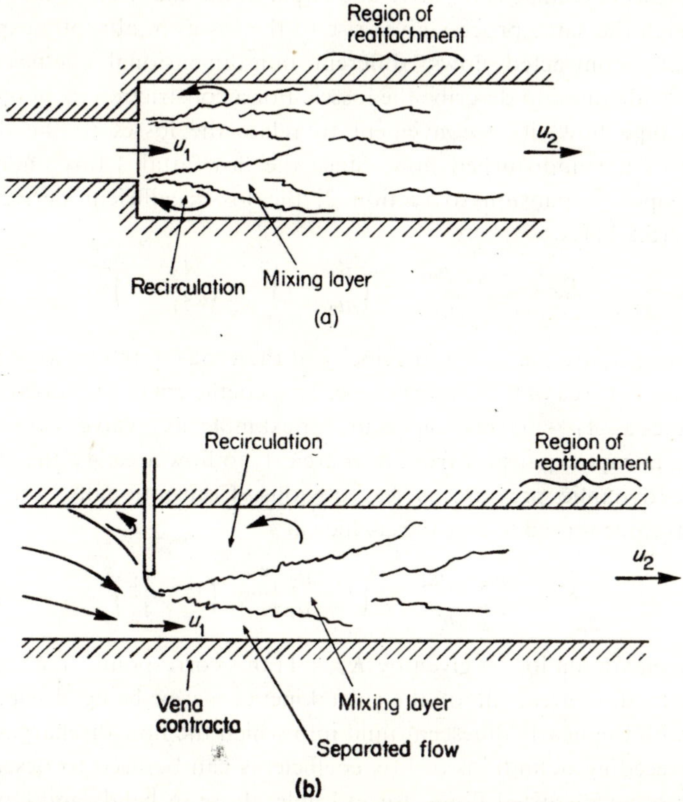

Figure 5.22 Examples of mixing and spreading flows (other examples are to be found in Figures 2.39, 4.13, 4.21 and 5.16). (a) Flow at an abrupt expansion in a duct. (b) Flow near a partly open gate valve

Two situations governed by these results are shown in Figure 5.22, the flow following a sudden expansion in the duct, and the contracting, then expanding flow at a partially closed valve. The flow through an orifice meter, shown in Figure 4.21, is essentially similar to that past a valve. The losses in the acceleration to the minimum flow area are small compared to those in the expansion. Hence the loss at a local contraction, such as a valve or orifice meter, is given with good accuracy by equation (5.5.11), with u_1 taken to be the maximum flow velocity at the vena contracta. The result (5.5.10) provides a fairly realistic estimate of the pressure rise from the vena contracta to the section at which the flow area achieves its original value. The net pressure change across the local contraction includes the fall in pressure from the undisturbed upstream flow to the vena contracta. Thus the net pressure change is negative and equal to the change in total pressure.

Although the same process gives rise to the losses in abruptly expanding and locally contracted flows, different non-dimensional coefficients are conventionally used to describe the losses. For a constriction in an otherwise uniform pipe flow, it is convenient to relate the losses to the dynamic pressure of the undisturbed flow. Since the flow attains this undisturbed velocity upon expansion to section 2, the loss coefficient derived from equation (5.5.11) is

$$K_c = \frac{p_{0_1} - p_{0_2}}{\frac{1}{2}\varrho_c u_2^2} = \left(\frac{u_1}{u_2} - 1\right)^2 = \left(\frac{A_2}{A_1} - 1\right)^2 \tag{5.5.12}$$

Here A_1 and u_1 are the area and velocity at the vena contracta, and A_2 is the cross-sectional area of the uniform pipe. This coefficient can attain arbitrarily large values as $A_1 \rightarrow 0$; this will occur, for example, as a valve is closed.

For an abrupt expansion from flow area A_1 to flow area A_2, the choice of reference dynamic pressure is less obvious. Conventionally, the higher, upstream value is used to give the coefficient

$$K_e = \frac{p_{0_1} - p_{0_2}}{\frac{1}{2}\varrho_c u_1^2} = \left(1 - \frac{u_2}{u_1}\right)^2 = \left(1 - \frac{A_1}{A_2}\right)^2 \tag{5.5.13}$$

Here the maximum loss is given by $K_e = 1$; this corresponds to a very large increase in flow area, all of the inlet kinetic energy being dissipated on mixing with the nearly quiescent fluid into which the flow discharges.

The preceding definitions of loss coefficients can be used to describe the losses in other distorted flows, for example, those in bends and elbows, at tee-junctions, and in more complex valves and meters. Even though the dissipation cannot be predicted, a local acceleration followed by inefficient diffusion can be expected to give rise to a loss which depends on the geometry of the flow boundaries and on some dynamic pressure characterizing the flow. Losses at a local distortion of a uniform pipe flow with mean velocity u can be described using the coefficient

$$K = \frac{p_{0_1} - p_{0_2}}{\frac{1}{2}\varrho_c u^2} \tag{5.5.14}$$

as in equation (5.5.12). Here, however, the subscript 1 refers to conditions in the pipe well upstream of the distortion. This coefficient can be given another interpretation. For constant-area, constant-density flow the velocity is the same well upstream and downstream of the localized irregularity in the flow. Then $u_1 = u_2$, and $p_{0_1} - p_{0_2} = p_1 - p_2$. Hence in this case

$$K = \frac{p_1 - p_2}{\frac{1}{2}\varrho_c u^2} \tag{5.5.15}$$

and may be interpreted as a pressure-drop coefficient.

The coefficient of equations (5.5.14, 15) will certainly depend on the geometry of the distorting section of the duct. It may also depend upon the Reynolds number of the flow. In general, the internal geometry of the flow depends on the location of the separation points within the channel, and their location is in turn dependent on the relative importance of viscous and inertia forces. Hence one must expect that the loss coefficient for a particular fitting will be of the form

$$K = K(\text{Re})$$

For flow at high Reynolds number the dependence is weak, as it is for the friction coefficient. Moreover, in fittings and valves with sharp internal corners the separation points are virtually fixed. For such fittings the inertial scaling of equations (5.5.14, 15) will account almost completely for the velocity dependence of the losses. For fittings, such as elbows, which have smoother internal contours, some dependence upon Reynolds number must be expected. For example, the losses in a 90°-bend are given approximately by

$$K = 1{\cdot}6f\frac{r}{d} + 0{\cdot}11\left(\frac{d}{r}\right)^{2\cdot5} + 2000f^{2\cdot5} \tag{5.5.16}$$

with r the bend radius, d the pipe diameter, and $f = 4\,c_f$ the friction factor for the pipe line. The friction factor depends on Reynolds number, as indicated in Figure 5.21; thus the loss coefficient itself is Reynolds-number dependent.

It is apparent that the losses at isolated fittings are equivalent, in their effect on the pressure of the flowing fluid, to the losses from friction distributed along the pipe wall. This equivalence can be made explicit in the following way. The frictional pressure drop in length L of uniform pipe is given by equations (5.5.4, 5) as

$$-\Delta p = p_1 - p_2 = 2\varrho_c u^2 \frac{L}{d} c_f = \frac{1}{2}\varrho_c u^2 \frac{L}{d} f \tag{5.5.17}$$

The pressure drop across a fitting in a uniform pipe is given by equation (5.5.15) as

$$-\Delta p = p_1 - p_2 = \tfrac{1}{2}\varrho_c u^2 K \tag{5.5.18}$$

where K is the loss coefficient for the fitting. Comparison of these two results gives

$$K = \frac{4L}{d} c_f = \frac{L}{d} f \tag{5.5.19}$$

This relationship is illustrated in Figure 5.23. It can be used to define either

(a) the loss coefficient K which is equivalent to length L of pipe of diameter d and friction coefficient c_f, or

(b) the length of pipe L of diameter d and friction coefficient c_f which is equivalent to a fitting with loss coefficient K.

The word 'equivalent' means simply 'giving the same dissipation'.

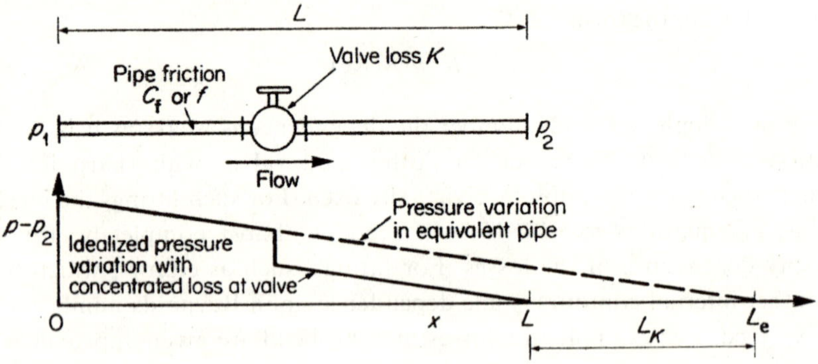

Figure 5.23 An example of the equivalence of specific losses and pipe friction

The relationship (5.5.19) is the basis of two methods of determining the losses in a pipe of uniform diameter.

Method I.

Find the loss coefficient equivalent to the actual run of the pipe

$$K_f = \frac{4L}{d} c_f \tag{5.5.20}$$

and add this to the sum of the loss coefficients for the fittings, to obtain the total loss coefficient

$$K_e = \Sigma K + K_f \tag{5.5.20}$$

The pressure drop follows from equation (5.5.18).

Method II.

Find the length of pipe equivalent to the sum of the loss coefficients for the fittings

$$L_K = \frac{d}{4c_f}\Sigma K \tag{5.5.21}$$

and add this to the length L of the pipe run, to obtain the equivalent total pipe length

$$L_e = L + L_K \tag{5.5.21}$$

The pressure drop follows from equation (5.5.17).

In most piping systems the sum of the losses at fittings is small compared to the frictional losses, that is, $L_K \ll L$. Hence the losses at fittings are sometimes termed *minor losses;* another name is *specific losses.* Of course, in a twisty pipe line with short runs of straight pipe, the 'minor losses' may turn out to be the greater source of dissipation. In such lines the problem of interference is encountered. It has tacitly been assumed that the fittings are separated by longish lengths of piping, so that each can be described by a unique loss coefficient. However, closely spaced fittings will interfere with one another, each influencing the flow within its neighbours, particularly that downstream. When this occurs, it is not possible to find the net dissipation simply by summing standard coefficients, and tests may be required to obtain a realistic estimate of the losses.

The relationship between equivalent length and loss coefficient (equation (5.5.19)) involves the Reynolds number, most obviously through the Reynolds-number dependence of the friction coefficient. The specific loss coefficient also depends on the Reynolds number and friction coefficient, since these determine the nature of the boundary layer of the flow entering the fitting. However, since the contribution of the so-called minor losses is in fact usually small, and since the friction coefficient itself is usually not known precisely, it is common practice to neglect the Reynolds-number dependence of the loss coefficient and the equivalent length of pipe. Table 5.3 gives some typical values of these empirical factors; they are related using equation (5.5.19) with $f = 4\,c_f = 0{\cdot}025$.

Several important aspects of pipeline flows have not been mentioned in this discussion. Nothing has been said about systems of pipes of differing diameters joined in series, or about pipes connected in parallel. Moreover, the important problem of friction in open-channel flow has only been touched on. Since these matters arise continually in practice, they are fully discussed in engineering handbooks. Although the calculations are necessarily more complicated, the methods of finding the losses in these more varied systems are in essence those developed in this chapter.

Table 5.3. Loss coefficients based on mean-flow dynamic pressure

Type of fitting	K	L_K/d
90° Standard elbow	0·75	30
Standard tee		
through crosspiece	0·5	20
through branch	1·5	60
Gate valve		
fully open	0·2	8
half open	5·5	220
Butterfly valve—open	0·5	20

Example 5.11. The following calculations are typical of those carried out thousands of times a day in engineering offices around the world. Two methods of obtaining the result will be illustrated, but many variants are in use.

What must be the pressure in a tank from which liquid ammonia (density 685 kg/m^3, viscosity 0·27 mN sec/m^2) flows at the rate of $\frac{1}{2}$ ft^3/sec through the pipe shown in Figure 5.24? The line is horizontal and within it fluid weight need not be taken into account. The pipe discharges freely to atmospheric pressure in another tank, the ammonia falling to the bottom of this second tank. The losses at entry, at an elbow in the line, and across a valve controlling the flow are shown. The entry loss arises from separation at the sharp-edged inlet to the pipe; see Figure 4.16(c). The pipe diameter is 2 in throughout, and the friction is specified by c_f = 0·006 on the basis of experience with similar installations.

Method I (equations (5.5.20)):

Since the mean velocity is uniform along the pipe, the flow area and density being constant, the loss coefficients may simply be summed to find their net effect: $\Sigma K = 4{\cdot}2$. The coefficient representing the pipe friction is $K_f = 4c_f L/d = 4 \times 0{\cdot}006 \times 600/2 = 7{\cdot}2$. Hence the net loss in total pressure in the line is given by

$$\frac{-\Delta p_0}{\frac{1}{2}\varrho_c u^2} = K_e = \Sigma K + K_f = 11{\cdot}4$$

This value does not give the change in pressure between the two tanks. Since the fluid is discharged at velocity u, the exit total pressure is $p_{0_e} = p_e + \frac{1}{2}\varrho_c u^2$, while $p_{0_i} = p_i$ relates the pressure and total pressure at inlet level within the first tank. Hence

$$p_i - p_e = p_{0_i} - p_{0_e} + \tfrac{1}{2}\varrho_c u^2$$

and

$$\frac{-\Delta p}{\frac{1}{2}\varrho_c u^2} = K_e + 1 = 12{\cdot}4$$

If the flow were efficiently diffused before discharge, some of the kinetic energy would be converted into displacement energy, the pressure would rise towards

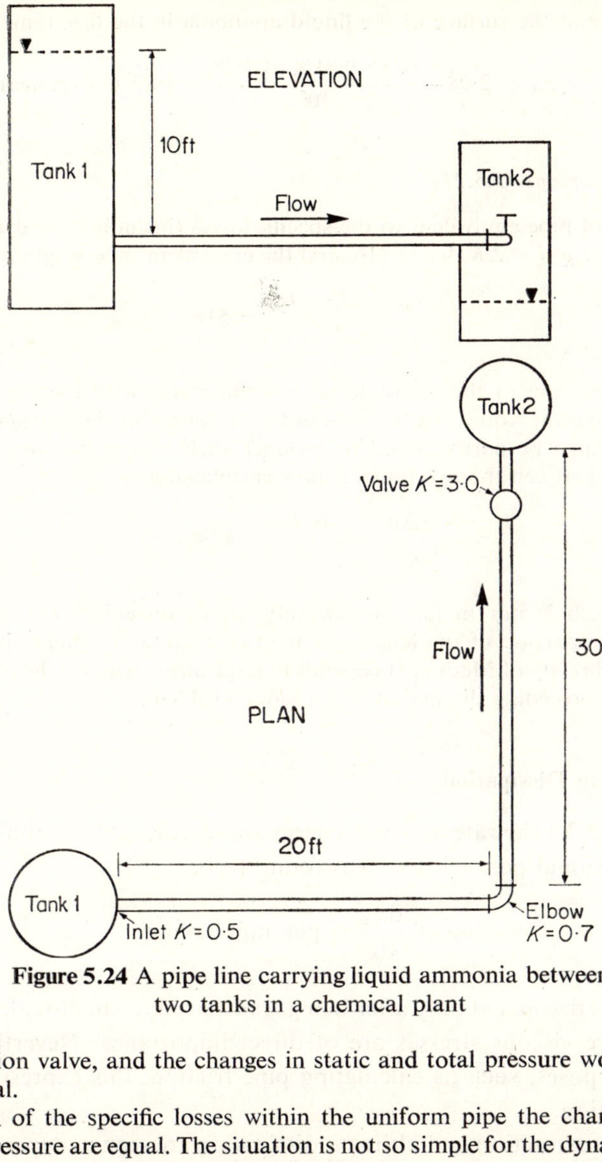

Figure 5.24 A pipe line carrying liquid ammonia between two tanks in a chemical plant

the stagnation valve, and the changes in static and total pressure would be more nearly equal.

For each of the specific losses within the uniform pipe the changes in static and total pressure are equal. The situation is not so simple for the dynamic pressure of the flow. The static pressure experiences this loss as soon as the fluid accelerates into the pipe, but the total pressure does not drop by this amount until after the fluid has left the pipe. Nevertheless, the kinetic energy 'thrown away' has the same effect on the resistance as would a specific loss with $K = 1$. If a diffuser is provided, it is possible to obtain $K < 1$ for the exit loss.

For the specified flow rate, $u = \dot{V}/A = 7{\cdot}0$ m/sec, and the dynamic pressure is $\frac{1}{2}\varrho_c u^2 = \frac{1}{2}\times 0{\cdot}685\times 7^2 = 16{\cdot}8$ kN/m². Hence the pressure at the pipe inlet must be

$$p_i - p_e = p_i - p_a = 12{\cdot}4\times 16{\cdot}8 = 208 \text{ kN/m}^2 \text{ (gauge)}$$

The pressure at the surface of the liquid ammonia in the first tank is given by

$$p_i - \varrho_c gh = 2{\cdot}08 - \frac{685 \times 9{\cdot}81 \times 10/3{\cdot}28}{10^5} = 1{\cdot}87 \text{ bar (gauge)}$$

Method II (equations (5.5.21)):

The length of pipe equivalent to the specific losses (including the dynamic pressure at exit) is $L_K/d = \Sigma K/4c_f = 216$, and the equivalent pipe length of the line is

$$\frac{L_e}{d} = \frac{L}{d} + \frac{L_K}{d} = 516$$

In this case the sum of the minor losses is comparable with the loss from pipe friction. However, it would not be without the considerable dissipation across the valve, which must be partly closed to produce such a large pressure drop. The pressure drop between the two tanks is now calculated as

$$\frac{-\Delta p}{\frac{1}{2}\varrho_c u^2} = \frac{4c_f L_e}{d} = 12{\cdot}4$$

This approach is not in fact significantly more convenient than Method I, although a comparison of the lengths of the two treatments might give that impression. The brevity of Method II depends in large measure upon the results available from the preceding discussion of the same problem.

5.5.3 Turbulent Dissipation

In Section 5.3.1 the rate at which energy was extracted by turbulence from a two-dimensional parallel flow was found to be

$$-\varrho_c \overline{u'v'} \frac{du_m}{dy} \quad \text{per unit volume} \tag{5.5.22}$$

This energy is dissipated only after being transferred to small-scale turbulent motion where viscous stresses are of direct importance. Nevertheless, for practical purposes, such as calculating pipe friction, this expression represents dissipation.

The energy extraction and ultimate dissipation are seen to depend upon an interaction of the mean velocity gradient, du_m/dy, and the turbulent mixing, represented by the mean stress, $-\varrho_c \overline{u'v'}$. This is quite different from the result for laminar parallel flow, where the dissipation is

$$\mu \left(\frac{du}{dy}\right)^2 \quad \text{per unit volume} \tag{5.5.23}$$

depending only on the velocity gradient and the passive viscosity coefficient.

This difference in the expressions for turbulent and laminar dissipation is the key to understanding the two kinds of turbulent dissipation which have been discussed in this chapter.

Figure 5.25 shows these two kinds of dissipative region, a *boundary layer* on the wall of a channel and a *mixing layer* in the body of the flow. For simplicity, two-dimensional channel flow is considered, and throughout this discussion a strip of the flow of unit breadth will be examined. The maximum velocity is arbitrarily taken to be $2u$, twice the mean velocity in the undisturbed uniform flow. Table 5.3 indicates that the dissipation at a half-open gate valve is equivalent to the wall friction on a few hundred diameters of uniform pipe flow. This large dissipation is accomplished, as indicated in Figure 5.25, within a length only a few times the channel breadth. This may seem puzzling, for the velocity gradient in the mixing layer is rather smaller than that in the boundary layer. The marked difference in the rates of dissipation must then depend on values of the mixing stress $-\varrho_c\overline{u'v'}$ for the two flows.

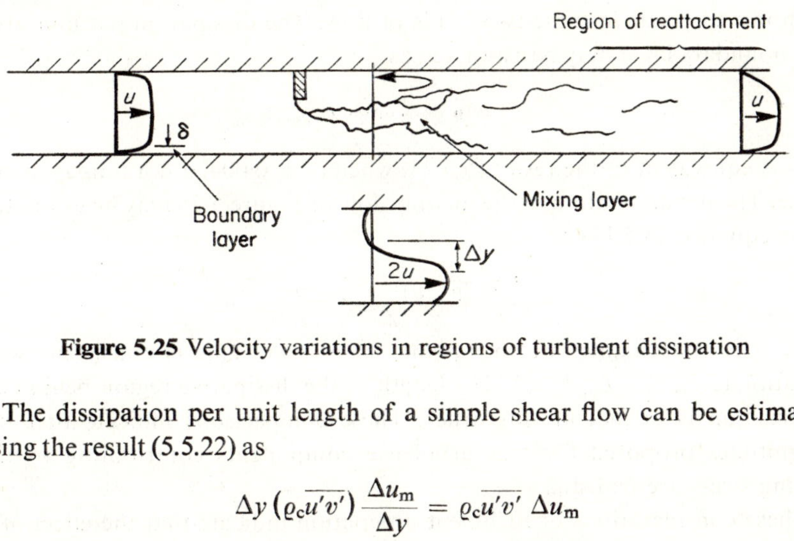

Figure 5.25 Velocity variations in regions of turbulent dissipation

The dissipation per unit length of a simple shear flow can be estimated using the result (5.5.22) as

$$\Delta y\,(\varrho_c\overline{u'v'})\,\frac{\Delta u_m}{\Delta y} = \varrho_c\overline{u'v'}\;\Delta u_m$$

where Δu_m is the change in mean velocity across the layer and Δy is its width. For the boundary layer we have $\Delta u_m \simeq u$, and may guess that the turbulence fluctuations contributing to the stress are something like

$$u' \sim \frac{\Delta u_m}{10} \simeq \frac{u}{10}, \qquad v' \sim \frac{\Delta u_m}{50} \simeq \frac{u}{50}$$

The component normal to the wall has been assigned a small value because motion in this direction is inhibited by the presence of the wall. Hence the dissipation per unit area of the boundary layer is around

$$\frac{\varrho_c u^3}{500} \tag{5.5.24}$$

For the mixing layer we have $\Delta u_m \simeq 2u$, and may guess that

$$u' \sim \frac{\Delta u_m}{8} \simeq \frac{u}{4}, \quad v' \sim \frac{\Delta u_m}{8} \simeq \frac{u}{4}$$

Here the turbulence is not inhibited by a nearby wall, and a larger mixing stress can be developed. Hence the dissipation (per unit breadth) for a mixing layer of length L_m is around

$$\frac{\varrho_c u^3 L_m}{8} \tag{5.5.25}$$

The estimates obtained above can be compared with the dissipation which is known to occur in these two kinds of flow. The dissipation per unit area of a boundary layer may be estimated as

$$u\tau_0 = \tfrac{1}{2}c_f\, \varrho_c\, u^3$$

This is equivalent to the result (5.5.24) when $c_f = 0{\cdot}004$, a not unreasonable value. The net dissipation in the mixing flow of Figure 5.25 may be estimated using equation (5.5.11) as

$$\tfrac{1}{2}\dot{m}(2u-u)^2 = \tfrac{1}{2}\varrho_c b u^3$$

Here b is the channel width, so that $\dot{m} = \varrho_c ub$. This is consistent with equation (5.5.25) if $L_m/b = 4$. The length of the dissipative region behind an obstruction is in fact of this order. These comparisons indicate that the magnitudes proposed for the turbulence components contributing to the mixing stress are realistic.

These considerations of turbulent dissipation indicate that the effect of a solid boundary on a turbulent flow is more complex than its effect on a laminar flow. In a laminar flow the dissipation is highest where the velocity gradient is highest, namely, at the wall. In a turbulent flow the stationary wall has a two-fold action; it establishes the velocity gradient necessary for dissipation but at the same time inhibits the equally important turbulent mixing. In a turbulent mixing layer, where this latter restraint is absent, very rapid dissipation can be achieved. Of course, even when the inhibition of mixing is taken into account, the dissipation in a turbulent boundary layer is very high compared to that in a laminar layer.

5.5.4 Flow under Gravity

For clarity, the role of fluid weight has been omitted from the discussion of dissipative flows. However, the weight of the fluid is of great importance for many liquid flows in piping systems, and we cannot end our consideration of pipe flows without seeing how it can be taken into account. The necessary analytical results are already available; they have only to be collected and interpreted.

One way of accounting for changes in elevation is suggested by the result

$$-\frac{\Delta(p_0+\varrho_c gz)}{\varrho_c} = \Delta e = c_v\,\Delta T = T_c\,\Delta s \tag{2.3.22}$$

This indicates that dissipation changes the sum $p_0+\varrho_c gz$, rather than the total pressure alone. Hence dissipative flows under gravity can be analysed using the results derived earlier in this chapter, with the total pressure and pressure replaced by the sums $p_0+\varrho_c gz$ and $p+\varrho_c gz$.

There is a second way of dealing with changes in elevation which makes more explicit use of the special features of constant-density flow under gravity. The energy equation for steady flow of a constant-density fluid has been written

$$\Delta\left[\frac{p-p_a}{\varrho_c g}+\frac{u^2}{2g}+z\right] = \Delta H_0 = -H_f \tag{2.3.23}$$

This defines the total head H_0 as the sum of the displacement, kinetic and potential energies, each represented as a vertical distance or head, and states that changes in the total head are equivalent to the head loss H_f. Another relevant result emerged during the discussion of the manometer in Section 2.5.1. The height of the surface in a manometer column was found to be the sum of the pressure head and the elevation at the point of measurement. This sum is the piezometric head

$$H_p = \frac{p-p_a}{\varrho_c g}+z \tag{2.5.5}$$

Note that the total and piezometric heads differ by the velocity head:

$$H_0 = H_p+\frac{u^2}{2g}$$

Hence the piezometric head provides a measure of the mechanical energy of the flow, once the local velocity head is added to it.

In the earlier discussions it was realized that

$$H_f = \frac{\Delta e}{g} > 0 \tag{2.3.24}$$

represented dissipation within the flowing fluid. Now methods are available to calculate the dissipation in typical pipeline flows and, from equations (5.5.17), the head loss is found to be

$$H_f = 4c_f \frac{L}{d}\frac{u^2}{2g} = f\frac{L}{d}\frac{u^2}{2g} \tag{5.5.26}$$

These results apply directly to pipe friction, but equations (5.5.21) allow specific losses to be represented in the same way. All of these inertia-scaled losses appear to be proportional to the velocity head $u^2/2g$. This is roughly true for many of the flows in engineering equipment, but it should be remembered that the friction and loss coefficients c_f and K may depend on the Reynolds number of the pipe flow.

The preceding results can be given an enlightening and useful graphical interpretation, best illustrated in terms of specific examples. Figure 5.26(a) shows water flowing from a tank with a free surface, through a straight inclined pipe, and discharging into the atmosphere at the end. It should be remembered that the results obtained above are forms of the energy equation for steady flow. They apply to this flow only if it is essentially steady, that is, only if the water level in the tank falls slowly.

Within the tank the fluid velocity is small and dissipation is negligible. Hence

$$H_{0_1} = \frac{p-p_a}{\varrho_c g} + z = \text{constant}$$

gives the total head within the tank and at entry to the pipe. Within the pipe dissipation occurs and the total head falls. Near the entry there is a concentrated loss, but once uniform flow is established, the rate of dissipation becomes uniform as well.

The line in the vertical plane which represents the variation of total head along the pipe is shown in Figure 5.26(a). This line is called the *energy gradient*. It is defined by equation (2.3.23) which gives

$$H_0 = H_{0_1} - H_f$$

For the linear portion, equations (5.5.26) give

$$H_0 = H_{0_1} - \left(K + \frac{4c_f x}{d}\right)\frac{u^2}{2g}$$

with x distance measured along the pipe from the entry, and K representing the additional loss near inlet.

Figure 5.26(a) also shows the *hydraulic gradient*, the line representing the

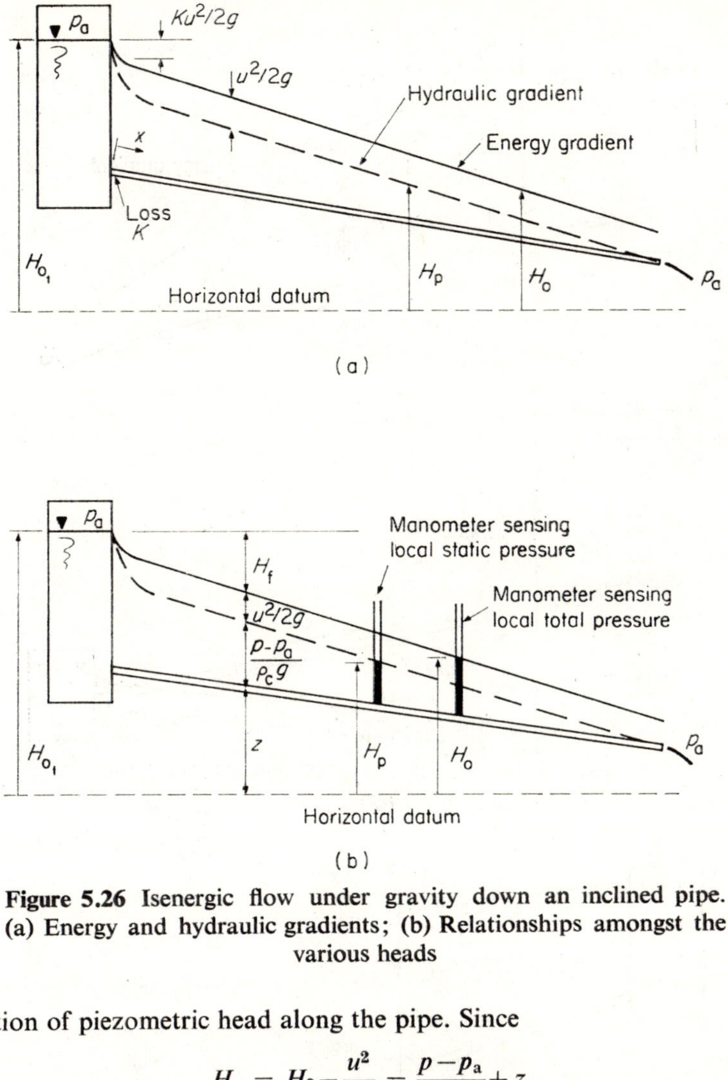

Figure 5.26 Isenergic flow under gravity down an inclined pipe. (a) Energy and hydraulic gradients; (b) Relationships amongst the various heads

variation of piezometric head along the pipe. Since

$$H_p = H_0 - \frac{u^2}{2g} = \frac{p - p_a}{\varrho_c g} + z$$

this line runs parallel to the energy gradient for the portion of the pipe where the flow is uniform. Note that it intersects the line of the pipe at the outlet end, where $p = p_a$. Near the entry the hydraulic gradient drops more abruptly than the energy gradient; the pressure experiences the additional 'loss' $u^2/2g$ as the fluid accelerates into the pipe.

The liquid in a manometer which senses the static pressure at a point in the pipe line will rise to the hydraulic gradient, as shown in Figure 5.26(b). In a manometer connected to a Pitot tube and sensing the total pressure

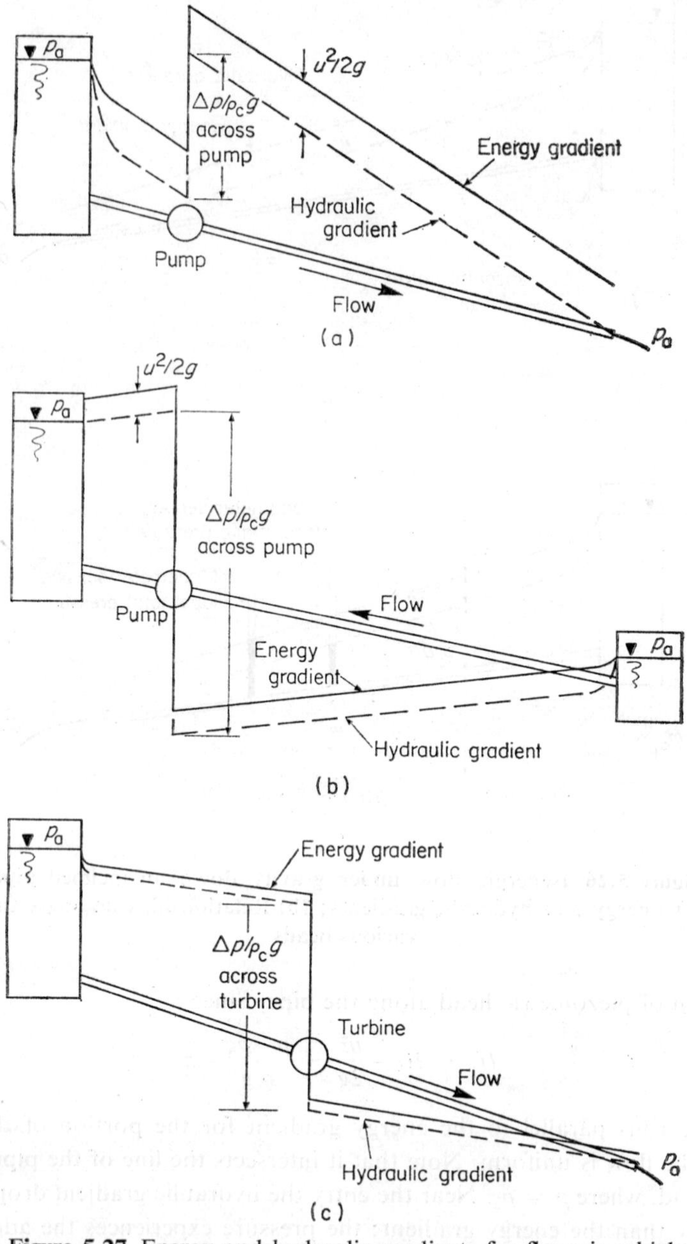

Figure 5.27 Energy and hydraulic gradients for flows in which work is done on and by the flowing fluid. (a) Downwards flow assisted by a pump; (b) Upwards flow under the action of a pump; (c) Downwards flow through a turbine

corresponding to the mean velocity of the flow, the liquid will rise to the energy gradient above the point at which the measurement is made. Figure 5.26(b) indicates also how the vertical distances representing kinetic energy $u^2/2g$, displacement or pressure energy $(p-p_a)/\varrho_c g$, potential energy z and dissipated energy H_f add graphically to give the initial total head H_{0_1}. The vertical distance between the hydraulic gradient and the line of the pipe represents the gauge pressure in the pipe, $p-p_a$. The pressure is above atmospheric when the pipe lies below the hydraulic gradient.

This graphical interpretation can be extended to cases in which work is done of the flowing fluid, so that its total head rises. Figure 5.27(a) shows the energy and hydraulic gradients which would result if a pump were installed to aid gravity in drawing water from the tank. Figure 5.27(b) indicates the energy and pressure distributions when the pump forces the liquid up the pipe into the tank. Finally, Figure 5.27(c) shows the energy gradient with a turbine installed in the pipe line and extracting mechanical energy from the flow. This represents the flow through a hydroelectric power station. The water flows from a reservoir to the turbine and is discharged into the river below the station with a finite velocity.

The graphical techniques illustrated here are useful as aids to computation, but are even more helpful in providing a clear picture of the interaction of the several energy components. Diagrams representing energy and hydraulic gradients can also be drawn for liquid flows in horizontal pipes and, with some modifications, for gas flows, in which changes in elevation are unimportant. However, it is for cases in which potential-energy changes are significant that the graphical methods prove particularly helpful.

Example 5.12. Figure 5.28 shows a pipe line carrying water from an outlying reservoir to a second reservoir near a city; the vertical scale is greatly exaggerated. The pipe, of diameter 4 ft, is 12 miles long and consists of three nearly straight sections, rising for the first half mile, then falling, and finally rising again. The pipe friction is specified by $f = 0{\cdot}02$. It is required to find the flow through the pipe and the maximum and minimum pressures in the line.

In this long, fairly straight line the velocity head and minor losses will be small in comparison with the uniformly distributed friction. Hence the energy and hydraulic gradients are nearly coincident. The pressure distribution can be found by drawing the straight hydraulic gradient between the levels H_{0_1} and H_{0_2} at the ends of the line. The maximum and minimum gauge pressures in the pipe could be read from a careful scale drawing of the system, as the maximum distance which the line of the pipe falls below the grade line and rises above it. Alternatively, they can be calculated once the velocity head is known.

The flow rate through the line is determined by

$$H_{0_1}-H_{0_2} = H_f = f\frac{L}{d}\frac{u^2}{2g}$$

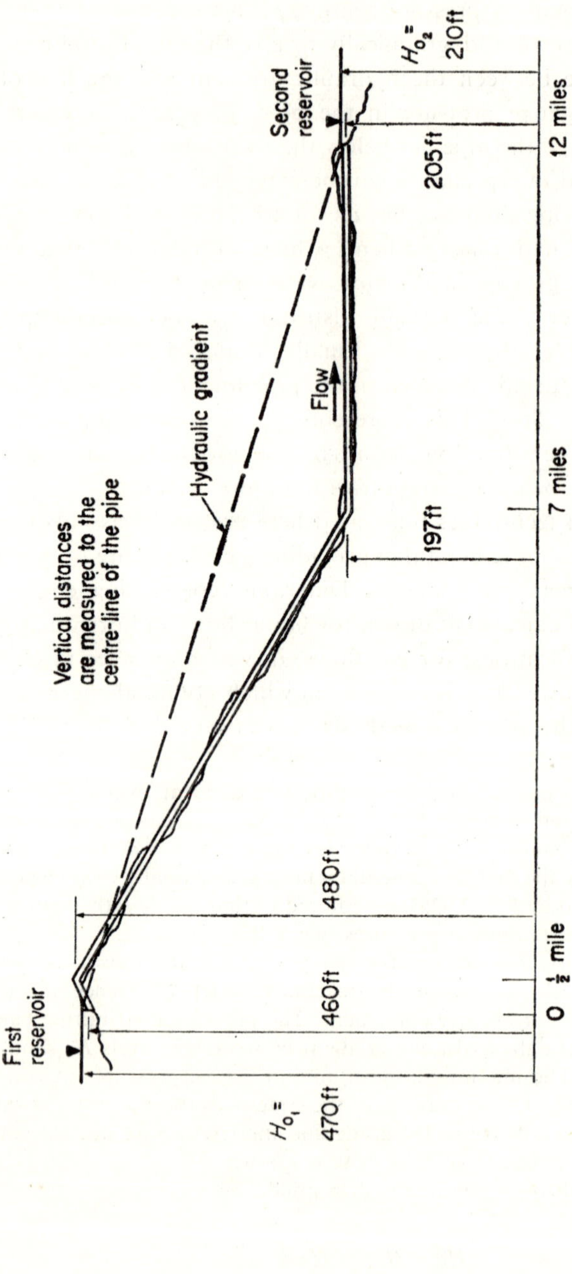

Figure 5.28 A pipe line between two reservoirs

Note that the elevations of the submerged entrance and exit are irrelevant, since the total head is uniform in each of the reservoirs. Then

$$\frac{u^2}{2g} = \frac{470-210}{0{\cdot}02\times 3\times 5280} = 0{\cdot}821 \text{ ft}$$

As was expected, the minor losses, which will be a few times the velocity head, are virtually negligible in this case. The velocity and flow rate are found to be

$$u = 7{\cdot}27 \text{ ft/sec} \quad \text{and} \quad \dot{V} = uA = 91{\cdot}4 \text{ ft}^3\text{/sec}$$

Since the velocity head is small, the variation in total head along the pipe is given by

$$H_0 = H_{0_1} - f\frac{x}{d}\frac{u^2}{2g} = H_{0_1} - (H_{0_1} - H_{0_2})\frac{x}{L}$$

and the pressure distribution by

$$\frac{p-p_a}{\varrho_c g} = H_0 - z$$

$$= (H_{0_1} - z) - (H_{0_1} - H_{0_2})\frac{x}{L}$$

For the situation shown in Figure 5.28

$$\frac{p-p_a}{\varrho_c g} = (470-z) - \frac{260}{12} x \text{ ft}$$

with z in ft and x in miles.

The positions of maximum and minimum pressures are evident from Figure 5.28. For $z = 480$ ft, $x = \frac{1}{2}$ mile

$$\frac{p-p_a}{\varrho_c g} = -10 - 10{\cdot}8 = -20{\cdot}8 \text{ ft}$$

For $z = 197$ ft, $x = 7$ miles

$$\frac{p-p_a}{\varrho_c g} = 273 - 152 = 121 \text{ ft}$$

Since the pipe radius is 2 ft, the minimum and maximum pressures are $-22{\cdot}8$ ft and 123 ft of water. A further adjustment of the order of 1 ft of water could be made to account for non-uniformity of losses along the pipe line.

These figures are vital criteria for the design of the pipe line. The pipe must be capable of withstanding both the inwards and outwards pressure differentials. Moreover, the magnitude of the former differential, considerably less than $p_a/\varrho_c g =$ 34 ft, indicates that a continuous flow is possible. If the absolute pressure were to drop too near the vapour pressure of the water, large quantities of air would evolve from the water and the syphon action would break down.

PROBLEMS FOR CHAPTER 5

Group A

5.1 A horizontal shaft of diameter 15 cm turns at 3000 rev/min in a partial journal bearing which surrounds only the bottom 180° of the shaft; the bearing is 20 cm wide. The load applied to the bearing is relatively small, so that the thickness of the lubricant film may be taken to be uniform at 0·1 mm, and side leakage may be neglected. The viscosity of the lubricant is 5 cP at the temperature at which the fluid is maintained.

(a) What is the torque required to overcome friction in the bearing?

(b) What power is dissipated in the bearing?

(c) At what rate must the lubricant be supplied to maintain the film?

(d) The specific heat of the lubricating oil is 0·6 kcal/kg °C, and its specific gravity is 0·9. Estimate the temperature rise which it experiences between entering the film and leaving it. Set down clearly any assumptions you make.

(e) Is the assumption of constant viscosity realistic?

5.2 In injecting a drug into the vein of a patient, a nurse applies a force of 1 lbf to the plunger of the syringe, whose barrel has cross-sectional area $\frac{1}{4}$ in^2. The pressure in the vein into which the fluid (mostly water) is injected is 9 in of water gauge. The hypodermic tube through which the fluid passes is 5 cm long and has inner diameter $\frac{1}{2}$ mm.

(a) Estimate the flow rate on the assumption of laminar flow.

(b) How long does it take to inject 10 cm^3 of fluid?

(c) Is the flow in fact likely to be laminar?

5.3 The constant 0·00231 appears in Callendar's equation written in the units of the SI steam tables. What are the units of this constant? What value does it adopt in a consistent set of units?

5.4 The speed of propagation of the crests of a train of water waves, the wave celerity c, is very little influenced by the viscosity of the water or by the air above the surface. Hence it may be expected to depend upon the density and specific weight of the fluid (ϱ and ϱg), on the wavelength λ, and on the depth of water d.

(a) Show that

$$\frac{c^2}{g\lambda} = f\left(\frac{d}{\lambda}\right)$$

(b) How will c depend on wavelength for very short ripples, for which the significant forces are dependent on the surface tension, $S = [F/L]$, and on fluid density?

5.5 Consider the dimensionless representation of the relationship between flow rate $\dot{V}$ and pressure rise Δp for a pump. The pump characteristic may be expected to involve the fluid density and viscosity, and also the speed of rotation N and rotor diameter D.

(a) Show that D and ND provide a suitable characteristic length and velocity for a family of similar pumps.

(b) Use these quantities to render the flow rate and pressure rise dimensionless, and write down the dimensionless flow law for the pump.

(c) Assuming the effect of Reynolds number to be negligible, and considering operation at the point of maximum efficiency, which corresponds to particular values of the flow and pressure parameters, find how much the flow rate and pressure rise change

(1) when the same pump runs at twice the speed, and
(2) when a geometrically similar pump of twice the size runs at the original speed.

5.6 For the flow discussed in Example 5.6, show that $\dot{m}/d\mu_0$ is a suitable Reynolds number for the duct ($\dot{m}$ is the mass flow through it and d its diameter). Hence, using the ratio $\dot{m}_m/\dot{m}_p$ obtained in Example 4.11, find the ratio of the Reynolds numbers for the model and prototype flows. Compare with that found in Example 5.6.

5.7 The drag of a wing-tip fuel tank for a light aeroplane is to be determined by model tests. The aeroplane is to cruise at 300 miles/hr at 15,000 ft. It is decided that the Reynolds number for the tests should be kept as close as possible to that for the prototype in this flight condition.

(a) For a one-third-scale model, what velocity would be required in a wind tunnel through which flows normal atmospheric air?

(b) For a one-tenth-scale model, what velocity would be required in a pressurized wind tunnel in which the pressure level is four atmospheres?

(c) For a one-tenth-scale model, what velocity would be required in a water tunnel?

(d) What fraction of the prototype drag are the drag forces generated in the tests (a), (b) and (c)?

5.8 Estimate the Reynolds number typical of each of the following flows, and state whether you would expect the flow to be turbulent:

(a) large and small blood vessels in the human body,
(b) the intake of a jet engine,
(c) the pipes of a household plumbing system,
(d) the ducts of the air-conditioning system of a building,
(e) the duct carrying water to the turbine of a hydroelectric power station,
(f) a tidal estuary,
(g) the gutter beside the road after a rainstorm,
(h) the tube supplying water to the washing spray for a car's windscreen, and
(i) the fuel line to an automobile engine.

5.9 Express equation (5.3.8) in terms of dimensionless groups. Is the result consistent with the more general (5.3.7)?

5.10 Assuming that the friction stress on the vertical side walls is the same as that on the channel bottom, make a more accurate estimate of the water depth for the situation described in Example 5.7.

5.11 In a domestic heating system a drawn copper tube of diameter 1 cm and length 30 m carries hot water to and from a radiator. The temperature is 80 °C in the outgoing line (15 m) and 50 °C in the return line (15 m).

(a) Taking the tube to be smooth-walled, and neglecting minor losses, prepare a plot showing how the pressure difference required to drive the flow depends on the flow rate.

(b) If heat is to be supplied to the radiator at the rate of 10,000 Btu/hr, what driving pressure is required? What power is dissipated in the line?

5.12 Consider the performance of a jet pump to which water is supplied through a central pipe of diameter 1 in with velocity 25 ft/sec. The surrounding pipe through which the pumped water flows has uniform diameter 3 in.

(a) Show that

$$\frac{\Delta p}{\frac{1}{2}\varrho_c u_j^2} = 2\frac{A_1}{A_2}\frac{A_j}{A_2}\left(1 - \frac{\dot{V}_1 A_j}{\dot{V}_j A_1}\right)^2$$

relates the pressure rise and the flow rate $\dot{V}_1$.

(b) Plot Δp against $\dot{V}_1$ for the range $\Delta p,\ \dot{V}_1 > 0$.

(c) What happens outside this range?

5.13 Calculate the shear stress, friction coefficient and friction factor for the flows of Examples 5.7 and 5.8.

5.14 Find the Reynolds number and friction stress on the wall of the pipe considered in Example 5.11.

5.15 An air-conditioning duct of cross-section 1 ft $\times$ 2 ft carries a flow of 2500 ft^3/min of air at normal temperature and pressure. The equivalent sand-grain roughness of the duct wall is 0·03 in. Find the pressure drop (expressed in inches of water gauge) over 100 ft of the ducting, using the result (5.5.9) to estimate the friction factor.

5.16 The air supporting a hovercraft flows from a pressurized bag around the craft periphery into the plenum chamber or air cushion beneath the craft through a series of holes, as shown in Figure 5.29.

(a) It is desired to express the drop in pressure between the bag and the cushion as $\Delta p = K(\frac{1}{2}\varrho u^2)$, where u is the hypothetical velocity at the orifice $(\dot{V}/A)$. Show

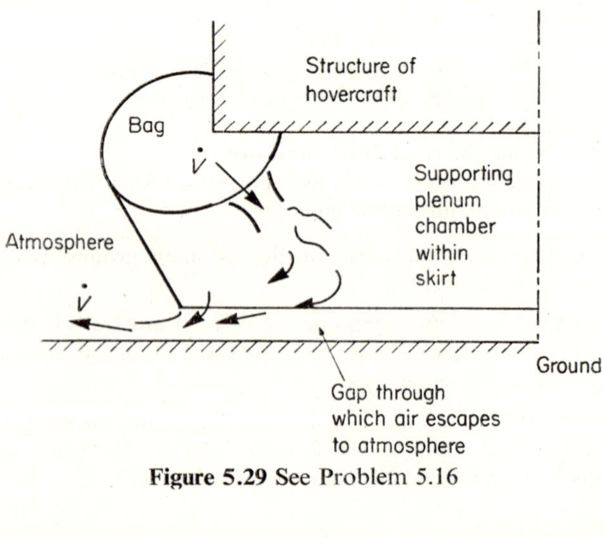

Figure 5.29 See Problem 5.16

that the loss coefficient must be taken to be $1/C^2$, where C is the contraction coefficient of the jet emerging from the holes.

(b) If the total hole area is 7 ft^2, and the contraction coefficient is 0·6, find the pressure drop for a volume flow of 5000 ft^3/min. The density of the air is little different from that in the atmosphere outside.

(c) The pressure difference between the cushion and the atmosphere is maintained by an essentially similar process. If the skirt perimeter is 40 ft, find the pressure difference between the plenum and the atmosphere when the gap between skirt hem and ground is 1 in. Use the same contraction coefficient.

(d) If the plenum plan area is 200 ft^2, what is the craft weight?

5.17 What is the loss coefficient, based on a higher velocity of 30 ft/sec, for the situation described in Problem 4.17?

5.18 Find the loss coefficients based on the inlet and outlet velocities for the flow discussed in Examples 2.4 and 2.9.

5.19 Reconsider the pipe flow of Example 5.11, changed only in that the line is no longer horizontal.

(a) What pressure difference is required to generate the flow of $\frac{1}{2}$ ft^3/sec if the 30-ft length of pipe following the elbow goes (1) vertically down, (2) vertically up, (3) upwards at an angle of 30°?

(b) What is the rate of dissipation in these three flows?

(c) Draw the line (3) to scale, and show the hydraulic and energy gradients along it.

5.20 At what rate is energy dissipated in the flows discussed in Examples 5.1, 5.2, 5.7, 5.8, 5.9, 5.11, and 5.12?

Group B

5.21 (a) Show that the formula for viscosity obtained in Example 5.1 becomes

$$\mu = \frac{L}{2\pi b u_1} \frac{\Delta r}{r^2}$$

when the gap between the cylinders is much less than the mean radius.

(b) How can this result be derived more easily?

(c) In Example 5.1 no correction has been made to account for non-uniformity of the flow near the bottom and top of the annular region occupied by the fluid. Will this result in a prediction of viscosity which is too high or too low?

5.22 A hydrostatic bearing operates without relative motion of the surfaces separated by the lubricant film. Consider the two-dimensional situation sketched in Figure 5.30. The fluid is supplied through the central slot and flows out through the films on either side.

(a) Show that the load supported (per unit length of bearing) is related to the rate at which fluid is supplied and to the film geometry by

$$F = \frac{6\mu L^2}{b^3} \dot{v}$$

where $\dot{v}$ is the total flow per unit length of the bearing.

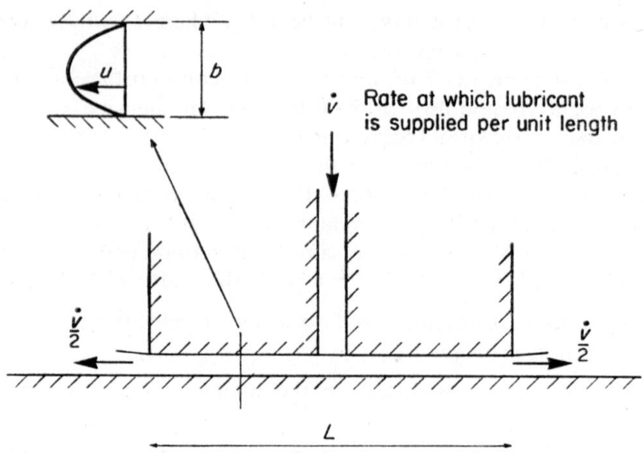

Figure 5.30 See Problem 5.22

(b) At what rate is energy dissipated in this flow?

(c) Pose two realistic cases, using oil and air as the supporting fluids, and calculate the volume flow and dissipation rate for them.

5.23 Equation (5.1.11) can be used to find the pressure distribution in a two-dimensional hydrodynamic bearing film. In such a bearing the lubricant film separates two solid surfaces which are in relative motion; for two-dimensional flow, $\dot{v}$ is uniform along the flow.

(a) Show that

$$\frac{d^2p}{dx^2} = \frac{6\mu U}{b^3}\frac{db}{dx}$$

at the point where $dp/dx = 0$.

(b) In order that a positive pressure be generated to hold the bearing surfaces apart, the lubricant must be drawn into a narrowing channel. Use the result (a) to support this statement.

5.24 Waves of period 3 sec wash up onto a beach. Just offshore, where the depth is nearly constant, this process results in a flow of water back and forth towards the beach. Taking the amplitude of the velocity variation to be 0·1 m/sec, use the results of Section 5.1.3 to estimate the thickness of the boundary layer on the bottom.

5.25 Multiplication of equation (5.1.12) by the velocity u gives an energy balance for a unidirectional, axisymmetric flow. Separate out the shear work and dissipation terms as in equation (5.1.23). For laminar flow, how is the dissipation distributed across the tube?

5.26 In Problem 5.4 the general form of the dependence of the celerity of water waves on wavelength and water depth is indicated. In shallow and deep water ($d/\lambda \to 0$ and ∞) simpler results can be found, on the basis of the observations that the speed is independent of wavelength in the former case, and of the water depth in the latter.

(a) How does celerity depend on wavelength and depth in the two limiting cases?

(b) Make a rough sketch of the function $f(d/\lambda)$ of Problem 5.4. In what ranges would you expect the simple results of (a) to apply?

(c) Use the results of (a) to find how the wave period depends on the wavelength in deep and shallow water.

(d) Compare these results with the wave speeds given in equations (3.2.7) and Section 4.3.4.

5.27 A smooth-crested weir at whose crest the flow becomes critical (see Figure 4.15) is used to meter the flow in an open channel. Apply dimensional analysis to the two-dimensional flow over the weir, assuming that the flow rate $\dot{V}$ depends on the breadth of the channel b, the elevation of the crest above the stream bed z_c, the depth of the water upstream d_1, and the acceleration of gravity g.

(a) Will any of these quantities necessarily be linked together in the flow law?

(b) How many dimensionless groups will appear in the law?

(c) Obtain the law in its general form, and show that this result is consistent with the theoretical prediction of Section 4.3.4.

5.28 The flow of heat from a heat-exchanger tube ($\dot{q}$ per unit area = [F/LT]) to the fluid passing along the tube may be supposed to depend upon the tube diameter d, the temperature difference between the body of the fluid and the tube wall ΔT, the thermal conductivity of the fluid k = [F/Tθ], the specific heat of the fluid c_p, and the diffusivities of momentum and heat in the fluid, ν and $\varkappa$, both with [L^2/T].

Use one of Methods II, III and IV to find the dimensionless groups involved in this relationship.

(a) Use k, ΔT, c_p, and ν to cancel the dimensions of each of the remaining quantities in turn.

(b) Use d, u, k, ΔT in the same way.

(c) Replace the conductivity k in the list of defining quantities by the fluid density ϱ, and use d, u, ϱ, ΔT to cancel the dimensions of the remaining quantities.

(d) Show that each set of results is equivalent to

$$\frac{\dot{q}}{\varrho c_p\, \Delta T\, u} = f\left(\frac{\nu}{\varkappa}, \frac{u^2}{c_p\, \Delta T}, \frac{ud}{\nu}\right)$$

(e) Try to give a physical interpretation to each of the groups of (a), (b), (c) and (d).

5.29 The equations (5.5.8, 9) can be used to estimate the Reynolds number at which a pipe becomes effectively rough.

(a) Show that this point is given by

$$\mathrm{Re} = e^{1\cdot 2}\frac{d}{k}\left[2{\cdot}5 \log_e \frac{d}{2k} + 4{\cdot}75\right]$$

(b) Find this Reynolds number for $k/d = 10^{-2}, 10^{-3}, 10^{-4}$. For the roughness distributions usually encountered, the transition from smooth-wall to rough-wall flow will extend over a range of Reynolds numbers from perhaps one-tenth to ten times that found here.

5.30 (a) Given the result (5.4.8), what rate of dissipation would you expect in the mixing of a flow which consisted initially of three parallel streams with differing velocities?

(b) Show that the rate of dissipation resulting from mixing within a flow across which the velocity is initially non-uniform is

$$\frac{\varrho_c}{2}\int_A u(u-\bar{u})^2 \, dA$$

where $\bar{u}$ is the ultimate uniform velocity.

5.31 It is proposed to replace the rectangular duct of Problem 5.15 by a smooth-walled tube which must carry the same air flow. The smooth-wall result (5.5.8) may be used to find the friction.

(a) What is the pressure drop over 100 ft of a round tube which just fits into the space available, that is, with a diameter of 1 ft?

(b) What is the pressure drop over 100 ft of a tube of elliptical section which just fits into the space available, that is, with major and minor axes of 2 ft and 1 ft?

(c) Comment on the practicality of the three alternative ducts.

5.32 (a) Find the Reynolds number and friction factor for the pipe flow described in Example 2.6.

(b) Compare with $f = 0{\cdot}24\ \mathrm{Re}^{-0{\cdot}173}$ which is valid for $10^5 < \mathrm{Re} < 10^7$. Does the prescribed pressure drop seem realistic?

(c) Make a more realistic estimate of the pressure at the bottom of the pipe, taking the inlet pressure and velocity to be unchanged.

5.33 Using the result (5.5.11), show that an abrupt expansion in a duct will diffuse a constant-density flow with efficiency

$$\eta_D = \frac{2}{1+A_2/A_1}$$

where A_1 and A_2 are the upstream and downstream flow areas. Use this result to estimate the limiting efficiency for Figure 4.18(b). It should always be possible to design a tapering diffuser which will do better than this.

5.34 Measurements show that the loss coefficient for a gate valve depends on the degree of closure in the following way:

	Fully open	3/4 open	1/2 open	1/4 open
K	0·2	1·2	5·6	24·0

See how these values compare with the predictions of equation (5.5.12), which gives the loss coefficient at a constriction in a uniform line. A meaningful comparison will require that account be taken of the continued contraction of the flow after passing beneath the sharp-lipped gate. Do this by assuming the minimum area of the flow to be 2/3 of the area of the opening.

5.35 (a) Draw an accurate scale diagram of the situation described in Example 5.12 and from it read the maximum and minimum pressures in the line.

(b) Imagine that a control valve is located in the line at the mid-point of one of the straight sections. Draw the hydraulic gradients for each of the three possible locations of the valve, for the case in which the pressure drop across it is 50 ft of water gauge. Note the maximum and minimum pressures.

(c) Where should the valve be located?

5.36 It is required to estimate the flow which can be carried away by the overflow system of a bath tub. The water surface must not be more than 180 cm above the outlet into the atmosphere. The water flows from the bath smoothly into a pipe of diameter 3 cm; this pipe is 2 m long and contains two elbows. The flow then expands abruptly into a straight pipe of diameter 4 cm and length 2 m, from which the water flows freely into the air. It may be assumed that the walls of the pipes are fully rough with relative roughness $k/d = 0{\cdot}02$; equation (5.5.9) gives the friction coefficient.

Group C

5.37 Consider the performance of an axisymmetric hydrostatic bearing, that is, one with a flat round bearing pad, supplied from a central inlet.

(a) Show that

$$F \propto \frac{\mu R^2}{b^3} \dot{V}$$

relates flow rate $\dot{V}$, load F, and bearing geometry (R and b are the pad radius and film thickness).

(b) Find the constant of proportionality by determining the pressure distribution and integrating it.

5.38 (a) Show that the pressure drop in laminar flow through a tapering tube ($r \propto x$) is given by

$$\Delta p = \frac{8\pi \dot{V} L}{3\mu(R_2 - R_1)} \left(\frac{1}{R_1^3} - \frac{1}{R_2^3} \right)$$

where L is the length of the tube, and R_1 and R_2 are the maximum and minimum radii.

(b) The pressure drop is the same whichever direction the flow takes. Does this mean that the flow is reversible?

(c) Show that the result of (a) reduces to the constant-area result as $R_2 \to R_1$.

(d) Under what conditions is the result of (a) likely to be applicable?

5.39 The general solution for axisymmetric flow (5.1.14) can be used to describe several kinds of flow, not merely the Poiseuille case (5.1.15). Consider now

(1) flow under pressure through an annulus with fixed inner and outer cylindrical walls (this kind of flow is found in certain heat exchangers); and

(2) flow under gravity down the outside of a cylindrical, vertical rod (this kind of flow occurs in certain mass-transfer and stirring processes).

(a) What boundary conditions should be applied at the inner and outer boundaries of these flows?

(b) For vertical flow under gravity, show that $-\varrho g$ should replace $\mathrm{d}p/\mathrm{d}x$ in equations (5.1.12, 13, 14).

(c) For the flows (1) and (2) obtain expressions for the velocity distribution and flow rate analogous to (5.1.15, 16).

(d) For the limit in which the flowing film is very thin, show that the results are equivalent to those for plane Poiseuille flow.

(e) Examine the other limit, that in which the inner radius approaches zero.

5.40 The non-steady solution (5.1.19) can be obtained fairly easily if one can postulate its structure, either through familiarity with solutions of equations like (5.1.18) or through an understanding of the physical situation.

(a) Show that a solution of the form

$$U \exp(ay)\cos(\omega t - by)$$

will be suitable if $2\mu ab = -\varrho\omega$ and $a^2 = b^2$.

(b) Put forward reasons for adopting the proposed form of solution.

(c) Show that equation (5.1.19) gives an appropriate solution.

5.41 A solution of the form (5.2.11) for equation (5.1.18) can be found by using a very powerful technique for the solution of partial differential equations—conversion to an ordinary differential equation. The result (5.2.11) shows that this can be done, since it indicates that there are solutions which depend on the single variable $\eta = y/(\nu t)^{1/2}$.

(a) Express the derivatives which occur in equation (5.1.18) in terms of $F(\eta) = u/U$, and hence show that the appropriate ordinary differential equation is

$$F'' = -\frac{\eta F'}{2}$$

where the primes indicate differentiation with respect to η.

(b) By integrating twice, show that the solution is of the form

$$F = A\int_0^\eta \exp\left(-\frac{s^2}{4}\right) \mathrm{d}s + B$$

This solution contains the error function, whose values are given in many sets of mathematical tables, and can therefore be evaluated easily.

(c) What values must the constants A and B have, if the motion is to satisfy the condition (5.2.10) and the requirement that the disturbance be small far from the moving plate?

(d) Plot u/U against η using error-function tables. How does the boundary-layer thickness grow in time?

5.42 The effective viscosity ε associated with turbulence is defined by

$$\varrho\varepsilon \frac{\mathrm{d}u_{\mathrm{m}}}{\mathrm{d}y} = \tau$$

This apparent kinematic viscosity is often referred to as the eddy viscosity. Sometimes, particularly in dealing with free-turbulent flows, such as wakes, jets and mixing layers, it is fairly realistic to take the eddy viscosity to be uniform across the flow. This is not so for channel flows, as is shown below.

(a) Using equations (5.1.6) and (5.3.13), show that

$$\varepsilon = Ku_{\mathrm{f}}y\left(1 - \frac{2y}{b}\right)$$

In what part of the channel does this result apply? Will ε be over-estimated or under-estimated in other parts of the flow? Make an estimate of ε_{max} in terms of K, u_f, and b.

(b) Show that $\varepsilon/\nu \simeq K\eta$ near the wall. Is this result consistent with the empirical observation that the viscous stresses are unimportant for $\eta > 70$?

5.43 (a) Use the results of the preceding problem to estimate the maximum eddy viscosity in the flows discussed in Examples 5.7 and 5.11.

(b) Find the ratio of eddy and kinematic viscosities for these two cases.

5.44 (a) Show that the rate at which energy is extracted from the mean flow ($\tau\, du_m/dy$) is given by τ_0^2/μ in the viscous sublayer, and by $(\tau_0^2/\mu)/K\eta$ in the fully turbulent part of the wall layer.

(b) Assume that the wall layer consists of two portions, a viscous layer and a logarithmic layer, the two meeting at η_m. Use the results of (a) to show that the total dissipation in the wall layer is

$$\tau_0 u_f \left[\eta_m + \frac{1}{K} \log_e \frac{\eta_c}{\eta_m} \right]$$

where $\eta_c = u_f b/2\nu$.

(c) By integrating the logarithmic law to find the volume flow, show that the dissipation per unit length of turbulent channel flow is

$$\frac{2\tau_0}{b} \dot{v} = 2\tau_0 u_f (2{\cdot}5 \log_e \eta_c + 3)$$

(d) Equate the results (b) and (c) to obtain an estimate of the thickness of the sublayer.

(e) Plot the hypothetical distribution of dissipation. Sketch a more realistic distribution.

5.45 The loss in total pressure in a conical diffuser, $-\Delta p_0$, may be supposed to depend upon the density of the fluid, the inlet and outlet velocities, u_1 and u_2, and the angle of expansion θ.

(a) Show that the losses can be represented by a law of the form

$$K' = f\left(\theta, \frac{A_2}{A_1}\right)$$

where

$$K' = \frac{-\Delta p_0}{\frac{1}{2}\varrho_c (u_1 - u_2)^2}$$

is the loss coefficient used in Figure 4.18(a).

(b) What is the value of this loss coefficient for an abrupt expansion, for which the diffuser efficiency has been seen in Problem 5.33 to be $2/(1+A_2/A_1)$?

(c) In the light of the result (b), comment on the fact that the performance of conical diffusers can be represented with fair accuracy by a single curve $K' = f(\theta)$, as shown in Figure 4.18(a). (Actually there is a range of expansion angles, between 30° and 100°, where the characteristics of diffusers with various expansion ratios A_2/A_1 do not collapse. Moreover, the losses for small values of θ depend on the wall friction. Nevertheless, the parameter K' is a useful tool in the presentation of diffuser performance.)

5.46 Two fully rough pipes run in parallel between two tanks, the pressures at each end thus being the same.

(a) Show that the ratio of the volume flows through these pipes is $(d_1/d_2)^5$.
(b) What are the ratios of friction stresses and Reynolds numbers?
(c) Find these ratios for smooth-walled pipes, for which $c_f \propto \mathrm{Re}^{-1/4}$.

5.47 In Example 2.12 the maximum flow through a cavitation test rig was determined on the assumption that friction in the supply line is negligible.

(a) Estimate the maximum flow taking into account the losses in the system: 27 m of pipe between sections 1 and 4, an elbow, a fully-open butterfly valve, and a diffuser with $A_2/A_1 = 2$, $\theta = 20°$.
(b) Is cavitation likely at the inlet section for the maximum flow condition?
(c) Draw the energy and hydraulic gradients for the line, plotting the heads in the vertical part of the line horizontally, and those in the horizontal part vertically.

5.48 A test programme for the development of a sodium-cooled nuclear reactor is to be specified. It is desired to use less exotic fluids for certain of the tests, so that the flow may be observed and measured more easily. Since the density of the heat-transfer fluid is fairly constant, tests at constant density without heat addition will be appropriate, so long as the Reynolds number can be represented with sufficient accuracy and the Mach number remains reasonably low. The maximum velocity in the sodium flow in the reactor is 20 ft/sec.

Tests are to be carried out on three components:

(1) the active core, which is 5 ft in diameter and 6 ft in height,
(2) one of the tube bundles making up the core, each of which contains several hundreds of fuel pins and is 2 ft in diameter;
(3) the outer flow path including the six heat exchangers, the whole being 40 ft in height and 40 ft in diameter; there is a degree of symmetry, so that only one third of the reactor need be modelled.

There are two, possibly conflicting geometrical requirements: it is desirable to introduce instruments into the flow passages, and the laboratory space available dictates that no one test rig extend beyond a fifty-foot cube.

(a) For various parts of the test programme, it is proposed to make use of

(1) atmospheric air,
(2) water at room temperature,
(3) carbon dioxide gas at around room temperature (say, 300 °K) at pressures up to 20 atm.

Comment on the utility of these fluids for tests on the three components. The properties of sodium in the reactor may be taken as: density 826 kg/m^3, viscosity 0·23 mN sec/m^2. For carbon dioxide at the test temperature, the viscosity is 15 μN sec/m^2.

(b) There is a free surface of sodium (with argon above to minimize chemical reactions there) at the top of the reactor (component 3 above), and certain of the tests must represent this. Hence water must be used for these tests. It is desirable to model the ratio V^2/gL, the Froude number relating inertia and gravity forces. Can this requirement be reconciled with Reynolds-number similarity?

6

Flow with Energy Addition

In Chapter 2 the concept of isenergic flow was introduced; subsequent chapters have concentrated on this important special case. Now the scope of the discussion will be broadened to include the effects of energy addition. Three methods are commonly used to add energy to a steadily flowing fluid: the burning of fuel in the stream, work interaction between the fluid and moving elements of machines such as pumps and turbines, and heat transfer between the fluid and adjacent solid surfaces. For the last two, the energy addition can be negative, the energy of the fluid being reduced. In the energy equation for steady, one-dimensional flow these three processes have the same effect—a non-zero term for the external energy addition. However, the response of the flow to this added energy depends also on the force applied to the fluid and on the dissipation involved in the process. Hence this discussion of energy addition to a one-dimensional flow has two fundamental aims:

(1) to find ways of specifying the energy flow for the three transfer mechanisms, and
(2) to determine the response of the flowing fluid to this energy addition.

Combustion differs in several important ways from the other two processes of energy transfer. Usually it involves a change in the mass flow of the stream, though only by a few per cent. More significant is the fundamental alteration in the nature of the fluid by contamination with combustion products and by a very large increase in temperature. The alteration is most dramatic in the actual region of combustion, with local departures from equilibrium and varied products of dissociation and ionization. Finally, combustion differs from the other modes of energy interchange in being restricted, almost always, to the particular case of a positive energy flow to a stream of air. This limited class of processes at present provides most of the energy used by mankind, through the burning of fossil fuels in furnaces, cylinders of internal combustion engines and combustion chambers of jet engines. The only significant alternative sources are nuclear power and hydroelectric power; the former is growing while the latter is declining in relative importance.

We shall not discuss the details of combustion in a flowing fluid; this would involve a study of the chemical reactions between fuels and oxidants and of the propagation of flame fronts into the unburnt reagents. Here we shall attempt only to estimate the magnitude of the energy addition which can be achieved by burning fuel and to predict the net changes in fluid properties which follow this energy release.

The second major means of energy addition is *work in flow machines*. The interaction of the fluid with the moving elements of pumps and turbines will be discussed in some detail in the next chapter. Here we shall consider the overall effect on the flow, treating the machine as a 'black box'. It is not enough to state merely the quantity of energy added or removed within the machine. The net change in the fluid properties between entry and exit depends also on the dissipation which occurs within the machine, for this determines the distribution of the transferred energy between mechanical energy and internal energy of the fluid. A specified input of energy to a water pump, for example, will normally act to increase the pressure of the water passing through the pump, but may, when the pump is operating very inefficiently, act only to increase slightly the temperature of the water. A variety of efficiencies are used to relate the useful output of a flow machine to the energy expended in operating it; these provide the additional information required to determine the response of the flow to the energy transfer. Efficiencies will be introduced for constant-density and variable-density flows through pumps, compressors and turbines, and the discussion will be extended to include propulsive devices such as propellers and jet engines.

The third mode of energy addition, *heat transfer*, will occupy the larger part of this chapter. There are two reasons for this emphasis: first, in many cases the energy transfer involves the interaction of fluid-mechanical and thermodynamic processes which is the main theme of this book; secondly, the energy transfer is often accomplished by passing the fluid through long tubes; here the flow is essentially one-dimensional and its development can be analysed using methods developed earlier. Typical heat-transfer processes are those within a boiler. Heat flows from the furnace gases (themselves heated by combustion) through the walls of the boiler tubes and into the water flowing within the tubes. The tube wall offers little resistance to the heat flow, and this resistance can in any case be predicted fairly easily. Thus the main problems of the heat transfer are associated with the fluid flows outside and inside the boiler tubes.

The transfer of energy between a stationary solid surface and the fluid flowing nearby is effected by small-scale mixing within the fluid—the same process which, through momentum transfer, gives rise to fluid friction. In most practical cases the mixing results primarily from turbulence. The methods developed in the preceding chapter on dissipative processes will be extended and used to discuss the energy transfers within some simple fluid flows. These results indicate the general nature of the relationship between heat transfer, friction and other characteristics of a flowing fluid. Thus they provide useful guides to the interpretation of empirical heat-transfer data. We shall also see what dimensional analysis has to say about the relationship between heat transfer and other characteristics of a flow system.

Finally the response of a flowing fluid to heat transfer will be considered. For a compressible fluid this is closely related to the response to friction, since the dynamics and thermodynamics of the fluid are closely coupled, as are the mechanisms of heat transfer and friction. The behaviour of a perfect gas will be given particular attention, because the analysis is fairly easy and because the results are of practical importance. The development of gas flows with constant flow area can be predicted for a variety of cases: with heat addition but without friction, with friction but without heat addition, and with the two combined to give isothermal flow. These results show that the flow of a highly compressible fluid within a system of pipes is in some respects markedly different from the constant-density pipe flows studied in the preceding chapter. The pressure variation depends not only on friction, but also on heat transfer; moreover, both heat addition and friction can lead to choking of the flow, even though the flow area remains constant.

6.1 THE ENERGY EQUATION FOR STEADY FLOW

6.1.1 General Results

The relationship between the specific energy of a flowing fluid and the energy supplied to it was obtained in Chapter 2 as the basis for the analysis of isenergic flow. Now it will be applied to a wider range of problems. Equation (2.3.7) gives

$$\Delta(e+pv+\tfrac{1}{2}u^2+gz) = w_e+q \tag{6.1.1}$$

relating the change in the sum of specific internal, displacement, kinetic and potential energies to the sum of the external work w_e per unit mass and the heat addition q per unit mass. Both these energy additions are reckoned as positive when they act to increase the total energy of the fluid. For a one-dimensional flow with mass flow $\dot{m}$, this result can be written

$$\dot{m}\Delta(e+pv+\tfrac{1}{2}u^2+gz) = \dot{W}_e+\dot{Q} \tag{6.1.2}$$

with $\dot{W}_e$ the rate at which external work is done on a section of the flow, and $\dot{Q}$ the rate at which heat flows into that section.

In general, an energy transfer to or from a flowing fluid will give rise to changes in each of the energy components. The relative size of these changes depends on the nature of the process and of the fluid. The relationship between the changes in internal energy and displacement energy is particularly interesting, because it is quite different for the two important cases of a gas and a constant-density fluid. For a perfect gas

$$\frac{\Delta e}{\Delta(pv)} = \frac{c_v}{R} = \frac{1}{\gamma-1}$$

This ratio lies in the range $\frac{3}{2}$ upwards; it is constant for any one perfect gas, being the same for any process that gas undergoes. For a constant-density fluid

$$\frac{\Delta e}{\Delta(pv)} = \varrho_c \frac{\Delta e}{\Delta p}$$

Moreover, as shown in Section 1.4.4, the internal energy and entropy are functions of temperature but not of pressure. Hence the ratio $\Delta e/\Delta(pv)$ can take on any value whatsoever. For example, $\Delta e/\Delta(pv) = 0$ in an isentropic process, for which the internal energy must also be constant. On the other hand, heat addition and heat removal at constant pressure will give $\Delta e/\Delta(pv) \to \pm\infty$.

The results of the preceding paragraph may be summarized as follows. Every energy addition to a gas is distributed in the same proportion between internal and displacement energies. For a constant-density substance this division of energy depends on the process. Since there are vital differences in the ways in which highly and slightly compressible substances absorb energy, it is necessary to use quite different forms of the energy equation in analysing flows of compressible and incompressible fluids. These special forms are restricted cases of the basic energy equation (6.1.1).

Constant-density Fluid

Introducing the constant density ϱ_c and noting that the internal energy and pressure are independent of one another, we rewrite equation (6.1.1) as

$$\Delta\left(\frac{p}{\varrho_c}+\frac{u^2}{2}+gz\right) = \Delta(gH_0) = w_e-(\Delta e-q) \tag{6.1.3}$$

where H_0 is the total head of the fluid,

$gH_0 = \dfrac{p}{\varrho_c}+\dfrac{u^2}{2}+gz$ is the total mechanical energy of the fluid, and $\Delta e - q$ represents the dissipation per unit mass.

The dissipation is the difference between the total rise in internal energy and that occasioned by heat addition. When applied to unit mass of a constant-density substance, equation (1.5.25) gives

$$\mathrm{d}e = T\,\mathrm{d}s = \mathrm{d}q + T\,\mathrm{d}P$$

where $\mathrm{d}P$ is the internal entropy production per unit mass. Then

$$\mathrm{d}e - \mathrm{d}q = T\,\mathrm{d}s - \mathrm{d}q = T\,\mathrm{d}P$$

This shows that dissipation is equivalent to entropy production within the fluid, and also, by virtue of the second law of thermodynamics (1.5.19), that the dissipation is always positive.

The dissipation can be expressed as a head loss, through

$$H_\mathrm{f} = \frac{\Delta e - q}{g} \tag{6.1.4}$$

a generalization of equation (2.3.24), which gives the head loss in isenergic flow. With the dissipation represented in this way, the energy equation can be expressed in terms of heads:

$$\Delta H_0 = \frac{w_\mathrm{e}}{g} - H_\mathrm{f} \tag{6.1.5}$$

where w_e/g is the work done per unit weight of fluid. The change in total head, a measure of the mechanical energy of the fluid, is seen to be the net result of the work done on the fluid and the dissipation occurring within the fluid. This last result is a generalization of equation (2.3.23).

Compressible Fluid

As was explained earlier when discussing isenergic flow, fluid weight usually has little influence on flows of highly dispersed, highly compressible substances. Moreover, since the changes in internal energy and displacement energy are of comparable magnitude, it is convenient to lump them together in the enthalpy, $h = e + pv$. The energy equation (6.1.1) now becomes

$$\Delta(h + \tfrac{1}{2}u^2) = \Delta h_0 = w_\mathrm{e} + q \tag{6.1.6}$$

with h_0 the total enthalpy of the fluid.

The change of state in any real energy-addition process must satisfy the second law of thermodynamics, which may be expressed as

$$T\,\mathrm{d}s - \mathrm{d}q = T\,\mathrm{d}P > 0$$

with $\mathrm{d}P$ the entropy production per unit mass. However, for a compressible fluid it is not possible to separate out distinct elements of mechanical and internal energy. Nor is it possible to distinguish a quantity of energy, the dissipation, which represents irreversibility in a flowing fluid. Hence the irreversibility which must be present in any real flow cannot be made explicit in the energy equation, as it has been in the constant-density result (6.1.5).

The application of the steady-flow energy equation will now be illustrated using some simple examples in which it is assumed that the energy inputs are known, so that the response of the fluid can be determined directly.

Example 6.1 A small windmill drives a pump located within a well, raising drinking water for cattle to a tank on the surface. The system is shown in Figure 6.1. An estimate of the rate at which water flows to the surface is required for the case in which the mill extracts energy from the wind at the rate of 1 hp. The water table is 50 ft below the surface of the water in the tank. The diameter of the pipe is 2 in, and its length is 65 ft. The useful work done on the water may be taken to

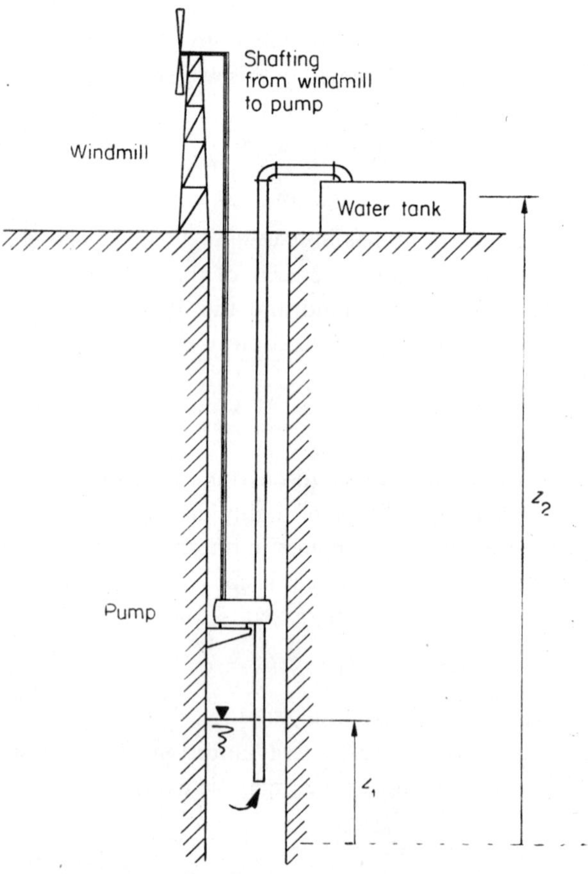

Figure 6.1 A windmill pumping water from a well

be 50 per cent of that supplied by the mill; the rest of the energy is dissipated in the system transmitting power to the pump and within the pump itself.

(a) Neglecting line losses

In making a first estimate of the flow rate, which may be accurate enough for some purposes, line losses in the pipe will be neglected. Consistent with this, the kinetic energy of the fluid will be neglected as well. Then equation (6.1.5) gives

$$\dot{m}(H_{0_2}-H_{0_1}) = \frac{\dot{W}_e}{g}$$

where H_{0_2} and H_{0_1} are the uniform total heads above and below the pump, and $\dot{W}_e$ is the useful power.

The total heads are most easily evaluated using values at the two free surfaces, where the pressures are virtually the same. Thus

$$H_{0_2}-H_{0_1} = z_2-z_1 = 50 \text{ ft}$$

and the flow rate is

$$\dot{m} = \frac{\dot{W}_e}{g\,\Delta H_0} = 5{\cdot}50 \text{ lb/sec}$$

The first estimate of the volume flow rate is then

$$\dot{V} = \frac{\dot{m}}{\varrho_c} = 317 \text{ ft}^3\text{/hr}$$

(b) With line losses

Using the preliminary estimate of the flow rate obtained above, the effect of losses in the pipeline can be determined. The friction factor will be taken to be $f = 0{\cdot}03$, and the minor losses to be four times the dynamic pressure. Then equation (5.5.26) gives the head loss in the line as

$$H_f = \left(\frac{fL}{d}+4\right)\frac{u^2}{2g} = \frac{15{\cdot}7u^2}{2g} = 4{\cdot}0 \text{ ft}$$

Now the head rise across the pump accounts not only for the increase in elevation of the water but also for the line losses:

$$H_{0_2}-H_{0_1} = z_2-z_1+H_f = 54 \text{ ft}$$

Then

$$\dot{V} = \frac{50}{54}\times 317 = 294 \text{ ft}^3\text{/hr}$$

provides an improved estimate of the flow rate. The correct value will be slightly greater than this but, in view of the approximate nature of the data, there is little point in refining the calculations.

(c) Positioning of the pump

One might wonder why the pump is installed within the well, for if it were on the surface maintenance would be easier and the system transmitting power from the windmill would be cheaper and more efficient. Unfortunately, the pump could not work at all if it were located on the surface of the ground. Once the absolute pressure on the suction side falls below the vapour pressure of the well water, the flow will break down. Since atmospheric pressure corresponds to $p_a/\varrho_c g = 34{\cdot}0$ ft of water gauge, the pump must certainly be located within this distance of the water level within the well. However, the fluid pressure will also fall as a result of line losses and acceleration at entry to the pump. Hence local cavitation may occur within the pump even when it is located well within 34 ft of the water surface below.

Suppose, for example, that the pump inlet is 20 ft above the water surface, and that the maximum velocity in the inlet is three times that of the flow in the pipe. Then the minimum pressure within the fluid may be estimated as

$$\frac{p}{\varrho_c g} = H_{0_1} - z - H_f(z) - \frac{9u^2}{2g}$$

$$= 34{\cdot}0 - 20{\cdot}0 - 1{\cdot}3 - 2{\cdot}3 = 10{\cdot}4 \text{ ft of water (abs)}$$

Here the losses in the pipe below the pump have been estimated as proportionate to those in the whole line, that is, as 20/65 of thehead loss. It total appears that this location of the pump would be satisfactory.

Example 6.2 In the process to be considered here the internal, potential, displacement and kinetic energy all change. Some of the changes are in fact small, but this may not be immediately apparent unless one has some experience with this kind of system.

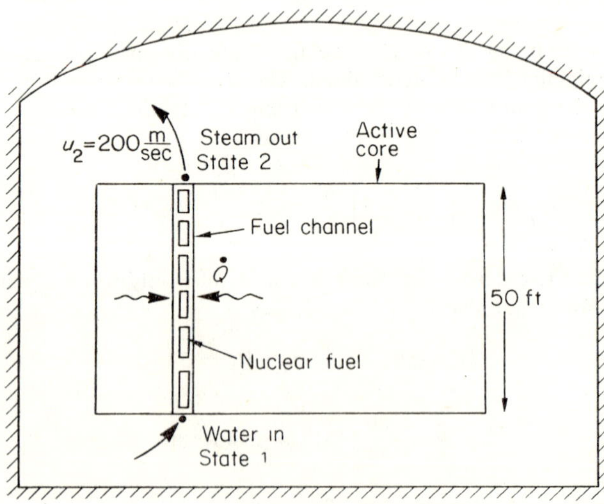

Figure 6.2 Flow through the core of a steam-generating nuclear reactor

Figure 6.2 shows one of the heat-exchange passages running vertically through the core of a steam-generating nuclear reactor. Water enters at 10 bar and 150 °C and emerges 50 ft higher at 9 bar and 600 °C, with a velocity of 200 m/sec. What is the energy input to the steam, if the flow area at outlet from the channel is 30 in^2?

It is not necessary to know anything about the complex process by which energy is added to the fluid as it passes through the core, nor about the flow passage or about the dissipation within it. The inlet and outlet conditions determine the energy input through the steady-flow energy equation, and since there is no external work

$$\dot{Q} = \dot{m}\left[(h_2-h_1)+g(z_2-z_1)+\tfrac{1}{2}(u_2^2-u_1^2)\right]$$

The mass flow can be determined from the outlet state:

$$\dot{m} = \frac{u_2 A_2}{v_2} = 200\times 30\times\frac{0{\cdot}0254^2}{0{\cdot}446} = 8{\cdot}68 \text{ kg/sec}$$

The enthalpy at outlet is found directly from steam tables: $h_2 = 3699$ kJ/kg. The determination of the inlet enthalpy requires more care, since the steam tables of Appendix IV give only the properties of water on the saturation line. Using the method of Example 1.16(b), we estimate the enthalpy of this nearly incompressible fluid as

$$h_1 = h_f + \frac{p_1 - p_s}{\varrho_c}$$

The appropriate saturation conditions are found by interpolation in the steam tables: for $T_1 = 150$ °C, $p_s = 4{\cdot}8$ bar, $h_f = 632$ kJ/kg. Then

$$h_1 = 632 + \frac{5{\cdot}2\times 10^5}{10^6} = 633 \text{ kJ/kg}$$

The change in specific enthalpy within the reactor core may now be calculated as

$$h_2 - h_1 = 3699 - 633 = 3066 \text{ kJ/kg}$$

The change in potential energy is

$$g(z_2 - z_1) = \frac{9{\cdot}81\times 50}{3{\cdot}28} = 150 \text{ J/kg}$$

which is negligible in the present context.

The kinetic energy at outlet is

$$\frac{1}{2}u_2^2 = \frac{1}{2}\times\frac{200^2}{10^3} = 20 \text{ kJ/kg}$$

just on the borderline of significance. The kinetic energy of the inflow is undoubtedly negligible, since the densities are in the ratio

$$\frac{\varrho_1}{\varrho_2} = \varrho_1 v_2 = 1000\times 0{\cdot}466 \simeq 500$$

When the preceding results are collected, we find the net rate of energy addition in the channel to be

$$\dot{Q} = \dot{m}\left(h_2 - h_1 + \tfrac{1}{2}u_2^2\right) = 8{\cdot}68\times(3066+20) = 26{,}800 \text{ kW}$$

Example 6.3 The energy principle can be generalized in a straightforward manner to treat systems into or out of which passes more than one steady flow. Consider part of a steam-raising plant, a feed-water heater in which waste steam is added to the boiler feed water before it is pumped to the boiler. This unit is shown schematically in Figure 6.3. The pressure within is maintained at 20 lbf/in^2 (abs). Water enters at 150 °F at the rate of 18,000 lb/hr, and steam enters at 300 °F at the rate of 2000 lb/hr. The pump drawing water from the heater supplies energy to the flow at the rate of 20 hp; the heat loss from the shell of the heater is estimated to be 200,000 Btu/hr. What is the temperature of the water emerging from the pump?

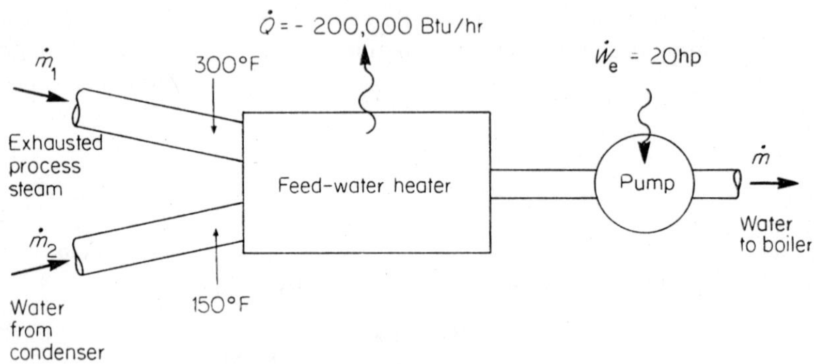

Figure 6.3 Flows into and out of a feed-water heater in a steam-raising plant

In Example 6.2 it was seen that potential and kinetic-energy changes are small in comparison with the large changes of enthalpy associated with phase change. Accordingly they will be neglected in the present analysis. It was also seen that at moderate pressures the enthalpy of liquid water is very nearly equal to that of water in equilibrium with steam at the same temperature. In other words, the enthalpy of water is very nearly a function of temperature alone. This fact will be used to estimate the enthalpy of the water entering and leaving the system.

The appropriate form of the energy equation is

$$\dot{m}h - \dot{m}_1 h_1 - \dot{m}_2 h_2 = \dot{W}_e + \dot{Q}$$

where $\dot{m} = 20{,}000$ lb/hr is the rate at which water leaves the pump with enthalpy h, and the subscripts 1 and 2 denote steam and water components of the inflow. The rate at which energy is supplied from outside the fluid is

$$\dot{W}_e + \dot{Q} = \frac{20\times 33{,}000\times 60}{778} - 200{,}000$$

$$= -149{,}000 \text{ Btu/hr}$$

Then, using steam-table values for 300 °F and 20 lbf/in² and for saturation at 150 °F, we find

$$h = \frac{\dot{m}_1}{\dot{m}} h_1 + \frac{\dot{m}_2}{\dot{m}} h_2 + w_e + q$$

$$= \frac{1192}{10} + 117{\cdot}9 \times \frac{9}{10} - \frac{149{,}000}{20{,}000}$$

$$= 119{\cdot}2 + 106{\cdot}1 - 7{\cdot}5 = 217{\cdot}8 \text{ Btu/lb}$$

The saturation temperature corresponding to this enthalpy is 249 °F, providing a good estimate of the temperature of the water leaving the pump.

6.1.2 Combustion

The analysis of combustion in a flowing fluid combines fluid mechanics and thermochemistry, and would find a natural place in this survey of the interactions between the dynamics and thermodynamics of a fluid flow. Regrettably, some topics must be abbreviated, and combustion, because it uses techniques not required elsewhere, is a natural choice. This topic is treated briefly, not because there is little to say about it, but because there is so much to say.

When fuel is burned within a flowing fluid, usually a flow of air, it has three effects:

(1) the temperature of the flowing fluid is raised, typically to local values in the order of 2500 °K in the region of combustion, and rather less after mixing;

(2) the mass flow is increased, unless before burning it contained both fuel and oxidant; and

(3) combustion products are generated, possibly of several species, depending on the nature of the fuel and on the completeness of combustion, but usually mostly carbon dioxide and/or water vapour.

A general idea of the effects of combustion can be obtained by neglecting the last two effects, and assuming that the burning of fuel merely increases the temperature of the air flow without contaminating it. For simplicity, this model will be used in most of our work.

It is important to realize the magnitude of the errors introduced by the idealizations suggested above. The actual change in mass flow is typically only a few per cent, as the air/fuel mass ratio is of the order 20:1 for a piston engine and 70:1 for a jet engine. The change in the characteristics of the fluid is more significant. For example, the ratio of specific heats in the gas passing through the turbine of a jet engine is usually around $\gamma = 1{\cdot}33$, instead of the value 1·4 in atmospheric air. The temperature of the air passing through the turbine is around 1000 °K, and the tables of Appendix II indicate that the change in specific-heat ratio can be ascribed to this high temperature. This suggests that departures of the heated air from the perfect-gas model are usually more important than contamination of the flow by combustion products. Even so, it sometimes proves convenient to

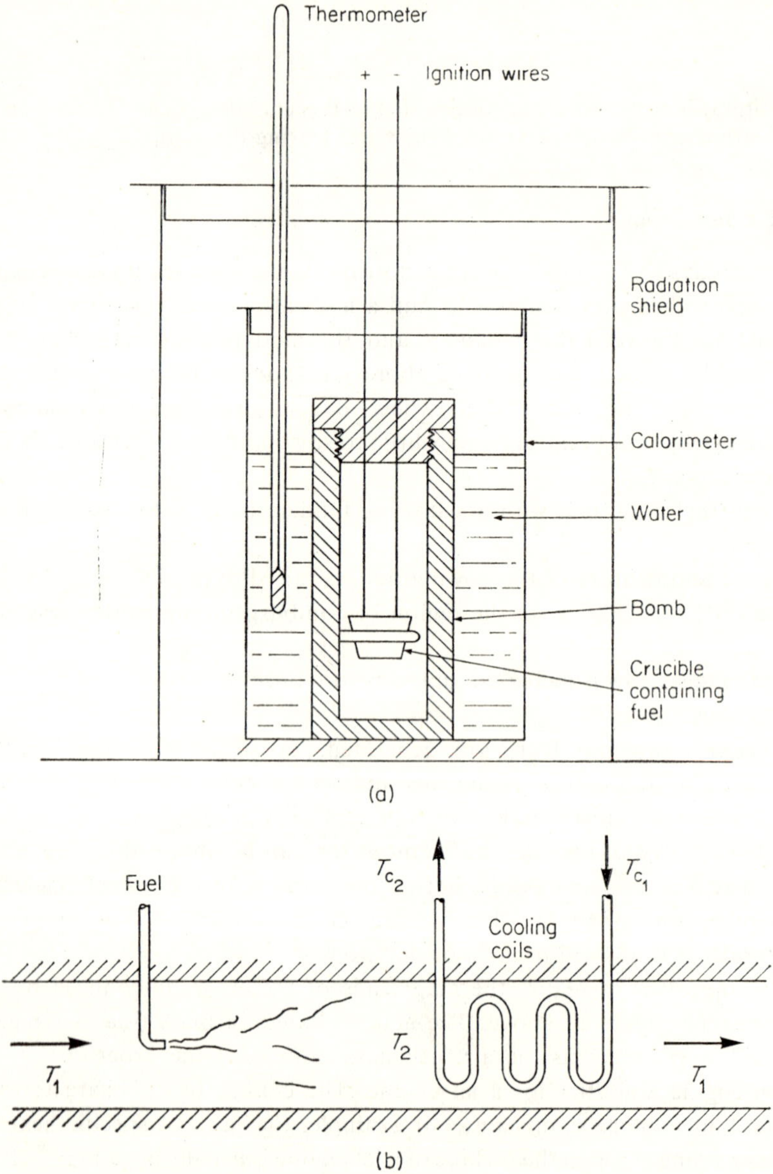

Figure 6.4 Methods of determining the calorific values of fuels. (a) Bomb calorimeter for solid and liquid fuels. (b) Flow calorimeter for gaseous fuels

introduce a further idealization and to use the perfect-gas model to make rough estimates of the response of an airflow to heating by combustion.

It is common practice to say that the burning of a fuel heats a stream of air. In fact, no heat need enter or leave the flow during combustion; the process is essentially an adiabatic conversion of the chemical energy of the reagents to macroscopic forms—internal energy measured by temperature, kinetic energy, and perhaps work done by the expanding fluid. In practice, there is often a significant radiation of energy from the flame to the walls of the combustion chamber; note that this actual heat flow is away from the region in which burning occurs. Nevertheless, the net effect of combustion is in many cases crudely equivalent to an addition of heat, even though the two processes differ in certain fundamental ways.

If we decide to treat combustion simply as a source of heat, it is still necessary to decide how much heat is introduced by the burning of a specified amount of fuel. The energy released when unit mass of a fuel is burned is termed its *heating value* (American usage) or *calorific value* (British usage). Quite different experimental techniques are used to determine this quantity for solid and liquid fuels, and for gaseous fuels. Figure 6.4 shows the two experimental situations. Each is designed to achieve complete burning: to this end an excess of oxygen is supplied. A consideration of the two kinds of experiment serves to introduce some of the essential ideas relating to combustion.

Bomb Calorimeter

To find the heating value of a solid or liquid fuel, it is usual to burn a sample within a rigid enclosure containing oxygen saturated with water.* The small ultimate temperature rise of the calorimeter indicates how much energy has been released. Since the volume of the gaseous system remains constant, the heat flow is equal to the fall in the internal energy of the reagents. Moreover, since the atmosphere is saturated with water, any additional water produced by the burning of hydrogen will be condensed, with the release of the latent heat of vaporization. The quantity determined in this way is called the higher (or gross) heating or calorific value ($HHV = HCV$).

Flow Calorimeter

Here a gaseous fuel is burned in a flow of atmospheric air, which is then cooled to its original temperature, the heat flow being used to measure the

* This method is also used to find the calorific values of foods.

heating value. Since the process takes place at constant pressure, the heat released must equal the fall in the enthalpy of the reagents. Water produced in the reaction will remain as vapour in the flow; it is unlikely that condensation will occur since the air contains a large amount of nitrogen which is not saturated with moisture. Hence in this case the latent heat of the water is not released. The quantity found in this way is termed the lower (or net) heating or calorific value ($LHV = LCV$).

Although one of the experiments described above gives the net change in internal energy of the reagents and the other the net change in their enthalpy, this difference proves to be relatively unimportant. Typically, the enthalpy and internal energy changes differ by less than one per cent. This implies that the heating value is only weakly dependent on the pressure at which burning occurs. Pressure has only a small influence on the internal energies of the substances involved and on the degree of completion of the reaction.

A more important difference between the two experiments lies in the final state of the water produced in the reaction. Consider the combustion of pure hydrogen:

$$H_2+\tfrac{1}{2}O_2 \rightarrow \mathrm{H_2O} \quad \text{(vapour)}$$

For the standard initial and final temperature of 25 °C = 77 °F, the enthalpy of the product is less than the sum of the enthalpies of the reagents by

$$LHV = LCV = 104{,}000 \text{ Btu/lb mole of hydrogen or water}$$

If the water is now condensed, there is a further enthalpy drop of

$$MW(h_g - h_f) = 18\times 1050 = 19{,}000 \text{ Btu/lb mole of water}$$

Here MW is the molecular weight of water and h_g and h_f are the steam-table values for vapour and liquid phases. Hence the net enthalpy drop is

$$HHV = HCV = 123{,}000 \text{ Btu/lb mole of hydrogen}$$

$$= 61{,}500 \text{ Btu/lb}$$

In this case the difference in the two heating values is 19/104 = 18 per cent. For fuels whose combustion products are not entirely water, the difference will be smaller. Even so, the distinction between LCV and HCV is usually large enough to be worth taking into account.

Given one of the heating values, the other can be found by adding or subtracting an appropriate latent heat. For example, the combustion of methane at 25 °C

$$\mathrm{CH_4 + 2\,O_2 \rightarrow CO_2 + 2\,H_2O} \text{ (vapour)}$$

gives

$$LHV = LCV = 802{,}000 \text{ kJ/kg mole of methane}$$

Condensation of the water releases the latent heat

$$2\,MW(h_g - h_f) = 2 \times 18 \times 2442$$
$$= 88{,}000 \text{ kJ/kg mole of methane}$$

Then

$$HHV = HCV = 890{,}000 \text{ kJ/kg mole of methane}$$
$$= 55{,}600 \text{ kJ/kg}$$

For the more usual fuels which are not pure substances, an analysis of the fuel and products of combustion indicates how much water is produced and makes possible the determination of the difference between the two calorific values. Some heating values obtained using these techniques are given in Table 6.1.

Table 6.1. Typical heating or calorific values of fuels for the standard temperature of 77 °F

	$HHV =$ HCV	$LHV =$ LCV	L/H (per cent)
Solid Fuels	(Btu/lb)		
Carbon	14,100	14,100	100
Anthracite coal	14,900	14,600	98
Wood	6,800	6,160	91
Liquid Fuels	(Btu/lb)		
Residual fuel (Bunker C)	18,100	17,200	95
Diesel oil	19,800	18,600	94
Kerosene	19,900	18,600	93
Motor spirit (gasoline or petrol)	20,200	18,800	93
Gaseous Fuels	(Btu/ft^3 at 1 atm)		
Natural gas	960	860	90
Coal gas	530	470	89
Hydrogen	313	264	84

How are experimentally determined heating values related to actual combustion processes? One obvious difference is in the final temperature. In the two experimental situations, this was very nearly the same as the initial temperature of the reagents. In practice, the work and heat extraction following combustion are seldom sufficient to reduce the exhaust temperature to the initial temperature. Hence the flow convects away a considerable portion of the energy which has been released, unless a recuperating system is provided, and this often proves undesirable on grounds of bulk, weight and expense.

Since the heating value is not very strongly dependent on temperature, an elevated outlet temperature does not necessarily indicate that the energy released differs significantly from the experimental heating value, but only that this energy has not been extracted from the flow. The chemical composition of the outlet gases, hot or otherwise, is more significant. For example, the burning of a hydrocarbon will, in general, give rise to carbon monoxide as well the carbon dioxide produced ideally. Since

$$C+O_2 \rightarrow CO_2 \quad \text{with} \quad CV = 394{,}000 \text{ kJ/kg mole}$$

and

$$C+\tfrac{1}{2}O_2 \rightarrow CO \quad \text{with} \quad CV = 111{,}000 \text{ kJ/kg mole}$$

the actual heating value is reduced. Such a difference in composition might arise from an elevated temperature of the exhausted gases, which could maintain a degree of dissociation in the products of combustion. More commonly it is the result of a basic difference in the two combustion processes. In practical combustion chambers, it usually proves impossible to carry the reaction as close to completion as in controlled experiments. Hence the experimental heating values give only a rough guide, an upper limit, to the energy release in normal combustion. Their main use is in providing a standard with which actual energy release can be compared.

There is some debate over the choice between higher and lower calorific values as the basis of comparison for real combustion processes. The argument hinges on the temperature of the combustion products when they are exhausted into the atmosphere, for this determines whether or not the water vapour can condense. For internal combustion engines, whose exhaust gases are usually quite hot, the lower value is generally accepted as providing the more meaningful standard. For steam plants, in which the gases are cooled by extensive heat exchangers before release, the choice is less obvious. It is possible, in principle, to approach the heat release associated with the higher calorific value. However, to do so would in many instances have undesirable consequences—a somewhat larger plant and the encouragement of corrosion in the outlet ducts. Hence even in this case the lower calorific value provides a realistic standard of comparison for the actual heat release. A widely applicable *efficiency of combustion* may then be defined as

$$\eta = \frac{\text{actual heat release per unit of fuel}}{LCV}$$

The main reasons why the combustion efficiency falls short of unity are heat loss from the combustion chamber and incomplete combustion. The primary requirement for complete combustion is that adequate oxygen be made available. It is also necessary that each bit of the fuel remain in contact with oxygen long enough and at a high enough temperature for the reaction to be completed. Thus the key to successful combustion is an appropriate mixing of fuel and air. Insufficient mixing will leave some of the fuel short of oxygen, while too rapid mixing can lower the temperature before combustion is complete. In practice, the mixing of the fuel with excess air, with a resulting reduction in temperature, is the main reason for incomplete combustion.

Example 6.4 Figure 6.5 shows the duct within a ram jet which has previously been discussed in Examples 2.2, 2.10, and 2.14. The air/fuel ratio is $\dot{m}_1/\dot{m}_f = 30$; the properties listed in the figure were found in Examples 2.2 and 2.10.

Let us determine the heat addition on the assumption that the air behaves like a perfect gas. The energy equation for steady flow (6.1.6) gives for this case

$$q = c_p(T_{0_2} - T_{0_1})$$

Since the velocities are specified, the heat addition can be found if the temperature rise is known.

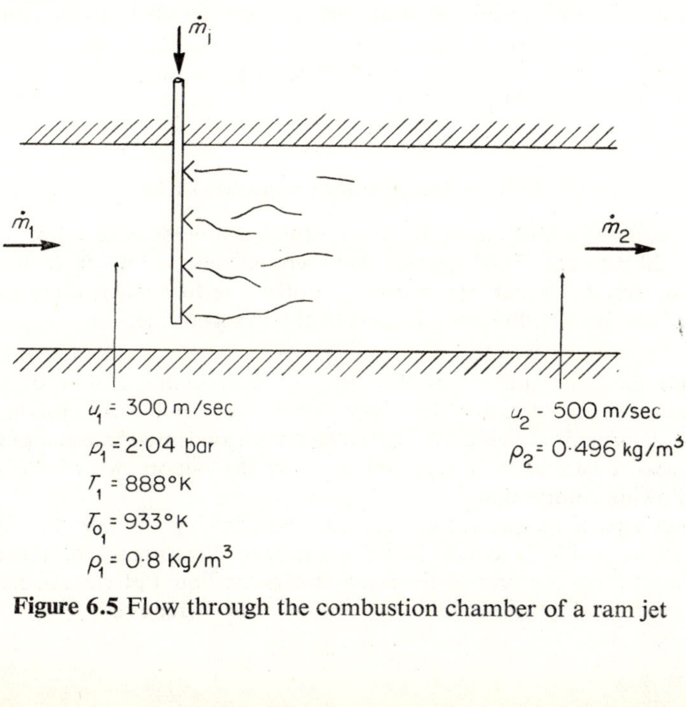

Figure 6.5 Flow through the combustion chamber of a ram jet

The first step is the determination of the pressure following combustion. If the injection system does not apply a drag to the flow, and if the injected fuel has no axial component of momentum (or if the two cancel), the momentum balance for the constant-area flow gives

$$p_2 - p_1 = \varrho_1 u_1^2 - \varrho_2 u_2^2$$
$$= 0{\cdot}72 - 1{\cdot}24 = -0{\cdot}52 \text{ bar}$$

Then

$$p_2 = 2{\cdot}04 - 0{\cdot}52 = 1{\cdot}52 \text{ bar}$$

Note that combustion produces a considerable drop in pressure by causing the fluid to expand and hence to accelerate.

The final temperature can now be calculated on the assumption that the gas constant R does not change; the data for air given in Appendix II indicate that this is a reasonable assumption. Thus

$$\frac{T_2}{T_1} = \frac{p_2}{p_1}\frac{\varrho_1}{\varrho_2} = \frac{1{\cdot}52}{2{\cdot}04} \times \frac{0{\cdot}8}{0{\cdot}496} = 1{\cdot}202$$

Then the temperature rise is $T_2 - T_1 = 179\ ^\circ\text{K}$

and the final temperature is $T_2 = 1067\ ^\circ\text{K}$

The heat addition will be estimated using the mean specific heat for the temperature range 888 °K to 1067 °K, namely, $c_p = 1{\cdot}135$ kJ/kg °K. Thus

$$q = c_p(T_2 - T_1) + \tfrac{1}{2}(u_2^2 - u_1^2)$$
$$= 203 + 80 = 283 \text{ kJ/kg of air}$$

Since the air/fuel ratio is 30, the heat released from the fuel can be estimated as

$$q\frac{\dot{m}}{\dot{m}_f} = 283 \times 30 = 8500 \text{ kJ/kg of fuel.}$$

Suppose that the fuel is kerosene with

$$LHV = 18{,}600 \text{ Btu/lb} = 43{,}300 \text{ kJ/kg}$$

Then the combustion efficiency for the postulated process is very low, only $\eta = 85/433 = 20$ per cent. This suggests that there is some serious fault in the combustion system. It should be possible to utilize the fuel much more efficiently, and to reduce the fuel flow to a fraction of the specified rate.

Example 6.5 The sudden condensation of water within a flow of steam—a condensation shock—has an effect very like that of combustion, the latent heat of the condensing fluid being transferred to the uncondensed fraction of the flow. This process illustrates in a very pointed way the importance of the phase of water following combustion.

Consider a particular case. Supercooled steam flowing between the blades of a steam turbine suddenly condenses; the pressure downstream of the shock is atmospheric; three per cent of the steam enters the liquid phase. An estimate of the change in total temperature of the vapour phase is required.

On the assumption that the condensed fluid is convected along with the vapour, the initial and final states are related by

$$h_1+\tfrac{1}{2}u_1^2 = yh_f+(1-y)h_g+\tfrac{1}{2}u_2^2$$

where y is the condensing fraction. Then

$$h_{0_2}-h_{0_1} = y(h_g-h_f) = yh_{fg}$$

where $h_{0_2} = h_g+\frac{1}{2}u_2^2$ and h_{fg} is the latent heat. For condensation to atmospheric pressure $h_{fg} = 970$ Btu/lb.

The specific heat of the vapour can be estimated by considering data for atmospheric pressure with 212 °F and 250 °F:

$$c_p = \frac{\Delta h}{\Delta T} = \frac{1169-1150}{250-212} = 0{\cdot}50 \text{ Btu/lb °R}$$

Then

$$T_{0_2}-T_{0_1} \simeq \frac{h_{0_2}-h_{0_1}}{c_p}$$

$$= \frac{y}{c_p} h_{fg} = 58 \text{ °F}$$

Condensation is seen to produce a marked change in the properties of the vapour.

From this point the analysis of the condensation shock is very like that of a normal shock in a gas. The final state is determined by the momentum and continuity conditions, $p+\varrho u^2 =$ constant and $\varrho u =$ constant. These requirements indicate that the pressure of the decelerated fluid will be higher than that before condensation.

6.2 EFFICIENCIES FOR FLOW MACHINES

6.2.1 The Hydraulic Efficiency of Pumps and Turbines

In the context of one-dimensional flow, the effect of a flow machine— pump, compressor, fan, turbine or hydraulic motor— is simply a change in the state of the fluid. The inlet and outlet states are each defined by two mean thermodynamic properties of the fluid, by the mean velocity (if the kinetic-energy change is significant), and by the elevation (if the potential-energy change is significant). Let us see what kind of information is required to relate these two states. The change in elevation is known from the geometry of the installation. If the flow rate is specified, the inlet and outlet velocities can usually be determined from the continuity relationship. The steady-flow energy equation relates the total energies at inlet and outlet. To these three constraints a fourth must be added to complete the specification of the change in state across the machine. This additional information usually takes the form of an efficiency which relates the actual work transfer to that in an ideal process.

In describing the efficiency of isenergic flow through a diffuser or nozzle (Section 4.4.2), we found it convenient to relate the actual process to an isentropic process between the same terminal pressures. A somewhat similar procedure will be used here; the efficiency will relate the actual work done (and actual change in total enthalpy) to the work done (and corresponding change in total enthalpy) in a hypothetical isentropic transition between the actual total pressures. Since the constant-density case is more straightforward, it will be considered first.

Constant-density Fluid

The energy equation written in the form

$$\Delta\left(\frac{p}{\varrho_c}+\frac{u^2}{2}+gz\right) = \Delta(gH_0) = w_e-(\Delta e-q) \tag{6.1.3}$$

leads directly to the definition of efficiencies for machines across which the density is nearly constant. Consider first a dissipation-free energy transfer which produces the same changes in total pressure and total mechanical energy. In this process $\Delta e-q = 0$; the internal energy changes, if at all, only as a result of heat addition. Equation (6.1.3) indicates that the actual change in total pressure is achieved with external work $\Delta(gH_0)$ per unit mass. For this ideal case

$$\Delta H_0 = \frac{w_e}{g}$$

This is a generalization of Bernoulli's equation for constant density (2.4.1). Now the total head can rise or fall even though there is no change in the entropy of the fluid.

In any realistic process the dissipation $\Delta e-q > 0$. Then for a pump or other device doing work on the fluid

$$w_e > \Delta(gH_0) > 0$$

For a turbine or other device absorbing work from the fluid

$$\Delta(gH_0) < w_e < 0$$

Hence the efficiency of a pump may be defined as

$$\eta_p = \frac{\Delta(gH_0)}{w_e} = \frac{\text{work absorbed ideally}}{\text{actual work}} \tag{6.2.1}$$

Hence

$$\Delta(gH_0) = \eta_p w_e \quad \text{and} \quad \Delta e - q = (1-\eta_p) w_e$$

These results show that the efficiency specifies the division of the total work input between mechanical energy and dissipation. For a turbine the efficiency is defined as

$$\eta_t = \frac{w_e}{\Delta(gH_0)} = \frac{\text{actual work}}{\text{work extracted ideally}} \tag{6.2.2}$$

Hence

$$-w_e = -\eta_t \, \Delta(gH_0) \quad \text{and} \quad \Delta e - q = -(1-\eta_t)\, \Delta(gH_0)$$

The efficiency is seen to specify the division of the fall in the mechanical energy of the fluid $-\Delta(gH_0)$ between work output $-w_e$ and dissipation $\Delta e - q$. The efficiencies defined above are referred to as *hydraulic efficiencies* to distinguish them from other efficiencies for flow machines which will be introduced later.

These definitions can be recast in alternative forms which prove convenient in particular applications. If the change in potential energy is negligible

$$\eta_p = \frac{1}{\eta_t} = \frac{\Delta p_0}{\varrho_c w_e} \tag{6.2.3}$$

relates the actual total-pressure change and the actual work transfer. In terms of powers

$$\eta_p = \frac{1}{\eta_t} = \frac{\dot{V}\, \Delta p_0}{\dot{W}_e} \tag{6.2.4}$$

This relates the change in the flux of mechanical energy and the work flow to or from the machine. We shall see later that these results apply to compressible fluids when the density change is not too large. Hence they prove useful in describing the performance of fans and blowers.

For cases in which the kinetic-energy change is small, either because the through-flow velocity is small or because the inlet and outlet velocities are nearly equal, the forms

$$\eta_p = \frac{1}{\eta_t} = \frac{\Delta(p + \varrho_c g z)}{\varrho_c w_e} = \frac{\Delta H_p}{w_e/g} \tag{6.2.5}$$

are useful. They relate the actual changes in pressure and piezometric head to the work transfer expressed in equivalent forms. The rise in piezometric head across a pump is referred to as the lift of the pump; this is the vertical distance by which the fluid might, in principle, be raised by the pump. Thus the pump efficiency is the ratio of the actual lift to that which could be achieved ideally.

Compressible Fluid

Here the energy transfer is determined by

$$\Delta(h+\tfrac{1}{2}u^2) = \Delta h_0 = w_e+q \tag{6.1.6}$$

As was seen earlier, for a gas the effects of heat addition and work transfer are essentially equivalent. Hence the heat addition must be clearly specified in any discussion of work transfer; it cannot simply be tucked away in a dissipation term, as was possible for constant density. We shall restrict the discussion to adiabatic work transfer. This is not unrealistic for turbomachines, which typically convey a large mass flow with little opportunity for a significant heat flow between machine and fluid. However, it is common practice to cool the cylinders of reciprocating compressors, and the results obtained for adiabatic flow cannot meaningfully be applied when this is done.

For an adiabatic work transfer

$$h_{0_2}-h_{0_1} = w_e$$

with $h_{0_2} = h(p_{0_2}, s_2)$ and $h_{0_1} = h(p_{0_1}, s_1)$. Since the energy transfer is related to the change in total enthalpy, it is convenient to define the ideal pattern process in terms of the actual total pressures p_{0_1} and p_{0_2}. As we saw earlier, the total temperatures and pressures can be measured directly using suitable probes; in some ways they are easier to measure than the corresponding static properties. The work transfer in the ideal process is

$$h_{0_i}-h_{0_1}$$

where $h_{0_i} = h(p_{0_2}, s_1)$ is the hypothetical total enthalpy. For a pump $h_{0_2} > h_{0_i}$; more energy has to be added in the actual process and the fluid emerges hotter. For a turbine as well $h_{0_2} > h_{0_i}$; less energy has been extracted in the machine than might have been ideally.

The efficiencies of machines which add and extract energy can be defined as

$$\eta_c = \frac{1}{\eta_t} = \frac{h_{0_i}-h_{0_1}}{h_{0_2}-h_{0_1}} \tag{6.2.6}$$

These efficiencies are referred to as *isentropic efficiencies* since they are based on an isentropic pattern process. They may also be described, like their special forms for constant density (6.2.1 to 6.2.5), as hydraulic efficiencies, since they describe the process within the fluid rather than the overall performance of the machine. The term *adiabatic efficiency* is also used; this refers to the restriction to adiabatic flow imposed when defining these efficiencies.

For the important case of a perfect gas, other forms for the efficiencies can be found. The derivation parallels that leading to the forms (4.4.13) for diffuser and nozzle efficiencies.

Since $\quad \Delta h_0 = c_p\,\Delta T_0 \quad$ and $\quad \dfrac{T_{0_2}}{T_{0_1}} = \left(\dfrac{p_{0_2}}{p_{0_1}}\right)^{\frac{\gamma-1}{\gamma}}$, $\quad$ we have

$$\eta_c = \frac{1}{\eta_t} = \frac{\left(\dfrac{p_{0_2}}{p_{0_1}}\right)^{\frac{\gamma-1}{\gamma}} - 1}{\dfrac{T_{0_2}}{T_{0_1}} - 1} \tag{6.2.7}$$

All of the total properties appearing here are those of the actual flow. The change in total pressure is given directly in terms of the work transfer as

$$\frac{p_{0_2}}{p_{0_1}} = \left(1 + \eta_c \frac{w_e}{h_{0_1}}\right)^{\frac{\gamma}{\gamma-1}} \tag{6.2.8}$$

This is a relationship of the kind postulated in equation (2.4.11). For fixed values of p_{0_1} and h_{0_1}

$$p_{0_2} = f(h_{0_2}, s_2)$$

since $w_e = h_{0_2} - h_{0_1}$ and $\eta_c = f(s_2)$. Here the total state of the fluid changes as a result of both energy addition and dissipation, unlike isenergic, isentropic flow for which the total properties are uniform along the flow.

The ideal case of isentropic, adiabatic flow is of some interest in its own right. Equation (1.5.14) gives for this case

$$\mathrm{d}h = T\,\mathrm{d}s + v\,\mathrm{d}p = \frac{\mathrm{d}p}{\varrho}$$

and equation (6.1.6) can be written

$$\Delta h_0 = \int_1^2 \frac{\mathrm{d}p}{\varrho} + \Delta\left(\frac{u^2}{2}\right) = w_e \tag{6.2.9}$$

This is a generalization of the Bernoulli result (2.3.29); the Bernoulli sum can now change as a result of energy transfer even though the entropy remains constant. Another way of expressing this relationship is to note that the total state is changed by the isentropic addition of energy, as indicated in equation (2.4.13).

Using the result (6.2.9) it is easy to show that the definitions for varying density reduce to those for constant density when the volume change in a compressible fluid is in fact small. Equation (6.2.9) indicates that the ideal work transfer is approximately

$$h_{0_i}-h_{0_1} \simeq \frac{\Delta p}{\varrho}+\Delta\left(\frac{u^2}{2}\right) \simeq \frac{\Delta(p+\frac{1}{2}\varrho u^2)}{\varrho} \simeq \frac{\Delta p_0}{\varrho}$$

where ϱ is the nearly constant density. Then equations (6.2.6) give

$$\eta_c = \frac{1}{\eta_t} = \frac{\Delta p_0}{\varrho w_e}$$

equivalent to equation (6.2.3). When the density change across the machine is very small, it does not much matter which density is used in the approximation $\Delta h = \Delta p/\varrho$; conventionally, the inlet value is adopted. For larger density changes, better results can be obtained by introducing the mean density $\varrho = \frac{1}{2}(\varrho_1+\varrho_2)$. Since certain of the error terms cancel when this value is used, this 'constant-density' approximation proves to be useful even when $\Delta\varrho/\varrho$ is of order unity.

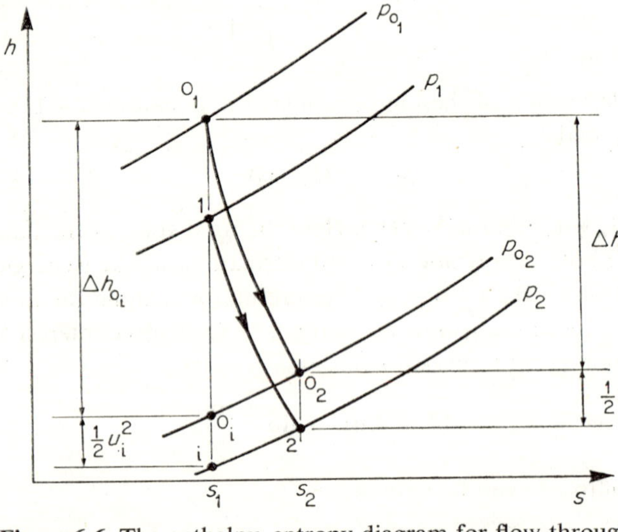

Figure 6.6 The enthalpy–entropy diagram for flow through a turbine, illustrating total-to-total and total-to-static efficiencies

Figure 6.6 shows the enthalpy changes related by the isentropic efficiency; since the enthalpy drops, this diagram relates to the case of flow through a turbine. Both the total and static pressure levels at inlet and outlet are shown. Sometimes the actual output of a turbine is compared with the enthalpy change $h_{0_1}-h_i$ rather than $h_{0_1}-h_{0_i}$, it being argued that a turbine could in principle decelerate the flow until the total pressure at outlet approached the static value. These two ideal changes are related by

$$h_{0_1}-h_i = h_{0_1}-h_{0_i}+\tfrac{1}{2}u_i^2$$

where u_i is the outlet velocity in the ideal process. For constant-density flow, the ideal outlet velocity equals the actual outlet velocity u_2, but for compressible flow the two will differ slightly. The turbine efficiency defined in terms of the static properties at outlet is

$$\eta_{ts} = \frac{h_{0_1} - h_{0_2}}{h_{0_1} - h_i} \tag{6.2.10}$$

or, for constant density

$$\eta_{ts} = \frac{-w_e}{-\Delta(gH_0) + \frac{1}{2}u_2^2} \tag{6.2.11}$$

These efficiencies are referred to as *total-to-static efficiencies* to distinguish them from those introduced earlier, which may be termed *total-to-total efficiencies*.

As noted above, in adopting a total-to-static efficiency one is in effect arguing that a turbine could be designed to reduce the fluid velocity to a negligible value before exhausting the fluid. This basis of assessment is reasonable when the kinetic energy of the exhausted fluid is in fact wasted. The kinetic energy of the water leaving the tailrace below a hydroelectric station is dissipated by mixing with the river water; here the total-to-static basis of assessment seems appropriate. On the other hand, the kinetic energy of the fluid leaving the turbine of a jet engine is utilized in thrust generation; here the total-to-total assessment seems more reasonable. The performance of pumps and compressors is usually assessed on a total-to-total basis.

It is useful to have an idea of the efficiencies which can be achieved in actual turbomachines. In large, carefully designed, carefully operated machines it is currently possible to achieve efficiencies in the order of 95 per cent in hydraulic and steam turbines, and a few per cent lower in pumps and compressors. The efficiency of machines in which the pressure increases tends to be lower, since flow separation is more likely in the diffusing channels within such machines.

The practically attainable efficiency depends to some extent on the size of the machine. The proportional losses within a class of similar turbomachines are found to vary roughly as $\mathrm{Re}^{-\frac{1}{5}}$, with Re a Reynolds number incorporating a typical dimension of the machine and a typical velocity of the flow within it. Hence the losses are a smaller proportion of the total energy transfer in a larger machine. This provides a justification for the recent increase in the size of electricity-generating steam turbines from around 60 MW output to around 500 MW output. Smaller flow machines suffer from other disadvantages: they are less likely to be carefully designed;

they are more likely to be part of a mass-produced assembly for which cost of production is the dominant design consideration; and they are more likely to be converted to functions for which they were not originally intended.

Example 6.6 In Example 2.5 the ideal operation of a pump and a compressor was considered. In each case the pressure rise was

$$\Delta p = 6240 \text{ lbf/ft}^2 = 100 \text{ ft of water gauge}$$

For these ideal pumping processes, it was found that

(a) for the water pump, $w_e = 100$ ft lbf/lb, and

(b) for the air compressor, $w_e = 47{,}400$ ft lbf/lb, with $p_1 = 2120$ lbf/ft^2, $T_1 = 530$ °R, and $\varrho_1 = 0{\cdot}075$ lb/ft^3.

Let us now consider the operation of these two machines with isentropic efficiencies of 80 per cent, each accepting fluid at the rate of 150 ft^3/min. The changes in velocity and elevation across the machines will again be taken to be negligible.

(a) Water Pump

Equation (6.2.5) gives

$$w_e = \frac{g\,\Delta H_p}{\eta_p} = 32{\cdot}2 \times 100/0{\cdot}80 \text{ ft lbf/slug}$$
$$= 125 \text{ ft lbf/lb}$$

Then

$$\dot{W}_e = \varrho_c \dot{V} w_e = 62{\cdot}4 \times 150 \times 125 \text{ ft lbf/min}$$
$$= 35{\cdot}5 \text{ hp}$$

Alternatively, equation (6.2.4) gives directly

$$\dot{W}_e = \frac{\dot{V}\,\Delta p}{\eta_p} = 150 \times \frac{6240}{0{\cdot}80} \text{ ft lbf/min}$$
$$= 35{\cdot}5 \text{ hp}$$

Twenty per cent of this power is dissipated within the pump, the rise in temperature being

$$\Delta T = (1-\eta_p)\frac{w_e}{c_v} = 0{\cdot}2 \times \frac{125}{778} = 0{\cdot}032 \text{ °F}$$

In the ideal process, the temperature is precisely constant, in the absence of heat transfer.

(b) Air Compressor

From equation (6.2.7) the work of compression is found to be

$$w_e = \Delta h = \frac{c_p T_1}{\eta_c}\left[\left(\frac{p_2}{p_1}\right)^{\frac{\gamma-1}{\gamma}} - 1\right]$$
$$= \frac{47{,}400}{0{\cdot}80} = 59{,}300 \text{ ft lbf/lb}$$

The power required is then

$$\dot{W}_e = \varrho_1 \dot{V}_1 w_e = 0{\cdot}075 \times 150 \times 59{,}300 \text{ ft lbf/min}$$
$$= 20{\cdot}2 \text{ hp}$$

In this case the temperature rise is

$$\Delta T = \frac{\Delta h}{c_p} = 318\ °\text{F}$$

greater by 20 per cent than the rise in ideal compression. However, it is not true to say that the additional energy required as a result of inefficiency has been wholly dissipated. In some circumstances it proves possible to convert some of the additional energy of the hotter gas into useful work.

In the process considered here, the pressure rises by a factor of four; obviously the constant-density approximation is inappropriate in its simplest form. It is of interest, however, to consider the utility of the constant-density form

$$\dot{W}_e = \frac{\dot{V}\,\Delta p}{\eta_c}$$

with $\dot{V}$ the volume flow based on the mean density across the compressor. The outlet density is $\varrho_2 = 8360/(53{\cdot}3 \times 848) = 0{\cdot}185$ lb/ft^3, and the mean density is 0·130 lb/ft^3. Hence the appropriate volume flow is

$$\dot{V} = \frac{0{\cdot}075}{0{\cdot}130} \times 150 = 86{\cdot}5 \text{ ft}^3\text{/min}$$

and

$$\dot{W}_e = \frac{86{\cdot}5 \times 6240}{0{\cdot}80} \text{ ft lbf/min} = 20{\cdot}4 \text{ hp}$$

The 'constant-density' result proves fairly accurate even though the density increases by a factor of $2\frac{1}{2}$ within the compressor.

Example 6.7 Let us now consider a case in which the kinetic energy of the fluid is important—the compressor of the jet engine of an aeroplane flying at high speed. Figure 6.7 shows the installation schematically. The aeroplane is flying at $M_1 = 0{\cdot}93$ at 10,000 m. The air reaches the compressor through a diffuser

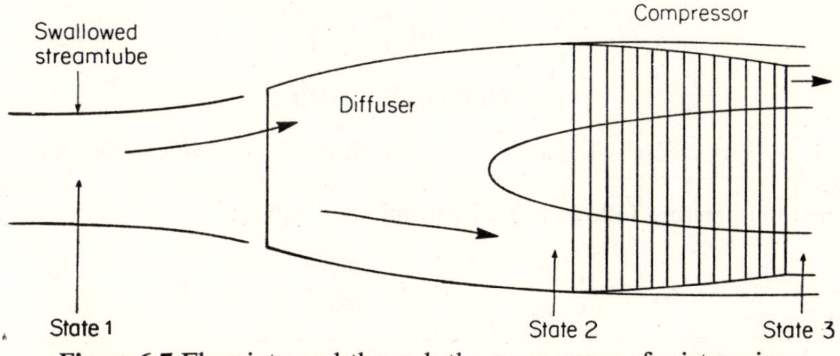

Figure 6.7 Flow into and through the compressor of a jet engine

of efficiency 80 per cent, the Mach number at inlet to the compressor being $M_2 = 0{\cdot}55$. If the isentropic efficiency of the compressor is 90 per cent, and the ratio of inlet and exit flow areas is $A_3/A_2 = \frac{1}{4}$, what are the properties of the air emerging from it for a total pressure ratio of $p_{0_3}/p_{0_2} = 11$?

We shall assume that the air behaves throughout like a perfect gas with $\gamma = 1{\cdot}4$.

State 1

The static properties of the air ahead of the aircraft are taken to have standard values:

$$p_1 = 0{\cdot}265 \text{ bar}, \qquad T_1 = 223 \text{ °K}$$

$$\varrho_1 = 0{\cdot}413 \text{ kg/m}^3, \qquad a_1 = 300 \text{ m/sec}$$

The total properties are found using isentropic flow tables:

$$p_{0_1} = \frac{0{\cdot}265}{0{\cdot}572} = 0{\cdot}463 \text{ bar}$$

$$T_{0_1} = \frac{223}{0{\cdot}853} = 262 \text{ °K}$$

State 2

In the isenergic flow within the diffuser the total temperature of the air remains constant:

$$T_{0_2} = T_{0_1} = 262 \text{ °K}$$

Knowing the two Mach numbers, we can find the temperature change using isentropic flow tables:

$$\frac{T_2}{T_1} = \frac{T_2/T_{0_2}}{T_1/T_{0_1}} = \frac{0{\cdot}943}{0{\cdot}853} = 1{\cdot}106$$

The pressure rise in the isenergic flow through the diffuser follows from equation (4.4.13):

$$\frac{p_2}{p_1} = \left[1 + \eta_D\left(\frac{T_2}{T_1} - 1\right)\right]^{\frac{\gamma}{\gamma-1}}$$

$$= (1{\cdot}0848)^{3{\cdot}5} = 1{\cdot}330$$

This is the ram compression achieved as the air decelerates on entering the intake duct.

The total pressure follows using isentropic flow tables:

$$p_{0_2} = \frac{p_{0_2}}{p_2}\frac{p_2}{p_1}p_1 = 1{\cdot}330 \times \frac{0{\cdot}265}{0{\cdot}814} = 0{\cdot}433 \text{ bar}$$

This is of course less than the initial value p_{0_1}.

The density at inlet to the compressor is

$$\varrho_2 = \frac{p_2}{p_1}\frac{T_1}{T_2}\varrho_1 = 1{\cdot}330 \times \frac{0{\cdot}413}{1{\cdot}106}$$

$$= 0{\cdot}496 \text{ kg/m}^3$$

and the velocity is

$$u_2 = M_2 a_2 = M_2 \left(\frac{T_2}{T_1}\right)^{\frac{1}{2}} a_1$$

$$= 0{\cdot}55 \times 1{\cdot}051 \times 300 = 174 \text{ m/sec}$$

State 3

The total pressure of the fluid emerging from the compressor is

$$p_{0_3} = \frac{p_{0_3}}{p_{0_2}} p_{0_2} = 11 \times 0{\cdot}433 = 4{\cdot}76 \text{ bar}$$

Equation (6.2.7) gives the total temperature as

$$T_{0_3} = T_{0_2}\left\{1 + \frac{1}{\eta_c}\left[\left(\frac{p_{0_3}}{p_{0_2}}\right)^{\frac{\gamma-1}{\gamma}} - 1\right]\right\}$$

$$= 262\left[1 + \frac{11^{0{\cdot}286} - 1}{0{\cdot}90}\right] = 549 \text{ °K}$$

The total density is then

$$\varrho_{0_3} = \frac{p_{0_3}}{RT_{0_3}} = \frac{476}{0{\cdot}287 \times 549} = 3{\cdot}02 \text{ kg/m}^3$$

This completes the determination of the total state. None of the static properties is known, nor is the outlet velocity. Hence an iterative calculation must be used to find the static outlet properties. We can start by assuming that they are approximately equal to the total properties at outlet. Then the continuity condition gives

$$u_3 \simeq \frac{\varrho_2}{\varrho_{0_3}}\frac{A_2}{A_3} u_2 = \frac{0{\cdot}496}{3{\cdot}02} \times 4 \times 174 = 114 \text{ m/sec}$$

and

$$a_3 \simeq (\gamma R T_{0_3})^{\frac{1}{2}} = (1{\cdot}4 \times 287 \times 549)^{\frac{1}{2}} = 470 \text{ m/sec}$$

Then

$$M_3 \simeq \frac{114}{470} = 0{\cdot}243$$

Improved estimates can now be found using isentropic flow tables:

$$\varrho_3 = 3{\cdot}02 \times 0{\cdot}971 = 2{\cdot}93 \text{ kg/m}^3$$

$$u_3 = \frac{3{\cdot}02}{2{\cdot}93} \times 114 = 118 \text{ m/sec}$$

$$a_3 = (0{\cdot}988)^{\frac{1}{2}} \times 470 = 467 \text{ m/sec}, \quad \text{and}$$

$$M_3 = \frac{118}{467} = 0{\cdot}253$$

6.2.2 Other Measures of Efficiency

So far we have considered only the efficiency of the work transfer between a flow machine and the fluid passing through it—the hydraulic efficiency. If we consider the operation of a complete energy-transfer system of which the machine is a part, a number of other measures of efficiency can be distinguished. Figure 6.8(a) shows a pumping installation, comprising an electric motor driving a water pump. Figure 6.8(b) shows the energy flows within this system. These are

P_1	the electrical power flowing to the motor,
$\dot{Q}_1$	the flow of heat from the motor to its surroundings,
P_2	the mechanical power transferred to the pump,
$\dot{Q}_2$	the flow of heat from the pump to its surroundings (other than the water flowing through it),
$\dot{W}_e = \dot{m}w_e$	the rate at which work is done on the fluid by the pump,
$\dot{Q} = \dot{m}q$	the heat flow from the pump to the fluid,
$\dot{m}\,\Delta(gH_0)$	the increase across the pump in the flux of mechanical energy, and
$\dot{m}\,\Delta e$	the increase across the pump in the flux of internal energy.

The heat flow $\dot{Q}_1$ results from mechanical and electrical losses within the electric motor. The heat flow $\dot{Q}_2+\dot{Q}$ represents mechanical losses within the pump itself, not within the fluid flowing through it. These losses arise from friction in bearings and seals, and between the rotating elements and the pump casing. Parasitic losses, such as those in cooling systems, may also be included.

Using these energy flows, the following efficiencies can be defined:

(1) for the electric motor $P_1 = P_2+\dot{Q}_1$ and the electrical efficiency is

$$\eta_e = \frac{P_2}{P_1}$$

(2) for the pump $P_2 = \dot{W}_e+\dot{Q}+\dot{Q}_2$ and the mechanical efficiency is

$$\eta_m = \frac{\dot{W}_e}{P_2}$$

(3) for the water $\dot{W}_e = \dot{m}[\Delta(gH_0)+(\Delta e-q)]$ and the hydraulic efficiency is

$$\eta = \frac{\dot{m}\,\Delta(gH_0)}{\dot{W}_e}$$

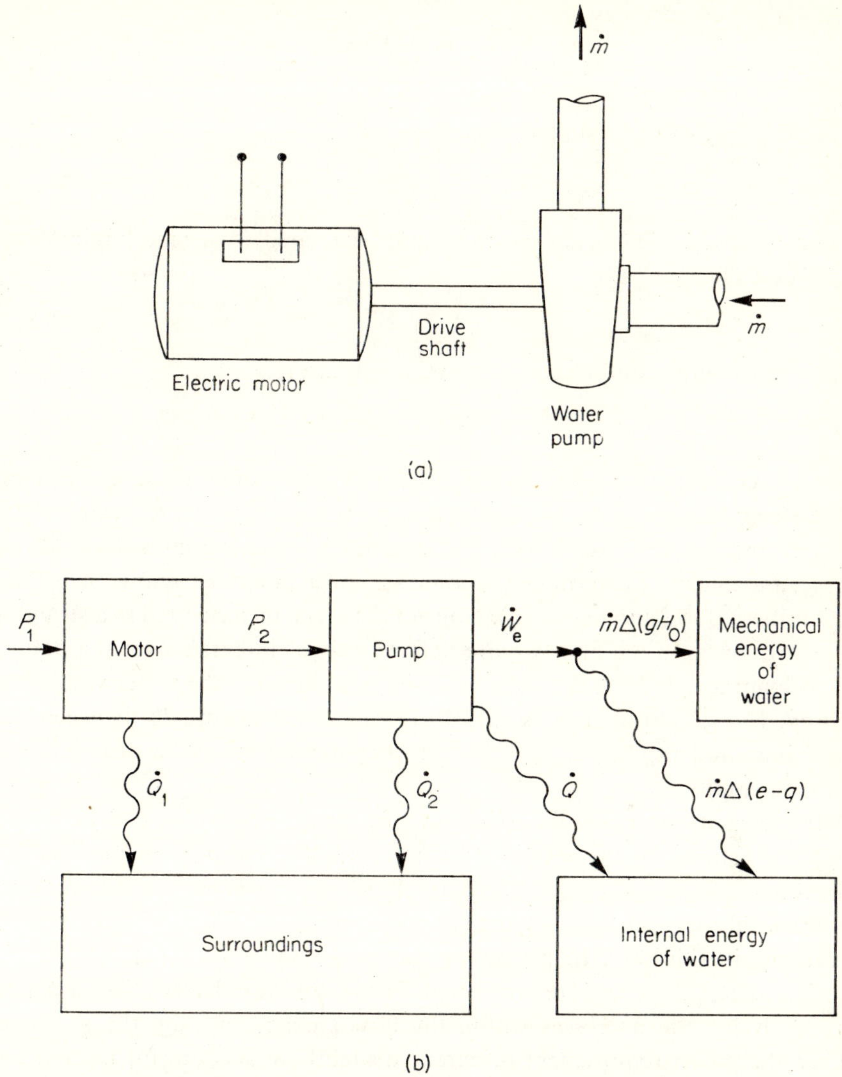

Figure 6.8 Energy flows in a water-pumping installation. (a) Plan view of the plant; (b) Distribution of the energy supplied to the plant

(4) for the pump and water $P_2 = \dot{m}[\Delta(gH_0)+\Delta e]+\dot{Q}_2$ and the overall efficiency is

$$\eta_0 = \frac{\dot{m}\,\Delta(gH_0)}{P_2}$$

(5) for the whole installation $P_1 = \dot{m}\,\Delta(gH_0)+\dot{Q}_1+\dot{Q}_2+\dot{m}\,\Delta e$ and the

installation efficiency is

$$\eta_i = \frac{\dot{m}\,\Delta(gH_0)}{P_1}$$

The efficiency of the installation is given by

$$\eta_i = \eta_e\eta_0$$

while the overall efficiency of the pump is related to its mechanical and hydraulic efficiencies by

$$\eta_0 = \eta_m\eta \tag{6.2.12}$$

The useful output and electrical power input are related by

$$\dot{m}\,\Delta(gH_0) = \eta_e\eta_m\eta P_1$$

Now consider how one might go about measuring these efficiencies. The energy flows P_1, P_2, $\dot{m}\,\Delta(gH_0)$ and $\dot{m}\,\Delta e$ can be measured using wattmeters, tachometers, weighing scales, piezometers and thermometers. Hence η_i, η_e, and η_0 can be determined. To find η_m and η one must know one of the quantities $\dot{Q}$, $\dot{W}_e$, and $\dot{m}(\Delta e - q)$. None of these can be measured in a straightforward manner; the first two are defined only as integrals over the pump/fluid interface, and the last only as an integral through the water within the pump. Hence, although it is easy to distinguish conceptually between the hydraulic and mechanical efficiencies, they cannot easily be distinguished experimentally.

In practice, the hydraulic efficiency of a pump or turbine can be found only by estimating the losses contributing to the mechanical inefficiency. Since these losses are, particularly for large machines, only a few per cent of the total energy flow, the hydraulic efficiency can be found with reasonable accuracy in this way. In predicting the performance of a machine at the design stage, this procedure is reversed. The hydraulic efficiency* is estimated by analysing the processes within the flow passage through the machine. Then the power requirement is increased slightly to account for mechanical losses.

Now consider the implications of the energy flows of Figure 6.8 for a machine conveying a compressible fluid. For this case it was necessary to limit consideration to a strictly adiabatic flow, in order to obtain an unequiv-

* It is common practice to break down the hydraulic efficiency further by introducing a volumetric efficiency, the ratio of the volume flow actually delivered by the pump to that passing through the impeller. The difference between these two flows is accounted for by leakage back past the impeller. No useful output is derived from the work done on this leakage flow.

ocal definition of the hydraulic efficiency. In practice, it is unlikely that the flow through a pump or turbine will be precisely adiabatic, if only because much of the energy dissipated within the mechanical parts of the machine will find its way into the flowing fluid. Hence for a compressible fluid the hydraulic efficiency is not only impossible to measure, but is impossible to define with precision. Nevertheless, in large machines these extraneous energy flows are only a few per cent of the total power, and adiabatic efficiencies can be used without serious error. Moreover, for machines of a particular class, say, axial-flow turbojet compressors, the energy balance will be roughly similar from machine to machine, and efficiencies calculated in the same way will provide a useful basis of comparison even if their absolute significance is uncertain.

Example 6.8 Figure 6.9 shows a hydroelectric generating station whose turbines are supplied from a reservoir 200 m above the tailrace. The efficiency of the electrical generator is 90 per cent. The total-to-static efficiency of the turbine is 92 per cent, with an exit velocity of 11 m/sec, and a mechanical efficiency of 98 per cent. In the flow to the turbine 5 per cent of the available energy (gH_{0_1} per unit mass) is dissipated. What is the overall efficiency of energy conversion of the system and how are the losses distributed through it?

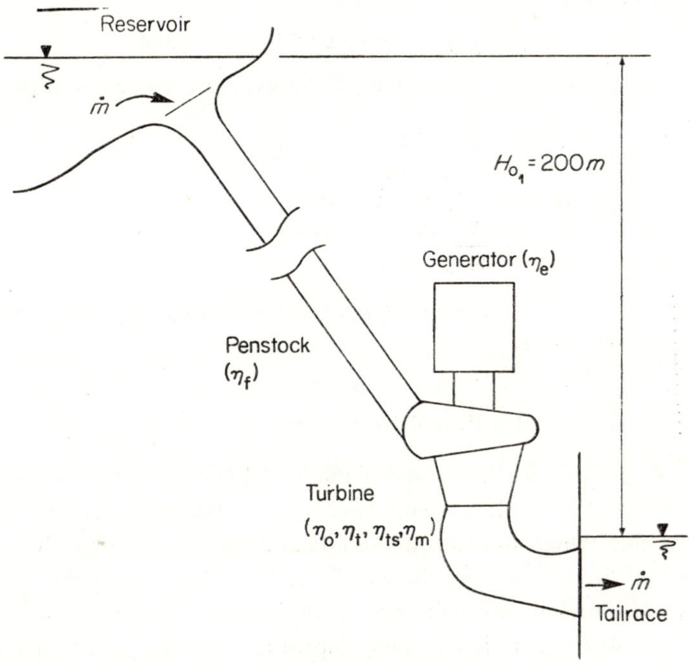

Figure 6.9 Efficiencies characterizing the components of a hydroelectric plant

The installation efficiency is

$$\eta_i = \eta_f \eta_{ts} \eta_m \eta_e$$
$$= 0{\cdot}95 \times 0{\cdot}92 \times 0{\cdot}98 \times 0{\cdot}90 = 77 \text{ per cent}$$

Here η_f is the efficiency of the intake.

The two hydraulic efficiencies are defined by equations (6.2.2, 11):

$$\eta_t = \frac{w_e}{\Delta(gH_0)} \quad \text{and} \quad \eta_{ts} = \frac{w_e}{\Delta(gH_0) - \frac{1}{2}u_2^2}$$

Then

$$\eta_t = \eta_{ts} \frac{\Delta(gH_0) - \frac{1}{2}u_2^2}{\Delta(gH_0)}$$

But

$$\Delta(gH_0) = \tfrac{1}{2}u_2^2 - \eta_f g H_{0_1}$$

when H_{0_1} is measured from the surface level at outlet, so that

$$\eta_t = \frac{\eta_{ts}}{1 - u_2^2/(2g\eta_f H_{0_1})}$$
$$= \frac{0{\cdot}92}{0{\cdot}968} = 95 \text{ per cent}$$

The overall efficiency of the turbine is then

$$\eta_0 = \eta_t \eta_m = 0{\cdot}98 \times 0{\cdot}95 = 93 \text{ per cent}$$

Hence we see that the dissipation of the available energy is distributed as follows:

5%	in the intake system
7%	in the turbine
3%	in the tailrace
9%	in the generator
24%	overall

As a result of errors in rounding-off, this total does not quite match the $\eta_i = 77$ per cent obtained above.

6.2.3 The Efficiency of a Propulsive System

Figure 6.10(a) shows the flows of mass and some of the flows of energy into and out of a generalized propulsive system. We shall confine our attention to devices which do not add significantly to the through-flow, that is, with the fuel flow, $\dot{m}_f \ll \dot{m}$, the air flow. This means that the results will not apply to a rocket, but will apply with fair accuracy to aircraft and ship propellers, and to turbojet, turbofan, propjet and ramjet engines as well.

Although the mass added to the flow by the fuel is taken to be negligible, the energy added is most certainly not. It is represented by the total energy

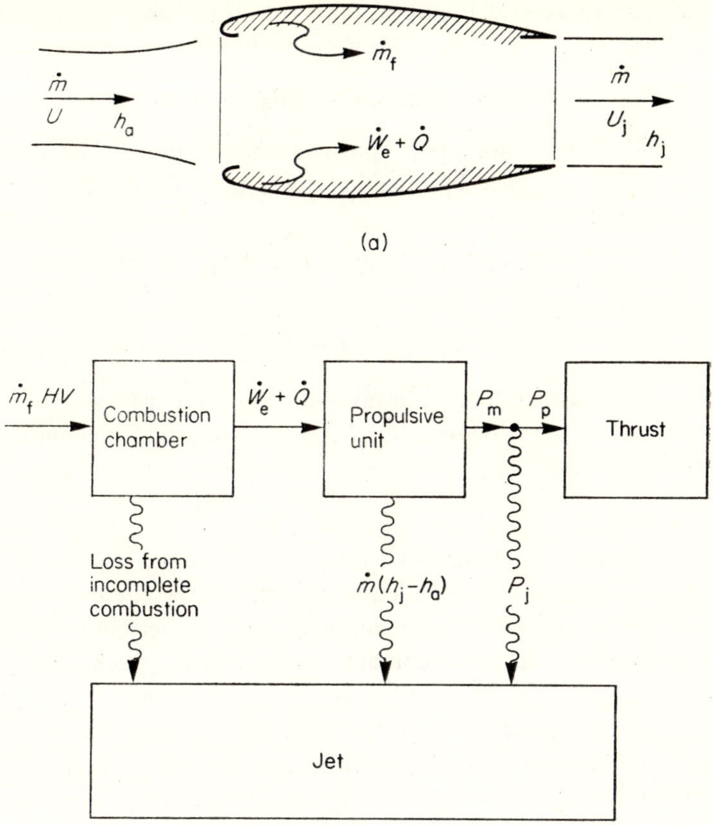

(b)

Figure 6.10 Energy flows in a generalized propulsive unit. (a) The propulsive unit as seen in relative axes, those in which the unit is stationary. (b) Distribution of the energy potentially available to the unit

input to the airflow, $\dot{W}_e + \dot{Q}$. In a jet engine the primary energy addition will be in the form of heat. The energy addition by a propeller contains a significant element of work. However, even in this case heat will be added to the air passing near the motor driving the propeller and through the cooling system for the motor.

Example 2.3 considered the momentum balance for this kind of idealized propulsive system. The net thrust was found to be

$$F = \dot{m}(U_j - U)$$

giving a propulsive power of

$$P_p = \dot{m}U(U_j - U) \tag{6.2.13}$$

This is the useful output of the system.

For the relative axes of Figure 6.10(a), the steady-flow energy equation can be applied to give the overall energy balance for the propulsive unit:

$$\dot{W}_e + \dot{Q} = \dot{m}(h_j - h_a) + \tfrac{1}{2}\dot{m}(U_j^2 - U^2)$$

In this frame of reference no propulsive work is done. The propulsive power can be distinguished by breaking down the kinetic-energy term into two components:

$$\tfrac{1}{2}\dot{m}(U_j^2 - U^2) = \dot{m}U(U_j - U) + \tfrac{1}{2}\dot{m}(U_j - U)^2$$

The second term

$$P_j = \tfrac{1}{2}\dot{m}(U_j - U)^2 \tag{6.2.14}$$

gives the flux of kinetic energy away from the unit, as seen in absolute axes, for which the jet velocity is $U_j - U$. Hence the total mechanical energy for the relative axes

$$P_m = P_p + P_j = \tfrac{1}{2}\dot{m}(U_j^2 - U^2) \tag{6.2.15}$$

can be interpreted for absolute axes as the sum of the propulsive power and the kinetic energy lost in the jet. For precision, the energy flow P_m will be referred to as 'organized' mechanical energy; the jet behind a realistic propulsive system will also contain 'unorganized' mechanical energy associated with swirling flow and non-uniform axial velocity.

These energy flows are shown in Figure 6.10(b); a rather similar diagram was used in Figure 6.8(b) to represent energy flows in a pumping plant. The mechanical-energy flows have been considered above. The other energy flows are:

$\dot{m}_f HV$ representing energy potentially available in the fuel expended, and

$\dot{m}(h_j - h_a)$ representing energy convected away in the hot fluid of the jet

Usually the lower heating value ($LHV = LCV$) of the fuel is used to determine the energy potentially available. In practice, some of the energy $\dot{m}_f LHV$ will, as a result of incomplete combustion, remain as chemical energy and will be convected away in the jet. When applied to actual propulsive systems, the term $\dot{m}(h_j - h_a)$ must include 'unorganized' mechanical energy as defined above.

The following measures of efficiency are used to describe the propulsive process:

(1) the combustion efficiency

$$\eta_c = \frac{\dot{W}_e + \dot{Q}}{\dot{m}_f HV}$$

(2) the efficiency of conversion to organized mechanical energy

$$\eta_m = \frac{P_m}{\dot{W}_e + \dot{Q}}$$

(3) the propulsive efficiency, the proportion of the organized mechanical energy converted to thrust power

$$\eta_p = \frac{P_p}{P_m}$$

(4) the thermal efficiency, the proportion of the available energy converted to mechanical form

$$\eta_t = \frac{P_m}{\dot{m}_f HV}$$

(5) the overall efficiency

$$\eta_0 = \frac{P_p}{\dot{m}_f HV}$$

These efficiencies are related by

$$\eta_0 = \eta_t \eta_p \quad \text{and} \quad \eta_t = \eta_c \eta_m$$

and the propulsive power is related to the energy expenditure by

$$P_p = \eta_c \eta_m \eta_p (\dot{m}_f HV)$$

Evidently efficient propulsion demands both efficient conversion of the energy of the fuel into mechanical energy, and efficient utilization of the mechanical energy for propulsion. The factors leading to a high thermal efficiency are complex, and vary from one species of propulsive plant to the next. The propulsive efficiency is rather easier to analyse, and the results are applicable to any species of plant. Equations (6.2.13, 14, 15) give

$$\eta_p = \frac{P_p}{P_m} = \frac{1}{1 + P_j/P_p} = \frac{2}{1 + U_j/U}$$

These results indicate that a high propulsive efficiency will be achieved when only a small proportion of the mechanical energy is lost in the jet, and that this condition will be satisfied when the rise in velocity $U_j - U$ is a small fraction of the forward speed. In other words, a specified thrust may be generated more efficiently by slightly accelerating a large mass flow than by greatly accelerating a small mass flow.

These conclusions would seem to indicate that a value of U_j/U close to unity should be sought. In practice, it proves undesirable to reduce U_j/U to values very near unity, for several reasons. For a specified speed and thrust, the propulsive unit becomes larger as $U_j - U$ is reduced, since the area of the captured flow is

$$A = \frac{\dot{m}}{\varrho_a U} = \frac{F}{\varrho_a U(U_j - U)}$$

The external drag on the unit will also increase. Finally, the effect of losses within the propulsive system becomes more important as the mass flow through it increases, and ultimately the energy requirement actually begins to rise as U_j/U is decreased further. The reason for this effect is that, roughly speaking, losses act to reduce the gross thrust $\dot{m}U_j$ and so produce a disproportionate change in the net thrust $\dot{m}(U_j - U)$ when U_j is close to U. Nevertheless, with the high efficiencies achieved for aircraft-engine components in recent years, it has proved possible to seek lower values of U_j/U in by-pass and fan-jet engines. The reduction in jet velocity has the added advantage of making the aircraft quieter.

Example 6.9 What are the various efficiencies for the propulsive plant of the supersonic aeroplane considered in Example 2.3(a)? For $U = 1500$ ft/sec, $P_p = 50{,}000$ hp and $T_a = 390$ °R, it was found that $\dot{m} = 11{\cdot}3$ slugs/sec, $U_j = 3120$ ft/sec, and $T_j = 1354$ °R. For this flight condition the engines consume kerosene at the rate of 8 lb/sec.

The various energy flows are (per unit of air flow):

$$\frac{P_p}{\dot{m}} = U(U_j - U) = 2{\cdot}43 \times 10^6 \text{ ft lbf/sl}$$

$$\frac{P_m}{\dot{m}} = \frac{1}{2}(U_j^2 - U^2) = 3{\cdot}74 \times 10^6 \text{ ft lbf/sl}$$

$$h_j - h_a = c_p(T_j - T_a)$$
$$= 6000 \times 964 = 5{\cdot}79 \times 10^6 \text{ ft lbf/sl}$$

(Here c_p has been given its normal atmospheric value.)

$$\frac{\dot{W}_e + \dot{Q}}{\dot{m}} = \frac{P_m}{\dot{m}} + (h_j - h_a)$$
$$= 9{\cdot}53 \times 10^6 \text{ ft lbf/sl}$$

$$\frac{\dot{m}_f HV}{\dot{m}} = \frac{8 \times 18{,}600 \times 778}{11{\cdot}3}$$
$$= 10{\cdot}23 \times 10^6 \text{ ft lbf/sl}$$

(Here the lower calorific value of kerosene has been taken from Table 6.1.)

The several efficiencies can now be found:

$$\eta_p = \frac{P_p}{P_m} = 65\%$$

$$\left.\begin{aligned} \eta_m &= \frac{P_m}{\dot{W}_e + \dot{Q}} = 39\% \\ \eta_c &= \frac{\dot{W}_e + \dot{Q}}{\dot{m}_f HV} = 93\% \end{aligned}\right\} \eta_t = \frac{P_m}{\dot{m}_f HV} = 37\%$$

$$\eta_0 = \frac{P_p}{\dot{m}_f HV} = 24\%$$

Another way of looking at the energy balance is to compare the various losses. As percentages of the available energy of the fuel these are

7% in incomplete combustion
13% in organized kinetic energy in the jet
56% in 'unorganized' energy in the jet, primarily an enthalpy rise
76% overall

Evidently the high temperature of the fluid in the jet, necessary in this kind of propulsive plant to reduce the density of the air and so increase its velocity to provide thrust, is the major source of inefficiency in the process.

6.3 HEAT TRANSFER

6.3.1 Modes of Heat Transfer

We have considered the effects of combustion and work transfer on a flowing fluid; now we come to the third mechanism of energy addition—heat transfer. Note that it is energy, not heat, which is actually transferred. The transfer, which is equivalent to a flow of heat, may be accomplished by one of several mechanisms, usually grouped into radiation, conduction and convection.

Radiation is the transfer of energy by electromagnetic waves. The energy transferred between two bodies is proportional to the difference of the fourth powers of their absolute temperatures and to the emissivity of the surfaces. Hence radiation is of particular importance in combustion chambers where large temperature differences exist. Incandescent particles of burning fuel are highly effective radiators, and radiation from them prevents flame temperatures from achieving the values which would be achieved in adiabatic flow.

Monatomic and diatomic gases are relatively passive with respect to radiation, but absorption and radiation are significant for gases with more complex molecules, in particular, CO_2 and H_2O. Hence radiation from combustion products can be considerable.

Radiation is not directly dependent on the fluid motion, though the flow pattern must of course influence the temperature distribution. Since the interaction between fluid flow and energy flow is relatively weak, this topic will not be considered further in this book.

Conduction is the transfer of energy by molecular activity in solids, in fluids at rest, and normal to the streamlines in moving fluids. In solids the transfer is achieved by processes such as atomic vibration; in gases, by molecular migration. The rate of heat flow is proportional to the temperature gradient within the material and to its thermal conductivity.

The transfer of energy by turbulent mixing across the mean streamlines of a turbulent flow may also be thought of as conduction. This transfer can be described using an effective or eddy conductivity. In the body of a turbulent flow this is usually much larger than the molecular conductivity, just as the eddy viscosity greatly exceeds the molecular viscosity.

Convection is the transfer of energy by macroscopic fluid motions. The rate of energy flow depends on the intrinsic energy of the moving fluid and on its velocity. In general, convection carries energy in the direction of the mean flow, but in a turbulent flow the small-scale convecting motions of the turbulence can produce a net energy flow normal to the mean streamlines. This transfer within a turbulent flow may be described either as turbulent conduction, or as a net convection resulting from the fluctuating turbulent motion.

Many of the processes described as convective heat transfer do in fact involve conduction as well. Consider, for example, the heat transfer from a hot sphere immersed in an air flow. A simple convective transfer of energy occurs only in the wake; the processes by which energy enters the flow from the hot surface are laminar conduction very near the surface and turbulent conduction farther away. Nevertheless, the term convective heat transfer is used to describe the whole flow. When the energy-convecting flow is established by the buoyancy of the heated fluid, as in the earth's atmosphere, the process is termed *free or natural convection*. When the fluid flow is essentially independent of the heat transfer, as in a hair dryer or in the flow over the wing of an aeroplane, the process is termed *forced convection*.

The boiler sketched in Figure 6.11 illustrates these modes of heat transfer.* Normally the circulation of the water and gases is forced by the pump and fan shown in the figure. However, if the fan ceased to provide a forced

* The flow passages of actual boilers are considerably more complicated than those shown. A somewhat idealized layout has been adopted to simplify this introductory discussion.

draught, some air flow would still be generated by natural convection. Through most of the heat exchanger, both on the water side and the gas side, convection (including small-scale turbulent convection) is the most important mechanism of energy transfer. However, the heat flow through the walls of the tubes and through the viscous layers on their inner and outer surfaces is accomplished by conduction. Radiation is of importance in the superheater section, where the temperatures are highest and the gases may contain specks of burning carbon.

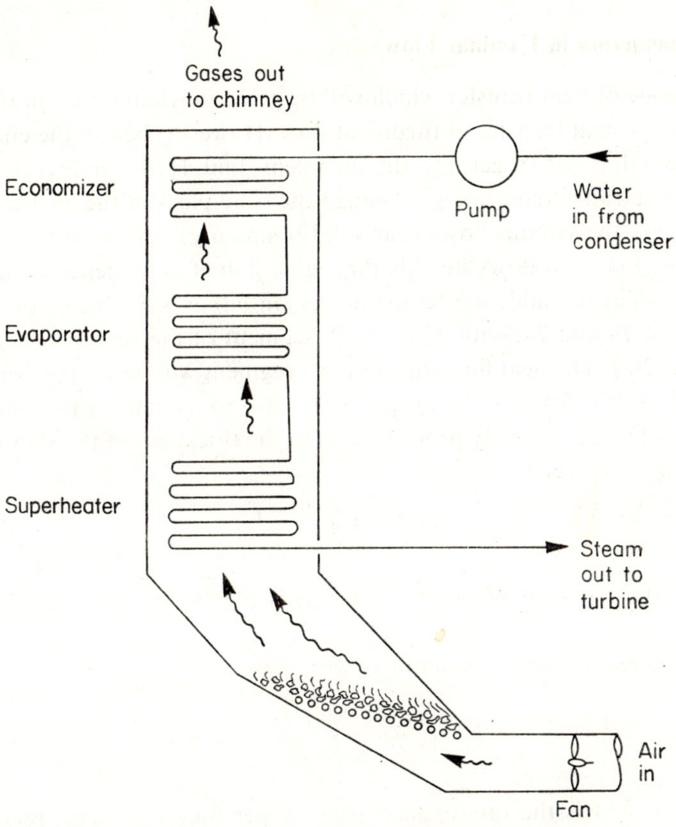

Figure 6.11 A schematic view of a furnace and boiler

Although heat transfer usually does not involve chemical changes in the flowing fluid, it often does involve phase changes. In the boiler of Figure 6.11, for example, the water is first heated as a liquid in the economizer, is then turned to steam in the evaporator, and finally flows as steam through the superheater tubes. (These three elements of the heat addition are clearly visible in Figure 1.30, as 2–a, a–b, b–3.) A phase change may also occur in the gas side of the flow through the boiler, for it is possible that water will condense out as the gases leave the economizer.

The three sections of the heat exchanger in a boiler differ markedly in their physical form, partly because they operate in very different environmental conditions, and partly because the heat-transfer characteristics of the fluid change from one section to the next. The heat-transfer properties of gases and liquids are somewhat different, and those of the two-phase boiling flow within the evaporator are quite different from either.

6.3.2 Conduction in Laminar Flow

The mode of heat transfer which will be given most attention in this book is convective heat transfer in turbulent flow. However, before the energy can be convected, it has to get into the turbulent fluid. Here conduction plays a vital part in transferring energy through the solid parts of the heat exchanger and through the viscous flows near solid boundaries.

Suppose that heat flows steadily through a slab of homogeneous stationary material, solid or fluid, the temperatures on either side being maintained constant at T_1 and T_2, with $T_1 > T_2$. A segment of the material is shown in Figure 6.12(a). The heat flow through this segment will be proportional to its plan area A. It is found to be proportional also to the temperature difference $T_1 - T_2$ and to be inversely proportional to the thickness of the slab t. Hence the heat flow is given by

$$\dot{Q} = -kA\,\frac{T_2 - T_1}{t}$$

where k, the constant of proportionality, is the *thermal conductivity* of the slab.

The last result can be generalized to

$$\dot{q} = -k\,\frac{\partial T}{\partial y} \tag{6.3.1}$$

where $\dot{q} = \dot{Q}/A$ is the rate of heat transfer per unit area. This relationship defines the local thermal conductivity in any material. In general, this property depends on the local temperature and pressure, but is independent of the temperature gradient. Appendix VII gives conductivities for some common fluids and for a number of solid materials. The dependence on temperature is particularly important for gases and vapours. The effect of pressure is small for most substances.

A partial derivative was used in equation (6.3.1) since that statement applies for non-steady, one-dimensional heat flows, for which $T = T(y, t)$.

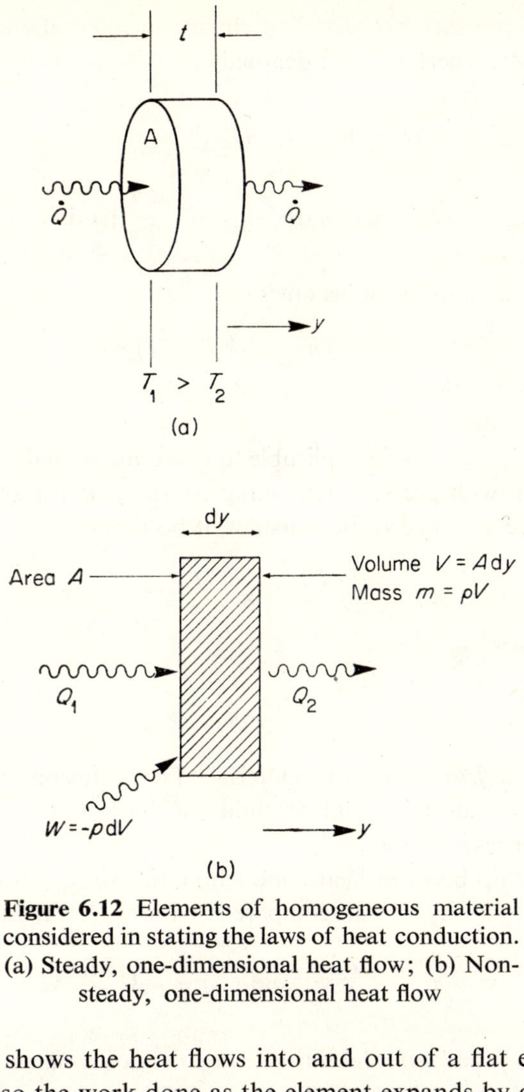

Figure 6.12 Elements of homogeneous material considered in stating the laws of heat conduction. (a) Steady, one-dimensional heat flow; (b) Non-steady, one-dimensional heat flow

Figure 6.12(b) shows the heat flows into and out of a flat element during time dt, and also the work done as the element expands by dV during that time. These energy flows are given by

$$Q_1 = -kA \frac{\partial T}{\partial y} \, \mathrm{d}t$$

$$Q_2 = -\left[k \frac{\partial T}{\partial y} + \mathrm{d}\left(k \frac{\partial T}{\partial y}\right)\right] A \, \mathrm{d}t$$

and

$$W = -p \, \mathrm{d}V = -p \, \mathrm{d}\left(\frac{m}{\varrho}\right) = -m \, \mathrm{d}\left(\frac{p}{\varrho}\right)$$

for a constant-pressure process. The change in internal energy during the process considered here may be denoted $\Delta E = m\,\mathrm{d}e$. Then the first law of thermodynamics gives

$$\Delta E = W + Q_1 - Q_2$$

or

$$m\,\mathrm{d}e = -m\,\mathrm{d}\left(\frac{p}{\varrho}\right) + \mathrm{d}\left(k\frac{\partial T}{\partial y}\right) A\,\mathrm{d}t$$

Since $m/A = \varrho\,\mathrm{d}y$, this result becomes

$$\varrho\frac{\partial h}{\partial t} = \frac{\partial(k\,\partial T/\partial y)}{\partial y}$$

in the limit as $\mathrm{d}t, \mathrm{d}y \to 0$.

The last result is generally applicable to one-dimensional heat flow, that is, for $T = T(y, t)$, with pressure remaining constant. If the specific heat and conductivity are assumed to be constant, it becomes

$$\frac{\partial T}{\partial t} = \varkappa\frac{\partial^2 T}{\partial y^2} \tag{6.3.2}$$

where

$$\varkappa = \frac{k}{\varrho c_p}$$

is the *thermal diffusivity* of the material. The diffusion equation (6.3.2) applies for any material—solid or fluid—so long as the assumptions of constant properties are valid.

The relationship between heat conduction and viscous friction in a fluid can be seen in two ways. First, compare equations (5.1.2) and (6.3.1):

$$\tau = \mu\frac{\partial u}{\partial y} \quad \text{and} \quad \dot{q} = -k\frac{\partial T}{\partial y}$$

The coefficients μ and k represent the rate at which momentum and energy are transported through the fluid by molecular activity; they are often grouped together using the name *transport properties*. The link between heat transfer and friction can also be established by comparing equation (6.3.2) with

$$\frac{\partial u}{\partial t} = \nu\frac{\partial^2 u}{\partial y^2} \tag{5.1.18}$$

which governs the transfer of momentum across a non-steady, parallel flow. The two equations have precisely the same form. Note that the kinematic viscosity ν is the diffusivity of momentum in the fluid.

The ratio of the two diffusivities in a fluid

$$\mathrm{Pr} = \frac{\nu}{\varkappa} = \frac{c_p\mu}{k} \tag{6.3.3}$$

is termed the *Prandtl number*. The Prandtl numbers for some common fluids are given in Appendix VII. For gases and vapours the Prandtl number remains constant over a considerable range of pressure and temperature, since the mechanisms of momentum and energy transfer are closely linked. The Prandtl number is a measure of the relative ease with which momentum and heat diffuse through a fluid. Thus it determines the relative thicknesses of the momentum and thermal boundary layers in laminar flow. For gases and vapours Pr $\simeq$ 1, and the two boundary layers have about the same thickness. For water Pr $\sim$ 10, indicating that heat conduction is relatively slow; while for liquid metals Pr $\sim$ 1/100, indicating that heat flows very rapidly through the fluid. These considerations do not apply to a turbulent flow, in which heat and momentum are transferred by the same mechanism of turbulent mixing, and $\mathrm{Pr_t} \simeq 1$ relates the two turbulent diffusivities.

Example 6.10 In a flowing fluid, heat transfers are determined not only by the temperature distribution at the boundaries, but by the dissipation occurring within the fluid. Consider, for example, the simplest kind of laminar motion: steady, plane Couette flow with constant properties c_p, ϱ, μ and k. In the flow shown in Figure 6.13, the dissipation is uniform at $\mu(U/b)^2$ per unit volume, as shown in equation (5.1.25). The temperature distribution will also be assumed to be invariant in time and in the x-direction; then $T = T(y)$ only. Hence, when the source term corresponding to dissipation is introduced, the conduction equation (6.3.2) becomes

$$k\frac{\mathrm{d}^2T}{\mathrm{d}y^2} + \mu\left(\frac{U}{b}\right)^2 = \varrho c_p\frac{\partial T}{\partial t} = 0$$

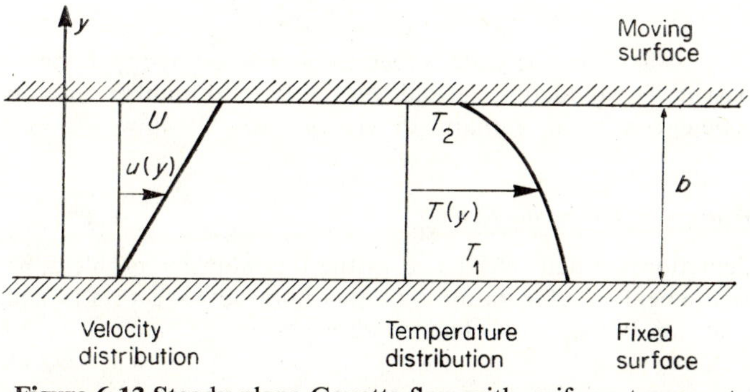

Figure 6.13 Steady plane Couette flow with uniform transport properties

Integrating twice, we find a parabolic temperature distribution:

$$T = -\frac{\mu}{k}\left(\frac{U}{b}\right)^2 \frac{y^2}{2} + Ay + B$$

The boundary conditions are

$$T = T_1 \quad \text{at} \quad y = 0$$
$$T = T_2 \quad \text{at} \quad y = b$$

Then the temperature distribution is

$$T = T_1 + (T_2 - T_1)\frac{y}{b} + \frac{\mu}{k}\frac{U^2}{2}\frac{y}{b}\left(1 - \frac{y}{b}\right)$$

Let us use this result to answer two questions:

(a) For a specified temperature difference $T_1 - T_2$, how is the heat-transfer rate influenced by the fluid flow?

(b) If the lower plate is insulated against heat flow, how is its temperature related to that of the upper plate?

(a) Heat transfer at the lower plate

The heat-transfer rate at $y = 0$ is given by

$$\dot{q} = -k\left(\frac{\partial T}{\partial y}\right)_0 = k\frac{T_1 - T_2}{b} - \frac{\mu U^2}{2b}$$

If the fluid were still, the heat transfer would be

$$\dot{q}_0 = k\frac{T_1 - T_2}{b}$$

Then

$$\frac{\dot{q}}{\dot{q}_0} = 1 - \frac{\mu}{k}\frac{U^2}{2(T_1 - T_2)}$$

$$= 1 - \text{Pr}\frac{U^2}{2c_p(T_1 - T_2)}$$

The heat flow away from the plate is seen to be reduced by the dissipation within the flow past it. The reduction depends on the Prandtl number, and on the ratio of the kinetic energy to the enthalpy difference across the flow.

(b) Temperature of the insulated plate

This calculation is often referred to as the thermometer problem, as it relates to the measurement of temperature by a thermometer beneath a boundary layer. For $\dot{q} = 0$ at $y = 0$, the result obtained above gives

$$T_{1_0} = T_2 + \text{Pr}\frac{U^2}{2c_p}$$

Since $T_{0_2} = T_2 + U^2/2c_p$ is the total temperature at the upper plate, this can be given as

$$\frac{T_{1_0} - T_2}{T_{0_2} - T_2} = \mathrm{Pr}$$

Comparison with equation (4.3.3) shows that Pr may be interpreted as the recovery factor for the lower plate, and T_{1_0} as the recovery temperature. Note that

$$T_{1_0} \gtrless T_{0_2} \quad \text{for} \quad \mathrm{Pr} \gtrless 1$$

The recovery factor for more realistic boundary layers is found to depend on Prandtl number in a rather similar manner, with $r \propto \mathrm{Pr}^{\frac{1}{2}}$ for a laminar layer, and $r \propto \mathrm{Pr}^{\frac{1}{3}}$ for a turbulent layer.

6.3.3 Heat Transfer across a Turbulent Boundary Layer

The transfer of energy within a turbulent flow cannot be related to the temperature gradient merely by introducing a constant of proportionality, as was done for a laminar flow in equation (6.3.1). The heat-transfer rate depends on the turbulent mixing within the fluid and not only on its thermal conductivity. In predicting the heat transfer within a turbulent flow, some measure of empiricism is always necessary, as it is in predicting the friction generated by turbulence. For flow passages of unusual shape, special tests are required to determine the heat-transfer characteristics. However, for parallel flow along a smooth wall it is possible to relate the heat transfer and momentum transfer in a fairly simple way. This relationship makes possible the use of empirical friction data to predict heat transfer.

The argument will be based on Figure 6.14, which shows the variations of the mean velocity and temperature across a turbulent boundary layer on a smooth wall. The viscous sublayer of thickness δ_s plays an important role in the discussion. In it the velocity variation is linear, friction being generated by the molecular viscosity. The temperature variation in the sublayer also arises from molecular activity; it too will be taken to be linear. Then in the sublayer

$$u = u_s \frac{y}{\delta_s} \quad \text{and} \quad T - T_w = (T_s - T_w)\frac{y}{\delta_s}$$

with u_s and T_s values at the edge of the layer. The stress and heat flow within the viscous sublayer are found using equations (5.3.11) and (6.3.1):

$$\tau_0 = \mu \frac{du}{dy} = \mu \frac{u_s}{\delta_s}$$

and (6.3.4)

$$\dot{q} = -k \frac{dT}{dy} = -k \frac{T_s - T_w}{\delta_s}$$

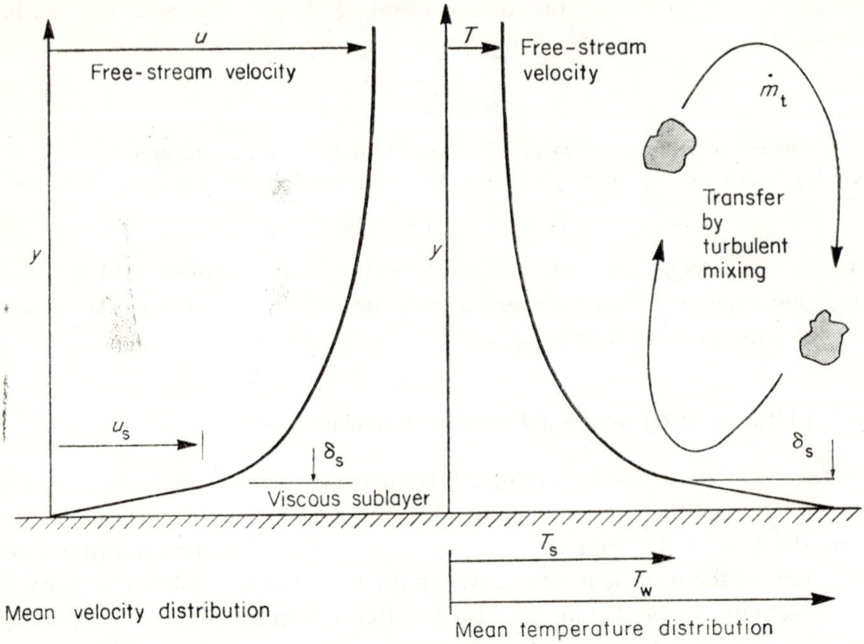

Figure 6.14 A simple model of the turbulent boundary layer on a smooth wall, illustrating the relationship between friction and heat transfer

Here $\dot{q}$ is the heat transfer rate per unit area, positive for energy flow away from the surface, that is, for heat transfer to the fluid.

Now consider the transfer of momentum and energy in the outer, turbulent part of the boundary layer. The turbulent mixing will generate net flows of momentum and energy across this layer. This process will be assumed to be equivalent to an interchange flow $\dot{m}_t$ which carries fluid from the edge of the sublayer to the core of the flow and then returns to the edge of the sublayer. The net effect of these two flows is to produce momentum and energy fluxes corresponding to the friction stress and heat transfer within the turbulent fluid. According to this model

$$\dot{m}_t(u_s - u) = -\tau_0$$

and (6.3.5)

$$\dot{m}_t c_p(T_s - T) = \dot{q}$$

Here u and T are free-stream values at the edge of the boundary layer. The validity of the second equation is limited not only by the crudeness of the model, but by the neglect of changes in kinetic energy and by the use of a constant specific heat. Moreover, an exact equivalence of the mechanisms of momentum and energy transfer has been assumed. For the purposes of this discussion we need not be able to predict the effective interchange rate $\dot{m}_t$; it is the relationship between momentum and energy transfers which is of interest.

More useful results can be obtained from equations (6.3.4, 5) by eliminating the quantities describing the viscous sublayer. These equations can be combined to eliminate δ_s and $\dot{m}_t$:

$$\frac{\dot{q}}{\tau_0} = \frac{k(T_w - T_s)}{\mu u_s}$$
$$= \frac{c_p(T_s - T)}{u - u_s}$$

The temperature at the edge of the sublayer can now be calculated in terms of T_w and T; when it is substituted back into the preceding equations, they give

$$\frac{\dot{q}}{\tau_0} = \frac{c_p(T_s - T)}{u - u_s} = \frac{c_p(T_w - T)}{u + (\mathrm{Pr} - 1)u_s} \tag{6.3.6}$$

Here $\mathrm{Pr} = c_p\mu/k$ is the Prandtl number of the fluid. Finally, the velocity at the edge of the sublayer can be eliminated. As suggested by Figure 5.14, the thickness can be estimated using $u_f\delta_s/\nu = 5$, with u_f the friction velocity. The first of equations (6.3.4) gives $\delta_s = \mu u_s/\tau_0 = \nu u_s/u_f^2$, and it follows that

$$u_s = 5u_f$$

Now equation (6.3.6) can be written

$$\frac{\dot{q}}{\tau_0} = \frac{c_p(T_w - T)}{u + 5(\mathrm{Pr} - 1)u_f}$$

independent of the characteristics of the sublayer.

When the friction coefficient $c_f = \tau_0/(\frac{1}{2}\varrho u^2) = 2u_f^2/u^2$ is introduced, the last result can be reduced to the dimensionless form

$$\mathrm{St} = \frac{\dot{q}}{\varrho u c_p(T_w - T)} = \frac{\frac{1}{2}c_f}{1 + 5(\mathrm{Pr} - 1)\left(\frac{1}{2}c_f\right)^{\frac{1}{2}}} \tag{6.3.7}$$

Here St is the *Stanton number*, one of several heat-transfer coefficients in common use. The Stanton number relates the heat flow from the wall to the flux of energy convected by the mean flow. Although the last result indicates the general nature of the relationship between heat transfer and friction, it is based on a highly simplified model of the boundary layer. This model neglects the existence of the buffer layer between the viscous and fully turbulent regions, and also assumes a similarity between the momentum and thermal boundary layers which will exist in fact only when Pr = 1. For fluids such as gases, with Prandtl numbers not too different from unity, it is found that the modified form

$$\mathrm{St} = \frac{\frac{1}{2}c_f}{1+7{\cdot}5\,\mathrm{Pr}^{-\frac{1}{6}}(\mathrm{Pr}-1)\left(\frac{1}{2}c_f\right)^{\frac{1}{2}}} \tag{6.3.8}$$

provides a better prediction of the heat transfer.

Equation (6.3.8) shows how closely heat transfer and friction are related. Indeed, for some purposes the approximate form

$$\mathrm{St} = \tfrac{1}{2}c_f \tag{6.3.9}$$

gives an adequate indication of the heat-transfer rate. This result is referred to as *Reynolds analogy* between heat transfer and friction. It is particularly useful for gas flows, where the Prandtl number is close to unity and the neglected term is small.

The dimensionless forms used above obscure an important aspect of the relationship between heat transfer and friction. From the dimensioned result (6.3.6) we have, to the approximation implicit in Reynolds analogy

$$\frac{\dot{q}}{\tau_0} = \frac{c_p\,\Delta T}{u} \tag{6.3.10}$$

This indicates that, although both friction and heat transfer increase with flow velocity, the former rises more rapidly. A more explicit comparison of the velocity dependence can be made by taking $c_f \propto \mathrm{Re}^{-\frac{1}{4}}$, appropriate to flow in a smooth tube at Reynolds numbers less than 100,000. Then, with factors other than the velocity remaining constant

$$\tau_0 = \tfrac{1}{2}c_f\,\varrho\,u^2 \propto u^{\frac{7}{4}}$$

and (6.3.10)

$$\dot{q} = \frac{c_p\,\Delta T\,\tau_0}{u} = \frac{1}{2}\,\varrho c_p\,\Delta T\,c_f\,u \propto u^{\frac{3}{4}}$$

These relationships point to the necessity of compromise in the design of heat exchangers. A specified heat transfer can be achieved in a large unit, with large heat-transfer area, large flow passages and low velocities, or in a smaller unit with reduced heat-transfer surface, smaller channels and higher velocities. A choice of the first kind will result in a bulky and expensive unit, while the second will entail higher friction, a larger pressure drop, and a greater requirement of pumping power.

Example 6.11 Although the result (6.3.8) provides a widely applicable relationship between turbulent heat transfer and friction, it is not in a form convenient for application to any one kind of flow. For a particular flow system and fluid, it is common practice to use an approximate expression of the form

$$\mathrm{Nu} = a\,\mathrm{Re}^b$$

where $\mathrm{Nu} = (\dot{q}d)/(k\,\Delta T)$ is the Nusselt number characterizing the heat-transfer process, d being a characteristic linear dimension of the system. The coefficients a and b are chosen to fit experimental results or to match the more general result (6.3.8) as closely as possible; the factor a, at least, will depend upon the Prandtl number of the fluid. The Nusselt number is used in preference to the Stanton number, since it contains only properties directly relevant to the heat transfer, leaving the Reynolds number to characterize the fluid flow.

To illustrate the method of obtaining a correlation of this kind, let us consider fully developed turbulent flow of air in a smooth-walled pipe, and develop a heat-transfer formula valid in the range $\mathrm{Re} = 5\times10^4$ to 10^6. For this case, $\mathrm{Pr} = 0{\cdot}7$, and the friction law is

$$\text{for} \quad \mathrm{Re} < 10^5 \qquad c_f = 0{\cdot}079\,\mathrm{Re}^{-\frac{1}{4}}$$

supplemented by

Re	c_f
2×10^5	0·00391
5×10^5	0·00329
10^6	0·00292

For $\mathrm{Pr} = 0{\cdot}7$, the result (6.3.8) reduces to

$$\mathrm{St} = \frac{c_f}{2\left[1 - 2{\cdot}4\left(\frac{1}{2}c_f\right)^{\frac{1}{2}}\right]}$$

Note that the Nusselt number based on pipe diameter is given by

$$\mathrm{Nu} = \mathrm{Pr}\,\mathrm{Re}\,\mathrm{St} = 0{\cdot}7\,\mathrm{Re}\,\mathrm{St}$$

Table 6.2 gives the relationship between Re and Nu defined by these results. When

Table 6.2. Correlation of Nusselt and Reynolds numbers for flow in smooth-walled pipes with $\mathrm{Pr} = 0{\cdot}7$

Re	c_f	St	Nu
5×10^4	0·00528	0·00301	105
10^5	0·00444	0·00250	175
2×10^5	0·00391	0·00219	307
5×10^5	0·00329	0·00182	637
10^6	0·00292	0·00161	1197

the calculated values are plotted in Figure 6.15, the approximate correlation

$$\mathrm{Nu} = 0{\cdot}016\ \mathrm{Re}^{0{\cdot}81}$$

is obtained. However, the data can be represented by other, slightly different pairs of coefficients.

This technique can be extended to obtain correlations of the form

$$\mathrm{Nu} = (a_0\ \mathrm{Pr}^c)\ \mathrm{Re}^b$$

Now the same value of b is used for the correlations for several values of Pr, and the several coefficients $a = a_0\ \mathrm{Pr}^c$ are then analysed to find suitable values of a_0 and c.

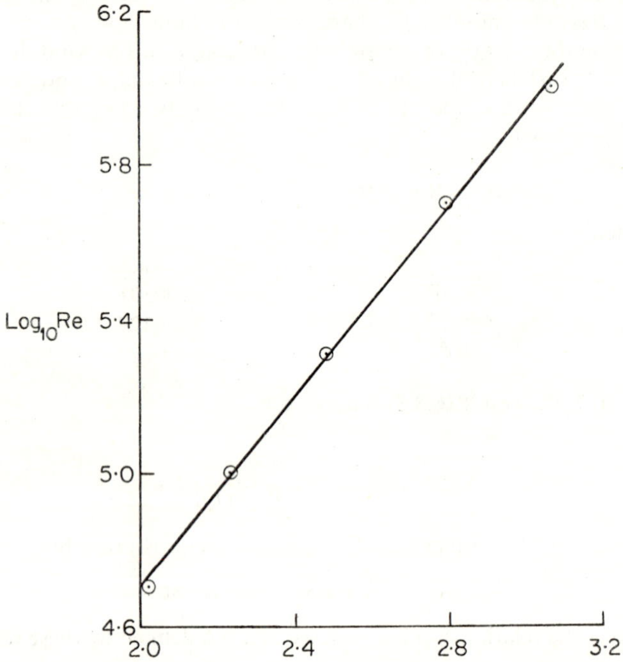

Figure 6.15 Relationship between Nusselt and Reynolds numbers for turbulent air flow in a smooth pipe. The straight line is the best power-law representation over the range of interest

6.3.4 Empirical Heat-transfer Coefficients

The preceding analysis has shown that it is possible to calculate the heat transfer for certain simple flow geometries, in particular, for flow along a smooth, plane wall. These results indicate the general nature of the heat-transfer laws for more complex heat-exchange systems, but cannot easily be generalized to give accurate numerical estimates of the heat flow. Hence the prediction of the heat transfer in a complex system depends upon empirical information relating to that particular system. We shall now see how this is specified, and how it is used in heat-transfer calculations.

To be specific, consider the heat exchanger shown in Figure 6.16(a). This consists of a series of hollow metal fins through which water flows and between which air passes. The mean temperature of the water, $T_f > T_{a_1}$, the initial temperature of the air; hence heat flows from it to the air through the walls of the fins. The difficulty of predicting the heat-transfer rate is suggested by Figure 6.16(b), which shows the distribution of time-mean

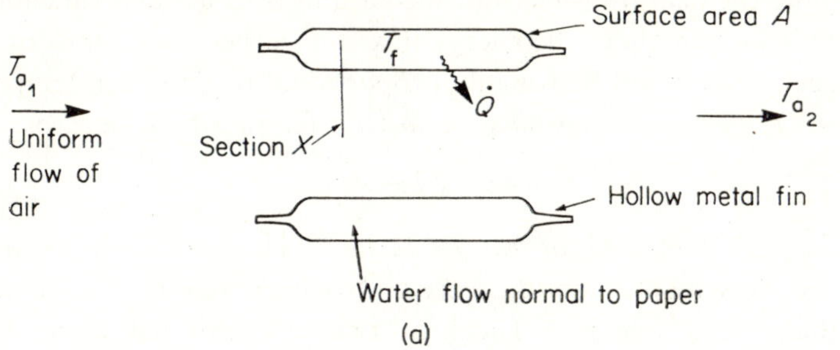

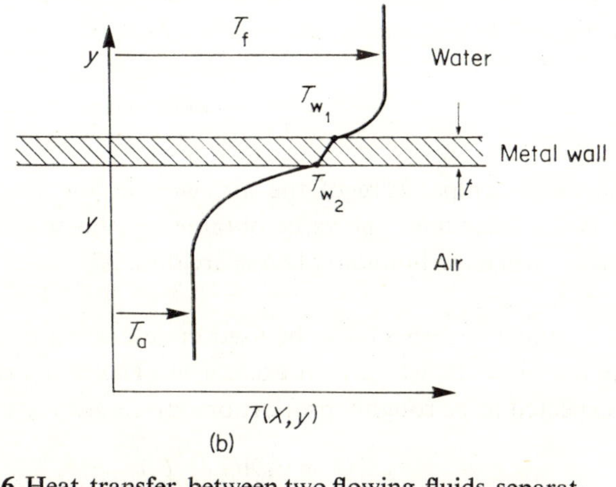

Figure 6.16 Heat transfer between two flowing fluids separated by a solid wall. (a) A representative heat exchanger made up of a number of parallel fins. (b) Distribution of mean temperature along the Section X shown in (a)

temperature (in general the flows will be turbulent) at Section X of Figure 6.16(a). The heat flow depends upon the heat-transfer characteristics of the wall and of the two wall layers, that in the liquid and that in the air. Moreover, the flow geometry and heat transfer will vary along the flow passage, and in a particularly complicated way near the leading and trailing edges of the fins.

Using existing knowledge of turbulent flows, it would be possible to make an estimate of the heat-transfer capabilities of the system of Figure 6.16. But even for this relatively simple geometry, the labour would be considerable and the results of uncertain accuracy. Hence the more usual approach in estimating the heat transfer from such a unit is to use experimental data. One thing is known about the energy flux: the earlier discussions of simple cases suggest that it will be roughly proportional to the mean temperature difference, $T_f - T_a$. Then the local heat-transfer rate may be expressed as

$$\dot{q} = C_h(T_f - T_a) \tag{6.3.11}$$

with $C_h = \dot{q}/\Delta T$ the local *heat-transfer factor*.* (This coefficient is sometimes denoted by the symbols h and U; here C_h has been used to avoid confusion with enthalpy and velocity.) The heat-transfer factor has dimensions of power/(area × temperature); the usual units are Btu/ft^2 hr °R and kW/m^2 °K.

In general, the factor C_h depends on the properties of the metal wall, on the properties of the two fluids, and on the characteristics of the two wall layers. It depends as well on the way in which the temperatures are defined. Within the fluid the time-mean temperature distribution is of the form $T(x, y)$. One way of obtaining a representative temperature for use in the relationship (6.3.11) is to integrate over the flow section. Thus, for example

$$T_a(x) = \frac{1}{b_a} \int_{b_a} T(x, y)\, dy \tag{6.3.12}$$

gives the mean temperature of the air, with b_a the width of the air gap between two similar fins. The value obtained in this way is termed the bulk temperature; when the boundary layers are thin, it is close to the temperature in the core flow.

A rather similar approach can be used in describing the performance of a complete fin, or of the whole heat exchanger. The total heat flow from a fin may be expected to be roughly proportional to its surface area. Then

$$\dot{Q} = C_h A(T_f - T_a) \tag{6.3.13}$$

defines the heat-transfer factor for the complete fin. This coefficient is in

* This is also called the surface coefficient of heat transfer.

some sense an average of the local heat-transfer factor over the surface of the fin. It will depend upon the properties of the two fluids, on their flow rates, and on the geometry of the flow passages. The way in which the temperatures are defined varies from situation to situation. In the case considered above, one might use the bulk temperature of the liquid

$$T_{\mathrm{f}} = \frac{1}{A_{\mathrm{f}}} \int_{A_{\mathrm{f}}} T(x, y)\, \mathrm{d}x\, \mathrm{d}y$$

(here A_{f} is the cross-sectional area within a fin), and the average of the upstream and downstream mean air temperatures

$$T_{\mathrm{a}} = \tfrac{1}{2}(T_{\mathrm{a}_1} + T_{\mathrm{a}_2})$$

For an isolated body exchanging heat with a uniform airstream, the temperature of the uniform flow would be selected as one of the relevant temperatures.

The local heat-transfer factor of equation (6.3.11) can be broken down into components representing the resistance of each of the elements through which the heat passes. For the case illustrated in Figure 6.16(b), we have two so-called film coefficients, one for the liquid and one for the air

$$C_{\mathrm{h}_1} = \frac{\dot{q}}{T_{\mathrm{f}} - T_{\mathrm{w}_1}} \quad \text{and} \quad C_{\mathrm{h}_2} = \frac{\dot{q}}{T_{\mathrm{w}_2} - T_{\mathrm{a}}}$$

and a factor for the intervening wall

$$C_{\mathrm{h_w}} = \frac{\dot{q}}{T_{\mathrm{w}_1} - T_{\mathrm{w}_2}} = \frac{k}{t}$$

This factor can be given an explicit form in terms of conductivity and thickness; the film coefficients must in general be determined empirically. By summing the temperature changes across each part of the heat flow path, we find

$$T_{\mathrm{f}} - T_{\mathrm{a}} = \dot{q}\left(\frac{1}{C_{\mathrm{h}_1}} + \frac{1}{C_{\mathrm{h_w}}} + \frac{1}{C_{\mathrm{h}_2}}\right) = \frac{\dot{q}}{C_{\mathrm{h}}}$$

with C_{h} the overall coefficient of equation (6.3.11). Hence the overall coefficient is given by

$$\frac{1}{C_{\mathrm{h}}} = \frac{1}{C_{\mathrm{h}_1}} + \frac{t}{k} + \frac{1}{C_{\mathrm{h}_2}}$$

Alternatively, one might define a thermal resistance

$$R = \frac{1}{C_{\mathrm{h}}} = \frac{\Delta T}{\dot{q}}$$

Then the overall resistance to heat flow through a series of components is

$$R = \Sigma R_i$$

the sum of the component resistances R_i. This analysis closely parallels that of a series electrical circuit.

Local and overall heat-transfer factors can be defined in a broadly similar manner for other flow geometries. Table 6.3 indicates the orders-of-magnitude typical of a number of common flow patterns and fluids. These serve to indicate which part of the heat flow path is likely to control the process. Consider, for example, furnace gases flowing across boiler tubes containing water. The values of C_h in Table 6.3 suggest that the resistance on the gas side will usually exceed that on the water side, particularly if the water is boiling. Hence it is worthwhile trying to increase the gas-side heat-transfer factor. This can be done by providing fins on the outside of the tubes to increase the effective heat-transfer area. Note also the very high heat-transfer factors associated with condensation; these explain how it is possible to reject very large amounts of energy in relatively small condensers.

Table 6.3. Typical values of heat-transfer factor C_h in Btu/ft^2 hr °R

Flow	Fluid	Range of C_h
Across tubes	Gas	2–10
	Liquid	10–100
In tubes	Gas	10–100
	Organic liquid	50–500
	Water	100–1000
	Liquid metal	1000–5000
Natural convection	Gas	0·1–100
	Liquid	10–500
Condensing	Liquid film	200–5000
	Liquid drops	5000–50,000
Boiling	Liquid/vapour	200–2000

Example 6.12 In a 'radiator' like that sketched in Figure 6.16(a), the surface area of the plates is 0·8 m^2. The walls are steel of thickness 0·16 in; the film coefficients for the water and air flows are 200 and 80, respectively. A scale deposited on the inside of the fins provides an additional resistance $R_f = 0{\cdot}001$ ft^2 hr °R/Btu; this is termed a fouling factor. The mean temperature difference between water and air is 40 °C. What power is 'dissipated' by the unit?

From Appendix VII, we find for steel $k = 25$ Btu/ft hr °R. Then $t/k = 1/1900 \simeq$ 0·0005 ft² hr °R/Btu is the resistance of the fin wall, and the overall resistance to the heat flow is

$$\frac{1}{C_h} = R_w + R_a + R_t + R_f$$

$$= 0{\cdot}0050 + 0{\cdot}0125 + 0{\cdot}0005 + 0{\cdot}0010$$

$$= 0{\cdot}0190 \text{ ft}^2 \text{ hr °R/Btu} = 3{\cdot}35 \text{ m}^2 \text{ °K/kW}$$

The heat-transfer rate is then

$$\dot{Q} = C_h A \, \Delta T = 0{\cdot}8 \times \frac{40}{3{\cdot}35} = 9{\cdot}55 \text{ kW}$$

The temperature drop across each part of the heat flow path is proportional to the resistance of that component. Then the temperature differentials for this system are

10·5 °C within the water

26·5 °C within the air

1 °C across the metal wall, and

2 °C across the scale

The resistance of the wall is of little importance in this case, and this is typical of most heat exchangers. In this situation the fouling factor is not very important either, although in practice it often does provide a significant resistance. Often the performance of a heat exchanger will deteriorate markedly as scale builds up on the heat-transfer surfaces. This scale must be removed in order to regain the original performance.

Example 6.13 The economizer of a boiler consists of tubes 100 ft long, with inner diameter 2 in and outer diameter 2·5 in, passing back and forth through the already fairly cool gases. Water flows through the tubes at 2 ft/sec, entering at 200 °F; the heat transfer within the tubes is specified by $\text{Nu} = 0{\cdot}023 \, \text{Re}^{0{\cdot}8} \, \text{Pr}^{0{\cdot}4}$. The gases flow past the tubes at 15 ft/sec, maintaining a nearly uniform temperature of 700 °F and density of 0·04 lb/ft³; the heat transfer within this flow is specified by $\text{Nu} = 0{\cdot}3 \, \text{Re}^{0{\cdot}6}$. In both gas and water correlations the properties are evaluated at the mean film temperature, that is, the average of the temperature at the wall and the temperature in the body of the flowing fluid.

The thermal resistance of the tube wall will be neglected. Moreover, the change in water temperature along the tubes will be neglected in the first instance; later this change will be taken into account. Also, since the temperature differential within the gas may be expected to exceed greatly that within the water, the properties of the water will be evaluated for the bulk temperature of 200 °F. However, the gas temperature changes from about 200 °F on the tube wall to 700 °F within the body of the flow. Hence the mean film temperature for the gas is 450 °F. The properties of air at this temperature will be used to describe the gases.

(a) Heat-transfer factors

Water side. The data of Appendix VII give

$$\nu = \frac{\mu}{\varrho} = 2{\cdot}9\times10^{-7}\ \text{m}^2/\text{sec} = 3{\cdot}1\times10^{-6}\ \text{ft}^2/\text{sec}$$

$$\text{Pr} = 1{\cdot}86$$

and

$$k = 0{\cdot}68\ W/\text{m}\ {}^\circ\text{K} = 0{\cdot}39\ \text{Btu/ft hr}\ {}^\circ\text{R}$$

Then

$$\text{Re} = \frac{ud_{\text{i}}}{\nu} = 107{,}000$$

$$\text{Nu} = 0{\cdot}023\ \text{Re}^{0\cdot8}\ \text{Pr}^{0\cdot4} = 310$$

and the heat-transfer factor for the water is

$$C_{\text{h}} = \frac{k\ \text{Nu}}{d_{\text{i}}} = 725\ \text{Btu/ft}^2\ \text{hr}\ {}^\circ\text{R}$$

Gas side. The data of Appendix VII for air give

$$\mu = 26{\cdot}7\times10^{-6}\ \text{N sec/m}^2 = 5{\cdot}58\times10^{-7}\ \text{lbf sec/ft}^2$$

and

$$k = 40{\cdot}4\ \text{mW/m}\ {}^\circ\text{K} = 0{\cdot}0234\ \text{Btu/ft hr}\ {}^\circ\text{R}$$

Then

$$\nu = \frac{\mu}{\varrho} = 4{\cdot}5\times10^{-4}\ \text{ft}^2/\text{sec}$$

$$\text{Re} = \frac{ud_0}{\nu} = 6940$$

$$\text{Nu} = 0{\cdot}3\ \text{Re}^{0\cdot6} = 60{\cdot}4$$

and the heat-transfer factor for the gas is

$$C_{\text{h}} = \frac{k\ \text{Nu}}{d_0} = 6{\cdot}79\ \text{Btu/ft}^2\ \text{hr}\ {}^\circ\text{R}$$

As anticipated, the resistance of the gas flow controls the heat transfer. The overall factor is about

$$C_{\text{h}} = 6{\cdot}7\ \text{Btu/ft}^2\ \text{hr}\ {}^\circ\text{R}$$

and the heat-transfer rate for one of the tubes with outer surface area $A = 65{\cdot}4\ \text{ft}^2$ is

$$\dot{Q} = C_{\text{h}}A\ \Delta T = 6{\cdot}7\times65{\cdot}4\times500 = 219{,}000\ \text{Btu/hr}$$

A greatly increased heat transfer could be achieved by adding fins on the outside of the tubes in order to reduce the gas-side resistance to heat flow.

(b) Effect of varying water temperature

The heat-transfer factor found above is sufficiently accurate to use in estimating the variation of the water temperature along the tube. The heat transfer into length dx of the tube and the change in water temperature in that length dT are related by

$$d\dot{Q} = \dot{m}c_p\, dT = C_h(P\, dx)[T_g - T(x)]$$

with P the tube perimeter, T_g the uniform bulk temperature of the gas, and $T(x)$ the varying bulk temperature of the water. Then the temperature distribution is determined by the differential equation

$$\frac{dT}{T-T_g} = -\frac{C_h P}{c_p \dot{m}}\, dx$$

which gives, when integrated

$$\frac{T-T_g}{T_i-T_g} = \exp\left[-\frac{C_h P}{c_p \dot{m}} x\right]$$

with T_i the inlet temperature of the water. The exit temperature T_f at the end of a tube of length L is given by

$$T_f - T_i = (T_g - T_i)\left[1 - \exp\left(-\frac{C_h A}{c_p \dot{m}}\right)\right]$$

where $A = PL$ is the outer area of the tube.

For the case considered here, the mass flow through the tube is $\dot{m} = 9800$ lb/hr. Then

$$\frac{C_h A}{c_p \dot{m}} = 0{\cdot}0447$$

and

$$T_f - T_i = 500 \times [1 - \exp(-0{\cdot}0447)] = 21{\cdot}8\ °F$$

Hence an improved estimate of the heat-transfer rate is

$$\dot{Q} = \dot{m}c_p(T_f - T_i) = 9800 \times 21{\cdot}8 = 214{,}000 \text{ Btu/hr}$$

(c) Mean heat-transfer temperature difference

Another way of accounting for the change of temperature along the tube is to introduce a suitable mean temperature for the water, T_m. The temperature differential which corresponds to the actual heat transfer is given by

$$\dot{Q} = \dot{m}c_p(T_f - T_i) = C_h A(T_g - T_m)$$

Using the results of (b), it is found to be

$$T_g - T_m = \frac{c_p \dot{m}}{C_h A}(T_f - T_i)$$

$$= \frac{T_i - T_f}{\log_e\left(\dfrac{T_g - T_f}{T_g - T_i}\right)}$$

For the case of present interest

$$T_g - T_m = -\frac{21{\cdot}8}{\log_e (478{\cdot}2/500)} = 489\ ^\circ\mathrm{F}$$

Here the logarithmic mean temperature difference equals the simple arithmetic mean; this is only the case when $(T_f - T_i)/T_i \ll 1$.

The heat-transfer rate can now be calculated as

$$\dot{Q} = C_h A(T_g - T_m) = 214{,}000\ \mathrm{Btu/hr}$$

the value obtained above.

Mean temperatures suitable for other forms of heat exchanger can be introduced by similar arguments.

6.3.5 Dimensional Analysis

Dimensional analysis is a valuable tool in discussing heat-transfer processes, provided that one has sufficient insight to sort out the many interrelated quantities into meaningful groups. The preceding discussion of heat transfer has put us in a position to do this.

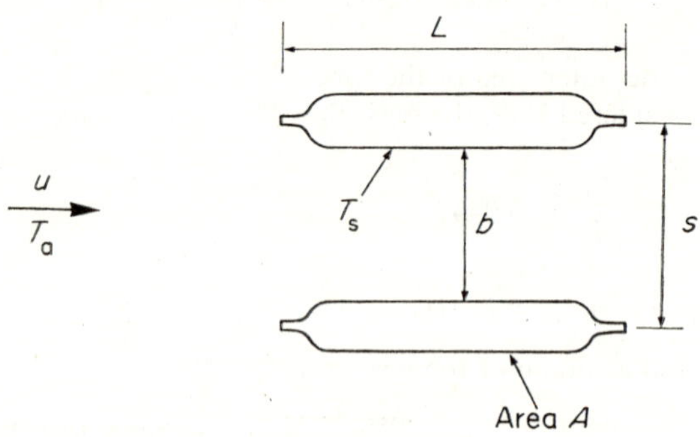

Figure 6.17 Quantities characterizing a simple heat-exchange process

For definiteness, we shall once again discuss a simple heat exchanger consisting of flat fins past which air is blown, as shown in Figure 6.17. Consider the heat transfer within the air, assuming the surface temperature to be uniform at T_s. The mean heat-transfer rate, $\dot{q} = \dot{Q}/A$, with A the total surface area of the fins, may be expected to depend upon:

$\Delta T = T_s - T_a$, the temperature difference producing the energy flow
u the velocity of the flow upstream
L, b, s the length, breadth and spacing of the passages between the fins
c_p, ϱ, k, μ, T_a, and a (the speed of sound), which are properties of the fluid upstream of the fins.

The buoyancy of the fluid is not taken into account, and the analysis relates to convective heat transfer only. Then

$$\dot{q} = f(\Delta T, u, L, b, s, c_p, \varrho, k, \mu, T_a, a)$$

Four independent dimensions are required to specify the twelve quantities linked in the last expression. Hence the pi-theorem indicates that a dimensionless representation will involve eight dimensionless groups. We have become familiar with the structure of a number of appropriate groups, and these can be selected without a formal analysis. It is necessary to check, however, that each group is independent of the others; alternatively, one of the Methods II, III and IV developed in Section 5.2.3 can be used, with a single set of four cancelling quantities applied to each of the remaining quantities. A suitable set of independent groups is

s/L and b/L—geometric ratios
$\Delta T/T_a$—a temperature ratio
$M = u/a$—the Mach number of the flow upstream
$\text{Re} = \varrho uL/\mu$—a Reynolds number for the flow
$\text{Pr} = c_p\mu/k$—the Prandtl number of the fluid
$\text{Ec} = u^2/(c_p\, \Delta T)$—the Eckert number
$\text{Nu} = \dot{q}L/k\, \Delta T = C_h L/k$—a Nusselt number

The Reynolds and Prandtl numbers have been discussed earlier. The Nusselt number relates the actual heat-transfer rate to a purely conductive heat-transfer rate which is characteristic of the system. Some comment is also required concerning the role of the Mach and Eckert numbers and of the temperature ratio $\Delta T/T_a$. If the flow velocity within the system is comparable with the speed of sound, changes in the Mach number will significantly alter the flow pattern and must be expected to change the heat transfer as well. The Eckert number has a somewhat similar structure to the Mach number. However, it may be significant even though the Mach number is small; as was seen in Example 6.10, the heat-transfer rate will be significantly altered when the temperature difference is comparable to $u^2/2c_p$. The temperature ratio $\Delta T/T_a$ will be important when the temperature change across the surface film is sufficient to produce significant changes in the fluid properties within the film. The role of this parameter can be accounted for in part by the use of properties evaluated at the mean film temperature rather than the free-stream temperature.

In a wide range of heat-transfer problems the role of the Mach and Eckert numbers is a minor one. Moreover, in many cases the use of mean film temperatures provides a reasonable correlation for different values of $\Delta T/T_a$.

With these restrictions the heat-transfer law may be given as

$$\mathrm{Nu} = f(\mathrm{Re}, \mathrm{Pr})$$

for a particular geometric configuration. In the case considered above, such a law would apply for particular values of s/L and b/L. Formulae of this type have been encountered in Examples 6.11 and 6.13. This formulation of the heat-transfer law is generally preferred, for the three groups represent, respectively, the heat transfer, the fluid motion and the fluid. However, it sometimes proves convenient to combine them into alternative dimensionless groups:

$\mathrm{Pe} = \mathrm{Re}\,\mathrm{Pr} = \varrho u c_p \dfrac{L}{k}$, the Péclet number, which characterizes the relative importance of convection and conduction, and

$\mathrm{St} = \dfrac{\mathrm{Nu}}{\mathrm{Pe}} = \dfrac{C_\mathrm{h}}{\varrho u c_p}$, the Stanton number (encountered earlier in Section 6.3.3), which relates the heat transfer and the convective capacity of the flow.

6.4 FLOW WITH DISSIPATION AND HEAT ADDITION

6.4.1 General One-dimensional Flow

It has been seen that the same mechanisms are responsible for the transfer of momentum and energy between a turbulent fluid and the solid boundaries past which it flows. Hence convective heat transfer will in general be associated with fluid friction. For a compressible fluid flow the inter-relationship is even closer, for there is a close coupling between the dynamic and thermodynamic response of the fluid. Thus heat addition will tend to expand the fluid, increasing its velocity and lowering its pressure. Friction, on the other hand, will tend to lower the pressure and density, increasing the velocity and reducing the temperature. (It will be seen later that the directions of certain of these changes are different when the flow is supersonic.) The remainder of this chapter will examine the development of a one-dimensional flow subjected to both heat addition and friction or, more generally, internal dissipation.

Figure 6.18 depicts an element of a continuous, steady, one-dimensional flow. The flow area changes gradually along the streamtube, giving the small net change $\mathrm{d}A$ across the element. The pressure, mean velocity and elevation of the streamtube axis also change slightly across the element. The net axial force on the fluid within the element is $\mathrm{d}F$, and the drag applied at the boundaries is $\mathrm{d}D$. Finally, there is a heat flow $\mathrm{d}\dot{Q} = \dot{m}\,\mathrm{d}q$ into the fluid.

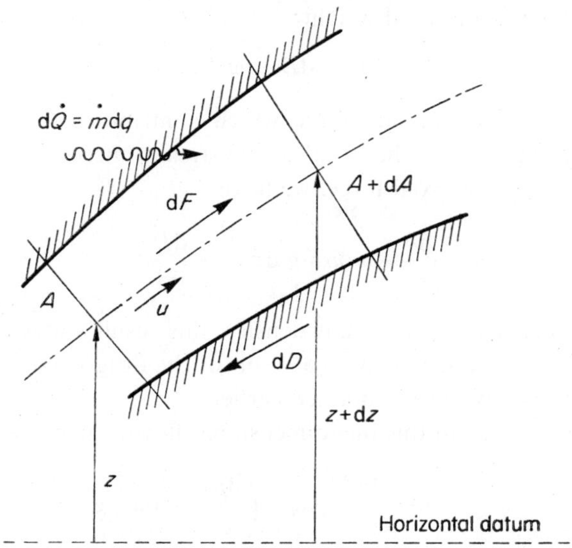

Figure 6.18 An element considered in setting down momentum relationships for generalized one-dimensional flow

No work transfer is considered, since seldom in practice is external work supplied gradually to a continuous steady flow. The drag forces at the boundaries do no work, nor do the friction stresses within the fluid do work on the fluid as a whole. These forces provide only dissipation, a redistribution of energy within the fluid. In the one-dimensional model, the dissipation is taken to be distributed uniformly over the cross-section, although this is far from the case in actual duct flows.

Let us see what information about this small change in state is provided by the three basic conservation principles and the second law of thermodynamics. The continuity and energy statements require little comment; in differential form they are

$$d(\varrho u A) = 0 \tag{6.4.1}$$

and

$$dh + u\,du + g\,dz = dq \tag{6.4.2}$$

The discussion of momentum is more involved. The momentum equations (2.2.12) give for a small change

$$dF_a = d(pA) + \varrho u A\,du \tag{6.4.3}$$

This expresses the element of the 'additional' force in terms of the fluid response. It can also be expressed as the sum of the axial components of pressure force, drag and fluid weight:

$$\mathrm{d}F_{\mathrm{a}} = p\,\mathrm{d}A - \mathrm{d}D - \varrho g A\,\mathrm{d}z \tag{6.4.4}$$

(The only element of pressure force which contributes to the additional force F_a is that applied by the 'wall' of the streamtube.) On equating these two expressions for the axial force, we have

$$\frac{\mathrm{d}p}{\varrho} + u\,\mathrm{d}u + g\,\mathrm{d}z = -\frac{\mathrm{d}D}{\varrho A} \tag{6.4.5}$$

For the particular case of constant density, this result states that the total mechanical energy changes only as a result of the drag applied to the fluid; equivalent results have been obtained earlier.

The entropy change in this one-dimensional flow is given by

$$T\,\mathrm{d}s = \mathrm{d}h - \frac{\mathrm{d}p}{\varrho} = \mathrm{d}q - \left(\frac{\mathrm{d}p}{\varrho} + u\,\mathrm{d}u + g\,\mathrm{d}z\right) \tag{6.4.6}$$

when use is made of the energy equation (6.4.2). The bracketed terms can be identified with the drag in equation (6.4.5), giving

$$T\,\mathrm{d}s = \mathrm{d}q + \frac{\mathrm{d}D}{\varrho A} \tag{6.4.7}$$

Here the entropy is seen to change as a result of heat addition and friction at the boundaries of the streamtube. The second law of thermodynamics, equation (1.5.19), implies that $\mathrm{d}D > 0$, the drag always opposing the flow, as shown in Figure 6.18. Let us compare these two contributions to the entropy change, using for simplicity Reynolds analogy (6.3.9) to relate heat transfer and friction. This gives

$$\frac{\dot{q}}{c_p\,\Delta T} = \frac{\tau_0}{u}$$

Introducing

$$\mathrm{d}q = \frac{\dot{q}\,P\,\mathrm{d}x}{\varrho u A} \quad \text{and} \quad \mathrm{d}D = \tau_0 P\,\mathrm{d}x$$

with P the perimeter of the streamtube, we obtain

$$\frac{dD/\varrho A}{\mathrm{d}q} = \frac{\tau_0 u}{\dot{q}} = \frac{u^2}{c_p\,\Delta T}$$

This is the Eckert number which emerged from the dimensional analysis of Section 6.3.5. It is seen here to be a measure of the relative importance of the entropy change due to friction and that due to heat addition.

The preceding discussion appears at first sight to provide a consistent picture of the response of a one-dimensional flow to energy addition and dissipation. Let us now consider its failings. Note first that the only source of dissipation appearing in equation (6.4.7) is boundary friction. We know that other flow processes lead to dissipation: shock waves and velocity redistribution across the flow. But where is the weakness in the development? The most obvious inadequacy is in the momentum relationship (6.4.3), which introduces the assumption that changes in the fluid momentum are proportional to a small change in a single velocity. Thus the large change occurring across a shock wave, and the redistribution in mixing and spreading flows are ruled out. In other words, the velocity profiles are assumed to be similar in shape and only slowly changing in magnitude. Hence the preceding results are applicable only to flows that develop very slowly, so that major changes in velocity profile do not occur over short distances. Apparently these results will apply to fully developed flows in ducts of constant section, to nozzle flows, and to flows in very slowly diffusing passages.

We shall now attempt to generalize the preceding analysis. Since the momentum statement (6.4.3) is restricted in its application, it cannot be used in predicting the development of a more general one-dimensional flow. The continuity and energy equations (6.4.1, 2) will be retained and also the relationship

$$T\,\mathrm{d}s = \mathrm{d}q - \left(\frac{\mathrm{d}p}{\varrho} + u\,\mathrm{d}u + g\,\mathrm{d}z\right) \tag{6.4.6}$$

between entropy and the other properties. The dissipation term cannot now be identified with the drag at the boundaries. However, we may express it as

$$\frac{\mathrm{d}p}{\varrho} + u\,\mathrm{d}u + g\,\mathrm{d}z = -\frac{\mathrm{d}D_e}{\varrho A}$$

with $\mathrm{d}D_e$ an element of effective drag, which represents all of the dissipation within the element of the flow. Alternatively, comparison of equations (1.5.25) and (6.4.6) shows that

$$\frac{\mathrm{d}p}{\varrho} + u\,\mathrm{d}u + g\,\mathrm{d}z = -T\,\mathrm{d}P$$

with $\mathrm{d}P = \mathrm{d}s - \mathrm{d}q/T > 0$ the production of entropy per unit mass. This entropy production is the net result of internal friction, mixing and heat transfer within the flowing fluid. Hence

$$T\,\mathrm{d}s = \mathrm{d}q + \frac{\mathrm{d}D_e}{\varrho A} = \mathrm{d}q + T\,\mathrm{d}P \tag{6.4.8}$$

gives the entropy change in this more general flow.

The result

$$dF_a = p\,dA - dD - \varrho g A\,dz \qquad (6.4.4)$$

still provides a way of calculating the force on an element of the flow. Now, however, $dD < dD_e$, since there will always be some internal dissipation additional to that associated with drag at the boundaries.

Collecting these results, we have

$$\begin{aligned} \frac{d\varrho}{\varrho} + \frac{du}{u} &= -\frac{dA}{A} \\ \frac{dp}{\varrho} + u\,du + g\,dz &= -\frac{dD_e}{\varrho A} = -T\,dP \qquad (6.4.9) \\ dh + u\,du + g\,dz &= dq \end{aligned}$$

These equations govern the response of the fluid to area change, internal dissipation and heat addition. They give the changes in the flow properties ϱ, p, h and u when the three kinds of perturbation are imposed on the flow. For specified distributions of flow area, dissipation and heat addition, and a particular equation of state, a step-by-step calculation gives the changes in state along the streamtube. For simple equations of state, notably the perfect-gas law, further analytical development is possible and the labour of fully numerical calculations can be avoided. However, even for a perfect gas analytical results can be obtained only for simple kinds of perturbation. Isentropic flow, considered earlier, is one such case. Others to be considered later are constant-area flows with heat addition but no friction, and with friction but no heat addition.

In principle, the results (6.4.9) can be applied to find the variation of fluid properties along a duct with any flow-area distribution. In practice, they can be applied meaningfully only to those flows whose velocity profiles remain nearly similar during development. Although we have omitted the momentum equation, which is the most sensitive to changes in velocity distribution, there remain difficulties in establishing suitable mean values for the retained equations. Moreover, while methods are available for estimating the friction dD acting on a section of a duct, there are fundamental difficulties in estimating the effective drag dD_e which represents the whole of the dissipation. The diffuser flow considered in Section 5.4.3 illustrates the problem. The dissipation seemingly associated with the changes in mean pressure and velocity does not occur wholly within the expanding section of the duct, but continues in the downstream parallel section. If one wished to construct a one-dimensional model of this development, it would be necessary to distribute the dissipation in a rather arbitrary way along the duct. It must be accepted that at the present time the development of many rapidly expanding flows cannot be predicted with confidence, but must be determined by experiment.

Example 6.14 To demonstrate the most general application of the results (6.4.9), a flow of steam in a diffuser will be considered. The calculation of the state of the steam leaving the diffuser proves to be a formidable task, and will not in fact be completed. However, it is instructive to make a start on it. The calculations will advance along the duct in small steps, the coefficients of equations (6.4.9) being estimated using the values already known.

Consider steam at 500 °C, 5 bar and 350 m/sec flowing into a diffusing tube which expands uniformly from 0·5 m^2 to 1·0 m^2 over a distance of 1·5 m. The effective coefficient of friction may be taken to be $c_f = 0{\cdot}005$, and the heat-transfer rate may be calculated using Reynolds analogy. The wall temperature is 400 °C.

A section of the duct of length $\Delta L = 50$ cm will be considered; then $\mathrm{d}A/A_1 = \frac{1}{3}$. The friction and heat-transfer area will be taken to be $A_w = \pi d_1\, \Delta L = 1{\cdot}25$ m^2. Then with $\varrho_1 = 1/0{\cdot}711$ kg/m^2 from steam tables

$$\frac{\mathrm{d}D_e}{A_1} = \tfrac{1}{2}\varrho_1 u_1^2 A_w \frac{c_f}{A_1} = 1{\cdot}076\ \mathrm{kN/m^2}$$

The heat-transfer factor for this section of the duct may be estimated as

$$C_h = \varrho_1 c_p u_1\, \mathrm{St} = \tfrac{1}{2}\varrho_1 c_p u_1 c_f = 2{\cdot}6\ \mathrm{kJ/m^2\ sec\ {}^\circ K}$$

The enthalpies for 5 bar and for 500 °C and 400 °C have been used to find c_p. The heat-transfer rate for the interval considered is

$$\mathrm{d}\dot{Q} = C_h A_w\, \Delta T = -2{\cdot}6\times 1{\cdot}25\times 100 = -326\ \mathrm{kW}$$

The mass flow is

$$\dot{m} = \varrho_1 u_1 A_1 = 350\times\frac{0{\cdot}5}{0{\cdot}711} = 248\ \mathrm{kg/sec}$$

and finally the heat addition per unit mass can be found:

$$\mathrm{d}q = \frac{\mathrm{d}\dot{Q}}{\dot{m}} = -\frac{326}{248} = -1{\cdot}313\ \mathrm{kJ/kg}$$

The equations (6.4.9) for the changes in properties over the interval now become

$$\begin{aligned} \mathrm{d}\varrho + 0{\cdot}00401\, \mathrm{d}u &= -0{\cdot}468 \\ \mathrm{d}p + 0{\cdot}492\, \mathrm{d}u &= -1{\cdot}076 \\ \mathrm{d}h + 0{\cdot}350\, \mathrm{d}u &= -1{\cdot}313 \end{aligned}$$

These equations can be solved only when the equation of state is known in sufficient detail. Callendar's equation provides a starting point for the iteration required to find the unknown state. For consistent units such as those used above, it gives

$$\mathrm{d}p - \frac{p_1}{\varrho_1}\, \mathrm{d}\varrho = 0{\cdot}231\ \varrho_1\, \mathrm{d}h$$

For the case considered above

$$dp - 356\, d\varrho = 0{\cdot}325\, dh$$

The preceding equations give $d\varrho$, dp and dh in terms of du. When these results are introduced into the last equation, we find

$$du = -158 \text{ m/sec}$$

It is of interest to consider the individual contributions to this change:

$$du\,(14{\cdot}26 - 4{\cdot}92 + 1{\cdot}14) = -1667 + 11 - 4$$

gives in turn the contributions from area change, friction and cooling. The combined effect of friction and heat transfer is less than 1 per cent of that of area change, and might be neglected if only a rough estimate of the final state were needed.

The other properties at the station 50 cm along the duct can now be estimated, for example

$$p_2 = p_1 + dp = 500 - 1{\cdot}1 + 77{\cdot}7 \text{ kN/m}^2 = 5{\cdot}77 \text{ bar}$$

The increment in length selected above, 50 cm, is too large for accurate calculations. If a digital computer is available to carry out the repetitive calculations, more precise results can be achieved by selecting a very short increment; the computer should be supplied with a better representation of the equation of state as well.

An alternative approach is to calculate the state of the steam at the end of an increment on the assumption of isentropic flow. An isentropic change is determined by the integral conductions s, h_0, $\dot{m}$ = constants, and can be calculated directly. The entropy change for the increment can then be estimated using equation (6.4.8). The resulting pattern of calculation is illustrated in Figure 6.19. The continuous curve is the actual process path, while the stepped path is that representing the proposed step-by-step calculation. This method will be successful only when the entropy changes are responsible for a small part of the change in state.

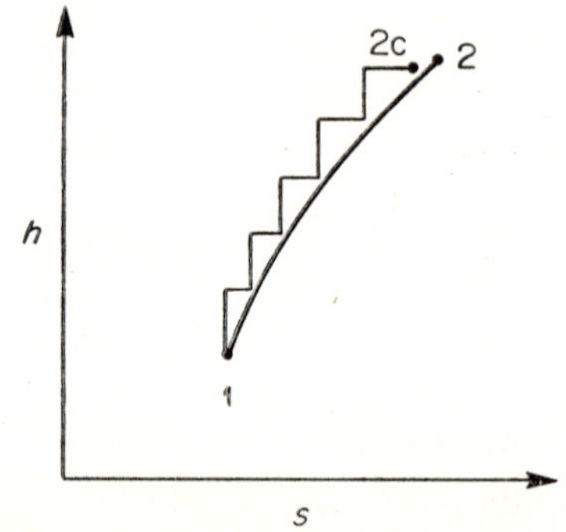

Figure 6.19 The actual process path for a diffuser flow and that obtained by accounting separately for area change and entropy change in alternating steps

It should be realized that the accuracy of any method of calculation is ultimately dependent not only on computing techniques, but on whether the friction and heat transfer can be specified in a realistic way.

6.4.2 Comparison of Energy Equations

There is another way of interpreting the results (6.4.9) which makes clearer the difference between the response to dissipation and heat addition of the constant-density flows which were considered earlier and the variable-density flows which will be considered henceforth. In Section 1.3.5, during the discussion of energy conservation for a system moving in a gravitational field, it was found that three energy equations could be distinguished: an overall equation

$$\Delta(TE) = \Delta(E+KE+PE) = W+Q+CE \tag{1.3.9}$$

a mechanical equation

$$\Delta(KE+PE) = W_m+CE_m \tag{1.3.10}$$

and a thermodynamic equation

$$\Delta E = W_t+Q+CE_t \tag{1.3.11}$$

The total work has been split into mechanical and thermodynamic components, as has the total convected energy. Let us now see what form these three equations take for the particular case of steady, one-dimensional flow.

Consider an elemental control mass; for this system the convected-energy terms are zero. The last of equations (6.4.9) provides the overall energy balance. Rewritten in the form (1.3.9) it is

$$\mathrm{d}(e+\tfrac{1}{2}u^2+gz) = -\mathrm{d}(pv)+\mathrm{d}q \tag{6.4.10}$$

The second of equations (6.4.9) provides the mechanical energy equation

$$\mathrm{d}(\tfrac{1}{2}u^2+gz) = -v\,\mathrm{d}p-T\,dP = -v\,\mathrm{d}p-\frac{\mathrm{d}D_e}{\varrho A} \tag{6.4.10}$$

Subtracting, we find the thermodynamic equation

$$\mathrm{d}e = -p\,\mathrm{d}v+\frac{\mathrm{d}D_e}{\varrho A}+\mathrm{d}q \tag{6.4.10}$$

Note that the dissipation does not appear in the overall equation; it serves merely to transfer energy from the mechanical equation to the thermodynamic equation. Note also that the overall displacement work $-\mathrm{d}(pv)$ is the sum of a mechanical work component $-v\,\mathrm{d}p$ representing work done against the pressure gradient, and a thermodynamic component $-p\,\mathrm{d}v$ representing work done in expansion.

For a constant-density fluid, the mechanical and thermodynamic energy equations simplify to

$$d(\tfrac{1}{2}u^2+gz) = -\frac{dp}{\varrho_c}-\frac{dD_e}{\varrho_c A}$$

and (6.4.11)

$$de = \frac{dD_e}{\varrho_c A}+dq$$

These equations are coupled only through the dissipation term. The addition of heat has no effect in the mechanical equation; nor do pressure and velocity changes appear in the thermodynamic equation. Hence the dynamics of a constant-density fluid can be analysed using only

$$d\left(\frac{p}{\varrho_c}+\tfrac{1}{2}u^2+gz\right) = -\frac{dD_e}{\varrho_c A}$$

without reference to the thermodynamics of the fluid. The density of a real liquid is slightly dependent on the temperature, and this may be important in establishing convection currents in very slow flows. Also, the viscosity of a liquid changes markedly with temperature, and this will be important in laminar flows. However, for a wide range of turbulent, high-velocity flows, these effects are of little importance. This has been tacitly assumed in a number of the preceding calculations.

Although the velocity distribution within a flowing liquid plays a vital role in the heat transfer into the liquid, the response of the liquid to this heat addition is virtually independent of the dynamics of the flow. Since the dissipated energy is usually very small in comparison with the heat capacity of the fluid, it produces very small changes in temperature and other properties. When heat transfer does occur, its effects usually swamp those of dissipation, and the second of equations (6.4.11) reduces to

$$de = dq$$

This characteristic was assumed in Example 6.13, in which no mention was made of the dissipation within the flow. In Example 6.10, on the other hand, the effect of dissipation on the heat transfer in a laminar flow was taken into account. Note, however, that the analysis was simplified by neglecting possible viscosity changes associated with the temperature variation across the flow.

The preceding remarks apply only to fluids with very small compressibility. The almost uncoupled equations (6.4.11) are never appropriate to gas flows. Even when the absolute changes in the pressure, density and temperature of a gas are small, they are closely linked. Hence we must use the results (6.4.10) which describe the coupling of the mechanical and thermodynamic response of the fluid. Since nearly all of our work on the development of flows subjected to friction and heat transfer has been based on the constant-density results (6.4.11), it will be necessary to look again at both problems and at the relationship between them.

6.4.3 The Rayleigh and Fanno Lines

It is not possible to find out much about the response of a fluid to a combination of area change, friction and heat transfer without specifying a particular equation of state. However, by restricting consideration to only one of these perturbations, one can discover a good deal about the response of compressible fluids in general. Isentropic flow with varying area has already been considered in Section 4.2.1. Frictionless, constant-area flow and isenergic, constant-area flow will now be discussed in general terms. Later the case of a perfect gas will be examined in detail.

When $dA = 0$ and $dD_e = 0$, the flow changing only as a result of heat addition, the first two of equations (6.4.9) can be integrated:

$$\varrho u = \frac{\dot{m}}{A}, \quad \text{a constant}$$

and (6.4.12)

$$p + \varrho u^2 = \frac{I}{A}, \quad \text{a constant}$$

The gravity term has been dropped as being unimportant for most flows of highly compressible fluids. This simplification will be retained throughout this chapter. The velocity can be eliminated from the results (6.4.12):

$$p + \frac{(\dot{m}/A)^2}{\varrho} = \frac{I}{A} \tag{6.4.13}$$

This equation defines a curve in the p, ϱ-plane. Indeed, since any other thermodynamic property can be expressed as a function of p and ϱ, this equation defines a unique relationship between every pair of thermodynamic properties. A curve specified in this way is called a *Rayleigh line;* its general form in the h, s-plane is indicated in Figure 6.20(a). The smooth curve corresponds to a single phase of the fluid.

The points on the Rayleigh line represent a family of flows with the same impulse function I and mass flow $\dot{m}$, but with differing total energy h_0. Example 6.4 deals with the transition between two such flows.* There the

* The mass flow is in fact not quite constant in that example.

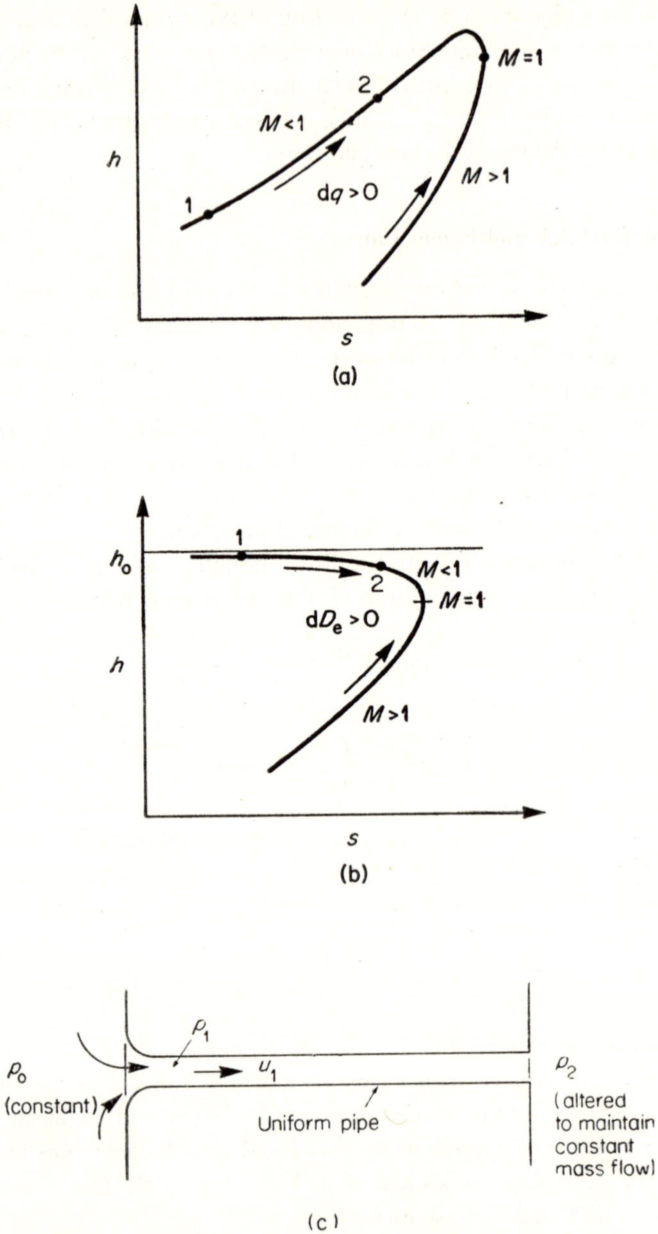

Figure 6.20 Characteristics of constant-area flows subjected to heat addition and friction. (a) The Rayleigh line for frictionless flow altered by heat addition. (b) The Fanno line for adiabatic flow altered by friction. (c) A situation illustrating choking as a result of friction and heat addition.

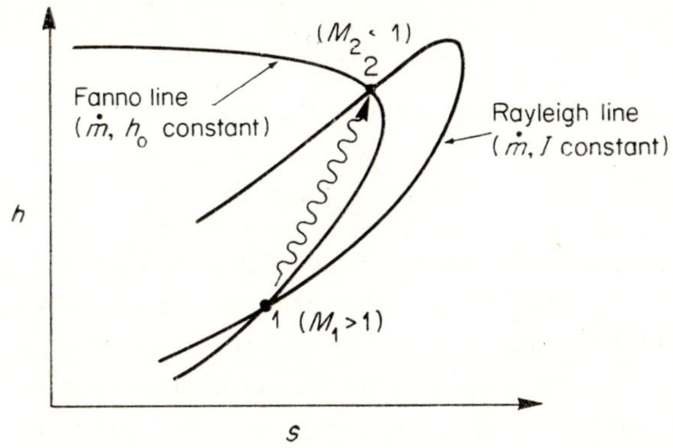

(d)

Figure 6.20 (d) Relationship of the normal-shock transition to the Rayleigh and Fanno lines

burning of fuel in the flow through a ram jet was found to raise the temperature, accelerate the fluid, lower its pressure and increase its entropy. The transition is like that from points 1 to 2 of Figure 6.20(a).

Since the addition of heat tends to increase the entropy, the state point will always move to the right along the Rayleigh line when heat flows into the fluid. As there is a maximum entropy for the curve sketched, there is apparently a limit to the heat which can be added to a fluid with a specified impulse function and mass flow. When differentiated, equation (6.4.13) gives

$$\frac{dp}{d\varrho} = \frac{(\dot{m}/A)^2}{\varrho^2} = u^2$$

for any point on a Rayleigh line. For the point of maximum entropy, $ds = 0$, and

$$\frac{dp}{d\varrho} = \left(\frac{\partial p}{\partial \varrho}\right)_s = u^2 = a^2$$

Hence $M = 1$ at the limiting point. It appears that heat addition will always move the flow towards the sonic condition. The sections of the Rayleigh line on either side of the sonic point represent subsonic and supersonic flows, the subsonic branch being that of higher enthalpy.

A rather similar discussion applies for isenergic, constant-area flow, altered only by friction. For $dq = 0$, $dA = 0$, the first and third of equations (6.4.9) give

$$\varrho u = \frac{\dot{m}}{A}, \quad \text{a constant} \tag{6.4.14}$$

and

$$h+\tfrac{1}{2}u^2 = h_0\,, \quad \text{a constant} \tag{6.4.14}$$

The Fanno line is obtained by eliminating the velocity:

$$h+\frac{(\dot{m}/A)^2}{2\varrho^2} = h_0 \tag{6.4.15}$$

The state postulate implies that a *Fanno line* is defined in each thermodynamic plane. The general form of the variation in the h, s-plane is indicated in Figure 6.20(b). Example 2.11 dealt with a transition along a Fanno curve. Steam flowing through a constriction in a whistle was found to have, on regaining the initial flow area, a reduced enthalpy and increased velocity and entropy. The transition is like that from points 1 to 2 of Figure 6.20(b).

Equation (6.4.7) shows that friction will always move the state point to the right along the Fanno curve, towards the condition of maximum entropy. At the point of maximum entropy

$$T\,\mathrm{d}s = \mathrm{d}h - \frac{\mathrm{d}p}{\varrho} = 0$$

But on the Fanno line

$$\frac{\mathrm{d}h}{\mathrm{d}\varrho} = \frac{(\dot{m}/A)^2}{\varrho^3} = \frac{u^2}{\varrho}$$

Then, for $\mathrm{d}s = 0$

$$\frac{\mathrm{d}p}{\mathrm{d}\varrho} = \left(\frac{\partial p}{\partial \varrho}\right)_s = u^2 = a^2$$

The limiting condition is once again that of sonic flow. Hence friction, which always increases the entropy, always moves the flow towards the sonic condition. Once again, subsonic flow corresponds to the part of the curve for which the enthalpy is higher.

The significance of these conclusions is best appreciated by considering a particular physical situation. Figure 6.20(c) shows a constant-area pipe leading from a reservoir where the conditions are maintained constant; the pressure at the end of the pipe is $p_2 < p_0$. In its flow along the pipe the fluid may be subjected to either friction or heat addition, but not both. Positive heat addition will move the state from point 1 to point 2 in Figure 6.20(a); friction will produce a similar change in Figure 6.20(b).

Consider a series of heated and frictional flows for which the mass flow is kept constant. For all of the frictional flows the inlet conditions (state 1) are the same, and

$$A,\ h_0,\ \dot{m} = \text{constants} \tag{6.4.16}$$

For all of the heated flows the inlet conditions are the same, and

$$A,\ I,\ \dot{m} = \text{constants}$$

However, the outlet conditions (state 2) will differ from flow to flow of each class. In general it will be necessary to reduce the pressure p_2 to maintain the mass flow constant as the heat addition or friction is increased.

For both the heated and frictional flows, there is a limit to the entropy change which can be accommodated. This limit is reached when the outlet state is sonic; different sonic states are generated by friction and heat addition, however. Nevertheless, there is nothing to stop us prescribing a heat addition greater than the limiting one, nor to stop us increasing the friction of the pipe beyond the limiting value. How does the flow respond if this is done?

Consider first adiabatic, frictional flow. If the reservoir conditions remain constant and the flow area is unchanged, only one of the constants of equations (6.4.16) can change—the mass flow. For the heated, frictionless flow the situation is generally similar, but here not only must the mass flow change but also the impulse function. These changes mean that the original Fanno and Rayleigh lines are no longer applicable. The new curves are defined by the condition of sonic flow at the outlet.

It was noted above that the mass flow can be kept constant, while the friction and heat addition increase, only by reducing the outlet pressure p_2. Hence, when added friction and heat addition cannot be accommodated in this way, the mass flow must be expected to fall. Thus both friction and positive heat addition have the same effect of *choking* the flow. A similar effect would be achieved in isentropic flow through the duct of Figure 6.20(c), if a valve in the line were slowly closed. Here too the mass flow could be maintained constant by downstream pressure reductions only as long as the sonic condition was not achieved at the valve.

The normal-shock transition can be given an interesting interpretation in terms of the Fanno and Rayleigh lines. The conditions (3.1.4) imply that the initial and final states must lie on both the Fanno and Rayleigh curves defined by the specified values of $\dot{m}/A$, I/A, and h_0. The two curves are shown in Figure 6.20(d); the shock carries the flow from the supersonic branch to the subsonic branch with an increase in entropy. It is not possible to show a path for the transition in this plot of equilibrium properties.

6.5 FLOW OF A PERFECT GAS

6.5.1 The General Case

This analysis will be based on equations (6.4.9) with the gravity terms omitted. Once again it will prove convenient to use the local Mach number as an index of the state of the flowing gas. For a perfect gas the speed of sound is given by

$$a^2 = \frac{\gamma p}{\varrho} = (\gamma-1)c_p T$$

Also $h = c_p T$ and $\mathrm{d}h = c_p\,\mathrm{d}T$. Using these results, we can recast equations (6.4.9) into the forms:

$$\begin{aligned} \frac{\mathrm{d}\varrho}{\varrho}+\frac{\mathrm{d}u}{u} &= -\frac{\mathrm{d}A}{A} \\ \frac{\mathrm{d}p}{p}+\gamma M^2\frac{\mathrm{d}u}{u} &= -\frac{\mathrm{d}D_e}{pA} \\ \frac{\mathrm{d}T}{T}+(\gamma-1)M^2\frac{\mathrm{d}u}{u} &= \frac{\mathrm{d}q}{h} \end{aligned} \tag{6.5.1}$$

The equation of state can be given in a comparable form as

$$\frac{\mathrm{d}p}{p} = \frac{\mathrm{d}\varrho}{\varrho}+\frac{\mathrm{d}T}{T} \tag{6.5.1}$$

In these equations the dimensionless factors $\mathrm{d}A/A$, $\mathrm{d}D_e/pA$ and $\mathrm{d}q/h$ prescribe the change in flow area, the dissipation and the heat addition. It will be noted that the response of the gas depends on the local Mach number and on its ratio of specific heats.

Explicit results for individual properties can be obtained by combining equations (6.5.1). If the property changes from the first three equations are introduced into the last of the set (6.5.1), we find

$$(1-M^2)\frac{\mathrm{d}u}{u} = -\frac{\mathrm{d}A}{A}+\frac{\mathrm{d}D_e}{pA}+\frac{\mathrm{d}q}{h} \tag{6.5.2}$$

The other changes follow immediately:

$$\begin{aligned} (1-M^2)\frac{\mathrm{d}\varrho}{\varrho} &= M^2\frac{\mathrm{d}A}{A}-\frac{\mathrm{d}D_e}{pA}-\frac{\mathrm{d}q}{h} \\ (1-M^2)\frac{\mathrm{d}p}{p} &= \gamma M^2\frac{\mathrm{d}A}{A}-[1+(\gamma-1)M^2]\frac{\mathrm{d}D_e}{pA}-\gamma M^2\frac{\mathrm{d}q}{h} \\ (1-M^2)\frac{\mathrm{d}T}{T} &= (\gamma-1)M^2\frac{\mathrm{d}A}{A}-(\gamma-1)M^2\frac{\mathrm{d}D_e}{pA} \\ &\quad +(1-\gamma M^2)\frac{\mathrm{d}q}{h} \end{aligned} \tag{6.5.2}$$

It is apparent that most kinds of perturbation have reversed effects on subsonic and supersonic flows. Earlier, equations (4.2.4) showed that this was so for the particular case of isentropic flow. Moreover, the reversal is consistent with the discussion of Rayleigh and Fanno flows in the preceding section.

Equations (6.5.2) provide the basis for a numerical integration giving the response to an arbitrary combination of friction, heating and area change. Since the coefficients depend upon the Mach number, its changes must also be found. For a perfect gas $\mathrm{d}M/M = \mathrm{d}u/u - \frac{1}{2}\,\mathrm{d}T/T$ leads to

$$(1-M^2)\frac{\mathrm{d}M}{M} = -\left(1+\frac{\gamma-1}{2}M^2\right)\frac{\mathrm{d}A}{A} + \left(1+\frac{\gamma-1}{2}M^2\right)\frac{\mathrm{d}D_e}{pA} + \frac{1}{2}(1+\gamma M^2)\frac{\mathrm{d}q}{h} \tag{6.5.3}$$

As was argued earlier when considering the Rayleigh and Fanno lines, both friction and positive heat addition tend to move the flow towards the sonic condition, as does a reduction in flow area.

It is also of interest to see how the total properties are affected by area change, dissipation and heat addition. The change in total temperature is given simply by

$$\frac{\mathrm{d}T_0}{T_0} = \frac{\mathrm{d}q}{h_0} \tag{6.5.4}$$

The change in total pressure can be found as follows. From equations (1.5.16) and (6.4.8) we have

$$\frac{\mathrm{d}s}{c_p} = \frac{\mathrm{d}T_0}{T_0} - \frac{\gamma-1}{\gamma}\frac{\mathrm{d}p_0}{p_0} = \frac{\mathrm{d}q}{h} + \frac{\gamma-1}{\gamma}\frac{\mathrm{d}D_e}{pA} \tag{6.5.4}$$

Then

$$\frac{\mathrm{d}p_0}{p_0} = -\frac{\mathrm{d}D_e}{pA} - \frac{\gamma}{\gamma-1}\left(\frac{\mathrm{d}q}{h} - \frac{\mathrm{d}q}{h_0}\right)$$

Introducing $h_0 = h\left(1+\frac{\gamma-1}{2}M^2\right)$, we obtain

$$\frac{\mathrm{d}p_0}{p_0} = -\frac{\mathrm{d}D_e}{pA} - \frac{\gamma M^2}{2}\frac{\mathrm{d}q}{h_0} = -\frac{\mathrm{d}D_e}{pA} - \frac{\gamma M^2}{2}\frac{\mathrm{d}T_0}{T_0} \tag{6.5.4}$$

As would be expected, it is found that changes in area have no direct influence on the total properties. Moreover, the effects of dissipation and heat addition are more straightforward than in equations (6.5.2); they are qualitatively the same for subsonic and supersonic flows.

Although the total temperature is changed only by heat addition, total pressure variations result from both dissipation and heat addition. A very considerable heat flow is required to produce significant changes in total pressure. Hence it is the drop in total pressure during combustion which is usually of interest; Example 6.4 illustrates this effect. Attempts have been made, in the so-called aerothermopressor, to increase the total pressure by an effective heat outflow. In this device, water is sprayed into a hot gas flow, its vaporization reducing the total temperature. Ideally, this would have the effect of compressing the flowing gas. In practice, it is found that additional dissipation seriously erodes the potential gain in total pressure.

Gas flows at low Mach number are of particular interest since they are most often encountered in practice. It has been seen in Section 2.4.1 and Example 4.1 that certain aspects of isenergic gas flows at low Mach number can be represented with adequate accuracy using the constant-density approximation. It is natural to ask whether this simplification can be used in a wider variety of calculations relating to low-speed gas flows. The equations (6.5.2, 3, 4) provide the means of answering this question. Table 6.4 gives the changes in the properties of a flowing gas with $M^2 \ll 1$ which result from area change, dissipation and heat addition*. It also gives the corresponding changes in the properties of a fluid of strictly constant density, with gravitational effects omitted. The latter results are obtained from equations (6.4.8, 9) and the continuity condition.

In Table 6.4 the degree of correspondence between the two sets of results decreases from top to bottom. Changes in total enthalpy and small changes in entropy are given correctly by the constant-density results. When the flow is isenergic, small changes in pressure and total pressure are correctly prescribed by the constant-density approximation. However, velocity changes are correctly given only when the flow is isentropic, that is, both adiabatic and dissipation-free. Finally, note that the temperature change in a gas flow is never correctly predicted by the constant-density approximation. Indeed, dissipation produces temperature changes of opposite sign in the two kinds of fluid. In summary, we may conclude that the constant-density model is most successful for isentropic flow at low Mach number, but even in that case the temperature change in the gas is grossly misrepresented.

Let us see how these conclusions apply to two constant-area gas flows, that in an insulated air line and that in a heat-exchanger tube. The pressure change in the air line will be given with fair accuracy by relating the local pressure gradient and local friction stress, but that within the heat-exchanger tube cannot always be calculated in this way. In neither case does the density in fact remain constant. In the insulated line the fluid expands as friction

* Certain of the terms containing M^2 have been retained to provide a more complete comparison.

Table 6.4. Comparison of the effects of area change, dissipation and heat addition on constant-density flow and perfect-gas flow at low Mach number

Change	$\varrho = \varrho_c$	$M^2 \ll 1$	Equivalent?
$\mathrm{d}h_0$	$\mathrm{d}q$	$\mathrm{d}q$	Always
$T\,\mathrm{d}s$	$\dfrac{\mathrm{d}D_e}{\varrho A} + \mathrm{d}q$	$\dfrac{\mathrm{d}D_e}{\varrho A} + \mathrm{d}q$	Always
$\dfrac{\mathrm{d}p}{\varrho}$	$u^2 \dfrac{\mathrm{d}A}{A} - \dfrac{\mathrm{d}D_e}{\varrho A}$	$u^2 \dfrac{\mathrm{d}A}{A} - \dfrac{\mathrm{d}D_e}{\varrho A} - u^2 \dfrac{\mathrm{d}q}{h}$	For $\mathrm{d}q = 0$
$\dfrac{\mathrm{d}p_0}{\varrho}$	$-\dfrac{\mathrm{d}D_e}{\varrho A}$	$-\dfrac{\mathrm{d}D_e}{\varrho A} - \frac{1}{2}u^2 \dfrac{\mathrm{d}q}{h}$	For $\mathrm{d}q = 0$
$\mathrm{d}h$	$u^2 \dfrac{\mathrm{d}A}{A} + \mathrm{d}q$	$u^2 \dfrac{\mathrm{d}A}{A} - u^2 \dfrac{\mathrm{d}D_e}{pA} + \mathrm{d}q$	For $\mathrm{d}D_e = 0$
$\dfrac{\mathrm{d}u}{u}$	$-\dfrac{\mathrm{d}A}{A}$	$-\dfrac{\mathrm{d}A}{A} + \dfrac{\mathrm{d}D_e}{pA} + \dfrac{\mathrm{d}q}{h}$	For $\mathrm{d}q = \mathrm{d}D_e = 0$
$\dfrac{\mathrm{d}\varrho}{\varrho}$	—	$M^2 \dfrac{\mathrm{d}A}{A} - \dfrac{\mathrm{d}D_e}{pA} - \dfrac{\mathrm{d}q}{h}$	Never
$\mathrm{d}e = c_v\,\mathrm{d}T$	$\dfrac{\mathrm{d}D_e}{\varrho A} + \mathrm{d}q$	$\dfrac{u^2}{\gamma} \dfrac{\mathrm{d}A}{A} - \dfrac{u^2}{\gamma} \dfrac{\mathrm{d}D_e}{pA} + \dfrac{\mathrm{d}q}{\gamma}$	Never
$\dfrac{\mathrm{d}M}{M}$	—	$-\dfrac{\mathrm{d}A}{A} + \dfrac{\mathrm{d}D_e}{pA} - \frac{1}{2} \dfrac{\mathrm{d}q}{h}$	Never

reduces its pressure, while in the heat-exchanger tube the density change will be predominantly the result of changes in temperature. In the adiabatic flow the temperature falls as the fluid accelerates, while in the heat-exchanger the temperature may rise or fall, depending on the direction of the heat transfer.

The preceding discussion relates to fluids which obey exactly the restrictions to constant density and perfect-gas behaviour. However, they are relevant to real fluids as well. Under most circumstances the constant-density results can be used to predict the behaviour of real liquids, while the perfect-gas results often provide an adequate representation of the flows of real gases and vapours. Moreover, the general conclusions may be used in deciding whether some aspect of a low-velocity vapour flow can be treated using the constant-density approximation.

6.5.2 Rayleigh Flow

There are two ways of analysing the constant-area, frictionless flow of a perfect gas. We might use special forms of the differential results (6.5.2, 3, 4). However, a more direct approach is available through the integral conditions

$$\varrho u = \dot{m}/A \quad \text{and} \quad p + \varrho u^2 = I/A \tag{6.4.12}$$

Although the equations (6.5.2, 3, 4) will not be used directly, they are relevant to the following discussion, since they show that changes in the fluid properties can be represented as functions of Mach number and the specific-heat ratio.

It is convenient to express the results in dimensionless form, as was done for isentropic flow in equations (4.2.7) to (4.2.11). Once again the local Mach number will be used as a parameter characterizing the development of the flow. However, the total state varies along the flow and does not now provide suitable reference properties. It is evident in Figure 6.20(a) that all the flows obeying equations (6.4.12) and with particular values of $\dot{m}/A$ and I/A will tend towards a unique sonic state when subjected to positive heat addition. This sonic state provides a convenient set of reference values for properties at other points along the Rayleigh line.

The second of equations (6.4.12) can be written

$$p + \varrho u^2 = p(1+\gamma M^2) = p^*(1+\gamma)$$

giving the non-dimensional result

$$\frac{p}{p^*} = \frac{1+\gamma}{1+\gamma M^2} \tag{6.5.5}$$

Using the special properties of a perfect gas and the continuity relation

$\varrho u = \varrho^* a^*$, we find

$$M = \frac{u}{a^*}\frac{a^*}{a} = \frac{\varrho^*}{\varrho}\left(\frac{T^*}{T}\right)^{\frac{1}{2}} = \frac{p^*}{p}\left(\frac{T}{T^*}\right)^{\frac{1}{2}}$$

and

$$\frac{T}{T^*} = \left(\frac{p}{p^*}\right)^2 M^2 \qquad (6.5.6)$$

This result applies to any constant-area flow of a perfect gas. For Rayleigh flow

$$\frac{T}{T^*} = \left[\frac{(1+\gamma)M}{1+\gamma M^2}\right]^2 \qquad (6.5.5)$$

Other dimensionless ratios follow directly from these results.

So far nothing has been said about the heat addition which gives rise to the change in state. It is related to the fluid properties by

$$c_p \Delta T_0 = q$$

a special case of the energy equation (6.1.6). The changes in total temperature are related to the local Mach number by

$$\frac{T_0}{T_0^*} = \frac{T}{T^*}\frac{T_0}{T}\frac{T^*}{T_0^*} = 2\frac{T}{T^*}\frac{\left(1+\frac{\gamma-1}{2}M^2\right)}{\gamma+1} \qquad (6.5.7)$$

with T/T^* as given above.

Tables of the functions of equations (6.5.5, 7) are given in Appendix V for $\gamma = 1{\cdot}4$. The column T_0/T_0^* is in a sense the key to these Rayleigh tables, since it is the change in T_0 that is related directly to the heat addition.

Figure 6.21 shows the variation of the total temperature ratio with Mach number. The transition between the two points shown corresponds to heat addition

$$q = c_p(T_{0_2} - T_{0_1}) \quad \text{per unit mass}$$

The maximum heat addition, starting from state 1, which is attainable without choking the flow is

$$q = c_p(T_0^* - T_{0_1}) \quad \text{per unit mass}$$

Note that little additional heat is absorbed as the Mach number approaches unity with state 1 fixed. Hence there is little point in accelerating the flow beyond a moderate Mach number, say, $M = 0{\cdot}7$.

It was argued earlier (in Section 6.4.3) that heat addition beyond the value $c_p(T_0^* - T_{0_1})$ will cause a reduction in mass flow. The reason for this fall in

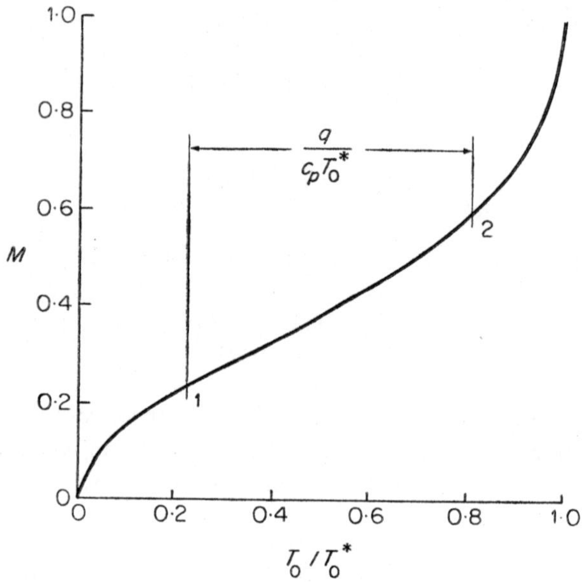

Figure 6.21 Relationship between Mach number and total temperature for Rayleigh flow of a perfect gas with $\gamma = 1{\cdot}4$

the flow rate can now be seen more clearly. For a choked flow of a perfect gas

$$\frac{\dot{m}}{A} = \varrho^* a^* \propto \frac{p^*}{T^{*\frac{1}{2}}}$$

It must be expected that additional heat addition will raise the sonic temperature and hence, by reducing the density, decrease the flow rate. This corresponds to a reduction in the initial Mach number M_1. Thus heat addition beyond the choking value will give rise to a movement of the initial state point 1 to the left along the curve of Figure 6.21.

Example 6.15 The layout of a jet engine is shown schematically in Figure 6.22. The part of the flow to be examined here is that between the compressor and turbine, the region in which the energy driving the engine is supplied by burning fuel in the air flow. The properties at entry to the combustion chamber are those found in Example 6.7, which dealt with the flow into the engine and through the compressor.

For a fuel with heating value 40,000 kJ/kg (this corresponds to kerosene with a combustion efficiency of 93 per cent) let us find

(a) the maximum air/fuel ratio possible without choking the flow, and
(b) the turbine inlet conditions for an air/fuel ratio of 65 : 1.

It will be assumed (rather unrealistically) that the air behaves like a perfect gas with $\gamma = 1{\cdot}4$ and $c_p = 1{\cdot}1$ kJ/kg °K.

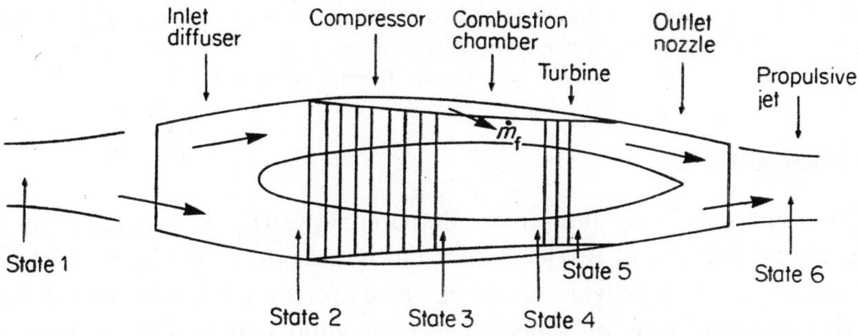

Figure 6.22 Schematic view of an aircraft jet engine

(a) Choking by heat addition

For $M_3 = 0{\cdot}253$, Rayleigh tables give $T_{0_3}/T_0^* = 0{\cdot}262$. For choking at state 4 $T_{0_4}/T_0^* = 1$. Then

$$T_{0_4} = \frac{T_{0_4}}{T_{0_3}} T_{0_3} = \frac{549}{0{\cdot}262} = 2097\ ^\circ\text{K}$$

and

$$T_{0_4} - T_{0_3} = 1548\ ^\circ\text{K}$$

Hence the air/fuel ratio is approximately

$$\frac{\dot{m}}{\dot{m}_f} = \frac{HV}{c_p \Delta T_0} = \frac{40{,}000}{1{\cdot}1 \times 1548} = 23$$

This ratio is achievable from the point of view of combustion, since there is adequate oxygen in air to consume this amount of kerosene. However, quite apart from the undesirable effects arising from the choking of the flow, the temperatures achieved are too high for continuous operation of the turbine.

(b) Air/fuel ratio of 65:1

In this case

$$T_{0_4} - T_{0_3} = \frac{\dot{m}_f HV}{\dot{m} c_p} = \frac{40{,}000}{1{\cdot}1 \times 65} = 560\ ^\circ\text{K}$$

Then

$$T_{0_4} = 560 + 549 = 1109\ ^\circ\text{K}$$

and

$$\frac{T_{0_4}}{T_0^*} = \frac{T_{0_4}}{T_{0_3}} \frac{T_{0_3}}{T_0^*} = \frac{1109}{549} \times 0{\cdot}262 = 0{\cdot}529$$

Rayleigh tables now give $M_4 = 0{\cdot}400$. The total pressure follows from Rayleigh tables as well:

$$p_{0_4} = \frac{p_{0_4}}{p_0^*} \frac{p_0^*}{p_{0_3}} p_{0_3} = \frac{1{\cdot}157}{1{\cdot}216} \times 4{\cdot}76 = 4{\cdot}53 \text{ bar}$$

Note that this considerable heat addition, 36 per cent of that required to choke the flow, is achieved with only a moderate increase in Mach number and only a small fall in total pressure ($p_{0_3} = 4{\cdot}76$ bar from Example 6.7).

6.5.3 Fanno Flow

The analysis of constant-area, adiabatic flow will be modelled on the preceding discussion of Rayleigh flow. Once again the properties of the limiting sonic state will provide reference values for dimensionless parameters. In this isenergic flow of a perfect gas the total temperature is constant. Hence the dependence of temperature on Mach number is given by

$$\frac{T}{T^*} = \frac{T_0}{T^*}\frac{T}{T_0} = \frac{\gamma+1}{2}\left(1+\frac{\gamma-1}{2}M^2\right)^{-1} \qquad (6.5.8)$$

The pressure variation follows from equation (6.5.6):

$$\frac{p}{p^*} = \frac{1}{M}\left(\frac{T}{T^*}\right)^{\frac{1}{2}} \qquad (6.5.8)$$

The velocity and density variations along the flow can be obtained from the perfect-gas law and continuity relationship. These functions are given in Appendix V in the Fanno tables for $\gamma = 1{\cdot}4$.

We have still to relate the changes in state to the retarding force which is applied to the flow. The analysis will be based on equation (6.5.3), specialized to the case of constant-area, adiabatic flow:

$$(1-M^2)\frac{\mathrm{d}M}{M} = \left(1+\frac{\gamma-1}{2}M^2\right)\frac{\mathrm{d}D}{pA} \qquad (6.5.9)$$

Since the element of drag $\mathrm{d}D/pA$ is dependent on the local velocity and Mach number, an immediate integration does not give an easily interpreted result. Before integrating we shall introduce measures of the drag which depend only weakly on the local velocity and Mach number. These are the friction coefficient and factor

$$c_f = \frac{f}{4} = \frac{\tau_0}{\frac{1}{2}\varrho u^2}$$

describing wall friction, and the drag coefficient

$$C_D = \frac{\mathrm{d}D}{\frac{1}{2}\varrho u^2 A}$$

describing the force on an isolated blockage in the flow. These two ways of characterizing the drag are essentially equivalent, since the same dependence on density and velocity is assumed in each case. It was seen in Section 5.5 that the losses from turbulent friction and from an isolated blockage are in fact roughly proportional to the local dynamic pressure of the flow.

For a perfect gas the dynamic pressure can be expressed as

$$\tfrac{1}{2}\varrho u^2 = \tfrac{1}{2}\gamma p M^2$$

Then the dimensionless measure of friction on a length dx of duct is given by

$$\frac{\mathrm{d}D}{pA} = \frac{\tau_0 P\,\mathrm{d}x}{pA} = \tfrac{1}{2}\gamma M^2 c_f \frac{P}{A}\,\mathrm{d}x$$

(here P is the duct perimeter) and the element of drag by

$$\frac{\mathrm{d}D}{pA} = \tfrac{1}{2}\gamma M^2 C_D$$

Since the two cases of distributed friction and an isolated blockage are equivalent, we shall confine further discussion to distributed friction. For a round pipe of diameter d, the element of frictional drag is

$$\frac{\mathrm{d}D}{pA} = \tfrac{1}{2}\gamma M^2 \frac{4c_f\,\mathrm{d}x}{d} = \tfrac{1}{2}\gamma M^2 \frac{f\,\mathrm{d}x}{d}$$

For a non-circular duct, d may be interpreted as the hydraulic diameter. When the last result is introduced into equation (6.5.9), we obtain

$$f\frac{\mathrm{d}x}{d} = \frac{(1-M^2)\,\mathrm{d}M^2}{\gamma M^4\left(1+\frac{\gamma-1}{2}M^2\right)}$$

Integration gives a relationship between the length of duct over which the friction is applied and the initial and final Mach numbers. For the purpose of tabulation, it is convenient to fix the final state by identifying it with the limiting sonic condition. The duct length then has the maximum value L_m consistent with the initial Mach number M. We have

$$\frac{1}{d}\int_0^{L_m} f\,\mathrm{d}x = \bar{f}\frac{L_m}{d} = \frac{1-M^2}{\gamma M^2} + \frac{\gamma+1}{2\gamma}\log_e\left[\frac{\tfrac{1}{2}(\gamma+1)M^2}{1+\frac{\gamma-1}{2}M^2}\right] \qquad (6.5.10)$$

Here $\bar{f}$ is the mean friction factor over the length L_m. Often sufficient accuracy is achieved by taking f to be constant. If it varies significantly, the duct can be divided into sections within which the variation is small.

The function of equation (6.5.10) is also given in the Fanno tables. It is the key to them, as it relates the frictional characteristics of the duct (or the drag coefficient, if appropriate) to the changes in properties along the flow.

The phenomenon of choking in adiabatic frictional flow is in many ways similar to choking as a result of heat addition. So long as the friction does not bring the flow to the sonic condition, it can be increased still further without necessitating a change in the initial state, although an alteration in the downstream conditions is required. To overcome the increase in the retarding force, the final pressure must fall. Once the flow is choked, the mass flow

$$\frac{\dot{m}}{A} = \varrho^* a^* \propto \frac{p^*}{T^{*\frac{1}{2}}}$$

cannot be maintained constant by adjustment of the downstream conditions. It will be reduced by additional friction, since this has the effect of reducing the final pressure p^*.

Example 6.16 Air is drawn from the atmosphere (barometric pressure 29·9 in of mercury) through 100 ft of straight pipe of diameter 2 in. Figure 6.23 shows the installation. The friction coefficient may be taken to be $c_f = 0{\cdot}006$ over the range of Reynolds numbers of interest. It is required to find the mass flow through the pipe when a suction of 10 in of mercury is maintained by a pump at the far end. The flow may be taken to be adiabatic.

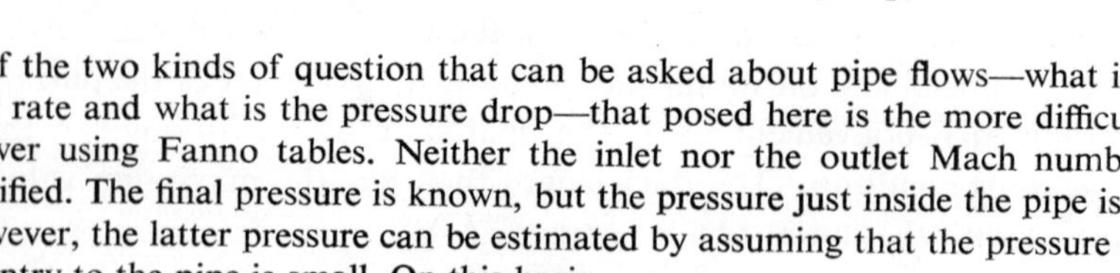

Figure 6.23 A pipe through which air is drawn by a pump at the far end

Of the two kinds of question that can be asked about pipe flows—what is the flow rate and what is the pressure drop—that posed here is the more difficult to answer using Fanno tables. Neither the inlet nor the outlet Mach number is specified. The final pressure is known, but the pressure just inside the pipe is not. However, the latter pressure can be estimated by assuming that the pressure drop on entry to the pipe is small. On this basis

$$\frac{p_2}{p_1} = \frac{29{\cdot}9 - 10{\cdot}0}{29{\cdot}9} = 0{\cdot}665$$

The relationship of the pipe length L to the Fanno tables may not be obvious at first sight. It is revealed by the following argument. The initial and final states (1 and 2) lie on the same Fanno line and have in common the same sonic condition. This sonic state could be reached from state 1 either after flow through length

L_{m_1} of pipe or after flow through length L to state 2, followed by length L_{m_2} of the same pipe. Hence these lengths are related by

$$L_{m_1} - L_{m_2} = L = 100 \text{ ft}$$

and the two states are connected by

$$\frac{fL_{m_1}}{d} - \frac{fL_{m_2}}{d} = \frac{fL}{d} = 14{\cdot}40$$

In Table 6.5 the pressure ratio p_2/p_1 is calculated for two assumed values of M_1. The flow with $p_2/p_1 = 0{\cdot}665$ lies between the two trials. Interpolation gives

$$M_1 = 0{\cdot}15 + \frac{0{\cdot}02 \times 0{\cdot}046}{0{\cdot}109} = 0{\cdot}158$$

Table 6.5. Isenergic duct flows

M_1	p_1/p^*	fL_{m_1}/d	fL_{m_2}/d	M_2	p_2/p^*	p_2/p_1
0·15	7·29	27·93	13·53	0·206	5·30	0·728
0·17	6·43	21·11	6·71	0·273	3·97	0·619

It is now possible to correct for the fall in pressure on entry to the pipe. It will be assumed that there are no losses at entry. Then the isentropic flow tables give

$$\frac{p_1}{p_0} = \frac{p_1}{p_a} = 0{\cdot}982 \quad \text{for} \quad M_1 = 0{\cdot}16$$

The pressure change along the pipe is now given by

$$\frac{p_2}{p_1} = \frac{p_2}{p_a}\frac{p_a}{p_1} = \frac{0{\cdot}665}{0{\cdot}982}$$

$$= 0{\cdot}678$$

The inlet Mach number can be estimated more accurately as

$$M_1 = 0{\cdot}15 + \frac{0{\cdot}02 \times 0{\cdot}059}{0{\cdot}109} = 0{\cdot}161$$

The changes in fluid properties along the pipe can now be found directly. For $M_1 = 0{\cdot}16$:

$$\frac{u_1}{u^*} = \frac{\varrho^*}{\varrho_1} = 0{\cdot}175, \quad \frac{T_1}{T^*} = 1{\cdot}194$$

and

$$\frac{fL_{m_1}}{d} = 24{\cdot}2$$

For $fL_{m_2}/d = 9{\cdot}8$:

$$\frac{u_2}{u^*} = \frac{\varrho^*}{\varrho_2} = 0{\cdot}246, \quad \frac{T_2}{T^*} = 1{\cdot}187$$

Then

$$\frac{u_2}{u_1} = \frac{\varrho_1}{\varrho_2} = 1{\cdot}41, \quad \frac{T_2}{T_1} = 0{\cdot}994$$

Even though the density, velocity and pressure change markedly along the pipe, the temperature change proves to be small. An examination of equations (6.5.8) shows why this must be so in a flow at low Mach number.

Although the density falls significantly along the line, it changes very little in the initial isentropic acceleration to $M_1 = 0{\cdot}161$. Hence the flow rate can be calculated with good accuracy as

$$\dot{m} = \varrho u A \simeq \varrho_a a_a M_1 A$$

With standard atmospheric values

$$\dot{m} = 1{\cdot}225 \times 340 \times 0{\cdot}161 \times 0{\cdot}00203 = 0{\cdot}136 \text{ kg/sec}$$
$$= 18{\cdot}0 \text{ lb/min}$$

6.5.4 Isothermal Flow

We have now developed methods of predicting the response of a flowing gas to one of area change, friction and heat addition. Analytical results can be obtained from the general results (6.5.2, 3, 4) for a number of cases in which the perturbations are combined. One of these is flow at constant temperature. It is of interest for two reasons. The flow in very long gas and air lines maintains a temperature only a little different from the environmental level. More generally, as the preceding example illustrated, the temperature changes only slightly in adiabatic flow at low Mach number. Such a flow may usefully be discussed using the simpler results for isothermal flow.

For constant-area flow at constant temperature, the last of equations (6.5.2) indicates that the heat transfer and friction must be related by

$$(1-\gamma M^2)\frac{dq}{h} = (\gamma-1)\, M^2 \frac{dD}{pA} = \frac{\gamma}{2}(\gamma-1)\, M^4 f \frac{dx}{d} \qquad (6.5.11)$$

Hence the Mach number variation (6.5.3) reduces to

$$f\frac{dx}{d} = \frac{1-\gamma M^2}{\gamma M^4}\, dM^2 \qquad (6.5.12)$$

Note that friction drives the flow not towards the sonic condition, but towards the state with $M = \gamma^{\frac{1}{2}}$, a slightly subsonic condition. The constant introduced on integrating equation (6.5.12) is evaluated in terms of this limiting state:

$$f\frac{L_m}{d} = \frac{1}{\gamma M^2} - 1 + \log_e(\gamma M^2) \qquad (6.5.13)$$

This gives the maximum length of duct in which isothermal flow can be maintained for an initial Mach number of M. Henceforth, the mean friction factor will be denoted simply by f.

The other fluid properties follow easily from the condition T = constant. Using the values for $M_t = \gamma^{-\frac{1}{2}}$ as references, we can write

$$\frac{p}{p_t} = \frac{\varrho}{\varrho_t} = \frac{u_t}{u} = \frac{M_t}{M} = \frac{1}{\gamma^{\frac{1}{2}} M} \qquad (6.5.14)$$

The results given above are much simpler than those for Fanno flow. They can be simplified even more for flow at low Mach number, for then only changes in the first term of equation (6.5.13) are significant. Using the result $p/p_t = 1/\gamma^{\frac{1}{2}} M$, we obtain

$$f\frac{L}{d} = \frac{p_1^2 - p_2^2}{p_t^2}$$

and

$$\left(\frac{p_2}{p_1}\right)^2 = 1 - f\frac{L}{d}\gamma M_1^2 \qquad (6.5.15)$$

It is possible to develop similar explicit relationships between other fluid properties and the duct length and friction factor.

The result (6.5.15) can also be obtained by the following simple line of argument. In Table 6.4 we note that, when the velocity is low enough and the heat transfer is small, pressure changes in a constant-area gas flow are dependent primarily on the wall friction. Then

$$-\mathrm{d}p = \frac{\mathrm{d}D}{A} = \frac{\varrho u^2}{2} f \frac{\mathrm{d}x}{d}$$

For a perfect gas

$$\varrho u^2 = \frac{(\dot{m}/A)^2}{\varrho} = \left(\frac{\dot{m}}{A}\right)^2 \frac{RT}{p}$$

and

$$-2p\,\mathrm{d}p = \left(\frac{\dot{m}}{A}\right)^2 RT \frac{f\mathrm{d}x}{d}$$

For constant area and temperature, integration between states 1 and 2 gives

$$p_1^2 - p_2^2 = \left(\frac{\dot{m}}{A}\right)^2 RT \frac{fL}{d} \qquad (6.5.16)$$

This is equivalent to equation (6.5.15). However, the second derivation does not reveal so clearly the conditions under which the formula is valid.

Example 6.17 In Example 6.16 it was noted that the temperature did not change greatly in the isenergic flow examined. Hence the isothermal results may be expected to give an accurate prediction of the flow rate.

Equation (6.5.16) can be applied directly:

$$\dot{m} = \left(\frac{fL}{d}\right)^{-\frac{1}{2}} A \left(\frac{p_1^2 - p_2^2}{RT}\right)^{\frac{1}{2}}$$

For the problem posed, with the pressure drop at entry neglected

$$p_1^2 - p_2^2 = 2120^2(1-0{\cdot}665^2) = 2{\cdot}50 \times 10^6 \text{ (lbf/ft}^2)^2$$

For standard atmospheric temperature

$$RT = 1716 \times 518 = 0{\cdot}889 \times 10^6 \text{ ft lbf/sl}$$

Also, from Example 6.16 we have $fL/d = 14{\cdot}40$. Then

$$\dot{m} = 14{\cdot}40^{-\frac{1}{2}} \times 0{\cdot}0218 \times \left(\frac{2{\cdot}50}{0{\cdot}889}\right)^{\frac{1}{2}} = 0{\cdot}00964 \text{ sl/sec} = 18{\cdot}6 \text{ lb/min}$$

With this information, it is possible to correct for the pressure drop at entry. The constant-density result $p_1 = p_a - \frac{1}{2}\varrho_a u_1^2$ would give a satisfactory estimate. However, it is not necessary to make this calculation, since the value $p_1/p_a = 0{\cdot}982$ is available from Example 6.16. Hence

$$p_1^2 - p_2^2 = 2080^2(1-0{\cdot}678^2) = 2{\cdot}34 \times 10^6 \text{ (lbf/ft}^2)^2$$

and an improved estimate of the mass flow is

$$\dot{m} = 18{\cdot}6 \times \left(\frac{2{\cdot}34}{2{\cdot}50}\right)^{\frac{1}{2}} = 18{\cdot}0 \text{ lb/min}$$

This is the value found in Example 6.16.

6.5.5 Thrust Calculations

In Section 4.2.3 during the study of isentropic flow the result

$$\frac{I}{I^*} = \frac{p}{p^*}\frac{A}{A^*}\frac{1+\gamma M^2}{1+\gamma} \tag{4.2.14}$$

was obtained. It is applicable to any one-dimensional flow of a perfect gas, but was applied earlier only to isentropic flows. For the constant-area flows considered here, $A/A^* = 1$, and the pressure ratios are:

for frictionless flow

$$\frac{p}{p^*} = \frac{1+\gamma}{1+\gamma M^2}$$

for isenergic flow

$$\frac{p}{p^*} = \frac{1}{M}\left(\frac{\gamma+1}{2}\right)^{\frac{1}{2}}\left(1+\frac{\gamma-1}{2}M^2\right)^{-\frac{1}{2}}$$

and for isothermal flow

$$\frac{p}{p^*} = \frac{1}{M}$$

Hence variations of the impulse function in these kinds of flow are given by: for frictionless flow

$$\frac{I}{I^*} = 1, \quad \text{as would be expected;}$$

for isenergic flow

$$\frac{I}{I^*} = \frac{1}{M}\frac{1+\gamma M^2}{[2(1+\gamma)]^{\frac{1}{2}}}\left(1+\frac{\gamma-1}{2}M^2\right)^{-\frac{1}{2}}$$

a function which is included in Fanno tables; and for isothermal flow

$$\frac{I}{I^*} = \frac{1+\gamma M^2}{M(1+\gamma)}$$

The method of application of these results in determining the force applied to a duct by the fluid is very similar to that for isentropic flow, illustrated in Example 4.8.

PROBLEMS FOR CHAPTER 6

Group A

6.1 Separate the following devices into two classes, those extracting energy from a natural source, and those transferring or converting it into other forms:
coal-fired furnace, steam condenser, water pump, turbine in a hydroelectric plant, turbine of a jet engine, windmill, plant converting crude oil into combustible components, steam turbine in a coal-burning electricity-generating plant, electrical generator, boiler fed by a coal-fired furnace, jet engine, automobile engine, automobile.

6.2 If the pump of Example 6.1 were stopped, the water would drain from it and the pipe into the well. The pump would have to be primed, that is, filled with water, before it could operate again. Why is this? How could the priming be carried out? How could this problem be avoided?

6.3 Water flows at 80,000 lb/hr through a heat exchanger, entering at 350 °F and leaving at 850 °F. The pressure may be taken to be 685 lbf/in^2 (gauge) throughout. What is the rate at which heat flows to the fluid? The method used in Example 6.2 may be used to estimate the enthalpy of the water at inlet.

6.4 The steam from the heat exchanger considered in the preceding problem passes through a turbine connected to an electrical generator. The steam emerges from the turbine at 2 lbf/in^2 (abs) and 94 per cent dryness. Estimate the electrical output of the generator, taking the gross efficiency of energy conversion to be 90 per cent.

6.5 Explain the trends in the ratios *LHV*/*HHV* given in Table 6.1.

6.6 Starting with the heating values for methane given in Section 6.1.2, find the heating values per ft^3 at atmospheric pressure. Compare with the values given in Table 6.1 for natural gas, which is largely composed of methane.

6.7 In a furnace coal (mostly carbon) is burned in air, but the combustion is incomplete so that 40 per cent of the carbon leaves as carbon monoxide, rather than carbon dioxide.

(a) Estimate the actual energy release, given that

$$C+O_2 \rightarrow CO_2 \quad \text{with} \quad CV = 32{,}800 \text{ kJ/kg}$$

$$CO+\tfrac{1}{2}O_2 \rightarrow CO_2 \quad \text{with} \quad CV = 10{,}200 \text{ kJ/kg}$$

(b) What is the combustion efficiency?

6.8 (a) For a pump and turbine handling a constant-density fluid, show that

$$\eta_p = \frac{1}{\eta_t} = \frac{\Delta H_0}{\Delta H_0 + H_f} = \frac{\Delta(gH_0)}{\Delta(gH_0) + c_v \,\Delta T}$$

for adiabatic flow.

(b) For a particular installation in which the flow may be taken to be adiabatic, the temperature rise is 0·041 °C and the pressure rise is 3 bar, the changes in kinetic and potential energies being negligible. What is the hydraulic efficiency of the pump?

6.9 (a) Show that, for a compressor which is cooled, the definition of isentropic efficiency could give a result greater than unity.

(b) How could an isentropic efficiency apparently greater than unity be achieved in a turbine?

6.10 (a) For an inlet-duct diameter of 1 m, what is the mass flow for the jet-engine intake considered in Example 6.7?

(b) What is the power requirement for the compressor?

6.11 Derive a result similar to equation (6.2.8) for a turbine.

6.12 What is the isentropic efficiency of the turbine considered in Problem 6.4?

6.13 Natural gas ($MW = 16, \gamma = 1{\cdot}29$) flows at the rate of 400 ft^3/sec with density 0·01 sl/ft^3 and temperature 70 °F through a pipe 36 in in diameter. An axial-flow compressor does work at the rate 2000 hp on this flow. An estimate of the pressure rise across the compressor is required; it may be assumed to have an efficiency of 82 per cent.

6.14 Show that the total-to-static efficiency of a turbine is always less than its total-to-total efficiency.

6.15 If water flows through the turbines of the hydroelectric station of Example 6.8 at the rate of 90 m^3/sec, find the electrical power generated and the rate of dissipation.

6.16 When the pump considered in Problem 6.8 carries a flow of 5 m^3/sec, the electrical power it absorbs is 3000 kW. The motor shaft turns at 1420 rev/min, and applies a torque of 1900 kgf m to the pump. Find

(a) the efficiency of the motor,

(b) the overall efficiency of the pump,

(c) the mechanical efficiency of the pump, on the assumption that there is no heat flow between pump and water.

6.17 (a) What is the propulsive efficiency of the ship propellers considered in Example 2.3(b)?

(b) At what rate is organized mechanical energy generated?

(c) If the unorganized mechanical energy of the wake is equal to the wasted component of the organized energy P_j, what is the shaft power supplied to the propellers?

(d) If the shaft power is 30 per cent of that potentially available in the fuel (Bunker C), at what rate is fuel consumed?

(e) How much fuel is required by the ship during an ocean crossing of 2000 miles?

6.18 Consider again the heat transfer across a Couette flow which was discussed in Example 6.10.

(a) Show that

$$\dot{q} = \frac{k(T_1 - T_{1_0})}{b}$$

(b) Express in words the condition which must be satisfied if heat is to flow away from the surface.

(c) How does this differ from the requirement when the fluid is still?

6.19 In Example 6.11 the relationship between heat transfer and friction is discussed for air flow in smooth pipes with Re $= 5\times 10^4$ to 10^6. Now a heat transfer law valid in the range Re $= 2\times 10^4$ to 10^5 is required.

(a) Use Reynolds analogy to find this law.

(b) How should the law be modified if $\text{Pr} \neq 0{\cdot}7$?

6.20 (a) What is the Stanton number for the water flow considered in Example 6.13?

(b) Use the result (6.3.8) to find the friction coefficient and the pressure drop along one of the heat transfer tubes.

6.21 If the mean temperature difference for the heat transfer of Problem 6.3 is 200 °F, estimate the heat transfer area required within the heat exchanger, taking

$$C_h = 8 \text{ Btu/ft}^2 \text{ hr } ^\circ\text{R}.$$

6.22 Liquid sodium is used as the heat transfer fluid within nuclear reactors of the fast-breeder type, where the active core is small and highly effective heat transfer is required. The following calculations indicate why liquid sodium proves useful in this situation.

Within the reactor core the temperature of the fluid rises from 400 °C to 600 °C. Fairly realistic mean values for the properties of sodium are those at 800 °K: $\varrho = 826$ kg/m^3, $c_p = 1{\cdot}26$ kJ/kg °K, $\mu = 0{\cdot}23$ mN sec/m^2, $k = 66{\cdot}3$ W/m °K. The flow velocity is 5 m/sec through a passage with equivalent diameter 1 cm.

(a) What are the diffusivities of momentum and heat for this fluid? What is the Prandtl number? How does the Prandtl number compare with those for more common heat-transfer fluids?

(b) What are the Reynolds and Péclet numbers for the flow?

(c) The Nusselt number is given approximately by

$$\mathrm{Nu} = 7 + 0{\cdot}025\,\mathrm{Pe}^{0{\cdot}8} \quad \text{for small Pr}$$

What are the Nusselt and Stanton numbers for the flow?

(d) What is the heat-transfer factor for this situation? How does this compare with those for other heat-transfer fluids?

(e) For a heat-transfer rate of $\dot{q} = 430$ kW/m^2, what is the mean temperature difference between the fluid and wall?

(f) In this kind of fluid molecular conduction is often found to be comparable to conduction effected by turbulent mixing. Why is this?

6.23 In any real flow, the element of actual drag will be less than the element of effective drag representing the net dissipation; that is, $\mathrm{d}D/\mathrm{d}D_e < 1$. Estimate the magnitude of this ratio for

(a) pipe flow,

(b) nozzle flow,

(c) a gradually expanding diffuser with $\theta = 2°$,

(d) a rapidly expanding diffuser with $\theta = 8°$, and

(e) a very rapidly expanding duct with separated flow, as shown in Figure 4.16(d).

6.24 Consider a perfect gas flowing in a constant-area duct.

(a) Explain why friction causes the fluid to accelerate in isenergic subsonic flow.

(b) Show that friction tends to 'cool' a subsonic isenergic flow. Why?

(c) Under what conditions does heat addition tend to reduce the temperature of the flowing fluid? Why does this occur? (The reason may be seen most clearly by examining the way in which the third of equations (6.5.1) is converted into the last of equations (6.5.2).)

6.25 Find the pressure and temperature following the combustion process of Example 6.15(b). What are the proportional falls in total and static pressure?

6.26 Steam enters a turbine through a duct 25 m in length and 10 cm in diameter. Its condition on entry to the duct is specified by $M = 0{\cdot}10$, 30 bar and 500 °C. The friction in the duct is specified by $c_f = 0{\cdot}01$, and the minor losses by $\Sigma K = 3{\cdot}5$. Assuming that the flow is adiabatic, and that the steam behaves like a perfect gas, find the temperature and pressure of the steam supplied to the turbine, using equations (6.5.2).

6.27 Natural gas ($MW = 16$, $\gamma = 1{\cdot}3$) is to be pumped for 50 miles through a buried pipeline of diameter 36 in; the friction coefficient of the pipe is $c_f = 0{\cdot}003$. Experience has shown that the temperature in such a line is around 70 °F. The pressure at outlet is 20 lbf/in^2 (gauge) and that at inlet is 100 lbf/in^2 (gauge).

(a) Find the flow in standard cubic feet per day, that is, the volume which the gas would occupy at 'standard' conditions, 70 °F and 1 atm.

(b) If a break occurred in the line so that its contents ultimately leaked away, what quantity of gas would be lost (in SCF)?

Group B

6.28 The windmill providing pumping power for the pump discussed in Example 6.1 is replaced by an electric motor. Water is now conveyed to the surface at the rate of 1000 ft^3/hr.

(a) Estimate the electrical power drawn by the motor, taking the overall efficiency of the installation to be 60 per cent.

(b) Where should the pump now be located?

6.29 A stoichiometric mixture is one containing just sufficient oxygen to consume completely the fuel supplied.

(a) What is the stoichiometric ratio for methane and air, by volume and by weight?

(b) Find the stoichiometric ratio by weight for a fuel composed mostly of heptane C_6H_{14}, for which

$$C_6H_{14}+9\tfrac{1}{2}\,O_2 \rightarrow 6\,CO_2+7\,H_2O$$

6.30 If methane and air are burnt at atmospheric pressure in the stoichiometric mixture, estimate the temperature of the resulting mixture, assuming perfect-gas behaviour of all constituents. As a result of dissociation, a much lower temperature will in fact be achieved. Take air to be 78 per cent N_2 and 22 per cent O_2 by volume and 76 per cent N_2 and 24 per cent O_2 by weight.

6.31 The difference between constant-volume and constant-pressure heating values can be estimated as

$$HV_p - HV_v = p(V_r - V_p)$$

where V_r and V_p are the volumes of reagents and products. For combustion at atmospheric pressure and 25 °C, find the difference for

(a) methane burned in air, for which

$$CH_4+2\,O_2 \rightarrow CO_2+2\,H_2O \quad \text{with} \quad LHV = 802{,}000 \text{ kJ/kg mol}$$

and

(b) octane vapour burned in air, for which

$$C_8H_{18}+12\tfrac{1}{2}\,O_2 \rightarrow 8\,CO_2+9\,H_2O \quad \text{with} \quad LHV = 5{,}120{,}000 \text{ kJ/kg mol}$$

6.32 A 2000-hp pump is installed in the pipeline of Example 5.12 to increase the flow of water between reservoirs. Taking the overall efficiency of the pump to be 85 per cent, estimate the new flow rate. An iterative solution will be required.

6.33 For a perfect gas, show that $\eta_c, \eta_t < 1$ corresponds to $\Delta s > 0$.

6.34 (a) Find the isentropic efficiencies for the turbine operations considered in Example 1.5(a) and (b).

(b) Is the latter case realistic? If the efficiency is in fact the same as that found for Example 1.5(a), what must be the condenser pressure?

(c) For a steam flow of 20,000 lb/hr, what is the power output for the two operating conditions?

6.35 The fluids emerging from the pump and compressor considered in Example 6.6 are used to drive an ideal hydraulic turbine and an ideal gas turbine within which the fluids are decelerated to rest at atmospheric pressure.

(a) Find the outputs from the two turbines.

(b) What would be the outputs if the pump and compressor had operated ideally themselves?

(c) What fraction of the additional energy absorbed by the fluids as a result of inefficient compression is recovered as work in the ideal hydraulic and gas turbines?

6.36 Tests of a number of pumps indicate that the maximum overall efficiency and design flow rate are related by

$\dot{V}$ (1000 gal/min)	10	3	1	0·5	0·2	0·1
η (%)	90	87	82	78	72	67

(a) Find an expression relating flow rate and efficiency, by taking the losses to be proportional to $\dot{V}^{-n}$.

(b) For what flow rate will the maximum efficiency be 95 per cent?

6.37 On leaving the combustion chamber, the flow within the ram jet considered in Example 6.4 is accelerated isentropically in a converging–diverging nozzle, and is then exhausted into the atmosphere at a pressure of 0·06 bar. The vehicle is flying at 1200 m/sec. For simplicity in the analysis, assume perfect-gas behaviour with $\gamma = 1{\cdot}4$, $c_p = 1{\cdot}0$ kJ/kg.

(a) What are the Mach number and total pressure following combustion?

(b) What are the Mach number and temperature on exit into the atmosphere?

(c) What is the velocity of the flow at outlet into the atmosphere?

(d) What is the thrust per unit mass?

(e) What is the propulsive efficiency of the unit?

(f) What is the overall efficiency of the unit? What would it be if combustion were perfect, and the fuel flow were reduced to achieve the same temperature rise during combustion?

6.38 It has been seen that the equations governing the non-steady diffusion of momentum across a parallel flow and the non-steady diffusion of heat through a stationary substance have the same structure. Hence for every non-steady parallel flow there is an analogous non-steady heat flow. How might heat flows analogous to the non-steady flows of Sections 5.1.3 and 5.2.5 be established?

6.39 Consider the heat transfer within laminar Couette flow which was discussed in Example 6.10.

(a) Express the result (a) in terms of the dimensionless groups of Section 6.3.5.

(b) Show that the net heat transfer from the flow is just sufficient to remove the dissipated energy.

6.40 The calculations of Example 6.12 relate to the operation of the heat exchanger when fairly new and when located in an air stream with temperature 20 °C. What will be the mean temperature of the water within the fins if the air temperature is

35 °C, the internal fouling factor has built up to 0·003, and lint has collected on the fins, reducing the heat-transfer factor for the airflow to 65 Btu/ft^2 hr °F?

6.41 Steam flows at 75,000 lb/hr through the superheater of a boiler. The heat-transfer area is 6000 ft^2 and $C_h = 7$ Btu/ft^2 hr °R. The gas from which the heat flows has a mean temperature of 600 °C. The steam is just on the saturation line on entering the superheater; its pressure is 685 lbf/in^2 (gauge) throughout.

Estimates are required for the temperature of the steam leaving the superheater, and for the rate of heat transfer. An iterative solution will be required; start by assuming that the initial temperature differential applies throughout the unit; for later calculations use the average temperature differential.

6.42 Apply dimensional analysis to convective heat transfer from a vertical plate. Take the heat transfer rate $\dot{Q}$ to depend on b and L, the breadth and height of the plate, on k, ν, $\varkappa$, $g\beta$, properties of the fluid (β is the coefficient of cubical expansion defined in equations (1.2.8)), and on the difference in temperature between the plate and the body of the fluid, ΔT.

(a) Show that suitable inter-related dimensionless groups are

$$\frac{\dot{Q}}{bk\,\Delta T}, \quad \frac{\nu}{\varkappa}, \quad \frac{L}{b} \quad \text{and} \quad \frac{g\beta\,\Delta T L^3}{\nu^2}$$

(b) The last of these groups is the Grashof number, Gr. Show that it plays the role of a Reynolds number in this situation.

(c) For a very broad plate, it is found that the heat-transfer law has the form

$$\text{Nu} = f(\text{PrGr})$$

Show that this result is consistent with (a).

6.43 The flame holder for the afterburner of an aircraft jet engine is tested on the ground by blowing air through a constant-area duct in which the system is mounted. The drag coefficient based on duct diameter is found to be $C_D = 0{\cdot}2$ in 'cold' tests. Are the changes in fluid properties resulting from this retardation large enough to be taken into account when determining the effect of heat addition? Use equations (6.5.2, 3) with $dq = dA = 0$ to estimate the effect on the flow. Consider the cases $M = 0{\cdot}3$ and $0{\cdot}9$. The gas flowing past the flame holder has

$$\gamma = 1{\cdot}33.$$

6.44 For the combustion process of Example 6.15(b), find the temperature rise in the flow, assuming that the fluid behaves like pure air, but taking into account the variation of the specific heat c_p with temperature.

6.45 (a) How much heat can be added to a supersonic airflow at $M = 1{\cdot}8$ and $T_0 = 300$ °C before it chokes?

(b) Suppose the flow considered above passed through a normal shock before the heat was added. How much heat could then be added before the sonic state was attained?

(c) The answers to (a) and (b) should be the same. Why?

6.46 Air flowing at $M = 1{\cdot}8$ passes through a normal shock, then flows isentropically through a converging–diverging nozzle. The ultimate flow area at the end of the nozzle is the same as the original flow area.

(a) Find the Mach number downstream of the shock and the final Mach number.

(b) Explain how the same state could be achieved by passing the fluid along a constant-area duct in which it is acted on by friction.

(c) Show the processes (a) and (b) on an h, s-plot. Mark in the sonic points on Fanno line and isentrope.

(d) Show that the friction force and force on the nozzle are the same.

6.47 Find a formula for the variation of p_0/p_0^* in Fanno flow of a perfect gas.

6.48 (a) What is the maximum flow which can be drawn through the air line considered in Example 6.16?

(b) Plot the pressure variation along the pipe for the case considered in (a). Sketch the pressure variation for the case considered in Example 6.16 on the same graph.

(c) Suppose that a maximum flow three times as great as that found in (a) were required. What is the least pipe diameter through which this amount of air can be drawn? (Take $c_f = 0{\cdot}006$ as before.)

6.49 Consider a perfect gas with $\gamma = 1{\cdot}4$ accelerating isentropically from rest into the rounded entrance of a pipe with uniform diameter and friction coefficient. The critical mass flow depends on the friction in the pipe, and not merely on the stagnation properties of the gas, as is the case for wholly isentropic flow.

(a) Show that the outlet temperature and velocity depend only on the reservoir conditions, for choked flow through a particular pipe.

(b) Use the result (4.3.12), together with isentropic and Fanno tables, to find how the critical flow depends upon friction. Plot the parameter $\dot{m}/(\varrho_0 a_0 A)$ against fL/d.

6.50 It is impossible to maintain isothermal, one-dimensional flow up to the limiting condition $M = \gamma^{-\frac{1}{2}}$. Use equation (6.5.11) to demonstrate this. What kind of flow is found very near the limiting condition?

6.51 (a) Show that equations (6.5.15, 16) are equivalent.

(b) Show that these results reduce to equations (5.5.17) as $\Delta p/p \to 0$.

(c) Find the proportional error in the constant-density result.

6.52 In an engineering handbook are given the following formulae for air flowing inside a round tube:

(1) the criterion for turbulent flow is

$$\dot{m}\left(\frac{T}{100}\right)^{-0{\cdot}68} d^{-1} > 2{\cdot}2$$

(2) if the flow is turbulent

$$C_h = 0{\cdot}24\left(\frac{T}{100}\right)^{0{\cdot}28} \dot{m}^{0{\cdot}8} d^{-1{\cdot}8}$$

and

$$\frac{\Delta p}{L} = (5{\cdot}7\times10^{-6})\left(\frac{T}{100}\right)^{1{\cdot}14} \dot{m}^{1{\cdot}8} d^{-4{\cdot}8} p^{-1}$$

where $\dot{m}$ is the mass flow in lb/hr
T is the absolute temperature in °R
d is the inner diameter in inches
L is the tube length in ft
Δp is the pressure drop in lbf/in^2
p is the pressure in lbf/in^2
C_h is the surface coefficient in Btu/ft^2 hr °R

(a) What temperature dependence for viscosity has been adopted in these formulae?
(b) Write the results in terms of Re, Nu, Pr and c_f.

Group C

6.53 The calculations of Problem 6.37 are unrealistic in two respects: the flow through the propulsion nozzle has been assumed isentropic, and a very large expansion in the flow area within the nozzle has been accepted.
(a) Taking the nozzle efficiency to be 80 per cent, find the exit state and efficiencies of propulsion.
(b) What are the exit areas (relative to that of the combustion chamber) for the ideal flow of Problem 6.37 and that of (a)?
(c) What thrust would be generated if the outlet area were only twice that of the combustion chamber and the nozzle flow were isentropic? By what fraction is the propulsive efficiency reduced?

6.54 Consider again the jet engine discussed in Examples 6.7 and 6.15, and shown in Figure 6.22. The efficiency of the turbine following the combustion chamber is 90 per cent and the efficiency of the nozzle through which the flow returns to the atmosphere (where the pressure is 0·265 bar) is 95 per cent.
(a) Find the state 5 at exit from the turbine. Use $\gamma = 1{\cdot}35$. Take the energy extracted by the turbine to be that absorbed by the compressor of Example 6.7, and select a suitable value for c_p.
(b) Find the state 6 at exit from the nozzle. Use $\gamma = 1{\cdot}38$ and a suitable value for c_p.
(c) What are the efficiencies characterizing this propulsive process?

6.55 Consider one-dimensional mass flow and one-dimensional conduction of heat in the same direction. By applying the overall energy equation to a fixed elemental volume of the flow, show that the energy equation governing the non-steady flow is

$$\frac{\partial(\varrho h_0)}{\partial t}+\frac{\partial}{\partial x}\left(u\varrho h_0-\frac{k}{c_p}\frac{\partial h}{\partial x}\right)=\frac{\partial p}{\partial t}$$

6.56 Consider a family of similar heat exchangers, with d the diameter of the heat-exchange tubes and L their length. It is required to find how the performance depends upon the choice of d and L, with the mass flow through the tubes, the temperature difference causing the heat transfer, and the density and other fluid properties all remaining constant. It may be assumed that the gas flow within the tubes controls the heat transfer, so that by Reynolds analogy $C_h \propto c_f u$ for any one fluid.

(a) Using equations (6.3.10), show that

$$\tau_0 \propto u^{\frac{7}{4}} d^{-\frac{1}{4}}, \quad \dot{Q} \propto u^{\frac{3}{4}} d^{\frac{3}{4}} L \quad \text{and} \quad \Delta p \propto u^{\frac{7}{4}} d^{-\frac{5}{4}} L$$

(b) Using the requirement of equality of mass flows, show that

$$\dot{Q} \propto L d^{-\frac{3}{4}} \quad \text{and} \quad \Delta p \propto L d^{-\frac{19}{4}}$$

(c) How does the ratio of the heat flow to the power expended in forcing the fluid through the exchanger vary with L and d?

(d) If the same heat output is required for two exchangers with $L_1/L_2 = 2$, what should be the ratio of the diameters of their heat-transfer tubes? What is the ratio of the pressure drops through them?

6.57 Nitrogen is to be compressed to a high pressure in a multistage centrifugal compressor. In the first three stages the pressure is raised from atmospheric to 5 atm with an overall isentropic efficiency of 70 per cent; the initial temperature is 15 °C. An intercooler is to be provided to reduce the temperature of the gas to 40 °C before it enters the next compressor stage. It is proposed to achieve the cooling by passing the gas through copper tubes immersed in flowing water whose temperature is 20 °C. The inner diameter of the tubes is 1 cm and the velocity through them is to be 5 m/sec. Only the heat transfer on the gas side need be taken into account; it is specified by $\text{Nu} = 0{\cdot}023\ \text{Re}^{0\cdot8}\ \text{Pr}^{0\cdot4}$, and the transport properties may be taken to be those of air.

Find the length of tube required and the pressure drop along it.

6.58 Steam flows vertically downwards through an array of parallel condenser tubes. The outer diameter of the tubes is 2 cm, and the distance between centres is 5 cm in the vertical and horizontal directions. The tubes of each row are located directly beneath those of the row above, so that condensate flows from the upper tubes onto the lower. Since the main component of the resistance to heat flow is the water film on the outside of the tubes, the thermal resistance increases for successive rows; it may be assumed to vary as $n^{\frac{1}{4}}$ down the bank, with n the number of the tube row from the top. The steam approaches the tube bank at 5 m/sec; it is saturated at 0·16 bar. The water temperature within the tubes is 20 °C.

(a) How many rows of tubes are required to condense the steam?

(b) At what rate is heat extracted from the steam per square metre of the plan area of the tube bank?

(c) How does the velocity of the steam vary within the bank, on the assumption that it contains no uncondensable gases?

6.59 At the end of Example 6.14 an approximate method is proposed for the determination of the development of a flow subjected to area change, friction and heat addition. The method of approximation is illustrated in Figure 6.19. Apply this technique to the 50 cm increment considered in Example 6.14.

6.60 For a perfect gas obtain the equations of the Rayleigh and Fanno lines in the form $f(h, s) = 0$, using the results (6.4.13, 15) and the canonical equation of state (1.5.18).

6.61 (a) For gas flow at low Mach number, show that

$$\frac{\mathrm{d}p}{p} = -\gamma M^2 \left[\frac{c_f}{2} + \frac{\Delta T}{T}\,\text{St}\right] \frac{P}{A}\,\mathrm{d}x$$

and

$$\frac{dp_0}{p_0} = -\frac{\gamma M^2}{2}\left[c_f + \frac{\Delta T}{T}\,\mathrm{St}\right]\frac{P}{A}\,dx$$

give the changes in pressure and total pressure along a constant-area duct, with $\Delta T = T_w - T$.

(b) If Reynolds analogy applies, show that

$$\frac{dp}{p} = -\frac{\gamma M^2}{2}\,\frac{T_w}{T}\,\frac{c_f P}{A}\,dx$$

(c) Can the pressure and total pressure be raised by the friction-linked heat flow considered here? Will these conclusions be changed for flows of fluids with Prandtl numbers much greater and much less than unity?

6.62 (a) For adiabatic flow of a perfect gas, show that

$$(1-M^2)\frac{dM}{M} = \left(1+\frac{\gamma-1}{2}M^2\right)\left(\frac{\gamma M^2}{2}\,\frac{f\,dx}{d} - \frac{dA}{A}\right)$$

(b) Show that, for specified variations of Mach number $M(x)$ and friction factor $f(x)$, the area variation $A(x)$ is given by a differential equation of simple form.

(c) An experimenter wishes to achieve a duct flow in which the Mach number is constant. If the friction factor is constant, show that he must arrange that

$$d \propto (x + \text{constant})$$

(d) What are the corresponding changes in velocity, density, temperature and pressure? Explain why the changes are related in this way.

6.63 Figure 6.24 shows air flowing isentropically into a constant-area nozzle in which heat is added to the flow. In the reservoir, $T_0 = 300$ °C.

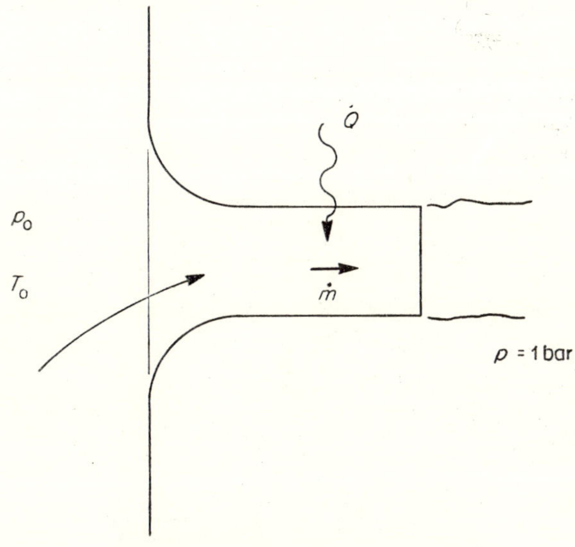

Figure 6.24 See Problem 6.63

(a) If $M = 0{\cdot}3$ before heat addition, and the flow is choked with $p = 1$ bar at the exit, what are the reservoir pressure p_0 and the heat addition q?

(b) By how much will the mass flow change if the rate of heat addition $(\dot{Q} = \dot{m}q)$ is increased by 50 per cent beyond the value found above, reservoir conditions being maintained constant? Note that q depends upon both $\dot{Q}$ and $\dot{m}$; hence a trial-and-error solution is required.

6.64 (a) Use Fanno tables to find the frictional force which the flow of Example 6.16 applies to the pipe. Consider only the parallel part of the pipe.

(b) Use the results of Section 6.5.5 to find the drag on the assumption of isothermal flow.

(c) Compare these results with $(p_1 - p_2)A$.

6.65 Consider the effect of friction and heat addition on the point in a duct at which the flow becomes sonic. It can no longer be expected that a transition to supersonic from subsonic flow will occur precisely at the geometrical throat.

(a) For adiabatic flow in a circular duct with $A = \frac{1}{4}\pi D^2$, show that continuous transition will occur for $\mathrm{d}D/\mathrm{d}x = \gamma c_f$.

(b) Show that heat addition in the absence of friction will give rise to sonic flow when $\mathrm{d}A/A = \mathrm{d}q/h$.

(c) Does sonic flow occur before or after the geometric throat when the flow is subjected to (1) friction, (2) heating and (3) cooling?

7

Flow Machines

In the preceding chapter the overall effect of a flow machine on the fluid passing through it was considered. Now we shall look at the processes which occur inside flow machines and give rise to the energy interchange. For simplicity, the flows discussed in this chapter will be taken to be adiabatic.

There are two widely used methods of doing work on or extracting work from a fluid.* The fluid can be contained briefly in a cylinder where it interacts with a moving piston, or it can be passed through an array of blades fastened to a rotating assembly. The processes in turbomachines are of particular interest here, since they often involve a close interaction of the thermodynamics and fluid mechanics of the flowing fluid.

The aims of this discussion are to examine the essential features of the energy transfer between a turbomachine and the fluid passing through it, and to reveal the primary constraints which apply to the design of turbomachines of various classes. Some of these constraints are: the necessity of minimizing dissipation within the machine; choking of the flow; cavitation in parts of the machine where the pressure is low; surging or instability of the flow through the machine; and centripetal stresses in the rotating assembly, coupled with the temperature of the fluid, which influences the strength of the rotor. In this introductory treatment it is not possible to say much about the construction of the many ingeniously contrived species of flow machines. Nor is it possible to discuss the methods of analysis used by designers of high-performance machines. The accurate prediction of performance requires modifications of the simple results to be developed here using empirical data obtained from earlier machines.

The basic tools for this analysis are the results obtained earlier for steady, one-dimensional flow. It can be argued that the flow through a turbomachine is neither one-dimensional nor steady. How then is it possible to use these simple results? The flows within machines whose rotors are closely enclosed

* This simplified classification does not take account of a class of rotary, positive-displacement machines which includes gear and screw pumps and the Wankel engine.

by a duct can usually be broken down into elements for which the one-dimensional model is fairly realistic; typically, the passage between each pair of blades can be treated in this way. However, the flow through an unenclosed rotor, such as a propeller or windmill, is less amenable to this kind of analysis; the fluid can pass around the ends of the blades and a complex three-dimensional flow pattern develops, with vortices spiralling downstream from the blade tips. We conclude that the one-dimensional results will be chiefly useful in analysing enclosed machines. Now consider the time dependence of the flow in a turbomachine. The flow near the rotating blades undoubtedly varies in time when viewed in absolute axes. However, a little upstream and downstream of a rotor with many blades the flow may be expected to be fairly steady. Moreover, viewed from relative axes, those fixed in the rotor, the flow through it usually appears to be steady. Hence, by switching back and forth between absolute and relative axes, we can use steady-flow results to analyse the flow through the whole machine.

The members of the family of turbomachines are extremely varied. One way of classifying them has been used above, the division into enclosed and unenclosed machines. It has been noted that the available methods of analysis can be applied directly only to the former class. Within it we shall consider two further kinds of division. The first is the distinction between *axial-flow* and *radial-flow* machines. We shall for simplicity assume that fluid passing through an axial-flow machine moves exactly along a constant-diameter cylinder, and that fluid passing through a radial machine moves exactly in a plane normal to the axis of rotation. In real machines these ideal motions are unlikely to occur; changes in density and velocity, and the exigencies of designing inlet and outlet passages, result in departures from the simple pattern. Moreover, complex secondary flows are set up in boundary layers within the machine. Finally, there exist mixed-flow machines in which the fluid motions are approximated by neither the axial nor the radial-flow pattern. Nevertheless, the ideal axial and radial-flow machines to be considered here reveal many of the essential features of real machines, and the results are applicable with fair accuracy to certain important classes of machines. The flow through most jet-engine compressors and steam turbines conforms reasonably well to the axial-flow model, while that in certain centrifugal pumps and hydraulic turbines is approximately radial.

Another kind of distinction between flow machines relates to the nature of the energy addition or extraction in the rotor. The term *impulse* machine is applied to one in which the kinetic energy changes while the pressure remains constant across the rotor. A *reaction* machine is one in which there is a significant pressure change across the rotor. The classification with respect to reaction has added significance, since the degree of reaction is closely linked to other characteristics of the machine, for example, the curvature of the blades.

We shall start by examining some of the basic features of the energy interchange between blades and fluid, using absolute and relative axes to describe the flow through a cascade of blades. These results will be applied in the first instance to a very simple impulse machine, and then to a variety of more realistic machines: impulse turbines, axial-flow turbines and pumps, and finally radial-flow machines. Next dimensional analysis will be applied to the whole family of turbomachines; it makes possible a compact presentation of the characteristics of the machines, and also provides additional means of classifying and comparing them. Finally, the rather special problem of cavitation in flow machines will be considered.

7.1 ENERGY CONVERSION IN A TURBOMACHINE

7.1.1 Absolute and Relative Axes

To understand the energy conversions within a turbomachine, it is necessary to look at the machine from two points of view. Absolute axes, those fixed with respect to the system with which work is exchanged, are used to discuss the net flow of energy. The consideration of relative axes, those fixed in the rotor, makes possible the use of steady-flow results in discussing the flow within the rotor and the changes in properties across it. The selection of appropriate absolute axes depends upon the circumstances of operation. For a turbine in a fixed plant, the earth provides an appropriate absolute frame of reference. On the other hand, the performance of the turbine of an aircraft jet engine would be discussed using absolute axes fixed in the aeroplane. However, if one wished to discuss the performance of the whole propulsive unit, an absolute frame in which the atmosphere is at rest would be appropriate, for it is in this frame that the useful effect of the engine is manifested.

The apparent geometry of the flow into and out of a rotor will depend on the choice of axes, absolute or relative. One must distinguish carefully between the two sets of velocities which in fact describe the same flow. *Absolute velocities* will be denoted by the symbol C and *relative velocities* by V; the difference between the two is the *rotor speed* U. These velocities are related by the vector equation

$$\mathbf{C} = \mathbf{U}+\mathbf{V} \tag{7.1.1}$$

Figure 7.1 illustrates the relationship between the velocities on the two sides of a cascade* of rotor blades. In the case shown, the fluid approaching the rotor has an absolute velocity inclined in the direction of rotor motion. However, the relative velocity is inclined in the opposite direction and the fluid is able to pass smoothly onto the moving blades. Note also that the change in the relative velocity within the blade row imposes a change on the absolute velocity as well. Actual flow machines display a wide variation in the geometry of the *velocity triangles* relating V, C and U. Some simple cases will be considered first, so that an understanding of the relationship between relative and absolute frames of reference can be built up.

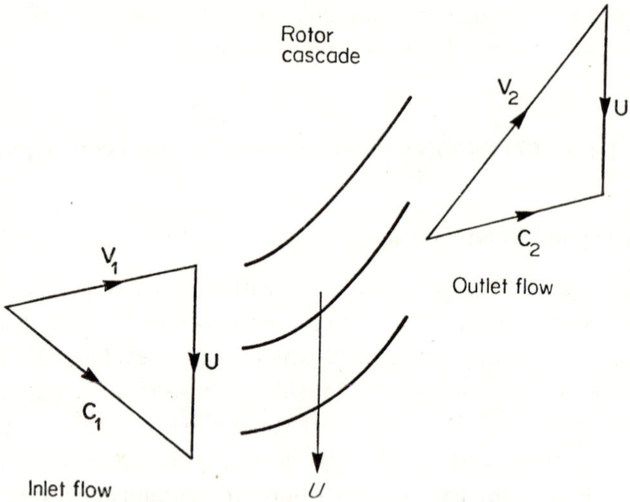

Figure 7.1 The use of velocity triangles to relate absolute velocities, relative velocities and rotor speed

Two limitations on this preliminary discussion of energy transfer are apparent from Figure 7.1. Note first that the rotor speed is the same on each side of the rotor. This is the case for an axial-flow machine, but not for radial flow, where a rotating frame of reference has to be selected to make the flow through the rotor appear steady. Hence the following discussion, which at some points assumes equal rotor speeds at inlet and outlet, does not have general application, but applies primarily to axial-flow machines.

* A hypothetical straight cascade, such as that considered here, is an infinite array of similar, parallel blades, through which the flow is two-dimensional. Real straight cascades are made up of a finite number of parallel blades; tests on them provide the aerodynamic characteristics of the blading. The flow through both theoretical and experimental cascades may be expected to reproduce certain of the leading features of that through the curving cascades of real turbomachines.

A more fundamental limitation which applies throughout this chapter is suggested by the skeletal form of the blades shown in Figure 7.1. The angles and deflexions referred to are those of the mean flow, not those of the blades. At the design point there will be a rough correspondence between the blade shape and the shape of the streamlines passing near the blades. Hence we can infer something about the shape of the blades which are associated with the specified inlet and outlet velocity triangles. However, this is at best an approximation, and can sometimes be seriously misleading. An accurate prediction of the deviation between flow angle and blade angle depends upon the availability of empirical information relating to similar blading.

If the flow through the cascade is taken to be adiabatic, and this is a good approximation for most turbomachines, equation (2.3.7) gives

$$(h_{0_2}+gz_2)-(h_{0_1}+gz_1) = w_e \tag{7.1.2}$$

relating the change in the state of the fluid to the work done on unit mass within the machine or part of it. Since actual flow machines are at most a few metres in height, the changes in gravitational potential energy within them are a small fraction of the overall energy conversion, except for low-head hydraulic turbines and pumps. Hence the subsequent discussion will be based on the results

$$h_{0_2}-h_{0_1} = w_e \tag{7.1.3}$$

and

$$\dot{m}(h_{0_2}-h_{0_1}) = \dot{W}_e \tag{7.1.4}$$

In Section 6.2 these relationships have been applied across whole machines. Now they will be applied to individual blade rows and flow passages within a machine.

For relative axes fixed in the moving rotor, $w_e = 0$, and

$$h_{0_2}-h_{0_1} = 0 \quad \text{with} \quad h_0 = h+\tfrac{1}{2}V^2 \tag{7.1.5}$$

For absolute axes fixed in the system with which work is exchanged, $w_e \neq 0$, and

$$h_{0_2}-h_{0_1} = w_e \quad \text{with} \quad h_0 = h+\tfrac{1}{2}C^2 \tag{7.1.6}$$

Since the static enthalpy is the same for both relative and absolute axes, these equations can be combined to give an alternative expression for the work transfer. Equation (7.1.5) implies that the enthalpy change across the rotor is

$$h_2-h_1 = \tfrac{1}{2}(V_1^2-V_2^2) \tag{7.1.7}$$

Then equation (7.1.6) gives

$$\begin{aligned} w_e &= \tfrac{1}{2}(C_2^2-C_1^2)+(h_2-h_1) \\ &= \tfrac{1}{2}(C_2^2-C_1^2)-\tfrac{1}{2}(V_2^2-V_1^2) \end{aligned} \tag{7.1.8}$$

This expresses the work in terms of the kinematics of the flow; given the velocity triangles at two points, one can find the work done between those points.

The preceding results apply to any fluid. For constant density more explicit relationships between the change in state and the work can be obtained. For this case equations (7.1.3, 4) can be written

$$\frac{p_{0_2}-p_{0_1}}{\varrho_c} = w_e - \Delta e \tag{7.1.9}$$

and

$$\dot{V}(p_{0_2}-p_{0_1}) = \dot{W}_e - \dot{m}\,\Delta e \tag{7.1.10}$$

For relative axes, $w_e = 0$, and

$$p_{0_2}-p_{0_1} = -\varrho_c\,\Delta e \quad \text{with} \quad p_0 = p+\tfrac{1}{2}\varrho_c V^2 \tag{7.1.11}$$

For absolute axes, $w_e \neq 0$, and

$$p_{0_2}-p_{0_1} = \varrho_c(w_e-\Delta e) \quad \text{with} \quad p_0 = p+\tfrac{1}{2}\varrho_c C^2 \tag{7.1.12}$$

Then the pressure change across the rotor is

$$p_2-p_1 = \tfrac{1}{2}\varrho_c(V_1^2-V_2^2)-\varrho_c\,\Delta e \tag{7.1.13}$$

Now equation (7.1.12) gives

$$\begin{aligned} w_e &= \tfrac{1}{2}(C_2^2-C_1^2)+\frac{p_2-p_1}{\varrho_c}+\Delta e \\ &= \tfrac{1}{2}(C_2^2-C_1^2)-\frac{1}{2}(V_2^2-V_1^2) \end{aligned}$$

as in equation (7.1.8).

The results (7.1.6, 12) for the absolute frame of reference are universally valid, subject to the validity of the assumptions of steady, one-dimensional, adiabatic flow. As was pointed out earlier, the results (7.1.5, 11) for relative axes are restricted in application to axial-flow machines, since centrifugal contributions are usually of considerable importance in radial and mixed-flow machines. However, the results (7.1.5, 11) are universally valid for stator rows, in which no work is done.

The neglect of potential-energy changes within a flow machine still leaves the possibility of augmenting the kinetic energy and enthalpy (and pressure) in different proportions. At one extreme is the impulse machine; across its rotor rows the pressure does not change, only the absolute kinetic energy being altered. In practice, most pure impulse machines are turbines, since large losses are usually involved in converting added kinetic energy into pressure and enthalpy. A machine across whose rotor blades the enthalpy change is large compared with the absolute kinetic-energy change is described as a high-reaction machine. The terms impulse and reaction refer in a general way to the source of the force applied to the blades; if the pressure is unchanged, the force may be thought of as arising from the impulse associated with the change in velocity of the fluid. On the other hand, the change in pressure across a high-reaction blade row may be said to produce a reaction force on the blades.

Even in an impulse machine there are pressure variations within the blading system and over the surface of each blade. The net force in the direction of rotor motion is responsible for the energy transfer, and the work done can be expressed as the product of this force and the rotor speed. The impulse applied to unit mass of the fluid can be reckoned as

$$C_{x_2} - C_{x_1} = V_{x_2} - V_{x_1}$$

with the subscript x denoting the component in the direction of rotor motion. Then

$$w_e = U(C_{x_2} - C_{x_1}) = U(V_{x_2} - V_{x_1})$$

These results have a very different appearance from equation (7.1.8). Nevertheless, the three expressions must be equivalent. It will be left as an exercise to show, using the properties of the velocity triangles of Figure 7.1, that they are.

7.1.2 An Elementary Machine

The first flow machine which will be studied in detail is not in fact a turbomachine, although it does possess many of the features of rotating machines. Figure 7.2 shows a simple impulse machine, a moving flat plate against which a water jet is directly normally. It is helpful to consider the operation of this device before looking at more realistic machines. The simple geometry provides an easy introduction to absolute and relative velocities, and the dynamics are correspondingly easy to understand.

In Figure 7.2(a) the flow is shown in the absolute frame of reference, that in which the nozzle producing the jet is stationary. The absolute jet velocity is C_1 and the speed of the plate is U. The vector equation (7.1.1) takes the simple algebraic form

$$C_1 = V_1 + U \qquad (7.1.14)$$

The relative velocities are shown in Figure 7.2(b); in these axes the plate

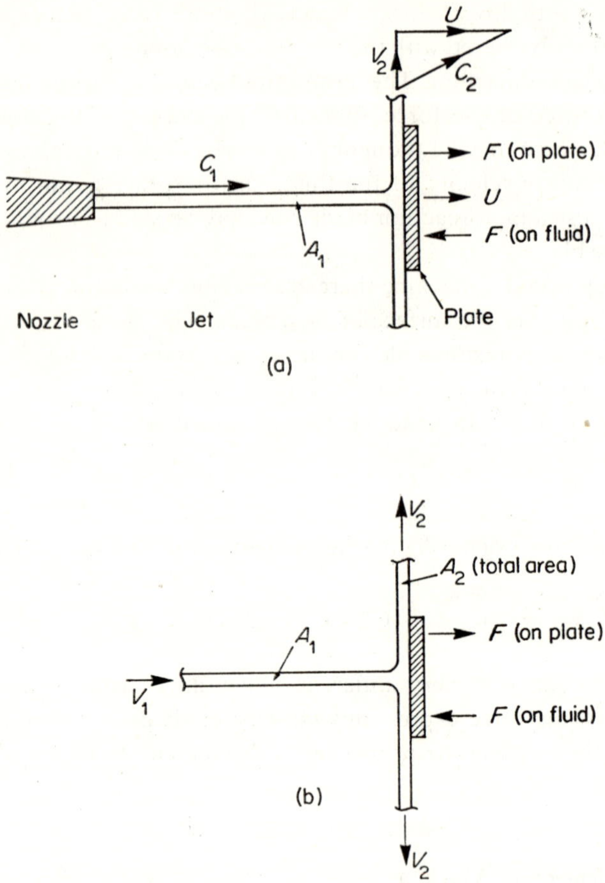

Figure 7.2 An elementary impulse turbine. (a) Absolute frame of reference; (b) Relative frame of reference

appears at rest. The relative velocity at exit is assumed to be along the plate. Hence the outflow velocities are related by

$$C_2^2 = V_2^2 + U^2 \tag{7.1.15}$$

This is a consequence of the vector addition (7.1.1).

The force F applied to the fluid (measured positive in the direction opposite to C_1 and U) is not altered by the change from absolute to relative axes. However, the energy interchange is modified. For the relative axes, no work is done on the fixed plate or on the fluid. In the absolute frame, work is done on the fluid at rate

$$\dot{W}_e = -FU \tag{7.1.16}$$

For $F > 0$, $U > 0$, $\dot{W}_e < 0$, corresponding to a loss of energy by the fluid, that is, to work done on the plate.

The three basic conservation principles will now be set down for the absolute and relative flows. By looking at the motion in these two ways one can demonstrate the equivalence of the two frames of reference, and also the greater simplicity of the analysis for relative axes. The flow will be taken to be lossless, one-dimensional, of constant density, and 'steady', that is, having C_1 and U constant. The pressure will be taken to be the same over the whole surface of the jet, though not of course over the whole plate. The calculations are directed towards obtaining expressions for the dynamic effects F and $\dot{W}_e$ in terms of the two velocities C_1 and U.

Absolute Axes

Consider a fixed control volume extending from the nozzle around the plate, and cutting normally across the outflow at the edge of the plate. The mass of fluid within this volume is continually increasing, as are the momentum and energy.

Relative Axes

Consider a control volume fixed relative to the plate, cutting through the jet at a fixed distance from the plate and across the outflow at the edge of the plate. The flow into and within this volume is steady.

Continuity

Absolute axes:

$$C_1A_1 = V_2A_2 + UA_1$$

The last term represents fluid 'stored' in the steadily lengthening jet. The outflow is reckoned as the flow area times the velocity component normal to that area.

Relative axes:

$$V_1A_1 = V_2A_2 \qquad (7.1.17)$$

Since $C_1 = V_1 + U$ the two continuity statements are equivalent.

Energy

Absolute axes:

$$\varrho_c C_1 A_1(\tfrac{1}{2}C_1^2) = \varrho_c V_2 A_2(\tfrac{1}{2}C_2^2) + \varrho_c U A_1(\tfrac{1}{2}C_1^2) - \dot{W}_e$$

In these axes work $-\dot{W}_e$ is done by the fluid; its magnitude cannot be calculated at this stage. The term before $\dot{W}_e$ represents energy stored in the lengthening jet.

Relative axes:

$$\tfrac{1}{2}V_1^2 = \tfrac{1}{2}V_2^2 \qquad (7.1.18)$$

In these axes the flow appears isenergic and steady. Since it is lossless, Bernoulli's equation applies. From (7.1.17) there follows immediately

$$A_1 = A_2$$

Momentum

$$F = \varrho_c C_1 A_1 C_1 - \varrho_c V_2 A_2 U - \varrho_c C_1 A_1 U$$

The first two terms give the fluxes of x-momentum into and out of the system. The last term gives the rate at which momentum is stored in the jet. By introducing $\dot{W}_e = -FU$ into the energy equation above, one can complete the analysis using only the absolute axes, finding $V_2 = C_1 - U$, then $A_1 = A_2$, and finally obtaining expressions for F and $\dot{W}_e$ in terms of U and C_1.

$$F = \varrho_c V_1^2 A_1 \qquad (7.1.19)$$

In this frame of reference there is no outflow of x-momentum, and no storage of momentum within the system. The power can now be calculated as $\dot{W}_e = -FU$. For this calculation of the actual work transfer, it is necessary to adopt the absolute frame. Finally, using equation (7.1.14) one can express the results for relative axes in terms of C_1 and U.

In the preceding calculations the approach using relative axes is undoubtedly the simpler. This is generally true for situations in which energy is transferred to and from a uniformly moving system. The best method is to discover what one can using relative axes in which the flow appears steady, turning to the absolute frame of reference only for information not available in the relative frame.

The energy imparted to the fluid is given by

$$w_e = -\frac{FU}{\dot{m}} \gtrless 0 \quad \text{for} \quad U \lessgtr 0$$

For $U < 0$, when the plate moves towards the nozzle, the machine considered acts as a pump. This can be seen in another way. The change in the specific energy of the fluid can be written with the help of equations (7.1.14, 15, 18) as

$$\tfrac{1}{2}C_2^2 - \tfrac{1}{2}C_1^2 = C_1^2 \frac{U}{C_1}\left(\frac{U}{C_1} - 1\right) > 0 \quad \text{for} \quad \frac{U}{C_1} < 0$$

$$< 0 \quad \text{for} \quad 1 > \frac{U}{C_1} > 0$$

As expected, the energy of the fluid is increased for $U < 0$. This behaviour is typical of hydraulic machines; a turbine running in reverse acts as a pump. However, unless the machine is specifically designed for this dual role, it is unlikely to be a very satisfactory pump. The device of Figure 7.2 is even less satisfactory as a pump than it is as a turbine, for there is no convenient way of storing and making use of the energy imparted to the fluid in the form of kinetic energy. Nevertheless, a pump of this species has found one application —in washing the outside of railway carriages. A cleaning solution is allowed to flow onto whirling blades which throw it against the sides of carriages moving slowly through the washing unit.

7.2 IMPULSE TURBINES

7.2.1 The Pelton Wheel

The first turbomachines which will be studied in detail are impulse turbines. These are in fact rather untypical of the wider class of flow machines. Nevertheless, being easy to analyse, they provide a convenient starting point.

Hydroelectric generating stations supplied from very high reservoirs, say, those 400 ft or more above the discharge level, have often been equipped with a very simple form of turbine—the Pelton wheel. The water is accelerated in nozzles and is discharged into the atmosphere in high-velocity jets. These jets strike against a series of cups or buckets fitted to the rim of a strongly constructed wheel, driving around the wheel and the generator coupled to it. The water leaves the blades with a much reduced absolute velocity, and drops into a tailrace beneath the turbine. Although fluid weight is the source of the energy driving the Pelton wheel, it does not significantly influence the flow within the nozzles and turbine.

Figure 7.3(a) shows the installation schematically; the velocities shown are absolute. Figure 7.3(b) shows the detail of the flow at a bucket, and indicates the relative velocities. The absolute and relative velocities are related by

$$C_1 = V_1 + U$$

and

$$C_2^2 = V_2^2 \sin^2\beta + (V_2\cos\beta - U)^2 \qquad (7.2.1)$$

These equations incorporate the assumption that the water approaches and leaves the buckets in a tangential plane. This cannot be precisely so, since the pans rotate through a finite angle while the jet plays against them, and there must be on average a small angle between C_1 and U.

The preceding study of a simpler machine suggested that analysis will be easier in terms of the relative axes, those moving with the instantaneous velocity of the cup on which the jet impinges. For lossless, one-dimensional, constant-density flow, the continuity, energy and momentum conditions are, for these axes

$$V_1 A_1 = V_2 A_2$$
$$\tfrac{1}{2}V_1^2 = \tfrac{1}{2}V_2^2 \qquad (7.2.2)$$

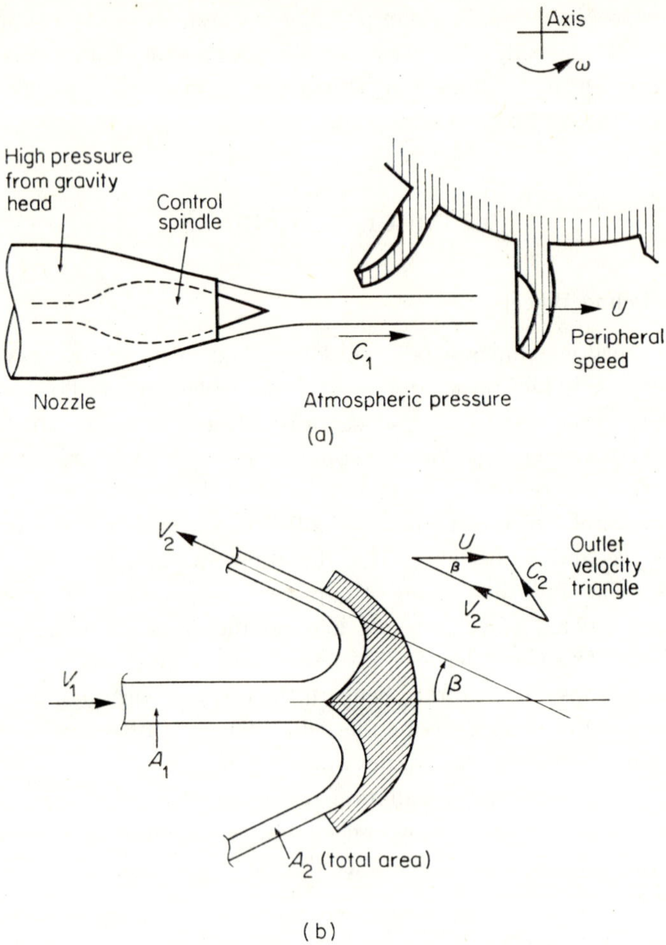

Figure 7.3 The Pelton wheel, a simple hydraulic turbine. (a) Inlet flow in absolute axes; (b) Section through a bucket fixed to the wheel as seen from relative axes

and

$$F = \varrho_c V_1 A_1 V_1 + \varrho_c V_2 A_2 V_2 \cos\beta \tag{7.2.2}$$

These give

$$V_2 = V_1, \quad A_1 = A_2$$

and

$$F = (1+\cos\beta)\varrho_c V_1^2 A_1 \tag{7.2.3}$$

To find the power extracted by the turbine it is necessary to look at the flow in the absolute frame of reference. This situation differs from that of Figure 7.2 in that no fluid is stored in the jet; ideally, the cups catch all of the jet flow as in succession they come into its path. Hence some of the time the jet impinges on more than one cup and the force found above is not the average force applied by the jet at the rim of the wheel. The net force can be found as follows. The impulse applied per unit mass reaching the wheel is

$$\frac{F}{\varrho_c V_1 A_1} = (1+\cos\beta)V_1$$

But all of the mass flow $\dot{m} = \varrho_c C_1 A_1$ ultimately impinges on one or other of the cups. Hence the net peripheral force is

$$(1+\cos\beta)\dot{m}V_1$$

and the net power extracted from the flow is

$$-\dot{W}_e = (1+\cos\beta)\,\dot{m}\,(C_1-U)U \qquad (7.2.4)$$

The last result indicates that no energy is extracted when the wheel is stalled ($U = 0$) or turning freely ($U = C_1$). The condition for maximum energy extraction from a given jet ($\dot{m}$, C_1 fixed) is $\mathrm{d}\dot{W}_e/\mathrm{d}U = 0$. This is satisfied for

$$U = \tfrac{1}{2}C_1$$

The kinematic significance of this result can be seen using the geometrical relations (7.2.1, 3). For any value of the peripheral speed

$$V_2 = V_1 = C_1 - U$$

For maximum energy extraction

$$V_2 = \tfrac{1}{2}C_1 = U$$

and

$$\left(\frac{C_2}{C_1}\right)^2 = \frac{1}{4}[\sin^2\beta+(\cos\beta-1)^2] = \frac{1}{2}(1-\cos\beta)$$

For small values of β the residual kinetic energy $\frac{1}{2}C_2^2$ is a very small fraction of the input energy $\frac{1}{2}C_1^2$; the fluid merely drops into the sump, its energy almost entirely extracted.

The efficiency of a Pelton wheel is usually defined as

$$\eta = -\frac{w_e}{\frac{1}{2}C_1^2} = \frac{\text{energy extracted}}{\text{kinetic energy input}} \qquad (7.2.5)$$

Note that this is a total-to-static efficiency, as defined in equation (6.2.11). For the lossless flow considered here, the total-to-total efficiency across the turbine rotor is 100 per cent. Using equation (7.2.4) one can express the total-to-static efficiency as

$$\eta = 2(1+\cos\beta)\left(1-\frac{U}{C_1}\right)\frac{U}{C_1} \leqslant \frac{1}{2}(1+\cos\beta) \qquad (7.2.6)$$

Maximum efficiency is achieved for $U/C_1 = \frac{1}{2}$; it can be made very close to 100 per cent by making β small. However, an attempt to reduce β too far is self-defeating, for the fluid leaving the bucket disturbs the jet approaching it and the resultant losses more than cancel out the expected gain.

In the preceding analysis it has been assumed that the flow is free of dissipation. A more realistic assessment of Pelton wheel performance can be made by introducing a *blade factor*

$$c = \frac{V_2}{V_1} \tag{7.2.7}$$

A value in the vicinity of 0·9 might be expected in practice. The impulse per unit mass is now altered to

$$V_1 + V_2 \cos\beta = V_1(1 + c\cos\beta)$$

and the power extracted and efficiency become

$$-\dot{W}_e = (1 + c\cos\beta)\,\dot{m}\,(C_1 - U)U$$

and

$$\eta = 2(1 + c\cos\beta)\left(1 - \frac{U}{C_1}\right)\frac{U}{C_1} \tag{7.2.8}$$

It appears that the best efficiency is still achieved with $U/C_1 = \frac{1}{2}$, although that optimum efficiency is reduced to $\frac{1}{2}(1 + c\cos\beta)$. For the realistic values of $c = 0{\cdot}90$ and $\beta = 20°$, the optimum efficiency is 92 per cent.

Example 7.1 A Pelton wheel driving a generator is supplied from a reservoir 200 m above the tailrace, the water flowing through a pipe of effective length 300 m; see Figure 6.9. Three jets of diameter 20 cm impinge on the cups at a radius of 75 cm. The total-to-static efficiency is 92 per cent, and the efficiency of conversion of the extracted energy into electrical energy is 90 per cent. It is required to find

(1) the electrical power delivered by the generator connected to the turbine, on the assumption that 95 per cent of the gravitational potential energy of the fluid is converted into kinetic energy in the jets, and

(2) the diameter of the inlet pipe which will make possible this efficiency of delivery with a pipe friction coefficient of $c_f = 0{\cdot}005$.

The rate at which potential energy is lost by the water passing through the turbine is $\dot{m}gH_{0_1}$, where H_{0_1} is the difference in elevation between reservoir and tailrace. This is the power which might, in principle, be converted into electrical energy. When account is taken of the losses in the conversion, the electrical output is given by

$$P_e = -\eta_g \dot{W}_e = \eta_g \eta_f \eta_{ts} \dot{m} g H_{0_1}$$

The three efficiencies have been specified, as has the driving head. The mass flow also depends on the efficiency with which the potential energy is converted into kinetic energy. Thus

$$C_1 = (2\eta_f g H_{0_1})^{\frac{1}{2}} = 61{\cdot}1 \text{ m/sec}$$

and the mass flow is

$$\dot{m} = \varrho_c C_1 A_1 = 5750 \text{ kg/sec}$$

Finally, the power generated is $P_e = 8900$ kW.

The dissipation per unit mass in the pipe leading to the turbine is given by equation (5.5.17) as

$$2c_f \frac{L_e}{d} u^2 = \frac{32}{\pi^2} c_f \frac{L_e}{d} \frac{\dot{V}^2}{d^4}$$

It is required that this be not more than $(1/20)gH_{0_1}^*$. Hence the required pipe diameter is given by

$$d^5 = \frac{640}{\pi^2} c_f \frac{L_e}{H_{0_1}} \frac{\dot{V}^2}{g} = 1{\cdot}64 \text{ m}^5$$

and $d = 1{\cdot}1$ m is the smallest pipe diameter consistent with the specified loss.

7.2.2 The Laval Turbine

Figure 7.4 shows a form of single-stage impulse turbine in which steam is the working fluid. The steam is accelerated to supersonic speed in the converging-diverging nozzle which directs it onto the moving blades. Since the pressure is the same on the two sides of the rotor, a uniform inflow is not required, and the flow usually consists of a few isolated jets somewhat like those of a Pelton wheel. However, the nozzles of a Laval turbine are usually located beside the rotor, rather than in its plane, since a steam jet travelling through air will spread and lose energy more rapidly than a water jet.

The analysis of the flow through a steam turbine is in general a formidable task. However, a single-stage impulse turbine is not too difficult to deal with, and further restriction to lossless, two-dimensional flow simplifies the analysis still more. This is the case which will be considered here. The state postulate is the key to the analysis. Two of the thermodynamic properties are unchanged across the rotor of an impulse turbine operating without losses:

$$p_1 = p_2 \quad \text{and} \quad s_1 = s_2 \tag{7.2.9}$$

Hence the other thermodynamic properties must be unchanged as well:

$$h_1 = h_2 \quad \text{and} \quad \varrho_1 = \varrho_2 \tag{7.2.9}$$

The difference in nozzle position makes the flow geometry more complicated than that of a Pelton wheel. Figure 7.4(c) shows the inlet and outlet

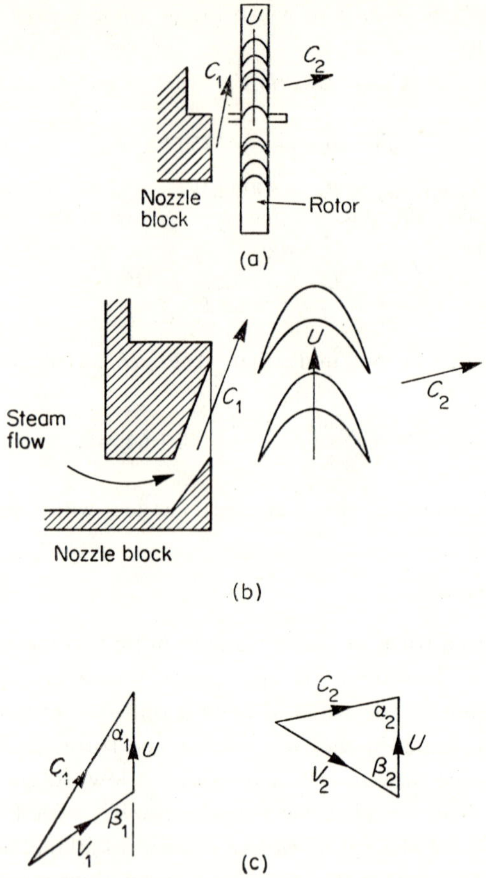

Figure 7.4 The Laval turbine, a simple steam turbine. (a) Absolute flow through the rotor; (b) Detail of nozzle and blades; (c) Inlet and outlet velocity triangles

velocity triangles for a Laval turbine. Since the enthalpy does not change across the rotor, equation (7.1.7) gives immediately

$$V_1 = V_2 \tag{7.2.9}$$

A number of interesting conclusions follow from the continuity requirement for two-dimensional flow:

$$\varrho_1 V_1 \sin \beta_1 = \varrho_2 V_2 \sin \beta_2 \tag{7.2.10}$$

The equalities (7.2.9) imply the further condition

$$\beta_1 = \beta_2 \tag{7.2.9}$$

The velocity triangles can now be combined in the simple manner of Figure 7.5. It is immediately apparent that the axial component of the velocity is the same on each side of the rotor:

$$C_a = V_1 \sin \beta_1 = V_2 \sin \beta_2 = C_1 \sin \alpha_1 = C_2 \sin \alpha_2 \qquad (7.2.11)$$

This conclusion can also be obtained from (7.2.10), with $\varrho_1 = \varrho_2$. Moreover, if we take C_1 and α_1 to be independent of the rotor speed

$$C_a = \text{constant for all } U$$

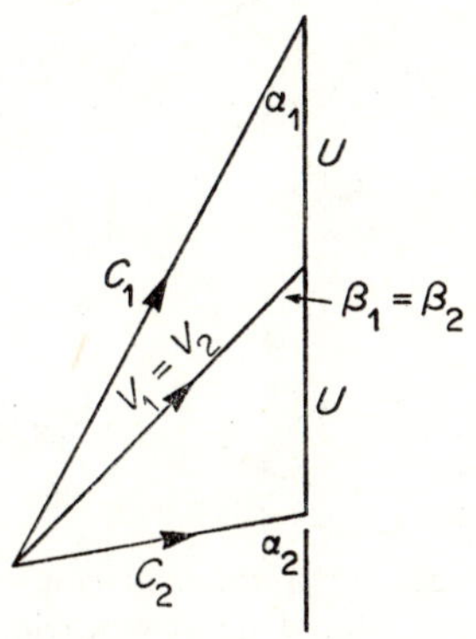

Figure 7.5 Velocity triangles for isentropic flow through a Laval turbine

So far the kinematics of the flow have been considered in isolation from the dynamics. Now the energy transformations will be considered. The work done by unit mass of the fluid is

$$-w_e = h_{0_1} - h_{0_2} = \tfrac{1}{2}C_1^2 - \tfrac{1}{2}C_2^2 \qquad (7.2.12)$$

For a specified inlet energy the work is greatest when the energy at outlet is least. This occurs for $\alpha_2 = 90°$, with $C_2 = C_a = C_1 \sin \alpha_1$. The maximum energy extraction is given by

$$-w_e = \tfrac{1}{2}C_1^2 \cos^2 \alpha_1 \qquad (7.2.13)$$

The corresponding rotor speed is seen from Figure 7.5 to be

$$U = \tfrac{1}{2}C_1 \cos \alpha_1 \qquad (7.2.14)$$

one-half of the absolute whirl component of the fluid incident on the rotor. Then the maximum energy extraction is

$$-w_e = 2U^2 \qquad (7.2.15)$$

The last result shows that the energy which can be extracted by a turbine is closely linked to the rotor speed. A similar result is obtained for the Pelton wheel, by setting $C_1 = 2U$ in equation (7.2.4). This points to the primary limitation on the performance of a turbine—centripetal stresses set up by its rotation. The nature of the restriction on the energy extraction is revealed by the following simple analysis.

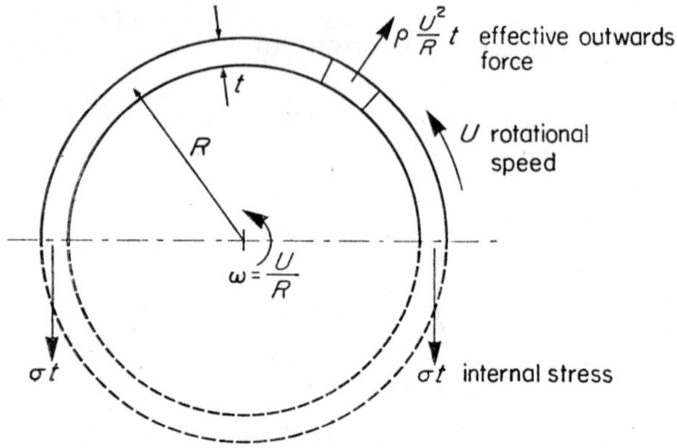

Figure 7.6 Forces on a rotating ring, illustrating the generation of centrifugal stresses in rotating machinery

Figure 7.6 shows a solid ring rotating about its axis of symmetry. The hoop stress and rotational speed can be related by an argument like that used in Example 3.11 to calculate the hoop stress resulting from internal pressure. The centripetal acceleration is equivalent to a pressure differential of $\varrho(U^2/R)t$, where ϱ is the density of the material of which the ring is made. Then

$$\frac{\varrho U^2 t}{R} 2R = 2\sigma t$$

relates the net force across a diametral plane to the hoop stress σ in the ring, and

$$U^2 = \frac{\sigma}{\varrho} \tag{7.2.16}$$

connects rotational speed, hoop stress and density. The permissible rotational speed for a particular material is determined by the allowable stress under the conditions of operation.

The result (7.2.16) cannot be applied directly to a turbine rotor. Some parts of the rotor do not carry a hoop stress. For example, the blades will be subjected to centripetal forces which must be borne by the ring or disc to which they are fixed. However, this added load can be compensated to some extent by the support given by a turbine wheel consisting of a solid disc. Another factor which determines the strength of a turbine rotor is the temperature of operation. Steam and gas turbines of necessity handle hot fluids, and the stresses in their rotors must be kept below the limit at which the heated material will creep outwards rapidly. Finally, some account must be taken of the other forces applied to the rotating assembly, particularly those applied by the fluid passing through the blading.

To get some feeling for the limitation imposed by equation (7.2.16), a particular case will be considered. For a rotor made of high-grade steel, an allowable stress of $\sigma = 140{,}000\ \text{lbf/in}^2$ and effective density of $20\ \text{sl/ft}^3$ might be achieved. For these values

$$U = \left(\frac{\sigma}{\varrho}\right)^{\frac{1}{2}} = 1000\ \text{ft/sec} \qquad (7.2.17)$$
$$= 300\ \text{m/sec}$$

Somewhat higher speeds can be achieved by the use of special alloys and by careful design to reduce the effective density. For this limiting speed, the energy which can be extracted in a single impulse stage is

$$-w_e = 2U^2 = 80\ \text{Btu/lb}$$
$$= 180\ \text{kJ/kg} \qquad (7.2.17)$$

The criterion found above has important implications for the several types of turbines in common use. If efficient operation is to be achieved in a steam-turbine plant, the temperature must fall by several hundred degrees across the turbine. A glance at the Mollier chart shows that the energy extraction associated with such a temperature change can be achieved only in a machine with many stages. On the other hand, the turbine of an aircraft jet engine extracts only the energy required to drive the compressor. In the jet engine considered in Examples 6.7 and 6.15 some 1000 kJ/kg are provided by combustion, but only some 300 kJ/kg are required to drive the compressor, the rest of the added energy being convected away in the propulsive jet. This energy requirement for the compressor can be satisfied by one or two turbine stages. Finally, consider the application of the limit (7.2.17) to a hydraulic turbine. The change in head corresponding to this energy is

$$\Delta z = -\frac{w_e}{g} = 18{,}000\ \text{m}$$

It appears that a single-stage turbine will be adequate for any terrestrial hydroelectric station.

Example 7.2 Consider a Laval turbine used to operate a centrifuge. The nozzles are set so that $\alpha_1 = 25°$. The temperature of the steam leaving the turbine, which operates at atmospheric pressure, is found to be 250 °F. For a flow through the turbine of 4000 lb/hr, the optimum performance is achieved for 9000 rev/min; the torque generated by the turbine is then 100 lbf ft. The specification of these externally measurable quantities makes it possible to discuss some of the processes within the turbine, on the assumption of lossless flow in rotor and nozzles.

Operation of the Rotor

The output of the turbine is given by

$$-\dot{W}_e = -\dot{m}w_e = T\omega$$

Then at the optimum operating point, $-\dot{W}_e = 171$ hp and the energy extracted from the steam is $-w_e = 109$ Btu/lb.

The rotor speed at this optimum operating condition is given by

$$U = \left(-\frac{w_e}{2}\right)^{\frac{1}{2}} = 1170 \text{ ft/sec}$$

The effective radius at which the jet acts is

$$\frac{U}{\omega} = 1{\cdot}24 \text{ ft}$$

and the acceleration at this radius is

$$U\omega = 1170 \times 943 \text{ ft/sec}^2 = 34{,}300 \text{ g}$$

Kinematics of the Relative Flow

The inlet absolute velocity is given by

$$C_1 = \frac{2U}{\cos \alpha_1} = 2570 \text{ ft/sec}$$

and the axial component by

$$C_1 \sin \alpha_1 = C_2 = 1085 \text{ ft/sec}$$

The specific volume of the outlet flow is found from steam tables to be $v_2 = 28{\cdot}4$ ft^3/lb. For ideal flow through an impulse turbine this will also be the specific volume at inlet: $v_1 = v_2$. Hence the effective flow area in the plane of the rotor is

$$\frac{\dot{m}v_2}{C_2} = 4{\cdot}18 \text{ in}^2$$

and the cross-sectional area of the jet is

$$A_1 = \frac{\dot{m}v_1}{C_1} = 1{\cdot}77 \text{ in}^2$$

It appears that a single jet will be adequate for the specified flow. The small size of the flow passage implies that the assumption of lossless flow is in fact quite unrealistic.

Kinematics of the Absolute Flow

From Figure 7.5 can be found

$$\tan \beta_1 = C_2/U = 2 \tan \alpha_1$$

Then

$$\tan \beta_1 = 2 \tan 25° = 0{\cdot}932$$

and

$$\beta_1 = \beta_2 = 43°$$

This gives an idea of the shape required for the rotor blades. They would have to be somewhat more sharply curved than the flow in order to produce the required mean deflexion.

The relative velocities are given by

$$V_1^2 = V_2^2 = U^2 + C_2^2 = \tfrac{1}{4} C_1^2 \cos^2 \alpha_1 + C_1^2 \sin^2 \alpha_1$$

Then

$$V_1 = \tfrac{1}{2} C_1 (1 + 3 \sin^2 \alpha_1)^{\frac{1}{2}} = 1660 \text{ ft/sec}$$

The speed of sound may be estimated on the assumption that the fluid behaves like a perfect gas. Since $T_2 = T_1$ for the assumed ideal operation

$$a_1 \simeq (\gamma R T_1)^{\frac{1}{2}} \simeq \left(1{\cdot}3 \times 710 \times 1545 \times \frac{32{\cdot}2}{18{\cdot}0}\right)^{\frac{1}{2}}$$
$$= 1600 \text{ ft/sec}$$

Then the Mach number of the inlet flow is

$$\frac{V_1}{a_1} \simeq 1$$

This value is too high for efficient operation, since locally supersonic relative flow would be likely to give rise to shocks and other dissipative mechanisms.

Conditions in the Nozzle and Reservoir

The absolute Mach number at inlet can be estimated as $C_1/a_1 = 2570/1600 = 1{\cdot}6$. On the assumption that the steam behaves like a perfect gas with $\gamma = 1{\cdot}3$, one can use isentropic tables to find

$$\frac{A_1}{A^*} = 1{\cdot}27, \quad \text{giving} \quad A^* = 1{\cdot}39 \text{ in}^2$$

the area of the nozzle throat. This small throat area again suggests that a single inlet nozzle will be sufficient.

The reservoir enthalpy can be determined using the steam-table value for $h_1 = h_2$:

$$h_{0_1} = h_1 + \tfrac{1}{2} C_1^2 = 1169 + 133 = 1302 \text{ Btu/lb}$$

Alternatively

$$h_{0_1} = h_{0_2} - w_e = h_2 + \tfrac{1}{2} C_2^2 - w_e = 1169 + 24 + 109 = 1302 \text{ Btu/lb}$$

For a constant-entropy change of state between reservoir and inlet states, the Mollier chart gives for the reservoir state

$$T_{0_1} = 540\ ^\circ\text{F} \quad \text{and} \quad p_{0_1} = 62 \text{ lbf/in}^2 \text{ (abs)}$$

The isentropic efficiency of the turbine is 100 per cent, the flow having been assumed isentropic in both nozzle and rotor. However, the energy-conversion efficiency for the process is very low. This efficiency will be calculated for two possible modes of operation.

(1) Suppose that the steam passing through the turbine runs to waste, the water in the boiler being made up by mains water at 70 °F, with $h_f \simeq 40$ Btu/lb. Then the heat added in the boiler would be

$$h_{0_1} - h_f = 1300 - 40 = 1260 \text{ Btu/lb}$$

and the energy-conversion efficiency can be reckoned as

$$\eta = \frac{109}{1260} = 8{\cdot}6 \text{ per cent}$$

(2) If the waste steam were condensed at atmospheric pressure and used again the make-up water would have a temperature of 212 °F, the condensing temperature, and $h_f \simeq 180$ Btu/lb. The energy added to the water on pumping it into the boiler is negligible. Then

$$h_{0_1} - h_f = 1300 - 180 = 1120 \text{ Btu/lb}$$

gives the heat which must be added in the boiler. The cycle efficiency is then

$$\eta = \frac{109}{1120} = 9{\cdot}7 \text{ per cent}$$

7.2.3 A Multistage Machine

In order to indicate some of the essential features of a multistage machine, one very simple case will be considered. Figure 7.7 shows an idealized two-stage impulse turbine, with the flow deflected through 180° in each blade row, and the rotors both moving at the same speed. The analysis which follows can be extended in a straightforward manner to an arbitrarily large number of stages; the two-stage machine suggests the characteristics of the whole class.

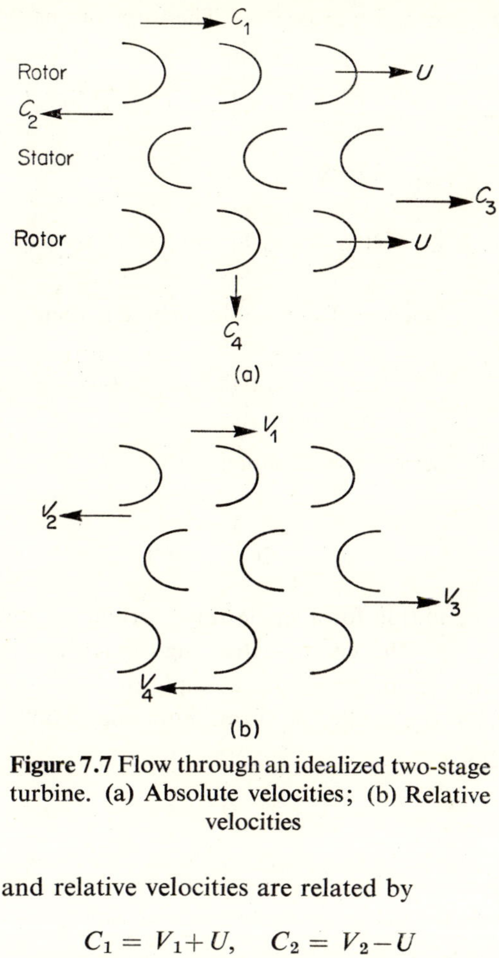

Figure 7.7 Flow through an idealized two-stage turbine. (a) Absolute velocities; (b) Relative velocities

The absolute and relative velocities are related by

$$C_1 = V_1 + U, \quad C_2 = V_2 - U$$
$$C_3 = V_3 + U, \quad C_4 = V_4 - U \tag{7.2.18}$$

With a blade factor c introduced to represent losses, the successive velocities are related by

$$V_2 = cV_1, \quad C_3 = cC_2, \quad V_4 = cV_3 \tag{7.2.18}$$

The relative velocities can now be expressed in terms of C_1 and U:

$$V_1 = C_1 - U$$
$$V_2 = c(C_1 - U)$$
$$V_3 = c^2 C_1 - (c^2 + c + 1) U$$
$$V_4 = cV_3$$

With the preceding kinematic relations available, it is an easy matter to discuss the energy-conversion processes. The force on the first rotor is given by

$$\frac{F_1}{\dot{m}} = C_1+C_2 = V_1+V_2 = (1+c)(C_1-U) \tag{7.2.19}$$

and that on the second rotor by

$$\frac{F_2}{\dot{m}} = C_3+C_4 = V_3+V_4 = (1+c)\,[c^2C_1-(c^2+c+1)\,U] \tag{7.2.19}$$

The total-to-static efficiency of the whole turbine is then

$$\eta = \frac{(F_1+F_2)\,U}{\frac{1}{2}\dot{m}\,C_1^2} = 2(1+c)\left[(1+c^2)-(2+c+c^2)\frac{U}{C_1}\right]\frac{U}{C_1} \tag{7.2.20}$$

The most efficient operation is obtained with

$$\frac{U}{C_1} = \frac{1+c^2}{2(2+c+c^2)} \tag{7.2.21}$$

The comparable condition for a single-stage turbine is given by equation (7.2.8) as $U/C_1 = \frac{1}{2}$. The energy-extracting capacities of single and two-stage machines can be compared most simply by considering the ideal case ($c = 1$, $\beta = 0°$), for the condition of optimum operation.

For a single stage:

$$\frac{U}{C_1} = \frac{1}{2} \quad \text{and} \quad -w_e = \frac{1}{2}C_1^2 = 2U^2$$

For two stages: (7.2.22)

$$\frac{U}{C_1} = \frac{1}{4} \quad \text{and} \quad -w_e = \frac{1}{2}C_1^2 = 8U^2$$

As might have been expected, the use of two stages greatly increases the energy which can be extracted for a particular rotor speed.

In reality there is a penalty to be paid when using this kind of multistage machine. This is most easily seen by looking at a particular case. For $c = 0{\cdot}80$, the best efficiency obtainable with a single stage is given by equation (7.2.8), with $\beta = 0°$, as $\eta = 90$ per cent for $U/C_1 = 0{\cdot}50$. For a two-stage machine, again with $C = 0{\cdot}80$, equation (7.2.21) gives the best operating speed as $U/C_1 = 0{\cdot}238$, for which equation (7.2.20) gives $\eta = 70{\cdot}3$ per cent. It is apparent that dissipation has little effect on the rotor speed for best operation, but does have a marked effect on the efficiency of the machine. This suggests that multistage machines will be successful only when it is possible to achieve high efficiencies for individual stages.

The addition of a second impulse stage has been found to increase abruptly both the output of the turbine and the losses within it. These two effects have the same root cause—the greatly increased (from U to $3U$) relative velocity at inlet. The fall in efficiency is not usually acceptable, and can be avoided by adopting reaction blading in which the inlet velocity is more nearly constant from stage to stage. As a result the energy transfers which can be achieved by multistage machines operating in comparable conditions are roughly proportional to the number of stages.

7.3 AXIAL-FLOW MACHINES

7.3.1 Work in a Straight Cascade

In many turbomachines the fluid moves along very nearly cylindrical surfaces concentric with the axis of rotation; this situation is shown in Figure 7.8. If the blade height $r_2 - r_1$ is small compared to the mean radius $\bar{r} = \frac{1}{2}(r_1 + r_2)$, the flow in a tangent plane will be essentially two-dimensional. It is usual to concentrate attention on the flow at the *pitchline*, that is, on the cylinder of radius $\bar{r}$. For $(r_2 - r_1)/\bar{r} \ll 1$, the pitchline flow is representative of the whole flow through the blades.

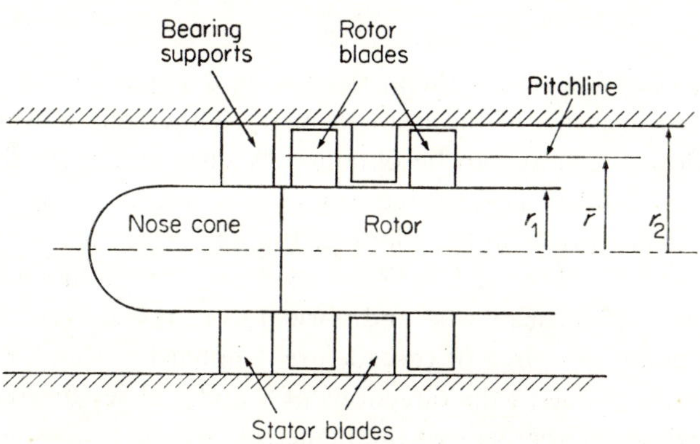

Figure 7.8 The basic geometry of an axial-flow machine

In a real machine the flow will depart from the two-dimensional pattern. The rotor speed varies from blade root to tip, according to the law $U = \omega r$; corresponding changes in blade shape are usually provided.* Moreover,

* The blade shape is also varied from tip to root in order that a desired variation in swirl velocity be achieved, for example, the free-vortex pattern, $C_x \propto r^{-1}$.

the boundary layers formed on the cylindrical walls of the duct must be expected to interact in a complex manner with the successive blade rows. Finally, the centripetal pressure field will give rise to radial flows in the boundary layers formed on the blades themselves. Hence the pitchline approximation is only the first step towards the prediction of the behaviour of real blading systems. It does, nevertheless, indicate the general features of the operation of axial-flow machines.

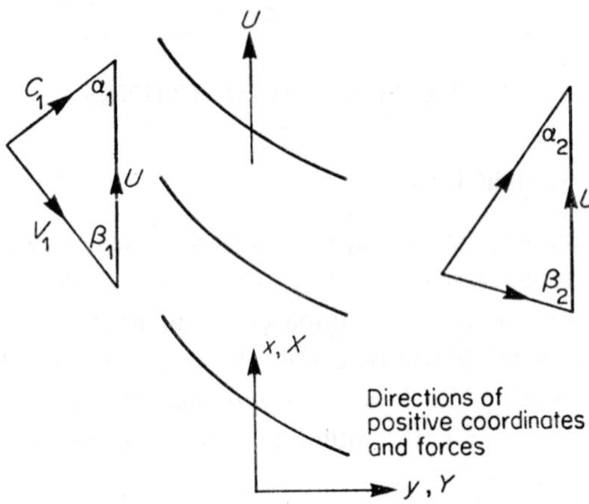

Figure 7.9 Notation for a straight cascade

Earlier, in equation (7.1.8), the work done on fluid passing through a rotor was expressed in terms of the flow geometry. Other relationships between work and flow geometry can be obtained by considering the forces acting on the moving blades. These forces are also of interest in determining the torque and axial thrust on the blading. Figure 7.9 defines the geometrical parameters which will be used in discussing them. The x and y-axes are taken in the direction of rotor motion and normal to it, respectively. The components of force on unit area of cascade are denoted by X and Y, along and normal to the cascade in the directions of x and y. They are related to the flow properties on either side of the blade row by

$$X = \dot{m}(C_{x_1} - C_{x_2}) = \dot{m}(V_{x_1} - V_{x_2}) \tag{7.3.1}$$

and

$$\begin{aligned} Y &= p_1 - p_2 + \dot{m}(C_{y_1} - C_{y_2}) \\ &= p_1 - p_2 + \dot{m}(V_{y_1} - V_{y_2}) \end{aligned} \tag{7.3.2}$$

Here $\dot{m}$ denotes the mass flow per unit area of the cascade:

$$\dot{m} = \varrho_1 C_{y_1} = \varrho_1 V_{y_1} = \varrho_2 C_{y_2} = \varrho_2 V_{y_2} \tag{7.3.3}$$

A family of geometric relations are also available:

$$\begin{aligned} C_{x_1} &= C_1 \cos\alpha_1 \qquad & C_{x_2} &= C_2 \cos\alpha_2 \\ V_{x_1} &= -V_1 \cos\beta_1 \qquad & V_{x_2} &= -V_2 \cos\beta_2 \end{aligned} \tag{7.3.4}$$

and

$$\begin{aligned} C_{y_1} &= V_{y_1} = C_1 \sin\alpha_1 = V_1 \sin\beta_1 \\ C_{y_2} &= V_{y_2} = C_2 \sin\alpha_2 = V_2 \sin\beta_2 \end{aligned} \tag{7.3.4}$$

The work done on the fluid is given by

$$w_e = -\frac{XU}{\dot{m}} = U(C_{x_2} - C_{x_1}) = U(V_{x_2} - V_{x_1}) \tag{7.3.5}$$

a purely kinematic result, as anticipated. It is apparent that the cascade operates as a pump, with $w_e > 0$, when

$$U\,\Delta C = U\,\Delta V > 0$$

that is, when the fluid is deflected in the direction of the rotor motion. Conversely, turbine operation, with $w_e < 0$, is obtained for

$$U\,\Delta C = U\,\Delta V < 0$$

the fluid being deflected in the direction opposite to the rotor motion.

One of the most useful pieces of information about a turbomachine is the relationship between the flow rate through it, its rotational speed and the energy transfer. The results (7.3.5) indicate the general features of the relationship, as is shown by the following considerations. As an approximation, the absolute angle α_1 of the flow into the cascade may be taken to be independent of the flow rate and machine speed; its direction is determined primarily by the geometry of the stator row or duct ahead of the rotor. Also, the relative angle β_2 of the flow leaving the rotor may be expected to vary only slightly with operating conditions, since it is primarily determined by the fixed geometry of the rotor blading. It will be assumed further that the axial component of the velocity is the same at inlet and outlet:

$$C_{y_1} = V_{y_1} = C_{y_2} = V_{y_2} = V_a = C_a \tag{7.3.6}$$

For constant-density flow these conditions are satisfied if the cross-sectional area of the annular space containing the blades is uniform. Moreover, equations (7.3.3) define a family of flows with varying density for which the conditions (7.3.6) apply. For all of these cases, the invariant angles α_1 and β_2 are related to the velocity components contributing to the work transfer (7.3.5) by

$$\frac{C_{x_1}}{C_a} = \cot\alpha_1 \quad \text{and} \quad \frac{V_{x_2}}{C_a} = \frac{-U + C_{x_2}}{C_a} = -\cot\beta_2$$

These results follow from the geometry of Figure 7.9. Now the work done by the rotor can be expressed as

$$w_e = U(C_{x_2} - C_{x_1}) = U[U - C_a(\cot\beta_2 + \cot\alpha_1)] \tag{7.3.7}$$

In non-dimensional form, this is

$$\frac{w_e}{U^2} = 1 - \frac{C_a}{U}(\cot\beta_2 + \cot\alpha_1) = 1 - \frac{fC_a}{U} \tag{7.3.8}$$

The preceding equations show that the relationship between the parameter specifying the work done (w_e/U^2) and that specifying the flow through the rotor (C_a/U) depends solely on the geometric parameter

$$f = \cot\alpha_1 + \cot\beta_2$$

For a machine with fixed blade angles, the parameter f should be nearly independent of the operating conditions, that is, of the flow rate and rotor speed. Figure 7.10 shows the relationship (7.3.8) for the three possible cases. The case $f > 0$ is of greatest importance. For turbines, this is obviously

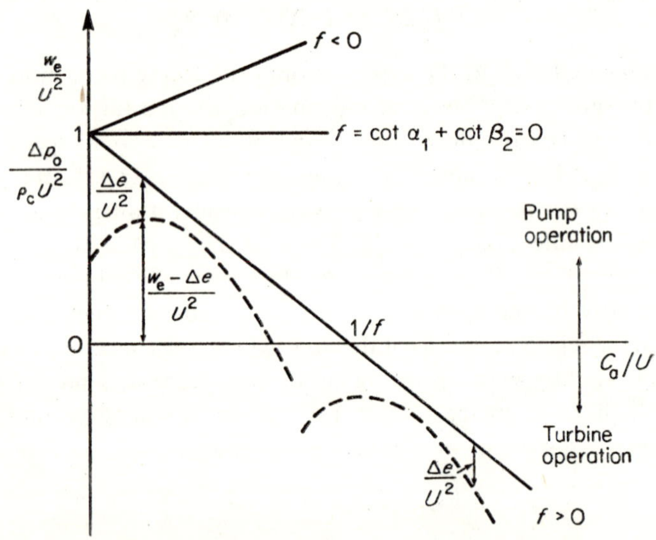

Figure 7.10 Dimensionless characteristics of single-stage axial-flow machines. The horizontal axis represents the flow rate through the machine, and the vertical axis represents the useful work done on the fluid passing through the machine. The solid lines give the ideal work done w_e/U^2 and are valid for any fluid. The dashed lines are valid only for a constant-density fluid, and give the useful work $\Delta p_0/\rho_c U^2$ with account taken of the dissipation $\Delta e/U^2$ within the machine

true, since only then can work be done by the fluid on the rotor. For pumps too a descending characteristic is desirable, since stable operation often requires that the pressure fall as the flow rate rises, a condition more likely to be achieved when $f > 0$. However, as will be seen shortly, the shape of the pressure-rise against flow-rate characteristic is strongly dependent on the losses in the machine, and a variety of pumps, corresponding to both $f \gtrless 0$, can be used.

The solid straight lines in Figure 7.10 represent work done on the fluid by the rotor. In a real pump not all of this work is recovered as a rise in the mechanical energy of the fluid. Conversely, in a real turbine, the fall in mechanical energy is greater than the work extracted. This is most easily seen for the case of constant-density, for which equation (7.1.12) gives

$$\frac{\Delta p_0}{\varrho_c} = w_e - \Delta e \quad \text{with} \quad \Delta e > 0$$

The dissipation Δe within a turbomachine of fixed geometry depends strongly on the ratio C_a/U. Only for a small range of the angles α_1 and β_1 will the flow pass smoothly onto the blades. Outside this range the flow separates from the leading edges of the blades, and the dissipation increases rapidly as the operating point moves away from the design point. Hence the dependence of $\Delta p_0/\varrho_c U^2$ on the dimensionless flow rate C_a/U can be expected to have the form shown in Figure 7.10 by dashed lines. The design condition, where $\Delta e/w_e$ is small, corresponds to a suitable matching of the relative inflow angles with the upstream shape of the blades.

It is apparent that a machine which is designed to operate as a pump will not usually be an efficient turbine, although it will become a turbine for large enough increases in flow rate or decreases in rotor speed. Note too that when the direction of flow through the machine is reversed the values of α_1 and β_2 will usually change as well, and one has in effect a different machine to consider. It is possible, however, to design machines which act as efficient pumps for flow in one direction and as efficient turbines for flow in the opposite direction. Use has been made of such machines in pumped-storage hydroelectric plants. When the demand for electricity is low, water is pumped into a reservoir; when the demand is high, water is allowed to flow back through the machines, now acting as turbines, to drive generators. A somewhat similar operation is found in the tidal-power plant recently brought into use in France. As the tide rises, the flow into an estuary is used to drive turbines; when the tide falls, the water passes backwards through the same machines, which again act as turbines to extract energy from the fluid a second time.

7.3.2 Losses in a Cascade

The preceding discussion has indicated in a general way the effects of internal dissipation on the performance of a turbomachine. Here the way in which losses arise and methods of describing them will be considered further.

The losses in a cascade of blades can be described in many ways. Three of these will be introduced and the relationships amongst them will be developed. The measures of dissipation to be considered are a loss coefficient for the cascade, a drag coefficient for the blades, and a diffuser efficiency for the passage between the blades. Only the constant-density case will be discussed. However, as was seen in Example 2.13, this simplification does not restrict the application of the results as much as might be expected at first sight. In many multistage machines the change in state across each blade row is relatively small, and the constant-density approximation provides an adequate description of the pressure and velocity changes in gas and vapour flows.

While only the component of force along the cascade contributes to the work done by a rotor, both components must be taken into account in discussing the losses within the blade row. For constant-density, two-dimensional flow, with the axial velocity constant, equation (7.3.2) reduces to

$$Y = p_1 - p_2 \tag{7.3.9}$$

Equation (7.1.13) now gives

$$\begin{aligned} Y &= \tfrac{1}{2}\varrho_c(V_2^2 - V_1^2) + \varrho_c\,\Delta e \\ &= \tfrac{1}{2}\varrho_c(V_{x_2}^2 - V_{x_1}^2) + \varrho_c\,\Delta e \end{aligned} \tag{7.3.9}$$

Thus the axial force is found to depend explicitly on the dissipation, which is given by equation (7.1.11) as

$$\Delta e = \frac{p_{0_1} - p_{0_2}}{\varrho_c} = -\frac{\Delta p_0}{\varrho_c} \tag{7.3.9}$$

Here the total pressures are those measured in the relative axes.

It was seen in equations (2.3.30) and again in Table 6.4 that the change in total pressure calculated using the constant-density approximation gives the dissipation in low-velocity flows of compressible fluids. Hence, if the total-pressure change is used to express the losses, the preceding results can be applied to any fluid, so long as the proportional change in density across the cascade is small. The fall in total pressure can be measured directly using a Pitot tube. Since much of the dissipation occurs in boundary layers which extend in thin wakes behind the blades, it is necessary to traverse the Pitot tube along the cascade in order to find the net dissipation. The losses in a cascade are often presented in terms of a dimensionless *loss coefficient*

$$\zeta = \frac{p_{0_1} - p_{0_2}}{\frac{1}{2}\varrho_c V_a^2} = -\frac{\Delta p_0}{\frac{1}{2}\varrho_c V_a^2} \tag{7.3.10}$$

The direction of the force acting on a cascade can be found using equations (7.3.1, 3, 9):

$$\frac{Y}{X} = -\frac{V_{x_2} + V_{x_1}}{V_{y_2} + V_{y_1}} - \frac{\Delta p_0}{X}$$

For lossless flow the force vector is normal to the mean-flow direction, as shown in Figure 7.11. Losses have the effect of tilting the force vector towards the rear of the cascade, by virtue of the additional axial component $-\Delta p_0$.

A more precise definition of the concept of the mean flow through a cascade makes it possible to set down the equations describing losses in a cascade in a more compact and more enlightening form. The geometric relations (7.3.4) show that

$$\frac{V_{x_2} + V_{x_1}}{V_{y_2} + V_{y_1}} = \frac{1}{2}\left(\frac{V_{x_2}}{V_{y_2}} + \frac{V_{x_1}}{V_{y_1}}\right)$$

$$= -\frac{1}{2}(\cot \beta_2 + \cot \beta_1)$$

The mean flow angle will be defined by

$$2 \cot \beta_m = \cot \beta_2 + \cot \beta_1 \tag{7.3.11}$$

Then the direction of the force vector is given by

$$\frac{Y}{X} = \cot \beta_m - \frac{\Delta p_0}{X}$$

The last result can also be used to express the losses in terms of the force components:

$$-\Delta p_0 = Y - X \cot \beta_m \tag{7.3.12}$$

The significance of this result becomes clearer when the forces are resolved into lift and drag components, respectively normal to and parallel to the mean flow. The force on an isolated aerofoil is conventionally described using these components; thus it is possible to relate cascade performance to that of the individual lifting surfaces which compose it. The geometry

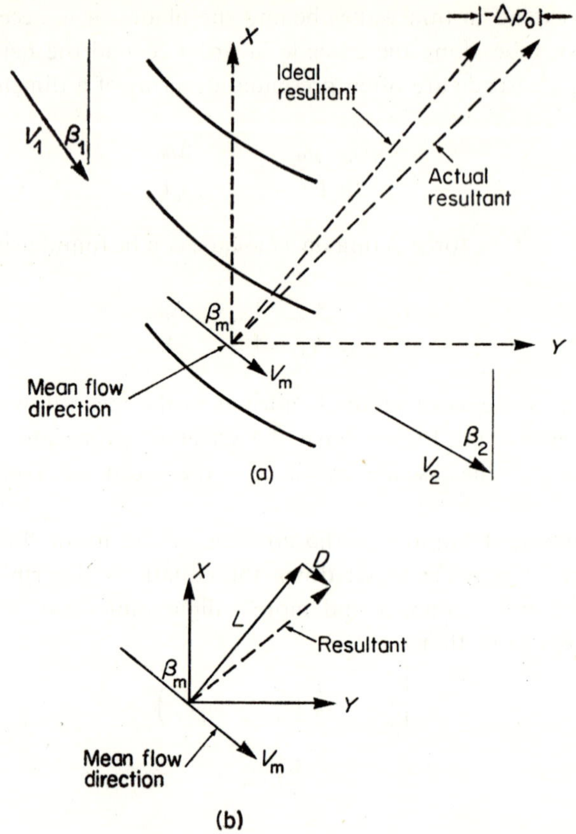

Figure 7.11 Geometry of the mean flow through a cascade and the forces acting on the cascade. (a) Effect of losses on the resultant force. (b) The lift and drag on the blades, and their relationship to the normal and transverse force components

of Figure 7.11(b) indicates that the lift and drag per unit area of the cascade are given by

$$L = X \sin \beta_m + Y \cos \beta_m \qquad (7.3.13)$$
$$D = -X \cos \beta_m + Y \sin \beta_m$$

Comparison with equation (7.3.12) gives immediately

$$D = -\Delta p_0 \sin \beta_m \qquad (7.3.13)$$

showing that the losses can be ascribed to the drag on the individual blades of the cascade.

The loss in a cascade with blades spaced at intervals s can be expressed in terms of a *drag coefficient* for an individual blade:

$$C_D = \frac{Ds}{\frac{1}{2}\varrho_c V_m^2 c}$$

Here $V_m = V_a/\sin\beta_m$ is the mean velocity of the flow past the blade, and c is the chord length of the blade, that is, the distance between its leading and trailing edges. The drag coefficient and loss coefficient (7.3.10) are related by

$$\begin{aligned} C_D &= -2\frac{s}{c}\frac{\Delta p_0}{\varrho_c V_m^2}\sin\beta_m \\ &= \frac{s}{c}\zeta\sin^3\beta_m \end{aligned} \tag{7.3.14}$$

A fluid passing through the blades of a compressor or pump experiences a rise in pressure and fall in relative velocity. Thus a compressor cascade acts as a diffuser, and we must expect that the losses in it will be larger than those in comparable turbine blading, in which the flow is accelerated. A *diffuser efficiency* can be used to describe the losses in a decelerating cascade, a convenient form for constant-density flow being given in equations (4.4.7):

$$\eta_D = \frac{1}{1-\Delta p_0/(p_2-p_1)}$$

For the situation considered here, equations (7.3.9) indicate that

$$\eta_D = \frac{1}{1+\Delta p_0/Y} \tag{7.3.15}$$

Equations (7.3.13) give

$$-\Delta p_0 = \frac{D}{\sin\beta_m} \quad \text{and} \quad Y = L\cos\beta_m + D\sin\beta_m$$

Then, since $|D/L| \ll 1$ in normal operation

$$\begin{aligned} \eta_D &\simeq 1+\frac{D}{L\sin\beta_m\cos\beta_m} \\ &= 1-\frac{2|D/L|}{\sin 2\beta_m} \end{aligned} \tag{7.3.15}$$

(Note that L is negative for the sign convention of Figure 7.11.)

The diffuser efficiency depends on the ratio $|D/L|$ characterizing the blades of the cascade. It also depends on the angle at which the blades are set, for in normal operation the mean flow angle β_m may be expected to be close to the mean blade angle. Thus it appears that it will be advantageous to set compressor blades at an angle near 45°. This prediction turns out to be correct in practice.

Example 7.3 A compressor cascade with 15 blades of height 20 cm and chord 10 cm is tested in a cascade tunnel, a wind tunnel specially designed for the study of cascade characteristics. The arrangement is shown in Figure 7.12. It is required to express the losses in terms of drag coefficient, loss coefficient and diffuser efficiency. The density will be taken to be constant at 1·225 kg/m^3, and the outlet velocity V_2 will be assumed to be uniform along the line of the cascade. (This last assumption neglects the existence of wakes behind the individual blades.)

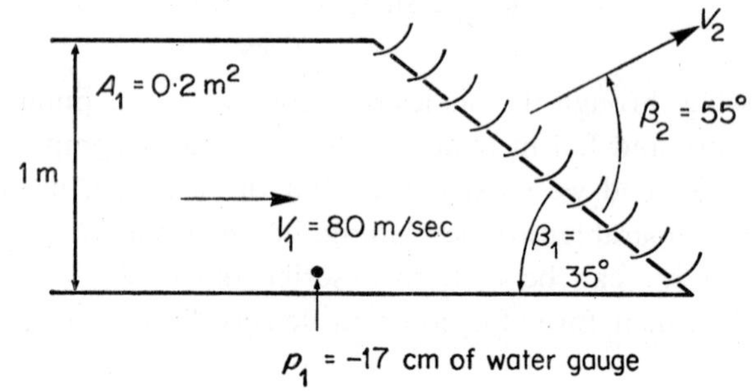

Figure 7.12 Details of a test in a cascade tunnel

Flow Geometry

The velocity component normal to the cascade is

$$V_a = V_1 \sin\beta_1 = V_2 \sin\beta_2 = 45{\cdot}8 \text{ m/sec}$$

and the outlet velocity is given by

$$V_2 = \frac{V_1 \sin\beta_1}{\sin\beta_2} = 56{\cdot}0 \text{ m/sec}$$

The mean flow direction is given by

$$\cot\beta_m = \tfrac{1}{2}(\cot 35° + \cot 55°) = 1{\cdot}064.$$

Then

$$\beta_m = 43{\cdot}2°$$

and the mean velocity is

$$V_m = \frac{V_a}{\sin\beta_m} = 67{\cdot}0 \text{ m/sec}$$

Finally, equations (7.3.4) give the velocity change across the cascade:

$$\begin{aligned} V_{x_2} - V_{x_1} &= -V_2 \cos\beta_2 + V_1 \cos\beta_1 \\ &= -V_a(\cot\beta_2 - \cot\beta_1) = 33{\cdot}2 \text{ m/sec} \end{aligned}$$

Forces on the Cascade

The flow per unit area of cascade is

$$\dot{m} = \varrho_c V_a = 1{\cdot}225 \times 45{\cdot}8 = 56{\cdot}2 \text{ kg/m}^2 \text{ sec}$$

The force along it is given by

$$X = \dot{m}(V_{x_1} - V_{x_2}) = -1870 \text{ N/m}^2$$

while that normal to it is

$$Y = p_1 - p_a = -0{\cdot}17 \times 1000 \times 9{\cdot}81 = -1670 \text{ N/m}^2$$

Losses

The dissipation is given by

$$-\frac{\Delta p_0}{\varrho_c} = \frac{p_1 - p_2}{\varrho_c} + \frac{1}{2}(V_1^2 - V_2^2)$$

$$= -1362 + 1629 = 267 \text{ J/kg}$$

In non-dimensional form this is

$$\zeta = -\frac{\Delta p_0}{\frac{1}{2}\varrho_c V_a^2} = \frac{2 \times 267}{(45{\cdot}8)^2} = 0{\cdot}254$$

For this cascade the spacing is

$$s = \frac{100/15}{\sin 35^\circ} = 11{\cdot}6 \text{ cm}$$

and the drag coefficient is

$$C_D = \frac{s}{c}\zeta \sin^3 \beta_m = 0{\cdot}095$$

Finally, the effective diffuser efficiency is

$$\eta_D = \frac{1}{1 + \Delta p_0 / Y} = 83{\cdot}6 \text{ per cent}$$

7.3.3 The Turbine Stage

Most turbomachines consist of alternating rows of moving rotor blades and fixed stator blades. Figure 7.8 illustrates this arrangement. Although only the rotors exchange work with the surroundings, the stators play a vital part in the operation of the machine. They guide the flow onto the rotor blades and induce suitable conversions between kinetic energy and enthalpy (or pressure). These energy conversions were discussed in general terms in Section 7.1.1; equations (7.1.5) to (7.1.13) specify them. Consideration will now be given to the flow geometry associated with a particular kind of conversion. It is necessary to distinguish between the two cases of turbine action and pumping action. The former case will be discussed first.

The blading system of Figure 7.13 is suitable for extracting energy from the fluid. This turbine stage may be thought of either as a complete single-stage turbine or as one of the stages of a multistage machine. The flow pattern shown is that for what is termed a *normal stage*, one which the fluid enters and leaves in the axial direction. The inlet and outlet absolute velocities will be taken to be equal in this normal stage:

$$C_1 = C_3 = C_a$$

This condition can be satisfied in a variety of flows, the most obvious being two-dimensional, constant-density flow.

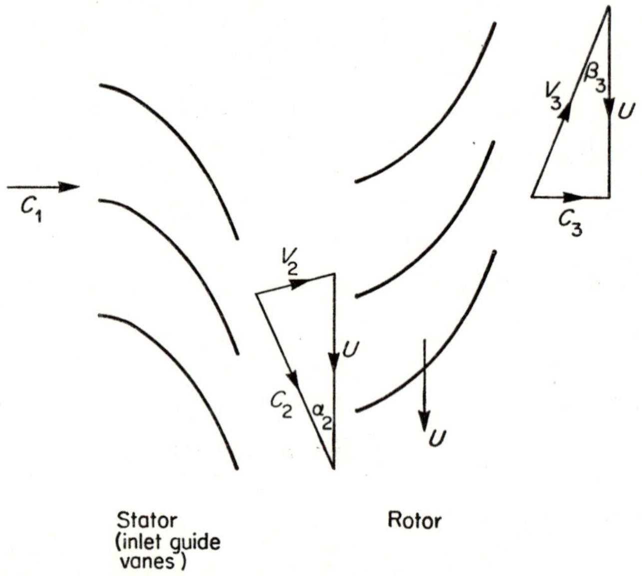

Figure 7.13 Flow geometry for a normal turbine stage

The enthalpy and total enthalpy changes within the turbine stage are given by equations (7.1.5, 6) and (7.3.5). They are shown below.

	STATOR	ROTOR	
Δh	$\frac{1}{2}(C_1^2 - C_2^2) < 0$	$\frac{1}{2}(V_2^2 - V_3^2) < 0$	
Δh_0	0	$U(V_{x_3} - V_{x_2}) < 0$	(7.3.16)

OVERALL (for normal stage, with $C_1 = C_3$):

$$\Delta h = \Delta h_0 = w_e < 0$$

For constant-density flow these energy conversions are more conveniently specified as pressure changes, using equations (7.1.11, 12) and (7.3.5).

	STATOR	ROTOR	
$\dfrac{\Delta p}{\varrho_c}$	$\frac{1}{2}(C_1^2-C_2^2)-\Delta e_1 < 0$	$\frac{1}{2}(V_2^2-V_3^2)-\Delta e_2 < 0$	
$\dfrac{\Delta p_0}{\varrho_c}$	$-\Delta e_1 < 0$	$U(V_{x_3}-V_{x_2})-\Delta e_2 < 0$	(7.3.17)

OVERALL (for a normal stage, with $C_1 = C_3$):

$$\frac{\Delta p}{\varrho_c} = \frac{\Delta p_0}{\varrho_c} = w_e - \Delta e_1 - \Delta e_2 < 0$$

It is convenient to have a means of specifying the distribution of the stage loading between the rotor and stator blades. The *degree of reaction* is commonly used for this purpose. It is defined either as

$$R_h = \frac{h_2 - h_3}{h_1 - h_3} \tag{7.3.18}$$

or as

$$R_p = \frac{p_2 - p_3}{p_1 - p_3} \tag{7.3.19}$$

The first definition will be used here, since it is more directly linked to the energy conversion. For a normal stage the results (7.3.16) give

$$R_h = \frac{V_2^2 - V_3^2}{2U(V_{x_3} - V_{x_2})}$$

For $V_{y_2} = V_{y_3} = V_a$, consistent with the restriction $C_1 = C_3 = C_a$, this reduces to

$$R_h = -\frac{V_{x_2} + V_{x_3}}{2U} = \frac{1}{2} - \frac{V_{x_2}}{2U} \tag{7.3.20}$$

since $V_{x_3} = -U$ for a normal stage.

The same result can be obtained for R_p, using equations (7.3.17), and with the additional requirement that the flow be lossless. This indicates that the two definitions of reaction are equivalent in certain circumstances, namely, when the density is nearly constant and the losses are small. Since in designing and operating flow machines one usually seeks to minimize losses, it may be expected that $R_p \simeq R_h$ in most realistic cases. Hence the degree of reaction defined in (7.3.18) may be thought of as roughly specifying the distribution of the pressure drop between rotor and stator.

Equation (7.3.20) shows that the blade shape and reaction are closely linked. For an impulse machine, with $R_h = 0$ (strictly, $R_p = 0$), equation (7.3.20) indicates that

$$V_{x_2} = U = -V_{x_3}$$

Impulse rotor blades must then be nearly symmetrical and highly scooped. However, for high reaction, with $R_h \simeq 1$, equation (7.3.20) shows that

$$V_{x_2} \simeq -U = V_{x_3}$$

In this case the rotor blades are nearly flat.

The degree of reaction also has important implications for energy transfer. Equation (7.3.5), applied to the rotor of the normal stage of Figure 7.13, gives

$$w_e = U(V_{x_3} - V_{x_2}) = -U(U + V_{x_2})$$

When combined with equation (7.3.20) this leads to

$$-\frac{w_e}{U^2} = 2(1 - R_h) \qquad (7.3.21)$$

For a specified rotor speed, the largest energy transfer is achieved with $R_h = 0$, that is, in an impulse machine. Hence the limiting value, $-w_e = 2U^2$, given in equations (7.2.17), applies more widely than appeared earlier. The energy extraction in a reaction turbine cannot approach this limit, let alone exceed it.

This analysis may seem to suggest that impulse blading should be used whenever possible, since its use leads to a reduction in the number of stages required to extract a specified amount of energy. However, it will be seen later that the losses associated with reaction blading are generally lower than those arising in impulse blades, where the relative velocities are high. Hence impulse blading is used only in rather special circumstances, where there is a special need to minimize the number of stages, or where it is desirable to minimize the pressure differential across a stage. Thus the first few stages of a multistage steam turbine are commonly of the impulse type, while the rest use reaction blading. Among hydraulic turbines, the Pelton wheel is used only for the highest supply heads, while propeller turbines with high-reaction blading are found in low-head hydroelectric stations.

Examples 7.4 Figure 7.14 shows a propeller-turbine installation across which the head falls by 20 m. The inlet guide vanes give the water a swirl before it passes through the turbine. The draught tube serves to decelerate the fluid before discharge, so that little power is lost in the form of kinetic energy at discharge. Although the flow before and after the rotor does not conform to the axial-flow pattern, that near the rotor does, and may be discussed using the results obtained above.

It is required to find the pressures just above and just below the turbine blades, on the assumption of lossless flow in the ducting, for the design operating condition: rotor speed, 40 rad/sec; absolute discharge in the axial direction; and relative outlet angle from the blades of $\beta_3 = 25°$ on the pitch circle, where $\bar{r} = 0{\cdot}625$ m.

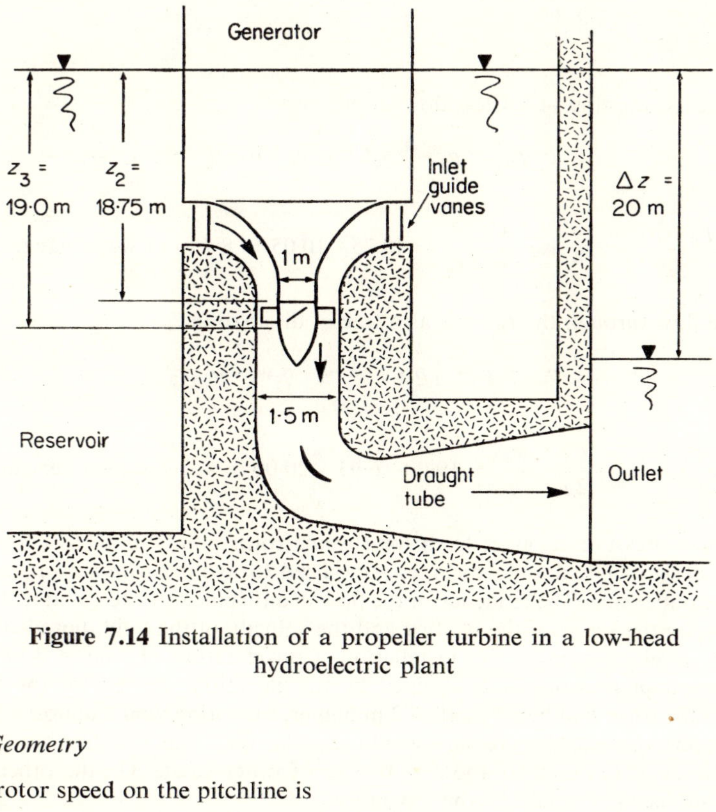

Figure 7.14 Installation of a propeller turbine in a low-head hydroelectric plant

Flow Geometry

The rotor speed on the pitchline is

$$U = \bar{r}\omega = 25{\cdot}0 \text{ m/sec}$$

and the axial velocity is

$$C_a = U \tan \beta_3 = 11{\cdot}66 \text{ m/sec}$$

For a normal stage $C_a/U = \tan \beta_3$ from Figure 7.13, and equation (7.3.8) can be written

$$\frac{w_e}{U^2} = 1 - \tan \beta_3 (\cot \beta_3 + \cot \alpha_2) = -\tan \beta_3 \cot \alpha_2$$

For the ideal case considered here, the energy extraction is, neglecting the small kinetic energy at outlet from the draft tube

$$-w_e = g\,\Delta z = 196{\cdot}2 \text{ J/kg}$$

and the absolute flow angle is given by

$$\cot\alpha_2 = -\frac{w_e}{U^2}\cot\beta_3 = 0{\cdot}674$$

Then $\alpha_2 = 56°$.

Pressures

On the assumption of lossless flow to the rotor

$$p_{0_2} = p_2+\tfrac{1}{2}\varrho_c C_2^2 = p_a+\varrho_c g z_2$$

Then

$$\frac{p_2-p_a}{\varrho_c g} = z_2-\frac{C_a^2}{2g\sin^2\alpha_2} = 18{\cdot}75-10{\cdot}05 = 8{\cdot}7 \text{ m of water gauge}$$

If the flow through the rotor is also free of dissipation

$$p_{0_3} = p_3+\tfrac{1}{2}\varrho_c C_3^2 = p_a+\varrho_c w_e+\varrho_c g z_3$$

and

$$\frac{p_3-p_a}{\varrho_c g} = z_3-\frac{C_a^2}{2g}+\frac{w_e}{g} = 19{\cdot}00-6{\cdot}93-20{\cdot}00 = -7{\cdot}9 \text{ m of water gauge}$$

The low pressure at outlet from the turbine rotor points to a basic limitation on the design of hydraulic turbines—the need to avoid *cavitation* on the low-pressure outlet side. The tendency of the flow in a turbine rotor to cavitate depends not only on the shape of the blading and the velocity of the fluid, but also on the pressure level inherent in the installation. Consideration of Figure 7.14 shows that the outlet pressure is determined by the elevation selected for the turbine rotor. If the rotor had been installed 2 m higher, cavitation would almost certainly have occurred, since the pressure would then be very near absolute zero, which corresponds at sea level to about $-10{\cdot}3$ m of water gauge. On the other hand, deeper submersion would inhibit cavitation.

7.3.4 The Compressor Stage

Figure 7.15 shows a blading system suitable for doing work on the fluid passing through it. This compressor stage may be thought of as a single-stage pump or blower or as part of a multistage compressor. Once again consideration will be given for the most part to a normal stage, with axially directed and equal velocities at inlet and outlet:

$$C_1 = C_3 = C_a$$

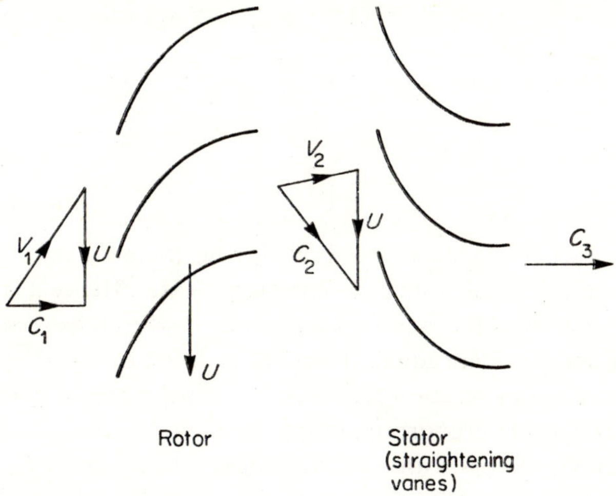

Figure 7.15 Flow geometry for a normal compressor stage

The energy transformations in the rotor and stator are given by equations (7.1.5, 6) and (7.3.5):

	ROTOR	STATOR
Δh	$\frac{1}{2}(V_1^2 - V_2^2) > 0$	$\frac{1}{2}(C_2^2 - C_3^2) > 0$
Δh_0	$U(V_{x_2} - V_{x_1}) > 0$	0

OVERALL (for a normal stage with $C_1 = C_3$):

$$\Delta h = \Delta h_0 = w_e > 0 \tag{7.3.22}$$

The *degree of reaction* will be defined as in equation (7.3.18), as the ratio of enthalpy change in the rotor to that in the whole stage. For the compressor stage

$$R_h = \frac{h_1 - h_2}{h_1 - h_3} \tag{7.3.23}$$

For a normal stage with $V_{y_1} = V_{y_2}$ and $V_{x_1} = -U$

$$R_h = \frac{V_1^2 - V_2^2}{2U(V_{x_2} - V_{x_1})} = \frac{1}{2} - \frac{V_{x_2}}{2U}$$

as in equation (7.3.20). It is again apparent that high-reaction machines have flat blades ($V_{x_2} \simeq V_{x_1}$). Note also that when $R_h = 50$ per cent the inlet and intermediate velocity triangles are similar; it may be expected that the rotor and stator blades will also be similar in shape.

The relationship between reaction and work transfer is obtained from

$$w_e = U(V_{x_2} - V_{x_1}) = U(V_{x_2} + U)$$

whence

$$\frac{w_e}{U^2} = 1 + \frac{V_{x_2}}{U} = 2(1 - R_h) \quad (7.3.24)$$

Since the pressure rises across a compressor row, there is a particular danger that separation will occur, with associated large losses. Hence it is reasonable to expect that the overall losses will be minimized when the pressure changes across stator and rotor are equal. It was seen earlier, on considering the two possible definitions of reaction (7.3.18, 19), that the pressure and enthalpy changes are roughly in proportion. It appears then that there is an advantage in designing compressors with a reaction near 50 per cent. For 50 per cent reaction, the work per stage is given by equation (7.3.24) as

$$w_e = U^2 \quad (7.3.25)$$

There is another reason for selecting a reaction near 50 per cent for a high-performance compressor. This is suggested by comparing the turbine and compressor stages shown in Figures 7.13 and 7.15. The high relative velocities in a turbine occur towards the trailing edges of the blades; in a compressor row they occur at the inlet edges. With high-velocity inlet flow of a gas, shock waves may form as the fluid accelerates over the leading edges of the blades. Separation and high losses would then result. For the stage as a whole these effects will be minimized with equal inlet velocities for stator and rotor. This again points to a stage with 50 per cent reaction and to similar rotor and stator blades.

A restriction on the *Mach number* of the relative flow ahead of a compressor row sets a limit on the energy transfer per stage. Suppose that the relative Mach number is to be kept within, say, $V_1/a_1 = 0{\cdot}7$, allowing a margin for local acceleration around the blades. Since the flow must have an axial component, this implies that for a normal stage the rotor speed must be somewhat less than $0{\cdot}7a_1$, say,

$$U = \frac{a_1}{2} \quad (7.3.26)$$

Equation (7.3.25) now indicates that the energy input for a stage with 50 per cent reaction will be in the order of

$$w_e = U^2 = \frac{a_1^2}{4} \quad (7.3.26)$$

For atmospheric air with $a_1 \simeq 340$ m/sec, equations (7.3.26) give

$$U = 170 \text{ m/sec} = 550 \text{ ft/sec}$$

and (7.3.27)

$$w_e = 30 \text{ kJ/kg} = 12 \text{ Btu/lb}$$

per stage. For air with $c_p \simeq 1$ kJ/kg °K, this corresponds to a temperature change of

$$\Delta T_0 = w_e/c_p = 30\ ^\circ\text{C} = 55\ ^\circ\text{F} \qquad (7.3.27)$$

per stage. Subject to this restriction, the properties of a gas cannot change by a large proportion across any one blade row of a compressor. Hence the constant-density approximation will provide a reasonably accurate representation of the flow through a single row. For multistage machines it is necessary to choose an appropriate mean density for each row.

The limits set out in equations (7.3.27) are more stringent than those (7.2.16, 17) imposed by the centripetal loading on the rotor. It is the approach to the sonic condition which limits the energy transfer per stage for a compressor, and not the stresses generated in the rotor. One reason for the relative freedom of turbines from the compressibility restriction has already been noted; the higher relative velocities occur at outlet from the blades. Since a turbine rotor usually moves in the same direction as the absolute inlet flow, a very high absolute velocity is required to generate a high relative inlet Mach number. This is illustrated by the calculations of Example 7.2. The turbine has a second advantage from this point of view, but not from that of strength, in dealing with high-temperature gases and vapours in which the sound speed is also high. For these reasons the maximum rotor speed for a turbine is usually determined, as was suggested earlier, by the strength of the rotor.

The discussion above relates to the pumping of gases and vapours. The pumping of liquids raises another problem—*cavitation.* Cavitation will usually occur when the absolute pressure drops locally to the liquid's vapour pressure. Since the pressure of the fluid within a turbomachine depends on the rotational speed, the velocity of the through-flow, and the energy transfer to or from the fluid, the commonly accepted requirement that cavitation be avoided imposes a rather complex restriction on the operation of the machine. This will be discussed at the end of the present chapter.

Example 7.5 Many of the preceding results depend upon the assumption that the fluid passes through a blade row with nearly constant density. Here this restriction will be relaxed, and account will be taken of density variations for the particular case of isentropic flow of a perfect gas. The flow shown in Figure 7.16 will be discussed on this basis. A row of blades of height 10 cm is fitted to a rotor

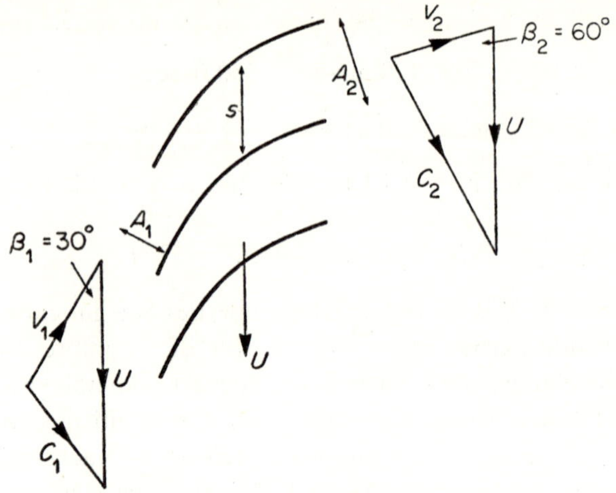

Figure 7.16 Flow through a compressor row

of inner diameter 80 cm, rotating at 4000 rev/min. The inlet pressure and temperature are 0·75 bar and 15 °C, and the relative velocity at inlet is 140 m/sec.

The isentropic tables are related to the flow geometry by the area ratio A/A^*. In this application it is necessary to express changes in flow area across the blade row in terms of the flow angles. For two-dimensional flow the relationship can be obtained by noting that the spacing between blades s can be expressed as

$$s = \frac{A_1}{\sin \beta_1} = \frac{A_2}{\sin \beta_2}$$

For the geometry of Figure 7.16

$$\frac{A_1}{A_2} = \frac{\sin 30°}{\sin 60°} = 0{\cdot}577$$

For the inlet conditions specified

$$M_1 = \frac{V_1}{a_1} = \frac{140}{340} = 0{\cdot}412$$

and isentropic tables for $\gamma = 1{\cdot}4$ give $A_1/A^* = 1{\cdot}553$.
Then

$$\frac{A_2}{A^*} = \frac{A_2}{A_1}\frac{A_1}{A^*} = \frac{1{\cdot}553}{0{\cdot}577} = 2{\cdot}69$$

and

$$M_2 = 0{\cdot}222$$

The relative outlet velocity can now be found:

$$V_2 = V_1 \frac{M_2}{M_1}\frac{a_2}{a_1} = V_1 \frac{M_2}{M_1}\left(\frac{T_2}{T_0}\right)^{\frac{1}{2}}\left(\frac{T_0}{T_1}\right)^{\frac{1}{2}}$$

$$= 140 \times \frac{0{\cdot}222}{0{\cdot}412}\left(\frac{0{\cdot}990}{0{\cdot}967}\right)^{\frac{1}{2}} = 76{\cdot}4 \text{ m/sec}$$

This leads to the energy transfer. The rotor speed on the pitchline is

$$U = 4000 \times 0{\cdot}45 \times \frac{2\pi}{60} = 188{\cdot}5 \text{ m/sec}$$

giving

$$w_e = U(V_1 \cos\beta_1 - V_2 \cos\beta_2) = 15{\cdot}6 \text{ kJ/kg}$$

Other properties of the outlet flow can be found in a straightforward manner. For instance, the outlet pressure is given by

$$p_2 = p_1 \frac{p_2}{p_0}\frac{p_0}{p_1} = 0{\cdot}75 \times \frac{0{\cdot}966}{0{\cdot}892} = 0{\cdot}812 \text{ bar}$$

and the inlet velocity for the following stator row by

$$C_2 = (V_2^2 + U^2 - 2UV_2 \cos\beta_2)^{\frac{1}{2}} = 164 \text{ m/sec}$$

7.4 RADIAL-FLOW MACHINES

7.4.1 Euler's Equation for a Turbomachine

Figure 7.17 defines the symbols which will be used to describe the axisymmetric flow through a cylindrical cascade of blades. Although some of the results apply more generally, the discussion will relate primarily to 'flat' rotors in which the flow is confined to a plane perpendicular to the rotor axis. The flow sketched is outwards from the axis, but the results apply, with suitable sign changes, to inwards flow as well.

The rotor speed is not now the same for the inner and outer edges of the rotor. The two values are related by

$$\frac{U_1}{r_1} = \frac{U_2}{r_2} = \omega \tag{7.4.1}$$

with ω the angular velocity of the rotor. The tangential and radial components of velocity will be denoted by the subscripts t and r. Thus, for example

$$C_1^2 = C_{t_1}^2 + C_{r_1}^2$$

and

$$V_1^2 = V_{t_1}^2 + V_{r_1}^2 \qquad (7.4.2)$$

The vector relationship (7.1.1) between absolute and relative velocities gives, for the two sides of the rotor

$$C_{t_1} = U_1 + V_{t_1} \quad \text{and} \quad C_{t_2} = U_2 + V_{t_2} \qquad (7.4.2)$$

The radial components are of course equal in pairs:

$$C_{r_1} = V_{r_1} \quad \text{and} \quad C_{r_2} = V_{r_2} \qquad (7.4.2)$$

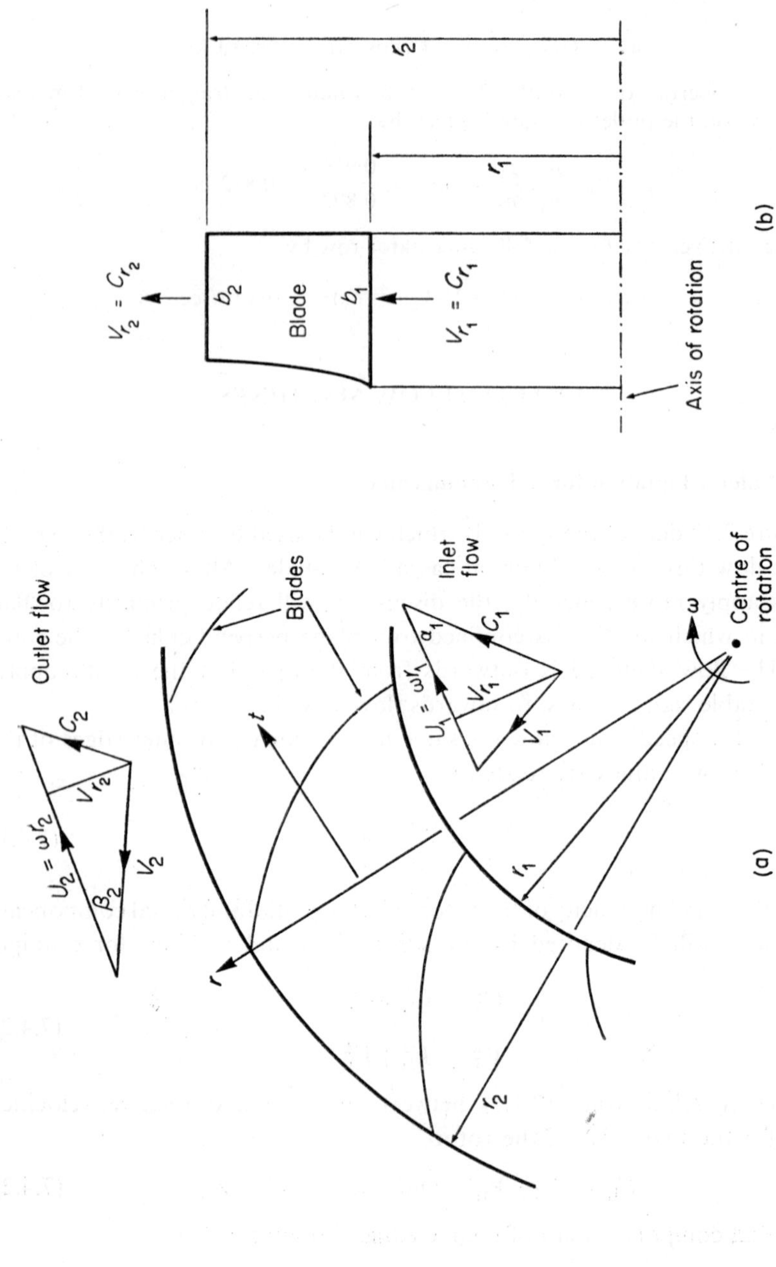

Figure 7.17 Geometry of the flow through a cylindrical cascade. (a) View in plane of rotation; (b) Section through the rotor

using values found in Example 7.4. Then taking $r = 1{\cdot}5$ m

$$C_t = \frac{\bar{r}}{r} C_{t_2} = \frac{0{\cdot}625 \times 7{\cdot}85}{1{\cdot}5} = 3{\cdot}27 \text{ m/sec}$$

The radial component at outlet from the guide vanes can be determined from the continuity requirement

$$2\pi r b C_r = A_a C_a$$

where $b = 20$ cm is the depth of the inlet flow passage, and A_a is the cross-sectional area of the annulus in which the blades rotate. Thus

$$C_r = \frac{A_a C_a}{2\pi r b} = 6{\cdot}08 \text{ m/sec}$$

and the outlet angle from the guide vanes is given by

$$\tan \alpha = \frac{C_r}{C_t} = \frac{6{\cdot}08}{3{\cdot}27} = 1{\cdot}86$$

giving

$$\alpha = 61{\cdot}7^\circ$$

Analysis of the Rotor

The flow through the rotor is

$$\dot{m} = \varrho_c A_a C_a = 11{,}450 \text{ kg/sec}$$

The torque applied to the rotor (and, with opposite direction, to the guide vanes) is

$$T = \dot{m} \bar{r} C_{t_2} = 56{,}200 \text{ N m}$$

It has been assumed that the angular momentum is uniform across the whole annulus, and is zero at outlet. The power extracted is then

$$-\dot{W}_e = T\omega = 56{\cdot}2 \times 40 = 2250 \text{ kW}$$

and the energy extracted per unit mass is

$$-w_e = -\frac{\dot{W}_e}{\dot{m}} = \frac{2250}{11{\cdot}45} = 196{\cdot}2 \text{ J/kg}$$

This is the value found in another way in Example 7.4.

7.4.2 Energy Conversions in Swirling Flow

The result (7.1.6) for absolute axes specifies the change of state which occurs in a fluid flowing through either a curved or straight rotor:

$$w_e = h_{0_2} - h_{0_1} = (h_2 - h_1) + \tfrac{1}{2}(C_2^2 - C_1^2) \tag{7.4.8}$$

However, the result (7.1.7) for relative axes will apply only when the rotation of the relative axes is negligible. The energy statement appropriate to a rotating frame of reference can be obtained by equating the energy transfer of equation (7.4.8) to that of Euler's equation (7.4.6).

From the geometric relations (7.4.2) it follows that

$$\begin{aligned} C^2 &= V_r^2+(V_t+U)^2 \\ &= V^2+U^2+2UV_t = V^2-U^2+2UC_t \end{aligned}$$

for both inlet and outlet flows. Thus equation (7.4.8) can be written

$$w_e = (h_2-h_1)+\tfrac{1}{2}(V_2^2-V_1^2)-\tfrac{1}{2}(U_2^2-U_1^2)+(U_2C_{t_2}-U_1C_{t_1})$$

Comparison with equations (7.4.6) now shows that

$$\begin{aligned} h_2-h_1 &= \tfrac{1}{2}(V_1^2-V_2^2)+\tfrac{1}{2}(U_2^2-U_1^2) \\ &= \tfrac{1}{2}(V_1^2-V_2^2)+\frac{\omega^2}{2}(r_2^2-r_1^2) \end{aligned} \qquad (7.4.9)$$

This is the relationship sought between enthalpy and velocity changes for relative axes in which no work is done. As would be expected, it is a generalization of equation (7.1.7), which applies for relative axes in steady translation relative to the absolute frame. The corresponding result for constant-density flow can be found by introducing

$$h_2-h_1 = \frac{p_2-p_1}{\varrho_c}+\Delta e$$

For a 'thin' rotor, one with $r_2 \simeq r_1$, the last term of equation (7.4.9) may be negligible and the result (7.1.7) is recovered. As would be anticipated, the cascade can then be treated as straight. However, it is more typical of radial-flow machines that $r_2/r_1 \sim 2$ to 3; hence the rotation-dependent term is often of great importance. Note that the 'centrifugal' contribution in equation (7.4.9) is positive for $r_2 > r_1$ and negative for $r_2 < r_1$. For outwards flow it contributes an increase in enthalpy and pressure, while for inwards flow it corresponds to a fall in the energy of the fluid. It is usually arranged that radial-flow pumps and compressors take advantage of this centrifugal energy addition by accepting fluid through a small-diameter 'eye' and ejecting it at the rotor periphery. Conversely, fluid usually passes inwards through a radial-flow turbine. In this way much greater energy transfers are achieved in centrifugal stages than in axial-flow stages operating in comparable conditions.

Equation (7.4.9) can be given another interpretation. It implies that

$$h+\frac{V^2}{2}-\frac{(\omega r)^2}{2} = \text{constant} \qquad (7.4.10)$$

for a flow which is isenergic for axes rotating with angular velocity ω. If the flow is isentropic as well, various forms of Bernoulli's equation can be obtained from this result. The case of constant density is simplest. For lossless and isentropic flow of a constant-density fluid

$$h = \frac{p}{\varrho_c} + e = \frac{p}{\varrho_c} + \text{constant}$$

giving

$$\frac{p}{\varrho_c} + \frac{V^2}{2} - \frac{(\omega r)^2}{2} = \text{constant} \tag{7.4.11}$$

as the appropriate Bernoulli relationship between pressure and relative velocity. This result could be used, for example, to find the variation of pressure in lossless flow through the rotor of a radial-flow machine.

7.4.3 The Centrifugal Pump or Blower

So far the operation of the various kinds of turbomachines has been considered in a rather general way. The intention has been to demonstrate the overall principles of operation, and no attempt has been made to provide a realistic prediction of performance. At this point we shall look at a particular kind of machine in more detail, in order to get some idea of the way in which it performs over a wide range of operating conditions. The single-stage centrifugal machine is chosen for this purpose, since it is reasonably easy to analyse, and is in very wide use. The fluid density will be taken to be constant throughout this discussion.

Figure 7.18 shows a simple form of centrifugal fan or pump.* This machine has no inlet guide vanes, and we may take

$$C_{t_1} = 0, \quad \alpha_1 = 90^\circ \quad \text{and} \quad \cot \alpha_1 = 0 \tag{7.4.12}$$

Considerable simplifications result in the formulae governing the operation of the pump when these results are introduced. However, before looking at the pump proper, it is appropriate to look further at the intake.

Equation (7.4.9) has shown that there is a 'centrifugal' energy addition to fluid moving outwards through a radial-flow pump; this energy addition is related to the change in rotor radius between inlet and outlet. However, a price must be paid for this additional energy, in the form of high velocities at

* In many pumps the inlet edges of the impeller blades extend into the impeller eye, the inlet section of the rotor being in some cases normal to the axis of rotation, so that the fluid enters in the axial direction. For such machines the details of the analysis must be altered, as they must for machines with inlet guide vanes.

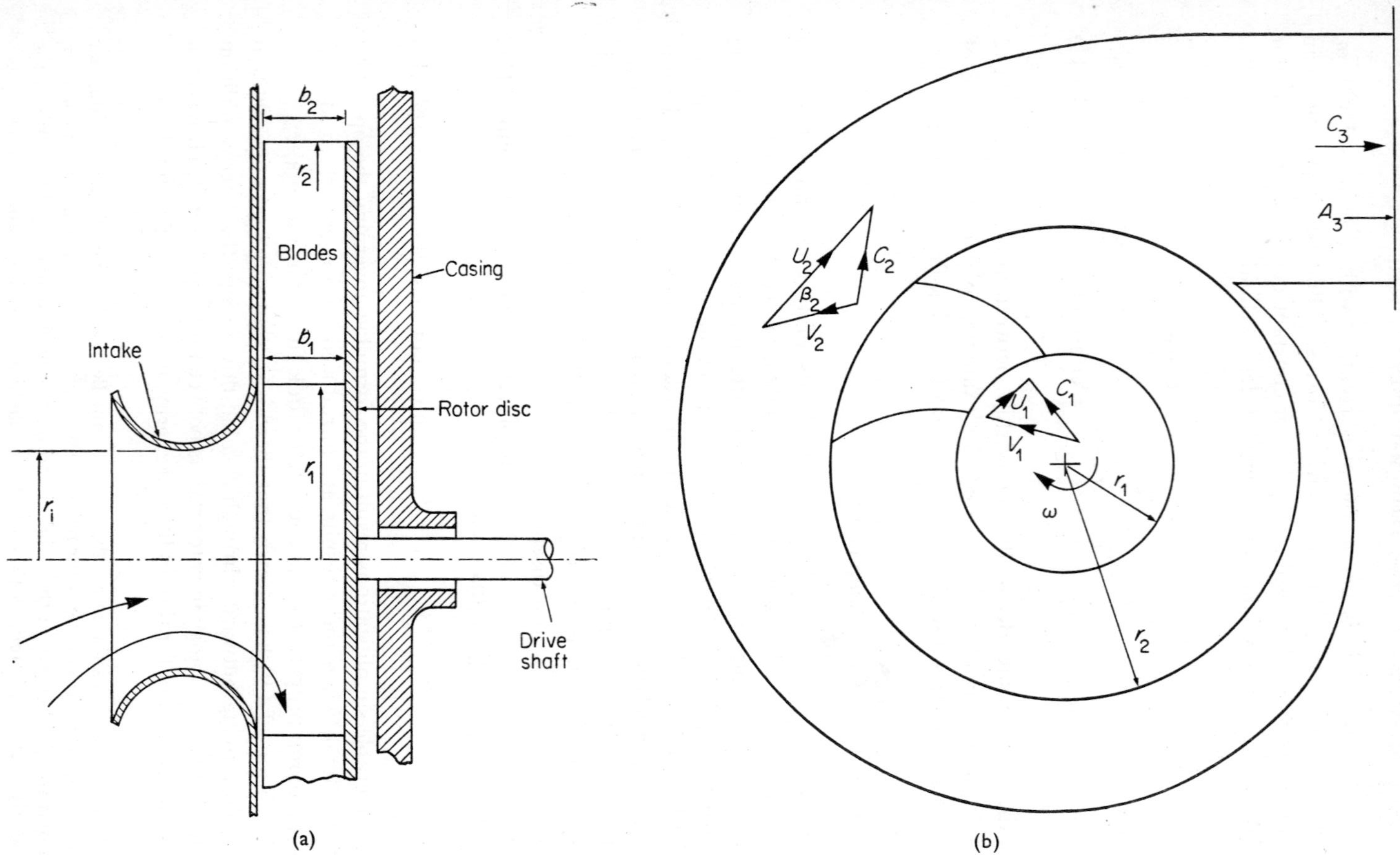

Figure 7.18 A simple single-stage blower. (a) Geometry of intake and impeller; (b) Geometry of the impeller and vaneless diffuser

the inlet eye. The velocity in the impeller eye vitally influences the performance of some machines. If the inlet velocity in a gas compressor becomes too high, shocks will form; in a pump handling a liquid, there is the danger of cavitation. In either case inefficient operation results when the inlet velocity exceeds a critical value. Setting aside these special problems, it is still desirable to minimize velocities, since the losses which inevitably occur in any real flow rise steeply with increasing velocity. Hence for any high-performance machine it is imperative to select an intake configuration which, for a given flow rate, achieves the minimum velocity in the inlet. This single requirement dictates many features of the inlet passages of a centrifugal pump or compressor. Here only the simple intake shown in Figure 7.18 will be examined, but similar considerations apply for more complex intakes and impellers.

For the pump of Figure 7.18 the absolute flow entering the impeller may be taken to be in the radial direction, with neither swirl nor axial component. Hence the relative velocity is given by

$$V_1^2 = C_1^2 + U_1^2 = \left(\frac{\dot{V}}{2\pi b_1 r_1}\right)^2 + (\omega r_1)^2$$

Consider now a family of geometrically similar intakes, for which $b_1/r_1 = c$, a constant; each member of the family rotates at the same speed ω and carries the same flow $\dot{V}$. For which member of this family is the relative velocity least? The expression

$$V_1^2 = \frac{\left(\dfrac{\dot{V}}{2\pi c}\right)^2}{r_1^4} + \omega^2 r_1^2$$

shows how this velocity depends on the inner radius of the impeller. With a large radius, the velocity is high because the rotor speed is high; with a small radius, it is high because the radial absolute velocity is large. Differentiation indicates that the minimum value of V_1 is achieved when

$$r_1^6 = 2\left(\frac{\dot{V}}{2\pi c\omega}\right)^2 \tag{7.4.13}$$

and

$$2C_1^2 = U_1^2 = \frac{2V_1^2}{3} \tag{7.4.14}$$

The inlet flow angle corresponding to this condition is given by

$$\beta_1 = \arctan\left(\frac{1}{\sqrt{2}}\right) \simeq 35^\circ \qquad (7.4.14)$$

There is little point in maintaining low relative velocities on the blades if high velocities occur in the intake duct. This has further implications for the intake geometry. Taking C_i to be the axial velocity in the eye of the intake of Figure 7.18, we have the constant-density continuity requirement

$$\pi r_i^2 C_i = 2\pi r_1 b_1 C_1$$

Then for the optimum inlet flow found above in equations (7.4.14)

$$\frac{r_1 b_1}{r_i^2} = \frac{C_i}{2C_1} = \frac{1}{2}\frac{C_i}{V_1}\frac{V_1}{C_1} = \frac{\sqrt{3}C_i}{2V_1}$$

If the velocities C_i and V_1 are to be roughly equal, it is necessary to have

$$\frac{r_1 b_1}{r_i^2} \simeq 0{\cdot}9$$

An intake geometry satisfying this criterion is

$$\frac{r_1}{r_i} \simeq \frac{3}{2} \quad \text{and} \quad \frac{b_1}{r_i} \simeq \frac{3}{5} \tag{7.4.15}$$

For such an intake, the assumption that the flow is radial and uniform on approaching the blades should be fairly realistic. This is the geometry sketched in Figure 7.18.

For the geometry suggested in equations (7.4.15), the geometric ratio appearing in equation (7.4.13) is

$$c = \frac{b_1}{r_1} \simeq \frac{2}{5}$$

Then the optimum radius given by equation (7.4.13) becomes

$$r_1 \simeq 0{\cdot}8\left(\frac{\dot{V}}{\omega}\right)^{\frac{1}{3}} \tag{7.4.16}$$

Other basic dimensions of the intake follow from equations (7.4.15). Although these considerations apply for a particular geometry and incorporate the assumption of uniform density, they illustrate in a general way the approach to the design of a wider range of intakes. An additional complication is encountered when pre-swirl vanes are incorporated in the pump. Inlet swirl vanes can be used to relax the restriction imposed by the inlet relative velocity. By giving the fluid a swirl in the direction of rotor motion, pre-swirl vanes allow a larger mass flow to pass into an impeller without raising the relative velocity at inlet to an undesirably high value.

Having seen how the fluid reaches the rotor, consider now the action of the impeller. Subject to the conditions (7.4.12), equations (7.4.6, 7) give

$$\frac{w_e}{U_2^2} = \frac{C_{t_2}}{U_2} = 1 - \frac{V_{r_2}}{U_2} \cot \beta_2 \qquad (7.4.17)$$

It has already been noted that stable operation is promoted by ensuring that w_e falls as V_{r_2} increases. This condition will be achieved when $\cot \beta_2 > 0$. Moreover, consideration of the velocity triangles of Figure 7.18 shows that a forwards-inclined outlet flow ($\beta_2 > 90°$, $\cot \beta_2 < 0$) is associated with a large absolute outlet velocity C_2 and with correspondingly large losses in the diffuser. This consideration too points to the desirability of having $\beta_2 < 90°$. In practice the outlet flow angles from centrifugal pumps usually lie in the range $\beta_2 = 20°$ to $30°$.

What does this imply for the blade geometry? Since centrifugal pumps, especially in the smaller sizes, have a rather small number of blades, the mean flow direction at rotor outlet is not exactly aligned with the blade tips. Hence the absolute whirl component C_{t_2} is not as large as it would be if the fluid followed the blades, and the relative flow angle β_2 is smaller than the angle of the blades at outlet. For radial blades, and even for some forwards-curved blades, the relative outlet flow will still be inclined backwards. Nevertheless, some degree of backwards curvature in the blades is required to achieve the flow outlet angles of $\beta_2 = 20°$ to $30°$ mentioned above.

In a realistic case some of the energy imparted to the fluid passing through the impeller will be dissipated within the impeller itself. The rise in the specific mechanical energy of the fluid is then, for absolute axes

$$\frac{p_{0_2} - p_{0_1}}{\varrho_c} = w_e - \Delta e \qquad (7.4.18)$$

Part of the dissipation arises from fluid friction and diffusion losses in the flow passages within the impeller. A second component is associated with separation of the flow from the inlet edges of the blades. Only over a narrow range of flow rates will the fluid pass smoothly onto the blades; the design operating point is chosen within this range. The separation losses* rise rapidly as the operating point moves away from this optimum range.

Finally, it is necessary to consider the diffuser lying beyond the impeller; here the fluid is decelerated and converted into a parallel flow which can be conveniently directed into a jet or pipeline. Vaneless diffusers are often used for small centrifugal pumps and blowers, the fluid being collected in a simple

* Sometimes referred to as losses due to 'shock', a rather misleading term, since compressibility and shock-wave formation do not usually play a part.

expanding scroll. This kind of diffuser has been analysed in Example 4.12. In more sophisticated machines, a row of stator vanes is provided to achieve a more rapid and carefully controlled diffusion; see Figure 2.22(b). If only the overall performance of the diffuser is of interest, the same analysis applies for both vaned and vaneless types. For constant-density flow the pressure rise in the diffuser is given by

$$\frac{p_3 - p_2}{\varrho_c} = \eta_D\left(\frac{C_2^2}{2} - \frac{C_3^2}{2}\right)$$

and the dissipation in the diffuser by

$$\frac{p_{0_2} - p_{0_3}}{\varrho_c} = (1 - \eta_D)\left(\frac{C_2^2}{2} - \frac{C_3^2}{2}\right) \tag{7.4.19}$$

The velocity at outflow from the diffuser is related to the impeller outlet flow by

$$C_3 = \frac{2\pi b_2 r_2}{A_3} V_{r_2} = \frac{A_2}{A_3} V_{r_2}$$

with A_3 the outlet flow area.

The results obtained above for the several components of the machine —intake, impeller and diffuser—will now be gathered together to provide an estimate of its operating characteristics. For simplicity, the flow in intake and impeller will be assumed to be dissipation-free, only losses in the diffuser being taken into account. Since a considerable part of the losses does occur in the diffuser in a real machine, this simplified analysis will give a general picture of the effect of losses on the performance of a centrifugal pump. The results are not, of course, accurate enough to serve as a basis for design. The assumption of lossless flow within the impeller is particularly doubtful, since separation at the inner edges of the blades must be expected when the pump operates well away from the design point.

If there is no dissipation within the rotor, equations (7.4.17,18) indicate that

$$w_e = \frac{p_{0_2} - p_{0_1}}{\varrho_c} = U_2 C_{t_2}$$

The only loss experienced by the fluid passing through the machine is given in equation (7.4.19). Then the useful work done by the machine is

$$\begin{aligned}\frac{p_{0_3} - p_{0_1}}{\varrho_c} &= w_e - \frac{p_{0_2} - p_{0_3}}{\varrho_c}\\ &= U_2 C_{t_2} - (1 - \eta_D)\left(\frac{C_2^2}{2} - \frac{C_3^2}{2}\right)\end{aligned} \tag{7.4.20}$$

It is convenient to express the preceding result in terms of dimensionless parameters. In equations (7.4.17) w_e/U_2^2 is given in terms of V_{r_2}/U_2 and β_2. The dissipation term of equation (7.4.20) can be represented in this way by introducing

$$\begin{aligned} C_2^2 &= V_{r_2}^2+(U_2-V_{r_2}\cot\beta_2)^2 \\ &= V_{r_2}^2\operatorname{cosec}^2\beta_2-2V_{r_2}U_2\cot\beta_2+U_2^2 \end{aligned}$$

Then, with the non-dimensional coefficients

$$\varphi_2 = \frac{V_{r_2}}{U_2} \quad \text{and} \quad \psi_{0_3} = \frac{p_{0_3}-p_{0_1}}{\varrho_c U_2^2}$$

representing the flow rate and rise in mechanical energy, equation (7.4.20) becomes

$$\begin{aligned} \psi_{0_3} &= 1-\varphi_2\cot\beta_2 \\ &\quad -\frac{1}{2}(1-\eta_D)\left[\varphi_2^2\operatorname{cosec}^2\beta_2-2\varphi_2\cot\beta_2+1-\left(\frac{A_2}{A_3}\right)^2\varphi_2^2\right] \\ &= \frac{1}{2}(1+\eta_D)-(2-\eta_D)\cot\beta_2\,\varphi_2 \\ &\quad -\frac{1}{2}(1-\eta_D)\left[\operatorname{cosec}^2\beta_2-\left(\frac{A_2}{A_3}\right)^2\right]\varphi_2^2 \end{aligned} \tag{7.4.21}$$

If the diffuser efficiency η_D and outlet angle β_2 are independent of the flow through the machine, this equation shows how the useful energy addition achieved by the machine depends on the rate at which fluid passes through it.

Other features of the operation can be found from the last result. The outlet pressure is specified by

$$\psi_3 = \frac{p_3-p_{0_1}}{\varrho_c U_2^2} = \psi_{0_3}-\frac{1}{2}\left(\frac{C_3}{U_2}\right)^2 = \psi_{0_3}-\frac{1}{2}\left(\frac{A_2}{A_3}\right)^2\varphi_2^2 \tag{7.4.22}$$

In the absence of losses up to the impeller periphery, the total-pressure rise to that point is specified by

$$\psi_{0_2} = \frac{p_{0_2}-p_{0_1}}{\varrho_c U_2^2} = \frac{w_e}{U_2^2} = 1-\varphi_2\cot\beta_2 \tag{7.4.23}$$

This is the maximum possible value of ψ_{0_3}, that obtained for $\eta_D = 1$ and $A_3 = \infty$. Finally, the pressure at the impeller periphery is specified by

$$\psi_2 = \frac{p_2-p_{0_1}}{\varrho_c U_2^2} = \psi_{0_2}-\frac{1}{2}\left(\frac{C_2}{U_2}\right)^2 \tag{7.4.24}$$

This is the pressure which would be attained in the absence of a diffuser. It corresponds to the least possible value of ψ_3, that for $\eta_D = 0$.

For backwards-inclined exit flow ($\beta_2 < 90°$) equations (7.4.21) and (7.4.22) indicate that the maximum energy addition (per unit mass) and pressure rise occur when there is no flow:

$$\psi_{0_3} = \psi_3 = \tfrac{1}{2}(1+\eta_D) \tag{7.4.25}$$

This pressure is often referred to as the shut-off head. Another important index of a machine's performance is the flow generated when the net pressure rise across it is zero. This is the operating condition for a blower which draws air from the atmosphere and exhausts it as a jet with the same atmospheric pressure. Equations (7.4.21, 22) give the condition determining this flow rate:

$$\frac{1}{2}(1+\eta_D)-(2-\eta_D)\cot\beta_2\,\varphi_2 - \frac{1}{2}\left[(1-\eta_D)\operatorname{cosec}^2\beta_2+\eta_D\left(\frac{A_2}{A_3}\right)^2\right]\varphi_2^2 = 0 \tag{7.4.26}$$

In the ideal case ($\eta_D \to 1$, $A_2/A_3 \to 0$) the flow rate for no pressure rise is given by

$$\varphi_2 = \tan\beta_2 \tag{7.4.27}$$

An actual maximum flow, though not expressible in a simple formula, will certainly be less than this.

The rate at which the rotor does work on the fluid passing through may be specified by the coefficient

$$\frac{\dot{W}_e}{\varrho_c U_2^3 A_2} = \frac{w_e}{U_2^2}\,\frac{V_{r_2}}{U_2} = \psi_{0_2}\varphi_2$$
$$= \varphi_2(1-\varphi_2\cot\beta_2) \tag{7.4.28}$$

Equations (7.4.17, 23) have been used in obtaining these results. The power $\dot{W}_e$ is the entire energy addition to the primary flow through the pump, the sum of the useful energy addition and the energy dissipated in the primary flow. However, additional dissipation occurs within the pump, but outside the primary flow; it results from friction in the bearings, rubbing within seals, shearing of fluid between rotor and casing, and other windage losses. Thus the power P transmitted to the pump is greater than $\dot{W}_e$ by an amount $\dot{W}_L$ which accounts for these frictional losses:

$$P = \dot{W}+\dot{W}_L$$

In order to form a general idea of the effect of these losses on pump performance, the frictional torque will be assumed to arise primarily from

Table 7.1 Characteristics of a centrifugal blower

φ_2	ψ_{0_2}	ψ_2	$\psi_{0_3} = \psi_3$	R_p	η_p	$\psi_{0_2}\varphi_2$	η_m	$\psi_{0_2}\varphi_2/\eta_m$	η_0
0	1	0·5	0·8	0·625	0·8	0	0	0·0278	0
0·05	0·95	0·397	0·729	0·545	0·767	0·0475	0·63	0·0753	0·483
0·1	0·9	0·290	0·656	0·442	0·729	0·09	0·765	0·1178	0·558
0·12	0·88	0·245	0·626	0·391	0·712	0·1058	0·792	0·1336	0·564
0·15	0·85	0·1775	0·581	0·306	0·684	0·1275	0·82	0·1553	0·561
0·2	0·8	0·060	0·504	0·119	0·630	0·16	0·851	0·1878	0·536
0·25	0·75	−0·0625	0·425	—	0·567	0·1875	0·87	0·2153	0·493
0·3	0·7	−0·19	0·344	—	0·492	0·21	0·883	0·2378	0·434
0·5	0·5	−0·75	0	—	0	0·25	0·90	0·2778	0

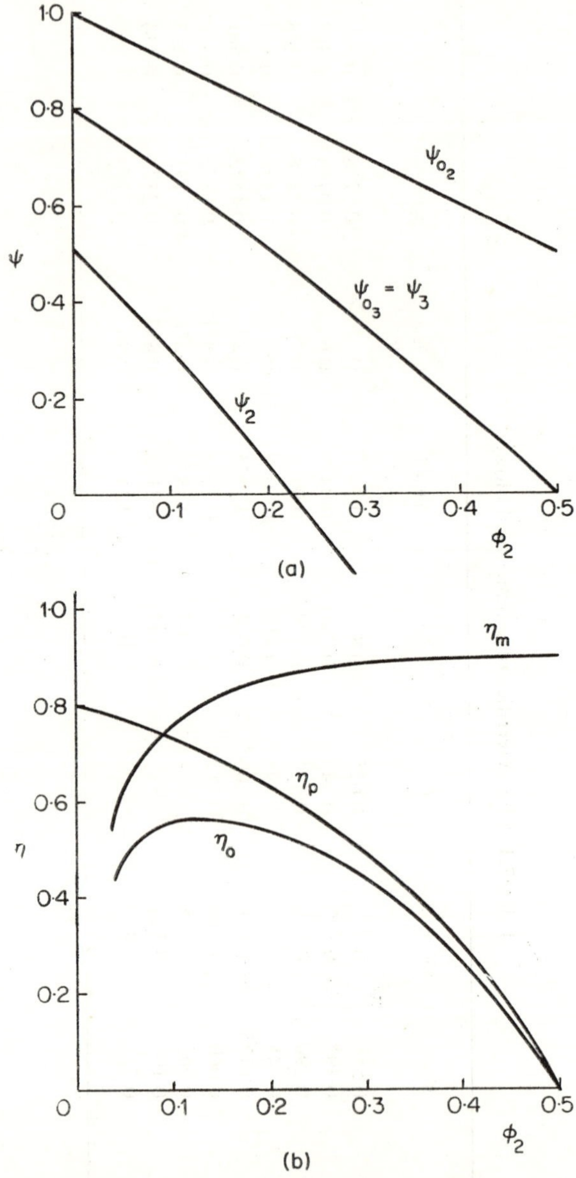

Figure 7.19 Dimensionless characteristics of the blower of Example 7.7. (a) Pressure coefficients against flow parameter; (b) Mechanical, hydraulic and overall efficiencies against flow parameter

with a power requirement of

$$P = 11{\cdot}7 \text{ ft lbf/sec} = 1/50 \text{ hp}$$

The flow for zero pressure rise is given by

$$\psi_3 = 0{\cdot}8 - 1{\cdot}4\varphi_2 - 0{\cdot}4\varphi_2^2 = 0$$

Then

$$\varphi_2 = 0{\cdot}5, \quad \text{and} \quad \dot{V}_{max} = 520 \text{ ft}^3\text{/min}$$

with a power requirement of

$$P = 116 \text{ ft lbf/sec} = 1/5 \text{ hp}$$

7.5 SIMILARITY LAWS

7.5.1 Flow with Constant Density

The preceding discussions have provided an overall view of the processes taking place within turbomachines and have introduced a few of the methods used in predicting their performance. Such results are required when designing a pump or turbine. However, a more commonly encountered problem is that of assessing the performance of an existing machine in order that its suitability for some particular duty may be determined. The selection of a pump or turbine and the prediction of its performance could be based on the analytical methods introduced earlier, but it is both simpler and more realistic to make use of measured characteristics. Dimensional analysis suggests the similarity laws governing a class of flow machines. These make possible a reduction in the number of measurements which are necessary and provide a compact presentation of the results.

The utility of a pump or compressor is measured by the flow it induces, the energy it imparts to the fluid, and the power expended in driving it. The performance of a turbine is measured by the useful power output, the flow through the machine, and the energy of the fluid supplied to it. To describe either operation a relationship is required amongst the flow rate $\dot{V}$, the power P transmitted to or from the machine, and the mechanical energy transferred to or extracted from the fluid. For a constant-density flow the useful work per unit mass is usually measured by the head H across the machine, defined by

$$gH = |w_e - \Delta e| = \left|\Delta\left(\frac{p_0}{\varrho_c} + gz\right)\right| \tag{7.5.1}$$

Comparison with equation (6.1.3) shows that H is equivalent to ΔH_0, the change in total head of the fluid. Hence the head H can be interpreted and measured in terms of the vertical heights in manometer columns filled with the fluid passing through the machine.

On what characteristics of machine and fluid do the relationships between $\dot{V}$, H and P depend? Obviously the fluid density and viscosity, ϱ_c and μ_c, are relevant, since they contribute to the stresses within the fluid. Many features of the geometry of a machine and of the flow inside influence its performance. However, if consideration is limited to a class of geometrically similar machines, a single length serves to specify a machine of the class, and a single velocity (in addition to the flow rate $\dot{V}$) serves to characterize the flow within. The impeller diameter D and angular velocity N of the rotor are commonly used for this purpose. Then the performance laws for a family of similar machines will involve the quantities

$$\dot{V},\ gH,\ P,\ \varrho_c,\ \mu_c,\ D \quad \text{and} \quad N$$

These laws are usually given in the forms

$$gH = f(\dot{V},\ \varrho_c,\ \mu_c,\ D,\ N)$$

and (7.5.2)

$$P = f(\dot{V},\ \varrho_c,\ \mu_c,\ D,\ N)$$

Other useful relationships can be obtained from these. The torque T applied to a pump and its overall efficiency η_0 are related to the quantities appearing in equations (7.5.2) by

$$T\omega = P = \varrho_c g \dot{V} H / \eta_0 \tag{7.5.3}$$

For a turbine

$$T\omega = P = \eta_0 \varrho_c g \dot{V} H \tag{7.5.4}$$

Hence from the laws (7.5.2) one can obtain

$$\eta_0 = f(\dot{V}, \text{etc.}) \quad \text{and} \quad T = f(\dot{V}, \text{etc.})$$

In equations (7.5.3, 4) ω denotes the rotational speed measured in rad/sec. The symbol N denotes the rotor speed in arbitrary units; in practice it is often measured in rev/min.

The dimensionless groups commonly used to express the laws (7.5.2) are

$$\varphi = \frac{\dot{V}}{ND^3} \quad \text{— flow coefficient}$$

$$\psi = \frac{gH}{N^2D^2} \quad \text{— head coefficient}$$

$$\Pi = \frac{P}{\varrho_c N^3 D^5} \quad \text{— power coefficient} \tag{7.5.5}$$

$$\text{Re} = \frac{\varrho_c N D^2}{\mu_c} \quad \text{— Reynolds number based on rotor characteristics}$$

Since each of the relationships (7.5.2) involves six quantities and three dimensions, these four parameters are sufficient to express the two laws in dimensionless form. The coefficients (7.5.5) have the general form of the dimensionless combinations which earlier emerged from the detailed analysis of flow machines. For example, the coefficients introduced during the discussion of the centrifugal pump are related to those of equations (7.5.5) by

$$\varphi_2 = \frac{V_{r_2}}{U_2} = \frac{2D}{\pi b_2}\varphi$$

and

$$\psi_{0_3} = \frac{p_{0_3} - p_{0_1}}{\varrho_c U_2^2} = 4\psi$$

Consistent units have been assumed in obtaining these relationships; the numerical factors will change if other units are used, for example, if N is measured in rev/min.

The performance laws (7.5.2) may be written in dimensionless form as

$$\psi = f(\varphi, \text{Re})$$

and

$$\Pi = f(\varphi, \text{Re}) \tag{7.5.6}$$

Since equations (7.5.3, 4) can be written

$$\Pi = \frac{\psi\varphi}{\eta_0} \quad \text{and} \quad \Pi = \eta_0\varphi\psi$$

the relationship

$$\eta_0 = f(\varphi, \text{Re}) \tag{7.5.6}$$

is implicit in these results. The forms (7.5.6) hold only for a family of similar machines, and of course only when the fluid density is essentially constant. Even for machines carrying liquids their application is restricted, since no account has been taken of the possibility of cavitation. Later, dimensionless groups which account for cavitation will be introduced, and the effects of cavitation on machine performance will be considered.

Equations (7.5.6) can also be generalized to account for variations in one or more geometric parameters. For example, the characteristics of a Kaplan turbine, a propeller turbine with pivoting blades, would take the form

$$\psi = f(\varphi, \text{Re}, \beta)$$

with β an angle specifying the blade incidence. Since changes in blade angle are associated with changes in the degree of reaction, a class of machines with basically similar geometry but varying reaction might be described by

$$\psi = f(\varphi, \text{Re}, \text{R}_\text{h}) \tag{7.5.7}$$

The pitchline analysis of axial-flow turbine and compressor stages has already provided laws of this form. Equations (7.3.21, 24) can be written

$$\frac{\mp w_e}{U^2} = \mp\frac{4w_e}{N^2D^2} = 2(1-R_h)$$

with D the pitchline diameter, and N measured in rad/sec. However

$$w_e = \frac{gH}{\eta_p} \quad \text{and} \quad -w_e = \eta_t gH$$

so that

$$\psi = \frac{1}{2}\eta_p(1-R_h) = \frac{1-R_h}{2\eta_t} \tag{7.5.8}$$

Since the hydraulic efficiency is dependent on Reynolds number and flow rate, these laws are of the form (7.5.7).

In many cases the Reynolds-number effect provides a rather small modification to a law of the form

$$\psi = f(\varphi) \quad \text{or} \quad \frac{gH}{N^2D^2} = f\left(\frac{\dot{V}}{ND^3}\right) \tag{7.5.9}$$

The correction is large enough to be of importance in designing machines, but can often be neglected if one seeks only to study the performance of existing machines under varying conditions. The relationship between the characteristics of two geometrically similar machines with characteristic diameters D_1 and D_2 operating at rotational speeds N_1 and N_2 is shown in Figure 7.20. For any two points at which ψ and φ are the same (these are termed corresponding points), we have

$$H_2 = H_1\left(\frac{N_2}{N_1}\right)^2\left(\frac{D_2}{D_1}\right)^2 \quad \text{and} \quad \dot{V}_2 = \dot{V}_1\frac{N_2}{N_1}\left(\frac{D_2}{D_1}\right)^3 \tag{7.5.10}$$

Equations (7.5.6) show that the power coefficient Π and the efficiency η_0 must also be the same at corresponding points.

So far, the relationships between H, $\dot{V}$, N and D have been thought of as performance characteristics of the form (7.5.2). One can equally well consider the relationships in the forms

$$N = f(\dot{V}, gH) \quad \text{and} \quad D = f(\dot{V}, gH)$$

The effect of viscosity, expressed by the Reynolds number, has been dropped for simplicity. The statements above can be expressed in dimensionless form

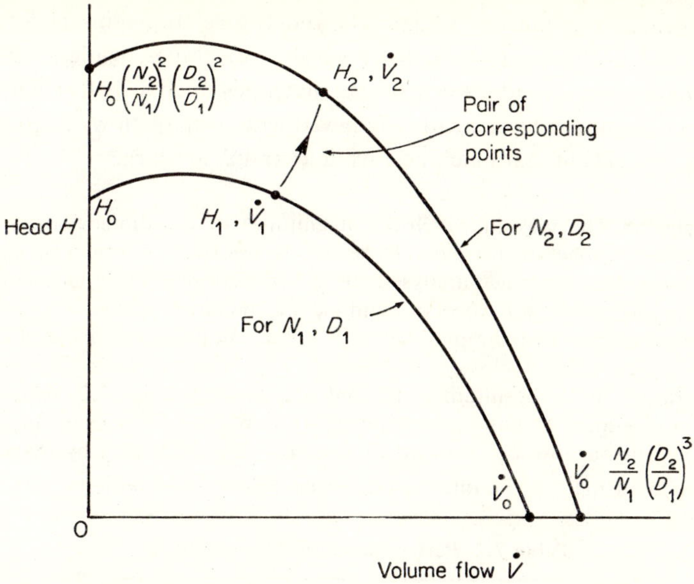

Figure 7.20 Relationship of performance characteristics for geometrically similar machines operating at different speeds

using the parameters

$$\Omega = \frac{N\dot{V}^{\frac{1}{2}}}{(gH)^{\frac{3}{4}}} \quad \text{— specific speed}$$

and (7.5.11)

$$\Delta = \frac{D(gH)^{\frac{1}{4}}}{\dot{V}^{\frac{1}{2}}} \quad \text{— specific diameter}$$

They are related to the flow and head coefficients by

$$\Delta = \psi^{\frac{1}{4}}\varphi^{-\frac{1}{2}} \quad \text{and} \quad \Omega = \varphi^{\frac{1}{2}}\psi^{-\frac{3}{4}} \tag{7.5.12}$$

The specific speed and diameter may be thought of as dimensionless machine characteristics, and their values for each class of turbomachine may be expected to fall within relatively narrow ranges. In particular, the value of the specific speed for maximum efficiency is often used as an index of machine type. (Note, however, that g is often dropped from the formula (7.5.11), while inconsistent units are used for the other quantities.) The nature of the relationship between machine form and the dimensionless characteristics can be seen as follows. For axial-flow machines, such as fans and propeller turbines, the rise in head H is usually small and the flow rate

$\dot{V}$ is large. The definitions (7.5.5, 11) and the relationships (7.5.12) then indicate that φ and Ω are large for axial-flow machines, while ψ and Δ are small. On the other hand for centrifugal machines, such as Pelton wheels and centrifugal pumps, a large head is achieved with a small flow. Then φ and Ω are small for centrifugal machines, while ψ and Δ are large.

Example 7.8 The way in which the magnitudes of the dimensionless characteristics relate to machine form can be seen most clearly by considering some particular cases. The machines analysed in earlier examples provide a convenient group for study: the Pelton wheel of Example 7.1, the Laval turbine of Example 7.2, the propeller turbine of Examples 7.4 and 7.6, the compressor cascade of Example 7.5, and the centrifugal blower of Example 7.7. The parameters characterizing their operation under optimum conditions are given in Table 7.2. Although the discussion of similarity has so far referred only to constant-density flow, it may for the present purpose be extended to cases of variable density by replacing gH by w_e and by using $\dot{V}_1$, the inlet value, as the characteristic volume flow.

Table 7.2 Performance of flow machines

Type	Fluid	$\dot{V}$ or $\dot{V}_1$	gH or w_e	N	D
Impulse turbine (Example 7.1)	Water 3 jets	5·75 m³/sec	1870 J/kg	40·7 rad/sec	1·5 m
(Impulse turbine Example 7.2)	Steam 1 jet	31·6 ft³/sec	2,730,000 ft lbf/sl	943 rad/sec	2·5 f
Propeller turbine (Examples 7.4, 6)	Water	11·45 m³/sec	196·2 J/kg	40 rad/sec	1·5 m
Axial compressor row (Example 7.5)	Air	19·8 m³/sec	15,600 J/kg	418 rad/sec	1 m
Centrifugal blower (Example 7.7)	Air	2·1 ft³/sec	6260 ft lbf/sl	298 rad/sec	$\frac{2}{3}$ ft

Most of the data of Table 7.2 were obtained in the examples quoted. For the impulse turbines D is the diameter at the centre of the jet; for the other machines it is the outer diameter of the rotor. The volume flows for the steam turbine and axial compressor are those at entry. For the Laval turbine, with a flow of 4000 lb/hr, and $v_1 = 28{\cdot}4$ ft³/lb

$$\dot{V}_1 = \frac{4000 \times 28{\cdot}4}{3600} = 31{\cdot}6 \text{ ft}^3/\text{sec}$$

For the axial-flow compressor, with inner and outer diameters of 0·8 and 1 m, and axial velocity of 70 m/sec

$$\dot{V}_1 = 70 \times \tfrac{1}{4}\pi \times (1 - 0{\cdot}64) = 19{\cdot}8 \text{ m}^3/\text{sec}$$

The corresponding values of the dimensionless parameters are shown in Table 7.3; the values shown are unit-free, being calculated from the consistent units of Table 7.2.

Table 7.3 Coefficients for machines of Table 7.2

Type	φ	ψ	Ω
Pelton wheel	0·042	0·50	0·35
Laval turbine	0·0021	0·49	0·079
Propeller turbine	0·085	0·055	2·6
Axial compressor	0·047	0·089	1·3
Centrifugal blower	0·024	0·16	0·60

These examples do not cover the whole range of machine types, but they do display the trends to be expected for the broad family of turbomachines. Axial-flow machines tend to have larger flow coefficients and specific speeds, and lower head coefficients.

7.5.2 Flow with Varying Density

In the example above, a start was made towards similarity laws for flow machines across which the density change is significant. The quantity w_e/N^2D^2 was used there as a generalization of the head coefficient gH/N^2D^2. However, for a pump $gH = w_e - \Delta e = \eta_p w_e$; only if the efficiency is unity does $w_e = gH$. A more consistent generalization of the head change is the isentropic change in enthalpy

$$\Delta h_{0i} = h_0(p_{0_2}, s_1) - h_0(p_{0_1}, s_1)$$

This quantity was used earlier in defining isentropic efficiencies for compres sors and turbines; see equations (6.2.6).

A more suitable extension of the first of equations (7.5.6) would therefore appear to be

$$\frac{\Delta h_{0_i}}{N^2D^2} = f\left(\frac{\dot{m}}{\varrho_1 N D^3}, \frac{\varrho_1 N D^2}{\mu_1}\right) \tag{7.5.12}$$

This statement proves adequate in many cases, but is in fact incomplete. Two factors have not been properly specified, the initial state of the fluid and its equation of state. The performance of the machine must obviously depend on them.

If the equation of state had a complicated form, the precise similarity laws would be correspondingly complex. For this reason, only those compressible fluids which behave rather like perfect gases will be considered. It has been seen that the thermodynamic response of a gas is vitally dependent on the ratio of its specific heats γ. Hence this parameter might be expected to play some part in the similarity laws for machines working with gases. In fact, changes in γ seem to have only a small effect on the flow laws and can usually be neglected. This observation suggests further that the precise equation of state of a vapour, such as steam, does not have much effect on the overall characteristics of a turbomachine. Of course, if a large fraction of the vapour were to condense within the machine, the effect could not be neglected.

The state of the fluid entering the machine is fixed when its velocity and two thermodynamic properties are specified; the viscosity has been seen to be a further relevant property. Convenient choices for the inlet properties are the density ϱ_1 and sound speed a_1; the density is already contained in the Reynolds number Re, and the sound speed can be introduced in a Mach number characterizing the flow, say ND/a_1. The inlet velocity is implicit in $\dot{m}$, ϱ_1, and D. Hence, for a particular machine shape

$$\Delta h_{0_i} = f(\dot{m}, \varrho_1, a_1, \mu_1, \gamma, N, D)$$

and equation (7.5.12) can be generalized to

$$\frac{\Delta h_{0_i}}{N^2 D^2} = f\left(\frac{\dot{m}}{\varrho_1 N D^3}, \frac{\varrho_1 N D^2}{\mu_1}, \gamma, \frac{ND}{a_1}\right) \tag{7.5.13}$$

Alternatively, one might choose to describe the inlet flow with stagnation properties. This procedure is perhaps more convenient, since the overall energy balance, $w_e = \Delta h_0$, involves stagnation states. An alternative to equation (7.5.13) is then

$$\frac{\Delta h_{0_i}}{N^2 D^2} = f\left(\frac{\dot{m}}{\varrho_{0_1} N D^3}, \frac{\varrho_{0_1} N D^2}{\mu_{0_1}}, \gamma, \frac{ND}{a_{0_1}}\right) \tag{7.5.14}$$

The laws (7.5.13, 14) were obtained in a direct attempt to generalize the constant-density forms (7.5.6). They suffer from the disadvantage that several of the quantities entering into them are not directly measurable. Pressures and temperatures are usually the most easily measured properties, and for this reason a similarity law involving them is more useful. Moreover, for a perfect gas the actual enthalpy change is given by

$$\Delta h_0 = c_p(T_{0_2} - T_{0_1}) \tag{7.5.15}$$

and the hypothetical enthalpy change of the isentropic pattern process is

$$\Delta h_{0_i} = c_p T_{0_1}\left[\left(\frac{p_{0_2}}{p_{0_1}}\right)^{\frac{\gamma-1}{\gamma}} - 1\right] \tag{7.5.15}$$

These simple relations provide further reasons for expressing performance in terms of total pressures and temperatures. Hence the usual similarity laws for machines handling gases take a very different form from those for constant density, even though they describe essentially similar operations.

If the initial state is specified by p_{0_1} and T_{0_1}, and the useful effect and total work implicitly by p_{0_2} and ΔT_0, the performance laws will have the forms

$$p_{0_2} = f(\dot{m}, p_{0_1}, T_{0_1}, \mu_{0_1}, \gamma, N, D)$$
$$\Delta T_0 = f(\dot{m}, p_{0_1}, T_{0_1}, \mu_{0_1}, \gamma, N, D)$$

The pressure ratio p_{0_2}/p_{0_1} is a convenient dimensionless measure of the useful transfer of energy, while $\Delta T_0/T_{0_1} = w_e/h_{0_1}$ is a measure of the net work done on the fluid. A mass-flow parameter like that used for a nozzle in equations (4.3.13, 14, 15), again provides a dimensionless measure of flow rate

$$\frac{\dot{m}}{\varrho_{0_1} a_{0_1} D^2} = \dot{m} \frac{(RT_{0_1}/\gamma)^{\frac{1}{2}}}{p_{0_1} D^2}$$

The performance laws expressed in terms of these parameters are

$$\begin{aligned} \frac{p_{0_2}}{p_{0_1}} &= f\left(\frac{\dot{m}\sqrt{RT_{0_1}}}{p_{0_1}D^2}, \frac{ND}{\sqrt{RT_{0_1}}}, \gamma, \text{Re}\right) \\ \frac{T_{0_2}-T_{0_1}}{T_{0_1}} &= f\left(\frac{\dot{m}\sqrt{RT_{0_1}}}{p_{0_1}D^2}, \frac{ND}{\sqrt{RT_{0_1}}}, \gamma, \text{Re}\right) \end{aligned} \tag{7.5.16}$$

Since the Reynolds number Re may be formed in several equally satisfactory ways, its structure has not been specified. The factor γ has been removed from two of the groups as it appears as an independent dimensionless group.

The results (7.5.16) are the commonly used generalizations of the constant-density performance laws (7.5.6). Note that the isentropic efficiencies for machines handling gases, given in equations (6.2.7), are expressed in terms of the ratios of equations (7.5.16). Hence these efficiencies are given by laws of similar structure, and can be calculated directly from the results above. Note also that the first of the results (7.5.16) is essentially equivalent to equation (7.5.14), the more straightforward generalization of the head vs flow-rate characteristic.

In practice any one compressor or turbine has usually to operate with only one gas or vapour. However, its speed of rotation may vary over a considerable range, and the fluid may be supplied to it with a variety of states. For example, the properties of the steam entering a turbine depend on the heat output of the boiler and on the way in which the feed-water pump operates, while the properties of the air entering an aircraft jet engine will depend on the altitude and speed of flight. The operation of a particular turbomachine is usually described by the simplified law

$$\frac{p_{0_2}}{p_{0_1}} = f\left(\frac{\dot{m}\sqrt{T_{0_1}}}{p_{0_1}}, \frac{N}{\sqrt{T_{0_1}}}\right) \tag{7.5.17}$$

For any one machine and any one gas, D, R and γ are constants; accordingly, they have been dropped from the similarity parameters. The resulting groups are, of course, no longer dimensionless.

The Reynolds number has not been included in equation (7.5.17). This

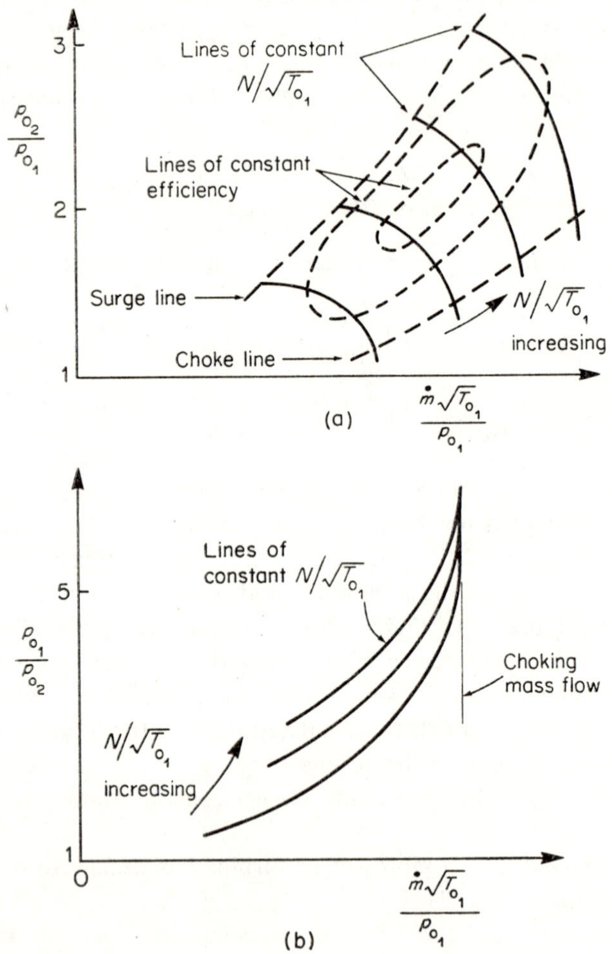

Figure 7.21 Characteristics of machines handling highly compressible fluids. (a) Compressor characteristics; (b) Turbine characteristics

may bc justified by arguing that, over the range of operation of a single machine, the Reynolds number will change by only a small factor at a particular operating point. Thus the major part of the dependence on Reynolds number is implicit in the parameters defining the operating point, $\dot{m}\sqrt{T_{0_1}}/p_{0_1}$ and $N/\sqrt{T_{0_1}}$. In other words, large Reynolds numbers correspond to large values of the velocity-determining parameters $\dot{m}$ and N, while the lower Reynolds numbers arise only when $\dot{m}$ and N are small.

The general forms of compressor and turbine characteristics (7.5.17) are shown in Figure 7.21. In both cases there is a tendency for the machine to choke as the mass-flow parameter increases. This parameter corresponds roughly to the inlet Mach number, and the Mach numbers within a machine tend to rise as the inlet Mach number is increased. Thus the phenomenon of *choking* is broadly analogous to the choking of a fixed nozzle, which was studied in Section 4.3.3. In some part of the machine the flow becomes sonic and no further change in Mach number distribution can occur, unless the rotor speed changes.

The dashed curve marked 'surge line' indicates the limit of stable operation of the compressor. An attempt to operate the machine at conditions corresponding to points beyond this line will lead to *surging*, severe oscillations of the flow within and through the compressor which are potentially destructive to the machine. This limit corresponds roughly to the points at which the characteristics (curves of constant $N/\sqrt{T_{0_1}}$) flatten out.

Example 7.9 The prediction of the change in the output of a turbine when the condition of the steam supplied is altered provides an interesting application of the preceding results.

Consider a turbine supplied with 10 lb/sec of steam at 100 lbf/in² (abs) and 400 °F, and exhausting into a condenser in which a vacuum of 26 in of mercury is maintained. The output of the machine is 3300 hp under these conditions. It is required to determine the flow rate and output when the inlet conditions are changed to 140 lbf/in² (abs) and 450 °F.

In these calculations it will be assumed that total and static properties are identical, that the efficiency of the turbine is the same at the two operating points, and that steam behaves like a perfect gas. Use will also be made of an empirical relationship generally found to be valid for multistage machines:

$$\dot{m}\frac{\sqrt{T_1}}{p_1} \propto 1 - \left(\frac{p_2}{p_1}\right)^2$$

This 'law of the ellipse' represents with fair accuracy the approach to choking when there is a large pressure ratio across the turbine.

For both of the operating conditions to be considered

$$p_2 = 3{\cdot}9 \text{ in of mercury (abs)} = 1{\cdot}9 \text{ lbf/in}^2 \text{ (abs)}$$

Then

$$\left(\frac{p_2}{p_1}\right)_a = \frac{1\cdot 9}{100} = 0\cdot 019$$

and

$$\left(\frac{p_2}{p_1}\right)_b = \frac{1\cdot 9}{140} = 0\cdot 014$$

Since the turbine is effectively choked in each case, the mass-flow parameter is virtually constant:

$$\left(\dot{m}\frac{\sqrt{T_1}}{p_1}\right)_a = \left(\dot{m}\frac{\sqrt{T_1}}{p_1}\right)_b$$

Hence

$$\dot{m}_b = \dot{m}_a\left(\frac{T_{1_a}}{T_{1_b}}\right)^{\frac{1}{2}}\frac{p_{1_b}}{p_{1_a}} = 10\times(860/910)^{\frac{1}{2}}\times 140/100 = 13\cdot 6 \text{ lb/sec}$$

The power delivered by the turbine is related to the ideal enthalpy drop by

$$\eta_0 = \eta_m\eta_t = \frac{P}{\dot{m}(h_1-h_i)}$$

If the overall efficiency is the same for the two supply conditions

$$P_a = P_a\frac{\dot{m}_b}{\dot{m}_a}\frac{(h_1-h_i)_b}{(h_1-h_i)_a}$$

Using values from the Mollier chart to evaluate the isentropic enthalpy changes, and the value $\dot{m}_b$ found above, we obtain

$$P_b = 3300\times 1\cdot 36\times\frac{1249-990}{1226-995} = 5000 \text{ hp}$$

7.6 CAVITATION

Cavitation is the name given to the boiling of a liquid as a result of a fall in pressure rather than heat addition. In Section 2.4.1 it was noted that cavitation could occur in isenergic, isentropic flow, the pressure reduction then being the result of fluid acceleration. We have also noted the possibility of cavitation in flow machines, in Example 6.1 on the inlet or suction side of a water pump, and in Example 7.4 on the outlet side of a hydraulic turbine. The discussion of cavitation within a flow machine is complicated by the fact that the absolute pressure at the point of cavitation is determined not only by the pressure level imposed on the system (for example, atmospheric pressure) and by the local velocity, but also by the elevation of the machine, the energy transfer within it, and the pressure changes associated with

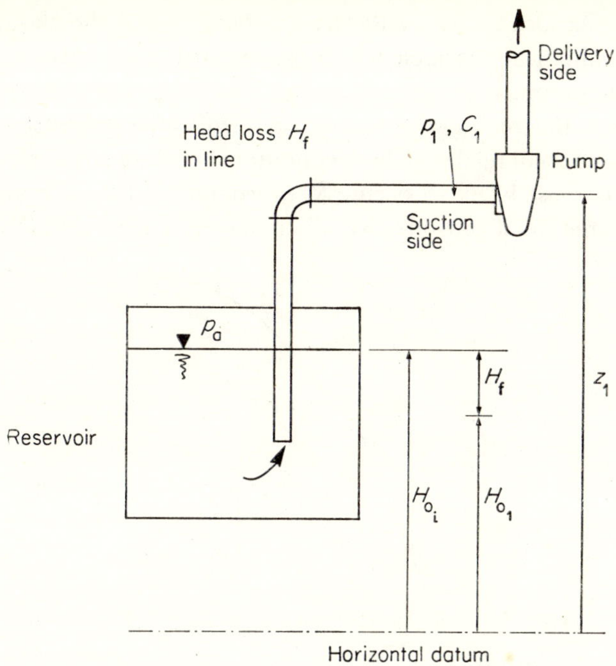

Figure 7.22 A water-pumping installation illustrating the factors influencing cavitation in a pump

dissipation. In the discussion which follows we shall see how these factors interact, and shall introduce parameters which can be used to describe the propensity of a hydraulic machine to cavitate.

Consider the pumping installation of Figure 7.22; water is drawn from a reservoir with a free surface by a pump located above the tank, the head loss between reservoir and pump inlet being H_f. On relating the reservoir conditions and the absolute flow at entry to the pump, we have

$$\frac{p_a}{\varrho_c}+gH_{0_i}=\frac{p_1}{\varrho_c}+\frac{C_1^2}{2}+gz_1+gH_f$$

with H_{0_i} the elevation of the free surface in the tank. The absolute total pressure at inlet is then

$$p_{0_1}=p_1+\frac{\varrho_c C_1^2}{2}=p_a-\varrho_c g(z_1-H_{0_i}+H_f) \tag{7.6.1}$$

This differs from the atmospheric pressure by virtue of the head loss H_f and the change in elevation $z_1-H_{0_i}$. The former will inevitably reduce the pressure at inlet to the pump, but the latter will augment the inlet pressure if the pump is located below the free surface. Hence the elevation of the pump is an important variable which the designer can control in attempting to avoid cavitation.

To see how the inlet total pressure is related to the propensity of a pump to cavitate, we shall relate the inlet conditions to those at the point at which cavitation is most likely to occur. At this point the kinetic energy must be large compared to the pressure rise which has been achieved in the machine.

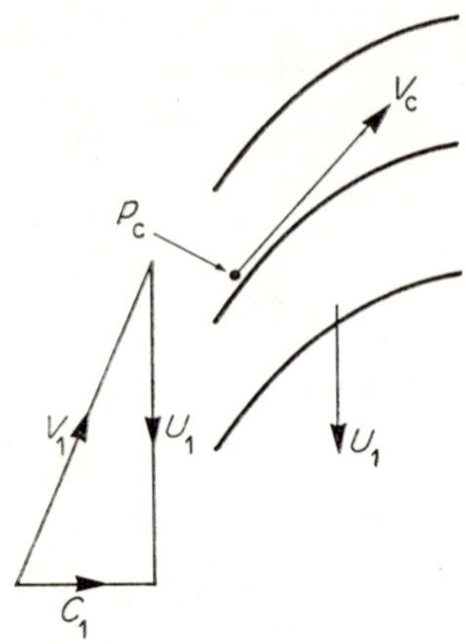

Figure 7.23 Flow from the intake of a pump to the point of cavitation inception as seen in relative axes

Hence cavitation is most likely in the stationary inlet of the pump (here no energy has been added, but the fluid may have undergone considerable acceleration), or just within the rotor (here the energy addition is small and even higher velocities are imposed). The result (7.6.1) provides the means of analysing cavitation in the intake. However, the second case, cavitation within the rotor, requires further consideration. Figure 7.23 shows for relative axes the flow approaching the rotor and the flow at the point of cavitation inception. Assuming lossless flow between the two points, and neglecting changes in potential energy, we have

$$p_1 + \frac{\varrho_c V_1^2}{2} = p_c + \frac{\varrho_c V_c^2}{2}$$

The likelihood of cavitation occurring is indicated by the difference between p_c and the vapour pressure of the liquid:

$$p_c - p_v = p_1 - p_v - \frac{\varrho_c}{2}(V_c^2 - V_1^2)$$

Using equation (7.6.1) we can convert this into

$$\frac{p_c - p_v}{\varrho_c} = \frac{p_{0_1} - p_v}{\varrho_c} - \frac{1}{2}(C_1^2 + V_c^2 - V_1^2) \tag{7.6.2}$$

The occurrence of cavitation is seen to depend on the difference between the total energy available at inlet and a kinetic-energy term determined by the rotational speed of the pump, the velocity of the flow into the pump, and the shape of the flow passages near the rotor inlet.

For a particular pump, operating at a specified speed and carrying a specified flow, the kinetic-energy term of equation (7.6.2) is fixed. Hence if cavitation is to be avoided, the available energy $(p_{0_1} - p_v)/\varrho_c$ must be greater than this dynamic term. This requirement is often expressed in terms of a related quantity, the *net positive suction head*

$$H_s = \frac{p_{0_1} - p_v}{\varrho_c g}$$

Equation (7.6.1) shows how this head can be calculated for a particular installation. The condition

$$H_s > \frac{C_1^2 + V_c^2 - V_1^2}{2g} \tag{7.6.3}$$

provides a rough indication of the conditions under which cavitation will not occur. Hence a pump which is designed to avoid cavitation is one capable of operating at a low value of the suction head H_s. Equation (7.6.1) indicates that H_s depends on the pressure imposed on the system (in the case of Figure 7.22, this is the atmospheric pressure), on the elevation of the pump, on the losses experienced by the fluid before reaching the pump, and of course on the vapour pressure. Hence the suction head H_s is primarily a quantity describing the system in which the pump is installed. However, the limiting value of H_s at which cavitation develops depends on the characteristics of the pump.

The criterion (7.6.3) assumes that the only fluid property relevant to the inception of cavitation is the difference between the local pressure and a unique vapour pressure of the pumped liquid. In fact, the surface tension and thermal conductivity of the liquid prove to be important, and also the presence of dissolved gases and impurities in the liquid and its composition in terms of mixed constituents. In petroleum oils, for example, bubble formation is slower than that in water, since the thermal conductivity is lower and the heat of vaporization cannot be supplied as rapidly. When the liquid contains dissolved gases, bubbles will appear at a pressure above the normal boiling point and, as was pointed out in Section 2.4.1, two forms of cavitation can be distinguished, gaseous and vaporous. Somewhat similar behaviour is exhibited by liquids composed of a number of distinct species, for example, petroleum products. Each of the components begins to vaporize at its own characteristic vapour pressure, so that the development of cavitation in the fluid as a whole is spread over a broad range of operating conditions. The discussion of cavitation which follows will concentrate on the fairly abrupt onset of cavitation, prescribed by the condition (7.6.3), which is characteristic of water. The conclusions are applicable in general terms to other liquids, but the details of the process must be expected to be different.

For a particular pump geometry, the kinetic-energy term of equation (7.6.2) depends on the absolute and relative velocities near the inlet to the pump. These in turn are defined by the rotor speed N, the flow rate $\dot{V}$, and the characteristic diameter of the machine D. Hence for a particular inlet shape

$$\tfrac{1}{2}(C_1^2+V_c^2-V_1^2) = f(N, \dot{V}, D)$$

Dimensional homogeneity requires that

$$\tfrac{1}{2}(C_1^2+V_c^2-V_1^2) = f(\varphi)\times(\dot{V}N^2)^{\frac{2}{3}}$$

with $\varphi = \dot{V}/ND^3$. the flow coefficient of equations (7.5.5). Subject to the limitations mentioned above, equation (7.6.2) suggests that cavitation will be avoided when

$$\frac{p_{0_1}-p_v}{\varrho_c} = gH_s > f(\varphi)\times(\dot{V}N^2)^{\frac{2}{3}} \tag{7.6.4}$$

The last result can be expressed compactly in terms of the *suction specific speed*

$$S = \frac{N\dot{V}^{\frac{1}{2}}}{(gH_s)^{\frac{3}{4}}} \tag{7.6.5}$$

Comparison with equations (7.5.11) shows that this has the structure of the specific speed for the machine, but with the suction head replacing the head across the machine. Now the criterion (7.6.3, 4) becomes

$$S < S_i = f(\varphi_i) \tag{7.6.6}$$

the subscript i denoting conditions at which cavitation is incipient. This result states that, for a given class of machines, the suction specific speed at which cavitation begins is determined by the flow coefficient; the converse is also true. So long as the suction specific speed does not pass this boundary, cavitation will not occur.

When it is possible that cavitation may occur, some quantity which specifies the inlet pressure must be added to the list of quantities determining the performance of a flow machine. The net positive suction head H_s is often used for this purpose. When it is added to the list of quantities in equations (7.5.2), an additional dimensionless group will appear in the non-dimensional laws for the machine. The suction specific speed is a suitable parameter incorporating H_s. Hence the affinity laws (7.5.6) are generalized to

$$\psi = f(\varphi, \mathrm{Re}, S)$$

and (7.6.7)

$$\eta_0 = f(\varphi, \mathrm{Re}, S)$$

The general form of the dependence on S for a conventional centrifugal pump is indicated in Figure 7.24. The main effects of cavitation are seen to be abrupt drops in head and efficiency once the limiting value of the flow coefficient is reached. For axial-flow machines, the reduction in performance is less abrupt, since the larger flow passages are not as rapidly filled by the developing cavity.

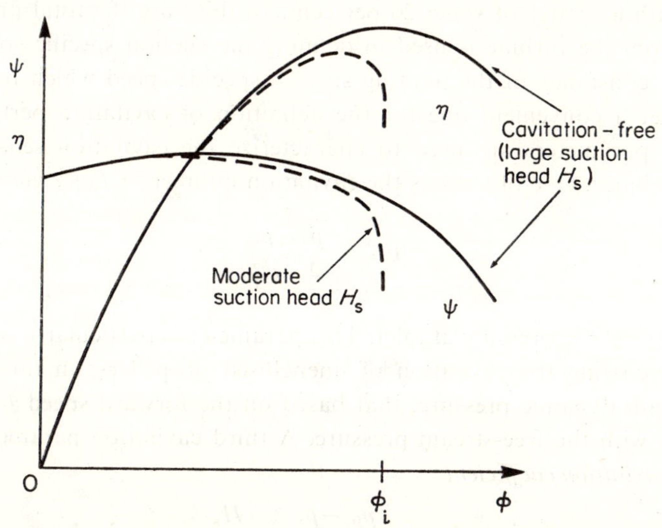

Figure 7.24 Dimensionless characteristics of a centrifugal pump, with and without cavitation

The fall in performance is one of the reasons for avoiding cavitation in flow machines. However, even more important problems are associated with the collapse of bubbles and cavities within the machine. This can produce serious vibration of the machine and damage to the flow passages and blades. Blades subjected to high-frequency shock loads arising from cavity collapse are often eroded away, and the resulting modification of the blade shape will give rise to further deterioration in machine performance.

As noted above, it is the geometry of the inlet passages of a water pump which determines the proneness to cavitation. Designers seeking to avoid cavitation have naturally selected geometries which delay its onset, subject to the requirement of achieving good overall performance. When the characteristics of a variety of pumps designed to avoid cavitation are compared, it is found that the limiting value of the suction specific speed does not vary over a very large range. For single-suction pumps (that is, pumps accepting fluid through only one inlet eye) it is found that, when N is measured in rad/sec

$$S < S_i = 3 \tag{7.6.8}$$

with a scatter of perhaps 20 per cent. Similarly, for turbines designed to avoid cavitation, it is found that

$$S < S_i = 4 \tag{7.6.9}$$

again with a scatter of some 20 per cent. In this case the total pressure at outlet from the turbine is used in defining the suction specific speed. It is the near constancy of the limiting suction specific speed which makes this parameter a convenient one for the definition of cavitation performance.

Other parameters are used to characterize the cavitation sensitivity of flow machines. One of these is the cavitation number or *Euler number*

$$\mathrm{Eu} = \frac{p_1 - p_v}{\frac{1}{2}\varrho_c C_1^2}$$

with p_1 the static pressure at inlet. This parameter is particularly convenient when discussing the cavitation of unenclosed propellers; in this case the free-stream dynamic pressure, that based on the forward speed U, is used, together with the free-stream pressure. A third cavitation parameter is the *Thoma cavitation coefficient*

$$\sigma = \frac{p_{0_1} - p_v}{\varrho_c g H} = \frac{H_s}{H}$$

with H the rise in head across the pump. Using equations (7.6.5) and (7.5.11), we see that

$$S = \frac{\Omega}{\sigma^{\frac{3}{4}}}$$

relates the suction specific speed, the Thoma coefficient and the specific speed of the machine. The criterion (7.6.8) can now be expressed as

$$\sigma > \sigma_i = \left(\frac{\Omega}{3}\right)^{\frac{4}{3}} \tag{7.6.10}$$

for cavitation-free operation.

The results (7.6.8, 9, 10) provide important constraints on the design of flow machines. They differ from the constraints discussed earlier—centrifugal stresses in turbines and relative Mach number in compressors—in being strongly dependent on the environmental conditions, and not merely on the characteristics of the machine and fluid. If the designer can arrange for a sufficiently high pressure at inlet to a pump or at outlet from a turbine (perhaps by lowering the machine to increase the gravity head), he can avoid cavitation. However, this is often impossible, and it is necessary to modify the design of the machine. The criteria (7.6.8, 9, 10) indicate what action is taken in conventional machines. If the volume flow, suction head and net head to be generated are fixed, the likelihood of cavitation will be reduced by reducing the rotor speed, with a consequent increase in machine size and reduction in head developed per stage. These changes imply that more stages may be required, or that a machine with a smaller specific speed must be selected. In any case, the machine will be bulkier and more expensive than it would be if cavitation had not to be taken into account.

In recent years pumps have been designed to operate continuously under conditions of cavitation. These are of two types, often referred to as *supercavitating pumps* and *inducers*. The former have compact blades rather like those of propellers; indeed, they were developed from the supercavitating propellers used in high-speed boats. The cavities formed on the blades collapse between and behind them; in this way serious cavitation damage is avoided. The kind of supercavitating machine referred to as an inducer has a quite different form. Here blades of very long chord and highly swept leading edge spiral around the rotor axis. The cavities collapse within the spiral blade passages, but the blade angle is initially very small and this delays the onset of cavitation and reduces its severity. Moreover, the massive construction of the rotor serves to absorb the energy of the collapse without leading to serious erosion. Inducers are sometimes used to supply fuel to rocket motors; in this application light weight is vital and a long life need not be achieved. Inducers are also being introduced in situations where a long life is required. For example, by placing an inducer in the impeller eye of a boiler feed pump (where cavitation is very important, since the inlet pressure is low and the vapour pressure high), one can raise the pressure at entry to the pump proper to allow a considerable increase in speed and a consequent reduction in the number of stages required in the pump.

Example 7.10 Is the turbine installation of Examples 7.4 and 7.6 likely to be satisfactory from the point of view of cavitation?

The total pressure at outlet from the turbine is, on the assumption of lossless flow, given by

$$\frac{p_{0_3}}{\varrho_c g} = \frac{p_a}{\varrho_c g} + \frac{w_e}{g} + z_3$$

$$= 10{\cdot}33 - 20{\cdot}00 + 19{\cdot}00 = 9{\cdot}33 \text{ m}$$

Steam tables give, for water at 20 °C, $p_v = 0{\cdot}023$ bar. Then

$$p_{0_3} - p_v = 91{\cdot}5 - 2{\cdot}3 = 89{\cdot}2 \text{ kN/m}^2$$

In the case to be considered, $\dot{V} = 11{\cdot}45$ m³/sec and $N = 40$ rad/sec are the values found earlier, so that

$$S = \frac{N\dot{V}^{\frac{1}{2}}}{[(p_{0_3} - p_v)/\varrho_c]^{\frac{3}{4}}} = 4{\cdot}7$$

Comparison with the criterion (7.6.9) suggests that cavitation is likely to occur within the turbine if it is installed and operated in the manner proposed.

PROBLEMS FOR CHAPTER 7

Group A

7.1 Laval turbines have often been used to drive cream separators. Why are they suitable in this application?

7.2 Consider the two-stage turbine discussed in Section 7.2.3.

(a) What is the ratio (F_2/F_1) of the energy extracted in the two stages for the optimum lossless operation?

(b) What is this ratio for optimum operation with blade factor 0·8?

7.3 In a paper mill, steam at 20 lbf/in² (abs) and 300 °F is used in the drying rolls. The steam is raised at a somewhat higher temperature and is first passed through a turbine which drives the paper-making machinery. Any residual power from the turbine is used to produce electricity for the paper mill. The dryer requires 40,000 lb/hr of steam; the machinery power is 2000 hp.

(a) Assuming ideal operation for the purpose of estimation, and taking the energy extraction to be that specified in equations (7.2.17, 22), find the quality of steam required for (1) a single-stage turbine, and (2) a two-stage turbine.

(b) Ideally, what power is available for electricity generation in cases (1) and (2)?

7.4 Use equation (7.3.7) to find the work done in the turbine discussed in Example 7.2. Take the values required from the results of the calculations carried out there.

7.5 Figure 7.10 suggests that the machine best suited for operation both as a pump and as a turbine without a change in the directions of flow and rotation will be such that $w_e = 0$ at the design point. Show that such a machine will have flat blades.

7.6 Show that the axial component of force on an impulse (constant-pressure) row is zero if the flow is lossless and two-dimensional. In what direction does the force act if there are losses?

7.7 Find the lift and drag forces acting on a blade of the cascade discussed in Example 7.3.

7.8 Accepting the energy-addition limit (7.3.27), estimate the number of axial-flow stages required for the compression process discussed in Example 6.7.

7.9 (a) For a normal turbine stage, show that

$$R_h = 1 - \frac{\tan\beta_3}{2\tan\alpha_2}$$

(b) What is the degree of reaction for the turbine of Example 7.4?

7.10 Carry out the calculations of Example 7.5 on the assumption that the density is constant at the inlet value. Find V_2, w_e and p_2. Compare the values with those obtained on the assumption of perfect-gas behaviour.

7.11 Consider the inlet and outlet flows of Example 7.5.

(a) Find the ratios of densities and of axial velocity components across the row.

(b) Draw the inlet and outlet velocity triangles superposed, that is, with a common side representing the rotor speed.

7.12 Construct an argument based on dimensional analysis to obtain the basic form of equation (7.4.16) without detailed calculations.

7.13 (a) Plot the characteristic (H in of water gauge against $\dot{V}$ ft^3/min) of the centrifugal blower analysed in Example 7.7.

(b) Using the curve of (a), obtain the characteristic for a rotor speed of 4000 rev/min.

7.14 A one-tenth-scale model of the turbine installation of Examples 7.4 and 7.6 is to be made. The water level will be scaled as well, in order that the inlet conditions be modelled correctly. It is required to estimate the flow through the model and the power the model turbine will generate at the design condition.

(a) Show that the requirement that the head coefficient be unchanged implies that the rotor speed must be increased by the factor $10^{1/2}$.

(b) Show that the flow will be decreased by the factor $10^{5/2}$, and the power by the factor $10^{7/2}$.

(c) Find the flow rate and power at the design point.

Group B

7.15 Consider the simple impulse machine discussed in Section 7.1.2. Show that the maximum power is extracted from the jet when $U = C_1/3$. What is the corresponding total-to-static efficiency? Why do these results differ from those for a Pelton wheel operating ideally?

7.16 In Example 7.1 the frictional dissipation in the flow from a reservoir to a Pelton wheel was considered, but no account was taken of the losses in the nozzles directing the fluid onto the turbine wheel. Taking the discharge coefficient of the nozzle to be 98·5 per cent, find the proportion of the ideal kinetic energy of the jet which is dissipated in acceleration from the pipe of diameter 1·1 m to the three jets of diameter 0·20 m.

7.17 Repeat the calculations of Problem 7.3, considering the same pair of energy extractions, but now taking

(1) isentropic efficiency of 60 per cent for the turbine,

(2) efficiency of transmission of turbine power to the machinery to be 80 per cent, and

(3) efficiency of conversion of turbine power to electrical power to be 70 per cent.

7.18 The requirement that the tensile stress in a blade fixed to a rotor not exceed the allowable value sometimes dictates the length of blade which can be used.

(a) For a blade of uniform section, extending between radii r_1 and r_2, show that the stress at the root is given by

$$\sigma = \tfrac{1}{2}\varrho\omega^2(r_2^2 - r_1^2)$$

(b) Show that the pitchline speed $U = \bar{r}\omega$ is limited to

$$U^2 = \frac{\sigma}{\varrho} \cdot \frac{\bar{r}}{r_2 - r_1}$$

(c) Use this result to estimate the root stress for the blades of Examples 7.4 and 7.5, taking the blades to be made of steel.

7.19 Show that equations (7.1.8) and (7.3.5) are equivalent, by manipulating the relationships between relative and absolute velocities.

7.20 (a) From one of equations (7.3.13) it is seen that

$$-\Delta p_0 = \frac{D}{\sin\beta_m}$$

Explain why the losses depend in this way on the angle of the blades.

(b) Explain also the way in which the diffuser efficiency (7.3.15) depends on β_m.

7.21 In order that the flow through a closed-circuit wind tunnel pass around the corners of the duct with small losses and with a fairly uniform velocity distribution, a row of turning vanes is fitted across each of the right-angled bends in the circuit. These vanes turn the flow through 90° without changing the mean velocity; thus they act as impulse blades. For a particular arrangement of blades with spacing/chord ratio of $s/l = 1$, the loss across the cascade is given by the profile loss co-efficient

$$\frac{p_{0_1} - p_{0_2}}{p_{0_2} - p_2} = 0{\cdot}07$$

(a) What are the loss coefficients defined in equations (5.5.14) and (7.3.10)? What is the comparable loss coefficient for a sharp corner without turning vanes?

(b) What is the drag coefficient of the blades?

(c) If such blades are used to turn an airflow in a duct of diameter 2 m, the flow speed being 100 m/sec, what are the forces on the whole cascade in directions normal to and along it? The air may be assumed to have standard sea-level properties.

(d) What are the lift and drag on the cascade of (c)?

7.22 In Examples 7.4 and 7.6 the flow on the pitch circle of a propeller turbine was considered. Now the conditions along the blades will be examined. It will be assumed that (1) C_a = constant across the duct, as well as across the rotor, (2) rC_{t_2} = (pitchline value) gives the variation of the swirl components before and after the rotor, and (3) $C_{t_3} = 0$ across the duct.

(a) What is the angle α_2 at blade root and tip?

(b) What is the angle β_3 at root and tip? Sketch a blade shape consistent with these values.

(c) What are the pressures just above and just below the rotor at blade root and tip?

(d) Show that the same amount of energy is extracted, no matter what the radius at which the fluid passes through the rotor.

7.23 Repeat the calculations of Example 7.5, taking the same data, but reversing the direction of rotor motion. Draw the velocity triangles, superposed, for this case.

7.24 Show that the limit (7.3.26) is equivalent to a limit on the proportional change in total temperature of

$$\frac{T_{0_2}-T_{0_1}}{T_{0_1}} \simeq \frac{1}{4}(\gamma-1)$$

7.25 In Example 7.5 the relationship $A_1/A_2 = \sin\beta_1/\sin\beta_2$ was obtained by purely geometric arguments. Show that it also follows from a combination of the continuity relations for

(1) the passage between a pair of blades, and

(2) a control volume with sides normal to the line of the cascade.

7.26 (a) Obtain from equation (7.4.10) the Bernoulli equation relating pressure and velocity changes in a perfect gas.

(b) In a centrifugal machine, the fluid moves through a flow passage such that $V = k/r$. Find an expression for the pressure variation using the constant-density result (7.4.11).

(c) Use the result of (b) to find the pressure variation within the rotor of the centrifugal blower of Example 7.7. Take $V_t/V_r = -\cot\beta_2 = -1$ throughout, so that

$$V = \sqrt{2}V_r = \frac{\dot{V}}{\sqrt{2}\,\pi rb}$$

7.27 (a) Use equations (7.4.15, 16) to find an inlet geometry appropriate to the blower specified in Example 7.7 when operating at maximum efficiency.

(b) Find the velocities in the eye of the intake and at inlet to the blades.

(c) How would you reconcile the geometry of (a) with the specified rotor depth, $b_2 = b_1 = 1$ in?

(d) Sketch a suitable intake for the blower.

7.28 Consider again the one-tenth-scale turbine model discussed in Problem 7.14.

(a) Show that the angle at which the flow approaches the rotor will be unchanged when the suggested scaling is used.

(b) Use the results of Example 7.4 to find the pressures just above and just below the rotor.

(c) What additional requirement must be satisfied if the cavitation performance of the installation is to be assessed in the experiments?

7.29 In Problem 6.13 an axial-flow compressor absorbing 2000 hp was proposed to pump natural gas of density 0·01 sl/ft^3 at the rate of 400 ft^3/sec along a line of diameter 36 in. Experience with such machines suggests the use of stages characterized by specific speed and diameter of $\Omega = 3$ and $\Delta = 1{\cdot}5$. The number of stages has still to be decided.

(a) Use the criterion (7.3.27) to find the minimum number of stages required.

(b) Find the rotational speed and diameter for one, two and four stages.

(c) Find the corresponding rotor speeds to see whether the centripetal forces are likely to be unrealistically high.

(d) What number of stages seems most suitable? What are the flow and head coefficients for this machine?

7.30 (a) For a propeller, show that the thrust coefficient

$$K_T = \frac{F}{\varrho_c N^2 D^2} = f(J, \text{Eu})$$

where $J = U/ND$ is the advance ratio of the screw, and Eu is the Euler number.

(b) For a model of diameter 1 ft, the 'shaft efficiency' and thrust coefficient are found to be 70 per cent and 0·16 when the forward speed is 25 ft/sec and the speed of rotation is 250 rev/min. For a forward speed of 30 kts, what is the rotational speed of a full-scale propeller of diameter 5 ft at which the advance ratio is reproduced? What are the corresponding thrust, shaft horsepower and torque?

(c) If the shaft of the prototype propeller is located 7 ft below the free surface, what pressure should be maintained at the centre-line of the test channel where the model shaft is located?

7.31 (a) Estimate the maximum speed at which the pump of Example 6.1 can operate without cavitating when it is located 20 feet above the water surface. Use the criterion (7.6.8).

(b) The specific speed of this pump is 0·5; what is the rotor diameter?

(c) What would be the answers to (a) and (b) if the pump were located 28 ft and 5 ft above the surface?

(d) What would be the answers to (a) and (b) for a flow of 1000 ft^3/hr, with the pump 20 ft above the water surface?

7.32 How is the Euler cavitation number related to the suction specific speed of a pump?

Group C

7.33 Consider the operation of a Pelton wheel under the condition $U = \frac{1}{2}C_1$ for maximum efficiency.

(a) Show that the energy dissipated in the flow over the rotor is $\frac{1}{4}(1-c^2)$ times the energy input, where c is the blade factor.

(b) Show that the fraction of the energy input lost in the tailrace is $\frac{1}{4}(1+c^2-2c\cos\beta)$.

(c) Find the total energy loss, and hence find the optimum total-to-static efficiency.

(d) Evaluate the two contributions to the loss for $\beta = 0°$ and $20°$, and for $c = 1{\cdot}0$ and $0{\cdot}8$; this will give rise to four cases.

7.34 Express the total-to-total efficiency of a Pelton wheel in terms of the nozzle efficiency η_N and the blade factor c.

7.35 Find the Carnot efficiency for the cycle of Example 7.2. Find also a Carnot-type efficiency based on the temperature at which most of the heat is accepted by the water, namely, the temperature of boiling at the reservoir pressure 60 lbf/in^2.

(a) Explain why the cycle efficiency found in Example 7.2 is so much lower than the Carnot efficiency.

(b) Discuss the utility of the Carnot efficiency as a measure of the potential utility of a steam plant.

(c) Discuss the feasibility of the various ways of trying to improve upon the efficiency of the Laval-turbine installation: exhausting to and condensing at a sub-atmospheric pressure, increasing the reservoir pressure, and increasing the temperature to which the steam is superheated. The operation discussed in Problems 7.3 and 7.17 is relevant to the present discussion.

7.36 Prepare a table similar to that of equations (7.3.17) for a pump handling a constant-density fluid.

7.37 In order to get a general idea of the effect of separation losses on the performance of a turbomachine, we may take the losses in the impeller to be proportional to $(\varphi-\varphi_d)^2$, where φ_d is the flow coefficient at the design condition, at which the flow passes smoothly onto the blades. Apply this method to the centrifugal blower discussed in Example 7.7, taking the losses in the impeller to be

$$20(\varphi-0{\cdot}12)^2$$

Since $\psi_{0_2} \neq w_e/U_2^2$ in this case, equations (7.4.21, 23, 24, 28, 30, 31) are altered.

(a) Obtain the formulae corresponding to equations (7.4.25, 26).

(b) Repeat the calculations of Table 7.1, now taking account of losses. Note that the first three columns all require the same adjustment. The column headed ψ_{0_2} in Table 7.1 now becomes w_e/U_2^2.

(c) Find the head, flow and power for (1) optimum efficiency, (2) no flow and (3) no net pressure rise.

7.38 A second simplification in Example 7.7 was the assumption of a constant direction for the relative outlet flow. In reality, the rotor does not impose as large an absolute whirl component as the ideal value, which is $C_t' = U - V_r \cot\beta'$. Here the subscripts 2 have been dropped; β' is the blade tip angle.

(a) One simple way of prescribing the actual whirl velocity is through a vane efficiency $\eta_v = C_t/C_t'$. A value around 0·7 is representative of typical rotor designs. Show that the actual flow angle is given by

$$\cot\beta = \frac{1-\eta_v}{\varphi} + \eta_v \cot\beta'$$

with $\varphi = V_r/U$, and that

$$w_e = \eta_v(U^2 - UV_r \cot\beta')$$

for swirl-free inlet flow.

(b) For a blade tip angle of 45°, and taking $\eta_v = 0{\cdot}7$, plot the flow angle β against the flow coefficient φ over the range of operation of the blower of Example 7.7.

(c) A second way of representing the departure from the ideal behaviour is the introduction of a factor B such that

$$w_e = BU^2 - UV_r \cot \beta'$$

for a pump with swirl-free inlet flow. Show that

$$\cot \beta = \frac{1-B}{\varphi} + \cot \beta'$$

It is possible to predict the factor B theoretically; $B \simeq 0{\cdot}7$ for a five-bladed pump with $\beta' = 45°$.

(d) For the blower considered in Example 7.7, show that the two methods of representing deviations from the ideal deflexion are nearly equivalent.

7.39 Figure 7.25 shows a simple form of turbine often used in lawn sprinklers. For simplicity, the jet may be taken to emerge perpendicular to both the central post and the horizontal tube. In this analysis all losses in the flow through the device will be neglected.

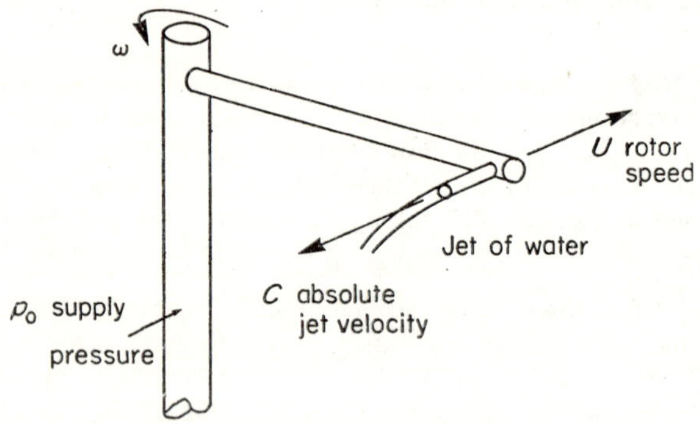

Figure 7.25 See Problem 7.39

(a) If the sprinkler does not rotate, show that the torque it applies to the central post is $2(p_0 - p_a)AR$, with A the area of the jet and R the radius at which the outlet is located.

(b) Show that, when the sprinkler rotates, the absolute velocity and rotor speed are related by

$$\frac{(C+U)^2}{2} - \frac{U^2}{2} = \frac{p_0 - p_a}{\varrho_c}$$

and that the work done on the rotor is given by $\varrho_c A(C+U)CU$.

(c) Discuss the efficiency of the device as (1) a turbine and (2) a water sprinkler.

Bibliography

Some books which look in more detail at matters discussed in the preceding chapters are listed below. For the most part these are not specialist works, but are textbooks or general introductions.

Thermodynamics

J. K. Roberts and A. R. Miller. *Heat and Thermodynamics*, Blackie and Son, London and Glasgow (1957)

H. C. Van Ness. *Understanding Thermodynamics*, McGraw-Hill, New York (1969)

M. W. Zemansky and H. C. Van Ness. *Basic Engineering Thermodynamics*, McGraw-Hill, New York (1966)

W. C. Reynolds. *Thermodynamics*, McGraw-Hill, New York (1965)

Fluid Mechanics

R. H. Sabersky and A. J. Acosta. *Fluid Flow*, Macmillan, New York (1964)

L. Prandtl. *The Essentials of Fluid Dynamics*, Blackie and Son, London and Glasgow (1952)

A. H. Shapiro. *Shape and Flow*, Anchor Books, Doubleday and Co., Garden City, New York (1961)

W. Kaufmann. *Fluid Mechanics*, McGraw-Hill, New York (1963)

Combined Treatments

G. F. C. Rogers and Y. R. Mayhew. *Engineering Thermodynamics: Work and Heat Transfer*, Longmans Green, London (1957)

N. H. Hall. *Thermodynamics of Fluid Flow*, Prentice-Hall, New York (1957)

J. M. Kay. *An Introduction to Fluid Mechanics and Heat Transfer*, Cambridge University Press, London (1963)

Gas and Aerodynamics

H. W. Liepmann and A. Roshko. *Elements of Gasdynamics*, Wiley, New York (1962)

A. H. Shapiro. *The Dynamics and Thermodynamics of Compressible Fluid Flow* (Two Volumes), Ronald Press, New York (1953)

T. von Karman. *Aerodynamics*, Cornell University Press, Ithaca, New York (1954)

A. M. Kuethe and J. D. Schetzer. *Foundations of Aerodynamics*, Wiley, New York (1950)

Heat Transfer

M. Fishenden and O. A. Saunders. *Heat Transfer*, Oxford University Press, London (1950)

W. H. Macadams. *Heat Transmission*, McGraw-Hill, New York (1954)

Flow Machines

S. L. Dixon. *Fluid Mechanics, Thermodynamics of Turbomachinery*, Pergamon Press, Oxford and London (1966)

A. Betz. *Introduction to the Theory of Flow Machines*, Pergamon Press, Oxford and London (1966)

G. T. Csanady. *Theory of Turbomachines*, McGraw-Hill, New York (1964)

H. Cohen and G. F. C. Rogers. *Gas Turbine Theory*, Longmans, Green, London (1951)

Special Topics

R. S. Scorer. *Natural Aerodynamics*, Pergamon Press, Oxford and London (1958)

O. G. Sutton. *Atmospheric Turbulence*, Methuen, London (1955)

O. G. Sutton. *Understanding Weather*, Penguin Books, Harmondsworth, Middlesex (1960)

N. Feather. *Vibrations and Waves*, Penguin Books, Harmondsworth, Middlesex (1964)

H. M. Morris. *Applied Hydraulics in Engineering*, Ronald Press, New York (1963)

G. H. Dury. *The Face of the Earth*, Penguin Books, Harmondsworth, Middlesex (1959)

A. B. Goodwin. *Power Hydraulics*, Cleaver-Hume Press, London (1963)

P. Bradshaw. *Experimental Fluid Mechanics*, Pergamon Press, Oxford and London (1964)

R. A. R. Tricker. *Bores, Breakers, Waves and Wakes*, Mills and Boon, London (1965)

J. Jeans. *An Introduction to the Kinetic Theory of Gases*, Cambridge University Press, London (1952)

Appendices

These tables are designed for use with the examples and problems of this book. They also serve to indicate the general nature of the data required for engineering calculations and the form in which it is presented. A variety of units has been used in these tables, since it is necessary to develop the ability to make use of data in whatever form they happen to be presented.

It may prove convenient to have a more extensive body of physical data available in a booklet separate from this book. A number of compact compilations of data are available, among which may be mentioned those of Mayhew and Rogers, published by Blackwell of Oxford: *Thermodynamic and Transport Properties of Fluids (SI Units)* and *Thermodynamic Properties of Fluids and Other Data (British Thermal Units)*.

APPENDIX I. UNITS AND CONVERSIONS

Mass

1 kg = 1 N sec^2/m = 2·205 lb = 0·0685 sl

1 sl = 1 lbf sec^2/ft = 32·174 lb

Length

1 m = 3·281 ft = 39·37 in

1 mile = 5280 ft = 1·609 km

1 nautical mile = 6076 ft

Volume

1 gal (British) = 277 in^3 = 4·55 litre
1 gal (American) = 231 in^3 = 3·79 litre
1 m^3 = 1000 litre

Density

1 kg/m^3 = 0·0624 lb/ft^3

Force

1 N = 1 kg m/sec^2 = 10^5 dyne = 0·2248 lbf
1 kgf = 9·807 N
1 metric ton = 1000 kgf
1 ton (British or long) = 2240 lbf
1 ton (American or short) = 2000 lbf

Pressure

1 bar = 10^5 N/m^2 = 10^6 dyne/cm^2 = 14·50 lbf/in^2
1 atm = 1·0133 bar = 14·70 lbf/in^2
= 760 mm mercury = 29·92 in mercury = 34·0 ft water

Energy

1 J = 1 N m = 10^7 erg
1 kJ = 0·9478 Btu = 737·6 ft lbf
1 kcal = 1000 cal = 4·187 kJ
1 Btu = 778·2 ft lbf = 0·252 kcal

Power

1 kW = 1000 N m/sec = 737·6 ft lbf/sec = 3413 Btu/hr
1 hp = 550 ft lbf/sec = 0·7457 kW = 2545 Btu/hr
1 metric hp = 0·736 kW

Temperature

1 °K = 1 °C = 1·8 °R = 1·8 °F
0 °C is equivalent to 32 °F, 273·15 °K and 491·68 °R

Specific Energy

1 kJ/kg = 0·4299 Btu/lb

Specific Heat

1 kJ/kg °K = 0·2388 Btu/lb °R

Viscosity

1 N sec/m² = 1 kg/m sec = 10 poise or dyne sec/cm²
= 0·0209 lbf sec/ft²

1 m²/sec = 10⁴ stoke or cm²/sec = 10·76 ft²/sec

Conductivity

1 kW/m °K = 577·8 Btu/ft hr °R

1 kW/m² °K = 176·1 Btu/ft² hr °R

Indication of multiples and submultiples of SI units

Factor	Prefix	Symbol	Example
10^6	mega	M	MW — megawatt
10^3	kilo	k	km — kilometre
10^{-1}	deci	d	dm — decimetre
10^{-2}	centi	c	cl — centilitre
10^{-3}	milli	m	mbar — millibar
10^{-6}	micro	μ	μm — micrometre or micron

APPENDIX II. PROPERTIES OF GASES

Universal gas constant, **R** = 8·314 kJ/kg mol °K
= 1545 ft lbf/lb mol °R

For air over a wide range of pressure and temperature,

R = 0·287 kJ/kg °K = 53·35 ft lbf/lb °R
= 1716 ft lbf/sl °R

Table 1. Characteristics of some gases

	MW	γ	a atm $\left(\dfrac{ft^3}{lb\ mol}\right)^2$	b ft^3/lb mol	% by weight in dry air
Air	28·96	1·40	344	0·585	100
Oxygen O_2	32·00	1·40	350	0·510	23·1
Nitrogen N_2	28·02	1·40	346	0·618	75·6
Argon A	39·95	1·67	346	0·517	1·3
Helium He	4·002	1·66	8·57	0·372	—
Hydrogen H_2	2·016	1·41	62·8	0·426	—
Water H_2O	18·02	1·30	1400	0·488	
Carbon dioxide CO_2	44·00	1·30	925	0·686	0·05
Carbon monoxide CO	28·00	1·40	381	0·639	—
Methane CH_4	16·03	1·32	581	0·686	—
Propane C_3H_8	44·09	1·12	2374	1·446	—

Note: The values of the specific heat ratio γ apply at normal temperatures.

Table 2. Thermal properties of air at low pressures

T °K	c_p kJ/kg °K	c_v kJ/kg °K	γ
200	1·003	0·715	1·401
300	1·005	0·718	1·400
400	1·014	0·726	1·395
500	1·030	0·742	1·387
600	1·051	0·764	1·376
1000	1·141	0·854	1·336
1500	1·211	0·924	1·311
2000	1·251	0·963	1·298
3000	1·295	1·008	1·285

APPENDIX III. PROPERTIES OF LIQUIDS AND SOLIDS

In these tables

- T is temperature in °C or °K as indicated
- v is specific volume in cm^3/gm mol
- β is the isobaric compressibility in $°K^{-1}$
- $\varkappa$ is the isothermal compressibility in cm^2/dyne
- c_p and c_v are the specific heats at constant pressure and constant volume in cal/gm mol °K
- γ is the ratio of specific heats

Table 1. Water at 1 atmosphere

Here T is temperature in °C
a is the speed of sound in m/sec
ϱ is the density in gm/cm³

T	ϱ	$\beta \times 10^6$	$\varkappa \times 10^{12}$	c_p	γ	a
0	1·000	− 67	50·8	4·22	1·00	1404
10	1·000	89	48·1	4·19	1·00	1448
20	0·998	208	45·8	4·18	1·01	1483
30	0·996	304	44·6	4·18	1·02	1510
40	0·992	390	43·8	4·18	1·03	1530
50	0·988	465	43·7	4·18	1·04	1544
60	0·983	522	43·7	4·18	1·05	1552
70	0·978	586	44·1	4·19	1·07	1555
80	0·972	643	44·8	4·20	1·08	1555

Table 2. Mercury at 0 °Centigrade

Here p is pressure in atm

p	v	$\beta \times 10^6$	$\varkappa \times 10^{12}$	c_p	c_v	γ
1	14·72	181	3·88	6·69	5·88	1·14
1,000	14·67	174	3·79	6·69	5·92	1·13
2,000	14·62	168	3·69	6·69	5·96	1·12
3,000	14·57	164	3·60	6·68	5·97	1·12
4,000	14·51	160	3·48	6·68	5·98	1·12
5,000	14·45	158	3·38	6·68	5·98	1·12
6,000	14·42	155	3·25	6·68	5·99	1·11
7,000	14·38	152	3·12	6·68	5·99	1·11

Table 3. Copper at 1 atmosphere

Here T is temperature in °K. Values in parentheses are extrapolated

T	v	$\beta \times 10^6$	$\varkappa \times 10^{12}$	c_p	c_v	γ
0	(7·0)	(0)	(0·710)	(0)	0	1·00
50	(7·002)	(11·5)	(0·712)	1·38	1·38	1·00
100	(7·008)	31·5	0·721	3·88	3·86	1·00
150	(7·018)	41·0	0·733	5·01	4·95	1·01
200	(7·029)	45·6	0·748	5·41	5·32	1·02
250	7·043	48·0	0·762	5·65	5·52	1·02
300	7·062	49·2	0·776	5·87	5·71	1·03
500	7·115	54·2	0·837	6·25	5·95	1·05
800	7·256	60·7	0·922	6·70	6·14	1·09
1200	7·452	69·7	1·030	7·34	6·34	1·16

Tables 2 and 3 have been extracted from *Basic Engineering Thermodynamics*, by M. W. Zemansky and H. C. Van Ness, published by the McGraw-Hill Book Co., New York (1966), with the permission of the authors and publishers.

APPENDIX IV. STEAM TABLES: PROPERTIES OF LIQUID WATER AND WATER VAPOUR

Table 1. Saturated vapour and water (SI units)

In this table
T is temperature in °C
p is pressure in bar
T_s, p_s are corresponding temperature and pressure
v_f and v_g are the specific volumes of liquid and vapour phases in m^3/kg
h_f and h_g are the specific enthalpies of liquid and vapour phases in kJ/kg
s_f and s_g are the specific entropies of liquid and vapour phases in kJ/kg °K

T	p_s	100 v_f	v_g	h_f	h_g	s_f	s_g
0·01	0·006112	0·1000	206·1	0	2500·8	0	9·155
1	0·006566	0·1000	192·6	4·2	2502·5	0·015	9·128
2	0·007054	0·1000	179·9	8·4	2504·3	0·031	9·102
4	0·008129	0·1000	157·3	16·8	2508·1	0·061	9·050
6	0·009346	0·1000	137·8	25·2	2511·8	0·091	8·999
8	0·01072	0·1000	121·0	33·6	2515·5	0·121	8·949
10	0·01227	0·1000	106·4	42·0	2519·2	0·151	8·900
12	0·01401	0·1001	93·83	50·4	2522·9	0·180	8·851
14	0·01597	0·1001	82·89	58·8	2526·6	0·210	8·804
16	0·01817	0·1001	73·38	67·1	2530·2	0·239	8·757
18	0·02063	0·1002	65·08	75·5	2533·9	0·268	8·712
20	0·02337	0·1002	57·84	83·9	2537·6	0·296	8·666
22	0·02642	0·1002	51·49	92·2	2541·2	0·325	8·622
24	0·02982	0·1003	45·92	100·6	2544·8	0·353	8·579
26	0·03360	0·1003	41·03	108·9	2548·4	0·381	8·536
28	0·03778	0·1004	36·73	117·3	2552·1	0·409	8·494
30	0·04242	0·1004	32·93	125·7	2555·7	0·436	8·452
32	0·04754	0·1005	29·57	134·0	2559·3	0·464	8·412
34	0·05318	0·1006	26·60	142·4	2562·9	0·491	8·372
36	0·05940	0·1006	23·97	150·7	2566·5	0·518	8·332
38	0·06624	0·1007	21·63	159·1	2570·1	0·545	8·294
40	0·07375	0·1008	19·55	167·5	2573·7	0·572	8·256
42	0·08198	0·1008	17·69	175·8	2577·2	0·599	8·219
44	0·09100	0·1009	16·03	184·2	2580·8	0·625	8·182
46	0·1009	0·1010	14·56	192·5	2584·3	0·651	8·145

continued

T	p_s	$100\ v_f$	v_g	h_f	h_g	s_f	s_g
48	0·1116	0·1011	13·23	200·9	2587·9	0·678	8·111
50	0·1233	0·1012	12·04	209·3	2591·4	0·704	8·075
55	0·1574	0·1015	9·578	230·2	2600·3	0·768	7·991
60	0·1992	0·1017	7·678	251·1	2609·0	0·831	7·909
65	0·2501	0·1020	6·201	272·0	2617·7	0·893	7·830
70	0·3116	0·1023	5·045	293·0	2626·3	0·955	7·755
75	0·3855	0·1026	4·133	313·9	2634·7	1·015	7·681
80	0·4736	0·1029	3·408	334·9	2643·2	1·075	7·611
85	0·5780	0·1032	2·828	355·9	2651·5	1·134	7·544
90	0·7011	0·1036	2·361	376·9	2659·7	1·192	7·478
95	0·8453	0·1040	1·982	398·0	2667·8	1·250	7·416
100	1·01325	0·1044	1·673	419·1	2675·8	1·307	7·355

p	T_s	v_g	h_f	h_g	s_f	s_g
0·006112	0·01	206·1	0	2501	0	9·155
0·01	7·0	129·2	29	2514	0·106	8·974
0·05	32·9	28·20	138	2561	0·476	8·394
0·10	45·8	14·67	192	2584	0·649	8·149
0·50	81·3	3·239	340	2645	1·091	7·593
0·75	91·8	2·217	384	2662	1·213	7·456
1·00	99·6	1·694	417	2675	1·303	7·359
1·50	111·4	1·159	467	2693	1·434	7·223
2·00	120·2	0·8856	505	2707	1·530	7·127
3·00	133·5	0·6057	561	2725	1·672	6·993
4·00	143·6	0·4623	605	2739	1·776	6·897
5·00	151·8	0·3748	640	2749	1·860	6·822
6·00	158·8	0·3156	670	2757	1·931	6·761
7·00	165·0	0·2728	697	2764	1·992	6·709
8·00	170·4	0·2403	721	2769	2·046	6·663
9·00	175·4	0·2149	743	2774	2·094	6·623
10	179·9	0·1944	763	2778	2·138	6·586
15	198·3	0·1317	845	2792	2·315	6·445
20	212·4	0·09957	909	2799	2·447	6·340
30	233·8	0·06665	1008	2803	2·645	6·186
40	250·3	0·04977	1087	2801	2·797	6·070
50	263·9	0·03944	1155	2794	2·921	5·973
60	275·6	0·03244	1214	2784	3·027	5·890
70	285·8	0·02737	1267	2772	3·122	5·814
80	295·0	0·02352	1317	2758	3·207	5·744
90	303·3	0·02048	1364	2743	3·286	5·679

continued

p	T_s	v_g	h_f	h_g	s_f	s_g
100	311·0	0·01802	1408	2725	3·360	5·615
110	318·0	0·01598	1450	2705	3·430	5·553
120	324·6	0·01426	1491	2685	3·496	5·493
130	330·8	0·01278	1531	2662	3·561	5·433
140	336·6	0·01149	1571	2638	3·623	5·373
150	342·1	0·01035	1610	2611	3·685	5·312
160	347·3	0·00932	1650	2582	3·746	5·248
170	352·3	0·00838	1690	2548	3·808	5·181
180	357·0	0·00751	1732	2510	3·872	5·108
190	361·4	0·00668	1777	2466	3·941	5·027
200	365·7	0·00585	1827	2411	4·014	4·928
210	369·8	0·00498	1889	2336	4·108	4·803
220	373·7	0·00368	2008	2178	4·289	4·552
221·2	374·15	0·00317	2084	2084	4·430	4·430

Table 2. Superheated steam (SI units)

In this table

T is temperature in °C
p is pressure in bar
v is specific volume in m^3/kg
v' is specific volume in $dm^3/kg = 10^{-3} \times m^3/kg$
h is specific enthalpy in kJ/kg
s is specific entropy in kJ/kg °K

p	T	50	100	150	200	250	300	400	500
0·006112	v	243·9	281·7	319·5	357·3	395·0	432·8	508·3	583·8
	h	2595	2689	2784	2880	2978	3077	3280	3489
	s	9·468	9·739	9·978	10·19	10·39	10·57	10·90	11·19
0·01	v	149·1	172·2	195·3	218·4	241·4	264·5	310·7	356·8
	h	2595	2689	2784	2880	2978	3077	3280	3489
	s	9·241	9·512	9·751	9·966	10·16	10·34	10·67	10·96
0·05	v	29·78	34·42	39·04	43·66	48·28	52·90	62·13	71·36
	h	2594	2688	2784	2880	2978	3077	3280	3489
	s	8·496	8·768	9·008	9·223	9·420	9·601	9·927	10·22
0·10	v	14·87	17·20	19·51	21·83	24·14	26·45	31·06	35·68
	h	2592	2688	2783	2880	2977	3077	3280	3489
	s	8·173	8·447	8·688	8·903	9·100	9·281	9·607	9·897

continued

p	*T*	50	100	150	200	250	300	400	500
0·50	*v*		3·420	3·890	4·356	4·821	5·284	6·209	7·134
	h		2683	2780	2878	2976	3076	3279	3489
	s		7·694	7·940	8·158	8·355	8·537	8·864	9·154
0·75	*v*		2·271	2·588	2·901	3·211	3·521	4·138	4·755
	h		2680	2779	2877	2975	3075	3279	3489
	s		7·500	7·750	7·969	8·167	8·349	8·676	8·967
1·0	*v*		1·696	1·937	2·173	2·406	2·639	3·103	3·565
	h		2676	2777	2876	2975	3075	3278	3488
	s		7·360	7.614	7·834	8·033	8·215	8·543	8·834
1·01325	*v*			1·912	2·145	2·375	2·604	3·062	3·519
	h			2777	2876	2975	3075	3278	3488
	s			7·608	7·828	8·027	8·209	8·537	8·828
1·5	*v*			1·286	1·445	1·601	1·757	2·067	2·376
	h			2773	2873	2973	3073	3277	3488
	s			7·420	7·643	7·843	8·027	8·355	8·646
2	*v*			·9602	1·081	1·199	1·316	1·549	1·781
	h			2770	2871	2971	3072	3277	3487
	s			7·280	7·507	7·708	7·892	8·221	8·513
3	*v*			·6342	·7166	·7965	·8754	1·031	1·187
	h			2762	2866	2968	3070	3275	3486
	s			7·078	7·312	7·517	7·702	8·032	8·324
4	*v*			·4710	·5345	·5953	·6549	·7725	·8893
	h			2753	2862	2965	3067	3274	3485
	s			6·929	7·172	7·379	7·566	7·898	8·191

p	*T*	200	250	300	350	400	450	500	600
5	*v*	·4252	·4745	·5226	·5701	·6172	·6641	·7108	·8040
	h	2857	2962	3065	3168	3272	3377	3484	3702
	s	7·060	7·271	7·460	7·633	7·793	7·944	8·087	8·351
6	*v*	·3522	·3940	·4344	·4743	·5136	·5528	·5919	·6697
	h	2851	2958	3062	3166	3270	3376	3483	3701
	s	6·968	7·182	7·373	7·546	7·707	7·858	8·001	8·267
7	*v*	·3001	·3364	·3714	·4058	·4397	·4734	·5069	·5737
	h	2846	2955	3060	3164	3269	3374	3482	3700
	s	6·888	7·106	7·298	7·473	7·634	7·786	7·929	8·195

continued

p	T	200	250	300	350	400	450	500	600
8	v	·2610	·2933	·3242	·3544	·3842	·4138	·4432	·5018
	h	2840	2951	3057	3162	3267	3373	3481	3699
	s	6·817	7·040	7·233	7·409	7·571	7·723	7·866	8·132
9	v	·2305	·2597	·2874	·3144	·3410	·3674	·3937	·4458
	h	2835	2948	3055	3160	3266	3372	3480	3699
	s	6·753	6·980	7·176	7·352	7·515	7·667	7·811	8·077
10	v	·2061	·2328	·2580	·2825	·3065	·3303	·3540	·4010
	h	2829	2944	3052	3158	3264	3370	3478	3698
	s	6·695	6·926	7·124	7·301	7·464	7·617	7·761	8·028
15	v	·1324	·1520	·1697	·1865	·2029	·2191	·2351	·2667
	h	2796	2925	3039	3148	3256	3364	3473	3694
	s	6·452	6·711	6·919	7·102	7·268	7·423	7·569	7·838
20	v		·1115	·1255	·1386	·1511	·1634	·1756	·1995
	h		2904	3025	3138	3248	3357	3467	3690
	s		6·547	6·768	6·957	7·126	7·283	7·431	7·701
30	v		·0706	·0812	·0905	·0993	·1078	·1161	·1324
	h		2858	2995	3117	3231	3343	3456	3682
	s		6·289	6·541	6·744	6·921	7·082	7·233	7·507
40	v			·0588	·0664	·0733	·0800	·0864	·0988
	h			2963	3094	3214	3330	3445	3674
	s			6·364	6·584	6·769	6·935	7·089	7·368
50	v			·0453	·0519	·0578	·0632	·0685	·0786
	h			2927	3070	3196	3316	3433	3666
	s			6·212	6·451	6·646	6·818	6·975	7·258
60	v			·0362	·0422	·0473	·0521	·0566	·0652
	h			2887	3045	3177	3301	3421	3657
	s			6·071	6·336	6·541	6·719	6·879	7·166
70	v			·0295	·0352	·0399	·0441	.0481	·0556
	h			2841	3018	3158	3287	3410	3649
	s			5·934	6·231	6·448	6·632	6·796	7·088
80	v			·0242	·0299	·0343	·0381	·0417	·0484
	h			2787	2990	3139	3272	3398	3641
	s			5·793	6·133	6·364	6·555	6·723	7·019
90	v				·0258	·0299	·0335	·0367	·0428
	h				2959	3118	3256	3385	3633
	s				6·039	6·286	6·484	6·657	6·958

continued

p	*T*	350	375	400	425	450	500	600
100	*v′*	22·41	24·53	26·39	28·12	29·72	32·75	38·31
	h	2926	3017	3097	3172	3241	3373	3624
	s	5·947	6·091	6·213	6·321	6·419	6·596	6·902
110	*v′*	19·60	21·69	23·50	25·14	26·66	29·49	34·65
	h	2889	2989	3075	3153	3225	3360	3616
	s	5·856	6·014	6·143	6·257	6·358	6·539	6·850
120	*v′*	17·19	19·31	21·07	22·65	24·10	26·77	31·59
	h	2849	2960	3052	3134	3209	3348	3607
	s	5·762	5·937	6·076	6·195	6·301	6·487	6·802
130	*v′*	15·09	17·26	19·01	20·53	21·93	24·47	29·01
	h	2804	2929	3028	3114	3192	3335	3599
	s	5·664	5·862	6·011	6·136	6·246	6·437	6·758
140	*v′*	13·21	15·48	17·22	18·72	20·06	22·50	26·79
	h	2753	2896	3003	3093	3175	3322	3590
	s	5·559	5·784	5·946	6·079	6·193	6·390	6·716
150	*v′*	11·46	13·91	15·66	17·14	18·44	20·78	24·87
	h	2693	2861	2977	3073	3157	3309	3581
	s	5·443	5·707	5·883	6·023	6·142	6·345	6·677
160	*v′*	9·76	12·48	14·27	15·73	17·02	19·28	23·19
	h	2617	2821	2949	3051	3139	3295	3573
	s	5·304	5·626	5·820	5·968	6·093	6·301	6·639
170	*v′*		11·17	13·03	14·49	15·76	17·96	21·71
	h		2778	2920	3028	3121	3281	3564
	s		5·541	5·756	5·914	6·044	6·260	6·603
180	*v′*		9·97	11·91	13·38	14·63	16·78	20·39
	h		2729	2888	3004	3102	3268	3555
	s		5·449	5·691	5·861	5·997	6·219	6·569
190	*v′*		8·82	10·89	12·38	13·62	15·72	19·21
	h		2674	2855	2980	3082	3254	3546
	s		5·348	5·625	5·807	5·950	6·180	6·536
200	*v′*		7·68	9·95	11·47	12·70	14·77	18·15
	h		2605	2819	2955	3062	3239	3537
	s		5·228	5·556	5·753	5·904	6·142	6·505
210	*v′*		6·50	9·08	10·64	11·87	13·90	17·19
	h		2500	2781	2928	3041	3225	3528
	s		5·050	5·484	5·699	5·859	6·105	6·474
220	*v′*		4·50	8·25	9·87	11·11	13·12	16·32
	h		2300	2738	2900	3020	3210	3519
	s		4·725	5·409	5·645	5·813	6·068	6·444

continued

p	T	350	375	400	425	450	500	600
221·2	v'	1·63	3·51	8·16	9·78	11·03	13·03	16·22
	h	1637	2139	2733	2896	3017	3208	3518
	s	3·708	4·515	5·398	5·638	5·807	6·064	6·441
250	v'	1·60	1·98	6·01	7·89	9·17	11·13	14·12
	h	1625	1850	2580	2807	2951	3165	3491
	s	3·682	4·026	5·142	5·474	5·677	5·962	6·361
300	v'	1·55	1·80	2·82	5·30	6·74	8·68	11·43
	h	1610	1791	2157	2614	2823	3084	3445
	s	3·645	3·933	4·482	5·157	5·444	5·795	6·234
350	v'	1·52	1·71	2·11	3·43	4·96	6·93	9·52
	h	1599	1762	1992	2375	2673	2998	3397
	s	3·614	3·875	4·219	4·776	5·197	5·633	6·120

The preceding tables using SI units have been extracted, with the permission of the Controller of Her Majesty's Stationery Office, from *Steam Tables 1964* published by H. M. S. O.

Table 3. Saturated water and vapour (British units)

In this table
p is pressure in lbf/in^2
T_s is the corresponding saturation temperature in °F
v_g is the specific volume of the vapour in ft^3/lb
h_f and h_g are the specific enthalpies of liquid and vapour in Btu/lb
s_f and s_g are the specific entropies of liquid and vapour in Btu/lb °R

p	T_s	v_g	h_f	h_g	s_f	s_g
0·0885	32·0	3306·0	0	1075·8	0	2·1877
0·5	79·6	641·4	47·6	1096·4	0·0924	2·0372
1·0	101·7	333·6	69·7	1106·0	0·1326	1·9782
2·0	126·1	173·7	94·0	1116·2	0·1749	1·9200
5·0	162·2	73·52	130·1	1131·1	0·2347	1·8441
7·5	179·9	50·29	147·9	1138·1	0·2629	1·8110
10·0	193·2	38·42	161·2	1143·3	0·2835	1·7876
14·696	212·0	26·80	180·1	1150·4	0·3120	1·7566
20	228·0	20·09	196·2	1156·3	0·3357	1·7319
30	250·3	13·75	218·8	1164·1	0·3680	1·6993
40	267·3	10·50	236·0	1169·7	0·3919	1·6763
60	292·7	7·175	262·1	1177·6	0·4270	1·6438

continued

p	T_s	v_g	h_f	h_g	s_f	s_g
80	312·0	5·472	282·0	1183·1	0·4531	1·6207
100	327·8	4·432	298·4	1187·2	0·4740	1·6026
150	358·4	3·015	330·5	1194·1	0·5138	1·5694
200	381·8	2·288	355·4	1198·4	0·5435	1·5453
250	401·0	1·844	376·0	1201·1	0·5675	1·5263
300	417·3	1·543	393·8	1202·8	0·5879	1·5104
400	444·6	1·161	424·0	1204·5	0·6214	1·4844
500	467·0	0·9278	449·4	1204·4	0·6487	1·4634
600	486·2	0·7698	471·6	1203·2	0·6720	1·4454
700	503·1	0·6554	491·5	1201·2	0·6925	1·4296
800	518·2	0·5687	509·7	1198·6	0·7108	1·4153
900	532·0	0·5006	526·6	1195·4	0·7275	1·4020
1000	544·6	0·4456	542·4	1191·8	0·7430	1·3897
1100	556·3	0·4001	557·4	1187·8	0·7575	1·3780
1200	567·2	0·3619	571·7	1183·4	0·7711	1·3667
1300	577·5	0·3293	585·4	1178·6	0·7840	1·3559
1400	587·1	0·3012	598·7	1173·4	0·7963	1·3454
1500	596·2	0·2765	611·6	1167·9	0·8082	1·3351
1600	604·9	0·2548	624·1	1162·1	0·8196	1·3249
1800	621·0	0·2179	648·3	1149·4	0·8412	1·3049
2000	635·8	0·1878	671·7	1135·1	0·8619	1·2849
2200	649·5	0·1625	694·8	1119·2	0·8820	1·2646
2400	662·1	0·1407	718·4	1101·1	0·9023	1·2434
2600	673·9	0·1213	743·0	1080·2	0·9232	1·2205
2800	685·0	0·1035	770·1	1054·8	0·9459	1·1946
3000	695·4	0·0858	802·5	1020·3	0·9731	1·1615
3206·2	705·4	0·0503	902·7	902·7	1·0580	1·0580

Table 4. Superheated steam (British units)

In this table

T is temperature in °F
p is pressure in lbf/in^2
v is specific volume in ft^3/lb
h is specific enthalpy in Btu/lb
s is specific entropy in Btu/lb °R

p	T	150	200	250	300	400	500	600	800	1000
1	v	362·6	392·6	422·5	452·3	512·0	571·6	631·2	750·4	869·5
	h	1127·9	1150·4	1173·1	1195·8	1241·7	1288·3	1335·7	1432·8	1533·5
	s	2·0156	2·0512	2·0843	2·1153	2·1720	2·2233	2·2702	2·3542	2·4283

continued

P	T	150	200	250	300	400	500	600	800	1000
	v		78·16	84·22	90·25	102·26	114·22	126·16	150·03	173·87
5	*h*		1148·8	1171·9	1195·0	1241·2	1288·0	1335·4	1432·7	1533·4
	s		1·8718	1·9056	1·9370	1.9942	2·0456	2·0927	2·1767	2·2509
	v		38·85	41·95	45·00	51·04	57·05	63·03	74·98	86·92
10	*h*		1146·6	1170·4	1193·9	1240·6	1287·5	1335·1	1432·5	1533·2
	s		1·7927	1·8274	1·8595	1·9172	1·9689	2·0160	2·1002	2·1744
	v			28·43	30·53	34·68	38·78	42·86	51·00	59·13
14·696	*h*			1169·0	1192·8	1239·9	1287·1	1334·8	1432·3	1533·1
	s			1·7834	1·8160	1·8743	1·9261	1·9734	2·0576	2·1319
	v			20·79	22·36	25·43	28·46	31·47	37·46	43·44
20	*h*			1167·3	1191·6	1239·2	1286·6	1334·4	1432·1	1533·0
	s			1·7476	1·7808	1·8396	1·8918	1·9392	2·0235	2·0978
	v				14·816	16·897	18·933	20·95	24·96	28·95
30	*h*				1189·3	1237·9	1285·7	1333·8	1431·7	1532·7
	s				1·7336	1·7937	1·8464	1·8940	1·9786	2·0530
	v				11·040	12·628	14·168	15·688	18·702	21·70
40	*h*				1186·8	1236·5	1284·8	1333·1	1431·3	1532·4
	s				1·6994	1·7608	1·8140	1·8619	1·9467	2·0212

p	T	350	400	450	500	600	700	800	1000	1200
	v	7·818	8·357	8·884	9·403	10·427	11·441	12·449	14·454	16·451
60	*h*	1208·2	1233·6	1258·5	1283·0	1331·8	1380·9	1430·5	1531·9	1636·6
	s	1·6830	1·7135	1·7416	1·7678	1·8162	1·8605	1·9015	1·9762	2·0434
	v	5·803	6·220	6·624	7·020	7·797	8·562	9·322	10·830	12·332
80	*h*	1204·3	1230·7	1256·1	1281·1	1330·5	1379·9	1429·7	1531·3	1636·2
	s	1·6475	1·6791	1·7079	1·7346	1·7836	1·8281	1·8694	1·9442	2·0115
	v	4·592	4·937	5·268	5·589	6·218	6·835	7·446	8·656	9·860
100	*h*	1200·1	1227·6	1253·7	1279·1	1329·1	1378·9	1428·9	1530·8	1635·7
	s	1·6188	1·6518	1·6813	1·7085	1·7581	1·8029	1·8443	1·9193	1·9867
	v		3·223	3·457	3·681	4·113	4·532	4·944	5·758	6·564
150	*h*		1219·4	1247·5	1274·1	1325·7	1376·3	1426·9	1529·4	1634·7
	s		1·5995	1·6312	1·6599	1·7109	1·7566	1·7984	1·8740	1·9416
	v		2·361	2·549	2·726	3·060	3·380	3·693	4·309	4·917
200	*h*		1210·3	1240·7	1268·9	1322·1	1373·6	1424·8	1528·0	1633·7
	s		1·5594	1·5937	1·6240	1·6767	1·7232	1·7655	1·8415	1·9094

continued

p	T	350	400	450	500	600	700	800	1000	1200
	v	2·151	2·291	2·427	2·688	2·942	3·192	3·439	3·928	4·413
250	*h*	1263·4	1291·4	1318·5	1371·0	1422·7	1474·5	1526·6	1632·7	1741·8
	s	1·5949	1·6233	1·6495	1·6969	1·7397	1·7793	1·8162	1·8843	1·9464
	v	1·7675	1·8891	2·005	2·227	2·442	2·652	2·859	3·269	3·674
300	*h*	1257·6	1286·8	1314·7	1368·3	1420·6	1472·8	1525·2	1631·7	1741·0
	s	1·5701	1·5998	1·6268	1.6751	1·7184	1·7582	1·7954	1·8638	1·9260

p	T	500	550	600	700	800	900	1000	1200	1400
	v	1·2851	1·3843	1·4770	1·6508	1·8161	1·9767	2·134	2·445	2·751
400	*h*	1245·1	1277·2	1306·9	1362·7	1416·4	1469·4	1522·4	1629·6	1739·5
	s	1·5281	1·5607	1·5894	1·6398	1·6842	1·7247	1·7623	1·8311	1·8936
	v	0·9927	1·0798	1·1591	1·3044	1·4405	1·5715	1·6996	1·9450	2·197
500	*h*	1231·3	1266·7	1298·6	1357·0	1412·1	1466·0	1519·6	1627·6	1737·9
	s	1·4919	1·5279	1·5588	1·6115	1·6571	1·6982	1·7363	1·8056	1·8683
	v	0·7947	0·8753	0·9463	1·0732	1·1899	1·3013	1·4096	1·6208	1·8279
600	*h*	1215·7	1255·5	1289·9	1351·1	1407·7	1462·5	1516·7	1625·5	1736·3
	s	1·4586	1·4990	1·5323	1·5875	1·6343	1·6762	1·7147	1·7846	1·8476
	v		0·7277	0·7934	0·9077	1·0108	1·1082	1·2024	1·3853	1·5641
700	*h*		1243·2	1280·6	1345·0	1403·2	1459·0	1513·9	1623·5	1734·8
	s		1·4722	1·5084	1·5665	1·6147	1·6573	1·6963	1·7666	1·8299
	v		0·6154	0·6779	0·7833	0·8763	0·9633	1·0470	1·2088	1·3662
800	*h*		1229·8	1270·7	1338·6	1398·6	1455·4	1511·0	1621·4	1733·2
	s		1·4467	1·4863	1·5476	1·5972	1·6407	1·6801	1·7510	1·8146
	v		0·5264	0·5873	0·6863	0·7716	0·8506	0·9262	1·0714	1·2124
900	*h*		1215·0	1260·1	1332·1	1393·9	1451·8	1508·1	1619·3	1731·6
	s		1·4216	1·4653	1·5303	1·5814	1·6257	1·6656	1·7371	1·8009
	v		0·4533	0·5140	0·6084	0·6878	0·7604	0·8294	0·9615	1·0893
1000	*h*		1198·3	1248·8	1325·3	1389·2	1448·2	1505·1	1617·3	1730·0
	s		1·3961	1·4450	1·5141	1·5670	1·6121	1·6525	1·7245	1·7886
	v			0·4532	0·5445	0·6191	0·6866	0·7503	0·8716	0·9885
1100	*h*			1236·7	1318·3	1384·3	1444·5	1502·2	1615·2	1728·4
	s			1·4251	1·4989	1·5535	1·5995	1·6405	1·7130	1·7775
	v			0·4016	0·4909	0·5617	0·6250	0·6843	0·7967	0·9046
1200	*h*			1223·5	1311·0	1379·3	1440·7	1499·2	1613·1	1726·9
	s			1·4052	1·4843	1·5409	1·5879	1·6293	1·7025	1·7672

continued

p	*T*	500	550	600	700	800	900	1000	1200	1400
	v			0·3569	0·4454	0·5131	0·5728	0·6284	0·7333	0·8336
1300	*h*			1209·0	1303·4	1374·3	1437·0	1496·2	1611·0	1725·3
	s			1·3850	1·4703	1·5290	1·5769	1·6190	1·6927	1·7577
	v			0·3174	0·4062	0·4714	0·5281	0·5805	0·6789	0·7727
1400	*h*			1193·0	1295·5	1369·1	1433·1	1493·2	1608·9	1723·7
	s			1·3639	1·4567	1·5177	1·5666	1·6093	1·6836	1·7489
	v			0·2815	0·3719	0·4352	0·4893	0·5390	0·6318	0·7199
1500	*h*			1174·5	1287·2	1363·8	1429·3	1490·1	1606·8	1722·1
	s			1·3412	1·4434	1·5068	1·5569	1·6001	1·6750	1·7406
	v				0·3417	0·4034	0·4553	0·5027	0·5906	0·6738
1600	*h*				1278·7	1358·4	1425·3	1487·0	1604·6	1720·5
	s				1·4303	1·4964	1·5476	1·5914	1·6669	1·7328
	v				0·2907	0·3502	0·3986	0·4421	0·5218	0·5968
1800	*h*				1260·3	1347·2	1417·4	1480·8	1600·4	1717·3
	s				1·4044	1·4765	1·5301	1·5752	1·6520	1·7185
	v				0·2489	0·3074	0·3532	0·3935	0·4668	0·5352
2000	*h*				1240·0	1335·5	1409·2	1474·5	1596·1	1714·1
	s				1·3783	1·4576	1·5139	1·5603	1·6384	1·7055
	v				0·2135	0·2721	0·3159	0·3538	0·4218	0·4849
2200	*h*				1217·4	1323·3	1400·8	1468·2	1591·8	1710·9
	s				1·3515	1·4393	1·4986	1·5465	1·6259	1·6937
	v				0·1828	0·2425	0·2848	0·3207	0·3843	0·4429
2400	*h*				1191·5	1310·3	1392·2	1461·7	1587·4	1707·7
	s				1·3228	1·4215	1·4842	1·5335	1·6143	1·6827
	v				0·1549	0·2172	0·2584	0·2926	0·3526	0·4074
2600	*h*				1160·6	1296·8	1383·4	1455·1	1583·1	1704·4
	s				1·2908	1·4040	1·4703	1·5212	1·6035	1·6725
	v				0·1281	0·1953	0·2356	0·2685	0·3254	0·3769
2800	*h*				1121·4	1282·4	1374·3	1448·5	1578·7	1701·2
	s				1·2524	1·3865	1·4569	1·5096	1·5933	1·6630
	v				0·0984	0·1760	0·2159	0·2476	0·3018	0·3505
3000	*h*				1060·7	1267·2	1365·0	1441·8	1574·3	1698·0
	s				1·1966	1·3690	1·4439	1·4984	1·5837	1·6540
	v					0·1583	0·1981	0·2288	0·2806	0·3267
3206·2	*h*					1250·5	1355·2	1434·7	1569·8	1694·6
	s					1·3508	1·4309	1·4874	1·5742	1·6452

The preceding tables using British units have been extracted, with the authors' permission, from *Thermodynamic Properties of Steam* by J. H. Keenan and F. G. Keyes, published by John Wiley & Sons, New York (1937).

Mollier Chart for steam; Sheet 1
(Sheets 2 and 3 follow)

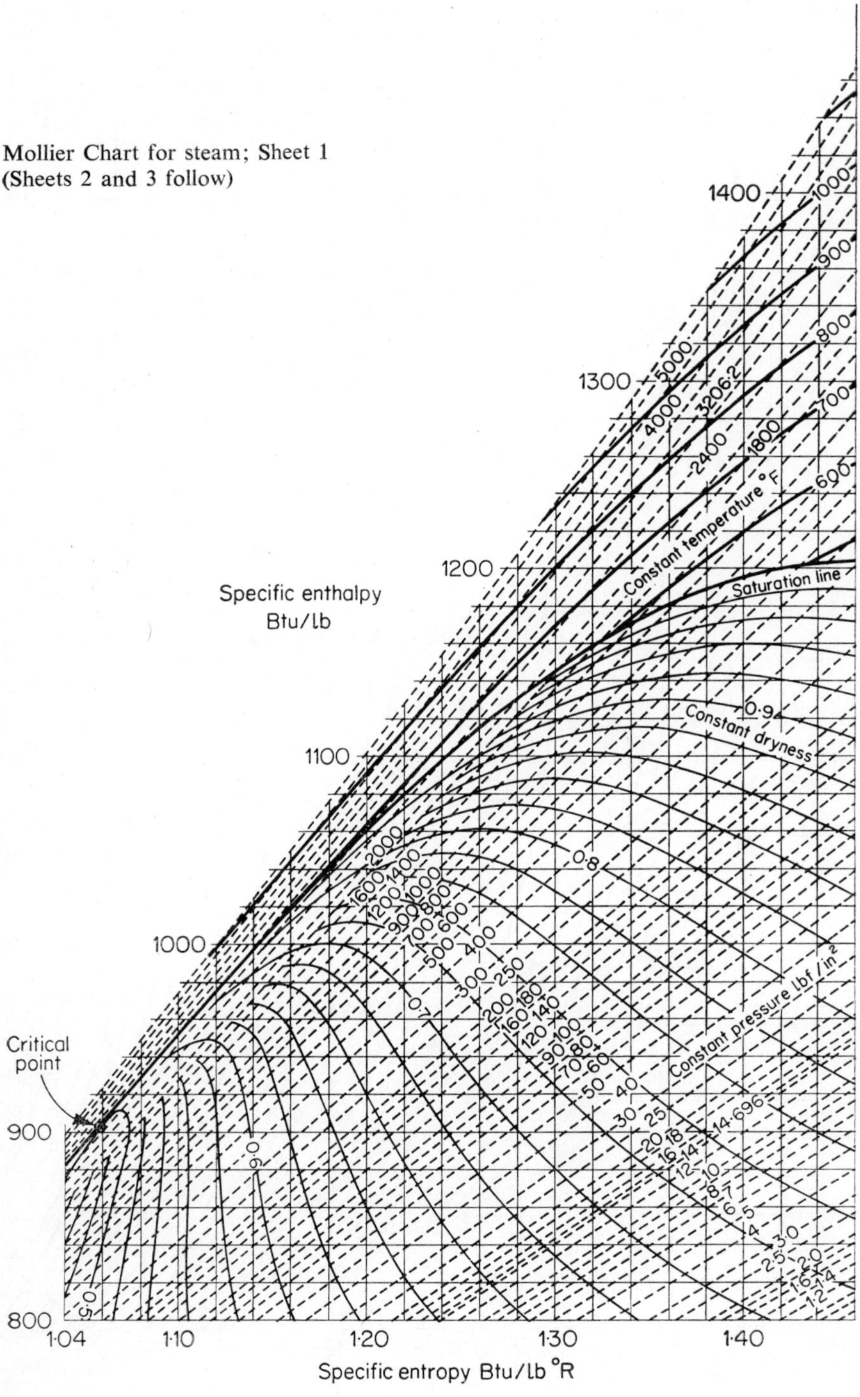

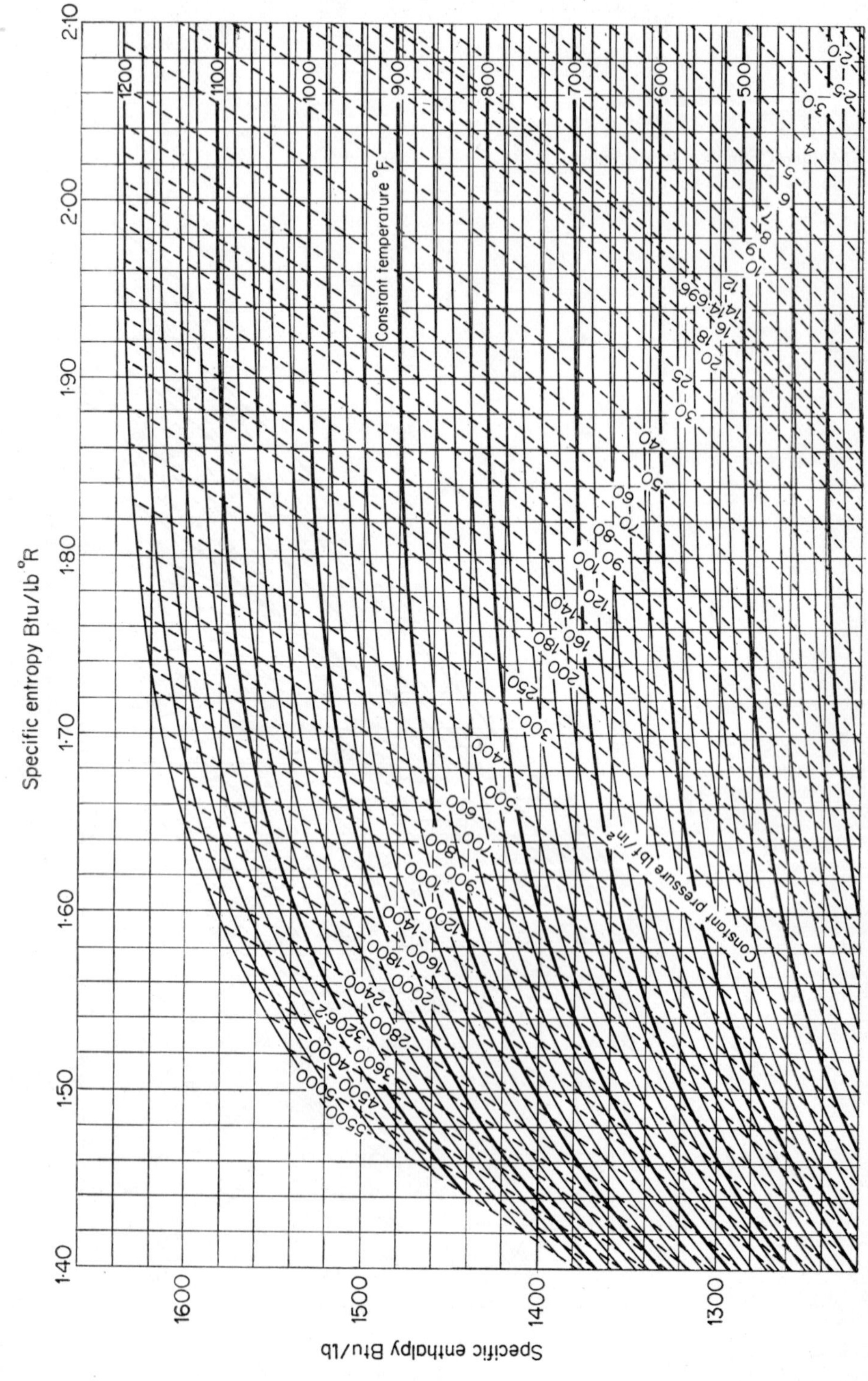
Specific entropy Btu/lb °R
Specific enthalpy Btu/lb
Constant temperature °F
Constant pressure lbf/in²

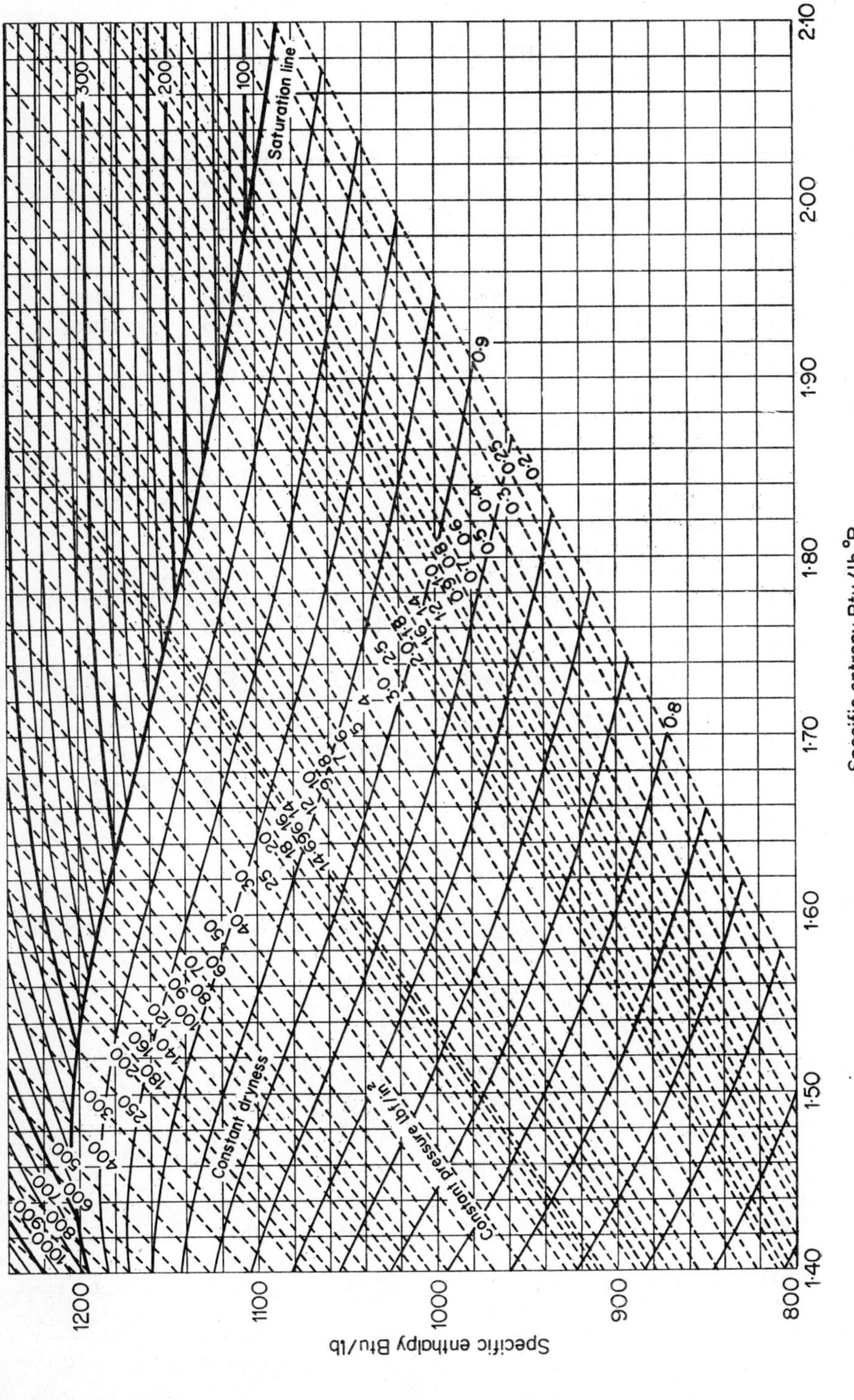
Saturation line
Constant dryness
Constant pressure lbf/in²
Specific entropy Btu/lb °R
Specific enthalpy Btu/lb
1·40
1·50
1·60
1·70
1·80
1·90
2·00
2·10
800
900
1000
1100
1200
0·9
0·8

APPENDIX V. TABLES FOR PERFECT-GAS FLOWS

Table 1. Normal shock for $\gamma = 1{\cdot}3$

M_1	M_2	p_2/p_1	ϱ_2/ϱ_1	T_2/T_1	p_{0_2}/p_{0_1}	p_{0_2}/p_1
1·00	1·000	1·000	1·000	1·000	1.000	1·832
1·05	0·953	1·116	1·088	1·026	1·000	1·941
1·10	0·911	1·237	1·178	1·051	0·999	2·058
1·15	0·874	1·364	1·269	1·075	0·997	2·183
1·20	0·840	1·497	1·362	1·100	0·993	2·316
1·25	0·810	1·636	1·456	1·124	0·987	2·46
1·30	0·783	1·780	1·551	1·148	0·979	2·61
1·35	0·758	1·930	1·646	1·172	0·968	2·76
1·40	0·735	2·085	1·742	1·197	0·956	2·92
1·45	0·714	2·246	1·838	1·222	0·942	3·09
1·50	0·694	2·41	1·935	1·247	0·926	3·27
1·55	0·676	2·59	2·031	1·273	0·908	3·45
1·60	0·660	2·76	2·127	1·299	0·889	3·64
1·65	0·645	2·95	2·223	1·326	0·868	3·83
1·70	0·630	3·14	2·318	1·353	0·847	4·03
1·75	0·617	3·33	2·41	1·380	0·824	4·24
1·80	0·605	3·53	2·51	1·408	0·800	4·45
1·85	0·593	3·74	2·60	1·437	0·776	4·67
1·90	0·583	3·95	2·69	1·467	0·751	4·90
1·95	0·572	4·17	2·79	1·497	0·726	5·13
2·00	0·563	4·39	2·88	1·527	0·701	5·37
2·05	0·554	4·62	2·96	1·558	0·675	5·62
2·10	0·546	4·86	3·05	1·590	0·650	5·87
2·15	0·538	5·10	3·14	1·623	0·625	6·12
2·20	0·530	5·34	3·23	1·656	0·600	6·39
2·25	0·523	5·59	3·31	1·690	0·576	6·66
2·30	0·516	5·85	3·39	1·725	0·552	6·93
2·35	0·510	6·11	3·47	1·760	0·528	7·22
2·40	0·504	6·38	3·55	1·796	0·505	7·50
2·45	0·498	6·66	3·63	1·832	0·483	7·80
2·50	0·493	6·94	3·71	1·869	0·461	8·10
2·55	0·488	7·22	3·79	1·907	0·440	8·41
2·60	0·483	7·51	3·86	1·946	0·420	8.72
2·65	0·478	7·81	3·93	1·985	0·400	9·04
2·70	0·474	8·11	4·01	2·03	0·381	9·36

continued

M_1	M_2	p_2/p_1	ϱ_2/ϱ_1	T_2/T_1	p_{0_2}/p_{0_1}	p_{0_2}/p_1
2·75	0·470	8·42	4·08	2·07	0·363	9·69
2·80	0·466	8·73	4·14	2·11	0·345	10·03
2·85	0·462	9·05	4·21	2·15	0·328	10·37
2·90	0·458	9·38	4·28	2·19	0·312	10·72
2·95	0·454	9·71	4·34	2·24	0·297	11·08
3·00	0·451	10·04	4·40	2·28	0·2820	11·44
3·50	0·424	13·72	4·96	2·76	0·1677	15·39
4·00	0·406	17·96	5·41	3·32	0·0993	19·96
4·50	0·393	22·76	5·77	3·95	0·0594	25·18
5·00	0·383	28·13	6·05	4·65	0·0361	30·90
6·00	0·370	40·6	6·47	6·27	0·01422	44·3
7·00	0·363	55·3	6·75	8·19	0·00610	60·1
8·00	0·357	72·2	6·94	10·40	0·00283	78·4
9·00	0·354	91·4	7·08	12·91	0·00140	99·1
10·0	0·351	112·9	7·19	15·71	0·00074	122·2

Table 2. Normal shock for $\gamma = 1{\cdot}4$

M_1	M_2	p_2/p_1	ϱ_2/ϱ_1	T_2/T_1	p_{0_2}/p_{0_1}	p_{0_2}/p_1
1·00	1·000	1·000	1·000	1·000	1·000	1·893
1·05	0·953	1·120	1·084	1·033	1·000	2·008
1·10	0·912	1·245	1·169	1·065	0·999	2·133
1·15	0·875	1·376	1·255	1·097	0·997	2·266
1·20	0·842	1·513	1·342	1·128	0·993	2·408
1·25	0·813	1·656	1·429	1·159	0·987	2·56
1·30	0·786	1·805	1·516	1·191	0·979	2·71
1·35	0·762	1·960	1·603	1·223	0·970	2·88
1·40	0·740	2·120	1·690	1·255	0·958	3·05
1·45	0·720	2·286	1·776	1·287	0·945	3·23
1·50	0·701	2·46	1·862	1·320	0·930	3·41
1·55	0·684	2·64	1·947	1·354	0·913	3·61
1·60	0·668	2·82	2·032	1·388	0·895	3·80
1·65	0·654	3·01	2·115	1·423	0·876	4·01
1·70	0·641	3·21	2·198	1·458	0·856	4·22
1·75	0·628	3·41	2·28	1·495	0·835	4·44
1·80	0·617	3·61	2·36	1·532	0·813	4·67
1·85	0·606	3·83	2·44	1·569	0·790	4·90
1·90	0·596	4·05	2·52	1·608	0·767	5·14
1·95	0·586	4·27	2·59	1·647	0·744	5·39

continued

M_1	M_2	p_2/p_1	ϱ_2/ϱ_1	T_2/T_1	p_{0_2}/p_{0_1}	p_{0_2}/p_1
2·00	0·577	4·50	2·67	1·688	0·721	5·64
2·05	0·569	4·74	2·74	1·729	0·698	5·90
2·10	0·561	4·98	2·81	1·770	0·674	6·17
2·15	0·554	5·23	2·88	1·813	0·651	6·44
2·20	0·547	5·48	2·95	1·857	0·628	6·72
2·25	0·541	5·74	3·02	1·901	0·606	7·00
2·30	0·534	6·01	3·08	1·947	0·583	7·29
2·35	0·529	6·28	3·15	1·993	0·561	7·59
2·40	0·523	6·55	3·21	2·040	0·540	7·90
2·45	0·518	6·84	3·27	2·089	0·519	8·21
2·50	0·513	7·13	3·33	2·14	0·499	8·53
2·55	0·508	7·42	3·39	2·19	0·479	8·85
2·60	0·504	7·72	3·45	2·24	0·460	9·18
2·65	0·500	8·03	3·50	2·29	0·442	9·52
2·70	0·496	8·34	3·56	2·34	0·424	9·86
2·75	0·492	8·66	3·61	2·40	0·406	10·21
2·80	0·488	8·98	3·66	2·45	0·389	10·57
2·85	0·485	9·31	3·71	2·51	0·373	10·93
2·90	0·481	9·65	3·76	2·56	0·358	11·30
2·95	0·478	9·99	3·81	2·62	0·343	11·68
3·00	0·475	10·33	3·85	2·68	0·328	12·06
3·50	0·451	14·13	4·26	3·32	0·213	16·24
4·00	0·435	18·50	4·57	4·05	0·1388	21·07
4·50	0·424	23·46	4·81	4·88	0·0917	26·54
5·00	0·415	29·00	5·00	5·80	0·0617	32·65
6·00	0·404	41·8	5·27	7·94	0·0297	46·8
7·00	0·397	57·0	5·44	10·47	0·01535	63·6
8·00	0·393	74·5	5·57	13·39	0·00849	82·9
9·00	0·390	94·3	5·65	16·69	0·00496	104·8
10·0	0·388	116·5	5·71	20·39	0·00304	129·2

Table 3. Isentropic flow for $\gamma = 1{\cdot}2$

M	p/p_0	ϱ/ϱ_0	T/T_0	A/A^*	I/I^*
0·00	1·000	1·000	1·000	∞	∞
0·05	0·999	0·999	1·000	11·86	9·562
0·10	0·994	0·995	0·999	5·95	4·822
0·15	0·987	0·989	0·998	4·00	3·260
0·20	0·976	0·980	0·996	3·03	2·493

continued

M	p/p_0	ϱ/ϱ_0	T/T_0	A/A^*	I/I^*
0·25	0·963	0·969	0·994	2·451	2·044
0·30	0·948	0·956	0·991	2·073	1·753
0·35	0·930	0·941	0·988	1·809	1·553
0·40	0·909	0·924	0·984	1·615	1·409
0·45	0·887	0·905	0·980	1·469	1·304
0·50	0·862	0·884	0·976	1·356	1·224
0·55	0·836	0·862	0·971	1·268	1·164
0·60	0·809	0·838	0·965	1·199	1·118
0·65	0·780	0·813	0·960	1·144	1·083
0·70	0·751	0·787	0·953	1·100	1·056
0·75	0·720	0·761	0·947	1·067	1·036
0·80	0·689	0·733	0·940	1·041	1·021
0·85	0·658	0·706	0·933	1·022	1·011
0·90	0·627	0·677	0·925	1·010	1·005
0·95	0·595	0·649	0·917	1·002	1·001
1·00	0·564	0·621	0·909	1·000	1·000
1·05	0·534	0·593	0·901	1·002	1·001
1·10	0·504	0·565	0·892	1·009	1·004
1·15	0·475	0·537	0·883	1·019	1·008
1·20	0·446	0·510	0·874	1·034	1·013
1·25	0·419	0·484	0·865	1·053	1·020
1·30	0·392	0·458	0·855	1·075	1·027
1·35	0·366	0·433	0·855	1·101	1·035
1·40	0·342	0·409	0·836	1·131	1·044
1·45	0·318	0·385	0·826	1·166	1·053
1·50	0·296	0·363	0·816	1·205	1·063
1·55	0·275	0·341	0·806	1·248	1·072
1·60	0·255	0·320	0·796	1·296	1·083
1·65	0·236	0·300	0·786	1·349	1·093
1·70	0·218	0·281	0·776	1·407	1·104
1·75	0·2013	0·2629	0·766	1·470	1·114
1·80	0·1856	0·2458	0·755	1·539	1·125
1·85	0·1710	0·2295	0·745	1·615	1·136
1·90	0·1573	0·2141	0.735	1·698	1·147
1·95	0·1446	0·1996	0·725	1·787	1·158
2·00	0·1328	0·1859	0·714	1·884	1·168
2·05	0·1218	0·1730	0·704	1·989	1·179
2·10	0·1117	0·1609	0·694	2·103	1·190
2·15	0·1023	0·1496	0·684	2·226	1·201
2·20	0·0936	0·1390	0·674	2·359	1·212

continued

M	p/p_0	ϱ/ϱ_0	T/T_0	A/A^*	I/I^*
2·25	0·0856	0·1290	0·664	2·50	1·222
2·30	0·0783	0·1197	0·654	2·66	1·232
2·35	0·0715	0·1110	0·644	2·83	1·242
2·40	0·0653	0·1029	0·635	3·01	1·252
2·45	0·0596	0·0953	0·625	3·21	1·262
2·50	0·0543	0·0883	0·615	3·24	1·272
2·55	0·0495	0·0817	0·606	3·65	1·282
2·60	0·0451	0·0756	0·597	3·90	1·291
2·65	0·0411	0·0700	0·588	4·17	1·300
2·70	0·0374	0·0647	0·578	4·46	1·309
2·75	0·0341	0·0599	0·569	4·77	1·318
2·80	0·0310	0·0553	0·561	5·10	1·327
2·85	0·0282	0·0512	0·552	5·47	1·335
2·90	0·0257	0·0473	0·543	5·86	1·343
2·95	0·0234	0·0437	0·535	6·28	1·352
3·00	0·02126	0·04039	0·526	6·74	1·360
3·50	0·00824	0·01834	0·449	13·76	1·434
4·00	0·00324	0·00842	0·385	28·4	1·493
4·50	0·00130	0·00395	0·331	58·0	1·541
5·00	0·00054	0·00190	0·286	116·3	1·580
6·00	$1{\cdot}06\times10^{-4}$	$4{\cdot}86\times10^{-4}$	0·2174	435	1·637
7·00	$2{\cdot}37\times10^{-5}$	$1{\cdot}40\times10^{-4}$	0·1695	1458	1·677
8·00	$6{\cdot}09\times10^{-6}$	$4{\cdot}51\times10^{-5}$	0·1351	4450	1·704
9·00	$1{\cdot}76\times10^{-6}$	$1{\cdot}60\times10^{-5}$	0·1099	12410	1·724
10·0	$5{\cdot}64\times10^{-7}$	$6{\cdot}21\times10^{-6}$	0·0909	31700	1·739

Table 4. Isentropic flow for $\gamma = 1{\cdot}3$

M	p/p_0	ϱ/ϱ_0	T/T_0	A/A^*	I/I^*
0·00	1·000	1·000	1·000	∞	∞
0·05	0·998	0·999	1·000	11·72	9·354
0·10	0·994	0·995	0·999	5·89	4·720
0·15	0·986	0·989	0·997	3·95	3·194
0·20	0·974	0·980	0·994	2·99	2·455
0·25	0·960	0·969	0·991	2·426	2·007
0·30	0·944	0·956	0·987	2·054	1·724
0·35	0·924	0·941	0·982	1·793	1·530
0·40	0·902	0·924	0·977	1·602	1·391
0·45	0·878	0·905	0·971	1·459	1·289

continued

M	p/p_0	ϱ/ϱ_0	T/T_0	A/A^*	I/I^*
0·50	0·853	0·885	0·964	1·348	1·213
0·55	0·825	0·863	0·957	1·261	1·155
0·60	0·796	0·839	0·949	1·193	1·111
0·65	0·766	0·815	0·940	1·139	1·078
0·70	0·735	0·790	0·932	1·097	1·052
0·75	0·704	0·763	0·922	1·064	1·034
0·80	0·672	0·737	0·912	1·040	1·020
0·85	0·640	0·710	0·902	1·021	1·010
0·90	0·608	0·682	0·892	1·009	1·004
0·95	0·577	0·655	0·881	1·002	1·001
1·00	0·546	0·628	0·870	1·000	1·000
1·05	0·515	0·600	0·858	1·002	1·001
1·10	0·485	0·574	0·846	1·008	1·003
1·15	0·457	0·547	0·835	1·018	1·007
1·20	0·429	0·521	0·822	1·032	1·012
1·25	0·402	0·496	0·810	1·050	1·018
1·30	0·376	0·471	0·798	1·070	1·024
1·35	0·351	0·447	0·785	1·095	1·031
1·40	0·327	0·424	0·773	1·123	1·039
1·45	0·305	0·401	0·760	1·154	1·047
1·50	0·284	0·379	0·748	1·189	1·055
1·55	0·264	0·359	0·735	1·228	1·063
1·60	0·245	0·339	0·723	1·271	1·072
1·65	0·227	0·319	0·710	1·318	1·081
1·70	0·210	0·301	0·698	1·369	1·090
1·75	0·1944	0·284	0·685	1·424	1·099
1·80	0·1797	0·267	0·673	1·484	1·108
1·85	0·1660	0·251	0·661	1·549	1·116
1·90	0·1533	0·236	0·649	1·618	1·125
1·95	0·1415	0·222	0·637	1·693	1·134
2·00	0·1305	0·2087	0·625	1·773	1·143
2·05	0·1203	0·1960	0·613	1·859	1·152
2·10	0·1108	0·1841	0·602	1·951	1·160
2·15	0·1020	0·1728	0·591	2·050	1·168
2·20	0·0939	0·1621	0·579	2·156	1·176
2·25	0·0865	0·1521	0·568	2·27	1·184
2·30	0·0796	0·1427	0·558	2·39	1·192
2·35	0·0732	0·1338	0·547	2·52	1·200
2·40	0·0673	0·1254	0·537	2·65	1·208
2·45	0·0619	0·1176	0·526	2·80	1·216

continued

M	p/p_0	ϱ/ϱ_0	T/T_0	A/A^*	I/I^*
2·50	0·0569	0·1103	0·516	2·95	1·223
2·55	0·0523	0·1034	0·506	3·12	1·230
2·60	0·0481	0·0969	0·497	3·30	1·237
2·65	0·0443	0·0909	0·487	3·48	1·244
2·70	0·0407	0·0852	0·478	3·68	1·250
2·75	0·0374	0·0799	0·469	3·89	1·257
2·80	0·0344	0·0749	0·460	4·12	1·264
2·85	0·0317	0·0702	0·451	4·35	1·270
2·90	0·0291	0·0659	0·442	4·61	1·276
2·95	0·0268	0·0618	0·434	4·88	1·282
3·00	0·02466	0·05796	0·426	5·16	1·288
3·50	0·01090	0·03092	0·352	9·11	1·338
4·00	0·00498	0·01692	0·294	15·94	1·378
4·50	0·00236	0·00954	0·248	27·39	1·409
5·00	0·00117	0·00555	0·211	45·95	1·433
6·00	$3{\cdot}21\times10^{-4}$	$2{\cdot}06\times10^{-3}$	0·1563	120·1	1·468
7·00	$1{\cdot}01\times10^{-4}$	$8{\cdot}47\times10^{-4}$	0·1198	285	1·491
8·00	$3{\cdot}60\times10^{-5}$	$3{\cdot}82\times10^{-4}$	0·0943	625	1·507
9·00	$1{\cdot}41\times10^{-5}$	$1{\cdot}86\times10^{-4}$	0·0761	1275	1.519
10·0	$6{\cdot}05\times10^{-6}$	$9{\cdot}68\times10^{-5}$	0·0625	2438	1·527

Table 5. Isentropic flow for $\gamma = 1{\cdot}4$

M	p/p_0	ϱ/ϱ_0	T/T_0	A/A^*	I/I^*
0·00	1·000	1·000	1·000	∞	∞
0·05	0·998	0·999	1·000	11·59	9·158
0·10	0·993	0·995	0·998	5·82	4·624
0·15	0·984	0·989	0·996	3·91	3·132
0·20	0·973	0·980	0·992	2·96	2·400
0·25	0·957	0·969	0·988	2·403	1·973
0·30	0·939	0·956	0·982	2·035	1·698
0.35	0·919	0·941	0·976	1·778	1·509
0·40	0·896	0·924	0·969	1·590	1·375
0·45	0·870	0·906	0·961	1·449	1·276
0·50	0·843	0·885	0·952	1·340	1·203
0·55	0·814	0·863	0·943	1·255	1·147
0·60	0·784	0·840	0·933	1·188	1·105
0·65	0·753	0·816	0·922	1·136	1·073
0·70	0·721	0·792	0·911	1·094	1·049

continued

M	p/p_0	ϱ/ϱ_0	T/T_0	A/A^*	I/I^*
0·75	0·689	0·766	0·899	1·062	1·031
0·80	0·656	0·740	0·887	1·038	1·019
0·85	0·624	0·714	0·874	1·021	1·010
0·90	0·591	0·687	0·860	1·009	1·004
0·95	0·559	0·660	0·847	1·002	1·001
1·00	0·528	0·634	0·833	1·000	1·000
1·05	0·498	0·608	0·819	1·002	1·001
1·10	0·468	0·582	0·805	1·008	1·003
1·15	0·440	0·556	0·791	1·017	1·006
1·20	0·412	0·531	0·776	1·030	1·011
1·25	0·386	0·507	0·762	1·047	1·016
1·30	0·361	0·483	0·747	1·066	1·022
1·35	0·337	0·460	0·733	1·089	1·028
1·40	0·314	0·437	0·718	1·115	1·035
1·45	0·293	0·416	0·704	1·144	1·042
1·50	0·272	0·395	0·690	1·176	1·049
1·55	0·253	0·375	0·675	1·212	1·056
1·60	0·235	0·356	0·661	1·250	1·063
1·65	0·218	0·337	0·647	1·292	1·071
1·70	0·203	0·320	0·634	1·338	1·079
1·75	0·1878	0·303	0·620	1·387	1·086
1·80	0·1740	0·287	0·607	1·439	1·094
1·85	0·1612	0·272	0·594	1·495	1·101
1·90	0·1492	0·257	0·581	1·555	1·108
1·95	0·1381	0·243	0·568	1·619	1·116
2·00	0·1278	0·2301	0·556	1·688	1·123
2·05	0·1182	0·2176	0·543	1·760	1·130
2·10	0·1094	0·2058	0·531	1·837	1·137
2·15	0·1011	0·1946	0·520	1·919	1·143
2·20	0·0935	0·1841	0·508	2·005	1·150
2·25	0·0865	0·1740	0·497	2·10	1·157
2·30	0·0800	0·1646	0·486	2·19	1·163
2·35	0·0740	0·1556	0·475	2·30	1·169
2·40	0·0684	0·1472	0·465	2·40	1·175
2·45	0·0633	0·1392	0·454	2·52	1·181
2·50	0·0585	0·1317	0·444	2·64	1·187
2·55	0·0542	0·1246	0·435	2·76	1·192
2·60	0·0501	0·1179	0·425	2·90	1·198
2·65	0·0464	0·1115	0·416	3·04	1·203
2·70	0·0430	0·1056	0·407	3·18	1·208

continued

M	p/p_0	ϱ/ϱ_0	T/T_0	A/A^*	I/I^*
2·75	0·0398	0·0999	0·398	3·34	1·213
2·80	0·0369	0·0946	0·389	3·50	1·218
2·85	0·0342	0·0896	0·381	3·67	1·223
2·90	0·0317	0·0849	0·373	3·85	1·228
2·95	0·0294	0·0804	0·365	4·04	1·232
3·00	0·02722	0·07623	0·357	4·23	1·237
3·50	0·01311	0·04523	0·290	6·79	1·274
4·00	0·00658	0·02766	0·238	10·72	1·303
4·50	0·00346	0·01745	0·198	16·56	1·325
5·00	0·00189	0·01134	0·167	25·00	1·342
6·00	$6{\cdot}33\times10^{-4}$	$5{\cdot}19\times10^{-3}$	0·1220	53·2	1·366
7·00	$2{\cdot}42\times10^{-4}$	$2{\cdot}61\times10^{-3}$	0·0926	104·1	1·381
8·00	$1{\cdot}02\times10^{-4}$	$1{\cdot}41\times10^{-3}$	0·0725	190·1	1·392
9·00	$4{\cdot}74\times10^{-5}$	$8{\cdot}15\times10^{-4}$	0·0581	327	1·399
10·0	$2{\cdot}36\times10^{-5}$	$4{\cdot}95\times10^{-4}$	0·0476	536	1·404

Table 6. Rayleigh flow for $\gamma = 1{\cdot}4$

M	T_0/T_0^*	T/T^*	p/p^*	p_0/p_0^*	u/u^*
0·00	0·000	0·000	2·40	1·268	0·0000
0·05	0·0119	0·0143	2·39	1·266	0·0060
0·10	0·0468	0·0560	2·37	1·259	0·0237
0·15	0·1020	0·1218	2·33	1·249	0·0524
0·20	0·1736	0·2066	2·27	1·235	0·0909
0·25	0·257	0·304	2·207	1·218	0·1379
0·30	0·347	0·409	2·131	1·199	0·1918
0·35	0·439	0·514	2·049	1·178	0·2510
0·40	0·529	0·615	1·961	1·157	0·3137
0·45	0·614	0·708	1·870	1·135	0·3787
0·50	0·691	0·790	1·778	1·114	0·444
0·55	0·760	0·860	1·686	1·094	0·510
0·60	0·819	0·917	1·596	1·075	0·574
0·65	0·868	0·961	1·508	1·058	0·637
0·70	0·909	0·993	1·424	1·043	0·698
0·75	0·940	1·014	1·343	1·030	0·755
0·80	0·964	1·025	1·266	1·019	0·810
0·85	0·981	1·029	1·193	1·011	0·862
0·90	0·992	1·025	1·125	1·005	0·911
0·95	0·998	1·015	1·060	1·001	0·957

continued

M	T_0/T_0^*	T/T^*	p/p^*	p_0/p_0^*	u/u^*
1·00	1·000	1·000	1·000	1·000	1·000
1·05	0·998	0·982	0·944	1·001	1·040
1·10	0·994	0·960	0·891	1·005	1·078
1·15	0·987	0·937	0·842	1·011	1·113
1·20	0·979	0·912	0·796	1·019	1·146
1·25	0·969	0·886	0·753	1·030	1·176
1·30	0·958	0·859	0·713	1·044	1·205
1·35	0·946	0·832	0·676	1·059	1·232
1·40	0·934	0·805	0·641	1·078	1·256
1·45	0·922	0·779	0·609	1·098	1·280
1·50	0·909	0·753	0·578	1·122	1·301
1·55	0·897	0·727	0·550	1·147	1·321
1·60	0·884	0·702	0·524	1·176	1·340
1·65	0·872	0·677	0·499	1·207	1·358
1·70	0·860	0·654	0·476	1·240	1·375
1·75	0·848	0·631	0·454	1·277	1·390
1·80	0·836	0·609	0·434	1·316	1·405
1·85	0·825	0·588	0·414	1·358	1·418
1·90	0·814	0·567	0·396	1·403	1·431
1·95	0·804	0·548	0·380	1·452	1·443
2·00	0·793	0·529	0·364	1·503	1·455
2·10	0·774	0·494	0·335	1·616	1·475
2·20	0·756	0·461	0·309	1·743	1·494
2·30	0·740	0·431	0·286	1·886	1·510
2·40	0·724	0·404	0·265	2·045	1·525
2·50	0·710	0·379	0·246	2·22	1·539
2·60	0·697	0·356	0·229	2·42	1·551
2·70	0·685	0·334	0·214	2·63	1·561
2·80	0·674	0·315	0·200	2·87	1·571
2·90	0·664	0·297	0·188	3·14	1·580
3·00	0·654	0·2803	0·1765	3·42	1·588
3·50	0·616	0·2142	0·1322	5·33	1·620
4·00	0·589	0·1683	0·1026	8·23	1·641
4·50	0·570	0·1354	0·0818	12·50	1·656
5·00	0·556	0·1111	0·0667	18·63	1·667
6·00	0·536	0·0785	0·0467	38·9	1·681
7·00	0·524	0·0583	0·0345	75·4	1·690
8·00	0·516	0·0449	0·0265	136·6	1·695
9·00	0·511	0·0357	0·0210	234	1·699
10·0	0·507	0·0290	0·0170	382	1·702

Table 7. Fanno flow for $\gamma = 1{\cdot}4$

M	T/T^*	p/p^*	p_0/p_0^*	u/u^*	I/I^*	fL_m/d
0·00	1·200	∞	∞	0·0000	∞	∞
0·01	1·200	109·5	57·9	0·0110	45·65	7134·40
0·02	1·200	54·8	28·9	0·0219	22·83	1778·45
0·03	1·200	36·5	19·30	0·0329	15·23	787·08
0·04	1·200	27·4	14·48	0·0438	11·44	440·35
0·05	1·199	21·90	11·59	0·0548	9·16	280·02
0·06	1·199	18·25	9·67	0·0657	7·64	193·03
0·07	1·199	15·64	8·29	0·0766	6·56	140·66
0·08	1·199	13·68	7·26	0·0876	5·75	106·72
0·09	1·198	12·16	6·46	0·0985	5·12	83·50
0·10	1·198	10·94	5·82	0·1094	4·62	66·92
0·11	1·197	9·95	5·30	0·1204	4·21	54·69
0·12	1·197	9·12	4·86	0·1313	3·87	45·41
0·13	1·196	8·41	4·50	0·1422	3·59	38·21
0·14	1·195	7·81	4·18	0·1531	3·34	32·51
0·15	1·195	7·29	3·91	0·1640	3·13	27·93
0·16	1·194	6·83	3·67	0·1748	2·95	24·20
0·17	1·193	6·43	3·46	0·1857	2·79	21·12
0·18	1·192	6·07	3·28	0·1965	2·64	18·54
0·19	1·191	5·74	3·11	0·2074	2·51	16·38
0·20	1·191	5·46	2·964	0·218	2·400	14·53
0·25	1·185	4·35	2·403	0·272	1·973	8·48
0·30	1·179	3·62	2·035	0·326	1·698	5·30
0·35	1·171	3·09	1·778	0·379	1·509	3·45
0·40	1·163	2·70	1·590	0·431	1·375	2·31
0·45	1·153	2·387	1·449	0·483	1·276	1·566
0·50	1·143	2·138	1·340	0·535	1·203	1·069
0·55	1·132	1·934	1·255	0·585	1·147	0·728
0·60	1·119	1·763	1·188	0·635	1·105	0·491
0·65	1·107	1·618	1·136	0·684	1·073	0·325
0·70	1·093	1·493	1·094	0·732	1·049	0·2081
0·75	1·079	1·385	1·062	0·779	1·031	0·1273
0·80	1·064	1·289	1·038	0·825	1·019	0·0723
0·85	1·048	1·205	1·021	0·870	1·010	0·0363
0·90	1·033	1·129	1·009	0·915	1·004	0·0145
0·95	1·017	1·061	1·002	0·958	1·001	0·0033
1·00	1·000	1·000	1·000	1·000	1·000	0·000
1·05	0·983	0·944	1·002	1·041	1·001	0·0027
1·10	0·966	0·894	1·008	1·081	1·003	0·0099
1·15	0·949	0·847	1·017	1·120	1·006	0·0205
1·20	0·932	0·804	1·030	1·158	1·011	0·0336

continued

M	T/T^*	p/p^*	p_0/p_0^*	u/u^*	I/I^*	fL_m/d
1·25	0·914	0·765	1·047	1·195	1·016	0·0486
1·30	0·897	0·728	1·066	1·231	1·022	0·0648
1·35	0·879	0·695	1·089	1·266	1·028	0·0820
1·40	0·862	0·663	1·115	1·300	1·035	0·0997
1·45	0·845	0·634	1·144	1·333	1·042	0·1178
1·50	0·828	0·606	1·176	1·365	1·049	0·1361
1·55	0·811	0·581	1·212	1·396	1·056	0·1543
1·60	0·794	0·557	1·250	1·425	1·063	0·1724
1·65	0·777	0·534	1·292	1·454	1·071	0·1902
1·70	0·760	0·513	1·338	1·483	1·079	0·2078
1·75	0·744	0·493	1·387	1·510	1·086	0·225
1·80	0·728	0·474	1·439	1·536	1·094	0·242
1·85	0·712	0·456	1·495	1·561	1·101	0·258
1·90	0·697	0·439	1·555	1·586	1·108	0·274
1·95	0·682	0·423	1·619	1·610	1·116	0·290
2·00	0·667	0·408	1·688	1·633	1·123	0·305
2·10	0·638	0·380	1·837	1·677	1·137	0·334
2·20	0·610	0·355	2·005	1·718	1·150	0·361
2·30	0·583	0·332	2·193	1·756	1·163	0·386
2·40	0·558	0·311	2·403	1·792	1·175	0·410
2·50	0·533	0·292	2·64	1·826	1·187	0·432
2·60	0·510	0·275	2·90	1·857	1·198	0·453
2·70	0·488	0·259	3·18	1·887	1·208	0·472
2·80	0·467	0·244	3·50	1·914	1·218	0·490
2·90	0·447	0·231	3·85	1·940	1·228	0·507
3·00	0·429	0·2182	4·23	1·964	1·237	0·522
3·50	0·348	0·1685	6·79	2·064	1·274	0·586
4·00	0·286	0·1336	10·72	2·138	1·303	0·633
4·50	0·238	0·1083	16·56	2·194	1·325	0·668
5·00	0·200	0·0894	25·00	2·236	1·342	0·694
6·00	0·1463	0·0638	53·2	2·30	1·366	0·730
7·00	0·1111	0·0476	104·1	2·33	1·381	0·753
8·00	0·0870	0·0369	190·1	2·36	1·392	0·768
9·00	0·0698	0·0294	327·2	2·38	1·399	0·779
10·0	0·0571	0·0239	535·9	2·39	1·404	0·787

APPENDIX VI. THE ICAO STANDARD ATMOSPHERE

z m	T °K	T °C	p bar	ϱ kg/m³	a m/sec
−1000	294·7	21·5	1·1393	1·347	344·1
0	288·2	15·0	1·0133	1·225	340·3
1000	281·7	8·5	0·8988	1·112	336·4
2000	275·2	2·0	0·7950	1·007	332·5
3000	268·7	− 4·5	0·7012	0·909	328·6
4000	262·2	−11·0	0·6166	0·819	324·6
5000	255·7	−17·5	0·5405	0·736	320·5
6000	249·2	−24·0	0·4722	0·660	316·5
7000	242·7	−30·5	0·4111	0·590	312·3
8000	236·2	−36·9	0·3565	0·526	308·1
9000	229·7	−43·4	0·3080	0·467	303·8
10000	223·3	−49·9	0·2650	0·414	299·5
11000	216·8	−56·4	0·2270	0·365	295·2
12000	216·7	−56·5	0·1940	0·312	295·1
13000			0·1658	0·267	
14000			0·1417	0·228	
15000			0·1211	0·1948	
16000			0·1035	0·1665	
17000			0·0885	0·1423	
18000			0·0757	0·1217	
19000			0·0647	0·1040	
20000	216·7	−56·5	0·0553	0·0889	295·1
22000	218·6	−54·6	0·0405	0·0645	296·4
24000	220·6	−52·6	0·0297	0·0469	297·7
26000	222·5	−50·6	0·0219	0·0343	299·1
28000	224·5	−48·6	0·0162	0·0251	300·4
30000	226·5	−46·6	0·0120	0·0184	301·7
32000	228·5	−44·7	0·0089	0·0136	303·0

APPENDIX VII. TRANSPORT PROPERTIES

Table 1. Air at moderate pressures

T °K	μ μN sec/m^2	k mW/m °K	c_p kJ/kg °K	Pr
200	13·29	18·09	1·003	0·736
300	18·46	26·24	1·005	0·707
400	22·86	33·65	1·014	0·688
500	26·70	40·41	1·030	0·680
600	30·17	46·61	1·051	0·680
800	36·24	57·74	1·099	0·690
1000	41·53	67·54	1·141	0·702
1200	46·26	76·40	1·175	0·711
1400	50·57	84·50	1·201	0·719
1600	54·57	91·99	1·221	0·724
1800	58·29	98·99	1·237	0·728

Table 2. Steam: vapour at saturation pressure

T °C	μ μN sec/m^2	k mW/m °K	c_p kJ/kg °K	Pr
0·01	8·84	15·85	1·86	1·04
20	9·56	17·39	1·87	1·03
40	10·28	18·97	1·89	1·02
60	11·00	20·55	1·91	1·02
80	11·73	22·23	1·95	1·03
100	12·45	24·01	2·01	1·04
150	14·3	29·2	2·29	1·12
200	16·2	35·9	2·91	1·31
250	18·2	44·9	3·94	1·60
300	20·7	60·2	6·18	2·12

Table 3. Water: liquid at saturation pressure

T °C	μ mN sec/m^2	k W/m °K	c_p kJ/kg °K	Pr
0·01	1·782	0·551	4·210	13·61
20	1·002	0·599	4·183	7·00
40	0·653	0·634	4·179	4·30
60	0·467	0·659	4·185	2·96
80	0·355	0·674	4·198	2·21
100	0·281	0·683	4·219	1·74
150	0·185	0·684	4·32	1·17
200	0·138	0·663	4·51	0·94
250	0·110	0·618	4·87	0·87
300	0·091	0·539	5·65	0·95

Table 4. Thermal conductivity of solids at 68 °F in Btu/ft hr °R

Substance	k
Commercial copper	200
Aluminium	118
Brass	50
Mild steel	25
Nickel steel	6
Concrete	0·7
Glass	0·5
Brickwork	0·5
Asbestos	0·1
Insulating material	0·03
For comparison	
Liquid mercury	5
Liquid water	0·3
Water vapour	0·01
Hydrogen	0·1
Air	0·015

Index